Handbook of Genetics and Biotechnology

nipa
Browse Subject
eBooks / eChapters / Articles
eBooks
Explore Now
Browse, Search, Read & Buy...
Subject Catalogues
eChapters
Publishers
Print Books
Forthcoming
eBooks
New Titles
scan for catalogue
NIPA
GENX
ONLINE RESOURCES
Publishing Books and Journals
Ebooks and e articles, Current Affairs
Reasoning, Logic and Aptitude
Competitive Examination Preparation
Language Learning and Development Programme
Document Quality Checker and Improvement Tool
Effective Public Speaking, Presentation and Interpersonal Skills
Online Programmes for Professional Development
Personality Development and Human Values
PAY USING
UPI
PayPal

Revised and expanded, 2nd Edition

Handbook of Genetics and Biotechnology

Editors

Rukum S. Tomar
Assistant Professor, Department of Biotechnology
Junagadh Agricultural University, Junagadh-362 001

Sunil V. Patel
Professor, Department of Biochemistry
Junagadh Agricultural University, Junagadh-362 001

Manoj V. Parakhia
Assistant Professr, Department of Biotechnology
Junagadh Agricultural University, Junagadh

B. A. Golakiya
Professor and Head, Department of Biotechnology
Junagadh Agricultural University, Junagadh

NEW INDIA PUBLISHING AGENCY
New Delhi – 110 034

NEW INDIA PUBLISHING AGENCY

101, Vikas Surya Plaza, CU Block, LSC Market
Pitam Pura, New Delhi 110 034, India
Phone: + 91 (11)27 34 17 17 Fax: + 91(11) 27 34 16 16
Email: info@nipabooks.com
Web: www.nipabooks.com

Feedback at feedbacks@nipabooks.com

ISBN : 978-93-83305-44-5

Composed and Designed by NIPA

Preface

Conceptual development in any science is a painful path. With the simple experiments on garden peas; Geoger Mandel laid down the foundation of Modern Genetics. Watson & Creek explored the molecular base of genetics; which started the era of biotechnology. Genetics and Biotechnology are the two side of the same coin. In fact we have been working in this science since last three sanctuaries. The factual knowledge has been distilled drop by drop. When it is placed at proper place the whole picture assumed on astonishing shape, color and texture. It becomes a master piece. Several hundreds of scientists have devoted their whole life behind both the branches of biological science. Of course whatever they have found it was with full-proof. If we want to work in this science to move it forward or to put it in use for the better life on this planate we need to install these innovations, ideas, processes and products and principles in our mind. Transgenic integration by plants is a natural phenomenon, so much so that we are still trying to figure out exactly how Mother Nature does it. Plant biotechnology has already made significant and positive contributions in plant breeding and development of resistant genotypes. It has the potential to be a powerful new tool for plant breeders, one that they will surely need in facing the challenges of rapid climate change, flood and drought, global warming, as well as the new pests and diseases that these changes may bring.

This text is devised for proper installing of biotechnology and genetics in the minds of those; whose are in the vicinity of these sciences. Past investigations forms the base for further leap forward in the particular direction. Continuous efforts and knowledge-recollection forms groves or passages to move ahead. We need to assimilate the previous know-how and knowledge on the first place before we initiate thought or working in the branches of our concerned. In a way this is a conditioning of the scientific minds. But it is essential because the passages which have been paved are resultant of long tiresome efforts within the frame of cause effect-relations. The subject of Biotechnology and Genetics is expanding its horizons. Most of the branches of biological science are mingling in this particular branch. The recent innovations are touching even to the psyche of mankind. Of course, we should not play God. Because we are still not wise enough to do so. We still want to understand origin, cause, purposes, interdependent, free will (if any?) in the life perpetuating on this planate. But one thing is sure that these two branches of the bioscience have drifted us nearer to God. Hope, this expanded text will illuminate the minds of all who have some interest in the biotechnology and genetics. This text will be of helpful to the students who want to pursue their career in the area of genetics and biotechnology since it contains the short of each topic followed by objective questions.

May 02, 2014 **Authors**
Junagadh

Contents

	Preface	*v*
1.	Introduction to Genetics and Biotechnology	1
2.	Cells, Its Development and Division	25
3.	Mendel and His Discoveries	121
4.	Extensions and Exceptions to Mendels Laws	147
5.	Genetic Mapping and Gene Regulation in Prokaryotes	177
6.	Genetic Mapping and Gene Regulation in Eukaryotes	197
7.	Genetic Transfer in Bacteria and Bacteriophages	223
8.	Sex Determination and Development	231
9.	Chromosomes	239
10.	Variation in Chromosome Structure and Number	267
11.	DNA Structure and Replication	273
12.	Gene Transcription and RNA Modification	395
13.	Translation of mRNA	403
14.	Chemistry of Gene	411
15.	Gene Mutation and DNA Repair Technology	511
16.	Recombination and Transposition at the Molecular Level	549
17.	Recombinant DNA Technology	563
18.	Genome Analysis and Its Uses	597

19. Functional Genomics, Proteomics and Bioinformatics 621

20. Genomics and the Future 637

21. Genetics of Cancer 645

22. Developmental Genetics 659

23. Quantitative Genetics 675

24. Population Genetics 687

25. Evolutionary Genetics 707

26. Relative Reading 721

27. Glossary 735

Chapter – 1

Introduction to Genetics and Biotechnology

A Overview

- Genetics is a branch of biology concerned with heredity and variation.
- Genes are the unit of inheritance and are composed of DNA.
- The effects of genes are noticeable at the molecular, cellular, family, and population levels.
- Genes are DNA sequences that instruct cells to produce particular proteins, which in turn determine traits.
- Genes can exist in more than one form and the variants of a gene arise by mutation.
- An allele is a variant of a gene that alters its expression in a detectable way. Alleles can be dominant or recessive.
- A polymorphism is a particular sequence of DNA that varies in one percent or more of the population.
- Single nucleotide polymorphisms are single base pair sites that differ among individuals.
- Chromosomes consist of genes and associated proteins.
- The human genome consists of 22 pairs of autosomes and one pair of sex chromosomes.

- A karyotype is the photographic catalog of the human chromosomes.
- Specialized cells and tissues arise by differentiation from the stem cells of the early embryo.
- Pedigree diagrams enable recessive and dominant traits to be followed through multiple generations of a family.
- A gene pool is the collection of alleles in a population.
- Comparative genomics is leading to a more detailed and subtle understanding of evolutionary relationships among species.

The Relationship between Genes and Traits

- Living cells are composed of biochemical.
- Each cell contains many different proteins that determine cellular structure and function.
- DNA stores the information for protein synthesis.
- The information within the DNA is accessed during the process of gene expression.
- The molecular expression of genes within cells leads to an organism's outwardly visible traits.
- Inherited differences in traits are due to genetic variation.
- Traits are governed by genes and by the environment.
- During sexual reproduction, genes are passed from parent to offspring.
- The genetic composition of a species evolves over the course of many generations.
- Mendelian traits are determined by single genes.
- Mendelian disorders are generally rare.
- Multifactorial traits are determined by multiple genes and environmental effects.
- Many common illnesses have a multifactorial basis.
- Genetic determinism is the point of view that the environment has little effect on gene expression.

Fields of Genetics

- Transmission genetics explores the inheritance patterns of traits as they are passed from parents to offspring.

- Molecular genetics seeks a biochemical understanding of the hereditary material.
- Population genetics is concerned with genetic variation and its role in evolution.
- Genetics is an experimental science.

Applications of Genetics

Establishing Identity-From Forensics to Rewriting History

- Genetic tools are used in diverse areas such as forensics, establishing paternity, agriculture, and health care.
- DNA typing can exclude an individual from being biologically related to someone else or from having committed a crime.

Health Care

- Genetic diseases differ from infectious diseases in that the recurrence risks are predictable, pre-symptomatic diagnosis using genetic testing is possible, and characteristic frequencies are observed in different populations.
- After identifying underlying genetic abnormalities (mutations), gene therapy can be used to treat and possibly prevent certain genetic disorders.

Agriculture

- Agriculture, both traditional and biotechnological, applies genetic principles.
- Traditional agriculture involves controlled breeding to select now combinations of inherited traits.
- Genetically modified (GM) foods have genes introduced from the genomes of other organisms.

What is Biotechnology?

- Biotechnology is the modern biology that comprises diverse and sophisticated techniques to exploit cellular processes of bacteria, plants and animals to obtain highly useful and novel products. In recent years, it has been proved to be a boon for modern agriculture and industry where its skills are applied to develop and design uniquely improved varieties of crops, new pharmaceutical products, plethora of chemicals, cosmetics, fertilizers, growth enhancers, processed foods, health care aids and environment-related substances.

- Biotechnology is generally divided into three sub fields - red, white and green biotechnology. Red biotechnology deals with genetically altered microorganisms that are used for producing products like insulin and vaccine for medical use. White biotechnology involves the creation of useful chemicals for the industrial sector through organisms like moulds or yeast. This form of biotechnology is also referred to as grey biotechnology. Green biotechnology, also known as agricultural biotechnology, deals with applications related to agriculture.
- Green biotechnology involves the manipulation of plants and animals to produce species that are more environments friendly and productive. Development of varieties of wheat that are disease resistant by cross breeding different types of wheat is an example of green biotechnology.

Traditional Breeding Compared to Biotechnology

- Genetically modified (GM) foods are becoming widely available.
- Genetic modification can mean adding a gene, enhancing an organism's own gene, or suppressing an organism's own gene (i.e. antisense technology).

Similar Steps, Different Degree of Precision

- Biotechnology entails genetic alteration of cells, growing multicellular organisms, breeding to obtain individuals that express the desired trait, and then field testing.
- Biotechnology differs from traditional plant breeding in that it is asexual, more precise, faster, and considers one trait at a time.

Government Regulation of Crops

- The U.S. government regulates GM crops the same way it does conventionally bred crops.
- The FDA evaluates the characteristics of a food, not its origin.
- The exoneration of Starlink corn was not as widely reported as the initial fears.

Biotechnology Provides Different Routes to Solving a Problem

- Mutant selection regenerates plants from cells grown in herbicide to uncover naturally occurring mutations.
- Transgenic technology is used to design an herbicide resistant crop.

Types of Plant Manipulations

Altering Plants at the Gene Level

- Methods of gene transfer include Ti plasmid in dicots and other methods in monocots.
- Genes in the nucleus or chloroplast can be manipulated.
- Many crop species have been engineered to be herbicide or pesticide resistant.
- The nutritional value of numerous crops has been enhanced.
- Root systems of engineered plants, through rhizosecretion, can be used for pharmaceutical production.
- Altering the levels of gene products (promotor enhancement; antisense knockouts) has been widely applied.

Altering Plants at the Cellular Level

- Fusing protoplasts from different types of plants can produce novel variants.
- An explant grows into an undifferentiated callus.
- Depending on the culture conditions the callus can yield plantlets, somatic embryos, or somaclonal variants.

Release of Genetically Modified Organisms to the Environment

Microcosm Experiments

- Genetically altered plants are tested in microcosms in the laboratory to see how they affect natural ecosystems.

Field Tests

- GM organisms are tested in the field to assess survival, fertility, containment, and interactions with other organisms.

Bioremediation

- Bioremediation is the intentional release of organisms that detoxify certain pollutants.
- The enzyme-catalyzed detoxification reactions may be a natural part of an organism's biochemistry or may be added transgenically.
- Organisms used in bioremediation help to detect or detoxify heavy metals, organic wastes or hidden explosives (land mines) in the environment.

Economic, Ecological, and Evolutionary Concerns

- The introduction of GM foods can have economic repercussions because they replace natural sources of substances.
- Measures must be taken to prevent pollen from transferring transgenes to other species.
- Herbicide- or pesticide- resistant crops can encourage persistence of this trait in weeds and pests.
- Use of GM foods varies internationally. Some countries ban GM foods. In the United States the percentage of GM crops produced is growing.

The Impact of Genomics

- The benefits and risks of GM crops are still hotly debated.
- Genomics aids in the identification of useful, as well as harmful, genes in plants.

Multiple Choice Questions

1. Which of the following is a correct definition of genetics?
   (a) The study of transmission of traits from parent to offspring
   (b) The study of genes and traits defined by genes
   (c) The study of DNA
   (d) The study of variation between members of a species
2. Which statement about nucleic acids is not correct?
   (a) They are macromolecules
   (b) They are organic molecules

(c) They are typically long, linear molecules

(d) None; all of these statements are correct

3. What is the main function of DNA?

(a) It stores information for protein synthesis

(b) It can be mutated

(c) It directs the process of protein synthesis

(d) It provides energy for the cell

4. Which statement is correct about Rieger syndrome in humans?

(a) The gene which, when mutant, causes this syndrome is located on the X chromosome

(b) This gene is required for production of an essential blood clotting factor, Factor XI

(c) It is located on the same chromosome as genes which, when mutant, cause Huntington disease and Hepatocellular carcinoma

(d) Rieger syndrome has not been demonstrated to have a genetic cause

5. Which of these might be a valid scientific reason for creating mice that glow green?

(a) We thought they might make popular pets

(b) It served as an effective demonstration that an extra gene could be incorporated into and expressed in most tissues of a model organism

(c) Glowing mice might be effective "flashlights" in search and rescue applications

(d) Glowing mice could be easily distinguished from unmanipulated animals

6. Which field of genetics is the oldest?

(a) Transmission genetics (b) Molecular genetics

(c) Population genetics (d) Developmental genetics

7. Which technique is most likely to be used by a molecular geneticist?

(a) Production of a new mutant allele of an interesting gene

(b) Establishing crosses between genetic variants

(c) Analysis of the amount of genetic variation in a population of individuals

(d) Identification of a previously unknown species

8. Morphs are an example of which level of genetic variation?

(a) Molecular (B) Cellular

(c) Organismal (d) All of the above

9. Which is an example of a study that might be done by a transmission geneticist?

(a) A comparison between many individuals indicates that the "short" variant of a particular gene seems to increase risk for alcoholism

(b) DNA sequencing allows geneticists to determine the difference between the short and long variants of a newly discovered gene

(c) Geneticists are able to clone and express a gene to produce human insulin

(d) A geneticist collects family histories in order to study the mechanism of transmission of a genetic disorder

10. In a particular breed of cats black (B) is dominant and white (b) is recessive. What are the chances of producing white kittens when crossing a Bb male with a bb female?

(a) 0% (b) 25%

(c) 50% (d) 100%

11. What are the chances of producing black kittens from the above cross?

(a) 0% (b) 25%

(c) 50% (d) 100%

12. Which of the following represents an individual that is homozygous for the dominant trait?

(a) bb (b) Bb

(c) BB (d) Aa

13. Which of the following represents an individual that is heterozygous for the dominant trait?

(a) bb (b) Bb

(c) BB (d) AA

14. Two common coat colors seen in Shetland sheepdogs (shelties) are sable and tricolor. Sable dogs B) are blonde and tricolored dogs (b) are black with a white collar and tan eyebrows. Sable is the dominant color; tricolor is recessive. A dog that is heterozygous for coat color (Bb) is called a tri-factored sable and is darker than a dog that is homozygous for sable (BB). What are the possible coat colors in a litter that results from a cross between a homozygous sable and a tri-factored sable?

(a) All sable puppies

(b) All tricolored puppies

(c) Both sable and tricolored puppies

(d) Sable and tri-factored sable puppies

15. What coat colors are possible from a cross between a sable and a tricolored dog?

(a) All sable puppies

(b) All tricolored puppies

(c) Both sable and tricolored puppies

(d) All tri-factored sable puppies

16. A tri-factored sheltie has a litter of puppies. Two of the pups are tricolored and the other two are tri-factored. What color was the father of the puppies?

(a) Sable (b) Tricolored

(c) Tri-factored (d) Either a or b

17. If the tri-factored sheltie had one sable puppy and three tri-factored puppies, then the father would have been what color?

(a) Sable (b) Tricolored

(c) Tri-factored (d) Either a or b

18. Which of the following is a result of nondisjunction of chromosomes during gamete formation?

(a) Cystic fibrosis

(b) Down syndrome

(c) Turner's syndrome

(d) Both b and c

19. A Barr body is a(n)
   (a) Inactivated X chromosome
   (b) Inactivated Y chromosome
   (c) An abnormality seen in the cells of persons with Down syndrome
   (d) Chromosome 21 trisomy complex

20. Mendel chose to work with garden pea plants because
   (a) They are small and easy to grow
   (b) They produce large numbers of offspring and mature quickly
   (c) The reproductive organs of the pea are enclosed within the flower and are self-pollinating
   (d) All of the above

21. In Mendel's experiment, a cross of a heterozygous purple-flowered plant (Ww) with a homozygous white-flowered plant (ww) will produce
   (a) 25% white-flowered offspring
   (b) 100% purple-flowered offspring
   (c) 50% homozygous purple-flowered offspring
   (d) 50% heterozygous purple-flowered offspring

22. In a dihybrid cross of a round, yellow pea that is heterozygous for both factors, with a wrinkled, yellow pea that is homozygous for the wrinkled factor and heterozygous for the color factor, we can predict
   (a) There will be no homozygous round peas
   (b) Half of the peas will be wrinkled
   (c) 25% of the peas will be green
   (d) a, b, and c are correct

23. The condition that results from one allele affecting more than one trait is
   (a) Incomplete dominance (b) Pleiotropy
   (c) Partial dominance (d) Epistasis

24. Which one of the following statements is not true about meiosis?
   (a) Meiosis occurs in reproductive cells
   (b) Meiosis results in four haploid daughter cells

(c) In meiosis, homologous pairs of chromosomes are pulled apart

(d) Meiosis occurs only in animals

25. Crossing over occurs during

(a) Prophase I: (b) Prophase II:

(c) Interphase I: (d) Interphase II:

26. Nondisjunction occurs during

(a) Mitosis (b) Meiosis I

(c) Meiosis II (d) Both b and c

27. Down syndrome results from

(a) The absence of one chromosome in position 21

(b) The presence of an extra chromosome in position 21

(c) The absence of both chromosomes in position 21

(d) Crossing over on the chromosomes in position 21

28. Which of the following is the result of nondisjunction of the sex chromosomes?

(a) Down syndrome (b) Barr body

(c) Turner syndrome (d) Phenylketonuria

29. Which one of the following genetic disorder is carried by a dominant allele?

(a) Huntington's disease (b) Sickle-cell anemia

(c) Tay-Sachs disease (d) Hemophilia

30. Males who inherit a XXY combination due to nondisjunction have the condition called

(a) Down syndrome (b) Klinefelter syndrome

(c) Turner syndrome (d) Hemophilia

31. The science of genetics is concerned with

(a) The transmission of traits

(b) The distribution of heritable characteristics in populations

(c) The expression of genetic information

(d) All of the above are correct

32. Who lower the energy needed to initiate a chemical reaction in a living cell?

(a) Enzymes (b) Nucleotides

(c) Carbohydrates (d) Proteins

33. Cytogenetics is related to all the following except

(a) Karyotype (b) Meiosis

(c) Chromosome behavior (d) Pedigree analysis

34. All of the following are examples of positive eugenic techniques except

(a) Selection of embryos free of disease-causing genes

(b) Outlawing reproduction by known criminals

(c) Allowing only Nobel prize winners to be egg or sperm donors for in-vitro fertilization

(d) Selection of egg or sperm cell with dominant genes for muscle strength and coordination

35. Identify the correct order of organization of genetic material, from largest to smallest.

(a) Genome, chromosome, gene, nucleotide

(b) Gene, chromosome, nucleotide, genome

(c) Chromosome, gene, genome, nucleotide

(d) Chromosome, genome, nucleotide, gene

36. Which is a significant source of genetic recombination during gamete production?

(a) Mutation (b) Crossing-over

(c) Nondisjunction (d) Controlled assortment

37. Human genes are not always readily produced by bacteria. Which of the following modifications can be useful?

(a) Inserting a cDNA sequence rather than the entire gene

(b) Inserting a synthetic DNA sequence if the coding region is very short

(c) Production of a hybrid or fusion protein

(d) All of these

(e) None of these

38. Prior to production of recombinant human insulin, diabetic patients could use insulin isolated from cattle. Which of the listed items would not be a drawback to this procedure

(a) Insulin from cattle is slightly different from human insulin, potentially leading to allergic reactions

(b) Members of some religions consider cattle sacred and thus are unable to use insulin isolated from cattle

(c) Infectious diseases may be transferred in the isolated insulin

(d) Despite the relatively stable supply of cattle pancreases from slaughter houses, insulin isolated from cattle is still relatively expensive and in somewhat limited supply

(e) None of these; all are drawbacks to use of cattle insulin

39. Bacteria that are unable to synthesize chemicals that allow formation of ice crystals have been engineered. Commercial fruit farmers can spray this bacterial strain on their fruit plants, with the idea that they will compete with unmodified bacteria and reduce frost damage. This is an example of:

(a) Bioremediation

(b) Biological control

(c) Biodegradation

(d) Bioeradication

40. Which of this is an example of biodegradation?

(a) A strain of bacteria chemically captures heavy metal ions into chemical compounds

(b) A population of microorganisms metabolized toluene from a massive leak of jet fuel in Hanahan, SC into carbon dioxide and other small molecules, reducing toluene levels from over 5,000 parts per billion to undetectable levels

(c) Naturally occurring microorganisms in Picatinny, NJ and other sites are able to dechlorinate common industrially produced contaminants such as trichloroethane, generating a nontoxic compound

(d) All of these are biodegradation

(e) None of these are biodegradation

41. In gene knockout technology:

(a) A gene is inserted by nonhomologous recombination

(b) A mutant gene is replaced by a functional copy

(c) A functional gene is replaced by a mutant copy

(d) A functional gene is inserted in addition to the mutant copy

42. To generate a gene knockout, which of the following is required?

(a) A method for screening to verify insertion of the cloned gene

(b) Regions of homology between the cloned gene and the target gene

(c) A method to verify that the ends of the cloned gene were removed due to homologous recombination rather than nonhomologous recombination

(d) All of these

(e) None of these

43. Some chimeric animals may not produce offspring carrying the gene replacement, while other chimeric animals with the same gene replacement can easily produce offspring carrying the gene replacement. Why?

(a) Mosaic animals are unable to produce offspring

(b) The gene replacement prevents correct gamete formation

(c) Some chimeras have a germline entirely composed of cells from the blastocyst rather than the modified cells

(d) Some chimeras have a germline entirely composed of modified cells rather than cells from the blastocyst

44. Which of the following explains why production of transgenic plants is easier than production of transgenic animals?

(a) Plant cells can grow in cell culture

(b) Plant cells have a lower number of potentially lethal genes

(c) Plant cells are totipotent

(d) Production of mutant plants poses less ethical dilemmas than production of mutant animals

(e) None of these

45. Which of the following groups of individuals could be considered to be clones?

(a) Fraternal twins

(b) Several strawberry plants that grew from runners from the same parent

(c) Littermates from an inbred strain of mice

(d) All of these

(e) None of these

46. What process causes shortening of telomeres during cell division?

(a) DNA synthesis

(b) Exonuclease removal of primers

(c) Ligation of DNA fragments

(d) Telomerase activity

47. Identify the best description of a stem cell.

(a) A cell that has differentiated into a specialized cell type

(b) A cell that is capable of division

(c) A cell that remains undifferentiated

(d) A cell that can grow in cell culture

48. Which of these is an example of an adult stem cell?

(a) A red blood cell

(b) A neuron

(c) A cell from a blastocyst

(d) Skin cells that produce new epithelial layers

49. Which of these uses of transgenic technology is not directly relevant to agricultural applications?

(a) Production of a pharmaceutical protein in goat's milk

(b) Production of salmon that grow all year round instead of just in warm weather

(c) Production of soybean plants that are resistant to herbicides

(d) Production of "super mice" that carry the human growth hormone gene

50. Which of these is an accurate description of gene therapy?
    (a) A form of site directed mutagenesis
    (b) Introduction of cloned genes into living cells to cure disease
    (c) Cloning of a disease gene in order to understand the disease
    (d) All of these
    (d) None of these

51. Diseases amenable to treatment by gene therapy:
    (a) Must be relatively mild
    (b) Must be caused by a defect in cells that are clinically accessible
    (c) Must have other types of treatments available in case gene therapy fails
    (d) All of these
    (e) None of these

Fill in the blanks

1. A gene ____________
    (a) Can be described as a unit of heredity
    (b) Contains information to produce a particular protein
    (c) Affects traits at the cellular level
    (d) All of the above

2. Haploid cells in the human body ____________
    (a) Are a typical example of human cells
    (b) Are caused by allelic variation
    (c) Are gametes, which allow sexual reproduction
    (d) Contain 46 chromosomes

3. In order to ensure he had true-breeding plants for his experiments Mendel ____________
    (a) Cross-fertilized each variety with another
    (b) Let each variety self-fertilize for several generations
    (c) Removed the female parts of the flowers to prevent cross-fertilization

(d) Removed the male parts of the flowers to prevent cross-fertilization
(e) Used numerous varieties of garden peas

4. When Mendel crossed two varieties of true-breeding plants the F1 generation were _____
(a) Heterozygous
(b) Of a phenotype midway between the parent plants
(c) Phenotypically similar to the parent plant with the dominant trait
(d) Both a and c

5. An Rh-negative mother pregnant with an Rh-positive child in her first pregnancy will _______________
(a) Produce "anti-Rh" antibodies that will cause clotting of her fetuses blood cells
(b) Produce "anti-Rh" antibodies that will cause clotting of any future Rh-positive fetuses blood cells
(c) Experience a life-threatening allergic response
(d) Develop tolerance to the Rh-antigen

6. The significance of Morgan's experiments was that _______________
(a) It presented the first clear evidence that genes determining Mendelian traits do reside on chromosomes
(b) It illustrated the effects of polygenic traits on the inheritance of sex-linked traits
(c) Chromosomal reduction through meiosis was proven
(d) A method to control economically important fruit flies resulted

7. Sickle-cell anemia __________
(a) Poor blood circulation due to abnormal hemoglobin molecules
(b) Brain fails to develop in infancy
(c) Failure of blood to clot
(d) Failure of chloride ion transport mechanism

8. Cystic fibrosis is _____________
(a) Poor blood circulation due to abnormal hemoglobin molecules
(b) Brain fails to develop in infancy

(c) Failure of blood to clot

(d) Failure of chloride ion transport mechanism

9. Hemophilia is ___________

(a) Poor blood circulation due to abnormal hemoglobin molecules

(b) Brain fails to develop in infancy

(c) Failure of blood to clot

(d) Failure of chloride ion transport mechanism

10. Phenylketonuria is ___________

(a) Poor blood circulation due to abnormal hemoglobin molecules

(b) Brain fails to develop in infancy

(c) Failure of blood to clot

(d) Failure of chloride ion transport mechanism

11. Hemophilia involving a lack of blood protein factors VIII and IX are ___________

(a) A dominant trait that usually only develops in individuals in their 30s

(b) A recessive trait inherited on autosomes

(c) A sex-linked trait

(d) Common in royal families due to several mutations in their genetic material

12. Ultrasound is ___________

(a) Cells are removed and grown in culture

(b) Technique that can be used as early as eight weeks in pregnancy and yields rapid results

(c) Used to examine fetus for major abnormalities

(d) None of the above

13. Amniocentesis is ___________

(a) Cells are removed and grown in culture

(b) Technique that can be used as early as eight weeks in pregnancy and yields rapid results

(c) Used to examine fetus for major abnormalities

14. Chorionic villi sampling means__________

(a) Cells are removed and grown in culture

(b) Technique that can be used as early as eight weeks in pregnancy and yields rapid results

(c) Used to examine fetus for major abnormalities

(d) None of the above

15. Garden peas were a good subject for Mendel's genetic studies because __________

(a) They are available in many varieties

(b) They are small, easy to grow plants

(c) It is easy to control their fertilization

(d) All of the above

16. To conduct genetic experiments it is necessary to begin with subjects that are __________

(a) Haploid (b) Heterozygous

(c) True-breeding (d) Previously unstudied

17. During your plant research project, you cross a blue-flowered variety with a yellow-flowered variety. The F1 generation consists of all yellow-flowered plants. The yellow-flowered plants are probably __________

(A) Recessive (b) Dominant

(c) Incomplete dominant (d) Hybrid

18. In the experiment mentioned above the purple-flowered plants are probably __________

(a) Recessive (b) Dominant

(c) Incomplete dominant (d) Hybrids

19. A cross between two individuals results in a ratio of 9331 of four possible phenotypes. This is an example of a(n) __________

(a) Monohybrid cross

(b) Dihybrid cross

(c) Testcross

(d) Incomplete dominance

20. When a trait is controlled by the interaction of the products of two genes it is referred to as ___________

(a) Dominance (b) Pleiotropy

(c) Epistasis (d) Incomplete dominance

21. The tri-factored shelties mentioned earlier are an example of ___________

(a) Dominance (b) Pleiotropy

(c) Incomplete dominance (d) Aneuploidy

22. A DNA molecule that is complexed with proteins into a rodlike structure is called a(n) __

(a) Chromophore (b) Chromosome

(c) Genophore (d) Genosome

23. The end result of meiosis in human males is ___________

(a) Two diploid sperm cells

(b) Four diploid sperm cells

(c) Two haploid sperm cells

(d) Four haploid sperm cells

24. The gap in Darwin's theory of evolution was his inability to ___________

(a) Observe variation

(b) Explain the basis of evolution

(c) Explain migration

(d) Observe large populations

25. ___________ is (are) a significant source of genetic recombination during gamete production.

(a) Mutation (b) Crossing-over

(c) Nondisjunction (d) Controlled assortment

True/False

1. When doing dihybrid crosses Mendel found the F2 generation showed a 931 ratio of phenotypes.

(a) True (b) False

2. Mendel's second law is often stated as genes located on different chromosomes are inherited independently each other.

   (a) True (b) False

3. The human genome is distributed over 24 chromosomes.

   (a) True (b) False

4. Development of genetic technologies are often rejected at first even though they may be used to create great advances in medicine, agriculture, and many other fields.

   (a) True (b) False

5. Gene expression can be defined as the process of transcription and translation.

   (a) True (b) False

6. DNA is the most important macromolecule for correct cellular function.

   (a) True (b) False

7. The major difference between different cells in the same organism is that they have different DNA

   (a) True (b) False

8. All sexually reproducing organisms are diploid

   (a) True (b) False

9. Genetic composition of a species is stable over time.

   (a) True (b) False

10. Most traits are controlled by genetic variation.

    (a) True (b) False

11. When doing dihybrid crosses Mendel found the F2 generation showed a 931 ratio of phenotypes.

    (a) True (b) False

12. Mendel's second law is often stated as genes located on different chromosomes are inherited independently each other.

    (a) True (b) False

13. In polygeny, one gene affects many traits.

    (a) True (b) False

14. In pleiotrophy, one gene affects many traits.

(a) True (b) False

15. Mendel called the factors that determined the inherited characteristics "genes."

(a) True (b) False

16. Phenotype describes the actual alleles received from both parents.

(a) True (b) False

17. A chromosome is a single strand of DNA in a compact formation.

(a) True (b) False

18. Sperm and egg cells have a diploid number of chromosomes.

(a) True (b) False

19. Accidental changes in genes are called mutations.

(a) True (b) False

20. Since hemophilia is a recessive allele, this condition will not show up in individuals carrying just one allele with this trait.

(a) True (b) False

21. A picture of an individual's particular array of chromosomes is called a pedigree.

(a) True (b) False

22. Individuals who are heterozygous for sickle-cell anemia receive a benefit by not being as susceptible to malaria

(a) True (b) False

23. Gene therapy by gene addition is best able to treat diseases in which a gene function is missing.

(a) True (b) False

24. Recombinant microorganisms are fundamentally different from nonrecombinant strains.

(a) True (b) False

25. Transgenic organisms contain recombinant DNA integrated into their genome

(a) True (b) False

26. The ability to grow embryonic stem cells in the laboratory has made gene replacement technology in mammals feasible.

(a) True (b) False

27. Knockout mice can serve as model organisms for study of human disease.

(a) True (b) False

Answers

Multiple Choice Question

1	2	3	4	5	6	7	8	9	10	11	12	13	14	15
B	C	A	C	B	A	A	D	D	C	C	C	B	D	D
16	17	18	19	20	21	22	23	24	25	26	27	28	29	30
C	C	D	A	D	D	D	B	B	A	D	B	C	A	B
31	32	33	34	35	36	37	38	39	40	41	42	43	44	45
D	A	D	B	A	B	D	C	B	B	C	D	C	C	B
46	47	48	49	50	51									
B	C	D	D	B	B									

Fill in the blanks

1	2	3	4	5	6	7	8	9	10	11	12	13	14	15
D	C	B	D	B	A	A	D	C	B	C	C	A	B	D
16	17	18	19	20	21	22	23	24	25					
C	B	A	B	C	C	B	D	B	D					

True/False

1	2	3	4	5	6	7	8	9	10	11	12	13	14	15
B	A	A	A	A	B	B	B	B	B	B	A	B	A	B
16	17	18	19	20	21	22	23	24	25	26	27			
B	A	B	A	B	B	A	A	B	A	A	A			

❑❑❑

Chapter – 2

Cells, Its Development and Division

Cells are the Working Units of Life

- With the possible exception of viruses, every form of life on Earth either is a cell or composed of cells. All cells come into existence through the activity of other cells.

All Cells are either Prokaryotic or Eukaryotic

- All cells can be classified as prokaryotic or eukaryotic. All plants, animals, fungi, and protists are either single eukaryotic cells or are composed of eukaryotic cells. Prokaryotic cells are either bacteria or archaea. Eukaryotic cells have most of their DNA contained in a membrane-lined nucleus, whereas prokaryotic cells do not have a nucleus. Eukaryotic cells also have more specialized structures, called organelles, than do prokaryotic cells. All prokaryotes are single-celled, whereas many eukaryotes are multi-celled.

The Eukaryotic Cell

- There are five principal components to the eukaryotic animal cell: the nucleus, other organelles, the cytosol, the cytoskeleton, and the plasma membrane. Organelles are "tiny organs" within the cell that carry out specialized functions, such as energy transfer and material recycling. The cytosol is the fluid in which these organelles are immersed (The cytoplasm is the region of the cell inside the plasma membrane but outside the nucleus.) The cytoskeleton is composed of several groups of proteins that give the cell support and facilitate transportation of cellular elements. The plasma membrane is the chemically active outer boundary of the animal cell. Plant cells have a plasma membrane that has, outside of it, a cell wall.

A Tour of the Animal Cell: Along the Protein Production Path

- Information for the construction of proteins is contained in the DNA located in the cell nucleus. This information is copied onto informational "tape" of messenger RNA that departs the cell nucleus through nuclear pores and goes to the sites of protein synthesis, structures called ribosomes. Many ribosomes first migrate to, and then embed in, a series of membrane sacs called the rough endoplasmic reticulum. Polypeptide (or amino acid) chains that will become proteins are dropped from ribosomes into the internal spaces of the rough endoplasmic reticulum, where these chains fold up and undergo editing. Some ribosomes are not embedded in rough endoplasmic reticulum but instead remain free-standing in the cytosol.
- Materials move from one structure to another in the cell via the endomembrane system, in which a piece of membrane, with proteins or other materials inside, can bud off from one organelle, move through the cell, and then fuse with another membrane-lined structure. Membrane-lined structures that carry cellular materials are called transport vesicles.
- Transport vesicles move proteins that are being produced from the rough endoplasmic reticulum to the Golgi complex, where the proteins are processed further and marked for shipment to appropriate cellular locations.

Outside the Protein Production Path: Other Cell Structures

- The smooth endoplasmic reticulum is a network of membranes that functions to synthesize lipids and to detoxify potentially harmful substances. Lysosomes are organelles that break down material coming into the cell (such as food) or worn-out cellular structures, returning the components of these structures to the cytoplasm for further use. Lysosomes will hold onto, or expel outside the cell, material that cannot be recycled inside the cell. Mitochondria function to extract energy from food and to transform this energy into a chemical form the cell can use.

The Cytoskeleton: Internal Scaffolding

- Cells have within them a web of protein strands, called a cytoskeleton, that provides the cell with structure, facilitates the movement of materials inside the cell, facilitates cell movement, and helps cells capture prey.
- There are three principle types of cytoskeleton elements. Ordered by size, going from smallest to largest in diameter, they are microfilaments, intermediate filaments, and microtubules. Microfilaments are made of the protein actin. They help the cell move and capture prey. Actin works with the protein myosin to bring about muscle contraction. Intermediate filaments provide support and structure to the cell. Microtubules play a structural role in cells and facilitate the movement of materials inside the cell by serving as transport "rails."
- Cilia and flagella are extensions of cells composed of microtubules. Cilia extend from cells in great numbers, serving to move the cell or to move material around the cell. By contrast, one—or at most a few—flagella extend from cells that have them. The function of flagella is cell movement.

The Plant Cell

- Plant cells have almost all the structures same as found in animal cells. Plant cells also have a cell wall, a large central vacuole, and organelles called plastids, one variety of which is the chloroplasts that are the sites of photosynthesis. The cell wall gives the plant structural strength and helps to regulate the intake and retention of water.

Cell Communication

- Cells are able to communicate with each other through special structures: channels called plasmodesmata in plants and channels called gap junctions in animal cells.

The Components of Cells

Chemical Constituents of Cells

- The human body is composed of trillions of cells that share common features.

- Cells are constructed from numerous small molecules and macromolecules.
- The building blocks of cells include carbohydrates (simple sugars and polysaccharides), amino acids and proteins, lipids (fats and oils), and nucleic acids (nucleotides, DNA, RNA) .
- Enzymes are proteins that catalyze the multitude of biochemical reactions that occur in the cell.
- The proteome is the set of proteins that a particular cell type can produce.
- Expression of different subsets of genes in different cells (and therefore production of different proteomes) gives rise to over 250 specialized cell types.

Organelles

- Organelles represent the compartments (and unique microenvironments) in the cell and are involved in a variety of functions (division of labor).
- The nucleus (the storehouse of the majority of DNA in the cell) has a double membrane and nuclear pores, which allow macromolecular traffic in and out of the nucleus.
- The rough endoplasmic reticulum (ER), smooth ER, and Golgi body function as a membrane network for the synthesis of proteins and lipids that are targeted for delivery to the plasma membrane, organelles, or for secretion. Secretions bud off in vesicles and leave the cell.
- A mitochondrion has a double membrane whose inner folds carry enzymes that catalyze reactions that extract energy from nutrients.
- Lysosomes contain enzymes that degrade cellular debris.
- Peroxisomes house enzymes that detoxify certain substances, break down lipids, and synthesize bile acids.

The Cell Membrane

- The cell membrane (composed of a bilayer of phospholipids) and the underlying cytoskeleton (microtubules and microfilaments) give a cell its specific architecture. The proteins (or glycoproteins) residing in the cell membrane function as enzymes, signal transduction receptors, transport proteins and cell adhesion proteins.

The Cytoskeleton

- The major cytoskeleton components include microtubules (tubulin), microfilaments (actin), and intermediate filaments (various families of intermediate filament proteins).
- Microtubules also serve at the foundation of cilia and flagella. Spherocytosis is a heredity defect in the cytoskeleton of the red blood cell. An abnormal cytoskeletal protein, ankyrin, lies beneath the cell membrane, and causes the red blood cells to balloon out, blocking narrow blood vessels in organs.

Cell-Cell Interactions

Signal Transduction

- Cellular activities, including secretion, signal transduction, cell division, cell adhesion, and the function of the cytoskeleton, affects each other.
- In signal transduction, cell surface receptors get information from extracellular first messengers and pass the signals to second messengers inside the cell, which responds. Neurofibromatosis type I (NF1) is caused by faulty signal transduction. Nerve cells beneath the skin inappropriately transmit a growth factor signal, triggering cell division. A tumor forms.

Cell Adhesion

- Cell adhesion is a precise sequence of interactions between cell surface proteins that join cells.
- In inflammation, cell adhesion molecules (CAMs) guide white blood cells to injury sites.
- Leukocyte-adhesion deficiency and cancer are disorders that can result from abnormal cell adhesion.

Stem Cells and Cell Specialization

- Cellular differentiation requires differential gene expression and results in the formation of specialized cells with distinctive characteristics and functions (i.e. erythrocytes, lymphocytes, heart muscle cell, fat cell, nerve cell etc.).

- The zygote or fertilized egg is totipotent, capable of producing any cell type.
- Cells of the early embryo are pluripotent, meaning they can differentiate into some but not all types of cells.
- Later in development, pluripotent stem cells give rise to progenitor cells that are committed to a particular pathway.
- Stem cells are non-specialized cells that retain the potential to differentiate and enable a tissue to grow or repair itself.
- Regenerative medicine uses stem cells to replace or rejuvenate injured or diseased tissue.

Viruses and Prions-Not Cells, but Infectious

Viruses

- Viruses are infectious agents that disrupt normal cellular function.
- A virus contains genetic material in the form of nucleic acid (DNA or RNA) wrapped in a protein coat.
- Viruses use the metabolic and protein-synthetic machinery of cells to replicate.

A Prion - One Protein that Takes Two Forms

- Prions do not contain genetic material but are abnormally folded glycoproteins that cause a rare type of disease called spongiform encephalopathy.
- A normally folded form of the prion protein (PrP) naturally occurs in the brain of mammals.
- Prions cause disease by interacting with normal prion proteins and coaxing them to adopt the infectious (abnormally folded) form.

Cell Division and Death

- Cell division takes place because old cells die and because there are many instances in which an organism needs quantities of new cells above "replacement" level. Cell division includes the duplication of DNA (replication); the apportioning of the copied DNA into two quantities in a parent cell (mitosis); and the physical splitting of this parent cell into two daughter cells (cytokinesis). In DNA replication, the two strands of the double helix unwind, after which each single strand serves as a template for construction of a second, complementary strand of DNA) The result is a doubling of the original quantity of DNA.

DNA is Packaged in Chromosomes

- DNA comes packaged in units called chromosomes. These chromosomes are composed of DNA and its associated proteins—a combined chemical complex called chromatin. Chromosomes exist in an unduplicated state until such time as DNA replicates, prior to cell division. DNA replication results in chromosomes that are in duplicated state, meaning one chromosome composed of two identical sister chromatids.
- Chromosomes in human beings (and many other species) come in matched pairs, with one member of each pair inherited from the mother, and the other member of each pair inherited from the father. Such homologous chromosomes have closely matched sets of genes on them, though many of these genes are not identical. A given paternal chromosome may have genes that code, for example, for different hair or skin color than the counterpart genes on the homologous maternal chromosome. Human beings have 46 chromosomes—22 matched pairs and either a matched pair of X chromosomes (in females) or an X and a Y chromosome (in males).
- Cell division fits into the larger framework of the cell cycle, meaning a repeating pattern of growth, genetic replication, and cell division. The cell cycle has two main phases. The first is interphase, in which the cell carries out its work, grows, and duplicates its chromosomes in preparation for division. The second is mitotic phase, in which the duplicated chromosomes separates and the cell splits in two.

The Cell Cycle

- The cell cycle consists of interphase, when a cell is not dividing; During interphase, proteins, lipids and carbohydrates are produced in the G1 phase; DNA and proteins are made during S phase; and more proteins are produced in G2.
- Replicated chromosomes have two sister chromatids attached at their centromeres. In mitotic prophase, replicated chromosomes condense, a spindle forms and the nuclear membrane breaks down.

- In metaphase, chromosomes align down the center of the cell (equator or metaphase plate).
- In anaphase, centromeres part, one chromatid from each pair pulled to opposite ends of the cell. In telophase, the cell pinches in the middle (cytokinesis) and the two new cells separate.
- The cell cycle is tightly controlled and regulated at several "checkpoints."
- A cellular clock that limits the number of divisions is based on shrinking telomeres.
- Crowding, hormones and growth factors are extracellular influences on mitosis.
- Within cells, kinases and cyclins activate the genes whose products carry out mitosis.
- Non-dividing cells may become arrested during interphase and enter a quiescent phase (G0).

Apoptosis

- Mitosis (cell division) and apoptosis (cell death) are continuous processes that occur in a series of steps and are both initiated by signals in the extracellular environment.
- The balance between cell division and death maintains tissues in growth, development, and repair.
- In prenatal development coordination of these processes sculpts body form. After birth, mitosis and apoptosis protect and maintain the body.
- Disruption of the balance between cell division and cell death can lead to cancer or other disorders.

Mitosis and Cytokinesis

- There are four stages in mitosis: prophase, metaphase, anaphase and telophase. The essence of the process is that duplicated chromosomes line up along an equatorial plane of the parent cell, called the metaphase plate, with the sister chromatids that make up each duplicated chromosome lying on opposite sides of the plate. Attached to fibers called microtubules, the sister chromatids are then pulled apart, to opposite poles of the parent cell. Once cell division is complete, sister chromatids that once formed a single chromosome will reside in separate daughter cells, with each sister chromatid now functioning as a full-fledged chromosome.

- Cytokinesis in animal cells works through a ring of protein filaments that tightens at the middle of a dividing cell. Membranes on the portions of the cell being pinched together then fuse, resulting in two daughter cells.

Variations in Cell Division

- Because of their cell wall, plant cell must carry out cytokinesis differently from animal cell. The plant's solution is to grow new cell walls and plasma membranes near the metaphase plate, thus dividing the parent cell into two daughter cells. Prokaryotes such as bacteria employ a process called binary fission: They double their single, circular chromosome, with the two resulting chromosomes attaching to different sites on the plasma membrane. Then an outgrowth of plasma membrane and cell wall, called a septum, begins growing from opposite sides of the cell, in between the two chromosomes. When the two septum extensions join in the middle, they divide the one cell into two.

Meiosis

- In human beings, nearly all cells have paired sets of chromosomes, meaning these cells are diploid Meiosis is the process by which a single diploid cell divides to produce four haploid cells—cells that contain a single set of chromosomes.
- The haploid cells produced through meiosis are called gametes. Female gametes are eggs; male gametes are sperm. They are the reproductive cells of human beings and many other organisms.
- When the haploid sperm and haploid egg fuse, a diploid fertilized egg (or zygote) is produced, setting into development of a new of organism.

The Steps in Meiosis

- In meiosis, there is one round of chromosome replication, followed by two rounds of cell division. There are two primary stages to meiosis, meiosis I and meiosis II. In meiosis I, chromosome duplication is followed by a pairing of homologous chromosomes with one another, during which time they exchange reciprocal sections of themselves. Homologous chromosome pairs then line up at the metaphase plate – one member of each pair on one side of the plate, the other member on the other side.

- These homologous pairs are then separated, in the first round of cell division, into separate daughter cells. In meiosis II, the chromatids of the duplicated chromosomes are separated into separate daughter cells.

Significance of Meiosis?

- Meiosis generates genetic diversity by ensuring that the gametes it gives rise to will differ from one another. In this, it is unlike regular cell division, or mitosis, which produces daughter cells that are exact genetic copies of parent cells.
- Meiosis generates genetic diversity in two ways. First, in prophase I of meiosis, homologous chromosomes pair with each other and, in the process called crossing over or recombination, exchange reciprocal chromosomal segments with one another. Second, in metaphase I of meiosis, there is a random alignment or independent assortment of maternal and paternal chromosomes on either side of the metaphase plate. This chance alignment determines which daughter cell each chromosome will end up in.
- The genetic diversity brought by meiosis and sexual reproduction is responsible, to a significant extent, for the great diversity of life-forms seen in the living world today. Evolution is spurred by differences among offspring and meiosis and sexual reproduction ensure such differences. By contrast, asexual reproduction, as is seen in bacteria and other organisms, generally produces organisms that are exact genetic copies, or clones, of the parental organism.

Meiosis and Sex Outcome

- Human females have 23 matched pairs of chromosomes – 22 autosomes and two X chromosomes. Human males have 22 autosomes, one X chromosome and one Y chromosome. Each egg that a female produces has a single X chromosome in it. Each sperm that a male produces has either an X or a Y chromosome within it. If a sperm with a Y chromosome fertilizes an egg, the offspring will be male. If a sperm with an X chromosome fertilizes the egg, the offspring will be female.

Multiple Choice Questions

1. Which of the following statements concerning prokaryotes and eukaryotes is probably valid?
   (a) Prokaryotes are all unicellular, while eukaryotes are all multicellular
   (b) Mitochondria are about the same size as a prokaryotic cell
   (c) The nuclei of eukaryotes are descended from prokaryotes
   (d) Prokaryotes are not a very successful group of living organisms, while eukaryotes have been very successful
2. Which of the following cell structures is NOT an extension of the cell's cytoskeleton?
   (a) Flagella
   (b) Cilia
   (c) Pseudopodia
   (d) All of the above ARE extensions of the cell's cytoskeleton
3. Which of these organism usually requires oxygen to live?
   (a) Prokaryotes
   (b) Both prokaryotes and eukaryotes
   (c) Eukaryotes
   (d) None of the above
4. Which were probably the first living cells to appear on Earth?
   (a) Both prokaryotes and eukaryotes
   (b) Prokaryotes
   (c) Eukaryotes
   (d) None of the above
5. Which of these organism can be unicellular?
   (a) Both prokaryotes and eukaryotes
   (b) Prokaryotes
   (c) Eukaryotes
   (d) None of the above

6. Which of the following best summarizes the movement of a protein molecule from the time it is formed until it is secreted from the cell?
   (a) Rough ER —> ribosome —> transport vesicle —> mitochondria —> plasma membrane
   (b) ribosome —> rough ER —> mitochondria —> Golgi body —> transport vesicle —> plasma membrane
   (c) Ribosome —> rough ER —> Golgi body —> transport vesicle —> plasma membrane
   (d) Ribosome —> smooth ER —> nucleoli —> transport vesicle —> plasma membrane
7. Many antibiotics kill bacteria because they hamper the function of prokaryotic ribosomes. Which of the following cellular processes is directly altered by antibiotics?
   (a) Intracellular digestion
   (b) Production of mRNA
   (c) Synthesis of proteins
   (d) Tagging of proteins with molecular markers
8. The cells of the pancreas secrete large amounts of insulin into the bloodstream. Where these proteins are synthesized?
   (a) Lysosomes (b) Golgi apparatus
   (c) Smooth ER (d) Rough ER
9. The membrane of which organelle is perforated with pores that regulate the passage of large molecules such as mRNA?
   (a) Mitochondrion (b) Golgi apparatus
   (c) Nucleus (d) Rough ER
10. Which organelles are very prevalent in cells that secrete steroid sex hormones such as cells of the ovaries and testes?
    (a) Mitochondria (b) Smooth ER
    (c) Rough ER (d) Golgi bodies
11. The primary function of the nucleolus is to ...
    (a) Protect the DNA and produce lipids
    (b) Produce energy for the cell

(c) Protect the DNA and produce ribosomes

(d) Protect proteins and make them

12. The organelle that is responsible for sorting and shipping proteins to the appropriate places in the cell is the ...

(a) Rough endoplasmic reticulum (b) Golgi apparatus

(c) Lysosome (d) Mitochondrion

13. The organelle responsible for energy production is the ...

(a) Golgi apparatus

(b) Ribosome

(c) Smooth endoplasmic reticulum

(d) Mitochondrion

14. Which of the following organelle is NOT found in plant cells?

(a) Mitochondrion

(b) Nucleus

(c) Rough endoplasmic reticulum

(d) Lysosome

15. Which of the following cell structure is NOT found in animal cells?

(a) Cytoskeleton

(b) Smooth endoplasmic reticulum

(c) Ribosome

(d) Cell wall

16. Which of the following organelle contains digestive enzymes?

(a) Lysosome (b) Nucleus

(c) Chloroplast (d) Mitochondrion

17. What is the primary reason that most cells are small?

(a) To allow for faster movement of unicellular organisms

(b) To allow for rapid cell-to-cell communication

(c) So that unicellular organisms can live undetected

(d) To maximize the surface-area-to-volume ratio

18. Which of these molecules is a product of photosynthesis that occurs inside plant chloroplast?

(a) Oxygen (b) Water
(c) Methane (d) Carbon dioxide

19. Which of these cytoskeletal filament is primarily responsible for anchoring the organelles inside the cell's cytoplasm?

(a) Microtubules (b) Microfilaments
(c) Cilia (d) Intermediate filaments

20. In the following list, which is the smallest structure?

(a) Virus (d) Protein
(c) Bacterium (d) Plant cell

21. Which of the following statements concerning membrane protein is incorrect?

(a) They often facilitate diffusion of molecules across the membrane that otherwise would diffuse too slowly to be of use to the cell

(b) They may be receptor proteins that bind specific molecules from the surrounding solution, which triggers endocytosis (i.e., receptor-mediated endocytosis)

(c) They often require ATP to actively transport materials across the membrane against a concentration gradient

(d) They are usually not particular about what types of chemicals they will allow to cross the membrane

22. Refer to the illustration. Which molecule represents a glycoprotein that may function as a cellular ID tag, allowing cells of the immune system to recognize a cell as native to the body?

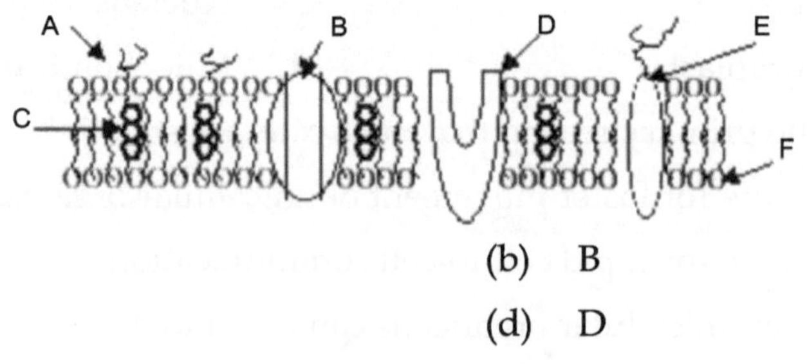

(a) A (b) B
(c) C (d) D
(e) E (f) F

23. Refer to the illustration. Which molecule may function in facilitated diffusion?

(a) A (b) B

(c) C (d) D

(e) E (f) F

24. Refer to the illustration. Which molecule is a polysaccharide attached to a phospholipid?

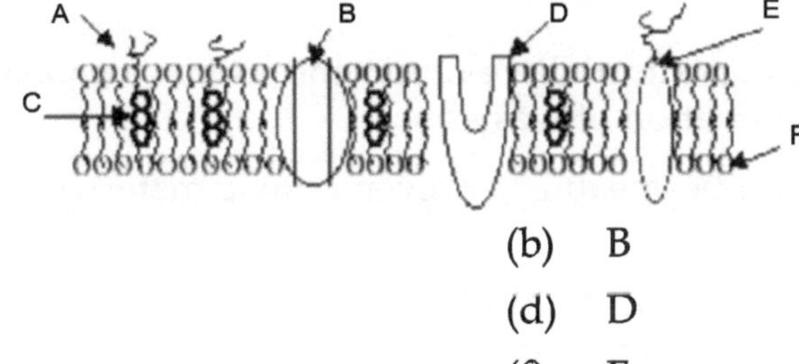

(a) A (b) B

(c) C (d) D

(e) E (f) F

25. Refer to the illustration. Which molecule might serve as a binding site for a hormone, thereby eliciting a response in the cell?

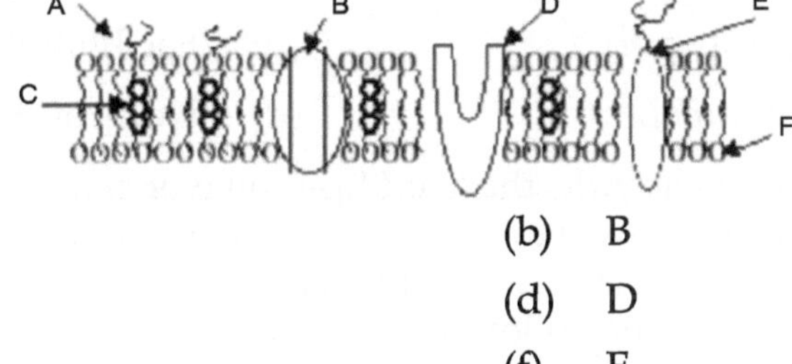

(a) A (b) B

(c) C (d) D

(e) E (f) F

26. Refer to the illustration. Which molecule works to keep the membrane at optimal fluidity?

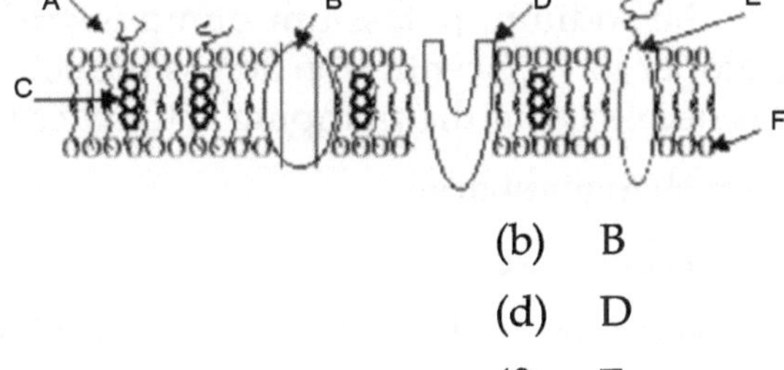

(a) A (b) B

(c) C (d) D

(e) E (f) F

27. Refer to the illustration. Which molecule is both hydrophobic and hydrophilic and aggregates as a bilayer to form the "fabric" of the membrane?

(a) A (b) B

(c) C (d) D

(e) E (f) F

28. What is the significance of selective permeability to biological membranes?

(a) Selective permeability permits the selective uptake of nutrients and the elimination of wastes

(b) Selective permeability prevents toxic materials from entering the cell

(c) Selective permeability allows cells to concentrate particular ions on either side of the membrane

(d) All of the above are correct

29. The movement of molecules across a membrane against their concentration gradient, at the expense of ATP energy, is called ...

(a) Active transport (b) Facilitated diffusion

(c) Passive transport (d) None of the above

30. Just as with most liquids, the fluid lipid bilayer can solidify and freeze at very cold temperature. However, there are certain kind of fish that live in very cold water. The cell membrane of these fish are likely to have more of which of the following membrane components that occupies warmer waters?

(a) Cholesterol (b) Phospholipids

(c) Glycoproteins (d) Transport proteins

31. The activity of the sodium/potassium pump is critically important in maintaining proper ion concentration across the cell membrane. Which of these step is necessary in the pumping activity of this protein?

(a) The pump changes shape

(b) The pump is energized with ATP

(c) Two potassium ions in the extracellular fluid bind with the pump

(d) All of the above are critical steps in the activity of the pump

32. Which of the following type of transport does NOT require the use of a transport protein?
    (a) Facilitated diffusion
    (b) Active transport
    (c) Both simple diffusion and facilitated diffusion
    (d) Simple diffusion

33. What do you think would happen to an animal cell that was placed into a drop of 100% pure water?
    (a) The cell will swell up and eventually burst
    (b) The cell will be unaffected by the pure water
    (d) The cell will shrivel and shrink down
    (d) None of the above

34. In the process of diffusion ...
    (a) Random motion of the molecules causes them to migrate toward each other
    (b) Molecules move from an area of greater concentration to an area of lesser concentration
    (c) Directed motion of the molecule causes them to migrate away from each other
    (d) Molecules move from an area of lesser concentration to an area of greater concentration

35. Materials that are too large to move out of the cell through either the lipid bilayer or a transport protein must be moved out through the process of ...
    (a) Endocytosis
    (b) Receptor-mediated endocytosis
    (c) Active transport
    (d) Exocytosis

36. Large food particles or perhaps whole organisms, can be engulfed and pulled into the cell by the process of ...
    (a) Pinocytosis
    (b) Active transport

(c) Receptor-mediated endocytosis

(d) Phagocytosis

37. Many of the proteins that are associated with the cell membrane are said to be "integral" proteins. An integral protein is defined as a protein that are ...

(a) Always used to transport material across the membrane

(b) Not associated with the hydrophobic portion of the lipid bilayer of the membrane

(c) Always a receptor protein

(d) Bound to the hydrophobic portion of the lipid bilayer of the membrane

38. Of the four primary functions of membrane proteins, which one does NOT require that the protein be in contact with the extracellular fluid?

(a) Communication proteins

(b) Structural proteins

(c) Transport proteins

(d) Recognition proteins

39. Which of the following molecules or ions would normally need to use a membrane protein in order to move through the cell membrane?

(a) H+ (b) O_2

(c) CO_2 (d) H_2O

40. Which of the following processes uses a difference in concentration (or concentration gradient) to transport molecules across the cell membrane? (level 2)

(a) Osmosis (b) Simple diffusion

(c) Facilitated diffusion (d) All of the above

41. Scientists have estimated that the earth is 4.6 billion years old how long have eukaryotes been on earth?

(a) 4.6 billion years (b) 3.5 billion years

(c) 2.5 billion years (d) 1.5 billion years

42. Plants are classified into which domain?

(a) Archaea (b) Plantae

(c) Eukarya (d) Bacteria

43. The endosymbiotic theory explains:

(a) The origin of all organelles in eukaryotic cell

(b) How bacterial cell can invade eukaryotic cell and cause disease

(c) How mitochondria and chloroplast originated from free-living cells

(d) How eukaryotic cells consume food

44. Which of the following components are found in animal cell membrane?

(a) Phopholipids

(b) Transmembrane proteins

(c) Cholesterol

(d) All of the above

45. If a cell's contents are in pure distilled water, which is hypotonic to the cell, the cell will probably

(a) Expand

(b) Become isotonic

(c) Burst

(d) Shrivel

46. Clathrin-coated pits are used to:

(a) Extrude bulk fluids from the cell

(b) Allow desired molecules directly into the cytoplasm

(c) Pass wastes in vesicles to the outside

(d) Pass information into adjacent cell

47. A lipid bilayer

(a) Permits water soluble molecules to pass through it

(b) Facilitates the passage of water soluble molecules through it

(c) Inhibits the passage of water soluble substances through it

(d) Actively transports water soluble molecules through it

48. By a mechanism called cotransport:

(a) Sugar moves down its concentration gradient while sodium moves against its concentration gradient

(b) Sugar and sodium move inward down their concentration gradient

(c) Sodium moves inward down its concentration gradient while sugar moves inward against its concentration gradient

(d) Sodium and sugar move outward against their concentration gradient

49. In each cycle, the sodium-potassium pumps transfer:

(a) By two sodium ions out of the cell, two potassium ions into the cell

(b) One sodium ion out of the cell and one potassium ion into the cell

(c) Three sodium ions out of the cell and two potassium ions into the cell

(d) Three sodium ions into the cell and two potassium ions out of the cell

50. In a lipid bilayer the lipids have:

(a) Their water repelling heads facing inward

(b) Their water repelling tails facing inward

(c) Their hydrogen bond-forming heads facing inward

(d) Their hydrogen bond-forming tails facing inward

51. The methods of membrane transport that don't require protein channels or carriers include:

(a) Osmosis (b) Diffusion

(c) Phagocytosis (d) All of the above

52. If an Elodea leaf is placed in a hypertonic (hyperosmotic) solution, what happens?

(a) Water moves out of the leaf

(b) Water moves into the leaf

(c) Water doesn't move in or out of the leaf

53. Which of the following statements refers to the cytoskeleton?

(a) Composed of actin, microtubules and intermediate filaments

(b) Gives cells their structure

(c) Involved in moving vesicles through a cell

(d) All of the above

54. Which process can move molecules from lower concentration to higher concentrations (against the concentration gradient)?

(a) Active transport

(b) Facilitated diffusion

(c) Osmosis

(d) Endocytosis

55. Where is cholesterol found in the cell?

(a) In the plasma membrane of all cells

(b) In the plasma membrane of animal cells

(c) In the cell wall of plant cells

(d) In Archaebacteria

56. Plants differ from animals in that plants have:

(a) Endoplasmic reticulum

(b) A central vacuole

(c) Golgi bodies

(d) Vesicles

57. What is the function of mitochondria?

(a) Organize all cellular activities

(b) Conduct oxidative metabolism to produce energy

(c) Produce energy via photosynthesis

(d) Site of protein synthesis

58. Prokaryotes have:

(a) Cytoplasm (b) Ribosomes

(c) DNA (d) All of the above

59. Which of the following is not a characteristic of prokaryotes?

(a) DNA (b) Cell membrane

(c) Cell wall (d) Endoplasmic reticulum

60. The term "nuclear envelope" is more correct than the term "nuclear membrane" because

(a) The enclosure has pores which membranes do not

(b) The enclosure is made up of two membranes

(c) The chemical composition is inconsistent with cellular membranes

(d) None of the above. The two terms are perfect synonyms

61. Proteins synthesized by the rough ER are

(a) For internal storage

(b) To build more membranes in the cell

(c) For internal regulation

(d) Exported from the cell

62. Plants differ from animals in that plants have

(a) An endoplasmic reticulum (b) A central vacuole

(c) Golgi complexes (d) Vesicles

63. Passage through pores in the nuclear envelope is restricted primarily to

(a) Proteins, RNA, and protein-RNA complexes

(b) Lipids and glycolipids

(c) DNA and RNA

(d) RNA and protein-carbohydrate complexes

64. In bacteria, some of the function of eukaryotic cells are performed by

(a) Lysosomes (b) Mitochondria

(c) Nucleoli (d) The plasma membrane

65. Glycoproteins and glycolipids assembled in Golgi bodies are packaged for distribution in

(a) Cisternae (b) Lysosomes

(c) Peroxisomes (d) Liposomes

66. Within chloroplast, light is captured by

(a) Grana within cisternae (b) Thylakoids within grana

(c) Cisternae within grana (d) Grana within thylakoids

67. With which of the following are basal bodies not associated?

(A) Animal cells (b) Centrioles

(c) Plant cells (d) Tubulin

68. Depolymerization of microtubule is inhibited by

(a) Kinesin (b) Dyneins

(c) Actin (d) Guanosine triphosphate

69. Bacteria may be propelled by

(a) Rotating thread-like flagellum

(b) Cilia

(c) Undulating 9+2 type flagellum

(d) Gel-sol changes in the cytoplasm

70. The bacterial cell wall is composed of

(a) A phospholipid matrix (b) A lipoprotein

(c) A polymer of sugars (d) Chitin

71. Unlike those of prokaryotes, eukaryotic cell wall are composed of

(a) A carbohydrate matrix cross-linked by short polypeptides

(b) Glycolipids and protein fibers

(c) Cellulose fibers embedded in a matrix

(d) Chitin

72. A gram-negative bacterium is enclosed by

(a) A single thick wall (b) A single thin wall

(c) A double thick wall (d) A double thin wall

73. The cytoplasm of a bacterium

(a) Is supported by the cytoskeleton

(b) Is supported by microtubules

(c) Is supported by keratin

(d) Has no internal support structure

74. Clusters of rRNA where ribosomes are assembled are called

(a) Nuclei (b) Cisternae

(c) Nucleoli (d) Golgi complexes

75. Mitochondrial enzymes for oxidative metabolism are

(a) On or within the surface of cristae

(b) Located on the outer membrane

(c) In the matrix

(d) Floating freely in intermembrane space

76. The smooth ER is especially abundant in cells that synthesize extensive amounts of

(A) Toxins (b) Proteins

(c) Enzymes (d) Lipids

77. Enzymes embedded in the membrane of the smooth ER

(a) Synthesize lipids

(b) May be used for detoxification

(c) Mostly are active only when associated with a membrane

(d) All of the above

78. Which of the following organelles is found in plant cells but not in animal cells?

(a) Ribosomes

(b) Mitochondria

(c) Peroxisomes

(d) None of the above

79. Most cells are very small. A typical eukaryotic cell, both plant and animal, will occur in which of the following size ranges?

(a) 1 mm to 100 μm (b) 100 μm to 10 μm

(c) 10 μm to 1 μm (d) 1 μm to 100 nm

80. Assuming all other factors to be the same, electron microscopes have greater resolving power than light microscopes because.

(a) The wavelength of electrons is much longer than the wavelength of visible light

(b) Electron microscopes have more lenses

(c) Because the beams in electron microscopes overlap creating a clearer picture

(d) Because the wavelengths in visible light are longer than with electrons

81. Bacterial flagella propel the cell by using

(a) A whipping-like motion

(b) Two flagella that move in opposite directions, like a flutter kick

(c) A rotating motion

(d) A flicking motion

82. Which characteristic do eukaryotic and prokaryotic flagella have in common?

(a) Chemical composition (b) Structure

(c) Location in the cell (d) Function

83. The cytoskeleton includes all of the following except

(a) Microtubules (b) Intermediate filaments

(c) Myosin filaments (d) Actin filaments

84. Ribosomes are found

(A) Only in the nucleus

(b) In the cytoplasm

(c) Attached to the smooth endoplasmic reticulum

(d) Only in eukaryotic cells

85. The Golgi apparatus is involved in

(a) Transporting proteins that are to be released from the cell

(b) Packaging proteins into vesicles

(c) Altering or modifying proteins

(d) All of the above

86. Who was the first microscopist to view an outline of cells?

(a) Janssen

(b) Robert Hooke

(c) Leeuwenhoek

(d) Matthias Schleiden

87. The cells of the Eukarya can grow larger than the cell of the Bacteria or the Archaea because:

(a) The Eukarya are more advanced

(b) The Eukarya have existed on Earth longer

(c) The Eukarya have membranous organelles

(d) None of the above

88. Bacteria that have cell walls with thick layers of peptidoglycan are:

(a) Gram-positive (B) Gram-negative

(c) Rod-shaped (d) Spirilla

89. In which way the Archaea are more similar to the Eukarya than to the Bacteria?

(a) Share similar surface molecules

(b) Same nucleic acid composition

(c) Archaea have histones with their genetic material.

(d) Shares the same habitat

90. Membrane-bound compartments in eukaryotic cells are called:

(a) Organelles (b) Tissues

(c) Organs (d) Cytoplasm

91. The structure in which proteins and RNA are made into ribosomes are called

(a) Nucleolus

(b) Nucleus

(c) Cytoplasm

(d) Rough endoplasmic reticulum

92. Membranous sac-like structures that are formed from the Golgi apparatus are called

(a) Mitochondria (b) Golgi complexes

(c) Vacuoles (d) Vesicles

93. In which of the following cell would you expect to find the most mitochondria?

(a) Brain (b) Stomach

(c) Liver (d) Muscle

94. The function of the Golgi apparatus is to

(a) Degrade and recycle worn-out cell parts

(b) Produce energy

(c) Process and modify specific proteins

(d) Store pigments and carbohydrates

95. Where are lysosomal enzymes stored in the cell?

(a) Golgi apparatus

(b) Rough endoplasmic reticulum

(c) Within the lysosomes

(d) Inner side of cell membrane

96. What is/are the function(s) of lysosomes?

(a) Engulf cellular debris

(b) Break down cholesterol

(c) Kill bacteria engulfed by white blood cells

(d) All of these are correct.

97. What plant organelle serves a function similar to that of animal lysosomes?

(a) Central vacuole (b) Chloroplasts

(c) Mitochondria (d) All of these are correct.

98. Peroxisomes function to

(a) Produce toxic compounds

(b) Concentrate acids

(c) Help the cell deal with toxins

(d) Help the cell defend against predators

99. What organelle temporarily stores or transports substances within the cell?

(a) Chloroplast (b) Lysosome

(c) Nucleus (d) Vesicle

100. Bacterial DNA is located in:

(a) The nucleus (b) The nucleoid region

(c) The nucleolus (d) The nucleoli

101. Which of the following is found both prokaryotic and eukaryotic cells?

(a) Mitochondria (b) A nucleus

(c) Chloroplasts (d) Ribosomes

102. The movement of chromosomes during mitosis is due to the:

(a) Shortening of microtubules to which they are attached

(b) Lengthening of microtubules to which they are attached

(c) Shortening of microfilaments to which they are attached

(d) Lengthening of microfilaments to which they are attached

103. 9 + 2 arrangement refers to the:

(a) Structure of prokaryotic flagella

(b) Structure of eukaryotic flagella

(c) Structure of eukaryotic cilia

(d) Both b and C

104. Human egg cells obtain nutrients while maturing when surrounding cells:

(a) Are ingested by the egg cell

(b) Secrete nutrients which the egg cell takes in by pinocytosis

(c) Lyse to release their contents

(d) Encase them in a coating of nutrients

105. Which of the following would be a problem for a cell if it were able to grow to twice its normal size?

(a) Water retention
(b) Salt balance
(c) Reproduction
(d) Intracellular communication

106. The cell theory states that

(a) Cells arise spontaneously
(b) Cells arise from preexisting cells
(c) Cells are the smallest unit of life
(d) Both b and C

107. Which portion of a phospholipid molecule would be facing the environment outside of the cell?

(a) The polar head
(b) The nonpolar head
(c) the polar tail
(d) The nonpolar tail

108. Which portion of a phospholipid molecule would be facing the interior of the cell?

(a) The polar head
(b) The nonpolar head
(c) The polar tail
(d) The nonpolar tail

109. Which portion of a phospholipid molecule is repelled by water?

(a) The polar head
(b) The nonpolar head
(c) The polar tail
(d) The nonpolar tail

110. Membrane proteins

(a) Exist in fixed positions within the plasma membrane
(b) Move freely within the plasma membrane
(c) Are involved in cell-to-cell recognition
(d) Are always large enough to span the entire width of the plasma membrane

111. How might viruses play a role in curing cystic fibrosis?

(a) Viruses could remove the defective gene from lung cells in persons with cystic fibrosis
(b) Viruses could repair the defective gene in lung cells in persons with cystic fibrosis

(c) Viruses could identify the defective gene for later removal

(d) Viruses could deliver a copy of the normal gene to lung cells in persons with cystic fibrosis

112. Which of the following exhibit the 9 + 2 structure of microtubules?

(a) Flagella

(b) Cilia

(c) Sensory hairs in the human ear

(d) All of the above

113. Centrioles may have originated as

(a) A unique arrangement of microtubules found in some prokaryotic species

(b) Protein fibers produced by ribosomes in some eukaryotic cells

(c) Protozoans that formed an endosymbiotic relationship with primitive aerobic prokaryotes

(d) Spirochetes that became involved in an endosymbiotic relationship

114. Access to the nucleus is regulated by

(a) Specialized proteins within the nuclear pore

(b) The rate of osmosis

(c) The amount of ATP available for active transport

(d) Hormones

115. A busy student pours herself a cup of coffee and adds a sugar cube. A moment later the phone rings so she doesn't have time to stir it. Ten minutes later she returns to her coffee, and when she tastes it, it is obvious that the sugar has dissolved and that the coffee is uniformly sweet. What happened?

(a) Osmosis has caused the sugar molecules to move down the concentration gradient until they were evenly distributed in the coffee

(b) Facilitated diffusion has occurred and has caused the sugar molecules to dissolve in the coffee

(c) Diffusion has occurred in the coffee causing the sugar to become evenly distributed

(d) Active transport has occurred using the heat in the coffee as a source of energy

116. The sugar molecules in the coffee mentioned above bounce and jostle each other because they have

(a) Kinetic energy (b) Potential energy

(c) Static energy (d) Random energy

117. Water molecules crossing a plasma membrane is due to

(a) Osmosis (b) Diffusion

(c) Facilitated diffusion (d) Active transport

118. Which of the following is a requirement for osmosis to occur?

(a) A semipermeable membrane

(b) A concentration gradient

(c) Sugar or salt molecules dissolved in water

(d) An energy source

119. What advantage would a plant cell have over an animal cell if both were placed in pure water?

(a) The plant cell would resist shrinking in pure water

(b) The plant cell would resist bursting in pure water

(c) The plant cell would reproduce at a faster rate in pure water

(d) Both a and c

120. The release of insulin from certain human cells is an example of

(a) Endocytosis (b) Exocytosis

(c) Phagocytosis (d) Pinocytosis

121. An amoeba is engulfing bacterial cell by surrounding them with a pseudopod. This is an example of

(a) Endocytosis (b) Exocytosis

(c) Phagocytosis (d) Both a and c

122. The removal of waste products from the amoeba is an example of

(a) Endocytosis (b) Exocytosis

(c) Phagocytosis (d) Pinocytosis

123. Diffusion of molecules across biological membrane that requires channel is called

(a) Active transport

(b) The sodium-potassium pump

(c) Facilitated diffusion

(d) Osmosis

124. The sodium-potassium pump functions to

(a) Remove sodium and potassium ions from the cell

(b) Pump sodium and potassium ions into the cell

(c) Pump potassium ions into the cell and sodium ions out of the cell

(D) Pump sodium ions into the cell and potassium ions out of the cell

125. A liver cell would detect the presence of insulin by

(a) Means of receptor proteins on its surface

(b) Means of receptor carbohydrates on its surface

(c) Producing insulin detection binding molecules

(d) The reduction of glucose in the external environment

126. Robert Hooke first described cells by looking at thin slices of

(a) Honeycomb (b) Skin

(c) Cork (d) Bacteria

127. Which of the following statements about the cell theory is true?

(a) All organisms are composed of several cells

(b) The smallest living thing is a cell

(c) Life continues to evolve anew with each new cell

(d) New cells arise spontaneously when conditions are right

128. What limits cell size?

(a) Surface-to-volume ratio

(b) Speed of movement of materials from the middle to the periphery of the cell

(c) The bigger the cell, the less access there is from the interior of the cell to the external environment

(d) All of the above limit cell size

129. Cystic fibrosis has all of the following characteristics except

(a) Is fatal if untreated

(b) Is a disease caused by a type of bacteria

(c) Is characterized by mucus buildup in pancreatic ducts

(d) Is associated with plasma membrane defects

130. Prokaryotic cells

(a) Have no membrane-bounded organelles

(b) Have a large nucleus containing hereditary information

(c) Comprise very complex individuals

(d) Are usually larger than eukaryotic cells

131. Eukaryotic cells

(a) Have membrane-bounded organelles

(b) Have a large nucleus containing hereditary information

(c) Are usually larger than prokaryotic cells

(d) All of the above

132. The cytoskeleton

(a) Synthesizes proteins

(b) Is composed of phospholipid molecules

(c) Confers strength and flexibility to the cell

(d) Serves as a hollow conduit or "pipeline" for materials throughout the cell

133. Which of the following structure contain microtubules?

(a) Cilia (b) Flagella

(c) Centrioles (d) All of the above

134. Chromosomes

(a) Are located in the nucleus

(b) Contain DNA

(c) Are only visible at certain times with a light microscope

(d) All of the above characterize chromosomes

135. Ribosomes

(a) Are synthesized in the inner region of the mitochondria

(b) Are located on the endoplasmic reticulum or float freely in the cytoplasm

(c) Pass through cilia

(d) Are involved in the synthesis of sugars

136. The endoplasmic reticulum

(a) Synthesizes minerals

(b) Synthesizes carbohydrates and lipids

(c) Functions as structural support for the cell

(d) Synthesizes ribosomes

137. Which of the following eukaryotic cellular organelles are made out of membranes?

(a) The endoplasmic reticulum

(b) Mitochondria

(c) The Golgi complex

(d) All of the above

138. The net movement of molecules down a concentration gradient is

(a) Osmosis (b) Osmotic pressure

(c) Diffusion (d) Exocytosis

139. Which of the following can move freely across a plasma membrane?

(a) Sugar (b) Protein

(c) Urea (d) Water

140. A red blood cell placed in pure water will burst because

(a) Osmotic pressure increases inside the cell

(b) The solution surrounding the cell is hypotonic

(c) The cell is hypertonic

(d) All of the above

141. Which one of the following terms is not related to a substance entering a cell?

(a) Endocytosis (b) Exocytosis

(c) Phagocytosis (d) Pinocytosis

142. Free movement of molecule in either direction through channels in the plasma membrane is

(a) Osmosis (b) Diffusion

(c) Active transport (d) Facilitated diffusion

143. Movement of molecules through channel in the plasma membrane is only one direction with the expenditure of energy is

(a) Osmosis (b) Diffusion

(c) Active transport (d) Facilitated diffusion

144. The sodium-potassium pump

(a) Allows sodium to move freely into the cell

(b) Can move molecules into the cell against the concentration gradient

(c) Is one of the many different kinds of active transport channels

(d) Removes potassium from the cell

145. Chemical messengers within the cell are

(a) Molecules used in metabolism

(b) With-in cell hormones

(c) Ions

(d) All of the above

146. During interphase chromosomes

(a) Are lined up on the equator

(b) Have separated into sister chromatids

(c) Are extended and in use

(d) Are coiled around a core of histone proteins

147. During the M phase of mitosis

(a) DNA is replicated

(b) Homologous chromosomes line up on the equator

(c) DNA is uncoiled and is in use

(d) Spindle fibers are bound to centromeres

148. Genes contain the information for which of the following processes?

(a) Constructing mRNA (b) Building polypeptides

(c) Constructing enzymes (d) All of the above

149. The relationship of blueprint to building is analogous to the relationship of

(a) A chromosome to the entire genome

(b) The genome in a fertilized egg to a newborn baby

(c) A gene to a chromosome

(d) A nucleotide to a gene

150. Which of the following statement concerning mitosis is incorrect?

(a) It occurs in somatic cells

(b) It splits up homologous chromosomes

(c) It occurs in both animal and plant cells

(d) It produces daughter cells with the same number of chromosomes as the original parent cell

151. The amount of DNA per nucleus was measured on several hundred cells from within the tip of an onion root. The amount of DNA ranged from 5 to 10 picograms. A cell with 5 picograms of DNA is probably in which of the following stages of the cell cycle?

(a) Interphase (G2) (b) Interphase (S)

(c) Anaphase (d) Interphase (G1)

152. The amount of DNA per nucleus was measured on several hundred cells from within the tip of an onion root. The amount of DNA ranged from 5 to 10 picograms. A cell with 8.3 picograms of DNA is probably in which of the following stages of the cell cycle?

(a) Interphase (G1) (b) Interphase (S)

(c) Interphase (G2) (d) Prophase

153. The following cell is in which stage of cell division?

(a) Prophase (b) Telophase

(c) Metaphase (d) Anaphase

154. The following cell is in which stage of cell division?

(a) Interphase (b) Anaphase

(c) Telophase (d) Prophase

155. The following cell is in which stage of cell division?

(a) Anaphase (b) Telophase

(c) Prophase (d) Interphase

156. Which of the following lists represents the correct order for the cell cycle?

(a) Prophase, anaphase, metaphase, telophase, interphase

(b) Prophase, interphase, metaphase, anaphase, telophase

(c) Interphase, metaphase, anaphase, telophase, prophase

(d) Metaphase, anaphase, telophase, interphase, prophase

157. The replication of DNA occurs during which portion of the interphase?

(a) G1 phase (b) G3 phase

(c) G2 phase (d) S phase

158. Spindle fibers first appear during which stage of mitosis?

(a) Metaphase (b) Prophase

(c) Anaphase (d) Telophase

159. Plant cell mitosis is basically like animal cell mitosis except that ...

(a) Plant cells do not have spindle fibers

(b) Plant cells do not replicate their DNA before mitosis

(c) Plant cells do not separate sister chromatids during anaphase

(d) Plant cells use a cell plate to separate daughter cells

160. Which of the following does NOT occur during anaphase of mitosis?

(a) Chromosomes are separated into identical sets of chromosomes

(b) Spindle fibers shorten as a result of the disassembly of the microtubules

(c) Sister chromatids are pulled apart

(d) Centrosomes separate and move to opposite poles of the cell

161. Which of the following does NOT occur during telophase of mitosis?

(a) Nuclear membrane begins to re-form

(b) Chromosomes become attached to spindle fibers

(c) Chromosomes relax and uncoil

(d) Mitotic spindle breaks down

162. During which stage of mitosis does the process of cytokinesis begin?

(a) Telophase (b) Anaphase

(c) Prophase (d) Metaphase

163. Which of the following statement concerning binary fission in a bacterial cell is NOT true?

(a) A bacterium uses spindle fibers to separate its two copies of DNA

(b) The replication of a bacterium's DNA is accomplished through the same DNA unwinding and replication mechanisms as those used in eukaryotes

(c) A bacterium pinches in between the two DNA copies to form the two daughter cells

(d) A bacterium produces an extra copy of its chromosome before it divides by binary fission

164. At the end of DNA replication ...

(a) An mRNA molecule is produced

(b) Two strands of DNA are produced

(c) A new protein is produced

(d) A new nucleotide is produced

165. Which of the following statements concerning homologous chromosomes, is a TRUE statement?

(a) The father donated one of each pair of homologous chromosomes, and the mother donated the other

(b) Homologous chromosomes are the same thing as sister chromatids

(c) Homologous chromosomes are always exactly alike

(d) The homologous pairs of chromosomes are created inside of a single cell after that cell completes the S phase of interphase

166. What is mitosis?

(a) The division of the cell and cytoplasm into halves

(b) The replication of nuclear material

(c) The process of nuclear division that reduces the number of chromosomes in the resulting cells by half

(d) Both A & B

167. Beginning with mitosis, what are the other parts of the cell cycle in order in which they occur?

(a) Cytokinesis, G1, S, G2,

(b) Cytokinesis, S, G1, S, G2

(c) S, cytokinesis, G1, S, G2,

(d) S, G1, S, G2

168. What occurs in the cell during the G2 phase?

(a) Growth processes and synthesis of compounds other than DNA

(b) DNA composing the chromosomes is duplicated

(c) Molecules and structures necessary for mitosis are synthesized

(d) Division of the cell and cytoplasm

169. Beginning with interphase, what are the phases of mitosis in order in which they occur?

(a) Telophase, anaphase, metaphase, prophase

(b) Anaphase, metaphase, prophase, telophase

(c) Prophase, anaphase, metaphase, telophase

(d) Prophase, metaphase, anaphase, telophase

170. In which phase do the chromosomes align on a plane in the center of the cell?

(a) Prophase

(b) Metaphase

(c) Anaphase

(d) Telophase

171. What occurs during anaphase?

(a) The mitotic spindle begins to form

(b) The chromosomes align on a plane in the center of the cell

(c) The sister chromatids separate

(d) The mitotic apparatus disassembles

172. What is the part of the chromosome that joins the two chromatids?

(a) Kinetochore (b) Centriole

(c) Centromere (d) Mitotic spindle

173. What is the structure in animal cells that divides the cytoplasm into two cells.

(a) Mitotic spindle (b) Centriole

(c) Cleavage furrow (d) Cell plate

174. What is a blastula?

(a) A bridge of microtubules produced by the centrioles that will pull the chromatids apart

(b) An early embryonic stage of a vertebrate and consists of a sphere of cells

(c) A microtubule organizing center that moves to opposite ends of the cell and pull the chromatids apart

(d) A localized area of rapid cell division

175. What is the term that describes a nucleus with two of each type of chromosome?

(a) Diploid (b) Blastula

(c) Meristem (d) Chromatid

176. The term "homologous chromosomes":

(a) Refers to replications of the same chromosome

(b) Is another name for sister chromatids

(c) Must be haploid

(d) Means a pair of chromosomes of the same kind, such as sex chromosomes

177. Cytoplasm in animal cells is divide during:

(a) Cytokinesis (b) G1

(c) G2 (d) Mitosis

178. Two cells exactly like the original cell result from:

(a) Gamete formation (b) Meiosis

(c) Mitosis (d) Sexual reproduction

179. A duplicated chromosome has how many chromatids?

(a) One
(b) Two
(c) Three
(d) Four

180. Each sperm cell of a dog contains 39 chromosomes. How many chromosomes are in each of the dog's somatic cells?

(a) 23
(b) 46
(c) 39
(d) 78

181. Which phase of mitosis is associated with aligning of chromosomes in the center of the cell?

(a) Interphase
(b) Prophase
(c) Metaphase
(d) Anaphase

182. During synapsis:

(a) Sister chromatids pair all along their length
(b) Sister chromatids pair at the centromeres
(c) Homologues repel each other except at the ends
(d) Homologues pair all along their length

183. Most multicellular eukaryotes form gametes by:

(a) Meiosis
(b) Oncogenesis
(c) Mitosis
(d) Binary fission

184. In sexually reproducing organisms, crossing-over results in:

(a) Independent assortment of genes
(b) A segregation of genes
(c) A higher proportion of dominant genes
(d) An increased genetic diversity in the population

185. The term "gametes" describes:

(a) Eggs and sperm cells
(b) All cells in the body
(c) All cells except egg and sperm cells
(d) None of the above

186. Chiasmata formation and crossing over occur during:

(a) Prophase of mitosis (b) Prophase I of meiosis

(c) Prophase II of meiosis (d) B and C

187. If one cell with 32 chromosomes divides by mitosis, how many cells are produced?

(a) One (b) Two

(c) Four (d) No way to know

188. If one cell with 32 chromosomes divides by mitosis, how many chromosomes does each new cell have?

(a) 64 (b) 48

(c) 32 (d) 16

189. If one cell with 32 chromosomes divides by meiosis, how many cells are produced?

(a) One (b) Two

(c) Four (d) No way to know

190. If one cell with 32 chromosomes divides by meiosis, how many chromosomes does each new cell have?

(a) 64 (b) 48

(c) 32 (d) 16

191. Which statement(s) is/are true regarding crossing over?

(a) Occurs between any two chromosomes in the nucleus of a cell

(b) Involve exchanges of segements of chromatids between two homologous chromosomes

(c) Is an important component of mitosis

(d) None of the above

192. Which statement describes an advantage of crossing over?

(a) Crossing over increases the genetic variability in a population

(b) Crossing over creates individuals that are identical copies (clones) of a parent

(c) Crossing over can occur during mitosis or meiosis

(d) None of the above

193. How do cell division and programmed cell death interact?

(a) Cell division occurs first, then programmed cell death, followed again by cell division and so forth

(b) Cell division resumes quickly before the last cell dies after programmed cell death

(c) Cell division and programmed cell death occur in balance so that a tissue does not shrink or grow too large

(d) None of these is correct

194. Interphase is divided into three main phases. Which of the following is not associated with interphase?

(a) G1 phase (b) G2 phase

(c) Synthesis (d) Mitosis

195. A cell can exit the cell cycle at G1 and enter at Go. What term best describes the Go phase?

(a) Quiescent (b) DNA replication

(c) Synapsis (d) Karyokinesis

196. Which cell component is produced in large quantities during the G2 phase?

(a) Chromosomes (b) Cell membranes

(c) Mitochondria (d) Endoplasmic reticulum

197. What is responsible for the equal distribution of replicated chromosomes into each of two daughter cells?

(a) Mitotic spindle (b) Random motion

(c) Centromeres (d) Chromatids

198. On average, how many times does an adult human cell divide?

(a) 35 to 63 times (b) 14 to 29 times

(c) 50 times (D) 70 times

199. What is the function of apoptosis?

(a) To weed out cells that grow uncontrollably

(b) To allow the immune system to distinguish self from non self

(c) To shape organs during development

(d) All of these are correct

200. The characteristics of cancer cells include

(a) Ability to produce signals to stimulate angiogenesis

(b) Uncontrollable growth

(c) Dedifferentiation

(d) All of the above are correct

201. Epidermal growth factor receptors

(a) Are now produced as a molecular therapy for the treatment of liver cancer

(b) Are used by cells to remove the "brake" the cell uses to restrain cell division

(c) Are overproduced in over 20% of breast cancers

(d) Amplify the signal in cytoplasm telling the cell to divide

202. Gametes

(a) Result from fusion of egg and sperm

(b) Another name for fertilization

(c) Egg and sperm

(d) Produces four haploid cells

203. Syngamy

(a) Results from fusion of egg and sperm

(b) Another name for fertilization

(c) Egg and sperm

(d) Produces four haploid cells

204. Meiosis

(a) Results from fusion of egg and sperm

(b) Another name for fertilization

(c) Egg and sperm

(d) Produces four haploid cells

205. A Zygote
- (a) Result from fusion of egg and sperm
- (b) Is another name for fertilization
- (c) Is composed of egg and sperm
- (d) Produces four haploid cells

206. Prophase I:
- (a) Chromosomes gather at the two poles of the cell
- (b) Crossing over occurs
- (c) Chromosomes align on the central plane
- (d) One version of each chromosome moves to a pole of the cell

207. Metaphase I:
- (a) Chromosomes gather at the two poles of the cell
- (b) Crossing over occurs
- (c) Chromosomes align on the central plane
- (d) One version of each chromosome moves to a pole of the cell

208. Anaphase I:
- (a) Chromosomes gather at the two poles of the cell
- (b) Crossing over occurs
- (c) Chromosomes align on the central plane
- (d) One version of each chromosome moves to a pole of the cell

209. Telophase I:
- (a) Chromosomes gather at the two poles of the cell
- (b) Crossing over occurs
- (c) Chromosomes align on the central plane
- (d) one version of each chromosome moves to a pole of the cell

210. The structure of eukaryotic chromosome is in part maintained by
- (a) Covalent bonding between the strands of the double helix
- (b) Ionic bonding between the strands of the double helix

(c) Winding the DNA molecule around a core of histone protein
(d) The centromere
(e) Microtubules present in the cytoplasm

211. Sister chromatids are

(a) Pairs of heterozygous chromosomes
(b) Only seen in prokaryotic cells
(c) Formed only during meiosis
(d) Duplicated DNA held together by a centromere

212. DNA binds with histone proteins because

(a) The combination of the two creates a positively charged structure
(b) The positively charged histone is attracted to the negatively charged DNA
(c) The combination of the two creates a complex with no net charge
(d) That is the only way that DNA replication can be stimulated

213. Human skin cells are

(a) Dead when mature
(b) Haploid
(c) Diploid
(d) Capable of becoming germ cells

214. While viewing stained onion root tip cells you notice that distinct chromosomes cannot be observed in most of the cells. This is because most of the cells

(a) Were destroyed while preparing the slide
(b) Must be in interphase
(c) Are in metaphase
(d) Must be undergoing meiosis

215. The S phase of mitosis involves synthesis of

(a) Cellular products for export from the cell
(b) Cellular membranes and organelles

(c) DNA

(d) The cell wall in plant cells

216. The majority of the cell cycle is taken up by

(a) Interphase (b) Prophase

(c) Metaphase (d) Anaphase

217. The nuclear envelope reforms in the daughter cells during

(a) Interphase (b) Prophase

(c) Metaphase (d) Telophase

218. The shortest stage of mitosis is

(a) Interphase (b) Prophase

(c) Metaphase (d) Anaphase

219. A fertilized egg is called a(n)

(a) Meiotic cell (b) Gamete

(c) Ovum (d) Zygote

220. Alternation of generations refers to

(a) A event that only occurs in plant life cycles

(b) The alternation of a haploid and diploid generation in a life cycle

(c) The lack of chromosome division between meiotic divisions

(d) The dependence of one generation on the next for provision of nutrients

221. Crossing over takes place during

(a) Anaphase of mitosis (b) Telophase of mitosis

(c) Meiosis I (d) Meiosis II

222. Genetic variation in a population is ensured by

(a) Crossing over during meiosis

(b) Random fertilization

(c) Independent assortment

(d) All of the above

223. Which of the following is not related to mitosis?

(a) Germ cells (b) Chromosomes

(c) DNA replication (d) Somatic cells

224. Which phase of mitosis is associated with chromosomes aligned at the center of the cell and centromeres replicating?

(a) Interphase (b) Prophase

(c) Metaphase (d) Anaphase

225. Which phase of mitosis is associated with formation of the nuclear envelope?

(a) Prophase (b) Metaphase

(c) Anaphase (d) Telophase

226. Which phase of mitosis is associated with separation of chromatids?

(a) Interphase (b) Prophase

(c) Anaphase (d) Telophase

227. During cell division, what structure forms as a network of protein cables?

(a) Spindle (b) Aster

(c) Centromere (d) Chromatid

228. Which one of the following statement is not true about meiosis?

(a) Meiosis occurs in reproductive cells

(b) Meiosis results in four haploid daughter cells

(c) In meiosis, homologous pairs of chromosomes are pulled apart

(d) Meiosis occurs only in animals

229. Crossing over occurs during

(a) Prophase I (b) Prophase II

(c) Interphase I (d) Interphase II

230. In the complex cell cycle, cell division preparation continues with the replication of mitochondria and chromosome condensation during which phase?

(a) G1 (b) G2

(c) M (d) C

231. Chromosomes exist in somatic cells as nearly identical copies of each other called

(a) Analogous chromosomes (b) Sister chromatids

(c) Homologous chromosomes (d) Centromeres

232. A growth disorder in cells that starts when an apparently normal cell begins to grow in an uncontrolled and invasive way is called

(a) A DNA aberration (b) Chromosomal condensation

(c) Syngamy (d) Cancer

233. In meiosis

(a) The chromosomes do not replicate between the two nuclear divisions

(b) The chromosome number is halved

(c) The second meiotic division is similar to the process of mitosis

(d) All of the above occur

234. What gene that serves as "quality control" relative to cell division is called the "guardian angel gene"?

(a) EGF (b) p 53

(c) Ras (d) E1B

235. The G1 checkpoint is the time when

(a) Sister chromatids separate and move to opposite poles of the cell

(b) The decision is made to divide, delay or enter a resting stage

(c) Cytokinesis begins

(d) Cancerous cells begin unchecked division

236. The cell cycle of eukaryotes is accessed and controlled by feedback

(a) At three checkpoints

(b) Provided by specialized cells of the immune system

(c) At the G1 checkpoint. If all is well, mitosis continues unchecked to completion

(d) During the S phase

237. How are meiosis and mitosis similar?

(a) Both produce daughter cells with the exact same genetic makeup as the original parent cell

(b) Both are preceded by duplication of DNA

(c) Both always produce haploid daughter cells

(d) Both occur in all living organisms

238. During which type(s) of cell division does a ring of contractile microfilaments form a cleavage furrow around the equator of the cell?

(a) Meiosis II and mitosis only

(b) Mitosis and meiosis I and II

(c) Meiosis I only

(d) Meiosis I and mitosis only

239. During which type(s) of cell division does the DNA begin to condense into visible chromosomes, the nuclear membrane breaks apart into small vesicles, nucleoli disappear, and chromosomes are composed of two chromatids?

(a) Meiosis II only (b) Meiosis I only

(c) Mitosis and meiosis I and II (d) Meiosis I and mitosis only

240. During which type(s) of cell division can chromosomes be composed of two chromatids, and each chromosome is lined up on the equatorial plane of the cell?

(a) Meiosis I only (b) Meiosis II and mitosis only

(c) Meiosis I and mitosis only (d) Mitosis and meiosis I and II

241. Which of the following is a type of sexual reproduction rather than asexual reproduction?

(a) Regeneration (b) Vegetative reproduction

(c) Self-fertilizing flowers (d) Binary fission

242. Which of the following is the primary reason why polar bodies are produced in the cytokinetic process of egg formation?

(a) To reduce the number of chromosomes in half for each of the gametes

(b) To make it easier for the cleavage furrow to divide up the cytoplasmic material

(c) To preserve as much of the cytoplasmic nutrients as possible for the egg and any potential zygotes

(d) To reject chromosomes that have mutations on them by not allowing them to be placed into the egg

243. During which type(s) of cell division do chromosomes that were composed of two chromatids become chromosomes composed of a single chromatid?

(a) Mitosis and meiosis I and II (b) Meiosis I and mitosis only

(c) Meiosis I only (d) Meiosis II and mitosis only

244. During which type(s) of cell division do tetrads form and crossing over occur?

(a) Mitosis and meiosis I and II (b) Meiosis I only

(c) Meiosis II and mitosis only (d) Meiosis I and mitosis only

245. Which type(s) of cell division is immediately preceded by replication of DNA?

(a) Meiosis II only (b) Meiosis I only

(c) Meiosis II and mitosis only (d) Meiosis I and mitosis only

246. Which of the following statement about meiosis is true?

(a) Meiosis produces haploid cells only after the meiosis II process has been completed

(b) Meiosis produces gametes that have chromosomes, each with two chromatids

(c) Meiosis produces haploid gametes in which each chromosome has a homologue

(d) Meiosis shuffles alleles and creates chromosomes that are genetically different from the parent's chromosomes

247. Which of the following statement concerning gametes is NOT correct?

(a) Gametes are produced after the second meiotic division.

(b) Gametes are diploid

(c) Gametes are haploid

(d) Gametes are said to be 1n

248. The important process of crossing over occurs during which stage of meiosis?

(a) Prophase II (b) Anaphase I

(c) Anaphase II (d) Prophase I

249. A couple has three children, and all of them are boys. When they get pregnant a fourth time, they are hopeful that this one is a girl. What is the chance that this child will be a girl?

(a) 1/1 (a sure thing) (b) 1/4

(c) 1/3 (d) 1/2

250. The Y chromosome can be used to trace a person's lineage many generations into the past. What characteristic of the Y chromosome allows this to be accomplished?

(a) The Y chromosome engages in only a tiny bit of crossing over

(b) All sperm cells carry the Y chromosome

(c) The Y chromosome is extremely large

(d) The Y chromosome is carried exclusively in females

251. At which stage during meiosis are sister chromatids separated into two different groups of chromosomes?

(a) During anaphase I (b) During anaphase II

(c) During metaphase I (d) During metaphase II

252. A certain type of diploid organism contains 58 chromosomes. How many chromosomes would you expect a gamete of this organism to contain?

(a) 29 chromosomes (b) 58 chromosomes

(c) 116 chromosomes (d) 23 chromosomes

253. One of the critical function of meiosis is the mixing of genetic material. This is important for producing variety in the offspring. What are the two phases of meiosis where this mixing takes place?

(a) Prophase II and metaphase II

(b) Metaphase I and telophase I

(c) Metaphase II and telophase II

(d) Prophase I and metaphase I

254. Which of the following definition for crossing over is accurate?

(a) Genes on chromosome #1 are traded for genes on chromosome #2

(b) Genes on the upper arm of one sister chromatid of chromosome #1 are traded for genes on the C)lower arm of the other sister chromatid of chromosome #1

(c) Genes on one sister chromatid of the paternal chromosome are traded for the same genes on the other sister chromatid of the paternal chromosome

(d) Genes on the paternal chromosome are traded for the same genes on the maternal chromosome

255. Which of the following statement concerning human oogenesis is correct?

(a) The complete meiotic division that produces an ovum (egg) does not finish until after fertilization

(b) The process of oogenesis begins within the female fetus while it is still in its mother's womb

(c) Oogonia can reproduce mitotically within an adult female's ovary to produce more oogonia

(d) Both A and B are correct

256. Which of the following statement concerning human spermatogenesis is INCORRECT?

(a) Spermatogonia can develop directly into spermatids without any meiotic divisions

(b) Spermatogonia are able to produce both primary spermatocytes and other spermatogonia

(c) Spermatogonia are the only cells in the human body that possess flagella

(d) None of the above

257. How many tetrads are found in meiotic prophase in an organism with a diploid number of 12?

(a) 4 (b) 6

(c) 18 (d) 24

258. When does DNA synthesis occur during the cell cycle?

(a) G1 (b) G2

(c) G0 (d) S

259. Human sperm and egg formation (gametes) differ from one another in which of the following way(s)?

(a) Timing of gamete formation

(b) Polar body formation

(c) Amount of cytoplasm in each daughter cell formed

(d) All of these statements refer to differences between human sperm and egg formation

260. Which feature of a eukaryotic cell is responsible for breaking down food energy into useable cellular energy in the form of ATP, a process called cellular respiration?

(a) Endoplasmic reticulum (b) Nucleus

(c) Mitochondrion (d) Chloroplast

261. Which of the following cell structures are directly involved with the making of proteins?

(a) Golgi Body and smooth endoplasmic reticulum

(b) Lysosomes and the nucleus

(c) Rough endoplasmic reticulum and ribosomes

(d) Chloroplasts and mitochondria

262. Which of the following cell feature is not wrapped in a membrane?

(a) Nucleus (b) Centrioles

(c) Endoplasmic reticulum (d) Lysosome

263. Which of the following groups of cellular components are ALL involved in protein production, processing, or modification?

(a) Ribosomes, Golgi Body, and endoplasmic reticulum

(b) Nucleus, mitochondria, chloroplasts

(c) Mitochondria, endoplasmic reticulum, and Golgi Body

(d) Centrioles, chloroplasts, and ribosomes

264. Where are the main instructions for the making of proteins kept inside a cell?

(a) Endoplasmic reticulum (b) Nucleus

(c) Cell membrane (d) Mitochondria

265. Which of the following is a reasonable estimate of the percentage of time spent in mitosis in a typical human cell cycle?

(a) 25% (b) 98%

(c) 50% (d) 5%

266. What are the primary differences often observed for mitosis in plant relative to animal cells?

(a) Plants typically skip the anaphase step in mitosis

(b) Animal cells spend longer in metaphase and lack the spindle fibers seen in plant mitosis

(c) Plants differ in details of telophase and some do not use centrioles to form the spindle system

(e) Plants spend much longer in both metaphase and anaphase than do animal cells

267. During which of the following stage of the cell cycle, individual chromosomes are not visible within cells?

(a) Prometaphase (b) Telophase

(c) Anaphase (d) Metaphase

268. If the p53 gene in a cell is mutated, which of the following situations may occur?

(a) The G0 checkpoint won't work properly

(b) Cells are marked for apoptosis

(c) Cells cannot pass the G1/S checkpoint

(d) Cells with damaged DNA may proliferate in an uncontrolled manner

269. Which of the following is NOT true about homologous chromosomes?

(a) Homologous chromosomes carry the same sequence of gene sites (loci)

(b) One chromosome is of maternal origin while the other is of paternal origin

(c) Homologous chromosomes are identical in length

(d) One of each pair of homologous chromosomes is produced from the other in S phase

270. Which of the following is NOT a difference between mitosis and meiosis?

(a) Mitosis produces two cells while meiosis produces 4 cells

(b) Mitosis produces genetically identical cells whereas meiosis produces genetically different cells

(c) Mitosis produces either haploid or diploid cells (depending on the original cell) whereas meiosis always produces haploid cells

(d) Mitosis begins with a 2n cell while meiosis begins with a 1n cell

271. The checkpoint that requires a cell to be of adequate size in order to move forward is:

(a) G1/S checkpoint

(b) G0 checkpoint

(c) The M checkpoint

(d) Mid metaphase checkpoint

272. The p53 gene product (the protein made from this gene) is involved in:

(a) The length of time a cell spends in interphase

(b) The M checkpoint

(c) Bringing cells out of G0

(d) Marking cells for programmed cell death

273. If the p53 gene in a cell is mutated, which of the following situations may occur?

(a) Cells cannot pass the M checkpoint, no matter what

(b) The G0 checkpoint won't work properly

(c) Cells with damaged DNA may proliferate in an uncontrolled manner

(d) Cells cannot pass the G1/S checkpoint

274. During which phase of meiosis are homologs most closely paired?

(a) Pachytene (b) Zygotene

(c) Leptotene (d) Diplotene

275. During metaphase II for an organism with a diploid number of 4, how many chromatids are aligned in the equatorial plane of a cell?

(a) 6 (b) 4

(c) 10 (d) 2

276. If an organism has a haploid number of 3, how many tetrad structures will align on the equatorial plane in metaphase I of meiosis?

(a) 3 (b) 2

(c) 1 (d) 6

277. In the random segregation of dyads that is the basis for the Mendelian principle of independent assortment, how many possible combinations are there for a diploid number?

(a) 12 (b) 2

(c) 4 (d) 8

278. Which of the following is characteristic of leptonema?

(a) It is the time when crossing over takes place, though this will not be apparent until later stages

(b) It is the stage when the previous doubling of chromosomes first becomes visible

(c) It corresponds to the first or reductional division in meiosis

(d) It is the period when the chromatin begins to condense

279. The initial alignment of chromosomes called rough pairing occurs in which part of prophase I?

(a) Leptonema (b) Pachynema

(c) Diakinesis (d) Zygonema

280. Because DNA polymerase III can only act from 5' to 3', continuous strand growth can be achieved only along one of the template strands (Leading strand) and strand growth along the other strand must occur discontinuously resulting in the production of a series of short sections of new DNA called :

(a) Replicon fragments (b) Okazaki fragments

(c) Klenow fragments (d) None of the above

281. The enzyme responsible for initiating the unwinding of double-stranded DNA (eliminating supercoiling) by nicking a single strand of the DNA molecule is :

(a) Ligase (b) Topoisomerase

(c) Gyrase (d) Helicase

282. A repeating DNA sequence at the end of chromosomes that prevents them from losing base pair sequences at their ends and from fusing together is called :

(a) A Telomere (b) A Telomerase

(c) A replicon (d) None of the above.

283. The enzyme that accomplishes the unwinding of the original double stranded DNA molecule, once supercoiling has been eliminated, by breaking the hydrogen bonds that hold the two strands together is :

(a) Helicase (b) Topoisomerase

(c) Primase (d) DNA Polymerase II

284. What is meiosis?

(a) Double sets of chromosomes

(b) Single sets of chromosomes

(c) Process of nuclear division that reduces the number of chromosomes in the resulting cells by half

(d) Nuclear replication in eukaryotic cells

285. What is the difference between diploid and haploid cells?

(a) Diploid cells have two sets of chromosomes and haploid cells have only one

(b) Diploid cells have one set of chromosomes and haploid cells have two sets

(c) Diploid cells don't go through meiosis and haploid cells do

(d) Diploid cells don't have homologous chromosomes and haploid cells do

286. At the end of oogenesis, how many functioning secondary oocytes are produced?

(a) 1 (b) 2

(c) 4 (d) Millions

287. Germ-line cells

(a) Produce gametes
(b) Are haploid
(c) Usually undergo mitosis
(d) Are special somatic cells

288. During synapsis

(a) Sister chromatids pair all along their length
(b) Sister chromatids pair at the centromeres
(c) Homologues repel each other except at the ends
(d) Homologues pair all along their length

289. Which of the following is not a unique feature of meiosis?

(a) Synapsis
(b) Homologous recombination
(c) Reduction division
(d) Diakinesis

290. Crossing-over can occur between homologues during

(a) Zygotene
(b) Pachytene
(c) Leptotene
(d) Diplotene

291. Terminal chiasmata are characteristic of

(a) Anaphase I
(b) Prophase II
(c) Metaphase II
(d) Metaphase I

292. In anaphase I

(a) Sister chromatids move toward opposite poles
(b) Homologous chromosomes move toward opposite poles
(c) Terminal chiasmata move toward opposite poles
(d) Homologues move toward the same pole

293. Most multicellular eukaryotes form gametes by

(a) Parthenogenesis
(b) Meiosis
(c) Oncogenesis
(d) Mitosis

294. After fertilization, the resulting zygote begins to divide by

(a) Syngamy
(b) Binary fission
(c) Mitosis
(d) Synapsis

295. The zygote of unicellular organisms

(a) Always divides by mitosis

(b) May immediately engage in syngamy

(c) Always divides by meiosis

(d) May divide by meiosis

296. Organisms subject to a chromosome damaging environment would benefit from

(a) Parthenogenesis (b) Binary fission

(c) A haploid adult (d) Syngamy

297. The crossing over process is believed to have originated from

(a) Recombinant DNA (b) Genetic engineering

(c) Mutation (d) Gene repair

298. When crossing over is complete in prophase I, sister chromatids

(a) Are fused together by the synaptonemal complex

(b) Are attached by their ends to the nuclear envelope

(c) Are held together near their common centromeres

(d) Drift away from each other

299. In plants, haploid cells

(a) Divide by meiosis

(b) Must immediately undergo syngamy

(c) Will cross over

(d) Divide by mitosis

300. In telophase I

(a) Homologous chromosomes cluster at the poles

(b) Identical chromatids cluster at the poles

(c) Non-identical chromatids cluster at the poles

(d) Separate, but identical chromatids cluster at the poles

301. Unlike gametes, body cells are called

(a) Somatic (b) Haploid

(c) Semantic (d) Synergic

302. The presence of a chiasma indicates that two chromatids

(a) Are about to form a synapsis (b) Are haploid

(c) Are identical (d) Have exchanged parts

303. Sexual reproduction favors

(a) Genetic stability (b) Highly successful species

(c) Stable populations (d) Genetic diversity

304. Unique to meiosis is the failure of chromosomes to replicate between

(a) Telophase I and prophase II

(b) Prophase I and metaphase I

(c) Prophase II and telophase II

(d) Telophase II and interphase

305. Which of the following is not a form of asexual reproduction?

(a) Fragmentation (b) budding

(c) Parthenogenesis (d) Binary fission

306. Which of the following are mismatched

(a) haploid-n (n) Somatic cells-2n

(c) Zygote-n (d) Sperm cell-n

307. Chiasmata formation and crossing over occur during:

(a) Prophase of mitosis (b) Prophase I of meiosis

(c) Prophase II of meiosis (d) b and c, but not A

308. If a cell with 32 chromosomes divides by meiosis, how many chromosomes will each nucleus contain at telophase I? (Assume cytokinesis has occurred).

(a) 64 (b) 48

(c) 32 (d) 16

309. Following the same cell in the previous question, how many chromosomes will each nucleus contain at telophase II?

(a) 64 (b) 48

(c) 32 (d) 16

310. Which one of the following statements is not true about meiosis?

(a) Meiosis occurs in reproductive cells

(b) Meiosis results in four haploid daughter cells

(c) In meiosis, chromosomes do not exchange genetic material

(d) In meiosis, homologous pairs of chromosomes are pulled apart

311. Crossing over occurs during:

(a) Prophase I (b) Prophase II

(c) Interphase I (d) Interphase II

312. Evolution of sexual reproduction may have occurred based on all of the following, The except

(a) As a means to keep animals within a population in close proximity

(b) As a means to correct damage to the double strands of the DNA

(c) Through independent assortment, offspring have new combinations of genes that can be beneficial

(d) Through crossing over, there is an unlimited amount of genetic variability in the population

313. When is binary fission the best method of cell division?

(a) When the environment is not changing

(b) When the environment is changing rapidly

(c) During the rainy season

(d) During the warm season

314. Sexually reproducing plants undergo alternation of generations. Which stage produces the gametes?

(a) Diploid sporophyte stage

(b) Diploid zygote

(c) Haploid gametophyte stage

(d) Haploid spores

315. What is an example of sex without reproduction?

(a) Bacterial conjugation (d) Parthenogenesis

(c) Binary fission (d) Mitosis

316. Which of the two meiotic divisions halves the chromosome number?

(a) First division (b) Second division

(c) Both (d) None

317. What process(es) increase(s) diversity in a population?

(a) Sexual reproduction (b) Crossing over

(c) Independent assortment (d) All of these are correct

318. At what stage of meiosis do the homologous chromosomes separate?

(a) Prophase I (b) Anaphase I

(c) Anaphase II (d) Telophase II

319. During what stage do the individual replicated chromosomes align down the center of the cell?

(a) Metaphase I (b) Telophase I

(c) Metaphase II (d) Telophase II

320. Cell types that develop into sperm cells during spermatogenesis (from early to late) are:

(a) Spermatogonium-primary spermatocyte-secondary spermatocyte-spermatid

(b) Primary spermatocyte-secondary spermatocyte-spermatid-spermatogonium

(c) Spermatogonium-spermatid-secondary spermatocyte-primary spermatocyte

(d) Primary spermatocyte-secondary spermatocyte-spermatogonium-spermatid

321. Where does meiosis occur in human males?

(a) In the scrotum (b) Vas deferens

(c) Seminiferous tubules (d) All of these are correct.

322. Cytokinesis is the division of

(a) A cell (b) A nucleus

(c) Cytoplasm (d) A chromosome

323. The genetic material in the nucleus is located in the

(a) Nucleolus (b) Nuclear membrane

(c) Nuclear sap (d) Chromosomes

324. Which of the following does mitosis normally accomplish?

(a) Production of two identical daughter cells

(b) Production of two nuclei with identical genetic content

(c) Precise division of the cytoplasm and its distribution to two daughter cells

(d) Reproduction of mitochondria and chloroplasts

325. Where does the duplication of chromosomes occur?

(a) Interphase (b) Prophase

(c) Metaphase (d) anaphase

326. In what phase does the movement of individual chromosomes towards opposite poles occur?

(a) Interphase (b) Prophase

(c) Metaphase (d) Anaphase

327. In what phase does the arrangement of chromosomes along the equator of the cell occur?

(a) Interphase (b) Prophase

(c) Metaphase (d) Anaphase

328. In what phase does the disappearance of nuclear membrane and nucleoli occur?

(a) Interphase (b) Prophase

(c) Metaphase (d) Anaphase

329. In what phase does the DNA become visible strands?

(a) Interphase (b) Prophase

(c) Metaphase (d) Anaphase

330. The two chromatids of a chromosome separate from each other during

(a) Prophase (b) Metaphase

(c) Anaphase (d) Interphase

331. Which of the following are typical of both mitosis and of the first division of meiosis?

(a) Spindle fibers form

(b) Two nuclei form as a result of the division

(c) All of the above

(d) None of the above

332. Which of the following is characteristic of the first division of meiosis but not mitosis?

(a) Chromosomes are duplicated prior to division

(b) Nuclear membrane and nucleolus disappear in prophase

(c) Sister chromatids are attached to each other at the centromere

(d) Homologous chromosomes pair

333. Which of the following are ways that meiosis differs from mitosis?

(a) In meiosis, the chromosome number is reduced

(b) In meiosis, the daughter cells are genetically different from the parent cell

(c) In meiosis, at least some of the daughter cells differ genetically from each other

(d) All of the above

334. An advantage of sexual reproduction over asexual reproduction is that it allows protection for the offspring.

(a) More offspring to be produced

(b) More genetic variation among the offspring

(c) All of the above

(d) None of the above

335. Which of the following is true of mitosis in a diploid cell?

(a) It results in 2 haploid daughter cells

(b) It involves transcription of the cell's DNA

(c) It results in daughter cells who are genetically identical, but physically smaller, than the original cell

(d) All of the above

336. From here on in, anything is fair game. Ready? Is crossing over (swapping of segments of DNA between homologous chromosomes) seen with mitosis?

(a) Yes, it is a normal part of mitosis

(b) No, crossing over is not associated with mitosis

(c) Yes, but it only occurs in fetal cells

(d) None of those above

337. During which stage does DNA replication occur?

(a) Prophase (b) Anaphase

(c) Metaphase (d) None of those above

338. Which stage of mitosis is seen in the pictured cell?

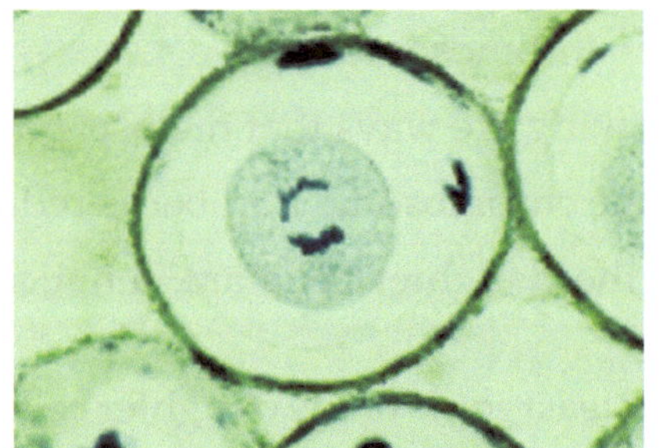

(a) Anaphase (b) Prophase

(c) Telophase (d) Interphase

339. At what phase in the cell cycle does DNA replication occur?

(a) G1 (b) S

(c) G0 (d) M

340. Which of the following is NOT correct?

(a) Mitosis is also known as karyokinesis

(b) Cytokinesis is a part of mitosis

(c) Metaphase occurs before anaphase

(d) None of these

341. What is the function of centrioles?

(a) To cause cytokinesis via the concentric shortening of microtubules

(b) They are required for DNA replication

(c) They are the structure which hold the 2 daughter chromatids together at the centromere

(d) They produce microtubules for the chromosomes to migrate along as they move to opposite poles of the cell

342. What stage of mitosis is in this cell?

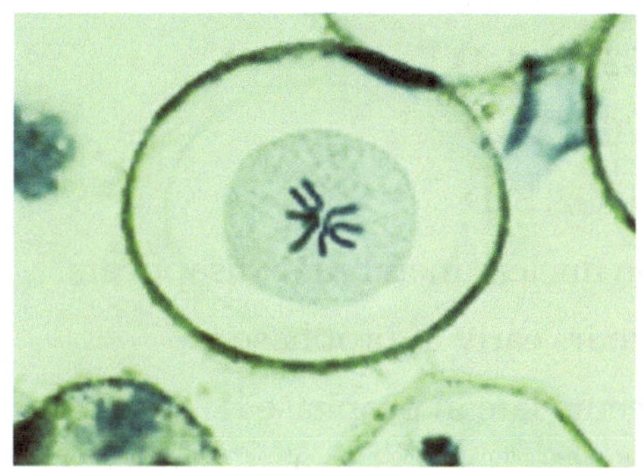

(a) Anaphase (b) Telophase

(c) Prophase (d) None of the above stages

343. Which answer gives the stages of mitosis in proper chronological order?

(a) Anaphase, prophase, metaphase, telophase

(b) Metaphase, anaphase, telophase, prophase

(c) Prophase, metaphase, anaphase, telophase

(d) Interphase, prophase, metaphase, anaphase, telophase

344. This is a picture of a cell in telophase.

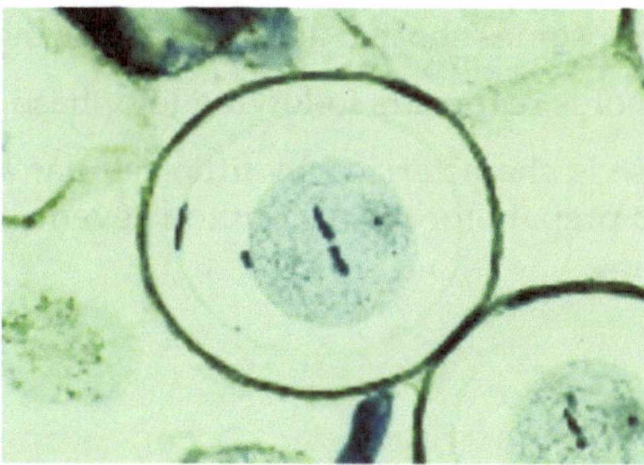

(a) Correct

(b) No, this is a cell in anaphase

(c) No, this is a cell in interphase

(d) None of the above choices

345. Which is the correct sequence of stages in the cell cycle of a terminally differentiated cell?

(a) G0, G1, S, G2, M, D

(b) GO, G1, G2, S, M, D

(c) G0, S, G1, G2, M, D

(d) G0, G1, M, G2, S, D

346. In prophase, the nuclear membrane disappears.

(a) True, it occurs early in prophase

(b) True, it occurs late in prophase

(c) False, it is not associated with prophase

(d) None

347. Match the terms with the appropriate stages in the answer: Migration, Shortening and Thickening, Cytokinesis, Prophase.

(a) Telophase, Anaphase, Prophase, centrioles forming

(b) Anaphase, Prophase, Metaphase, microtubules

(c) Anaphase, Prophase, Telophase, centrioles forming

(d) Metaphase, Anaphase, Telophase, microtubules

348. Which of the following is correct?

(a) The chromosomes shorten and thicken during prophase

(b) The nucleolus reappears following telophase

(c) Interphase is characterized by little cellular activity, as the cell is resting to prepare for the next mitotic event

(d) All of the above

349. Place the pictures in the correct chronological order.

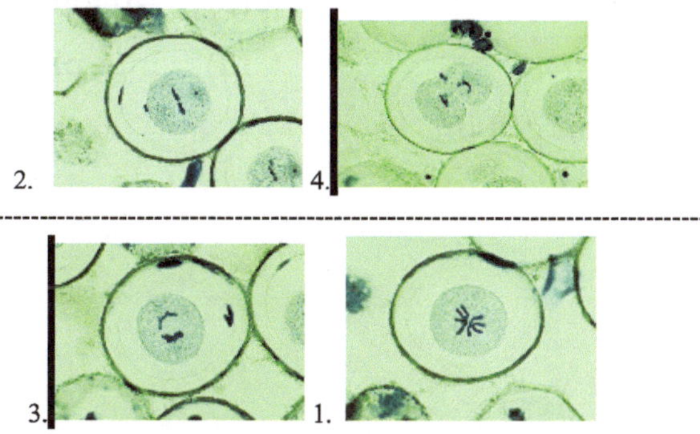

(a) 1, 2, 3, 4 (b) 2, 3, 1, 4
(c) 4, 1, 3, 2 (d) 3, 4, 1, 2

350. Which stage is seen in this cell?

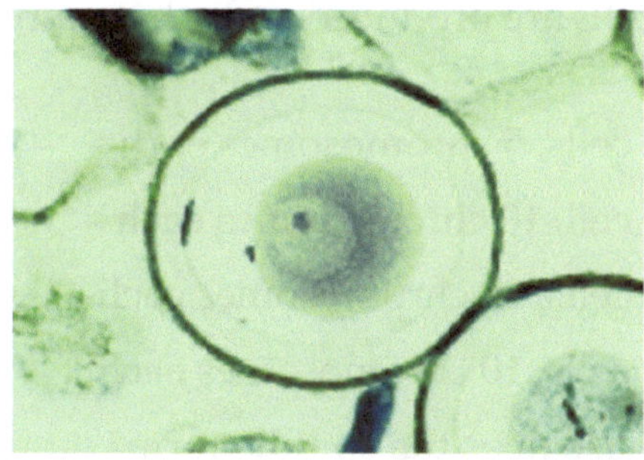

(a) Prophase (b) Anaphase
(c) Metaphase (d) Interphase

351. What phase is this cell in?

(a) Prophase (b) Metaphase
(c) Anaphase (d) Telopase

352. During which phase do chromosomes become visible?

(a) Telophase (b) Anaphase

(c) Prophase (d) Interphase

353. Which phase is this cell in?

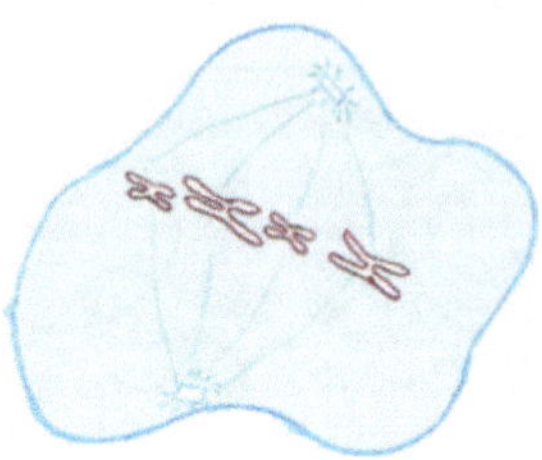

(a) Prophase (b) Telophase

(c) Metaphase (d) Interphase

354. A cell with 10 chromosomes undergoes mitosis and cell division. How many daughter cells are produced and what number of chromosomes do they have

(a) 2 daughter cells, 5 chromosomes each

(b) 1 daughter cell, 10 chromosomes each

(c) 4 daughter cells, 20 chromosomes each

(d) 2 daughter cells, 10 chromosomes each

355. During which phase does the spindle form?

(a) Prophase (b) Interphase

(c) Metaphase (d) Anaphase

356. Which phase is the cell in?

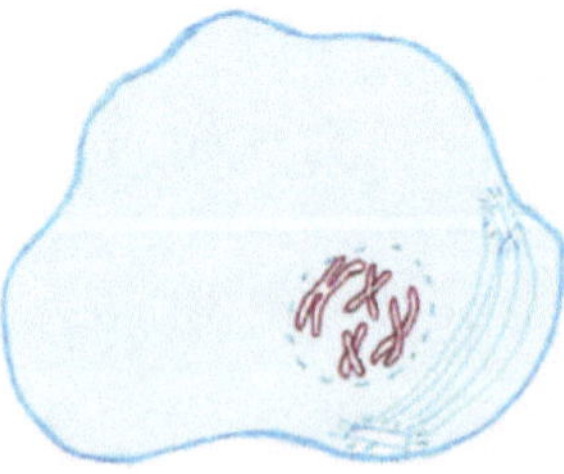

(a) Propase (b) Telophase

(c) Interphase (d) Metaphase

357. The process where the cytoplasm divides and forms two new cells is called:

(a) Mitosis (b) Cytokinesis
(c) Karyotype (d) Spindle formation

358. What phase is the cell in?

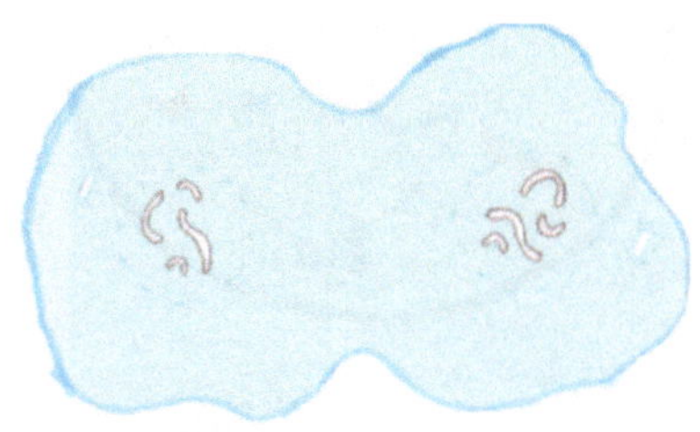

(a) Anaphase (b) Interphase
(c) Telophase (d) Prophase

359. Cytokinesis begins in:

(a) Interphase (b) Metaphase
(c) Telophase (d) Prophase

360. What is depicted in the illustration

(a) Mitosis of an animal cell (b) Cytokinesis of an animal cell
(c) Mitosis of a plant cell (d) Cytokinesis of a plant cell

361. The process of mitosis ensures that:

(a) Each new cell is genetically different from its parent
(b) Each new cell receives the proper number of chromosomes
(c) Cells will divide at the appropriate time
(d) DNA is replicated without errors

362. The image illustrates what phase of mitosis

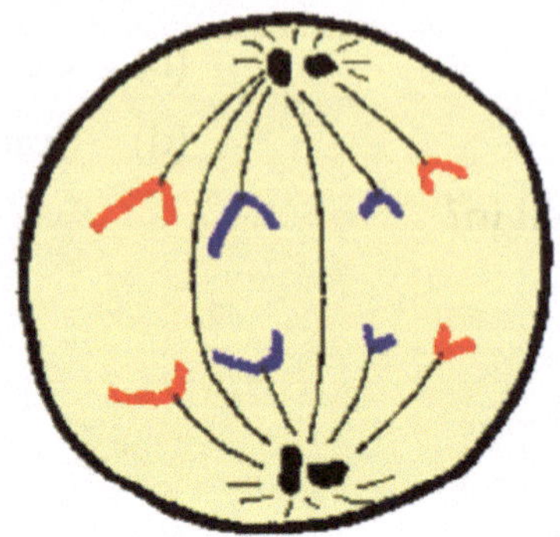

(a) Anaphase (b) Teleophase
(c) Metaphase (d) Prophase

363. Which of the following is NOT part of mitosis
(a) Prophase (b) Metaphase
(c) Telophase (d) Interphase

364. Which of the following is NOT part of the chromosome?
(a) Kinetochore (b) Chromatid
(c) Centromere (d) Spindle

365. A cell that has 20 chromosomes undergoes mitosis. Which of the following is true?
(a) Two daughter cells will be created, each have 20 chromosomes
(b) Two daughter cells will be created, each have 40 chromosomes
(c) 4 daughter cells will be created, each having 10 chromosomes
(d) 2 daughter cells will be created, each having 10 chromosomes

366. A spindle forms during which phase?
(a) G2 (b) Interphase
(c) Prophase (d) Metaphase

367. Compared to the X chromosome, the Y chromosome is:
(a) Much larger (b) Much smaller
(c) More twisted (d) Inherited more often

368. Which of the following can be determined from a karyotype?

(a) The sex of the individual

(b) Whether the individual has Down Syndrome

(c) The number of chromosomes present

(d) All of these

369. Most cells spend their lives in:

(a) Prophase (b) Metaphase

(c) Interphase (d) Telophase

370. Cytokinesis begins during which phase?

(a) Telophase (b) Synthesis phase

(c) Anaphase (d) Metaphase

371. First one! Let's make it simple to get you going: Are meiosis and mitosis synonymous (do they describe the same process)?

(a) Yes they are

(b) Yes, but it's only called meiosis when it occurs in the testes or ovaries

(c) No, these words describe different processes

(d) No, except on the third Sunday in a month whose Julian date is the first decimals in pi or has 7 as a factor

372. As before, anything is fair game from here on in. Ready? In meiosis the chromosome number is reduced to the haploid number in a diploid animal.

(a) Yes, and this is a normal part of mitosis, too

(b) Yes

(c) No, it remains at the diploid number

(d) None of those above

373. During which stage does DNA replication occur?

(a) Prophase I

(b) Prophase II

(c) Metaphase I

(d) None of those above

374. Which of the following is true of the labelled structures from this diploid animal?

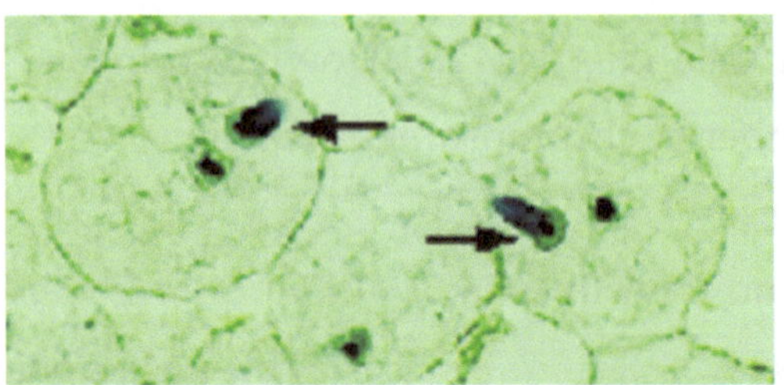

(a) They are currently within a cell in the ovary

(b) They have a haploid chromosome number

(c) They are female pronuclei

(d) They have a diploid chromosome number

375. In which organs does meiosis occur?

(a) The thyroid and parathyroid

(b) The ovary and testis

(c) The testis and seminiferous tubules

(d) The ovary and Graafian follicle

376. Which of the following is NOT correct?

(a) The nuclear membrane breaks down during Metaphase I

(b) Prophase I can be subdivided into five stages

(c) Following Telophase II, a single primary spermatocyte will have produced four spermatids

(d) None of these

377. What is mean by the term, "pairing of homologous chromosomes?"

(a) It indicates that two different chromosomes (different genes on each) come together

(b) It indicates that two identical chromosomes (same genes with the same DNA sequences in each gene) come together

(c) It indicates that two similar chromosomes (same genes, but with potentially different DNA sequences in each gene) come together

(d) It is irrelevant, as this does not occur as part of meiosis

378. In which stage of meiosis does crossing over occur?

(a) Anaphase I
(b) Metaphase II
(c) Zygotene
(d) Prophase I

379. Which meiotic division is seen to be occurring within this Ascaris cell?

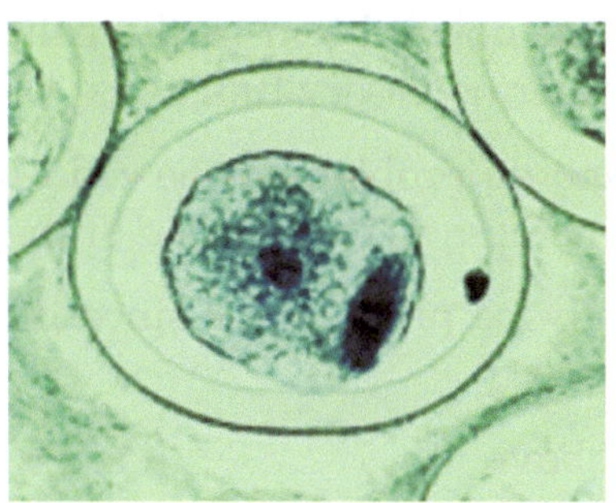

(a) The first
(b) The second
(c) The third
(d) It's impossible to tell from this picture

380. Which is the correct statement?

(a) The first meiotic division is quicker than the second
(b) Some stages of meiosis occur in the ovum following the fusion of the male and female pronuclei
(c) Secondary spermatocytes have completed the first meiotic division
(d) All of the above choices

381. Identify the true statement regarding polar bodies.

(a) They contain genetic material and can undergo Telophase II
(b) They are seen with both spermatogenesis and oogensis
(c) There are two polar bodies following the completion of meiosis associated with oogenesis in all species
(d) They are associated only with mitosis

382. Place the stages of Prophase I in the correct order.

(a) Diakinesis, diplotene, leptotene, pachytene, zygotene
(b) Diplotene, leptotene, pachytene, zygotene, diakinesis

(c) Leptotene, pachytene, diakinesis, diplotene, zygotene

(d) Leptotene, zygotene, pachytene, diplotene, diakinesis

383. A cell is exposed to a substance that prevents it from dividing. The cell becomes larger and larger. This situation:

(a) Should present no problem to the cell since it can continue to perform all other necessary functions

(b) Should present no problem to the cell because the surface area of the cell will increase as the volume of the cell increases

(c) Will eventually be problematic since the surface area to volume ratio is decreasing

(d) None of the above

384. Which of these structures would never be found in an animal cell?

(a) Endoplasmic reticulum (b) Nucleus

(c) Central vacuole (d) Cell membrane

385. The ability of some bacteria to cause disease relates to the presence of both:

(a) Pili and capsules

(b) Cytoplasm and a nucleoid region

(c) Ribosomes and DNA

(d) The cell membrane and plasmids

386. A scientist discovers an unidentified unicellular organism. To identify it as eukaryotic she must determine if it has:

(a) Ribosomes (b) Cytoplasm

(c) DNA (d) Mitochondria

387. Which of the following organelles are concerned with energy production?

(a) Plastids and mitochondria

(b) Endoplasmic reticulum and the Golgi complex

(c) Mitochondria and chloroplasts

(d) Lysosomes and chloroplasts

388. Chromosomes are made of

(a) DNA (b) Genes

(c) Both A and B (d) None of the above

389. Because individuals have different combinations of traits,

(a) There is variation in nature (b) Everyone is the same

(c) No traits are visible (d) All of the above

390. Meiosis

(a) Is the same as mitosis

(b) Is the process that forms eggs and sperm

(c) Is required for the body to make more skin cells

(d) Occurs in heart cells

391. The joining of an egg and a sperm (in humans) produces

(a) A zygote with 23 chromosomes

(b) A zygote with 46 chromosomes

(c) A zygote with 1/2 the full set of chromosomes

(d) None of the above

392. Meiosis

(a) Occurs in the ovary to produce 1 egg and 3 polar bodies

(b) Occurs in the testes to produce 4 sperm

(c) Forms sex cells with 23 chromosomes (in humans)

(d) All of the above

Use the following diagram to answer this question:

393. How many chromosomes are in the original cell?

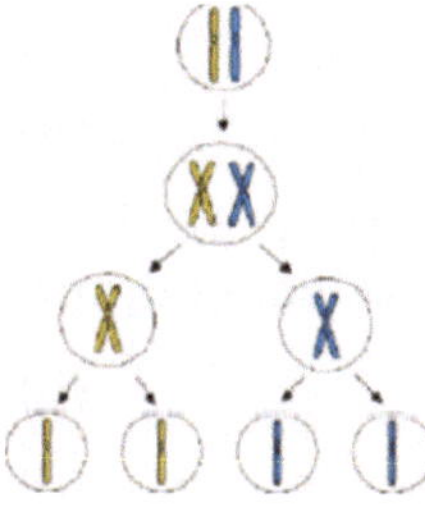

(a) 1 (b) 2

(c) 3 (d) 4

394. What happens to the number of chromosomes before the cell divides for the first time?

(a) The number of chromosomes stays the same

(b) The number of chromosomes becomes less

(c) The number of chromosomes doubles

(d) None of the above

395. What happens to the number of chromosomes before the cell divides for the first time?

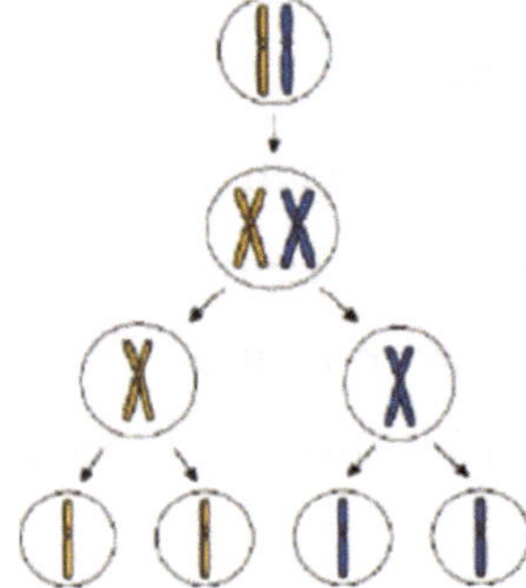

(a) The number of chromosomes stays the same

(b) The number of chromosomes becomes less

(c) The number of chromosomes doubles

(d) None of the above

Fill In the blanks

1. A gram-positive bacterium is stained ______________ by the gram stain

(a) Pink (b) Purple

(c) Blue (d) Green

2. The endosymbiont theory proposes that chloroplasts and mitochondria evolved from ________ and were engulfed by larger ________.

(a) Archaea; eukarya (b) Bacteria; archaea

(c) Bacteria; eukarya (d) Eukarya; bacteria

3. __________ only occur(s) in the gonads to produce gametes.

(a) Mitosis (b) Meiosis

(c) Mitosis and meiosis (d) Sporogony

4. ____________ are genes that produce proteins that stimulate cell division. Damages to these genes can cause overproduction of these proteins, leading to cancer.

(a) Proto-oncogenes (b) Tumor-suppressing genes

(c) Chiasmata (d) None

5. The sister chromatids move toward the poles in ____________ .

(a) Anaphase (b) Interphase

(c) Metaphase (d) Prophas

6. Prokaryotes reproduce via __________.

(a) Binary fission (b) Meiosis

(c) Syngamy (d) All of the above

7. During __________ the nuclear membrane begins to reform and the nucleolus reappears.

(a) Prophase (b) Metaphase

(c) Anaphase (d) Telophase

8. Unlike gametes, body cells are called _________cells.

(a) Somatic (B) Haploid

(c) Semantic (d) Synergic

9. __________hold homologous chromosomes together during prophase I of meiosis.

(a) Chiasmata (b) Centrioles

(c) Centromeres (d) Metaphase plates

10. The process where two gametes fuse is called ________.

(a) Zygote

(b) Syngamy

(c) Fertilization

(d) B and C are both correct

11. In anaphase I of meiosis, _________ are separated; in anaphse II of meiosis, _________ are separated.

(a) Homologous chromosomes, non-homologous chromosomes

(b) Homologous chromosomes, sister chromatids

(c) Sister chromatids, homologous chromosomes

(d) Homologous chromatids, sister chromosomes

12. The type of programmed cell death that is a part is normal development is called _______________.

(a) Mitosis (b) Apoptosis

(c) Cytokinesis (d) Angiogenesis

13. During prophase, chromosomes _________________.

(a) Align at the center of the cell

(b) Condense and become visible

(c) Separate at the centromeres

(d) Begin to replicate

14. The phase in which chromosomes separate and move to opposite poles is ___________.

(a) Prophase (b) Anaphase

(c) Metaphase (d) Telophase

15. When plant cells construct a new cell wall that separates two newly-formed cells, the first structure to form is the ______________.

(a) Cell plate (b) Cleavage furrow

(c) Phragmoplast (d) Centriole

16. A "checkpoint" that controls a cell's fate during a specific point within the cell cycle is actually made up of ________________.

(a) A group of interacting proteins

(b) The mitotic spindle

(c) Telomeres

(d) Actin and tubulin

17. In cells of humans and other vertebrates, chromosomes ______________.

(a) Have telomeres at their ends

(b) Lose 50 to 200 DNA nucleotides with each cell division

(c) Shorten each time the cell divides

(d) All of these are correct

18. Molecules that act locally to stimulate mitosis are called ______________.

(a) Growth factors (b) Hormones

(c) Stimulants (d) Kinases

19. Levels of ____________ fluctuate in the cell and form the checkpoint that operates in G1 to determine whether or not a cell divides, while ____________, always present in the cell, bind to cyclin molecules before cell division.

(a) Epidermal growth factor; necrosis factor

(b) Hormones; growth factors

(c) Cyclins; kinases

(d) Kinases; growth factors

20. Many tissues retain the ability to renew themselves and to repair damage because they contain a few ______________.

(a) Stem cells (b) Molecules of telomerase

(c) Static cell populations (d) None of these is correct

21. Apoptosis begins when ______________.

(a) The cell is injured

(b) A "death receptor" on the cell membrane receives the signal to die

(c) Caspases become activated

(d) Killer enzymes tear up the cytoskeleton

22. The term for a mass of cells growing out of control is ______________.

(a) A benign tumor

(b) A malignant tumor

(c) Metastasis

(d) All of the above are correct

23. Inactivation of ______________ is one of the steps leading to the development of cancer.
(a) Tumor suppressor genes (b) Oncogenes
(c) Growth factors (d) Stem cells

24. Mutations ____________ are associated with many cancers.
(a) Preventing the initiation of metaphase
(b) Disabling key elements of the G1 checkpoint
(c) Of the cell membrane
(d) Both a and c

25. Every nucleus in a eukaryotic organism contains the same amount of DNA, with the exception of nuclei in the _________.
(a) Epidermal cells (b) Muscle cells
(c) Gametes (d) Neurons

26. All of the following are cellular organelles in a eukaryotic cell except the _________.
(a) Mitochondria (b) Endoplasmic reticulum
(c) Plasma membrane (d) Nucleus

27. The two primary divisions of the cell cycle are __.
(a) Interphase and telophase (b) Metaphase and G1
(c) Telophase and prophase (d) Mitosis and interphase

28. A cell with damaged DNA can lead to cancer if it is allowed to divide. In the normally functioning cell cycle such cells are prevented from division by __________.
(a) The product of mutated p53 genes
(b) The M checkpoint
(c) The selectivity of the kinetochores
(d) The G1/S and G2/M checkpoints

29. Cells that are metabolically active but not destined to proliferate are said to be in _____ phase.
(a) S (b) G0
(c) G2 (d) G1

30. The checkpoint that requires a cell to be of adequate size in order to move forward is:

(a) The M checkpoint (b) G1/S checkpoint

(c) Mid metaphase checkpoint (d) G0 checkpoint

31. The p53 gene product (the protein made from this gene) is involved in:

(a) The length of time a cell spends in interphase

(b) Bringing cells out of G0

(c) The length of time a cell spends in prophase

(d) Marking cells for programmed cell death

32. The most important difference between chromosome behavior in mitosis and meiosis is that __.

(a) There are two divisions of chromatids in mitosis but only one in meiosis

(b) Meiosis occurs in all somatic cells but mitosis can occur only in special reproductive cells

(c) Crossing over plays a critical role in mitosis but only a minor one in meiosis

(d) In mitosis paternal and maternal members of a homolog are autonomous but in meiosis they synapse

33. The first division in meiosis is known as the ______ division.

(a) Additional (b) Primary

(c) Pairing (d) Reductional

34. In the second division of meiosis each _____ splits to form two _____.

(a) Homologous pair, dyads (b) Dyad, monads

(c) Chromosome, chromatids (d) Sister chromatid, homologs

35. A cell with damaged DNA can lead to cancer if it is allowed to divide. In the normally functioning cell cycle such cells are prevented from division by ___.

(a) cdc mutations

(b) The G1/S and G2/M checkpoints

(c) The M checkpoint

(d) The selectivity of the kinetochores

36. Cells that are metabolically active but not destined to proliferate are said to be in _____ phase.

(a) Metaphase (b) G0

(c) G2 (d) G1

37. The occurrence of __ is evidence for crossing over in an earlier stage of meiosis.

(a) Bivalents (b) Chromomeres

(c) Chiasmata (d) Synapsis

38. Crossing over refers to __.

(a) The onset of homology search in the leptotene stage of meiosis I

(B) The random segregation of dyads in metaphase I of meiosis

(c) The formation of lateral elements between paired homologs during zygonema

(d) Genetic exchange between nonsister chromatids during meiosis

39. Before the Hershey-Chase experiment, many scientists believed that ____ carried the hereditary information.

(a) Proteins (b) Phages

(c) Helicase (d) The plasma membrane

40. In RNA, _____ codon(s) translate to ______ amino acid(s)

(a) 1, 1 (b) 3, 1

(c) 3, 3 (d) 1, 20

41. The sugar in RNA is _____, the sugar in DNA is _______

(a) Deoxyribose, ribose (b) Ribose, deoxyribose

(c) Ribose, phosphate (d) Ribose, uracil

42. The science of genetics is concerned with _________.

(a) The origin of heritable differences

(b) The transmission of traits

(c) The distribution of heritable characteristics in populations

(d) All of the above are correct.

43. _________ lower the energy needed to initiate a chemical reaction in a living cell.

(a) Enzymes (b) Nucleotides

(c) Carbohydrates (d) Proteins

44. Cytogenetics is related to all the following except _________.

(a) Karyotype (b) Mitosis

(c) Meiosis (d) Pedigree analysis

45. All of the following are examples of positive eugenics techniques except ________.

(a) Selection of embryos free of disease-causing genes

(b) Outlawing reproduction by known criminals

(c) Allowing only Nobel prize winners to be egg or sperm donors for *in-vitro* fertilization

(D) selection of egg or sperm cell with dominant genes for muscle strength and coordination

46. The gap in Darwin's theory of evolution was his inability to _________.

(a) Observe variation

(b) Explain the basis of evolution

(c) Explain migration

(d) Observe large populations

47. Homologous chromosomes have the same genes at the same ____.

(a) Allele (b) Loci

(c) Synapsis (d) Diploid

48. During which phase of meiosis does crossing-over usually occur?

(a) Prophase I (b) Prophase II

(c) Telophase I (d) Telophase II

49. Alternate states of a gene are called ___.

(a) Allele (b) Loci

(c) Synapsis (d) Diploid

50. During which phase of meiosis can it be said that reduction division occurs?

(a) Prophase I (b) Prophase II

(c) Anaphase I (d) Anaphase II

51. After telophase I, the daughter cells are ___ with ___ chromatid(s).

(a) Diploid, two (b) Diploid, one

(c) Haploid, two (d) Haploid, one

52. Gametes are reproductive cells with ___ nuclei resulting from ____.

(a) Diploid, mitosis (b) Haploid, mitosis

(c) Diploid, meiosis (d) Haploid, meiosis

53. In human males, gamete arise through _____, which occurs in the _____.

(a) Gametogenesis, seminiferous tubules

(b) Spermatogenesis, seminiferous tubules

(c) Oogenesis, seminiferous tubules

(d) Oogenesis, Graafian follicle

54. Of the following meiotic phases, ____ is most similar in behavior to the mitotic phase of the same name.

(a) Prophase II (b) Anaphase I

(c) Metaphase I (d) None of the above

55. __________ distinguishes prophase from mitotic prophase.

(a) The number of chromatids per chromosome

(b) Synapsis

(c) The number of homologues

(d) Terminal chiasmata

56. If a zygote has 4 chromosomes, the somatic cells formed from it have _______________ chromosomes.

(a) 4 (b) 8

(c) 2 (d) 1

57. In anaphase I, the ____________ are pulled apart.

(a) Chiasmata (b) Centromeres

(c) Kinetochores (d) Sister chromatids

58. If gametes have 8 chromosomes, the cell resulting from syngamy will have ________ chromosomes.

(a) 4 (b) 8

(c) 16 (d) 2

59. Binary fission is ___________.

(a) Meiotic division of a cell

(b) A process by which amoebae reproduce sexually

(c) Division of one cell into two

(d) The most complex form of cell division

60. A cell having one set of chromosomes is said to be ___________.

(a) Haploid (b) Diploid

(c) Triploid (d) Polyploid

61. Meiosis is a form of cell division that ____________.

(a) Is common to asexual life cycles

(b) Does not require fertilization to follow

(c) Prevents evolution

(d) Halves the genetic material

62. The sac of enzymes contained in a sperm cell enabling penetration into the egg is the __________.

(b) Protamine (b) Acrosome

(c) Spermatocyte (d) Midpiece

63. The ____________ stores spermatozoa until ejaculation.

(a) Bulbourethral gland (b) Epididymis

(c) Urethra (d) Prostate gland

64. Male sex hormones are secreted by __________.

(a) Spermatogonia (b) Secondary spermatocytes

(c) Sustentacular cells (d) Interstitial cells

65. Egg-making in the female is called ____________.
   (a) Spermatogenesis
   (b) Oogenesis
   (c) Fertilization
   (d) Embryo transfer

66. In the female, small cells that form as a result of meiosis that contain little cytoplasm are ____________.
   (a) Ova
   (b) Oogonia
   (c) Polar bodies
   (d) Secondary oocytes

67. In humans, approximately how many oocytes develop to maturity from puberty to menopause and are released during ovulation?
   (a) 100
   (b) 400
   (c) 1,000
   (d) 10,000

68. Homologous chromosomes are not necessarily identical in their __________.
   (a) Overall size
   (b) Centromere placement
   (c) Arm ratio
   (d) Alleles

69. According to the Denver Classification, a chromosome with its centromere located very close to, but not on, the end of the chromosome is __________.
   (a) Telocentric
   (b) Acrocentric
   (c) Metacentric
   (d) Subcentric

70. During meiosis __________.
   (a) The chromosome number is doubled
   (b) Chromosomes are not involved
   (c) Two identical cells are produced
   (d) The chromosome number is reduced by half

71. Meiosis involves _______ division(s) of a nucleus.
   (a) One
   (b) Two
   (c) Four
   (d) Eight

72. As cell volume increases, it takes ____________ time for nutrients to diffuse into a cell.
   (a) More
   (b) Less
   (c) The same
   (d) None

73. Oxidative metabolism is carried out ____ of mitochondriA)

(a) In the intermembrane space

(b) On the surface of the inner membrane

(c) In the inside of the outer membrane

(d) In the matrix

74. Ribosomes are made up of ____subunits.

(a) 0 (They are whole.) (b) 2

(c) 3 (d) 4

75. The rough ER is so named because it has an abundance of ____________ on it.

(a) Mitochondria (b) Lysosomes

(c) Golgi bodies (d) Ribosomes

True/False

1. The microtubules of cilia and flagella are organized in a characteristic 9 + 2 pattern, and they slide past one another.

(a) True (b) False

2. Proteins lacking a signal peptide sequence will probably be secreted from the cell.

(a) True (b) False

3. According to the endosymbiotic theory, the infoldings and specializations of the plasma membrane led to the evolution of the endomembrane system.

(a) True (b) False

4. True or false: Mitochondria are inherited from the female parent.

(a) True (b) False

5. A cell that measures 200 μm in diameter compared to a cell that measures 20μm will have 100 times the surface area but 1000 times the volume.

(a) True (b) False

6. Because they have chloroplasts for energy production, plant cells lack mitochondria.

   (a) True (b) False

7. The cell theory was firmly established in the nineteenth century and has not changed since.

   (a) True (b) False

8. True or false: Genetic evidence does not substantiate the endosymbiont theory.

   (a) True (b) False

9. The nucleus envelope is a bilayered structure which has numerous pores.

   (a) True (b) False

10. The gelatinous fluid inside the nucleus is called nucleoplasm.

    (a) True (b) False

11. Prior to replication the DNA in the nucleus is in the form of a tangled mass called chromatin.

    (a) True (b) False

12. Blood cells placed in pure water will burst because there is a net movement of water out of the cell.

    (a) True (b) False

13. Diffusion is the movement of water molecules across a semi-permeable membrane which continues until equilibrium has been reached.

    (a) True (b) False

14. The nucleus is surrounded by two layers of membrane.

    (a) True (b) False

15. Detoxification of potentially dangerous oxidizing moelcules occurs within the lysosome.

    (a) True (b) False

16. The Golgi complex synthesizes proteins.

    (a) True (b) False

17. Energy for the cell is generated by the mitochondrion.

    (a) True (b) False

18. Chloroplasts are found only in plants.

(a) True (b) False

19. Cell vacuoles are comprised of empty space.

(a) True (b) False

20. Cell walls in plants are composed of cellulose.

(a) True (b) False

21. The chemical reactions of photosynthesis occur on the membranes of the mitochondrion.

(a) True (b) False

22. Mitochondria are a kind of bacterial cell.

(a) True (b) False

23. Chloroplasts are a kind of bacterial cell.

(a) True (b) False

24. Some cell surface proteins respond to chemical signals called hormones.

(a) True (b) False

25. The term "simple cell cycle" is a reference to the production of two identical diploid eukaryotic cells.

(a) True (b) False

26. Binary fission begins after DNA replication is completed.

(a) True (b) False

27. The spindle is a network of protein cables extending from centrioles at each pole of the cell during mitosis.

(a) True (b) False

28. Two diploid daughter cells are the result of mitosis.

(a) True (b) False

29. p53 is the strain of virus most often responsible for causing cells to become cancerous.

(a) True (b) False

30. A disabled G1 check point is associated with many cancers.

(a) True (b) False

31. Chromosome replication takes place during interphase.
(a) True (b) False

32. Cytokinesis takes place during metaphase.
(a) True (b) False

33. Normal human cells can go on dividing indefinitely.
(a) True (b) False

34. Following telophase in animal cells, a cell plate begins to form,
(a) True (b) False

35. Bacteria divide through a procell called binary fission.
(a) True (b) False

36. A chromosome is a single, long DNA molecule in a compact formation.
(a) True (b) False

37. Sperm and egg cells have a diploid number of chromosomes.
(a) True (b) False

38. The chromosome number in species increases as you climb the evolutionary tree, plants having lower numbers and animals having higher numbers.
(a) True (b) False

39. A tumor-suppressor protein Rb blocks the activity of another protein called E2F.
(a) True (b) False

40. Whatever the force that led to sexual reproduction, its evolutionary consequences have been profound—no genetic process generates diversity more quickly.
(a) True (b) False

41. DNA replicates during the S phase of metaphase.
(a) True (b) False

42. Cancer can be caused by chemicals that mutate DNA
(a) True (b) False

43. In meiosis, two haploid cells—the egg and the sperm—unite to form a diploid cell called a zygote.

(a) True (b) False

44. Cytokinesis always follows shortly after karyokinesis.

(a) True (b) False

45. Meiotic and mitotic cycles differ in the number of chromatids per chromosome in prophase.

(a) True (b) False

46. Asexual growth is synonymous with mitosis.

(a) True (b) False

47. Crossing over typically occurs between two chromatids of the same duplex chromosome.

(a) True (b) False

48. Sperm cells can ferry cocaine to an ovum during fertilization.

(a) True (b) False

49. In humans, the X and the Y chromosomes are the autosomes.

(a) True (b) False

50. Traits are passed down from offspring to parents.

(a) True (b) False

51. The science of genetics includes topics such as cloning, gene therapy, and DNA fingerprinting.

(a) True (b) False

52. All traits can be seen.

(a) True (b) False

53. All genes are the same length.

(a) True (b) False

54. A gene codes for a particular protein molecule which will result in a specific trait.

(a) True (b) False

Answers

Multiple Choice Question

1	2	3	4	5	6	7	8	9	10	11	12	13	14	15
B	D	C	B	A	C	C	D	C	B	C	B	D	D	D
16	17	18	19	20	21	22	23	24	25	26	27	28	29	30
A	D	A	D	B	D	E	B	A	D	C	F	D	A	D
31	32	33	34	35	36	37	38	39	40	41	42	43	44	45
D	D	A	B	D	D	D	B	A	D	D	C	C	D	A
46	47	48	49	50	51	52	53	54	55	56	57	58	59	60
D	A	B	C	C	D	C	D	C	C	B	D	D	D	B
61	62	63	64	65	66	67	68	69	70	71	72	73	74	75
D	B	A	D	A	B	C	D	A	C	C	D	D	C	A
76	77	78	79	80	81	82	83	84	85	86	87	88	89	90
D	D	D	B	D	C	D	C	B	D	B	C	A	C	A
91	92	93	94	95	96	97	98	99	100	101	102	103	104	105
A	D	D	C	C	D	A	C	D	B	D	A	D	B	D
106	107	108	109	110	111	112	113	114	115	116	117	118	119	120
D	A	A	D	A	D	D	D	A	C	A	A	A	B	B
121	122	123	124	125	126	127	128	129	130	131	132	133	134	135
D	B	C	C	A	C	B	D	B	A	D	C	D	D	B
136	137	138	139	140	141	142	143	144	145	146	147	148	149	150
D	D	C	D	D	B	D	C	B	D	C	D	D	B	B
151	152	153	154	155	156	157	158	159	160	161	162	163	164	165
D	B	C	B	D	D	D	B	D	D	B	A	A	B	A
166	167	168	169	170	171	172	173	174	175	176	177	178	179	180
B	A	C	D	B	C	C	C	B	A	D	A	C	B	D
181	182	183	184	185	186	187	188	189	190	191	192	193	194	195
C	D	A	D	A	B	B	C	C	D	B	A	C	D	A
196	197	198	199	200	201	202	203	204	205	206	207	208	209	210

B	A	B	D	D	C	C	B	D	A	B	C	D	A	C
211	212	213	214	215	216	217	218	219	220	221	222	223	224	225
D	C	C	B	C	A	D	D	D	B	C	D	A	C	D
226	227	228	229	230	231	232	233	234	235	236	237	238	239	240
C	A	D	A	B	C	D	D	B	B	A	B	B	D	B
241	242	243	244	245	246	247	248	249	250	251	252	253	254	255
C	C	D	B	D	C	B	D	D	A	B	A	D	D	D
256	257	258	259	260	261	262	263	264	265	266	267	268	269	270
A	B	D	D	C	C	B	A	B	D	C	A	D	D	D
271	272	273	274	275	276	277	278	279	280	281	282	283	284	285
A	D	C	A	B	A	D	D	D	B	B	A	A	C	A
286	287	288	289	290	291	292	293	294	295	296	297	298	299	300
A	A	D	D	B	D	B	B	C	D	D	D	C	D	C
301	302	303	304	305	306	307	308	309	310	311	312	313	314	315
A	D	D	A	B	C	D	D	D	C	A	A	A	C	A
316	317	318	319	320	321	322	323	324	325	326	327	328	329	330
A	D	B	C	A	C	C	D	B	A	D	C	B	B	C
331	332	333	334	335	336	337	338	339	340	341	342	343	344	345
A	D	D	B	C	B	D	A	B	D	D	C	C	D	A
346	347	348	349	350	351	352	353	354	355	356	357	358	359	360
B	C	A	C	D	D	C	C	D	A	A	D	C	C	B
361	362	363	364	365	366	367	368	369	370	371	372	373	374	375
B	A	D	D	A	C	B	D	C	A	C	D	D	B	B
376	377	378	379	380	381	382	383	384	385	386	387	388	389	390
A	C	D	B	C	A	D	C	C	A	D	C	C	A	B
391	392	393	394	395										
B	D	B	A	A										

Fill in the blanks

1	2	3	4	5	6	7	8	9	10	11	12	13	14	15
B	B	B	A	A	A	D	A	A	D	B	B	B	B	C
16	17	18	19	20	21	22	23	24	25	26	27	28	29	30
A	D	A	C	A	B	B	A	B	C	C	D	D	B	B
31	32	33	34	35	36	37	38	39	40	41	42	43	44	45
D	D	D	B	B	B	C	D	A	A	B	D	A	D	B
46	47	48	49	50	51	52	53	54	55	56	57	58	59	60
B	B	A	A	C	C	D	B	A	B	A	A	B	C	A
61	62	63	64	65	66	67	68	69	70	71	72	73	74	75
D	B	B	D	B	C	B	D	B	D	B	C	B	B	D

True/False

1	2	3	4	5	6	7	8	9	10	11	12	13	14	15
A	B	B	A	A	B	B	B	A	A	A	B	B	A	B
16	17	18	19	20	21	22	23	24	25	26	27	28	29	30
B	A	B	B	A	B	B	B	A	B	A	A	A	B	B
31	32	33	34	35	36	37	38	39	40	41	42	43	44	45
A	B	B	B	A	A	B	B	A	A	B	A	B	B	B
46	47	48	49	50	51	52	53	54						
B	B	A	B	B	A	B	B	A						

❑❑❑

Chapter – 3

Mendel and His Discoveries

About Mendel

- Born in what is now the Czeck Republic.
- Grew up in an agricultural environment.
- Became a priest-teacher at a local monastery.
- Took courses in botany, physics, and mathematics at the University of Vienna.
- Gregor Mendel was the first person to comprehend some of the most basic principles of genetics. He reached these understandings in the mid-nineteenth century; working in what is now the Czech Republic and using as his experimental subjects a species of garden pea, Pisum sativum..
- Carried out "hybridization" experiments with the common garden pea.
- Mendel's hypotheses became the laws of inheritance in modern genetics.
- Mendel's work re-discovered by Correns, DeVries, and Tschermak in 1900. Period of modern "Mendelian" genetics begin.

The Experimental Subject: Pisum sativum

- Mendel took seven characters in his plant—such attributes as seed color and texture. In his plants, each of these characters came in two varieties or traits, one of them dominant, the other recessive. His experiments involved breeding pea plants, starting with plants that had a given set of traits and then observing which of those traits showed up in succeeding generations.

- The physical functioning, bodily characteristics, or actions of an organism are its phenotype. In Mendel's plant, purple flowers were one phenotype and white flowers were another. Phenotypes in any organism are in significant part determined by that organism's genotype, meaning its genetic makeup. Mendel realized that the phenotypes in his plants were being controlled by what we would today call their genotypes.

Starting the Experiments

- Mendel realized that it was possible for organisms to have identical phenotypes—for all his pea plants to have yellow seeds, for example—and yet to have differing underlying genotypes.
- One of Mendel's central insights was that the basic unit of genetics are material elements that, in his pea plants, came in pairs. These elements, today called genes, come in alternative forms called alleles. One member of an allele pair resides on one chromosome, while the other allele resides on a second chromosome that is homologous to the first.
- Another of Mendel's insight was that genes retain their character through many generations, rather than being "blended" together. Genes that coded for green pea color, for example, were retained in their existing form over many generations.

Terms and Tools to Follow Segregating Genes

- A homozygous individual possesses two identical alleles (i.e. AA or aA). A heterozygous individual possesses two different alleles (i.e. AA).
- Phenotype is the outward expression of a trait (i.e. tall vs. short; blue eyes vs. brown eyes).
- Genotype is the actual genetic makeup of the individual (i.e. AA, Aa, aA).
- Wild type refers to the most common form. A mutant is a variant that has under gone a mutation (change in the DNA).

- The physical nature of meiosis (chromosome behavior) explains the law of segregation. The law of segregation states that inherited "characters" (alleles) separate during meiosis, so that each offspring receives one copy of each allele from each parent.
- Punnett squares are a convenient method for diagramming a genetic cross. Inspection of the square gives you the genotypic and phenotypic results and ratios.
- The genotypic ratio for a monohybrid cross is 1:2:1, and the phenotypic ratio is 3:1.
- A test cross reveals the presence of recessive genes in an individual with an unknown genotype by crossing them with an individual homozygous recessive for the genes in question.

Another Generation for Mendel

- A third insight of Mendel's was that alleles separate prior to the formation of gametes. Though Mendel did not know it, the physical basis for this is that the alleles he was observing resided on homologous chromosomes, which always separate in meiosis. An organism that has two identical alleles of a gene for a given character is said to be homozygous for that character. An organism that has differing alleles for a character is said to be heterozygous for that character. Dominant means: expressed in the heterozygous condition (as with the yellow color of peas, expressed in the heterozygous Yy condition). Recessive means: not expressed in the heterozygous condition (as with the green color of peas in the Yy condition).

Crosses Involving Two Characters

- Mendel observed that the genes for the different characters he studied were passed on independently of one another. This was of because the genes for these characters resided on separate, non-homologous chromosomes. The physical basis for what he found is the independent assortment of chromosomes during meiosis.

Following the Inheritance of Two Genes-Independent Assortment

- Mendel's law of independent assortment considers genes transmitted on different chromosomes.
- The phenotypic ratio of 9:3:3:1 of a dihybrid cross indicates that a gene on one chromosome does not influence transmission of a gene on a different chromosome.

- In meiosis, random assortment of maternally and paternally derived chromosomes results in gametes that have different combinations of genes.
- Punnett squares and probability are used to predict recurrence of more than one trait.

Pedigree Analysis

- Pedigree charts depict family relationships and transmission of inherited traits.
- Squares represent males and circles represent females.
- Horizontal lines indicate parents, vertical lines show generations and elevated horizontal lines depict siblings.
- Symbols for heterozygotes are half-shaded, and for individuals with a particular phenotype, completely shaded.
- Pedigrees can reveal mode of inheritance, and can include molecular information, carrier status, and input from other genes and the environment.
- Interpretation of pedigrees can be complicated by lack of information, adoption, children born out of wedlock, assisted reproductive technologies (i.e. artificial insemination), lack of penetrance.

Reception of Mendel's Ideas

- Gregor Mendel published his work, but the significance of it was never recognized in his lifetime. It was only rediscovered 16 years after his death, in 1900.

Incomplete Dominance

- Not all inheritance works through the principles Mendel perceived in his peas. Incomplete dominance operates when neither allele for a given gene is completely dominant with the result that heterozygous genotypes can yield an intermediate phenotype (such as pink snapdragons).

Lessons from Blood Types: Codominance

- In some instances, differing alleles of the same gene will have independent effects, rather than one allele being dominant over the other. Such is the case with the genes that code for the human blood proteins, A and B. Neither of the alleles for these proteins are dominant over the other. Rather, a person can have a genotype that produces either or both of the proteins, or neither of them.

Multiple Alleles and Polygenic Inheritance

- Human beings and many other species can have no more than two alleles for a given gene, each allele residing on a separate, homologous chromosome. Many allelic variants of a gene can, however, exist in a population. Most traits in living things are governed not by one gene but by many, with these genes often having several allelic variants. The term for such genetic influence is polygenic inheritance, meaning the inheritance of a genetic character that is determined by the interaction of multiple genes, with each gene having a small additive effect on the character.
- Polygenic inheritance produces continuous variation in phenotypes, meaning there are no fixed increment of difference between individuals. Human skin, for example, comes in a range of colors in which one color shades imperceptibly into the next.
- The traits produced in polygenic inheritance tend to manifest in bell-curve distributions, in which most individuals display near-average trait values, rather than extreme trait values. Gene interactions and gene-environment interactions are so complex in polygenic inheritance that predictions about phenotypes are a matter of probability, not certainty.

Genes and Environment

- The effects of genes can vary greatly in accordance with the environment in which the genes are expressed an organism's genotype and environment interact to produce that organism's phenotype.

One Gene, Several Effects: Pleiotropy

- Genes work in an interrelated fashion, such that a single gene is likely to have multiple effects. Pleiotropy is a phenomenon in which one gene has many effects.

Multiple Choice Questions

1. Alternate form of a gene that govern the expression of the same trait and that occur at the same positions on homologous chromosomes are called.
   (a) mRNA (b) Autosomes
   (c) Chromatids (d) Alleles

2. A diploid cell with four total chromosomes is heterozygous at 10 different genes. Do these genes assort independently during meiosis?
   (a) No
   (b) Not enough information is given to determine if the genes assort independently
   (c) Yes
   (d) Can't say

3. In Mendel's garden peas, the tallness (T) trait is dominant over the dwarf condition (t), round seed (R) is dominant to wrinkled seeds (r), purple flowers (P) are dominant to white flowers (p), and one-pod seeds (S) are dominant to three-pod seeds (s). These genes are all located on separate chromosomes. Suppose a plant heterozygous for each gene is self-fertilized, and 1,000 offspring are produced. How many would Mendel have expected to be dwarf with purple flowers and wrinkled one-pod seeds?
   (a) 35 (b) 1,000
   (c) 562 (d) 9

4. For codominant genes, the standard procedure of using capital and lowercase letters for the different alleles is inadequate. For codominant genes, a superscript is used for example, the human blood type A allele is indicated as "IA" and the Type B allele is indicated as "IB." The recessive allele that is responsible for Type O blood is still given the lowercase letter "i." In a family of six children, two are blood Type O, two are blood Type AB, one is blood Type B, and another Type A. What are the possible genotypes of the parents?
   (a) IBi x IAIB (b) IAi x IBi
   (c) IAIB x ii (d) IAIB x IAIB

5. The fact that all seven of the garden pea traits studied by Mendel obeyed the principle of independent assortment means that
   (a) Garden pea plants produce gametes with seven chromosomes
   (b) Seven pairs of alleles determining these traits behave as if each is on separate chromosome pairs
   (c) Seven pairs of alleles determining these traits are on the same pair of homologous chromosomes
   (d) Garden pea plants have seven pairs of chromosomes
6. In an individual with genotype AaBbCcDDEe, what proportion of the gametes produced will have an "A" allele?
   (a) 1/2 (b) 1/4
   (c) 3/4 (d) 9/16
7. Assuming no gene interactions, a cross between two heterozygous individuals (AaBb) at two independently assorting genes will produce an expected phenotypic ratio of ...
   (a) 15:1 (b) 3:1
   (c) 9:3:3:1 (d) 1:2:1
8. A cross between two heterozygous individuals (AaBb) at two independently assorting incompletely dominant genes will produce an expected phenotypic ratio of ...
   (a) 9:3:3:1 (b) 1:1:1:1
   (c) 1:2:2:4:1:2:1:2:1 (d) 15:1
9. You have become a renowned marine biologist, and in your travels you discover a species of fish that is diploid and reproduces exactly like other sexually reproducing animals with the exception that this animal is hermaphroditic (it has both male and female reproductive organs and can self-fertilize). You mate two fish, and then raise one of their offspring and allow it to self-fertilize, producing hundreds of offspring. Which of the following would be true of the offspring of this individual?
   (a) They would have completely different genes from their parent, but would be phenotypically identical
   (b) They would have the same alleles as their parent, but they would be shuffled into different combinations, resulting in phenotypically different offspring

(c) They would be phenotypically identical to their parent

(d) None

10. In another species of flowering plant, the tallness (T) trait is dominant over the dwarf condition (t), and red flowers (R) are dominant to white flowers (r). Suppose a plant heterozygous for both genes are crossed with an individual that is homozygous recessive for each trait. The cross produces a 1:1:1:1 phenotypic ratio. Are these two genes located on separate homologous pairs?

(a) No, if the two genes were on separate chromosomes, then we would expect a 9:3:3:1 ratio

(b) Yes, the cross TtRr x ttrr would give us an expected 1:1:1:1 phenotypic ratio if the genes assorted independently

(c) No, if the two genes were on separate chromosomes, then we would expect something other than a 1:1:1:1 phenotypic ratio

(d) None

11. For Mendel's pea plants, tall (T) was dominant over dwarf (t), and purple flowers (P) were dominant over white flowers (p). Assuming that these genes would sort independently, how many different types of gametes would be produced in a plant with the genotype TtPP?

(a) 4 (b) 2

(c) 8 (d) 3

12. For Mendel's pea plants, tall (T) was dominant over dwarf (t), and purple flowers (P) were dominant over white flowers (p). A plant with genotype the ttPp would have which phenotype?

(a) Tall plant with white flowers

(b) Dwarf plant with white flowers

(c) Dwarf plant with purple flowers

(d) Tall plant with purple flowers

13. For Mendel's pea plants, tall (T) was dominant over dwarf (t), and purple flowers (P) were dominant over white flowers (p). If a plant with the genotype ttPp were crossed with a plant that was Ttpp, the phenotypic ratio of the offspring would be

(a) 9:3:3:1 (b) 9:1

(c) 1:1:1:1 (d) 3:1

14. For Mendel's pea plants, tall (T) was dominant over dwarf (t), and purple flowers (P) were dominant over white flowers (p). A plant that had a phenotype of dwarf plant and white flowers had which genotype?

(a) ttPP (b) TtPp

(c) ttpp (d) TTpp

15. For Mendel's pea plants, tall (T) was dominant over dwarf (t), and purple flowers (P) were dominant over white flowers (p). A plant that had a phenotype of tall plant and purple flowers must have had which genotype?

(a) The genotype cannot be deduced from the phenotype

(b) TtPP

(c) TTPp

(d) TtPp

16. A mother who is heterozygous for blood Type A states that a man who is heterozygous for blood Type B is the father of her baby. What blood type must the baby have if the man is indeed the baby's father?

(a) Type AB

(b) Type A

(c) Type B

(d) All the above are possible blood types for the baby

17. On their wedding day, a man and wife decide that they want to have three daughters. What is the probability that this will actually occur?

(a) 1/4 (b) 1/8

(c) 1/16 (d) 1/2

18. For the Punnett square below, the genotype of parent 1 is ______ while that of parent 2 is ______.

parent 2

parent I		
	Aa	Aa
	aa	aa

(a) Aa / Aa (b) AA / aa

(c) aa / aa (d) Aa / aa

19. A plant with orange flowers is crossed with another plant with orange flowers. The offspring show the phenotypic ratio of one red flower:two orange flowers:one yellow flower. These plants are displaying a classic example of ... (level 1) [Hint]

(a) Pleiotropy (b) Incomplete dominance
(c) Codominance (d) Multiple alleles

20. A trait, such as skin color, that is coded for by more than one gene is an example of

(a) Pleiotropy (b) Incomplete dominance
(c) Polygenic inheritance (d) Codominance

21. Mendel's unit factors in pairs are most accurately known to be ________.

(a) Two maternal chromatids
(b) Two alleles on nonhomologous chromosomes
(c) Two alleles on a paternal and a maternal homologue
(d) Identical alleles on homologous chromosomes

22. Fetal cells may be removed along with fluid from the womb by a process known as _____.

(a) Karyotyping (b) Uterine tapping
(c) Amniocentesis (d) Sonogramming

23. An organism has a diploid number of 12. On the basis of segregation and independent assortment, how many different types of gametes can be formed?

(a) 1 (b) 2
(c) 12 (d) 64

24. Normal parents have a Tay-Sachs child Tay-Sachs is a recessive condition. What is the probability that their next child will be normal?

(a) 1.0 (100%) (b) 0.25 (25%)
(c) 0.75 (75%) (d) 0.67 (67%)

25. If you have a dominant disorder in a diploid organism, what two genotypes would express the same phenotype?

(a) Homozygous dominant and homozygous recessive
(b) Homozygous dominant and heterozygous

(c) Homozygous recessive and heterozygous

(d) None

26. What type of experiment can be done to differentiate between a homozygous dominant and a heterozygous genotype for a dominant trait?

(a) Chi-square test (b) Test cross

(c) Punnett square (d) Human pedigree analysis

27. How did Mendel's work support the chromosomal theory of inheritance?

(a) The behavior of chromosomes during meiosis (as observed under a microscope) correlated with Mendel's principles of inheritance

(b) Mendel's work matched with the continuous variation theory

(c) Mendel's statistical evidence of gene segregation indicated the involvement of chromosomes

(d) The traits that Mendel studied were known to lie on chromosomes

28. How would the dwarf trait that disappeared in the first filial (F1) generation be described?

(a) Dominant (b) Haploid

(c) Heterozygous (d) Recessive

29. If a home garden is planted with all hybrid, round peas, what proportions would be observed in the progeny? (The wrinkled trait is recessive).

(a) All round (b) 1/4 round and 3/4 wrinkled

(c) 3/4 round and 1/4 wrinkled (d) All wrinkled

30. A plant with the genotype Dd will produce gametes containing which alleles?

(a) D only (b) D or d

(c) DD only (d) DD, Dd and Dd gametes

31. A test cross is designed to determine the genotype of a parent with the dominant phenotype. A test cross produces progeny that are ½ tall and ½ dwarf. What is the genotype of the tall parent?

(a) dd (b) Dd

(c) D (d) DD

32. If a plant that is heterozygous for purple flower is crossed with a plant that is white, what proportion of the progeny will have white flowers? (Purple is dominant).

(a) 75% (b) 100%

(c) 50% (d) 25%

33. In the following test cross [AaBb X aabb], what are the dimensions of the Punnett square used to determine the genotypes of the next generation?

(a) 4 X 1 (b) 2 X 4

(c) 4 X 2 (d) 2 X 2

34. What genotype is present most often among the progeny (F2) of a dihybrid cross (AaBb X AaBb)?

(a) AAbb (b) AaBb

(c) AABB (d) AABb

35. In the F1 dihybrid cross (WwGg X WwGg), what is the probability of obtaining an individual that is round, green, and true-breeding? (G = Yellow seed, g = green seed; W = round seed, w = wrinkled)

(a) 9/16 (b) 1/2

(c) 1/16 (d) 3/16

36. How many different types of gametes can be formed by individuals of the genotype AaBbcc?

(a) AaBbcc (b) Aa, Bb, cc

(c) ABc, Abc, aBc, abc (d) ABc, Abc, aBc

37. How many different types of gametes can be formed by individuals of the genotype AaBB?

(a) Aa, BB (b) A, a, B, B

(c) AaBB (d) AB, aB

38. A pea plant that is true breeding (homozygous) for green round seeds is crossed with a plant that has yellow wrinkled seeds. The F1 progeny all have round green seeds. Predict the results of test crossing these F1 plants.

(a) 1:3:1:3 (b) 9:3:3:1

(c) 1:0:0:0 (d) 1:1:1:1

39. In the following list of genotypes, how many are examples of heterozygosity, and which ones are they?

(a) AA, Aa, Gg, bb, BB, Bb, GG, Ww, ww, aa

(b) 6 heterozygotes: AA, bb, BB, GG, ww, aa

(c) 3 heterozygotes: bb, ww, aa

(d) 4 heterozygotes: Aa, Gg, Bb, Ww

40. Using the forked line method, determine the phenotypic ratios of the trihybrid cross AaBbCc X AaBBCC. What proportion of the progeny have the dominant phenotype for A, B, and C?

(a) 2/16 (or 1/8) (b) 1/16

(c) 9/16 (d) 12/16 (or 3/4)

41. Using the forked line method, determine the phenotypic ratios of the trihybrid cross AaBbCc X AaBBCC. What proportion of the progeny are aaBBCC?

(a) 12/16 (or 3/4) (b) 1/16

(c) 8/16 (or 1/2) (d) 9/16

42. Assume that the probability of giving birth to a baby girl is 1/2 and to a baby boy is also 1/2. (It's not quite. More boys are born than girls.) If a family has two girls, what is the probability that the next child will be a boy?

(a) 3/16 (b) 1/2

(c) 1/4 (d) 1/8

43. A couple learn that both are heterozygous (carriers) for the same cystic fibrosis gene. What is the probability that their first child will have cystic fibrosis? What is the probability that their second child will have cystic fibrosis?

(a) 1/4 and 3/4 (b) 1/4 and 1/4

(c) 3/4 and 3/4 (d) 1/4 and 1/16

44. In the last problem you determined that the probability of a couple having a child with cystic fibrosis was 1/4. What is the probability that the couple's first child will be a normal boy?

(a) 1/16 (b) 1/2

(c) 1/8 (d) 3/8

45. Red color in carnations is codominant. Heterozygotes are pink. Curly petals are dominant to straight petals. A pink plant that is heterozygous for curly is self pollinated. What is the probability of pink curly progeny?

(a) 1/8 (b) 1/16

(c) 3/16 (d) 3/8

46. By using Pascal's triangle from your text, expand the binomial (a + b)5. What is the coefficient for the term a4b?

(a) 4 (b) 1

(c) 10 (d) 5

47. In a family of 5, what is the probability of having 5 boys?

(a) 1/16 (b) 1/4

(c) 1/8 (d) 1/32

48. In a family of five, what is the probability of having 3 boys and 2 girls?

(a) 6/16 (b) 1/32

(c) 10/32 (d) 5/32

49. A couple is heterozygous for cystic fibrosis. They would like to have three children. What is the probability that all children will be normal?

(a) 3/4 (b) 27/64

(c) 1/8 (d) 1/16

50. A couple is heterozygous for cystic fibrosis. They would like to have three children. What is the probability that two children will be normal and one will have cystic fibrosis?

(a) 3/16 (b) 1/8

(c) 3/4 (d) 27/64

51. A couple is heterozygous for cystic fibrosis. If they have three children, what is the possibility that AT LEAST one child will have cystic fibrosis?

(a) 37/64 (b) 27/64

(c) 9/64 (d) 1/64

52. Pedigree analysis is especially useful in determining the mode of inheritance in organisms:

(a) With greater than 500 chromosomes

(b) That reproduce asexually

(c) With very few offspring produced in each generation and with long generation times

(d) That are haploid

53. In addition to the reasons addressed in the previous question, pedigree analysis is essential for studies of inheritance in humans because of:

(a) The exceptionally large size of the human genome

(b) The great expense of experimenting with humans

(c) Ethical constraints on human experimentation

(d) The number of chromosomes that a diploid human has

54. The technical term for the individual in a pedigree who has the phenotype of interest and is the starting point for undertaking the pedigree analysis is:

(a) Progeny (b) Proband

(c) Grandparent (d) Victim

55. Although pedigree analysis does not provide the same degree of certainty about modes of inheritance as do designed, controlled, experimental crosses, confidence in the hypotheses generated by pedigree analysis is increased:

(a) When you only examine families with large numbers of children

(b) By examining a large number of traits in the same pedigree

(c) When you only examine two generations in a pedigree

(d) When you examine multiple independent pedigrees for the same trait that are consistent with the hypothesized mode of inheritance

56. A trait that is present in one parent but not present in the second generation is characteristic of a(n):

(a) Autosomal dominant (b) Y-linked trait

(c) Autosomal recessive trait (d) Quantitative trait

57. What is the best hypothesis for the mode of transmission of a trait where one parent expresses the trait, the other parent does not, and half of the children (regardless of sex) express the trait?

(a) X-linked dominant (b) Y-linked dominant

(c) X-linked recessive (d) Autosomal dominant

58. In the previous question where one parent and half of the children (regardless of sex) express the trait, what is the most likely genotype of the parent expressing the trait?

(a) Hemizygous (b) XX

(c) homozygous (d) Heterozygous

59. What is the best hypothesis for the mode of transmission of a trait where neither parent expresses the trait but one half of the sons do?

(a) Rare autosomal recessive

(b) A new mutation in each of the sons expressing the trait

(c) Rare X-linked dominant

(d) Rare X-linked recessive

60. If the pattern of trait appearance in a large number of pedigrees is not consistent with any of the typical modes of inheritance, the trait likely:

(a) Is polygenic

(b) Has an environmental component to its expression

(c) Both a and b are possible

(d) Is the result of recurrent mutations

61. A limitation of pedigree analysis of human traits is:

(a) Missing information about some members of the family

(b) The relatively small number of children ("sample sizes") per generation

(c) Both a and b are limitations

(d) None of the above; pedigree analysis is as robust a method as controlled experiments

62. A mutation of an X-linked, recessive allele is lethal. Male fetuses that inherit the allele and female fetuses that are homozygous recessive abort before birth. What is the probability that a child born to a female who is heterozygous for the mutation will have one copy of the allele?

(a) 3/4 (b) 1/3

(c) 2/3 (d) 1/4

63. What type of inheritance is shown in the pedigree given above?

(a) Autosomal dominant (b) Sex-linked dominant

(c) Y-linked trait (d) Sex-linked recessive

64. Which of the following genotypes will have a pigmented aleurone?

(a) AaCcRrC′c′ (b) AACCRRC′C′

(c) aaccrrc′c′ (d) aaccRrc′c′

65. A chromosome alteration resulting in the removal of part of a chromosome is known as

(a) Inversion (b) Duplication

(c) Translocation (d) Deletion

66. A person with the sex chromosomes would be what gender and have how many Barr bodies: XXXY

(a) Male, 3

(b) The gender of the person cannot be determined

(c) Male, 2

(d) Female, 3

67. What is the genetic content and number of cells at the end of meiosis I and meiosis II?

(a) 2 cells, diploid; 4 cells, haploid

(b) 2 cells, haploid; 4 cells, diploid

(c) 2 cells, diploid; 4 cells, diploid

(d) 2 cells, haploid; 4 cells, haploid

68. Human body (somatic) cells contain how many chromosomes?

(a) 46 (b) 11.5

(c) 69 (d) 92

69. Crossing over is

(a) Making an RNA copy of a DNA strand

(b) Independent assortment of chromosomes

(c) The exchange of homologous portions of nonsister chromatids

(d) The movement of genetic material from one chromosome to a nonhomologous chromosome

70. The genome of chimpanzees consists of 48 chromosomes (2n = 48). For the following question, assume that no mutation occurred during DNA replication.How many DNA molecules are in the genome of a chimpanzee cell before it begins meiosis?

(a) 96 (b) 48

(c) 24 (d) 12

71. An X-linked recessive allele is responsible for a human disease whose symptoms include loss of teeth and loss of sweat glands. When males have the disease, they are toothless and have no sweat glands.Females who are heterozygous are often missing some teeth and have some areas of skin lacking sweat glands. Which of the following best explains why heterozygous females express the trait in this manner?

(a) Epistasis

(b) X-inactivation

(c) Pleiotropy

(d) Incomplete dominance

72. If C and D are linked genes with a crossover frequency of 50%, which of the following is true for sperm of the individual with the genotype CcDd?

(a) 1/4 CD + 1/4 Cd + cD + 1/4 cd

(b) 1/4 CCDD + 1/2 CcDd + 1/4 ccdd

(c) 1/2 CD + 1/2 cd

(d) All CcDd

73. The above diagram shows a distribution of skin colors within a population of humans. What kind of inheritance does skin color represent?

(a) Sex-linked inheritance

(b) Pleiotropic inheritance

(c) Incomplete dominance

(d) Polygenic inheritance

74. A man who has type B blood and a woman who has type A blood could have children of which of the following phenotypes?

(a) AB or O (b) A, B or O

(c) A, B, AB or O (d) A or B only

75. Sometimes, one gene pair will interact to control the expression of a second gene pair (for example, coat color in Labrador retrievers). This gene interaction is called

(a) Epistasis

(b) Complete dominance

(c) Incomplete dominance

(d) Second pair control

76. Every genetic defect teaches a lesson about what is needed for proper function. Defects connected with parental imprinting teach us that proper function requires

(a) Not only the correct alleles, but also the expression of those alleles at the right time

(b) Two alleles of each gene

(c) That the "genetic load" of harmful alleles cannot exceed the Hayflick limit

(d) A functional allele from the mother and a functional allele from the father

77. Each cell in an individual with Down Syndrome contains how many chromosomes?

(a) 45 (b) 24

(c) 3 (d) 47

78. Human hemophilia A is inherited as a sex-linked recessive. If a hemophiliac male mates with a normal, noncarrier female, the offspring will be

(a) 1/2 normal males and 1/2 carrier females

(b) All hemophiliacs

(c) 1/2 hemophiliac males and 1/2 carrier females

(d) 1/2 hemophiliac males and 1/2 normal females

79. An AABb individual is mated with another AABb individual. The possible number of genetically different kinds of offspring is

(a) 2 (b) 4

(c) 1 (d) 3

80. Tom's brother suffers from phenylketonuria, a recessive disorder. The brother's parents do not have PKU. What are the chances that Tom, who is normal for this trait, is a carrier of PKU?

(a) 2/3 (b) 3/4

(c) 1/4 (d) 1/3

81. During anaphase II

(a) The cell is haploid

(b) Sister chromatid separates and migrates toward opposite poles

(c) Homologues separates and migrate toward opposite poles

(d) Nuclei re-form

82. Andalusian chickens with the genotype CBCB are black, those with the genotype CWCW are white and those with the genotype CBCW are gray. What is the relationship between the CB and CW alleles?

(a) The relationship between the alleles is one of incomplete dominance

(b) CB is recessive to CW

(c) CW is dominant to CB

(d) CB and CW are co-dominant

83. For the parents in the previous question, what is the probability that their next child will have black chicken?

(a) 0 (b) ¼

(c) ½ (d) 1

84. For these parents, what is the probability that their next child will be homozygous dominant for white chicken?

(a) 0 (b) ¼

(c) ½ (d) 1

85. In a cross between two pea plants - Pp x Pp:

(a) Pp is the phenotype (b) Pp is the genotype

(c) The plants are dihybrids (d) None

86. In the same cross between two pea plants -Pp x Pp:

(a) Both parents are homozygotes

(b) Both parents are heterozygotes

(c) One parent is a heterozygote, one parent is a homozygote

(d) None

87. In the same cross between two pea plants - Pp x Pp:

(a) These parent plants have a dominant phenotype

(b) These parent plants have a recessive phenotype

(c) The phenotype of these parent plants cannot be determined by the information given

(d) None

88. When Mendel crossed pure-breeding purple (PP) and white flowered pea plants (pp), the offspring were:

(a) All purple (b) All white

(c) Half purple and half white (d) 75% purple, 25% white

89. When heterozygous purple pea plants (Pp) were crossed, the offspring were:

(a) All purple (b) All white

(c) Half purple and half white (d) 75% purple, 25% white

90. A testcross is a mating of an individual expressing a dominant phenotype with:

(a) Another individual of dominant phenotype

(b) A homozygous recessive individual

(c) Itself

(d) A homozygous dominant individual

91. In garden peas, height is determined by a single gene. Tall is dominant to short. If two heterozygous plants are crossed, which proportion of the offspring will be short?

(a) 3/4 (b) 2/3

(c) 1/2 (d) 1/3

92. In garden peas, height is determined by a single gene. Tall is dominant to short. If two heterozygous plants are crossed, which proportion of the offspring will be heterozygous?

(a) 3/4 (b) 2/3

(c) 1/2 (d) 1/3

93. When a pollen from a flower on one pea plant is transferred to a flower on another pea plant, this is:

(a) Self-pollination (b) Cross-pollination

(c) The F1 generation (d) None of the above

94. In peas, yellow seed color is dominant to green. Y indicates the dominant seed color and y indicates the recessive color. What will be the expected phenotype ratio of offspring in a cross between two heterozygous yellow peas?

(a) 3 yellow: 1 green

(b) 1 yellow : 2 yellow-green : 1 green

(c) 3 YY : 1 yy

(d) 1 YY : 2 Yy : 1 yy

95. In peas, yellow seed color is dominant to green. Y indicates the dominant seed color and y indicates the recessive color. What will be the expected genotype ratio of offspring in a cross between two heterozygous yellow peas?

(a) 3 yellow: 1 green

(b) 1 yellow : 2 yellow-green : 1 green

(c) 3 YY : 1 yy

(d) 1 YY : 2 Yy : 1 yy

96. In tobacco plants, the allele for tall plants (T) is dominant relative to the dwarf allele (t). What are the genotypes of the parents that produce 3/4 tall plants and 1/4 dwarf plants among their offspring?

(a) TT x tt (b) TT x Tt

(c) Tt x Tt (d) T x t

97. PTC is a harmless chemical that some people can taste and others cannot. The taster allele "T" is dominant and the non-taster allele "t" is recessive. If two people heterozygous for the taster character have children together, what are the possibility that their child will be a non-taster?

(a) 0% (b) 25%

(c) 50% (d) 75%

98. If a yellow guinea pig is crossed with a white guinea pig, the offspring are cream colored. What is the simplest explanation for this?

(a) Incomplete dominance (b) Epistasis

(c) Pleiotropy (d) All of the above

99. If flower color exhibits incomplete dominance, what is the phenotypic ratio of the offspring of monohybrid parents - Rr x Rr?

(a) 1 red : 2 pink : 1 white (b) 3 red : 1 white

(c) 1 RR : 2 Rr : 1 rr (d) 3 RR : 1 rr

100. Which statement is NOT true regarding polygenetic characters?

(a) Height and weight are examples of polygenetic characters

(b) Two or more genes act together to determine an individual's phenotype

(c) The presence of one gene masks the expression of another gene

(d) Most characters are the result of polygenes

101. An individual has attached ear lobes, a recessive phenotype. Which statement is true?

(a) This individual has two alleles that are heterozygous

(b) This individual has two alleles that are homozygous

(c) We cannot determine whether these alleles are heterozygous or homozygous based on the given information

(d) None

102. In corn, when dominant allele A and dominant allele B are present, purple color is produced. A cross of AaBb x AaBb would yield how many purple offspring out of 16?

(a) 9 (b) 7

(c) 3 (d) 1

103. In pepper plants, red peppers (R) are dominant over yellow peppers (r) and tallness (T) is dominant over shortness (t). In the cross between the two pepper plants - RRTT x rrtt, what will the phenotypes of the offspring be?

(a) Red and short (b) Red and tall

(c) Yellow and short (d) Yellow and tall

Fill In the blanks

1. The term phenotype refers to the __________ of an individual.
   (a) Breeding pattern
   (b) Number of chromosomes
   (c) Number of genes
   (d) Appearance of a characteristic or trait
2. A human pedigree may give us information on __________.
   (a) How a gene is inherited
   (b) The type of mutation that caused a specific genetic trait
   (c) How the mutation occurred
   (d) The protein product of a mutated gene
3. Mendel's cross of the true-breeding round and wrinkled parents produced all round progeny peas in the F1 generation. This is the P1 generation of a ________.
   (a) Test cross (b) Monohybrid cross
   (c) Genetic cross (d) Self cross
4. Mendel's plants in the parental or P1 generation were true-breeding. The phrase 'true-breeding' means that these peas came from plants that are ________.
   (a) Homozygous (b) Heterozygous
   (c) Crossed (d) Haploid
5. A pea plant that is heterozygous for the dwarf gene (Dd) will be _______.
   (a) Dead (b) Short
   (c) Extra tall (d) Tall
6. A plant that has the genotype DD is said to be ________.
   (a) Mutant
   (b) Heterozygous
   (c) Homozygous
   (d) Polyploid

7. Mendel also studied other pea traits. He crossed two true breeding plants, one with purple flowers and one with white flowers. All of the progeny had purple flowers. The purple phenotype is said to be ________.

(a) Dominant (b) Haploid
(c) Mutant (d) Diploid

8. Assume tall (T) is dominant to dwarf (t). If a homozygous dominant individual is crossed with a homozygous dwarf, the offspring will ______.

(a) All be intermediate in height (b) Be half tall and half dwarf
(c) Be 3/4 tall and 1/4 dwarf (d) All be short

9. Michelle and Keith are apparently normal, but their daughter was born with alkaptonuria, an inherited metabolic disorder. If alkaptonuria is like most other human hereditary disorders, the probability of their next child being born with alkaptonuria is ______

(a) 2/3 (b) 0
(c) 3/4 (d) 1/4

10. Fetal cells may be removed along with fluid from the womb by a process known as _____.

(a) Karyotyping (b) Uterine tapping
(c) Amniocentesis (d) Sonogramming

11. Flower color in snapdragon is an example of incomplete dominance. If a red-flowered plant is crossed with a white-flowered plant, the F1 generation has pink flowers. If a pink-flowered plant is crossed with a pink-flowered plant, the progeny plants will be __________.

(a) 50% pink and 50% red
(b) 25% red, 50% pink, 25% white
(c) 100% red
(d) 25% white and 75% red

12. In corn, pigment production is dependent upon the expression of two pigment genes. If the plant doesn't express one of the two genes, no pigment is produced. When one gene can interfere with the expression of another gene, this is called __________.

(a) Pleiotropy (b) Epistasis
(c) Incomplete dominance (d) None

13. People with cystic fibrosis have overly sticky mucus, salty sweat, and liver and pancreatic problems. When one gene affects multiple phenotypes, this is _______.

(a) Pleiotropy (b) Epistasis

(c) Incomplete dominance (d) None

Answers

Multiple Choice Question

1	2	3	4	5	6	7	8	9	10	11	12	13	14	15
D	A	A	B	B	A	C	C	B	B	B	C	C	C	A
16	17	18	19	20	21	22	23	24	25	26	27	28	29	30
D	B	D	B	C	C	C	D	C	B	B	A	D	C	B
31	32	33	34	35	36	37	38	39	40	41	42	43	44	45
B	C	A	B	C	C	D	D	D	D	B	B	B	D	D
46	47	48	49	50	51	52	53	54	55	56	57	58	59	60
D	D	C	B	D	A	C	C	B	D	C	D	D	D	C
61	62	63	64	65	66	67	68	69	70	71	72	73	74	75
C	B	A	D	D	C	A	A	C	B	B	A	D	C	A
76	77	78	79	80	81	82	83	84	85	86	87	88	89	90
D	D	A	D	A	B	A	B	A	A	A	A	B	D	C
91	92	93	94	95	96	97	98	99	100	101	102	103		
B	A	D	C	B	A	A	C	B	B	B	B	A		

Fill in the blanks

1	2	3	4	5	6	7	8	9	10	11	12	13
A	A	B	A	D	C	A	B	D	C	B	B	A

❑❑❑

Chapter – 4

Extensions and Exceptions to Mendels Laws

When Gene Expression Appears to Alter Mendelian Ratios

Lethal Allele Combinations

- Us recessive lethal alleles eliminaters a progeny class.

Multiple Alleles

- A gene can have more than two alleles, but a diploid individual only has one or two of them.
- Different allele combinations can produce different phenotypes and different severities of symptoms.

Different Dominance Relationships

- Incomplete dominance of an allele produces a phenotype in the heterozygote that is intermediate between that of either homozygote.
- Codominant alleles are both expressed in a heterozygote.

Epistasis-When One Gene Affects Expression of Another

- In epistasis, one gene masks the effect of another. An example of epistasis in humans is the Bombay phenotype.

- The Bombay phenotype results in O type blood for individuals homozygous recessive for the recessive "h" allele.

Penetrance and Expressivity

- Genotypes vary in penetrance (percent of individuals affected) and expressivity (severity of symptoms).
- Penetrance and variable expression are not well understood biochemically and are probably due to the complex biochemical environment all genes function in.

Pleiotropy-One Gene, Many Effects

- Effect is pleiotropic

Phenocopies-When It's Not in the Genes

- A trait caused by the environment but resembling a genetic trait or occurring in certain family members is a phenocopy.

Genetic Heterogeneity-More than One Way to Inherit a Trait

- Genetic heterogeneity occurs when different genes (or alleles) cause the same phenotype.

Maternal Inheritance and Mitochondrial Genes

- Only females transmit mitochondrial genes, in the oocyte cytoplasm. In maternal inheritance, a trait passes from females to offspring of both sexes, but not from males.
- Mitochondrial DNA does not cross over, mutates faster than nuclear DNA and is present in many copies per cell.
- The genotype of the mother determines the phenotype of the offspring for maternal effect genes.
- Female gametes receive gene products from the mother that affect early developmental stages of the embryo.

Mitochondrial Disorders

- A variety of disorders, many causing muscle weakness and fatigue, result from mutations in mitochondrial genes.

- Ooplasmic transfer has been used to over come mitochondrial disease in in-vitro conceived children.

Heteroplasmy Complicates Mitochondrial Inheritance

- In heteroplasmy, a mutation is present in some mitochondria DNA/ gene only.

Mitochondrial DNA Studies Clarify the Past

- Mitochondrial DNA can be used for forensic pedigree and phylogenetic analysis.

Epigenetic Inheritance

- Dosage compensation is necessary to ensure genetic equality between the sexes.
- Dosage compensation occurs in female mammals by the random inactivation of one X chromosome.
- In adult female mammals, one X chromosome has been permanently inactivated.
- X-inactivation in mammals depends on the Xic locus and the XIST gene.
- The expression of an imprinted gene depends on the sex of the parent from which the gene was inherited.
- The imprinting of genes and chromosomes is a molecular marking process that involves DNA methylation.

Extranuclear Inheritance

- Mitochondria and chloroplast contains circular chromosomes with many genes.
- Extranuclear inheritance produces non-Mendelian results in reciprocal crosses.
- Studies in yeast and Chlamydomonas provided genetic evidence for extranuclear inheritance of mitochondria and chloroplasts.
- The pattern of inheritance of mitochondria and plastids varies among different species.
- A few rare human diseases are caused by mitochondrial mutations.

- Extranuclear genomes of mitochondria and chloroplasts evolved from an endosymbiotic relationship.
- Eukaryotic cells occasionally contain symbiotic infective particles.

Linkage

Linkage was Discovered in Pea Plants

- Linkage was discovered in peas when a dihybrid cross did not show a typical 9:3:3:1 ratio, but had an excess of parental types.
- Genes on the same chromosome are linked, and have different patterns of inheritance than the unlinked genes Mendel observed.
- The further apart two genes are on as same chromosome, the greater the probability that crossing over will occur between them.

Linkage Maps

- Genetic linkage maps are based on the observation that the likelihood of a crossover between two linked genes are directly proportional to the distance between them.
- Linkage maps are built by tracking transmission of two traits at a time, and then determining the percent recombination between them.

Examples of Linked Genes in Humans

- Linkage data from pooled human pedigrees can be used to develop a human linkage map.
- Linkage disequilibrium (LD) in the human genome reveals non-randomly distributed sequences where recombination occurs infrequently.

The Evolution of Gene Mapping

- Computer analysis of linkage data allows the computation of a LOD score (logarithm of the odds). A LOD score of 3 or greater signifies linkage between markers.
- Haplotypes are a tightly linked series of markers that appear to be inherited as a single unit.
- Markers may be protein coding alleles, variable restriction sites (RFLPs), variable short repeated sequences (VNTRs), or single nucleotide polymorphisms (SNPs).

Gene Interactions

- A cross involving a two-gene interaction can produce a 9:3:3:1 ratio in offspring when four distinct phenotypes are produced.
- A cross involving a two-gene interaction can produce a 9:7 ratio when both genes are epistatic to each other.
- Bridges observed an 8:4:3:1 ratio because the cream-eye allele can modify the eosin-eye allele, but not the red or white alleles.

Multiple Choice Question

1. Which of the following were key reasons for Mendel's success in elucidating his genetic laws?
   (a) Mendel had a strong understanding of the chemical nature of cells and knew DNA was passed on between generations
   (b) Mendel carefully established true breeding lines and used them in hybridization experiments carried out over multiple generations
   (c) Mendel's use of peas as a model organism and his experience with plant breeding in his father's orchard
   (d) Mendel was able to dedicate the efforts of a large research team and arranged to have very large test plots available for growing the numbers of plants required
2. Which is NOT a characteristic that made peas useful as a model system for Mendel's studies?
   (a) They are diploid and produce haploid gametes
   (b) They have a complex phenotype with much genetic variability
   (c) They are relatively small and easy to grow
   (d) They produce edible fruit
3. How might the ability of pea plants to self fertilize be a problem for genetic experiments?
   (a) Self fertile plants tend to be weaker and require more care
   (b) Self fertile plants are not able to produce viable seeds due to inbreeding
   (c) Stamens must be removed from self fertile plants to control fertilization
   (d) Pistils of self fertile plants are unable to accept pollen from other plants

4. Which description of plants from a true breeding strain is correct?

(a) They only have one allele of the gene that controls the trait of interest

(b) They carry two different alleles for the gene that controls the trait of interest

(c) They can only reproduce by self fertilization

(d) They all have dominant alleles for the trait of interest

5. Drosophila females are capable of storing sperm and using it to fertilize eggs long after mating. How might this present a problem for a geneticist?

(a) In order to design a specific cross, females must be prevented from mating with any other strain of males

(b) This allows Drosophila females to "select" sperm from the stronger male

(c) Individual offspring from these females may have more than one father

(d) This should not present a problem for a geneticist; it is merely a reproductive advantage to the fly if the population size is low

6. In single factor crosses between two true breeding strains of tree frog, a scientist discovers that plain green tree frogs and tree frogs with a red 'racing stripe' always produce offspring that have red racing stripes. Why?

(a) Stripes are dominant to plain green coloration

(b) Stripes must allow the frog to hide more effectively in the leaves

(c) Tree frogs with stripes are often eaten by birds

(d) Tree frogs with stripes must be more common than plain green ones

7. Which of the following most accurately restates Mendel's law of segregation?

(a) Genes are inherited in discrete units from one generation to the next

(b) Genes can exist in different forms, known as alleles

(c) Homologous chromosomes separates during gamete formation

(d) Genes on different chromosomes are inherited independently

8. You capture several interesting insects from a population in your yard. By inbreeding, you develop true breeding strains that differ only in the shape of their wings - one strain has round wings while the other strain has oval wings. How might you demonstrate that Mendel's principles apply to your insects?
   (a) Set up a monohybrid cross and determine if your F2 results are predicted by a 3:1 ratio based on the law of segregation
   (b) Set up a dihybrid cross and determine if your F2 results are predicted by a 9:3:3:1 ratio based on the law of independent assortment
   (c) Set a test cross to demonstrate that your traits breed true
   (d) Set up a two point test cross to demonstrate that your genes are linked

9. You set up a monohybrid cross between true breeding round winged and true breeding oval winged insects. Your F1 offspring have round wings, and you collect the following numbers of F2: Round wings: 302; Oval wings: 86. Which of the following statements is NOT correct?
   (a) The critical value for this chi square calculation is 7.815
   (b) The chi square value for this calculation is 1.663
   (c) Mendel's rules accurately predict the outcome of this cross
   (d) None of these; all are correct statements

10. I know that black coat color in mice is dominant to albino. Black mice get a higher price at the pet store so I am disappointed when I cross a black mouse with an albino mouse and get 6 albino and 2 black mice in the litter. I expected all black. What mistake did I make?
   (a) I assumed that black alleles were more common than white
   (b) I assumed that white alleles were more common than black
   (c) I assumed that my black mouse was true-breeding
   (d) I assumed that my white mouse was true-breeding

11. I do a chi square analysis of the offspring from my cross of a black mouse and albino mouse. If the black mouse was heterozygous I should have expected a 1:1 ratio, and I got 6 albino and 2 black mice. What is my chi square value?
   (a) 0.25 (b) 0.5
   (c) 1 (d) 2

12. While studying for your genetics exam, a fellow student comments that the albino coat color gene in mice must be recessive. What mistake did they make?
    (a) The gene that controls the coat color character comes in both dominant and recessive forms
    (b) The black coat color gene is actually dominant
    (c) The albino coat color gene is actually dominant
    (d) Coat color in mice cannot be easily described in dominant/recessive relationships
13. Which is the most correct description of a dominant trait?
    (a) The most common trait in the population
    (b) The trait that is seen in a heterozygous organism (hybrid)
    (c) A trait that reappears in the F2 of a monohybrid cross
    (d) A trait that confers a selective advantage to the organism and thus increases its frequency of reproduction
14. If you have true breeding strains that differ only in a single character, how could you determine which trait for that character is dominant?
    (a) Create another true breeding strain for another trait for that character
    (b) Create a monohybrid for that character and see which trait is present
    (c) Create a dihybrid and observe the phenotype
    (d) You cannot easily determine which trait is dominant
15. You cross true breeding lizards that are either green or albino and discover that all the offsprings are striped green and white. Which is the most likely result if you cross two striped lizards?
    (a) All striped offspring (b) 3 striped: 1 albino
    (c) 3 green: 1 albino (d) 1 green: 2 striped: 1 albino
16. If you performed a dihybrid cross and found that your results were significantly different from the 9:3:3:1 ratio, what would you suspect?
    (a) You miscounted your results
    (b) You had too small a sample size and need to repeat the experiment

(c) You picked two genes that are linked and thus do not assort independently

(d) You picked two genes that are linked and thus do not segregate during gamete formation

17. Pedigrees for human genetic traits are important because:

(a) Humans tend to have small numbers of offspring

(b) Matings between humans cannot be established by experimenters

(c) Knowledge about family history can help determine probabilities for individuals

(d) All of the above

18. Laurie met her husband Jim at a support group for families dealing with PKU (phenylketonureA), a recessive genetic disorder. Laurie and Jim both have one sister who is affected with PKU and no other family history of the disorder. They come to you as a genetic counselor to ask what their risk is of having an affected child. What do you tell them?

(a) They must both be heterozygotes so their risk of having an affected child is ¼

(b) They might both be heterozygotes so they should get tested to see if they carry the disease allele or not

(c) Since neither one of them is affected, they are not at risk of having an affected child

(d) You can't give them any firm answers because you don't know their genotypes

19. Which of the following statements about a pedigree would be most definitive in determining a recessive pattern of inheritance?

(a) Two affected individuals have three affected children

(b) Two unaffected individuals have one affected and two unaffected children

(c) There are four instances where at least one affected parent has an affected child

(d) Two known heterozygotes have two affected and one unaffected child

20. Statistical tests:

(a) Are useful to distinguish between competing hypotheses to explain a set of data

(b) Allow determination of how well a hypothesis predicts a set of experimental results

(c) Can provide a quantifiable measure of goodness of fit

(d) B and C only

21. You identify a previously unidentified plant species and decide to study its genetics. You establish true breeding strains for leaf shape (longer than wide vs equal length and width) and flower color (white vs. yellow). When you perform a dihybrid cross with these strains, you find that your results are very different than those predicted by a 9:3:3:1 ratio. You calculate a chi square and get a p value of less than 0.0001. Which of the following statements is most accurate based on this observation?

(a) Your plant species is not governed by Mendel's laws

(b) You need to test a different hypothesis, such as close linkage of these genes

(c) You must have designed your crosses wrong, so you need to throw the experiment out

(d) You have discovered a new mechanism of sexual reproduction

22. The genotype of an individual is not always reflected in the phenotype. This may be due to which of the following?

(a) Epistasis

(b) Dominance

(c) Incomplete dominance

(d) All the answers suggest ways in which the genotype may not be observed in the phenotype

23. A father with myotonic dystrophy has three daughters who are all carriers of the mutant allele and two sons who are unaffected noncarriers. The three daughters have six sons of which four are affected and two are not, and four daughters of which two are carriers and two are not. What type of mutation is suggested from this description?

(a) Autosomal dominant (b) Autosomal recessive

(c) Sex-linked dominant (d) Sex-linked recessive

24. A gene is identified which confers great athletic ability. A family of 10 children are identified in which all 10 children carry this gene. However, only three of the children exhibit great athletic ability, four have slightly better than average athletic ability, and three have average athletic ability. What does this tell you about the expressivity of this gene?

(a) 100% expressivity

(b) <100% expressivity

(c) 1% expressivity

(d) You cannot tell anything about the expressivity from this information

25. A recessive mutation was discovered in a small island population (1,000 people). The mutation was observed in the heterozygous state, but never in the homozygous state. What is a possible explanation for this observation?

(a) Only individuals with opposite alleles at this locus reproduced

(b) The recessive allele is lethal in a homozygous state

(c) The heterozygous state is "fixed" in this population

(d) The population size is too small to give an explanation for this phenomenon

26. What phenotypes and phenotypic ratios would you expect in a test cross of a pink flowered and a red flowered snapdragon?

(a) 1 red : 1 pink

(b) 3 pink: 1 red

(c) 3 red: 1 pink

(d) 1 white: 2 pink: 1 red

27. In a paternity suit a man disputes that he is the parent of a child although the mother of the child says that he is. The man has type O blood. The child also has type O blood, and her mother has type B blood. Could he be the father?

(a) Yes, because the allele for type O is dominant

(b) Can't determine if he might or might not be

(c) No, because B is dominant to O

(d) Yes, because the mother could be heterozygous

28. A female is heterozygous for the recessive X-linked gene for Lesch-Nyhan syndrome, a recessive gene that leads to mental retardation, self mutilation, and early death. What proportion of her daughters will be "carriers" for the trait?

(a) 25% (b) 50%

(c) 0% (d) 90%

29. If an X-linked disorder is normally lethal before the onset of reproduction, which of the following is true?

(a) Heterozygous females pass the disorder to half of their daughters, all of which express the trait

(b) The disorder occurs only in males

(c) Males are always heterozygous carriers but do not express the disorder

(d) The disorder is twice as likely in males as in females

30. A mutation in which most of the protein coding sequence of the gene is removed is most likely to be which type of allele?

(a) Lethal (b) Dominant

(c) Loss of function (d) Sex-limited

31. The phenomenon where half the "normal" level of functional protein is not enough to generate a normal phenotype is called haploinsufficiency. Which of the following situations demonstrates haploinsufficiency?

(a) Persons with type AB blood display both A and B antigens on their red blood cells

(b) Heterozygous and homozygous flowers in Fig 4.1 are both purple

(c) "Small eye" in mice is caused by loss of one copy of the Pax6 gene

(d) A mutation in a protein causes it to become much more active than usual

32. Which of the following statements about a lethal allele is NOT correct?

(a) Lethal alleles are always recessive

(b) Lethal alleles may have a late age of onset

(c) Lethal alleles may be caused by mutations in essential genes

(d) Lethal alleles may affect one individual differently than another

33. Sarah has chronic bronchitis and frequently ends up hospitalized with pneumonia. She has no family history of cystic fibrosis, a serious recessive genetic disorder which affects the respiratory and digestive tracts. She and her husband, Greg, have one child after undergoing fertility treatments to overcome Greg's infertility. They are surprised when the child is diagnosed with cystic fibrosis and the genetic counselor tells them that they each also have the condition, which causes their bronchitis and infertility. Which of the following explanations would be a reasonable explanation of their situation?

(a) There are several different alleles of cystic fibrosis, which cause differing forms of the disease

(b) Although the underlying disease is genetic, the severity of the disorder is influenced by environmental variation

(c) Cystic fibrosis displays variable expressivity

(d) All of the above

34. A semilethal allele is an example of which of the following extensions of Mendelian genetics?

(a) Incomplete dominance (b) Incomplete penetrance

(c) Variable expressivity (d) Overdominance

35. In a rare blood type referred to as the Bombay phenotype, individuals are unable to attach the A or B antigens to their red blood cells. Individuals that are homozygous recessive for gene "H" have the Bombay phenotype and their blood type is type "O" regardless of their ABO genotype. This is an example of what extension of Mendelian inheritance?

(a) Multiple alleles (b) Overdominance

(c) Environmental influence (d) Epistasis

36. Which of the following examples would NOT be consistent with a gene dosage effect?

(a) Males with one normal allele for the X linked gene dystrophin are unaffected, while females (who shut off one of their two X chromosomes) with one normal allele for dystrophin may be mildly affected with muscular dystrophy

(b) Flower color in four-o-clocks seems to be incompletely dominant

(c) In fruit flies, homozygotes for an allele that produces a small amount of functional protein are severely abnormal but survive to adulthood, while homozygotes for an allele that produces no protein at all die as embryos

(d) In bees, males normally are haploid while females normally are diploid

37. Which of the following is the best example of overdominance?

(a) Two lines of true breeding white parents produce purple flowered offspring

(b) Two lines of true breeding tomatoes that are susceptible to verticillium wilt (a fungal infection) produce offspring that are resistant to verticillium wilt

(c) Two true breeding lines, one tall and one short, produce offspring of medium height

(d) Two true breeding lines, one tall and one short, produce all tall offspring

38. The recessive mutant allele that causes cystic fibrosis is much more frequent in Caucasians of eastern European descent than in other populations. Some scientists believe heterozygotes must have had a survival advantage during plagues such as cholera that occasionally swept through this population. What concept does this illustrate?

(a) Multiple alleles (b) Overdominance

(c) Environmental influence (d) Epistasis

39. You find a pink geranium in your flowerbed of red geraniums. Seeds from self fertilization of this plant produce ¼ red plants, ½ pink plants, and ¼ white plants. When considering the visible phenotype, which explanation is most likely?

(a) The soil in which the pink plant was growing was deficient in a critical nutrient

(b) The pink plant had a new mutation, creating an allele to which the wild type "red" is incompletely dominant

(c) The pink plant had a new mutation, creating an allele which is incompletely dominant to red

(d) The pink plant had a new mutation, creating an allele which is codominant to "red"

40. What are the possible offspring from a type A mother and a type AB father?

(a) A, B, AB, and O (b) A, B, and AB

(c) A and AB (d) A only

41. In soybeans, resistance to sudden death syndrome (SDS) is inherited as a dominant condition. In some cases, individuals that are not resistant to SDS (but whose parents were) can pass the trait on to their progeny. This is an example of which of the following?

(a) Overdominance (b) Incomplete penetrance

(c) Incomplete dominance (d) Variable expressivity

42. Neurofibromatosis is a condition in humans which is generally inherited in a dominant fashion. Assuming this disorder is 90% penetrant, what is the chance that a single child born to a father with neurofibromatosis and a mother known not to carry the mutation will have the disorder?

(a) 90% (b) 50%

(c) 45% (d) 5%

43. In a group of fruit flies homozygous for an allele that causes a reduced number of thoracic bristles, you notice that the number of bristles varies from about half the normal number to only a few bristles. This is an example of which of the following?

(a) Incomplete dominance (b) Overdominance

(c) Incomplete penetrance (d) Variable expressivity

44. You are trying to develop true breeding blue hydrangeas, but the plants in one corner of your plot consistently turn pink regardless of your mating strategy. Which of the following would you suspect?

(a) Incomplete penetrance

(b) Variable expressivity

(c) Multiple genes affecting this trait

(d) Environmental effects

45. A trait that appears to be dominant in one sex but recessive in the other is called:

(a) A sex limited trait (b) A sex influenced trait

(c) A sex linked trait (d) None of the above

46. Which of the following could NOT be used to describe a sex limited trait:
    (a) This term refers to a trait controlled by a gene located on the X or Y chromosomes
    (b) This is a trait that occurs more frequently in one sex than another
    (c) This is an extreme example of a sex influenced trait
    (d) This trait is likely to be a secondary sex characteristic
47. What ratio of phenotypes would you expect if you crossed a heterozygous Pea comb chicken to a heterozygous Rose comb chicken?
    (a) 9 walnut: 3 rose: 3 pea: 1 single
    (b) 1 walnut: 1 rose: 1 pea: 1 single
    (c) 3 walnut: 1 single
    (d) 3 rose: 1 pea
48. Which of the following events is LEAST likely to occur in a cross involving epistasis by both genes?
    (a) A monohybrid has a different phenotype from either of the two parents
    (b) Two lines of true breeding white parents produce purple flowered offspring
    (c) A 9:3:3:1 ratio of F2 phenotypes is seen
    (d) All of these are likely to be seen in epistatic interactions
49. If you establish a dihybrid cross and find an unexpected phenotypic ratio in the F2, which of the following might you suspect?
    (a) Epistasis
    (b) Multiple genes affecting your trait of interest
    (c) Linkage
    (d) All of these
50. Which of the following is NOT true of a maternal effect gene?
    (a) It is located in the nuclear DNA
    (b) Maternal genotype affects offspring phenotype
    (c) It may control deposition of material into oocytes
    (d) It must be located on the X chromosome

51. Which of the following statements about nurse cells are true?

(a) They are haploid

(b) They are able to express both alleles present in the female

(b) They are precursors of the embryo

(b) Both A and B

52. Which of the following statements regarding epigenetic inheritance is FALSE?

(a) Epigenetic inheritance can be reset during gametogenesis

(b) Epigenetic inheritance can temporarily affect an individual

(c) Epigenetic inheritance does not involve a change in DNA sequence

(d) A and C

53. In dosage compensation:

(a) An X chromosome can be inactivated

(b) Gene expression can be increased from a single X chromosome

(c) Gene expression can be reduced from a pair of X chromosomes

(d) All of the above

54. Which of the following statements regarding X inactivation in mammals is FALSE?

(a) The process is entirely random

(b) X inactivation may occasionally occur in males

(c) This process ensures a homogenous phenotype in heterozygotes

(d) X inactivation occurs early in embryonic development

55. You attempt to replicate the experiment shown in Fig 7.6, but an incompetent lab assistant inoculated the flasks in step 3 from entire plates rather than single clones. What result would you expect?

(a) The results should be the same as seen in Fig 7.6

(b) You would see only fast or slow alleles in every sample

(c) You would see both fast and slow alleles in every sample

(d) The results would support the Lyon hypothesis

56. A female mouse has been identified in which the vast majority of active X chromosomes were inherited from her father. Which of the following statements may explain this?

(a) The TsiX allele inherited from her father is particularly weak

(b) The Xce allele inherited from her father is particularly weak

(c) The Xic region is missing from the chromosome inherited from her father

(d) The Xist allele inherited from her father is particularly strong

57. A mutation in Xist which prevents binding to proteins that promote compaction would interfere with which stage of X inactivation?

(a) Initiation (b) Spreading

(c) Maintenance (d) None of these

58. Monoallelic expression of a gene:

(a) Occurs when one copy of the gene has been deleted

(b) Occurs when one copy of the gene carries a mutation

(c) Occurs when the gene is imprinted

(d) Is not seen in mammals

59. A mouse heterozygous at the Igf-2 locus is dwarf and has 50% dwarf progeny when mated to a homozygous Igf-2m animal. Which statement about this animal is FALSE?

(a) This mouse must be a male

(b) The Igf-2m allele in this mouse came from its mother

(c) This mouse expresses abnormal Igf-2 protein

(d) None of these (all statements are true)

60. Imprinted genes:

(a) Provide an example of epigenetic inheritance

(b) Are near differentially methylated regions

(c) Are only transmitted by one parent

(d) A and B

61. You develop an inbred strain of mice that display one of two distinct syndromes. What evidence might lead you to believe that these mice are mutant for a gene or genes affected by imprinting?

(a) The two syndromes occur randomly in the colony of mice

(b) All mice of the same litter have the same syndrome

(c) All mice have the same syndrome as their mother

(d) All mice have the same syndrome as their father

62. A pattern of transmission where all offspring have the same phenotype as their mother is consistent with which type of non-Mendelian inheritance?

(a) Maternal effect (b) Genomic imprinting

(c) Dosage compensation (d) Extranuclear inheritance

63. The gene dosage effect which affects eye color in fruit flies is most likely due to which phenomenon?

(a) Maternal effect (b) Genomic imprinting

(c) Dosage compensation (d) Extranuclear inheritance

64. Both X-linked genes and extranuclear inheritance produce different results in reciprocal crosses. How can you distinguish between them?

(a) Extranuclear traits can be transmitted from an unaffected mother to an affected son

(b) Extranuclear traits can be transmitted from an affected mother to an affected offspring

(c) Extranuclear traits cannot be transmitted from father to son

(d) None of these

65. For a trait known to be caused by a mitochondrial gene variant, there may be an occasional female that can transmit different phenotypes to her progeny on an apparently random basis. What situation might you suspect?

(a) She is heterozygous for this trait

(b) She displays heteroplasmy for this trait

(c) She displays a low rate of mutation

(d) She is the product of a somatic mutation

66. In *S. cerevisiae*, mutations which cause a segregational petite phenotype:
   (a) Are located in mitochondrial genes
   (b) Cause mitochondria to have better than normal function
   (c) Are transmitted according to Mendelian rules
   (d) Prevent growth under all conditions

67. Neutral petite mutations:
   (a) Are an example of extranuclear inheritance
   (b) Always produce petite progeny when mated to wild type
   (c) Might be expected to produce normal progeny when mated to a suppressive petite
   (d) All of the above

68. Mitochondrial mutations in humans tend to affect some tissues more than others. Which is the most likely explanation?
   (a) Some human tissues do not have mitochondria
   (b) Some human tissues have higher energy demands than others
   (c) Some human tissues can synthesize large amounts of ATP in the absence of mitochondrial function
   (d) Heteroplasmy and subsequent segregation often leads to a variegated phenotype

69. Which piece of evidence provides the strongest support for the endosymbiosis theory?
   (a) All eukaryotes have mitochondrion, but not all have chloroplasts
   (b) Genes in chloroplasts and mitochondrion have a high degree of similarity to bacterial genes
   (c) Some genes which control mitochondrial or chloroplast function are located in the nucleus
   (d) All of these

70. Symbiotic infective particles:
   (a) Tend to be inherited in an extranuclear pattern
   (b) Have their own DNA
   (c) Appear to be similar to viruses in function
   (d) All of these

71. Moth pigmentation can be maternally influenced If a+ is the allele for pigmentation, a mother with genotype a+/a and a father with genotype aa will produce

(a) All pigmented larvae

(b) All unpigmented larvae

(c) A 3 pigmented: 1 unpigmented ratio in the larvae

(d) A 1 pigmented: 1 unpigmented ratio in the larvae

72. As adults, what percentage of the larvae in question 2 will be pigmented?

(a) 100% (b) 75%

(c) 50% (d) 25%

73. Inheritance of variegation in corn is cytoplasmic because

(a) Pollen grains do not carry chloroplasts

(b) Neither gamete carries chloroplasts

(c) Both gametes carry chloroplasts

(d) Coloration is environmentally induced

74. Autogamy in heterozygous killer Paramecium results in

(a) All killer (b) All sensitive

(c) 3 killer: 1 sensitive ratio (d) 1 killer: 1 sensitive ratio

75. What are the possible phenotypes and genotypes of the female parent of a sinistral coiled snail?

(a) Sinistral, DD (b) Sinistral, Dd

(c) Dextral, DD (d) Sinistral or dextral, dd

76. A cross is made between a petite yeast strain and a wild-type strain. After meiosis, the ratio of petite: wild type is 1:1. The mutation is of a

(a) Nuclear gene

(b) Mitochondria gene and is of the neutral type

(c) Mitochondrial gene and is of the suppressive type

(d) Chloroplast gene

77. Prader-Willi syndrome and Angelman syndrome are both associated with deletions of the same bands from the long arm of chromosome 15.

However, the physical expression associated with each disorder is very different. This is an example of

(a) Imprinting (b) Nuclear inheritance

(c) Mendelian inheritance (d) Cytoplasmic inheritance

78. Which statement concerning the mitochondrial genome in humans is untrue?

(a) There are few noncoding regions

(b) There are no introns

(c) The DNA is linear

(d) tRNA genes are located in this genome

79. If the stigma on a green corn plant (genotype IjIj) is pollinated by pollen from a variegated plant (genotype ijij), and the F1 are selfed, the F2 will be

(a) All variegated

(b) 50% green and 50% variegated

(c) 75% green and 25% variegated

(d) 25% green and 75% variegated

Fill in the blanks

1. A gene that is not on a sex chromosome is termed __.

(a) Autosomal (b) X-linked

(c) Y-linked (d) Chromosomal

2. In dealing with an X-linked trait, a male will express X chromosome alleles directly in the phenotype because the Y chromosome lacks homology with the X-linked gene. This condition is referred to as being __.

(a) Heterozygous (b) Monozygous

(c) Parazygous (d) Hemizygous

3. A female is diagnosed with hemophilia, an X-linked recessive gene disorder characterized by failure of blood to clot. In checking her medical history, the doctor will most likely discover that ___.

(a) Her brother has hemophilia

(b) Her father's mother had hemophilia

(c) Her mother had hemophilia, her father was normal

(d) Her father had hemophilia, her mother was normal

4. For a male to have X-linked hemophilia, ___.

(a) His mother must have hemophilia

(b) His father's mother must have hemophilia

(c) His mother's father must have hemophilia

(d) His mother must be normal

5. Alleles at one locus masking or blocking the expression of alleles at another locus is an example of ___.

(a) Incomplete dominance

(b) Epistasis

(c) Quantitative inheritance

(d) A dominant gene

6. Individuals with the Bombay phenotype in the ABO blood group system ___.

(a) Are functionally (phenotypically) type O

(b) May have the alleles for A or B antigens

(c) Don't have any antigens at all in their red blood cells

(d) a and b above

7. In the human ABO blood group system ___.

(a) Alleles for A and B antigens are codominant

(b) The allele producing the type O phenotype is recessive

(c) The allele producing the O phenotype is incompletely dominant

(d) a and b above

8. Mendelian crosses involving phenotypes resulting from alleles at two segregating loci ___.

(a) Will have the expected Mendelian dihybrid phenotypic ratio

(b) Will have a ratio typical of a trihibrid

(c) Will have variant Mendelian genotypic ratios

(d) Will have the expected Mendelian dihybrid genotypic ratio

9. Gene interaction in the development of phenotypes caused by two genes ________.

(a) Can only be identified by examining the metabolic pathways generating the phenotype

(b) Will result in 9:3:3:1 F2 phenotypic ratios

(c) Tends to render all but one allele recessive

(d) May result in variations of the expected Mendelian dihybrid phenotypic ratios

10. A ratio of 12:3:1 is most characteristic of a cross involving ________.

(a) 1 gene pair (b) 2 gene pairs

(c) 3 gene pairs (d) 4 gene pairs

11. A mother who is blood type AB has a child who is AB also. A potential father is blood type O. A well-informed geneticist concludes that ________.

(a) He cannot be the father

(b) He might be the father, but it is unlikely

(c) He is very likely to be the father

(d) He or any other male of blood type O could be the father

12. Genetic anticipation refers to the phenomenon which shows ________.

(a) Dominant disorders that affect the victim later in life

(b) An increase in severity of symptoms and a decrease in the age of onset as it is passed through the family

(c) A disorder that affects only individuals from a specific ethnic group

(d) An increase in the number of mutations on a single chromosome over time

13. Some mutations have been found that are expressed differently depending on the sex of the parent the gene is passed through. This is called ________.

(a) Sex-linked inheritance (b) Penetrance

(c) Imprinting (d) Heterozygous

14. Some mutations have been found that are expressed differently depending on the sex of the parent the gene is passed through. This is called __________.
    (a) Sex-linked inheritance (b) Penetrance
    (c) Imprinting (d) Heterozygous
15. Heterozygotes of single locus traits with codominant alleles ___.
    (a) Have the phenotype of both alleles
    (b) Have a phenotype that is not clearly like either homozygote's phenotype
    (c) Have a 1:2:1 phenotypic ratio in Mendelian crosses
    (d) a and c above
16. In incomplete dominance, heterozygotes have a phenotype that is ___.
    (a) Not clearly like either homozygous phenotype
    (b) Indistinguishable from the homozygous phenotypes
    (c) More environmentally influenced than the phenotypes of the homozygotes
    (d) Characteristic of both homozygous phenotypes
17. The phenotypic ratio expected from a Mendelian monohybrid cross involving incompletely dominant alleles ___.
    (a) Is 1:6:1
    (b) Is the same as the genotypic ratio
    (c) Cannot be predicted
    (d) Is 1:1

True/False

1. Mendel had extensive training in biology and heredity prior to his groundbreaking work.
    (a) True (b) False
2. Fusion of gametes is called fertilization and this creates a diploid individual.
    (a) True (b) False

3. Cross fertilization is the typical mode of reproduction for pea plants.
   (a) True (b) False
4. The probability that two heterozygotes produce an affected and a heterozygous offspring is ¾. Keep in mind that each offspring is considered a separate event.
   (a) True (b) False
5. A chi square value lower than the critical value demonstrates that the hypothesis was correct.
   (a) True (b) False
6. Critical values for the chi square test are larger when there are more degrees of freedom because each class of data is expected to show a small amount of variation.
   (a) True (b) False
7. The p values given on the chi square chart indicate probability of that much variation occurring by random chance alone.
   (a) True (b) False
8. A high chi square value always means that our hypothesis is wrong.
   (a) True (b) False
9. A chi square value below the critical value means that our hypothesis is correct.
   (a) True (b) False
10. Lethal alleles violate the rules of Mendelian inheritance.
    (a) True (b) False
11. Mutations may be recessive because cells can increase the amount of functional protein produced from their remaining normal allele.
    (a) True (b) False
12. Essential genes produce proteins that are always required by the cell.
    (a) True (b) False
13. Sex limited inheritance refers to traits that are located on the X or Y chromosome.
    (a) True (b) False

14. For alleles that display incomplete dominance, the phenotypic ratio expected in the F2 generation of a monohybrid cross would be the same as the genotypic ratio.

(a) True (b) False

15. For genes which have multiple alleles, the relationships between those alleles can be a variety of types of dominant/recessive relationships.

(a) True (b) False

16. In some cases, the type of dominance relationship observed depends on how the phenotype is defined.

(a) True (b) False

17. You would expect individuals with type B blood to produce the anti-B antibody.

(a) True (b) False

18. Genes that show a dosage effect must be located on the X chromosome.

(a) True (b) False

19. A couple, both of whom display the recessive trait of albinism, are recruited for a study on the genetics of albinism. Their first child is normally pigmented, causing the scientists conducting the study to accuse the mother of infidelity and throw them out of the study. True or false, the scientists in this study are certain to have correctly accused the mother.

(a) True (b) False

20. In maternal effect inheritance, the phenotype of the mother determines the genotype of the offspring.

(a) True (b) False

21. Gene products produced by nurse cells are controlled by both alleles of the female.

(a) True (b) False

22. Any individual having only one copy of a gene is the same as any other individual having only one copy of a gene.

(a) True (b) False

23. TsiX would be expected to play a role in the spreading phase of X inactivation.

(a) True (b) False

24. Dosage compensation is not required for all X linked genes in mammals.

(a) True (b) False

25. Genomic imprinting is an example of epigenetic inheritance.

(a) True (b) False

26. The mitochondrial genome encodes all proteins required to carry out oxidative phosphorylation.

(a) True (b) False

27. In Chlamydomonas, mitochondria are inherited from one parent while chloroplasts are inherited from the other.

(a) True (b) False

28. Plant cells are the only cells capable of photosynthesis.

(a) True (b) False

Answers

Multiple Choice Question

1	2	3	4	5	6	7	8	9	10	11	12	13	14	15
1	2	3	4	5	6	7	8	9	10	11	12	13	14	15
B	D	C	A	A	A	C	A	A	C	D	A	B	B	D
16	17	18	19	20	21	22	23	24	25	26	27	28	29	30
C	D	B	B	D	B	B	D	B	B	A	D	B	B	C
31	32	33	34	35	36	37	38	39	40	41	42	43	44	45
C	A	D	B	D	D	B	B	B	B	B	C	D	D	B
46	47	48	49	50	51	52	53	54	55	56	57	58	59	60
A	B	C	B	D	B	B	D	C	C	B	B	C	B	D
61	62	63	64	65	66	67	68	69	70	71	72	73	74	75
B	D	C	A	B	C	A	B	B	D	A	C	A	B	B
76	77	78	79											
A	A	C	C											

Fill in the blanks

1	2	3	4	5	6	7	8	9	10	11	12	13	14	15
A	D	D	C	B	D	D	A	D	B	B	B	A	C	D
16	17													
A	B													

True/False

1	2	3	4	5	6	7	8	9	10	11	12	13	14	15
B	A	B	B	B	A	A	B	B	B	A	B	B	A	A
16	17	18	19	20	21	22	23	24	25	26	27	28		
A	B	B	B	B	A	B	B	A	A	B	A	B		

❑❑❑

Chapter – 5

Genetic Mapping and Gene Regulation in Prokaryotes

Genetic Transfer and Mapping in Bacteria

- Bacteria can transfer genetic material during conjugation
- Hfr strains contain an f factor integrated into the bacterial chromosome
- Conjugation experiments can be used to map genes along the *E. coli* chromosome
- A detailed genetic map of the *E. coli* chromosome has been obtained from many conjugation studies
- Bacteriophages can also transfer genetic material from one bacterial cell to another in the process of transduction
- Cotransduction can be used to map genes that are within 2 minutes of each other
- Bacteria can also transfer genetic material by transformation
- Horizontal gene transfer is the transfer of genes among different species

Intragenic Mapping in Bacteriophages

- Mutations in viral genes can alter plaque morphology
- A complementation test can reveal if mutations are in the same gene or in different genes

- Intragenic maps were constructed using data from a recombinational analysis of mutants within the rii region
- Deletion mapping can be used to localize many rii mutations to specific regions in the riia or riib genes
- Intragenic mapping experiments provided insight into the relationship between traits and molecular genetics

Transcriptional Regulation

- The phenomenon of enzyme adaptation is due to the synthesis of cellular proteins
- The lac operon encodes proteins that are involved in lactose metabolism
- The lac operon is regulated by a repressor protein
- Bacterial conjugation experiments provide evidence that the lacI gene encodes a repressor protein
- The lac operon is also regulated by an activator protein
- Further studies have revealed that the lac operon has three operator sites for the lac repressor
- The ara operon can be regulated positively or negatively by the same regulatory protein
- The trp operon is regulated by a repressor protein and also by attenuation
- Repressible operons usually encode anabolic enzymes, and inducible operons encode catabolic enzymes

Translational and Posttranslational Regulation

- Repressor proteins and antisense RNA can inhibit translation
- Posttranslational regulation can occur via feedback inhibition and covalent modifications

Gene Regulation In The Bacteriophage Life Cycle

- Phage l can follow a lytic or lysogenic life cycle
- During the lysogenic cycle, the cII/cIII complex activates expression of the l repressor

- The lytic cycle depends on the action of the cro protein
- Cellular proteases influence the choice between the lytic and lysogenic cycle. The O_R region provides a genetic switch between the lytic and lysogenic cycles

Multiple Choice Question

1. How is genetic transfer different from sexual reproduction?
   (a) In genetic transfer, genetic material is transferred from one individual to another
   (b) In sexual reproduction, genetic material from two individuals is mixed equally
   (c) In both processes, the individual donating genetic material retains their original complement of genetic material
   (d) None of these
2. Which method of gene transfer involves direct contact between the bacteria?
   (a) Conjugation (b) Transduction
   (c) Transformation (d) All of these
   (e) None of these
3. How might the results of Bernard Davis's U tube experiment have been different if each strain were only mutant for a single gene?
   (a) There would be no difference
   (b) Transformation by small pieces of DNA could restore gene function
   (c) Conjugation would not be required to change the phenotype of the bacterial strains
   (d) B and C
4. Which of the following is not a protein or group of proteins required for conjugation to occur?
   (a) The relaxosome
   (b) Pilin protein
   (c) The origin of transfer
   (d) Coupling protein

5. Approximately how many minutes would it take for an Hfr strain with an origin between cheA and pabB (oriented toward pabB) take to transfer galE?

   (a) 17 minutes (b) 23 minutes

   (c) 77 minutes (d) 117 minutes

6. Which of the following is a description of generalized transduction?

   (a) Inaccurate excision of prophage can occasionally remove a small portion of the bacterial chromosome

   (b) Random pieces of bacterial DNA can become incorporated into a phage coat

   (c) Pieces of bacterial DNA may be released into the culture medium when a cell is lysed

   (d) All of these

7. What is the best explanation for why Bernard Davis didn't observe generalized transduction in his U tube experiments?

   (a) He used genes that could not be packaged into a bacteriophage head

   (b) He used bacterial strains that could not fit through the filter

   (c) He used bacterial strains that did not host a prophage

   (d) He used virulent bacteriophage

8. A researcher would like to map the location of galE and trpA genes in a new species of bacterium that appears to be closely related to *E. coli*. He decides to use cotransduction, and generates appropriate donor and recipient strains. He is disappointed when cotransduction is not seen in his experiement. What is the most reasonable explanation for this situation?

   (a) His new bacterial species does not have galE or trpA genes

   (b) His new bacterial species cannot survive galE or trpA mutation

   (c) These two genes are too far apart to be mapped by cotransduction

   (d) These two genes are too close together to be mapped by cotransduction

9. Which of the following is NOT a critical function of competence factors?
   (a) Transportation of DNA fragments into bacterial cells
   (b) Binding of DNA to a cell surface receptor
   (c) Degradation of one strand of the transforming DNA
   (d) Degradation of the bacterial chromosome in the presence of single standed DNA

10. How might the function of extracellular endonucleases help protect the bacterium from infection by viruses?
   (a) They would prevent the virus from binding to the outside of the bacterial cell
   (b) They would destroy the viral coat, so it cannot inject its genetic material
   (c) They would cut the viral DNA into small pieces
   (d) All of the above

11. What do transduction, transformation, and conjugation have in common?
   (a) All require presence of an outside factor to facilitate gene transfer
   (b) In all three processes, DNA is transferred as a single stranded molecule
   (c) They all transfer large pieces of DNA into recipient cells
   (d) All of the above

12. Which type(s) of genetic transfer lead to incorporation of new DNA into the bacterial chromosome? I. Conjugation mediated by F factor; II. Hfr mediated conjugation; III. Transduction; IV. Transformation
   (a) I and II only (b) III and IV only
   (c) II, III, and IV (d) I, II, III, and IV

13. How would you expect a mutation in a gene encoding a tail fiber protein would affect the plaque phenotype of T4?
   (a) It could lead to very large plaques
   (b) It would be unlikely to have an affect on plaque formation
   (c) It could lead to smaller than normal plaques
   (d) A and B

14. In Benzer's intragenic mapping experiments, what event was required to allow production of infectious phage from rII mutants?

(a) rII mutant strains with mutations in different genes

(b) Homologous recombination between phage

(c) Crossover within the mutant gene, between the mutations

(d) Crossover at one end of the mutant gene, outside the mutation

15. In deletion mapping, mutations can be localized to a region of the gene because:

(a) The mutation fails to complement other mutations in the gene

(b) The mutation fails to complement a deletion of a known portion of the gene

(c) The gene sequence can be determined

(d) A and B

16. A key finding of intragenic mapping was:

(a) That genes consist of indivisible particles, like beads on a string

(b) That genes do not change over time

(c) That genes can be modified by crossing over and mutation

(d) All of these

17. Which of the following would you expect to find in an inducible system?

(a) A repressor protein, which is bound to DNA in absence of any other factor

(b) A repressor protein, which is bound to DNA in the presence of a corepressor

(c) An activator protein, which is bound to DNA in absence of any other factor

(d) An activator protein, which is bound to DNA only in the absence of an inhibitor

18. Which of the following would be an example of enzyme adaptation?

(a) Enzymes required for glucose utilization are constitutively expressed

(b) Addition of phosphate groups to a protein can activate it

(c) Certain enzymes are made up of multiple subunits

(d) Enzymes for tryptophan synthesis are not made in the presence of tryptophan

19. What effect would you expect if gene expression of the lac operon were completely repressed?

(a) The cell would be more efficient without "wasting" the energy required for the low level of acZ, lacY, and lacA gene expression

(b) Allolactose would accumulate within the cell and become toxic

(c) Lactose would not be converted into the inducer and the operon could not be induced

(d) All of these

20. Which of the following genes is not a structural gene of the lac operon?

(a) lacA (b) lac I

(c) lacY (d) lacZ

21. The most convincing evidence that lacI is a diffusible protein is:

(a) A merozygote with only one functional lacI gene is able to regulate both copies of the lac operon

(b) A merozygote with only one functional lacO is unable to regulate both copies of the lac operon

(c) The lacI gene is located very close to the lac operon

(d) Mutants lacking lacI function do not respond to changes in lactose concentration

22. Which of the following elements, when nonfunctional, would lead to a reduction in the total amount of functional lacZ produced by a bacterial cell?

(a) The operator (b) The lacI gene

(c) The CAP site (d) The lacA gene

23. Which of these describes a similarity between the lac and ara operons?

(a) Expression of structural genes does not occur at high glucose concentrations

(b) lacI and araI both encode trans acting factors

(c) Binding of a regulatory protein is prevented in the presence of an inducer

(d) All of these are similarities

24. What is the best explanation for the stem loop arrangement seen in the absence of translation of the trpL gene?

(a) Region 4 binds to region 3 before region 2 has been transcribed

(b) Region 1 is unable to bind to region 3 or 4

(c) The 1:2 stem loop has more base pairs than a 2:3 stem loop

(d) The ribosome prevents region 1 from binding to any other region

25. Why does the ribosome stall on region 1 during attenuation?

(a) Translation of this region requires tRNAtrp

(b) High levels of tryptophan interfere with ribosome function

(c) Low levels of tryptophan lead to intrinsic termination

(d) Binding of the repressor protein prevents further translation

26. How might antisense RNA affect translation?

(a) It can interfere with ribosome binding by blocking binding sites

(b) It can enhance ribosome binding by modifying binding sites

(c) It can prevent identification of the start codon by the ribosome

(d) It can interfere with the normal secondary structure of the mRNA molecule

27. How does the expression of Q protein lead to the lytic cycle?

(a) The Q protein functions as an antiterminator

(b) The Q protein stimulates production of the coat proteins for assembly of new phage

(c) The Q protein interacts with RNA polymerase

(d) All of these

28. Presence of cellular proteases that degrade cII:

(a) Prevents activation of the lytic cycle

(b) Is typical under poor growth conditions

(c) Allows high levels of expression of the λ repressor protein.

(d) None of these

29. Transcription is turned off by

(a) Induction (b) Repression

(c) Activation (d) All of the above

30. In regulation by repression

(a) A sugar, such as lactose, acts as an inducer and combines with the repressor to prevent transcription

(b) An inducer activates the activator so that it binds to DNA and prevents transcription

(c) An amino acid activates the repressor so that the repressor binds to the operator and prevents transcription

(d) An amino acid binds to the operator, blocking the repressor, allowing transcription to proceed

31. When both glucose and lactose are present

(a) Cyclic AMP is high so transcription occurs

(b) The lac repressor binds with the lactose and transcription occurs

(c) RNA polymerase is able to bind to the operator so transcription occurs

(d) Transcription is turned off

32. Which is NOT a DNA-binding domain?

(a) Leucine zipper

(b) Helix-turn-helix

(c) Zinc finger

(d) Intron-exon-intron

(e) All of the above are domains that bind to DNA

33. If a normal (wild type) repressor is added *in vitro*, which DNA fails to bind the repressor?

(a) I+ O+ Z+ (b) I+ Oc Z+

(c) I- O+ Z+ (d) I+ O+ Z-

34. The lac repressor is a

(a) RNA molecule
(b) Lipid molecule
(c) DNA molecule
(d) Protein molecule

35. If a normal repressor is added *in vitro,* in which of the genotypes does the repressor fail to bind to the operator?

(a) I+ O^c Z+
(b) I+ O+ Z+
(c) I+ O+ Z-
(d) I- O+ Z+

36. A mutation in the I gene directly affects

(a) The repressor
(b) The operator region
(c) β-galactosidase
(d) The promoter

37. What would be the effect of a mutation in the operator that blocked the lac repressor binding?

(a) The lacZYA genes would not be expressed
(b) The lacZYA genes would be repressed by lactose
(c) The genes would be inducible by lactose
(d) The lacZYA genes would be expressed constitutively

38. What would be the effect of a mutation in the lacI gene that blocked binding of the lac repressor to the operator?

(a) The lacZYA genes would be repressed by lactose
(b) The genes would be inducible by lactose
(c) The lacZYA genes would be expressed constitutively
(d) The lacZYA genes would not be expressed

39. What would be the effect of a mutation in the lacI gene that blocks the binding of lactose to the repressor?

(a) The genes would be inducible by lactose
(b) The lacZYA genes would be repressed by lactose
(c) The lacZYA genes would not be expressed
(d) The lacZYA genes would be expressed constitutively

40. Genes that are always transcribed are called:

(a) Transposons
(b) Constitutive genes
(c) Operator genes
(d) Repressor genes

41. The catabolite-activating protein is involved in

(a) Repressing the expression of the lac operon when glucose is present

(b) Activating the expression of the lac operon when glucose is present

(c) Repressing the lac operon when lactose is present

(d) Activating transcription when glucose is present

42. The role of glucose in catabolite repression is to

(a) Increase the activity of cAMP by inhibiting the lac operon

(b) Inhibit the activity of the operon causing an increase in cAMP levels in the cell

(c) Increase the activity of adenyl cyclase, causing the decline of cAMP in the cell

(d) Inhibit the activity of adenyl cyclase, causing a decline in the level of cAMP in the cell

43. Which of the following are examples of cis acting genes?

(a) lacI, lacY, lacZ (b) lacP, lacO

(c) lacY, lacZ, lacA (d) lacI

44. A regulatory site within the leader sequence of the tryptophan operon that controls this operon by diminishing transcription is called:

(a) Upregulator (b) Transcriptor

(c) Attenuator (d) Repressor

45. In the absence of tryptophan:

(a) The repressor is active and does not bind to the promoter. Transcription does not occur

(b) The repressor binds to the co-repressor. Transcription is initiated

(c) The repressor is inactive and does not bind to the promoter. Transcription is initiated.

(d) The trp operon is repressed and transcription occurs

46. In the lac and trp operators, the repressor is said to be an allosteric molecule because:

(a) It stimulates the mutation of these operons

(b) Interaction with another molecule causes a conformational change in the repressor's active site

(c) It diminishes the transcription of the related gene structure

(d) It induces the production of lactose or tryptophan

47. Which of the following is shared by both prokaryotes and eukaryotes?

(a) Promoters (b) Introns

(c) 3'-poly A tails (d) RNA splicing

48. Which is not a DNA-binding domain?

(a) Leucine zipper (b) Intron-exon-intron

(c) Zinc finger (d) Helix-turn-helix

49. What is the name for the transcripts that ultimately get converted to the form that is translated?

(a) Mature RNA (b) mRNA

(c) Small nuclear RNA (d) Heterogeneous nuclear RNA

50. Which of the following sequences contains a eukaryotic promoter?

(a) CATAGATACCCGGACC (b) GTCCTATATAAATGGGC

(c) CGAGAGATTTTGCCGCA (d) GCATCCCGGTACGCGAC

51. Which of the following are post-transcriptional modifications involved in the control of gene expression?

(a) mRNA processing (b) Methylation

(c) Polyadenylation (d) All of the above

Fill in the blanks

1. The end result of F factor mediated conjugation is __________

(a) Is that both strains are F+

(b) Involves transfer of the entire bacterial chromosome

(c) Converts the recipient strain to F+ and the donor to F-

(d) a and b

2. Transformation is a sexual process by which __________

(a) A phage transfers bacterial DNA between cells

(b) Bacterial DNA is transferred between cells during "mating"

(c) Bacterial DNA from the environment is incorporated into the host

(d) Bacterial DNA with another species is exchanged

3. A wild-type of E.coli is transformed by DNA from a trp-try-strain. The following auxotrophic transformants are scored: tyr-= 11; trp-tyr-=80. What is the co-transfer index for trp and tyr? __

(a) 0.09 (b) 0.11

(c) 0.20 (d) 0.88

4. In the above question (number 2) the two genes under study are________

(a) Closely linked (b) Moderately linked

(c) Distantly separated (d) Unlinked

5. A prototroph (D+E+F+) is used to make transducing phages that then infect D-E-F- cells. The resulting data: D+E+F+=35; D+E+F-=60; D+E-F+=2; D+E-F-=210 (Total=307). _______ is the correct gene order.

(a) DEF (b) DFE

(c) EDF (d) FDE

6. In the above problem__________is the relative cotransduction frequency for genes D and E?

(a) 0.12 (b) 0.31

(c) 0.71 (d) 0.88

7. In the above problem_________ is the relative cotransduction frequency for genes D and F?

(a) 0.12 (b) 0.31

(c) 0.71 (d) 0.88

8. In the above problem__________is the relative cotransduction frequency for genes E and F?

(a) 0.12

(b) 0.31

(c) 0.71

(d) Should not be calculated

9. Given the following results of an interrupted mating experiment, give the gene order_________. Time=0, no genes transferred; time= 5 mins,

genes A and C present; time=10 mins, genes A and C present; time=15 mins, genes A,C, and B present.

(a) ABC (b) ACB

(c) CAB (d) Two of these are possible

10. A wild-type strain of *B. Subtilis* is transformed by DNA from a strain that cannot grow on galactose and also needs biotin for growth. Transformants are isolated and determined to have the following genotypes: gal-bio-=17; gal-bio+=4; gal+bio-=9. ____ is the relative co-occurance of these two loci?

(a) 0.133 (b) 0.30

(c) 0.43 (d) 0.57

11. A bacterial strain of the genotype a-b- would be ________

(a) Be unable to grow on minimal medium

(b) Be able to grow on complete medium

(c) Be unable to grow on minimal medium with "a" added

(d) All of the above

12. DNA methylation may be a significant mode of genetic regulation in eukaryotes. Methylation regulates gene activity by _________.

(a) Adding methyl groups to the cytosine of CG doublets in DNA

(b) Altering RNA polymerase activity by methylation

(c) Altering translational activity, especially of highly methylated tRNAs

(d) Changing DNA-DNA hydrogen bonding

13. The trp operon is _________.

(a) Repressible and under negative control

(b) Inducible and under negative control

(c) Inducible and under positive control

(d) Repressible and under positive control

14. A mutation in the I gene of the lac operon directly affects _________.

(a) B-galactosidase (b) The repressor

(c) The promoter (d) Permease

15. Enhancer elements are found ___.
    (a) Within coding sequences
    (b) Downstream from the start site
    (c) Upstream from the start site
    (d) In any of the above regions
16. Regarding enhancers and transcription factors (TFs) and their effect on gene expression, ___.
    (a) Both are trans-activating
    (b) Enhancers are trans-activating, while TFs are cis-activating
    (c) Enhancers are cis-activating, while TFs are trans-activating
    (d) Both are cis-activating
17. Posttranscriptional regulation occurs in ___.
    (a) RNA synthesis (b) The synthesis of proteins
    (c) The synthesis of DNA (d) The processing of mRNA
18. Translational regulation occurs in ___.
    (a) The mitochondria (b) The nucleus
    (c) The cytoplasm (d) The geology apparatus
19. Transcription factors are ___.
    (a) Regulatory DNA sequences that bind to the promoter site
    (b) Regulatory bases that bind to the promoter
    (c) Regulatory sequences that bind to a protein
    (d) Regulatory proteins that bind to a promoter site
20. Transcriptional regulation ___
    (a) Is highly efficient at completely preventing transcription
    (b) Allows the cell to only produce proteins that are needed at the time
    (c) Can be induced by a repressor protein
    (d) All of these
21. Genes within an operon is ___
    (a) Tend to be regulated by a common regulatory mechanism
    (b) Are generally involved in the same biochemical pathway

(c) Are expressed as a polycistronic RNA

(d) All of the above

22. The inducer for the lac operon is ________________

(a) Lactose

(b) Allolactose

(c) β-galactosidase

(d) Galactose

23. Attenuation of the trp operon is ___________

(a) Occurs when transcription is complete before translation begins

(b) Is mediated by the trp repressor protein

(c) Occurs in the presence of high levels of tryptophan

(d) All of the above

24. A translational regulatory protein ______________

(a) Binds to DNA and prevents translation

(b) Binds to mRNA and prevents translation

(c) Binds to rRNA and prevents translation

(d) Binds to tRNA and prevents translation

25. An allosteric enzyme is ____________

(a) Has two different binding sites

(b) Is regulated by binding of a regulatory molecule somewhere other than the active site

(c) May be regulated by feedback inhibition

(d) All of these

26. After phage λ initially enters a bacterial cell, activation of the early promoters _______

(a) Directs synthesis of an antiterminator protein

(b) Prevents accumulation of the cII/cIII complex

(c) Is controlled by levels of nutrients in the cell

(d) Leads directly to lysis of the cell

27. Functions of the λ repressor protein include all of the following except ____________

(a) Inhibition of the PR promoter for the cII gene, preventing activation of the PRE promoter

(b) Activation of high levels of its own expression

(c) Inhibition of the PL promoter, preventing activation of the lytic cycle

(d) Binding to OR and OL

28. PR and PRM is ______________

(a) Activate transcription of the same genes under different regulation

(b) Prevent establishment of the lytic cycle when activated

(c) Stimulate transcription in opposite orientations on the phage λ DNa

(d) All of these.

29. In regulation by induction is ______________

(a) An inducer inactivates the repressor so transcription can proceed

(b) An inducer inactivates the repressor so transcription is stopped

(c) An inducer activates the repressor so transcription can proceed

(d) An inducer activates the repressor so transcription is stopped

30. When glucose is present as ____________

(a) Cyclic AMP is high, the catabolite activator protein (CAP) binds to the activator binding site, and transcription of lactose is turned on

(b) Cyclic AMP is low, CAP binds to the site activator binding site, and transcription of lactose is turned on

(c) Cyclic AMP is high, CAP does not bind to the activator binding site, and transcription of lactose is turned on

(d) Cyclic AMP is low, CAP does not bind to the activator binding site, and transcription is turned off

31. The lac repressor ______________

(a) Binds to the operator and prevents transcription

(b) Binds to the CAP site and prevents transcription

(c) Binds to the CAP site and facilitates transcription

(d) Binds to the operator and facilitates transcription

True/False

1. Conjugation involves direct transfer of genetic material from one bacterium to another.

   (a) True (b) False

2. Gene transfer by conjugation converts a strain that cannot act as a donor to a strain that can initiate conjugation.

   (a) True (b) False

3. Hfr strains of bacteria contain F factors that include some genes from the bacterial chromosome.

   (a) True (b) False

4. According to Fig 6.7, argR and argG are expected to be transferred about 2 minutes apart in interrupted mating experiments.

   (a) True (b) False

5. Virulent phages generally follow the lysogenic cycle when infecting bacteria.

   (a) True (b) False

6. In generalized transduction, random bits of the bacterial chromosome are packaged in a phage head.

   (a) True (b) False

7. According to figure 6.7, you would expect that hipA and purL genes could frequently be cotransduced.

   (a) True (b) False

8. Homologous recombination is a mechanism used by bacteria to incorporate new DNA into their chromosome, if it is similar to an existing sequence.

   (a) True (b) False

9. Horizontal gene transfer may be mediated via conjugation but not transduction or transformation.

   (a) True (b) False

10. Bacteriophage may be able to infect some strains of bacteria but not others.

    (a) True (b) False

11. Mutations which complement each other are in the same genE)

    (a) True (b) False

12. Deletion mapping is more efficient but less detailed than fine structure mapping of genes.

    (a) True (b) False

13. In the absence of glucose, when lactose is present it combines with the repressor, allowing RNA polymerase to carry on transcription.

    (a) True (b) False

14. When both glucose and lactose are absent, transcription occurs.

    (a) True (b) False

15. Negative control occurs when a repressor is inactivated by an inducer.

    (a) True (b) False

16. An activator must react with its substrate before it can bind to DNA.

    (a) True (b) False

17. Catabolite repression requires an inducer and an activator protein.

    (a) True (b) False

18. The cro protein and λ repressor proteins bind to the same sites within OR

    (a) True (b) False

19. If regulation of a gene involves a repressor protein, it is an inducible gene.

    (a) True (b) False

20. The lac repressor protein can repress the operon by binding to any one of the three binding sites in the operator.

    (a) True (b) False

21. Induction is the process by which phage λ enters the lysogenic life cycle.

    (a) True (b) False

22. A gene which is usually turned down but can be activated is under negative control.

    (a) True (b) False

23. All of the genes involved in functioning of the lac operon are inducible.

    (a) True (b) False

Answers

Multiple Choice Question

1	2	3	4	5	6	7	8	9	10	11	12	13	14	15
B	A	D	C	B	B	C	C	D	C	D	C	C	C	B
16	17	18	19	20	21	22	23	24	25	26	27	28	29	30
C	A	D	C	B	A	C	A	C	A	A	D	D	D	C
31	32	33	34	35	36	37	38	39	40	41	42	43	44	45
D	D	B	D	A	A	D	C	C	B	A	D	B	C	C
46	47	48	49	50	51									
B	A	B	D	B	D									

Fill in the blanks

1	2	3	4	5	6	7	8	9	10	11	12	13	14	15
A	C	D	A	A	B	A	D	D	D	D	A	A	B	D
16	17	18	19	20	21	22	23	24	25	26	27	28	29	30
C	D	C	D	B	D	B	C	B	D	A	B	C	A	D
31														
A														

True/False

1	2	3	4	5	6	7	8	9	10	11	12	13	14	15
A	B	B	A	B	A	B	A	B	A	B	A	A	B	A
16	17	18	19	20	21	22	23							
A	A	A	B	B	B	A	B							

❑❑❑

Chapter – 6

Genetic Mapping and Gene Regulation in Eukaryotes

Linkage and Crossing Over

- Crossing over may produce recombinant phenotypes.
- Bateson and Punnett discovered two traits that did not assort independently.
- Morgan provided evidence for the linkage of several X-linked genes and proposed that crossing over between X chromosomes can occur.
- A chi square analysis can be used to distinguish between linkage and independent assortment.
- Creighton and McClintock correlated crossing over that produced new combinations of alleles with the exchange of homologous chromosomes.
- Crossing over occasionally occurs during mitosis.

Genetic Mapping in Plants and Animals

- The frequency of recombination between two genes can be correlated with their map distance along a chromosome.
- Alfred Sturtevant used the frequency of crossing over in dihybrid crosses to produce the first genetic map.
- Trihybrid crosses can be used to determine the order and distance between linked genes.

- Interference can influence the number of double crossovers that occur in a short region.

Genetic Mapping in Haploid Eukaryotes

- Ordered tetrad analysis can be used to map the distance between a gene and the centromere.
- Unordered tetrad analysis can be used to map genes in dihybrid crosses.

Regulation VIA Transcription Factors

- Structural features of regulatory transcription factors allow them to bind to DNA
- Regulatory transcription factors recognize response elements that function as enhancers or silencers
- Regulatory transcription factors may exert their effects through TFIID and mediator
- The function of regulatory transcription factor proteins can be modulated in three ways
- Steroid hormones exert their effects by binding to a regulatory transcription factor and controlling the transcription of nearby genes
- The CREB protein is an example of a regulatory transcription factor modulated by protein-protein interaction and covalent modification

Changes in Chromatin Structure

- Gene accessibility can be controlled by changes in chromatin structure
- DNaseI sensitivity can be used to study changes in chromatin structure
- Chromatin packing and nucleosome location are altered during globin gene expression
- Transcriptional activators recruit chromatin-remodeling enzymes to the promoter region
- DNA methylation inhibits gene transcription
- DNA methylation is heritable

Regulation of RNA Processing and Translation

- Alternative splicing regulates which exons occur in an RNA transcript, allowing different proteins to be made from the same structural gene.
- The nucleotide sequence of RNA can be modified by RNA editing.
- The stability of mRNA influences mRNA concentration.
- Double-stranded RNA can mediate mRNA degradation.
- Phosphorylation of ribosomal initiation factors can alter the rate of translation.
- The regulation of iron assimilation is an example of the regulatory effect of RNA-binding proteins on translation.

Multiple Choice Question

1. You cross a truebreeding normal corn plant with a true breeding plant with variegated seed coat and colorless endosperm. You suspect the genes are linked, so you cross one of the resulting normal offspring to a true breeding variegated, colorless plant. What are your expected results if the genes are completely linked?

(a) 1 normal: 1 variegated: 1 colorless: 1 variegated and colorless

(b) 9 normal: 3 variegated: 3 colorless: 1 variegated and colorless

(c) 3 normal: 1 variegated and colorless

(d) 3 variegated and colorless: 1 normal

2. For the cross in the previous question, you get the following results. Calculate a chi square value using an expectation of independent assortment and determine if the genes are likely to be linked or not. Normal 350; variegated 63; colorless 57; variegated and colorless 330.

(a) The chi square value is greater than the critical value; these genes are likely to be linked

(b) The chi square value is greater than the critical value; these genes are unlikely to be linked

(c) The chi square value is less than the critical value; these genes are likely to be linked

(d) The chi square value cannot help us predict linkage

3. From the results in the previous question, calculate a map distance between these two genes.

   (a) 0.15 map units (b) 0.85 map units

   (c) 1.5 map units (d) 15 map units

4. How did Morgan demonstrate crossing over?

   (a) A fly that received only y+w+m+ and ywm chromosomes from her parents was able to transmit different combinations of alleles to her offspring

   (b) A fly with a gray body and red eyes (wild type) was able to produce offspring with gray bodies and white eyes

   (c) True breeding yellow, white, miniature flies always had yellow, white, miniature offspring

   (d) None of these examples demonstrate crossing over

5. What was unique about the chromosomes studied by Creighton and McClintock?

   (a) They carried different alleles for two linked genes

   (b) They were unable to cross over

   (c) These chromosomes were unable to synapse during meiosis I

   (d) Physical structures of the chromosomes allowed them to be distinguished from each other

6. Which of the following would be a reasonable way to demonstrate that twin spots are caused by mitotic recombination?

   (a) Collect a large number of flies with twin spots and verify that in all cases, both halves of the twin spot are the same size

   (b) Collect a large number of petunias with twin spots and verify that they are able to pass this phenotype to their progeny

   (c) Design cytologically distinct chromosomes that must recombine to cause the twin spot; analyze chromosomes from the spots with a microscope

   (d) Try to clone new individuals from the cells in the twin spots and analyze their phenotype

7. You begin to study a novel plant species and discover that this diploid plant has 16 chromosomes. How many linkage groups would you expect to find?

   (a) 16

   (b) 8

   (c) 4

   (d) It is impossible to tell from the information provided

8. Calculate map distances between genes A, B, and C based on all pairwise dihybrid crosses. When you perform the trihybrid cross to verify your results, you discover that, despite analysis of a very large number of progeny, you have only about half as many double crossover progeny as you expect. Which explanation is the most reasonable?

   (a) Your dihybrid cross data are an underestimate of the distance between the more widely separated genes

   (b) You are seeing an example of interference

   (c) You are seeing an example of random sampling error

   (d) None of these

9. In a dihybrid cross involving Bar eye and little fly genes, what frequency of recombinant progeny do you expect?

   (a) 2.1 (b) 11.1

   (c) 57.0 (d) 68.1

10. What percentage of double crossover progeny would you expect in a trihybrid cross involving aristaless, dumpy, and black? (Assume no interference.)

   (a) 0.05% (b) 5%

   (c) 13% (d) 48.5%

11. You set up the following dihybrid mapping cross in fruit flies. P e+e+ roro cq cq x ee ro+ro+ cq+cq+. After backcrossing F1 males to ee roro cq cq females, you get the following results: gray body, rough eyes, claret eyes 576; ebony body, smooth, red eyes 564. How can you explain this result?

   (a) The genes are not linked

   (b) Crossing over does not occur in male Drosophila

(c) These genes are X linked

(d) A and B only

12. What is a convenient way to identify gene order in a trihybrid mapping cross?

(a) Look for double crossover progeny and identify the gene that was flipped relative to parentals

(b) Look for the largest class of single crossovers and the two genes are the ones on the ends

(c) Look for the smallest class of single crossovers to identify the genes which are closest together

(d) A and B only

13. Why is tetrad analysis useful for studying genetics?

(a) Cells within a tetrad are haploid so genotype is directly reflected in the phenotype

(b) Cells within a tetrad reflect the events of meiosis

(c) Analysis of simple systems can elucidate general rules that apply more broadly

(d) All of the above

14. Formation of the following pattern in an ordered octad is due to what event during meiosis? Pattern: AAaaaaAA

(a) No crossing over and normal segregation in meiosis I

(b) Crossing over and normal segregation during meiosis I

(c) Crossing over and segregation during meiosis II

(d) No crossing over and segregation during meiosis II

15. You set up a mapping cross involving two genes in Aspergillus and determine the following percentages of tetrad types: parental ditype 51%, nonparental ditype 49%. What can you say about your genes?

(a) They are closely linked

(b) They are unlinked

(c) They are lethal

(d) They are not expressed in the phenotype

16. You set up a mapping cross involving your favorite gene (YFG) in Neurospora and determine the following octad types: 4:4 arrangement 83; 2:4:2 arrangement 7; 2:2:2:2 arrangement 10. What is the distance between your gene and the centromere?

(a) 7 cm (b) 10 cm

(c) 17 cm (d) 83 cm

17. Consider three recessive alleles (a, b, and c) in a hypothetical organism. Assume that a female heterozygous for all three traits is testcrossed. Which of the data sets below would indicate that all three genes are linked on the same chromosome?

(a) a+ b+ c+ = 378; a b c = 375; a+ b c = 370; a b+ c+ = 372; a+ b+ c+ = 370; a b c = 379; a+ b c+ = 380; a b+ c = 376

(b) a+ b+ c+ = 940; a b c = 932; a+ b c = 282; a b+ c+ = 286; a+ b+ c+ = 245; a b c = 240; a+ b c+ = 40; a b+ c = 35

(c) a+ b+ c+ = 940; a b c = 932; a+ b c = 125; a b+ c+ = 130; a+ b+ c+ = 938; a b c = 943; a+ b c+ = 132; a b+ c = 135

(d) a+ b+ c+ = 300; a b c = 900; a+ b c = 300; a b+ c+ = 300; a+ b+ c+ = 300; a b c = 300; a+ b c+ = 300; a b+ c = 2

18. Consider three recessive alleles (a, b, and c) in a hypothetical organism. Assume that a female heterozygous for all three traits is testcrossed. Which of the data sets below would indicate that all three genes are independently assorting?

(a) a+ b+ c+ = 378; a b c = 375; a+ b c = 370; a b+ c+ = 372; a+ b+ c = 370; a b c+ = 379; a+ b c+ = 380; a b+ c = 376

(b) a+ b+ c+ = 940; a b c = 932; a+ b c = 282; a b+ c+ = 286; a+ b+ c = 245; a b c+ = 240; a+ b c+ = 40; a b+ c = 35

(c) a+ b+ c+ = 940; a b c = 932; a+ b c = 125; a b+ c+ = 130; a+ b+ c = 938; a b c+ = 943; a+ b c+ = 132; a b+ c = 135

(d) a+ b+ c+ = 300; a b c = 900; a+ b c = 300; a b+ c+ = 300; a+ b+ c = 300; a b c+ = 300; a+ b c+ = 300; a b+ c = 2

19. Consider three recessive alleles (a, b, and c) in a hypothetical organism. Assume that a female heterozygous for all three traits is testcrossed. Which of the data sets below would indicate that two of the genes are linked and the third is independently assorting?

(a) a+ b+ c+ = 378; a b c = 375; a+ b c = 370; a b+ c+ = 372; a+ b+ c = 370; a b c+ = 379; a+ b c+ = 380; a b+ c = 376

(b) a+ b+ c+ = 940; a b c = 932; a+ b c = 282; a b+ c+ = 286; a+ b+ c = 245; a b c+ = 240; a+ b c+ = 40; a b+ c = 35

(c) a+ b+ c+ = 940; a b c = 932; a+ b c = 125; a b+ c+ = 130; a+ b+ c = 938; a b c+ = 943; a+ b c+ = 132; a b+ c = 135

(d) a+ b+ c+ = 300; a b c = 900; a+ b c = 300; a b+ c+ = 300; a+ b+ c = 300; a b c+ = 300; a+ b c+ = 300; a b+ c = 2

20. Consider three recessive alleles (e, g, and k) in a hypothetical organism. Assume that a female heterozygous for all three traits is testcrossed and the following offspring result: e+ g+ k+ = 471; e+ g+ k = 20; e+ g k = 141; e g+ k+ = 142; e+ g k+ = 122; e g+ k = 121; e g k = 466; e g k+ = 17. Assuming the three genes are linked, what is the proper gene order?

(a) e g k (b) k g e

(c) g e k (d) e k g

21. What is the allelic arrangement in the F1 heterozygote?

(a) Cis (b) Trans

(c) Cis and trans (d) Axial

22. You cross a mutant strain of Neurospora to a wild-type strain and recover 960 asci with the 4:4 first-division segregation pattern and a total of 1040 asci with the second-division segregation pattern (2:2:2:2 and 2:4:2). What is the distance between the gene under study and the centromere?

(a) 26 (b) 46

(c) 52 (d) 92

23. Gene regulation in eukaryotes may be influenced by:

(a) Environmental conditions (b) Developmental stage

(c) Tissue or cell type (d) All of these

24. DNA sites bound by regulatory transcription factors:

(a) Are trans acting factors

(b) Are cis acting factors

(c) Are promoters

(d) Have similar sequences in all cells

25. A homodimer:

(a) Is a protein with a helix-loop-helix DNA binding motif

(b) Is a protein made of two unique subunits

(c) Is a pair of identical proteins

(d) Typically prevents DNA binding

26. In comparing effects of regulatory transcription factors that affect TFIID and those that affect mediator:

(a) TFIID binding prevents transcription while mediator binding allows transcription

(b) Mediator allows binding of RNA polymerase but transcription can be prevented by transcription factors binding to silencers

(c) Transcription factors which bind to enhancers stimulate binding of both proteins

(d) Transcription factors which bind to silencers prevent binding of both proteins

27. In the absence of glucocorticoid:

(a) No glucocorticoid receptor is present

(b) Glucocorticoid receptor is present in an active form

(c) Glucocorticoid receptor dimerizes to prevent glucocorticoid binding

(d) None of these

28. GREs:

(a) Are located near promoters of many different genes

(b) Are activated in response to increased glucocorticoid levels

(c) Are bound by a dimerized regulatory transcription factor

(d) All of these

29. Extracellular signaling molecules:

(a) Are usually able to directly affect transcription

(b) Are required to enter the cytoplasm and bind a receptor molecule

(c) Usually bind plasma membrane receptors which then pass along the signal

(d) Are unable to affect transcription

30. Which of the following pieces of DNA would you expect to be most susceptible to DNase-I digestion?

(a) A 30nm fiber

(b) An 11nm fiber

(c) An actively transcribed gene

(d) "Naked" DNA, which is not bound to histones

31. You want to perform a DNase-I protection experiment to see where binding sites for regulatory transcription factors for the mouse calmodulin gene are located. What would you use for a probe?

(a) The calmodulin gene coding sequence

(b) A DNA fragment including about 100bp on either side of the promoter

(c) A DNA fragment containing several thousand bp upstream of the promoter, one small piece at a time

(d) Any of these would be useful probes

32. Housekeeping genes:

(a) Are constitutively expressed

(b) Have highly methylated CpG islands

(c) Encode proteins required only in specific tissues

(d) All of these

33. Methylation of CpG islands:

(a) Enhances binding of regulatory transcription factors

(b) Prevents activation of enhancers

(c) Prevents binding of chromatin remodeling proteins

(d) Interferes directly with RNA polymerase binding

34. The best explanation for interference of DNA methylation with transcription factor binding is:

(a) Methyl-CpG binding protein competes with the transcription factor for binding to methylated DNA

(b) Methylation of the CpG island modifies the shape of the transcription factor

(c) Methylation of the CpG island modifies the shape of the response element to which the transcription factor normally binds

(d) Methylation prevents dimerization of DNA binding proteins

35. In RNA editing, the guide RNA:

(a) Must have the same sequence as the end result of the edited RNA

(b) Can control the insertion or deletion of uracils

(c) Can change a uracil to another base

(d) Can interact with many different RNAs

36. RNA editing:

(a) Is certainly a very rare event

(b) Occurs in widely varied species

(c) Probably has only minor effects on RNA function

(d) All of these

37. The stability of a particular mRNA can be enhanced by:

(a) Presence of destabilizing elements

(b) A short poly-A tail

(c) Presence of short bits of antisense RNA

(d) All of the above

38. Which of these would be the best evidence for a functional role of RNA interference in resistance to viral infection?

(a) Cells resistant to double stranded RNA viruses are also likely to be resistant to single stranded RNA viruses

(b) Cells resistant to DNA viruses tend not to also be resistant to RNA viruses

(c) Cells resistant to a single stranded RNA virus tend to be resistant to DNA viruses

(d) Cells resistant to one double stranded RNA virus tend to be resistant to all double stranded RNA viruses

39. Receptor tyrosine kinases are transmembrane proteins that are activated by binding of an extracellular signal and subsequent phosphorylation of the receptor. Control of the function of receptor tyrosine kinases primarily occurs at what level?

(a) Transcription

(b) RNA processing

(c) Translation

(d) Posttranslational modification

40. Changes in function of which of the following proteins by posttranslational modification (such as phosphorylation or acetylation) would be likely to, in turn, affect the expression of specific genes?

(a) TFIID (b) eIF4F

(c) Histones (d) IRP

41. Which of the following statements of the eukaryotic chromosome is false?

(a) The chromosomal DNA is not complexed with protein

(b) The chromosome is uninemic

(c) Dark bands are due to tight coiling of the DNA

(d) Each end of the chromosome is referred to as a telomere

42. Which type(s) of chromatin in eukaryotic chromosomes is (are) transcriptionally active?

(a) Intercalary heterochromatin

(a) Constitutive heterochromatin

(c) Euchromatin

(d) a and b

43. Which of the following statements of the eukaryotic chromosome is false?

(a) The chromosomal DNA is not complexed with protein

(b) The chromosome is uninemic

(c) Dark bands are due to tight coiling of the DNA

(d) Each end of the chromosome is referred to as a telomere

44. A disease that inactivates the X-ray repair system is

(a) Wilm's tumors (b) Ataxia telangiectasia

(c) Retinoblastoma (d) HIV

45. A homeotic mutation is one in which

(a) Some proteins are made from mRNA transcribed by the mother

(b) One cell type follows the developmental path of another

(c) The X-ray repair system is inactivated

(d) Programmed cell death occurs

46. Genomic imprinting refers to the fact that

(a) Some proteins are made from mRNA transcribed by the mother

(b) One cell type follows the developmental path of another

(c) The X-ray repair system is inactivated

(d) Gene activity depends upon whether the gene is of maternal or paternal origin

47. In which of the stages in the formation of an immunoglobulin molecule is diversity not generated?

(a) Choice of which subunits to combine

(b) Choice of how to combine the subunit genes

(c) De novo generation of diversity at junctions

(d) All of the above generate diversity

48. Linked genes:

(a) Are located near each other on the same chromosome

(b) Violate the law of independent assortment

(c) Segregate together during meiosis

(d) All of the above

(e) None of the above

49. You cross a true breeding normal corn plant with a true breeding plant with variegated seed coat and colorless endosperm. You suspect the genes are linked, so you cross one of the resulting normal offspring to a true breeding variegated, colorless plant. What are your expected results if the genes are completely linked?

(a) 1 normal: 1 variegated: 1 colorless: 1 variegated and colorless

(b) 9 normal: 3 variegated: 3 colorless: 1 variegated and colorless

(c) 3 normal: 1 variegated and colorless

(d) 3 variegated and colorless: 1 normal

50. For the cross in the previous question, you get the following results. Calculate a chi square value using an expectation of independent assortment and determine if the genes are likely to be linked or not. Normal 350; variegated 63; colorless 57; variegated and colorless 330.

(a) The chi square value is greater than the critical value; these genes are likely to be linked

(b) The chi square value is greater than the critical value; these genes are unlikely to be linked

(c) The chi square value is less than the critical value; these genes are likely to be linked

(d) The chi square value cannot help us predict linkage

51. From the results in the previous question, calculate a map distance between these two genes.

(a) 0.15 map units (b) 0.85 map units

(c) 1.5 map units (d) 15 map units

52. How did Morgan demonstrate crossing over?

(a) A fly that received only y+w+m+ and ywm chromosomes from her parents was able to transmit different combinations of alleles to her offspring

(b) A fly with a gray body and red eyes (wild type) was able to produce offspring with gray bodies and white eyes

(c) True breeding yellow, white, miniature flies always had yellow, white, miniature offspring

(d) None of these examples demonstrate crossing over

53. What was unique about the chromosomes studied by Creighton and McClintock?

(a) They carried different alleles for two linked genes

(b) They were unable to cross over

(c) These chromosomes were unable to synapse during meiosis I

(d) Physical structures of the chromosomes allowed them to be distinguished from each other

54. Which of the following would be a reasonable way to demonstrate that twin spots are caused by mitotic recombination?

(a) Collect a large number of flies with twin spots and verify that in all cases, both halves of the twin spot are the same size

(b) Collect a large number of petunias with twin spots and verify that they are able to pass this phenotype to their progeny

(c) Design cytologically distinct chromosomes that must recombine to cause the twin spot; analyze chromosomes from the spots with a microscope

(d) Try to clone new individuals from the cells in the twin spots and analyze their phenotype

55. You begin to study a novel plant species and discover that this diploid plant has 16 chromosomes. How many linkage groups would you expect to find?

(a) 16

(b) 8

(c) 4

(d) It is impossible to tell from the information provided

56. You calculate map distances between genes A, B, and C based on all pairwise dihybrid crosses. When you perform the trihybrid cross to verify your results, you discover that, despite analysis of a very large number of progeny, you have only about half as many double crossover progeny as you expect. Which explanation is the most reasonable?

(a) Your dihybrid cross data are an underestimate of the distance between the more widely separated genes

(b) You are seeing an example of interference

(c) You are seeing an example of random sampling error

(d) None of these

57. You set up the following dihybrid mapping cross in fruit flies. P e+e+ roro cq cq x ee ro+ro+ cq+cq+. After backcrossing F1 males to ee roro cq cq females, you get the following results: gray body, rough eyes, claret eyes 576; ebony body, smooth, red eyes 564. How can you explain this result?

(a) The genes are not linked

(b) Crossing over does not occur in male *Drosophila*

(c) These genes are X linked
(d) A and B only
(e) None of the above

58. What is a convenient way to identify gene order in a trihybrid mapping cross?

(a) Look for double crossover progeny and identify the gene that was flipped relative to parentals
(b) Look for the largest class of single crossovers and the two genes are the ones on the ends
(c) Look for the smallest class of single crossovers to identify the genes which are closest together
(d) A and B only
(e) All of the above

59. Why is tetrad analysis useful for studying genetics?

(a) Cells within a tetrad are haploid so genotype is directly reflected in the phenotype
(b) Cells within a tetrad reflect the events of meiosis
(c) Analysis of simple systems can elucidate general rules that apply more broadly
(d) All of the above
(e) None of the above

60. Tetrad analysis automatically includes the centromere as an additional point in mapping crosses, because

(a) Crossovers that are not between a gene and the centromere do not affect the relationship of that gene to the chromosome during division
(b) Crossovers that are between a gene and the centromere cannot be identified in this system
(c) When studying crossovers between genes, the position of the centromere is irrelevant
(d) All of the above
(e) None of the above

61. Formation of the following pattern in an ordered octad is due to what event during meiosis? Pattern: AAaaaaAA

(a) No crossing over and normal segregation in meiosis I

(b) Crossing over and normal segregation during meiosis I

(c) Crossing over and segregation during meiosis II

(d) No crossing over and segregation during meiosis II

62. In an unordered tetrad with the parental ditype:

(a) All spores have the same genotype

(b) All spores have the genotype of one of the parental strains

(c) All four spores have different genotypes

63. You set up a mapping cross involving two genes in *Aspergillus* and determine the following percentages of tetrad types: parental ditype 51%, nonparental ditype 49%. What can you say about your genes?

(a) They are closely linked

(b) They are unlinked

(c) They are lethal

(d) They are not expressed in the phenotype

64. You set up a mapping cross involving your favorite gene (YFG) in *Neurospora* and determine the following octad types: 4:4 arrangement 83; 2:4:2 arrangement 7; 2:2:2:2 arrangement 10. What is the distance between your gene and the centromere?

(a) 7 cm (b) 10 cm

(c) 17 cm (d) 83 cm

65. You set up a mapping cross involving two genes of interest in Saccharomyces and determine the following tetrad types: 8 nonparental ditype; 37 tetratype; 80 parental ditype. Calculate the most accurate map distance between these two genes.

(a) 18.7mu (b) 32.3mu

(c) 64.6mu (d) None of these is accurate

Fill in the blanks

1. Linked genes_________
   (a) Are located near each other on the same chromosome
   (b) Violate the law of independent assortment
   (c) Segregate together during meiosis
   (d) All of the above

2. Tetrad analysis automatically includes the centromere as an additional point in mapping crosses, because __________
   (a) Crossovers that are not between a gene and the centromere do not affect the relationship of that gene to the chromosome during division
   (b) Crossovers that are between a gene and the centromere cannot be identified in this system
   (c) When studying crossovers between genes, the position of the centromere is irrelevant
   (d) All of the above

3. In an unordered tetrad with the parental ditype__________
   (a) All spores have the same genotype
   (b) All spores have the genotype of one of the parental strains
   (c) All four spores have different genotypes
   (d) All spores are diploid in a parental ditype, the spores are of both parental types in a 2:2 ratio

4. You set up a mapping cross involving two genes of interest in Saccharomyces and determine the following tetrad types: 8 nonparental ditype; 37 tetratype; 80 parental ditype. Calculate the most accurate map distance between these two genes______
   (a) 18.7mu (b) 32.3mu
   (c) 64.6mu (d) None of these is accurate

5. Consider two recessive alleles (a and b) in a hypothetical organism. Given the following data, the offspring of a test-crossed dihybrid, determine

the ________distance between the two genes (in map units). a+ b+ = 206; a+ b = 704; a b+ = 736; a b = 254.

(a) 12.7 (b) 15.3
(c) 24.2 (d) 28

6. Given the gene sequence A B C D E F, crossing over should occur most frequently between ________

(a) A - F (b) A - C
(c) A - D (d) A - E

7. If crossover interference is 100% then __________

(A) The coefficient of coincidence is 100% and the expected number of double crossovers is observed

(b) The coefficient of coincidence is 0% and the expected number of double crossovers is observed

(c) The coefficient of coincidence is 100% and no double crossovers are observed

(d) The coefficient of coincidence is 0% and no double crossovers are observed

8. It is estimated that _____ % of the DNA in a eukaryote is "junk DNA."

(a) Less than 10% (b) 25%
(c) 50% (d) Over 90%

9. The cot measure of DNA-DNA hybridization reveals that ___% of the eukaryotic chromosome consists of satellite DNA while ___% consists of unique DNA and ___% is intermediately repetitively.

(a) 10, 15, 75 (b) 15, 10, 75
(c) 10, 75, 15 (d) 15, 75, 10

10. The cancer ______ results from a defective tumor suppressor gene

(a) Burkitt's lymphoma (b) Retinoblastoma
(c) Ataxia telangiectasia (d) Xeroderma pigmentosum

11. The cancer ______ results from a chromosomal rearrangement.

(a) Burkitt's lymphoma (b) Retinoblastoma
(c) Ataxia telangiectasia (d) Xeroderma pigmentosum

12. Avoidable substances in the environment and the diet are estimated to cause ____% of all cancers.

(a) 10-20 (b) 20-50

(c) 50-60 (d) 80-90

13. In order for CREB to activate transcription ___________

(a) It must act as a heterodimer

(b) It must be phosphorylated

(c) It must bind to cAMP

(d) It must bind to RNA polymerase

14. Gene amplification is different from other alterations in DNA structure because_______

(a) It is the only alteration that involves loss of DNA

(b) It is the only alteration that involves addition of DNA

(c) It is the only alteration that involves changes to the chemical structure of the DNA

(d) It is the only alteration that involves changes in chromatin structure

15. Two different proteins produced by alternative splicing are most likely to__________

(a) Have completely different functions

(b) Have identical functions

(c) Have similar functions under different forms of regulation

(d) Have different functions under similar forms of regulation

16. Control of alternative splicing involves_______________

(a) Repressors, which cause exon skipping

(b) Enhancers, which increase recognition of weak splice sites

(c) Splicing factors, which modulate spliceosome function

(d) All of these

17. C-bands consist primarily of ____________

(a) Constitutive heterochromatin (b) Intercalary heterochromatin

(c) Intercalary euchromatin (d) Euchromatin

18. The width of a nucleosome (Å) is approximately ________

(a) 1 (b) 11

(c) 110 (d) 300

19. Lampbrush chromosomes ________

(a) Appear after heat shock

(b) Are found in amphibian oocytes

(c) Form secondary buoyant-density bands

(d) Represent transcriptionally inactive regions of DNA

20. At G2 of interphase, a eukaryotic chromosome contains ________

(a) Many DNA copies

(b) Two circles of DNA

(c) Two linear DNA molecules

(d) One circle of DNA

21. The fact that any nucleus can give rise to any, and therefore all cell types is called ________

(a) Apotosis (b) Totipotency

(c) Genomic imprinting (d) Morphogenesis

22. Transposons that control gene expression in corn were discovered by ________

(a) Morgan (b) Brenner

(c) McClintock (d) Rous

23. If a nucleus from an intestinal cell of Xenopus is injected into an enucleated egg, the most significant result will be ________

(a) An intestinal cell

(b) Normal development to adult

(c) An indeterminate amphibian

(d) A carrot embryoid

True/False

1. Every chromosome in a diploid cell comprises a linkage group.
   (a) True (b) False
2. Crossing over involves the physical exchange of DNA between chromatids.
   (a) True (b) False
3. If two genes are linked, a double heterozygote will only be able to produce two possible types of gametes.
   (a) True (b) False
4. Crossing over only occurs within a bivalent.
   (a) True (b) False
5. Parental phenotypes in the offspring are the most common in a mapping cross.
   (a) True (b) False
6. Morgan was able to demonstrate linkage of genes to chromosomes because he studied X linked genes.
   (a) True (b) False
7. Chi square analysis can help distinguish between two competing hypotheses.
   (a) True (b) False
8. Creighton and McClintock's important discovery was that they were able to demonstrate linkage between a translocation and a neighboring gene.
   (a) True (b) False
9. Twin spots are only possible in doubly heterozygous individuals.
   (a) True (b) False
10. In linkage mapping, one map unit is equal to one % crossover frequency.
    (a) True (b) False
11. Trihybrid mapping crosses are more accurate than dihybrid mapping crosses.
    (a) True (b) False

12. You can predict the number of double crossovers in a region accurately if you have a detailed genetic map.

(a) True (b) False

13. Unordered tetrad analysis allows determination of the specific events of meiosis I and II.

(a) True (b) False

14. Unordered tetrad analysis is useful for determining recombination frequencies.

(a) True (b) False

15. Double crossovers can occur between multiple chromatids.

(a) True (b) False

16. DNA in an open configuration is more likely to be transcribed than DNA in a closed configuration.

(a) True (b) False

17. A CpG island consists of about 2000 copies of the CG dinucleotide.

(a) True (b) False

18. A transcription factor is a protein that affects the ability of RNA polymerase to transcribe a gene.

(a) True (b) False

19. Response elements may be located within the gene itself.

(a) True (b) False

20. Transcription factors, if present, will always bind their appropriate response elements.

(a) True (b) False

21. A single RNA molecule can encode several different proteins.

(a) True (b) False

22. Methylation of DNA is a heritable trait.

(a) True (b) False

23. The length of the poly-A tail has a dramatic effect on RNA stability.

(a) True (b) False

24. Every chromosome in a diploid cell comprises a linkage group.

(a) True (b) False

25. Crossing over involves the physical exchange of DNA between chromatids.

(a) True (b) False

26. If two genes are linked, a double heterozygote will only be able to produce two possible types of gametes.

(a) True (b) False

27. Crossing over only occurs within a bivalent

(a) True (b) False

28. Parental phenotypes in the offspring are the most common in a mapping cross.

(a) True (b) False

29. Morgan was able to demonstrate linkage of genes to chromosomes because he studied X linked genes.

(a) True (b) False

30. Chi square analysis can help distinguish between two competing hypotheses

(a) True (b) False

31. Creighton and McClintock's important discovery was that they were able to demonstrate linkage between a translocation and a neighboring gene.

(a) True (b) False

32. Twin spots are only possible in doubly heterozygous individuals.

(a) True (b) False

33. In linkage mapping, one map unit is equal to one % crossover frequency.

(a) True (b) False

34. Trihybrid mapping crosses are more accurate than dihybrid mapping crosses.

(a) True (b) False

35. You can predict the number of double crossovers in a region accurately if you have a detailed genetic map.

(a) True (b) False

36. Unordered tetrad analysis allows determination of the specific events of meiosis I and II.

(a) True (b) False

37. Unordered tetrad analysis is useful for determining recombination frequencies.

(a) True (b) False

38. Double crossovers can occur between multiple chromatids.

(a) True (b) False

Answers

Multiple Choice Question

1	2	3	4	5	6	7	8	9	10	11	12	13	14	15
C	A	A	A	D	C	B	B	B	B	B	D	D	C	B
16	17	18	19	20	21	22	23	24	25	26	27	28	29	30
C	B	A	C	D	A	A	D	B	C	B	D	D	C	D
31	32	33	34	35	36	37	38	39	40	41	42	43	44	45
C	A	B	C	B	B	C	D	D	D	B	C	B	B	B
46	47	48	49	50	51	52	53	54	55	56	57	58	59	60
D	D	D	C	A	A	A	D	C	B	B	B	D	D	A
61	62	63	64	65										
C	B	B	C	B										

Fill in the blanks

1	2	3	4	5	6	7	8	9	10	11	12	13	14	15
D	A	B	B	C	A	D	D	C	B	A	D	B	B	C
16	17	18	19	20	21	22	23							
D	A	C	B	C	B	C	B							

True/False

1	2	3	4	5	6	7	8	9	10	11	12	13	14	15
B	A	B	B	A	A	A	B	A	A	A	B	B	A	A
16	17	18	19	20	21	22	23	24	25	26	27	28	29	30
A	B	A	A	B	A	A	A	B	A	B	B	A	A	A
31	32	33	34	35	36	37	38							
B	A	A	A	B	B	A	A							

❑❑❑

Chapter – 7

Genetic Transfer in Bacteria and Bacteriophages

Genetic Transfer and Mapping in Bacteria a

- Bacteria can transfer genetic material during conjugation
- Hfr strains contain an F factor integrated into the bacterial chromosome
- Hfr strains can transfer a portion of the bacterial chromosome to recipient cells
- Conjugation experiments can map genes along the *E. coli* chromosome
- A genetic map of the *E. coli* chromosome has been obtained from many conjugation studies
- Bacteria may contain different types of plasmids
- Bacteriophages transfer genetic material from one bacterial cell to another via transduction
- Cotransduction can be used to map genes that are within 2 minutes of each other
- Bacteria can also transfer genetic material by transformation
- Bacteria may acquire new genes by horizontal gene transfer

Intragenic Mapping in Bacteriophages

- Mutations in viral genes can alter plaque morphology

- A complementation test can reveal if mutations are in the same gene or in different genes
- Intragenic maps were constructed using data from a recombinational analysis of mutants within the rII region
- Deletion mapping can be used to localize many rII mutations to specific regions in the rIIA or rIIB genes
- Intragenic mapping experiments provided insight into the relationship between traits and molecular genetics

Multiple Choice Question

1. How is genetic transfer different from sexual reproduction?
   (a) In genetic transfer, genetic material is transferred from one individual to another
   (b) In sexual reproduction, genetic material from two individuals is mixed equally
   (c) In both processes, the individual donating genetic material retains their original complement of genetic material
   (d) None of these
2. Which method of gene transfer involves direct contact between the bacteria?
   (a) Conjugation (b) Transduction
   (c) Transformation (d) All of these
   (e) None of these
3. How might the results of Bernard Davis's U tube experiment have been different if each strain were only mutant for a single gene?
   (a) There would be no difference
   (b) Transformation by small pieces of DNA could restore gene function
   (c) Conjugation would not be required to change the phenotype of the bacterial strains
   (d) B and C
   (e) None of the above

4. The end result of F factor mediated conjugation:
   (a) Is that both strains are F+
   (b) Involves transfer of the entire bacterial chromosome
   (c) Converts the recipient strain to F+ and the donor to F-
   (d) A and B
   (e) None of the above

5. Which of the following is not a protein or group of proteins required for conjugation to occur?
   (a) The relaxosome
   (b) Pilin protein
   (c) The origin of transfer
   (d) Coupling protein
   (e) None of the above

6. Which of the following is a description of generalized transduction?
   (a) Inaccurate excision of prophage can occasionally remove a small portion of the bacterial chromosome
   (b) Random pieces of bacterial DNA can become incorporated into a phage coat
   (c) Pieces of bacterial DNA may be released into the culture medium when a cell is lysed
   (e) All of these
   (e) None of the above

7. What is the best explanation for why Bernard Davis didn't observe generalized transduction in his U tube experiments?
   (a) He used genes that could not be packaged into a bacteriophage head
   (b) He used bacterial strains that could not fit through the filter
   (c) He used bacterial strains that did not host a prophage
   (d) He used virulent bacteriophage

8. A researcher would like to map the location of galE and trpA genes in a new species of bacterium that appears to be closely related to *E. coli*. He decides to use cotransduction, and generates appropriate donor and

recipient strains. He is disappointed when cotransduction is not seen in his experiment. What is the most reasonable explanation for this situation?

(a) His new bacterial species does not have galE or trpA genes

(b) His new bacterial species cannot survive galE or trpA mutation

(c) These two genes are too far apart to be mapped by cotransduction

(d) These two genes are too close together to be mapped by cotransduction

9. Which of the following is NOT a critical function of competence factors?

(a) Transportation of DNA fragments into bacterial cells

(b) Binding of DNA to a cell surface receptor

(c) Degradation of one strand of the transforming DNA

(d) Degradation of the bacterial chromosome in the presence of single standed DNA

10. How might the function of extracellular endonucleases help protect the bacterium from infection by viruses?

(a) They would prevent the virus from binding to the outside of the bacterial cell

(b) They would destroy the viral coat, so it cannot inject its genetic material

(c) They would cut the viral DNA into small pieces

(d) All of the above

(e) None of the above

11. What do transduction, transformation, and conjugation have in common?

(a) All require presence of an outside factor to facilitate gene transfer

(b) In all three processes, DNA is transferred as a single stranded molecule

(c) They all transfer large pieces of DNA into recipient cells

(d) All of the above

(e) None of the above

12. Which type(s) of genetic transfer lead to incorporation of new DNA into the bacterial chromosome? I. Conjugation mediated by F factorII. Hfr mediated conjugationIII. TransductionIV. Transformation

(a) I and II only

(b) III and IV only

(c) II, III, and IV

(d) I, II, III, and IV

(e) None of these

13. How would you expect a mutation in a gene encoding a tail fiber protein would affect the plaque phenotype of T4?

(a) It could lead to very large plaques

(b) It would be unlikely to have an affect on plaque formation

(c) It could lead to smaller than normal plaques

(d) A and B

(e) None of the above

14. In deletion mapping, mutations can be localized to a region of the gene because:

(a) The mutation fails to complement other mutations in the gene

(B) The mutation fails to complement a deletion of a known portion of the gene

(c) The gene sequence can be determined

(d) A and B

(e) None of the above

15. A key finding of intragenic mapping was:

(a) That genes consist of indivisible particles, like beads on a string

(b) That genes do not change over time

(c) That genes can be modified by crossing over and mutation

(d) All of these

(e) None of these

True/False

1. Conjugation involves direct transfer of genetic material from one bacterium to another.

   (a) True (b) False

2. Gene transfer by conjugation converts a strain that cannot act as a donor to a strain that can initiate conjugation.

   (a) True (b) False

3. Hfr strains of bacteria contain F factors that include some genes from the bacterial chromosome.

   (a) True (b) False

4. Virulent phages generally follow the lysogenic cycle when infecting bacteria.

   (a) True (b) False

5. In generalized transduction, random bits of the bacterial chromosome are packaged in a phage head.

   (a) True (b) False

6. Homologous recombination is a mechanism used by bacteria to incorporate new DNA into their chromosome, if it is similar to an existing sequence.

   (a) True (b) False

7. Horizontal gene transfer may be mediated via conjugation but not transduction or transformation.

   (a) True (b) False

8. Bacteriophage may be able to infect some strains of bacteria but not others.

   (a) True (b) False

9. Mutations which complement each other are in the same gene.

   (a) True (b) False

10. Deletion mapping is more efficient but less detailed than fine structure mapping of genes.

    (a) True (b) False

Answers

Multiple Choice Question

1	2	3	4	5	6	7	8	9	10	11	12	13	14	15
B	A	D	A	C	B	C	C	D	C	E	C	C	B	C

True/False

1	2	3	4	5	6	7	8	9	10
A	B	B	B	A	A	B	A	B	A

Chapter – 8

Sex Determination and Development

Sex Chromosomes

- Sex can be considered at chromosomal, gonadal, phenotypic, and gender identity levels.
- Human females are homogametic (XX) and males are heterogametic (XY).
- The Y chromosome contains few identified genes, while the larger X chromosome contains several thousand genes. Pseudoautosomal regions at both tips of the Y chromosome contain genes that have counterparts on the X chromosome.
- The male determining gene, SRY, has been found on the Y chromosome.

The Phenotype Forms

- The SRY gene is a single copy gene unique to the Y chromosome.
- The SRY protein triggers a cascade of gene action that initiates development of male features while suppressing development of female features.
- Factors necessary for male development include Anti-Müllerian Hormone, testosterone, and dihydrotestosterone (DHT).
- In the absence of these male factors, the embryo follows the female developmental pathway.

- Pseudohermaphrodites are chromosomally one sex, but look phenotypically like the other sex.

Gender Identity-Is Homosexuality Inherited?

- Evidence is accumulating that genes contribute to homosexuality.

Traits Inherited on Sex Chromosomes

- Y-linked traits are passed on the Y chromosome, and X-linked traits on the X
- Males are hemizygous for X-linked traits, expressing genes on the X chromosomE) Females express homozygous recessive alleles on the X chromosomE)

X-Linked Recessive Inheritance

- X-linked recessive traits pass from carrier mothers to sons with a 50% probability, and can be passed from hemizygous men to their daughters, if the phenotype is not severE) There is no father to son transmission of X-linked traits.

X-Linked Dominant Inheritance

- X-linked dominant conditions are expressed in both males and females.
- These conditions are generally more severe in males.

X Inactivation Equalizes the Sexes

- Early in female development, the maternal or paternal X chromosome is turned off in each cell.
- X inactivation compensates for differences between males in females in the number of gene copies on the X chromosome.
- The XIST gene encodes RNA that inactivates these genes.
- The inactivated X takes up stain and appears as a Barr body.
- A female may express a sex-linked trait if the mutant allele is on the active X in affected tissues.
- The Lyon hypothesis proposed that the Barr body was the inactive X chromosomE)

Gender Effects on Phenotype

Sex-limited Traits

- A sex-limited trait (which may be autosomal or sex-linked) affects body parts or functions present in only one gender.

Sex-influenced Traits

- A sex-linflunced trait (which may be autosomal or sex-linked) affects body parts or functions present in only one gender like man patterned baldness, bearded, etc.

Multiple Choice Question

1. An abnormal human of the karyotype 49, XXXYY will form how many Barr bodics?

(a) 3 (b) 4

(c) 2 (d) 5

(e) 1

2. Which of the following defines the pseudoautosomal region?

(a) Homologous regions of the X and Y chromosomes

(b) Hemizygous regions of the X and Y chromosomes

(c) Heterozygous regions of autosomal chromosomes

(d) Homozygous regions of autosomal chromosomes

(e) Regions of the X and Y chromosomes that don't recombine during meiosis

3. Klinefelter's syndrome and Turner's syndrome are related in which of the following ways?

(a) Result from nondisjunction

(b) Involve an unusual number of X or Y chromosomes

(c) Can be identified by a karyotype

(d) Have 44 autosomes

(e) All of these answers describe ways in which Klinefelter's syndrome and Turner's syndrome are related

4. An individual is diploid with an XXY sex chromosome complement. It would be

(a) A male fly, male human being

(b) A male fly, female human being

(c) A female fly, male human being

(d) A female fly, female human being

(e) An intersex fly, male human being

5. A woman whose husband worked at the Chernobyl nuclear reactor gives birth to a hemophiliac son.

(a) She should blame the reactor accident because the radiation caused the hemophilia

(b) She should not blame the accident because she carried the hemophilia allele

(c) Her husband should sue for divorce because it cannot be his child

(d) Further genetic tests should be done to determine who is at fault

(e) Hemophilia is environmentally induced, not genetic

6. The probability that a male inherits his Y chromosome from his maternal grandfather is

(a) 0 (b) 0.25

(c) 0.50 (d) 0.75

(e) 1.0

7. White-eyes is a sex-linked recessive trait in Drosophila Which of the crosses below will yield all males of one phenotype and all females of the opposite phenotype?

(a) Homozygous white female x white male

(b) Homozygous white female x wild-type male

(c) Heterozygous wild female x white male

(d) Homozygous wild-type female x white male

(e) Homozygous wild-type female x wild-type male

8. In moths, males are homogametic Spotted wings (s) are recessive to the wild-type (+) and sex-linked. A wild-type male is mated to a spotted female, and all the F1 are wild-type. What is the genotype of an F1 female?

   (a) + / s (b) + / +
   (c) + / W (d) s / W
   (e) s / s

9. Some people with polydactyly have an extra toe; others have an extra finger. This situation is an example of a trait that is

   (a) Dosage compensated (b) Variably expressed
   (c) Sex-limited (d) Sex-influenced
   (e) Pseudoautosomal

10. How many Barr bodies would a cell from an XXYY human being have?

   (a) None (b) One
   (c) Two (d) Three
   (e) None

11. How many Barr bodies would a cell from an XXYY fruit fly have?

   (a) None (b) One
   (c) Two (d) Three
   (e) Four

Fill in the blanks

1. In Drosophila, sex is determined by __________.

   (a) The ratio of the number of X chromosomes to the number of autosomes
   (b) The ratio of the number of Y chromosomes to the number of autosomes
   (c) The ratio of the number of X chromosomes to the number of sets of autosomes
   (d) The presence or absence of the Y chromosome
   (e) The ratio of the number of Y chromosomes to the number of sets of autosomes

2. The primary sex ratio in humans ________.
   (a) Is possibly affected by the lower mass of the Y chromosome
   (b) Is probably about 1.2 to 1.6
   (c) May vary in different ethnic groups according to secondary sex ratio findings
   (d) Is higher than the secondary sex ratio
   (e) All answers are true
3. The effect of Barr body formation in mammals is that ________.
   (a) Females are mosaics for heterozygous X-linked genes
   (b) Males are mosaics for heterozygous autosomal genes
   (c) Males are mosaics for hemizygous X-linked genes
   (d) Females are mosaics for homozygous X-linked genes
   (e) Females are mosaics for heterozygous autosomal genes
4. The gene which confers the primary male sex characteristics was discovered by observing ________.
   (a) XX males who had part of the Y chromosome translocated onto one of the X chromosomes
   (b) XY females whose Y chromosome was missing the SRY region
   (c) That XX mouse embryos transformed with the mouse SRY gene developed as male mice
   (d) All of these answers describe observations leading to the discovery of the gene conferring primary male sex characteristics in humans
   (e) None
5. In maize, there are mutant genes that confer male sterility to all of the offspring of the female parent carrying this genotype. This genetic information is not found on any of the chromosomes that make up the maize genome. This type of inheritance is called ________.
   (a) Extrachromosomal inheritance
   (b) Transgenic inheritance
   (c) Pseudoinheritance

(d) Biparental inheritance

(e) none

6. A ____ trait is passed only from father to sons.

(a) Autosomal dominant

(b) Autosomal recessive

(c) Sex-linked dominant

(d) Sex-linked recessive

(e) Y-linked

7. A ____ trait is passed from an affected father to all of his daughters and to none of his sons.

(a) Autosomal dominant

(b) Autosomal recessive

(c) Sex-linked dominant

(d) Sex-linked recessive

(e) Y-linked

Answers

Multiple Choice Question

1	2	3	4	5	6	7	8	9	10	11
C	A	E	C	B	A	B	C	B	B	A

Fill in the blanks

1	2	3	4	5	6	7
C	E	A	D	A	E	C

❑❑❑

Chapter – 9

Chromosomes

Portrait of a Chromosome

- Cytogenetics is the study of chromosome abnormalities and associated effects on health or other traits.
- Excess or deficient genetic material can cause syndromes or end prenatal development.
- Chromosomes consist of highly coiled DNA, RNA, histones, and non-histone proteins.

Telomeres and Centromeres are Essential

- Telomeres consist of repeat sequences and protect chromosome tips.
- A centromere is a constricted site where spindle fibers attach during cell division.
- Centromeres are regions of repeated DNA bound to centromere-associated proteins.

Karyotypes are Chromosomal Charts

- Karyotypes are charts that display chromosomes in size order.
- Chromosomes are numbered from largest to smallest, 1 through 22, plus X and Y.
- Chromosomes are distinguished by size, centromere location, differential staining, and DNA probes.

- Five human chromosomes (13, 14, 15, 21, and 22) have satellites, which carry repeats of genes encoding ribosomal components.
- Translocations result in exchanges of material between two chromosomes.
- Some cancers arise from translocations.

Visualizing Chromosomes

Obtaining Cells for Chromosome Study

- Any cell with a nucleus can be used to obtain chromosomes to prepare a karyotype.
- Fetal karyotypes are constructed from cells obtained by amniocentesis, chorionic villus sampling, and by fetal cell sorting from maternal blood.
- Fetal karyotypes are prepared for patients with advanced maternal age, repeated pregnancy losses, and increased risk of a chromosomal anomaly as indicated by a maternal serum marker test or family history.

Preparing Cells for Chromosome Observation

- To obtain chromosomes for karyotyping, cells are halted in metaphase, broken open on a glass slide, and the chromosomes spread over the surface.
- Chromosomes are stained (or exposed to fluorescent DNA probes), identified, and arranged in order of size and centromere location.
- Chromosomal shorthand summarizes the total number of chromosomes, types of sex chromosomes, and type of aberration present.
- Ideograms are maps of the distinguishing features of individual chromosomes.
- Fluorescent in situ hybridization (FISH) combines chromosome specific probes and fluorescent dyes to "paint" chromosomes (or label specific gene loci).

Variation in Chromosome Number

Polyploidy

- A euploid cell has a normal chromosome number.
- Polyploid cells have extra chromosome sets.
- Polyploids result from the fertilization of an oocyte by two sperm or after fertilization involving a diploid gamete.
- Abnormal chromosome number is a common cause of spontaneous abortion.

Aneuploidy

- An aneuploid cell lacks a chromosome or has an extra one and can result from meiotic nondisjunction.
- Individuals with trisomies are more likely to survive than those with monosomies. Sex chromosome aneuploidy is less severe than autosomal aneuploidy.
- The most common autosomal aneuploids are trisomies 13, 18 and 21. Sex chromosome aneuploids include XO (Turner syndrome), XYY (Klinefelter syndrome), triplo-X females, and XYY males.
- Aneuploidy causes an imbalance in gene expression that is often detrimental to the phenotype of the individual
- Aneuploidy in humans causes abnormal phenotypes

Euploidy

- Variations in euploidy occur naturally in a few animal species
- Variations in euploidy can occur in certain tissues within an animal
- Variations in euploidy are common in plants

Natural And Experimental Ways to Produce Variations in Chromosome Number

- Meiotic nondisjunction can produce aneuploidy or polyploidy
- Mitotic nondisjunction or chromosome loss can produce a patch of tissue with an altered chromosome number
- Changes in euploidy can occur by autopolyploidy, alloploidy, and allopolyploidy

- Allodiploids are often sterile, but allotetraploids are more likely to be fertile
- Experimental treatments can promote polyploidy
- Cell fusion techniques can be used to make hybrid plants
- Monoploids produced in agricultural and genetic research can be used to create homozygous and hybrid strains

Variation in Chromosome Structure

- Natural variation exists in chromosome structure
- Mutations can alter chromosome structure
- The loss of genetic material in a deficiency tends to be detrimental to an organism

Deletions and Duplications

- Deletions and duplications can result when translocations or inversions disrupt pairing of chromosomes in meiosis.
- Deficiencies can be detected using cytological, genetic, and molecular techniques
- Duplications tend to be less harmful than deletions
- A gene duplications produced the bar-eye phenotype in Drosophila
- Duplications provide additional material for gene evolution, sometimes leading to the formation of gene families

Translocations

- In a Robertsonian translocation, a piece of one chromosome breaks off and attaches to another chromosome.
- In a reciprocal translocation, two nonhomologous chromosomes exchange parts.
- A translocation that deletes, duplicates, or disrupts a gene can harm health.
- Translocation carriers may have a normal phenotype may have affected children.
- Unbalanced translocations usually have detrimental phenotypic effects

- Individuals with balanced translocations may produce abnormal gametes due to the segregation of chromosomes

Inversions

- Inversions result when part of a chromosome flips, and may affect health.
- Inversions often occur without phenotypic consequences
- Inversion heterozygotes may produce abnormal chromosomes due to crossing over

Isochromosomes and Ring Chromosomes

- An isochromosome has two identical arms and therefore introduces duplications and deletions.
- Isochromosomes arise in meiosis when the centromere splits in the wrong plane.
- Ring chromosomes arise when telomeres are lost, leaving sticky ends that close and form rings.
- Ring chromosomes can produce symptoms when they result in additional genetic material.

Uniparental Disomy-Two Genetic Contributions from One Parent

- Uniparental disomy (UPD) results when two chromosomes or chromosome parts are inherited from the same parent. UPD therefore doubles part of one parent's contribution.
- UPD can arise from a trisomy and subsequent chromosome loss, or from two nondisjunction events.
- UPD can cause disease if it creates a homozygous recessive condition associated with an illness.
- UPD can also cause disease by disrupting genomic imprinting.

Viral Genomes

- Viral genomes are relatively small and can be composed of RNA or DNA
- Viral genomes are packaged into the capsid in an assembly process

Bacterial Chromosomes

- Bacterial chromosomes contain circular DNA molecules in which a few thousand gene sequences are interspersed with other functionally important sequences
- The formation of chromosomal loops helps make the bacterial chromosome more compact
- DNA supercoiling further compacts the bacterial chromosome
- Chromosome function is influenced by DNA supercoiling

Eukaryotic Chromosomes

- The sizes of eukaryotic genomes vary substantially
- Eukaryotic chromosomes contain many functionally important sequences including genes, origins of replication, centromeres, and telomeres
- The genome of eukaryotes contains sequences that are unique, moderately repetitive, and highly repetitive
- Highly repetitive DNA can be separated from the rest of the chromosomal DNA by equilibrium density centrifugation
- Sequence complexity can be evaluated in a renaturation experiment
- Eukaryotic chromatin must be compacted to fit within the cell
- Linear DNA wraps around histone proteins to form nucleosomes, the repeating structural unit of chromatin
- The repeating nucleosome structure is revealed by digestion of the linker region
- Nucleosomes become closely associated to form a 30 nm fiber
- Chromosomes are further compacted by anchoring the 30 nm fiber into radial loop domains along the nuclear matrix
- Condensin and cohesin promote the formation of metaphase chromosomes

Multiple Choice Question

1. Which of the following descriptions of chromosomes is not correctly matched?
   (a) Metacentric – chromosome arms are almost equal in size
   (b) Submetacentric – chromosome arms are slightly different in size
   (c) Acrocentric – chromosome arms are identical in size
   (d) Telocentric – there is only one chromosome arm
   (e) None
2. Which of the following is not a type of chromosomal mutation?
   (a) Duplication
   (b) Inversion
   (c) Translocation
   (d) All of these are chromosomal mutations
   (e) None of these are chromosomal mutations
3. A duplication is:
   (a) The presence of an extra copy of an existing chromosome
   (b) The presence of an entirely new piece of chromosomal material
   (c) The presence of an extra copy of a piece of chromosomal material
   (d) The result of chromosomal material exchanged between two chromosomes
   (e) A piece of genetic material which has been moved to a new chromosome
4. Which explanation for the loss of a terminal piece of a chromosome makes the most sense?
   (a) Only pieces of DNA with centromeres are correctly sorted into daughter nuclei during mitosis
   (b) Only pieces of DNA with telomeres are correctly sorted into daughter nuclei during mitosis
   (c) This piece of DNA is recognized as foreign by the cell
   (d) A and B
   (e) None of these

5. Which technique can be used to quickly identify large deficiencies?
   (a) Karyotype analysis
   (b) Genetic studies
   (c) Pedigree analysis
   (d) All of these
   (e) None of these
6. Why might gene duplication allow for generation of more genetic variability?
   (a) The second copy of the gene could be expressed under different regulation
   (b) The second copy of the gene could accumulate mutations which would be lethal if only one copy were present
   (c) The second copy of the gene could be deleted without effect on the organism
   (d) A and B
   (e) None of the above
7. Which statement regarding the human globin genes is likely to be correct?
   (a) Myoglobin is probably more similar to α-globin than is β-globin
   (b) α-globin and β-globin are probably more similar to each other than they are to myoglobin
   (c) The globin genes in mice are probably more similar to those of humans than are globin genes of chimpanzees
   (d) All of these
   (e) None of these
8. Which of the following chromosomal rearrangements would you expect to have the least phenotypic effect on the organism?
   (a) A paracentric inversion
   (b) A paracentric duplication
   (c) A deficiency
   (d) A duplication
   (e) None of these
9. Why are unaffected carriers of inversions likely to produce genetically abnormal progeny?
   (a) The inversion chromosome is more likely to be placed in a gamete than the normal chromosome
   (b) The inversion chromosome is unable to accomplish synapsis with the normal chromosome during meiosis

(c) Crossovers cannot occur between inversion and normal chromosomes

(d) Crossovers between the inversion and normal chromosomes lead to chromosomes with deletions, deficiencies, or abnormal structure

(e) None of these

10. What do deficiencies, inversions, and translocations have in common?

(a) They all involve loss of genetic material

(b) They all contain normal amounts of genetic material in unusual arrangements

(c) They all can be caused by DNA breakage and inaccurate repair

(d) They all involve extra genetic material in a karyotype

(e) None of these

11. What do inversions and balanced translocations have in common?

(a) They both contain a normal amount of DNA

(b) They both tend to have mild or no phenotypic consequences

(c) They both tend to cause abnormal gamete production

(d) All of these

(e) None of these

12. Which of the following is an example of a euploid individual?

(a) A D. melanogaster embryo with 7 chromosomes

(b) A human with 47 chromosomes

(c) A plant of a new wheat species with twice the usual number of chromosomes

(d) None of these

(e) All of these

13. If X inactivation in humans shuts down extra X chromosomes, why do XXY and XXX individuals show any phenotypic effects?

(a) X inactivation can only shut down one X per cell

(b) The extra Barr body takes up physical space in the cell

(c) Some X linked genes escape inactivation

(d) All of these

(e) None of these

14. Which of the following might cause nondisjunction?

(a) Failure of synapsis in meiosis I

(b) Chromosome breakage

(c) Lack of synapsis in mitosis

(d) Centromere duplication

(e) None of these

15. An allotetraploid species:

(a) Contains a full diploid set from each of two related species

(b) Is likely to be able to produce viable gametes

(c) May be agriculturally important

(d) All of these

(e) None of these

16. If a woman has a hemophilia carrier mother and a father who has hemophilia, what is the probability of the woman having hemophilia?

(a) 25% (b) 50%

(c) 75% (d) 100%

(e) None of these

17. Aneuploidy

(a) Results from nondisjunction

(b) Occurs when there is an abnormal number of chromosomes

(c) Generally causes fetuses to stop developing before birth

(d) All of these are correct

(e) None of these

18. Females with delayed or no sexual development but who have normal intelligence most likely suffer from

(a) Turner syndrome (b) Triplo-X

(c) Klinefelter syndrome (d) Hormonal abnormalities

(e) None of these

19. Consider the following phenotype: tall and thin, menstrual irregularity, slightly lower IQ. With which syndrome are these characterstics consistent?

(a) Trisomy18 (b) XXY

(c) XXX (d) XO

(e) None of these

20. Individuals having sex chromosomes XXY would have which disorder?

(a) Turner syndrome (b) Marfan syndrome

(c) Down syndrome (d) Klinefelter syndrome

(e) None of these

21. What type of smaller-scale chromosome abnormality causes Cri du chat syndrome?

(a) Deletion (b) Duplication

(c) Inversion (d) Translocation

(e) None of these

22. What type of smaller-scale chromosome abnormality occurs when the short arms of two nonhomologs break and join the two long arms making a new, large chromosome?

(a) Deletion (b) Duplication

(c) Robertsonian translocation (d) Reciprocal translocation

(e) None of these

23. Which of the following is an example of a genome?

(a) The F factor

(b) A homologous pair of chromosomes

(c) All the DNA contained in an embryonic plant

(d) All the DNA contained in a human gamete

(e) None of these

24. Why are viruses not considered to be living organisms?

(a) They rely on host cells for all energy requiring activities

(b) They do not contain genetic material

(c) They have a very primitive cellular structure
(d) All of the above
(e) None of the above

25. How might directed assembly be a benefit to the virus?
(a) It can utilize a wider array of building materials
(b) It can produce a more complex viral particle
(c) It requires less genes than self assembly
(d) It is not likely to be beneficial to the virus
(e) None of these

26. Scientists have identified a new bacterial species which typically has 6-10 nucleoids per cell. How many copies of the bacterial chromosome would you expect are present in each cell?
(a) 3-5
(b) 6-10
(c) 12-20
(d) There is not enough information here to determine
(e) None of these

27. Which type of gene sequence occupies the most space in the bacterial chromosome?
(a) Origins of replication
(b) Structural gene sequences
(c) Intergenic regions
(d) A and B equally
(e) None of the above

28. What are topoisomers?
(a) DNA structures that vary by a single nucleotide
(b) DNA structures that vary in length
(c) DNA structures that vary in conformation
(d) DNA structures that vary in orientation of the backbones
(d) None of these

29. Which of the following is an effect that might be caused by supercoiling in bacterial cells?

(a) Negative supercoiling can increase the tendency for strand separation in some regions

(b) Positive supercoiling can enhance gene activation

(c) Positive supercoiling can interfere with the process of DNA replication

(d) None of these

(e) All

30. Which of the following is a description of chromatin?

(a) All the genetic sequences contained by members of a particular species

(b) The DNA-protein complex which comprises eukaryotic chromosomes

(c) Repetitive sequences contained within the genome of an organism

(d) The protein coding sequences and their regulatory elements

(e) None of these

31. Which of the following structures of a eukaryotic chromosome is not primarily composed of DNA?

(a) Telomeres

(b) Origin of replication

(c) Kinetochore

(d) Centromere

(e) None of these

32. What is a major function of the telomeres?

(a) They prevent chromosomes from fusing with each other

(b) They prevent chromosomes from shrinking due to DNA loss during replication

(c) They allow chromosomes to segregate properly during cell division

(d) A and B

(e) None of the above

33. In fruit flies, a gene called bicoid controls embryonic development. What type of DNA would you expect this gene to be?

(a) Nonrepetitive sequences

(b) Moderately repetitive sequences

(c) Highly repetitive sequences

(d) Retroelements

(e) None of these

34. How do DNAs with a slow rate of renaturation compare to those with a high rate of renaturation?

(a) Slow renaturing DNAs are longer than fast renaturing DNAs

(b) Slow renaturing DNAs have more AT base pairs than fast renaturing DNAs

(c) Slow renaturing DNAs are less repetitive than fast renaturing DNAs

(d) Slow renaturing DNAs are less likely to contain protein coding sequences than fast renaturing DNAs

(e) None of these

35. Which of the following statements regarding a nucleosome is correct?

(a) It contains a single large protein which is connected to DNA by smaller linker proteins

(b) It contains approximately 146 base pairs of DNA

(c) It includes histone and nonhistone proteins

(d) All of these

(e) None of these

36. DNase protection experiments were used to support the beads on a string model of chromatin structure. Which of the following statements would apply to DNase protection experiments?

(a) DNase can cut DNA whether or not it is bound to proteins

(b) DNase can only cut highly condensed DNA

(c) Proteins which are tightly bound to DNA can prevent DNase digestion

(d) Proteins are digested by DNase

(e) None of these

37. Which of the following is a correct order of levels of chromatin compaction?

(a) Nucleosomes, 30 nm fibers, radial loop domains

(b) 30 nm fibers, nucleosomes, radial loop domains

(c) Nucleosomes, radial loop domains, 30 nm fibers

(d) Radial loop domains, nucleosomes, 30 nm fibers

(e) None of these

38. Which of the following is an example of constitutive heterochromatin?

(a) The Barr body

(b) A region of active gene expression

(c) Telomeres

(d) A and B

(e) None of these

39. What does the endosymbiotic hypothesis state?

(a) Eukaryotes evolved from prokaryotes

(b) Prokaryotes evolved from eukaryotes

(c) Mitochondria and chloroplasts are living prokaryotes

(d) Mitochondria and chloroplasts were once living prokaryotes

(e) Prokaryotes and eukaryotes show no relationship to one another

40. All of the following are true about mitochondrial DNA except ________.

(a) Mitochondrial DNA is self-replicating

(b) It is only passed through the female

(c) It contains no introns

(d) It has ribosomes more similar to prokaryotes than to eukaryotes

(e) It is only found in mammalian cells

41. If a small chromosome consists of 500,000,000 nucleotides, how many nucleosomes will attach to that chromosome?

(a) 250,000,000 (b) 2,500,000
(c) 500,000,000 (d) 1000
(e) None

42. Are histone proteins the only proteins associated with chromosomes?

(a) Yes (b) No
(c) It depends on the life cycle (d) None of these
(e) All

43. Nondisjunction during the second meiotic division will result in which of the following gametes?

(a) All monosomic gametes
(b) All trisomic gametes
(c) All haploid gametes
(d) Two diploid gametes and two haploid gametes
(e) Two haploid gametes, one diploid gamete, and one gamete missing a copy of one of the chromosomes

44. In humans, how many of our chromosomes have been observed in an aneuploid condition?

(a) X, Y, and chromosome 21
(b) Chromosomes 21, 18, and 13
(c) Only the sex chromosomes X and Y
(d) Only chromosome 21
(e) All the chromosomes

45. Which is NOT correct regarding nondisjunction?

(a) Occurs more frequently in human females over age 35
(b) May fail to separate maternal from paternal chromatids
(c) May fail to separate maternal chromatids from one another or paternal chromatids from one another
(d) May lead to Down syndrome
(e) Separates maternal from paternal chromatids

46. Which of the following supports the hypothesis that says "gene duplication is essential to evolution"?

(a) Gene families
(b) Intercalary deletions
(c) Inversions
(e) Fragile sites
(e) Genetic anticipation

47. Why would the transmission of Down syndrome due to a Robertsonian translocation be unaffected by the age of the mother?

(a) Mothers with a Robertsonian translocation never have children
(b) Down syndrome due to a Robertsonian translocation is not due to nondisjunction, which is affected by age
(c) Robertsonian translocations only occur in young mothers
(d) This statement is incorrect; the age of the mother does affect the transmission of Down syndrome due to Robertsonian translocations
(e) None of these

48. In which phase(s) of meiosis does non-disjunction occur?

(a) Interphase
(b) Prophase I only
(c) Anaphase I and Anaphase II
(d) Telophase II only
(e) Metaphase I and Metaphase II

49. A somatic cell in which there is not an exact multiple of the haploid chromosome number is called:

(a) Polyploidy
(b) Disomic
(c) Non-disjunctive
(d) Aneuploid
(e) Diploid

50. A disomic zygote arises from the fertilization of gametes in which:

(a) Both gametes are normal
(b) One gamete is normal and the other suffered non-disjunction in Meiosis I
(c) One gamete is normal and the other suffered non-disjunction in Meiosis II
(d) Both gametes suffered non-disjunction in Meiosis I
(e) Both gametes suffered non-disjunction in Meiosis II

51. When non-disjunction occurs in Meiosis I

(a) One of the products of Meiosis I has one of each of the replicated chromosome homologous pairs and the other has none

(b) One of the products of Meiosis I has a single unreplicated chromosome for every chromosome pair while the other has three copies (not all identical) of a particular chromosome

(c) Three of the products of Meiosis II contain too many chromosomes while one product is missing a chromosome for a particular set

(d) Two of the products of Meiosis II have both maternal and paternal chromosomes of a particular homologue set while the other two have none for that set

(e) Two of the products of Meiosis II have replicated chromosomes for a particular homologue set while the other two have none for that set

52. A trisomic zygote contains:

(a) One third of the normal chromosome number

(b) Three chromosomes for every unique chromosome type

(c) Three chromosomes

(d) A combination of chromosomes produced from three different possibilities

(e) An extra chromosome associated with a particular chromosome set

53. Down Syndrome or Trisomy 21 in humans results from non-disjunction in:

(a) Meiosis I of both gametes

(b) Meiosis I of either the male or female gamete (but not both)

(c) Either Meiosis I or Meiosis II in the male or female gamete (but not both)

(d) Meoisis II of both gametes

(e) Meiosis II of either the male or female gamete (both not both)

54. A monosomic zygote results in:

(a) The union of two gametes, both carrying one chromosome for each homologous pair

(b) The union of two normal gametes

(c) The union of two gametes that both suffered Meiosis II non-disjunction

(d) The union of two gametes that both suffered Meiosis I non-disjunction

(e) The union of a normal gamete with one missing a chromosome from the haploid set

55. An organism that contains more than two sets of chromosomes is said to be

(a) Disomic (b) Allopolyploid

(c) Tetraploid (d) Polyploid

(e) Amphidiploid

56. A cell that has only one set of chromosomes is said to be:

(a) Diploid (b) Tetraploid

(c) Allopolyploid (d) Haploid

(e) Polyploid

57. A plant resulting from the combination of chromosome sets from different species is said to be

(a) Autopolyploid (b) Allopolyploid

(c) Autotetraploid (d) Polyploid

(e) Autohybridoid

58. In chrysanthemum, the haploid chromosome number is 9. In tetraploid chrysanthemum species, the following number of chromosomes will be found:

(a) 18 (b) 24

(c) 36 (d) 9

(e) 72

59. Which of the following is an example of a genome?

(a) The F factor

(b) A homologous pair of chromosomes

(c) All the DNA contained in an embryonic plant

(d) All the DNA contained in a human gamete

60. Why are viruses not considered to be living organisms?

(a) They rely on host cells for all energy requiring activities

(b) They do not contain genetic material

(c) They have a very primitive cellular structure

(d) All of the above

(e) None of the above

61. How might directed assembly be a benefit to the virus?

(a) It can utilize a wider array of building materials

(b) It can produce a more complex viral particle

(c) It requires less genes than self assembly

(d) It is not likely to be beneficial to the virus

62. Scientists have identified a new bacterial species which typically has 6-10 nucleoids per cell. How many copies of the bacterial chromosome would you expect are present in each cell?

(a) 3-5

(b) 6-10

(c) 12-20

(d) There is not enough information here to determine

63. Which type of gene sequence occupies the most space in the bacterial chromosome?

(a) Origins of replication

(b) Structural gene sequences

(c) Intergenic regions

(d) A and B equally

(e) None of the above

64. What are topoisomers?

(a) DNA structures that vary by a single nucleotide

(b) DNA structures that vary in length

(c) DNA structures that vary in conformation

(d) DNA structures that vary in orientation of the backbones

Fill in the Blanks

1. DNA euchromatin is more loosely folded and includes areas ___________.

   (a) That encode protein (b) Near centromeres
   (c) At the tips of chromosomes (d) All of these are correct
   (e) None of these

2. A chromosome is ____________ if the centromere is located very close to a chromosome tip.

   (a) Metacentric (b) Submetacentric
   (c) Acrocentric (d) Telocentric
   (e) None of these

3. Genes that reside on the same chromosome are said to be __________.

   (a) Transposable (b) Satellite genes
   (c) Autosomal (d) Linked
   (e) All of these

4. When a dominant allele is linked to a recessive allele, the genes are said to be in ___________.

   (a) cis (b) trans
   (c) parallel (d) synchrony
   (e) All of these

5. An individual that has two different sex chromosomes is said to be the ____________.

   (a) Homogametic sex (b) Heterogametic sex
   (c) Dominant sex (d) Recessive sex
   (e) All of these

6. Calico cats are walking demonstrations of X inactivation because they have __________.

   (a) Orange patches only (b) Black patches only
   (c) Both orange and black patches (d) No coloration and are white
   (e) All

7. An inactivated X chromosome in the nucleus of a cell appears as a __________ in the microscope.

   (a) Normal X chromosome (b) Polyploid

   (c) Barr body (d) Monosomy

   (e) None

8. Viral genomes differ from eukaryotic genomes in all of the following ways except __________.

   (a) Nucleic acid may consist of DNA or RNA

   (b) Nucleic acid may be single stranded or double stranded

   (c) Ability to pack large amounts of DNA into a small volume

   (d) Few, if any, DNA-binding proteins

   (e) Inability to replicate its own DNA

9. Heterochromatin and euchromatin are both used to describe __________ and relative __________.

   (a) DNA structure/function

   (b) DNA structure/length

   (c) DNA function/protein products

   (d) Chromosome number/function

   (e) None

10. Transposable sequences fall into a unique group of repetitive DNA sequences that are identified by their ability to __________.

   (a) Interfere with DNA replication

   (b) Prevent proper chromosome segregation during meiosis

   (c) Move from place to place within the genome

   (d) Stop viruses from infecting a cell

   (e) Interfere with telomere function

11. Telomeres can best be characterized by the following statement: __________.

   (a) Made up of heterochromatin

   (b) Found at the ends of linear chromosomes

(c) Composed of short tandem repeats

(d) Highly conserved in evolution

(e) All these statements characterize telomeres

12. The histone protein that joins with H2B to form one tetramer subunit of a nucleosome is ___.

(a) H1 (b) H3

(c) H4 (d) H2A

(e) H2B

13. The histone protein that joins with H3 to form another tetramer subunit of a nucleosome is ___.

(a) H4 (b) H2B

(c) H3 (d) H1

(e) H2A

14. The histone protein that attaches to DNA strands between nucleosomes is ___.

(a) H2B (b) H2A

(c) H1 (d) H4

(e) H3

15. Chromosomes must condense to approximately 1/500th of their length for cell division. The first reduction is ___.

(a) Looping of solenoid fibers to form a 300 nm fiber

(b) Looping of 300 nm fibers

(c) Coiling around nucleosomes

(d) Forming a coiled solenoid fiber

(e) None

16. Data are consistent with a model that shows nucleosomes packing together to form a coil. This is called a ___.

(a) 30 nm fiber (b) Chromatid

(c) Chromosome (d) Solenoid

(e) None

17. Chromosomal aberrations may be characterized by all of the following except __________.

(a) Chromosome loss
(b) Chromosome duplication
(c) Translocations
(d) Inversions
(e) Frameshift mutations

18. The gain of one chromosome to a diploid genome results in __________, and the addition of a complete set of chromosomes to a diploid genome results in the formation of a(n) __________.

(a) Triploidy/autopolyploid
(b) Euploidy/triploid
(c) Polyploidy/haploid
(d) Allopolyploidy/tetraploid
(e) Aneuploidy/haploid

19. The difference between paracentric inversions and pericentric inversions is __________.

(a) The number of centromeres involved in the inversion
(b) The number of genes involved in the inversion
(c) That one suppresses crossovers and the other encourages crossovers
(d) The formation of dicentric bridges
(e) The placement of the centromere in the inversion

20. The normally cultivated cotton is a tetraploid (52 chromosomes). The probable ancestral haploid had _____ chromosomes.

(a) 52
(b) 4
(c) 104
(d) 14
(e) 26

True/False

1. Fragile X syndrome affects human males but not females.

(a) True
(b) False

2. Deletions tend to be less harmful than duplications.

(a) True
(b) False

3. Members of a gene family may be created by abnormal crossing over.
   (a) True (b) False
4. Inversions tend to have more serious phenotypic effects than other chromosomal rearrangements.
   (a) True (b) False
5. Alternate and adjacent-1 segregation are equally likely to produce nonviable gametes.
   (a) True (b) False
6. Aneuploidy is usually a lethal condition in humans.
   (a) True (b) False
7. While plants can often exist as polyploid species, animals cannot.
   (a) True (b) False
8. All of the cells within an individual are genetically identical.
   (a) True (b) False
9. An allodiploid can be created by fusion of diploid gametes.
   (a) True (b) False
10. Men with an extra Y chromosome very often are violent and aggressive.
    (a) True (b) False
11. The pea plant traits in Mendel's crosses were transmitted on different chromosomes.
    (a) True (b) False
12. The severity of Down syndrome rarely varies and if so, can be predicted from prenatal examinations.
    (a) True (b) False
13. Human sons receive X-linked traits only from their fathers.
    (a) True (b) False
14. Human polyploids are common and often survive for decades.
    (a) True (b) False

15. Although the likelihood of giving birth to a child with trisomy 21 increases with a woman's age, 40% of trisomy 21 cases result from non disjunction in the male.

(a) True (b) False

16. Viral genomes are composed of DNA.

(a) True (b) False

17. Directed assembly is a critical process for replication of TMV (tobacco mosaic virus).

(a) True (b) False

18. Bacterial and eukaryotic chromosomes each have a single origin of replication.

(a) True (b) False

19. Supercoiling generally prevents information within the DNA from being read.

(a) True (b) False

20. Eukaryotic genomes must be highly compacted to fit with the nucleus.

(a) True (b) False

21. Satellite DNAs in an individual have different AT/GC composition than the majority of DNA in that individual.

(a) True (b) False

22. Proteins bound to the DNA are constant over the life of the cell.

(a) True (b) False

23. Histone proteins bind to both the core DNA and the linker DNA.

(a) True (b) False

24. Compaction of DNA involves interaction with the nuclear matrix.

(a) True (b) False

25. Active genes are probably located in euchromatic regions of DNA.

(a) True (b) False

26. Condensin plays a critical role in the formation of metaphase chromosomes.

(a) True (b) False

27. Viral genomes are composed of DNA.

(a) True (b) False

28. Directed assembly is a critical process for replication of TMV (tobacco mosaic virus).

(a) True (b) False

29. Bacterial and eukaryotic chromosomes each have a single origin of replication.

(a) True (b) False

30. Supercoiling generally prevents information within the DNA from being read.

(a) True (b) False

31. Eukaryotic genomes must be highly compacted to fit with the nucleus.

(a) True (b) False

32. Proteins bound to the DNA are constant over the life of the cell.

(a) True (b) False

33. Histone proteins bind to both the core DNA and the linker DNA.

(a) True (b) False

34. Compaction of DNA involves interaction with the nuclear matrix.

(a) True (b) False

35. Active genes are probably located in euchromatic regions of DNA.

(a) True (b) False

36. Condensin plays a critical role in the formation of metaphase chromosomes.

(a) True (b) False

Answers

Multiple Choice Question

1	2	3	4	5	6	7	8	9	10	11	12	13	14	15
C	D	C	A	A	D	B	A	D	C	D	C	C	A	D
16	17	18	19	20	21	22	23	24	25	26	27	28	29	30
B	D	A	C	D	A	C	D	A	B	B	B	C	A	B
31	32	33	34	35	36	37	38	39	40	41	42	43	44	45
C	D	A	C	B	C	A	C	D	E	B	A	E	E	E
46	47	48	49	50	51	52	53	54	55	56	57	58	59	60
A	B	C	D	A	E	E	C	E	D	D	B	C	D	A
61	62	63	64											
B	B	B	C											

Fill in the Blanks

1	2	3	4	5	6	7	8	9	10	11	12	13	14	15
A	D	D	B	B	C	C	C	A	C	E	D	A	C	C
16	17	18	19	20										
D	E	A	A	D										

True/False

1	2	3	4	5	6	7	8	9	10	11	12	13	14	15
B	B	A	B	B	A	B	B	B	B	A	B	B	B	A
16	17	18	19	20	21	22	23	24	25	26	27	28	29	30
B	B	B	B	A	A	B	A	A	A	A	B	B	B	B
31	32	33	34	35	36									
A	B	B	A	A	A									

❑❑❑

Chapter – 10

Variation in Chromosome Structure and Number

Variation in Chromosome Structure

- Natural variation exists in chromosome structure
- Changes in chromosome structure include deletions, duplications, inversions, and translocations
- The loss of genetic material in a deletion tends to be detrimental to an organism
- Duplications tend to be less harmful than deletions
- Duplications provide additional material for gene evolution, sometimes leading to the formation of gene families
- Copy number variation is relatively common among members of the same species
- Comparative genomic hybridization is used to detect chromosome deletions and duplications
- Inversions often occur without phenotypic consequences
- Inversion heterozygotes may produce abnormal chromosomes due to crossing over
- Translocations involve exchanges between different chromosomes
- Individuals with reciprocal translocations may produce abnormal gametes due to the segregation of chromosomes

Variation in Chromosome Number

- Aneuploidy causes an imbalance in gene expression that is often detrimental to the phenotype of the individual
- Aneuploidy in humans causes abnormal phenotypes
- Variations in euploidy occur naturally in a few animal species
- Variations in euploidy can occur in certain tissues within an animal
- Variations in euploidy are common in plants

Natural and Experimental Ways to Produce Variations in Chromosome Number

- Meiotic nondisjunction can produce aneuploidy or polyploidy
- Mitotic nondisjunction or chromosome loss can produce a patch of tissue with an altered chromosome number
- Changes in euploidy can occur by autopolyploidy, alloploidy, and allopolyploidy
- Allodiploids are often sterile, but allotetraploids are more likely to be fertile
- Experimental treatments can promote polyploidy
- Cell fusion techniques can be used to make hybrid plants
- Monoploids produced in agricultural and genetic research can be used to make homozygous and hybrid strains

Multiple Choice Question

1. Which of the following descriptions of chromosomes is not correctly matched?

   (a) Metacentric – chromosome arms are almost equal in size

   (b) Submetacentric – chromosome arms are slightly different in size

   (c) Acrocentric – chromosome arms are identical in size

   (c) Telocentric – there is only one chromosome arm

2. Which of the following is not a type of chromosomal mutation?

   (a) Duplication

   (b) Inversion

(c) Translocation

(d) All of these are chromosomal mutations

(e) None of these are chromosomal mutations

3. A duplication is:

(a) The presence of an extra copy of an existing chromosome

(b) The presence of an entirely new piece of chromosomal material

(c) The presence of an extra copy of a piece of chromosomal material

(d) The result of chromosomal material exchanged between two chromosomes

(e) A piece of genetic material which has been moved to a new chromosome

4. Which explanation for the loss of a terminal piece of a chromosome makes the most sense?

(a) Only pieces of DNA with centromeres are correctly sorted into daughter nuclei during mitosis

(b) Only pieces of DNA with telomeres are correctly sorted into daughter nuclei during mitosis

(c) This piece of DNA is recognized as foreign by the cell

(d) A and B

(e) None of these

5. Which technique can be used to quickly identify large deficiencies?

(a) Karyotype analysis (b) Genetic studies

(c) Pedigree analysis (d) All of these

(e) None of these

6. Why might gene duplication allow for generation of more genetic variability?

(a) The second copy of the gene could be expressed under different regulation

(b) The second copy of the gene could accumulate mutations which would be lethal if only one copy were present

(c) The second copy of the gene could be deleted without effect on the organism

(d) A and B

(e) None of the above

7. Which statement regarding the human globin genes is likely to be correct?

(a) Myoglobin is probably more similar to α-globin than is β-globin

B) α-globin and β-globin are probably more similar to each other than they are to myoglobin

(c) The globin genes in mice are probably more similar to those of humans than are globin genes of chimpanzees

(d) All of these

(e) None of these

8. Which of the following chromosomal rearrangements would you expect to have the least phenotypic effect on the organism?

(a) A paracentric inversion

(b) A paracentric duplication

(c) A deficiency

(d) A duplication

9. Why are unaffected carriers of inversions likely to produce genetically abnormal progeny?

(a) The inversion chromosome is more likely to be placed in a gamete than the normal chromosome

(b) The inversion chromosome is unable to accomplish synapsis with the normal chromosome during meiosis

(c) Crossovers cannot occur between inversion and normal chromosomes

(d) Crossovers between the inversion and normal chromosomes lead to chromosomes with deletions, deficiencies, or abnormal structure

10. What do deficiencies, inversions, and translocations have in common?

(a) They all involve loss of genetic material

(b) They all contain normal amounts of genetic material in unusual arrangements

(c) They all can be caused by DNA breakage and inaccurate repair

(d) They all involve extra genetic material in a karyotype

11. What do inversions and balanced translocations have in common?

A) They both contain a normal amount of DNA

B) They both tend to have mild or no phenotypic consequences

C) They both tend to cause abnormal gamete production

D) All of these

E) None of these

12. Which of the following is an example of a euploid individual?

(a) A *D. melanogaster* embryo with 7 chromosomes

(b) A human with 47 chromosomes

(c) A plant of a new wheat species with twice the usual number of chromosomes

(d) None of these

13. If X inactivation in humans shuts down extra X chromosomes, why do XXY and XXX individuals show any phenotypic effects?

(a) X inactivation can only shut down one X per cell

(b) The extra Barr body takes up physical space in the cell

(c) Some X linked genes escape inactivation

(d) All of these

(e) None of these

14. Which of the following might cause nondisjunction?

(a) Failure of synapsis in meiosis I

(b) Chromosome breakage

(c) Lack of synapsis in mitosis

(d) Centromere duplication

15. An allotetraploid species:

(a) Contains a full diploid set from each of two related species

(b) Is likely to be able to produce viable gametes

(c) May be agriculturally important

(d) All of these

(e) None of these

True /False

1. Deletions tend to be less harmful than duplications.

(a) True (b) False

2. Members of a gene family may be created by abnormal crossing over

(a) True (b) False

3. Alternate and adjacent-1 segregation are equally likely to produce nonviable gametes.

(a) True (b) False

4. Aneuploidy is usually a lethal condition in humans.

(a) True (b) False

5. While plants can often exist as polyploid species, animals cannot.

(a) True (b) False

6. All of the cells within an individual are genetically identical.

(a) True (b) False

7. An allodiploid can be created by fusion of diploid gametes.

(a) True (b) False

Answers

Multiple Choice Question

1	2	3	4	5	6	7	8	9	10	11	12	13	14	15
C	D	C	A	A	D	B	A	D	C	D	C	C	A	D

True/False

1	2	3	4	5	6	7
B	A	B	A	B	A	B

❑❑❑

Chapter – 11

DNA Structure and Replication

Identification of DNA as the Genetic Material

- Experiments with pneumococcus suggested that DNA is the genetic material
- Hershey and Chase provided evidence that the genetic material injected into the bacterial cytoplasm is T2 phage DNA
- RNA functions as the genetic material in some viruses

Work Before Watson and Crick

- Nucleotides are the building blocks of nucleic acids
- Nucleotides are linked together to form a strand
- In 1869, Miescher first isolated DNA and named it "nuclein."
- In 1902, Archibald Garrod linked heredity to enzyme (protein) defects. In 1909 he published a book entitled "Inborn Errors of Metabolism."
- In 1928, Griffith identified a "transforming factor" that transmitted infectiousness in bacteria.
- In 1944, Avery, MacLeod and McCarty showed that the transforming factor was DNA)
- In 1950, Hershey and Chase, using their famous "blender" experiment, confirmed that DNA, not protein, is the genetic material.
- The work of chemist Phoebus Levene showed that DNA includes deoxyribose, nitrogenous bases, and phosphates, in equal proportions.

- Chargaff's data showed that the number of purines equals the number of pyrimidines.
- Using Rosalind Franklin's X-ray diffraction patterns, Watson and Crick deduced that DNA is a double helix.

Discovery: Watson and Crick

- Watson and Crick met in the early 1950s at Cambridge University in England and set about to decipher the structure of DNA) Their research was aided by the work of others, including Rosalind Franklin, who was using x-ray crystallography to learn about DNA's structure.
- James Watson and Francis Crick discovered the chemical structure of DNA in 1953. This event ushered in a new era in biology because it allowed researchers to understand some of the most fundamental processes in genetics.
- In trying to decipher the structure of DNA, Watson and Crick were performing work in molecular biology. This is the investigation of life at the level of its individual molecules. Molecular biology has grown greatly in importance since the 1950s.

The Components of DNA and Their Arrangement

- The DNA molecule is composed of building blocks called nucleotides, each of which consists of one sugar (deoxyribose), one phosphate group, and one of four bases: adenine, guanine, thymine, or cytosine (A, G, T, or C). The sugar and phosphate groups are linked together in a chain that forms the "handrails" of the DNA double helix. Bases then extend inward from the handrails, with base pairs joined to each other in the middle by hydrogen bonds. In this base pairing, A always pairs with T across the helix, while G always pairs with C)
- DNA is copied by means of each strand of DNA serving as a template for the synthesis of a new, complementary stranD) The DNA double helix first divides down the middle. Each A on an original strand then specifies a place for a T in a new strand, each G specifies a place for a C on the new strand, and so forth.
- Each double helix produced in replication is a combination of one parental strand of DNA and one newly synthesized complementary stranD) This is how life builds upon itself. A group of enzymes known as DNA polymerases are central to DNA replication; they move along the double helix, bonding together new nucleotides in complementary DNA strands.

- DNA can encode the information for the huge number of proteins utilized by living things because the sequence of bases along DNA's handrails can be laid out in an extremely varied manner. A collection of bases in one order encodes the information for one protein, while a different sequence of basis encodes the information for a different protein.
- The gene is a segment of DNA containing the information to specify a sequence of amino acids in a protein.
- The DNA double helix is a ladder of alternating deoxyribose and phosphate groups, with rungs formed by complementary base pairs.
- Adenine (A) and guanine (G) are purine bases; cytosine (C) and thymine (T) are pyrimidine bases. A pairs with T and G with C.
- The DNA double helix is directional, the two strands are anti-parallel.
- DNA is highly coiled and wound tightly about histones, forming nucleosomes, which wind into chromatin

DNA Replication

- Meselson and Stahl in 1957 showed that DNA replication was not conservative or dispersive, but semi-conservative. The two parental strands separate and each is a template for assembling new daughter strands.
- Existing DNA strands act as templates for the synthesis of new strands.
- Three different models were proposed that described the net result of DNA replication

Steps and Participants in DNA Replication

- Replication occurs simultaneously at several points on each human chromosome.
- At each initiation site, primase builds a short RNA primer. DNA polymerase builds a new strand from the 5' end of the DNA to the 3' enD) Bases in the template and complementary bases in the new strand hydrogen bond to form the new double helix.
- The new strand of DNA is checked for errors and the RNA primers removed and replaced with DNA bases. Ligase joins the sugar-phosphate backbone.

- DNA replication is discontinuous because DNA polymerase can't readily replicate strands that run in opposite directions.

PCR-Directing DNA Replication

- The polymerase chain reaction (PCR) can amplify selected DNA in a manner similar to the way DNA replication works in the nucleus.
- The PCR method involves repeated cycles of denaturation of double stranded DNA, annealing of primers, and DNA synthesis using heat stable DNA polymerases.

Bacterial DNA Replication

- Bacterial chromosomes contain a single origin of replication
- Replication is initiated by the binding of DnaA protein to the origin of replication
- DNA strand separation and the synthesis of RNA primers are necessary before daughter strands of DNA can be made
- DNA polymerases link nucleotides to synthesize the daughter strands
- DNA polymerase III is a processive enzyme that uses deoxynucleoside triphosphates
- Replication is terminated when the replication forks meet at the terminus sequences
- Certain enzymes of DNA replication bind to each other to form a complex
- The fidelity of DNA replication is ensured by proofreading mechanisms
- Bacterial DNA replication is coordinated with cell division
- DNA replication can be studied *in vitro*
- The isolation of mutants has been instrumental in our understanding of DNA replication

Eukaryotic DNA Replication

- Initiation occurs at multiple origins of replication on linear eukaryotic chromosomes
- Eukaryotes contain several different DNA polymerases

- Nucleosomes containing new histone proteins are quickly formed after DNA replication
- The ends of eukaryotic chromosomes are replicated by telomerase

Mutations in DNA Structure

- The error rate in DNA replication is very low, partly because repair enzymes are able to correct mistakes. When such mistakes are made and then not corrected, the result is a mutation: a permanent alteration in a cell's DNA base sequences.
- Most mutations have no effect on an organism, but when they do have an effect, it is generally negative. Cancers result from a line of cells that have undergone types of mutations that cause them to proliferate wildly.
- Some mutations come about in the body's germ-line cells, meaning cells that become eggs or sperm. Such mutations are heritable: They can be passed on from one generation to another. The gene for Huntington disease, which is expressed in nerve cells, is a heritable, mutated form of a normal gene.

Multiple Choice Question

1. With regard to their chromosomes, what is a major difference between prokaryotes and eukaryotes?
   (a) Eukaryotes are diploid while prokaryotes are haploid
   (b) Prokaryotes have DNA located in their cytoplasm, while all the DNA in eukaryotic cells is located inside a membrane bound nucleus
   (c) Eukaryotes have linear chromosomes while most prokaryotes contain a single circular chromsome
   (d) Prokaryotes compact their DNA to a high degree while eukaryotes maintain the chromosomes in an uncompacted state
2. Which of the following would be a valuable application of a karyotype?
   (a) Microscopic analysis could allow identification of unusual cellular structures
   (b) An individual could be matched to a forensic sample based on identical karyotypes

(c) A specific gene which causes a genetic disorder could be identified

(d) Genetic disorders caused by chromosomal rearrangements could be diagnosed

3. A locus is:

(a) One version of a gene

(b) The location of a gene along the chromosome

(c) The character the gene effects

(d) The mechanism of action of a gene

4. Cellular division is important for which of the following processes?

(a) Growth of a multicellular organism

(b) Growth of a unicellular organism

(c) Reproduction of a unicellular organism

(d) A and C only

5. Which of the following is a correct description for mitosis?

(a) Cell division of prokaryotic cells

(b) A sorting mechanism which accompanies cell division in eukaryotic cells

(c) A mechanism which reduces chromosome number from diploid to haploid during gamete formation

(d) The process whereby a eukaryotic cell is created from a prokaryote

6. What is the major difference between sexual and asexual reproduction?

(a) Sexual reproduction does not require cell division

(b) Asexual reproduction cannot occur in diploid species

(c) Asexual reproduction cannot occur in eukaryotes

(d) Asexual reproduction produces offspring identical to the parent

7. What is a restriction point?

(a) A point at which a cell is prevented from proceeding through the cell cycle

(b) A point at which a cell has become committed to proceeding through the cell cycle to cell division

(c) A point at which the cell density in a population has become too high and none of the cells are allowed to divide

(d) None of these

8. What change occurs in chromosome structure between G1 and G2 phases of interphase?

(a) By G2, they have become more tightly condensed

(b) They begin to be more actively transcribed in G2 than G1

(c) Chromosomes in G2 contain two linear pieces of DNA while those in G1 have only one

(d) Chromosomes in G2 have a centromere which was not present in G1

9. Which of the following pairs of structures has the same amount of DNA?

(a) An unreplicated chromosome and a chromatid

(b) A replicated chromosome and a sister chromatid

(c) A replicated chromosome and a bivalent

(d) An unreplicated chromosome and a bivalent

10. Which of the following events does not occur during prophase?

(a) Condensation of chromosomes

(b) Formation of the mitotic spindle

(c) Duplication of the DNA

(d) Fragmentation of the nuclear membrane

11. Where do the spindle fibers connect to the chromosomes?

(a) To the centromere (b) To the kinetochore

(c) To the centriole (d) To the centrosomes

12. What is the characteristic feature of metaphase?

(a) The chromosomes are aligned in the middle of the cell

(b) The chromosomes have been duplicated and consist of sister chromatids

(c) Kinetochores are connected to kinetochore microtubules

(d) Sister chromatids have been separated

13. Which event(s) create the most critical difference between mitosis and meiosis?

(a) Separation of sister chromatids

(b) Formation of bivalents and crossing over

(c) Replication of chromosomes and connection to the spindle

(d) Formation of the spindle

14. During which phase of meiosis are homologous chromosomes separated?

(a) Metaphase I (b) Metaphase II

(c) Anaphase I (d) Anaphase II

15. Which meiotic division more closely resembles mitosis?

(a) Meiosis I

(b) Meiosis II

(c) Both resemble mitosis closely

(d) Both have significant differences from mitosis

16. How are meiosis and gametogenesis linked in plants?

(a) Meiosis produces haploid gametes directly

(b) Meiosis produces haploid gametes that are then modified by the sporophyte to produce functional gametes

(c) Meiosis produces a haploid cell that divides by mitosis to form a gametophyte, which then produces gametes by mitosis

(d) They are not linked in any way

17. Which structure of the plant becomes the part of an apple that you eat?

(a) The embryo sac (b) The ovule

(c) The ovary (d) The megaspore

18. During which phase of meiosis does the segregation described by Mendel primarily occur?

(a) Meiosis I

(b) Meiosis II

(c) Segregation occurs equally in meiosis I and meiosis II

(d) Segregation does not occur during meiosis

19. How might segregation of genes affected by crossing over differ from that of most genes?

(a) It occurs in meiosis I instead of meiosis II

(b) It occurs in meiosis II instead of meiosis I

(c) It occurs twice

(d) It does not occur at all

20. Not considering crossing over, what gamete is not possible from the following diploid individual? Listed are alleles of five genes; members of homologues are separated by a commA) Chromosome 1: ABC , abc; Chromosome 2: Qr , qr.

(a) ABCqr (b) abcqr

(c) ABcqr (d) abcQr

21. Which of the following is not a mechanism for sex determination?

(a) The Z-W system (b) The X-O system

(c) The haplo-diploid system (d) The parthenogenetic system

22. How could you verify that a gene you suspect to be X linked is actually located on the X chromosome?

(a) Set up reciprocal crosses

(b) Set up a two point test cross with another gene known to be located on the X

(c) Set up crosses between the F1 and both parental types

(d) A and B only

23. You set up a cross between true breeding wild type male fruit flies and true breeding female flies with a yellow body color. You notice in the F1 generation that half the flies, specifically all the males, have yellow body color while the females are wild type (gray). Which of the following explanations is the most likely?

(a) Some of your females had already mated with yellow males before you collected them

(b) Some of your parental flies must have had a new mutation in their germ cells

(c) The mutation which causes yellow body color is on the X chromosome

(d) The mutation which causes yellow body color is on the Y chromosome

24. During which phase of meiosis does independent assortment occur?

(a) Metaphase I (b) Anaphase I

(c) Metaphase II (d) Anaphase II

25. You have a white eyed, male fruit fly with vestigial wings and a wild type female known to be heterozygous for both traits. You know that vestigial is located on an autosome. What is the probability that a mating between these two flies will generate a female offspring with vestigial wings and white eyes?

(a) 1/16 (b) 1/8

(c) ¼ (d) ½

26. In the ladder anology of the DNA molecule, the "rungs" of the ladder are:

(a) Sugars (b) Phosphates

(c) Paired nitrogenous bases (d) Joined sugar and phosphate

27. Which of the following best describes the arrangement of the sides of the DNA molecule?

(a) Twisted (b) Antiparallel

(c) Bonded (d) Alternating

28. If a DNA molecule is found to be composed of 40% thymine, what percentage of guanine would be expected

(a) 10% (b) 20%

(c) 40% (d) 80%

29. The enzymes that break hydrogen bonds and unwind DNA are:

(a) Primers (b) Forks

(c) Helicases (d) Polymerases

30. DNA replication results in:

(a) 2 completely new DNA molecules

(b) 2 DNA molecules that each contain a strand of the original

(c) 1 new DNA molecule, 1 old molecule is conserved
(d) 1 new molecule of RNA

31. During replication, what enzyme adds complimentary bases?
(a) Helicase (b) Synthesase
(c) Replicase (d) Polymerase

32. Transcription results in:
(a) An amino acid chain (b) Messenger RNA
(c) Complementary DNA (d) Okazaki fragments

33. How many different amino acids are there?
(a) 3 (b) 20
(c) 100 (d) An infinite number

34. A stretch of chromosome that codes for a trait can be called a(n):
(a) Chromatid (b) Replication fork
(c) Gene (d) Base-pair

35. Translation begins:
(a) At the replication fork (b) On the lagging strand
(c) At the start codon (d) In the nucleus

36. A nucleotide consists of:
(a) A phosphate and a base
(b) A phosphate, and a sugar
(c) A base and an amino acid
(d) A phosphate, a sugar, and a base

37. Which of the following is found on RNA but not DNA?
(a) Uracil (b) Deoxyribose
(c) Phosphate (d) Adenine

38. Which of the following is NOT a necessary component of translation?
(a) Anticodon (b) mRNA
(c) Ligase (d) Amino acid

39. Given the following DNA strand, which of the following is its complementary mRNA? G G A C T G A T T

(a) C C T G A C T A A (b) C C U G A C U A A

(c) G G A C T G A T T (d) T T A G T C A G G

40. Amino acids are joined together into a protein chain by which of the following?

(a) Transfer RNA (b) DNA polymerase

(c) Hydrogen bonds (d) Messenger RNA

41. The base pair rules states that:

(a) Replication is semiconservative

(b) A pairs with T, G pairs with C

(c) DNA is a double helix held together by hydrogen bonds

(d) A pairs with G, T pairs with C

42. Once transcription has been completed, which of the following is NOT necessary for protein synethesis to occur?

(a) tRNA (b) Ribosomes

(c) mRNA (d) DNA

43. Which site of the tRNA molecule binds to the mRNA molecule?

(a) Anticodon (b) Codon

(c) Amino acid (d) 5 prime end

44. How many different codons are possible?

(b) 3 (b) 20

(c) 64 (d) An infinite number

45. DNA is a polymer of:

(a) Nucleosides

(b) Fatty acids

(c) Deoxyribose sugars connected by phosphodiester bonds

(d) Nucleotides

46. Which of the following are pyrimidines?

(a) Adenine and cytosine (b) Adenine and guanine

(c) Adenine and thymine (d) Cytosine and thymine

47. A nucleic acid was analyzed and found to contain 32 percent A, 18 percent G, 17 percent C, and 33 percent T. The nucleic acid must be:

(a) Single-stranded RNA (b) Single-stranded DNA

(c) Double-stranded RNA (d) Double-stranded DNA

48. A nucleic acid was analyzed and found to contain 37 percent A, 16 percent G, 22 percent C, and 25 percent T. The nucleic acid must be:

(a) Single-stranded RNA

(b) Single-stranded DNA

(c) Double-stranded RNA

(d) Double-stranded DNA

49. For double-stranded DNA, consider the following base ratios:

1. A/G
2. C/T
3. C/G
4. (A+C)/(G+T)
5. (A+G)/(C+T)
6. (A+T)/(G+C)

Which of those ratios always equals 1?

(a) 1 and 2 (b) 4 and 6

(c) 3, 4, and 5 (d) 1, 4, and 5

50. The two polynucleotide chains in DNA are:

(a) Semidiscontinuous (b) Semiconservative

(c) Parallel (d) Antiparallel

51. In the Meselson-Stahl experiment, which model of DNA replication was eliminated by the analysis of DNA isolated from bacteria one replication cycle after shifting from 15N to 14N medium?

(a) Semiconservative

(b) Conservative

(c) Dispersive

(d) All DNA was of intermediate density

52. In the semiconservative replication of DNA, progeny DNA molecules consist of:

(a) One-half of the molecules with two parental strands and one-half of the molecules with two new strands

(b) All molecules with interspersed parental and new segments

(c) All molecules with one parental and one new strand

(d) All molecules with two new strands

53. Which of these will complete the mRNA strand matched to DNA?

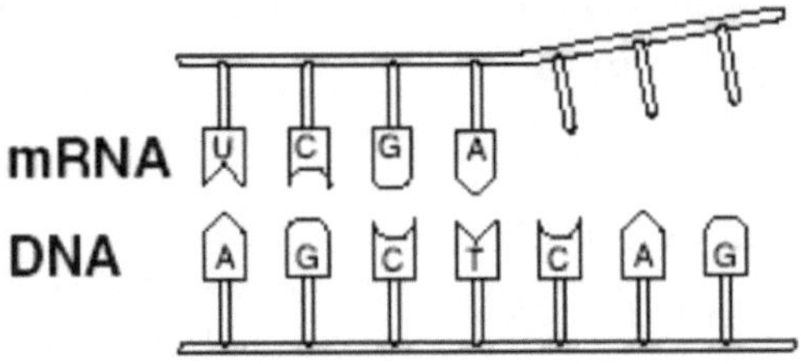

(a) AUG (b) GUC

(c) CAG (d) UAC

54. The parts of DNA that provide the code for proteins are the—

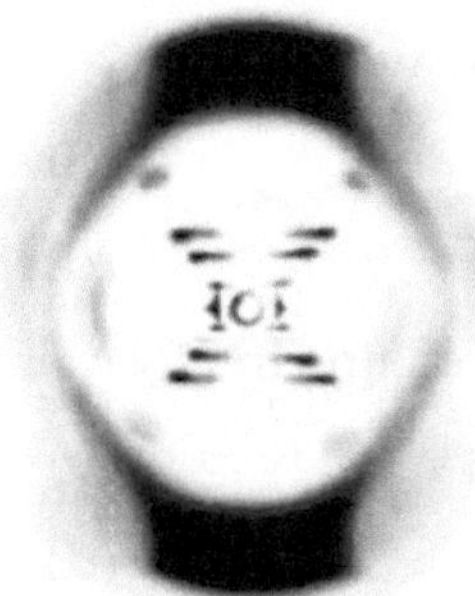

(a) Nitrogenous bases (b) Sugars

(c) Hydrogen bonds (d) Phosphates

55. The picture shows an x-ray diffraction of DNA. The x-ray diffraction of DNA led to the idea that DNA-

(a) Contains paired bases (b) Can copy itself

(c) Is a double helix (d) Is a very long molecule

56. The Human Genome Project was begun in 1988 by scientists from 13 nations as a worldwide effort to understand the sequencing of all of the DNA in the human body. What is one potential scientific benefit of this research?

(a) It will help explain cultural differences

(b) It helps to classify man most accurately in the animal kingdom

(c) It will help find the genes responsible for many diseases

(d) It will create communication between research centers

57. The most accurate way of determining the degree of relationship between species is probably to compare their –

(a) Fossil records (b) Body structures

(c) Stages of development (d) Nucleic acid sequences

58. Name this structure.

(a) Deoxynucleic acid (b) Deoxyribonucleic acid

(c) Denatured ribonucleic acid (d) Deoxoribonuclear acid

59. The process by which RNA is made from DNA is called:

(a) Synthesis (b) Translation

(c) Transcription (d) Replication

60. Adenine always pairs with:

(a) Cytosine (b) Thymine

(c) Guanine (d) Ribose

61. The rungs of the DNA ladder are made of

(a) Base pairs (b) Phosphates and hydrogen

(c) Glucose and sugars (d) Sugars and phosphates

62. The DNA molecule is held together by:

(a) Hydrogen bonds (b) Glucose

(c) Magnetism (d) Glue

63. The process by which a DNA molecule makes a copy of itself is called:

(a) Replication (b) Synthesis

(c) Transcription (d) Translation

64. A gene is:

(a) A segment of DNA that codes for a protein

(b) A set of homologous chromosomes

(c) A type of pants

(d) A molecule within the DNA

65. The twisted ladder shape of the DNA molecule is called a:

(a) Deoxyribose flip (b) Double helix

(c) Transcription (d) Hydrogen twist

66. The sugar found in the DNA molecule is:

(a) Equal (b) Phosphate

(c) Deoxyribose (d) Ribose

67. Which of the following takes the genetic code to the cytoplasm:

(a) Deoxyribose (b) DNA

(c) mRNA (d) tRNA

68. The three nucleotide sequence on RNA is called a:

(a) tRNA (b) Codon

(c) Triplet (d) Gene

69. Each 3 nucleotides codes for:

(a) 1 amino acid (b) 3 amino acids

(c) 1 protein (d) 3 proteins

70. The sugar found in RNA is

(a) Uracil (b) Deoxyribose

(c) Helicase (d) Ribose

71. DNA is called the "blueprint of life" because:

(a) It is like a fingerprint

(b) It has a blue color

(c) It contains the plans for building an organism

(d) It can relay messages to other molecules

72. The two men who discovered the structure of DNA were:

(a) Frederick and alvers (b) Watson and crick

(c) Berkeley and fry (d) Darwin and lamarke

73. Amino acids are the building blocks of:

(a) DNA and RNA (b) Lipids

(c) Proteins (d) Carbohydrates

74. The largest molecules in our bodies are:

(a) Nucleic acids (d) Chromosomes

(c) Proteins (d) Amino acids

75. Which of the following statements is true?

(a) Each DNA molecule is a single gene

(b) DNA molecules are in the shape of a double helix

(c) mRNA molecules are in the shape of a double helix

76. The human body is composed mostly of:

(a) Calcium, iron, phosphorus, and zinc

(b) Oxygen, hydrogen, calcium, and iron

(c) Oxygen, nitrogen, hydrogen, and carbon

77. The process of making RNA from DNA is:

(a) Transcription (b) Translation

(c) Conjunction (d). Peptide synthesis

78. The process of making proteins from RNA is:

(a) Transcription (b) Translation

(c) Conjunction (d) Peptide synthesis

79. Which of the following is NOT one of the bases in a DNA molecule?

(a) Adenine (b) Cytosine

(c) Uracil (d) Thymine

80. Cytosine will form a base pair only with:

(a) Cytosine (b) Adenine

(c) Thymine (d) Guanine

81. What is a nucleotide?

(a) A phosphate, five-carbon sugar, and a nitrogen base

(b) A group of linked amino acids

(c) One half of a DNA molecule

(d) A protein, a sugar, and a phosphate

82. The weak bonds between complementary nitrogen bases involve:

(a) RNA (b) DNA

(c) Hydrogen (d) Five-carbon sugars

83. DNA assumes a form termed what?

(a) A triple-linked chain (b) A six sided figure

(c) A step ladder (d) A double helix

84. DNA may be described functionally as:

(a) An enzyme

(b) A source for RNA production

(c) An arrangement of mRNA molecules

(d) A sink for disposal of RNA

85. Where is the messenger RNA synthesized in a cell?

(a) Nucleus

(b) Mitochondria

(c) Ribosome

(d) Endoplasmic reticulum

86. Which of the following correctly describes messenger RNA?

(a) Single stranded

(b) Uracil replaces thymine

(c) Many sugar to phosphate bonds occur

(d) All of the above

87. Which of the following are correct types of RNA?

(a) mRNA (n) qRNA

(c) tRNA (d) a and c

88. The directions used in protein synthesis are provided by:

(a) mRNA (b) Centromere

(c) RNA in the ribosomes (d) tRNA

89. The coding of amino acids by multiple sets of nucleotides is referred to as:

(a) Base pairing specificity

(b) Hydrogen bonding

(c) Chemical coding

(d) Triplet coding

90. What is the role of ribosomes in protein synthesis?

(a) They provide a source of amino acids

(b) They provide a site for transfer RNAs to link to messenger RNAs

(c) They translate the basic DNA code using transfer RNA

(d) They carry the proteins to their site of action

91. Which of the following is NOT a major grouping of proteins in cells?

(a) Enzymes (b) Antibodies

(c) Energy storage units (d) Structural units

92. Arrange the following proteins in the proper order in which they participate in DNA replication. 1 = Primase; 2 = Helicase; 3 = Single-strand binding proteins; 4 = DNA polymerase

(a) 1, 2, 3, 4 (b) 1, 3, 2, 4

(c) 2, 3, 1, 4 (d) 2, 3, 4, 1

(e) 2, 4, 3, 1

93. What is the function of the enzyme primase during DNA replication?

(a) To unwind the double helix to prime it for replication

(b) To prime DNA for replication by preventing the formation of knots

(c) To add a DNA primer to enable replication of DNA

(d) To add a RNA primer to enable replication of DNA

(e) To correct errors made during DNA replication

94. The short DNA segments formed on the discontinuously replicated strand are called

(a) Kawasaki fragments (b) Nagasaki fragments

(c) Miyazaki fragments (d) Okazaki fragments

(e) Suzuki fragments

95. Which of the following are limitations of DNA polymerase?

(a) It can only add bases to the exposed 5' end of a pre-existing strand

(b) It can only replicate the leading strand

(c) It can only replicate the lagging strand

(d) All of the above

(e) None of the above

96. Which of the following statements about DNA replication is TRUE?

(a) The leading strand is replicated continuously, while the lagging strand is replicated discontinuously

(b) The leading strand is replicated discontinuously, while the lagging strand is replicated continuously

(c) Both the leading and lagging strands are replicated continuously

(d) Both the leading and lagging strands are replicated discontinuously

97. Translation is the synthesis of

(a) mRNA from DNA (b) mRNA from proteins

(c) Proteins from DNA (d) Proteins from mRNA

98. Which of the following occurs as the ribosome shifts down the mRNA by a distance of three nucleotides?

(a) The tRNA that was in the A site moves into the E site

(b) The tRNA that was in the P site moves into the A site

(c) The tRNA that was in the E site moves into the P site

(d) The tRNA that was in the P site moves into the E site

99. Which of the following are limitations of DNA polymerase?

(a) It can only add bases to the exposed 5' end of a pre-existing strand

(b) It can only replicate the leading strand

(c) It can only replicate the lagging strand

(d) None of the above

100. Which of the following statements about DNA replication is TRUE?

(a) The leading strand is replicated continuously, while the lagging strand is replicated discontinuously

(b) The leading strand is replicated discontinuously, while the lagging strand is replicated continuously

(c) Both the leading and lagging strands are replicated continuously

(d) Each of the leading and lagging strands is replicated both continuously and discontinuously

101. The segment of the DNA molecule where messenger RNA synthesis begins is called the...

(a) Promoter region (b) Sigma factor

(c) Transcription terminator (d) Template

102. The complimentary messenger RNA strand that would be synthesized from the DNA base sequence of CTGAC would be...

(a) GACTG (b) UGACU

(c) AGTUG (d) GACUG

103. The name of the structure that causes the synthesis of RNA to cease is called the...

(a) Promoter region (b) Sigma factor

(c) Transcription terminator (d) Polymerase

104. Synthesis of mRNA is

(a) In the 5' to 3' direction with new nucleotides being added to the 5' end of the mRNA molecule

(b) In the 3' to 5' direction with new nucleotides being added to the 5' end of the mRNA molecule

(c) In the 5' to 3' direction with new nucleotides being added to the 3' end of the mRNA molecule

(d) In the 3' to 5' direction with new nucleotides being added to the 3' end of the mRNA molecule

105. Sigma factor is

(a) A subunit of mRNA

(b) A subunit of RNA polymerase

(c) A subunit of DNA

(d) An enzyme needed for transcription to proceed

106. Transcription begins when RNA polymerase binds to the

(a) Terminator on DNA (b) Initiator on DNA

(c) Promoter on DNA (d) Promoter on RNA

107. The bases of RNA are the same as those of DNA with the exception that RNA contains

(a) Cysteine instead of cytosine

(b) Uracil instead of thymine

(c) Cytosine instead of guanine

(d) Uracil instead of adenine

108. Which one of the following is not a type of RNA?

(a) nRNA (nuclear RNA) (b) mRNA (messenger RNA)

(c) rRNA (ribosomal RNA) (d) tRNA (transfer RNA)

109. Each amino acid in a protein is specified by

(a) Several genes (b) A promoter

(c) An mRNA molecule (d) A codon

110. The bases of RNA are the same as those of DNA with the exception that RNA contains

(a) Cysteine instead of cytosine

(b) Uracil instead of thymine

(c) Cytosine instead of guanine

(d) Uracil instead of adenine

111. Which one of the following is not a type of RNA?

(a) nRNA (nuclear RNA) (b) mRNA (messenger RNA)

(c) rRNA (ribosomal RNA) (d) tRNA (transfer RNA)

112. Each amino acid in a protein is specified by

(a) Several genes (b) A promoter

(c) An mRNA molecule (d) A codon

113. The TATA box in eukaryotes is a

(a) Core promoter (b) - 35 sequence

(c) - 10 sequence (d) 5' cap

114. The site where RNA polymerase attaches to the DNA molecule to start the formation of RNA is called a(n)

(a) Promoter (b) Exon

(c) Intron (d) GC hairpin

115. When mRNA leaves the cell's nucleus, it next becomes associated with

(a) Proteins (b) A ribosome

(c) tRNA (d) RNA polymerase

116. If an mRNA codon reads UAC, its complementary anticodon will be

(a) TUC (b) ATG

(c) AUG (d) CAG

117. The nucleotide sequences on DNA that actually have information encoding a sequence of amino acids are

(a) Introns (b) Exons

(c) UAA (d) UGA

118. Which of the following statements is correct about prokaryotic gene expression?

(a) Prokaryotic mRNAs must have introns spliced out

(b) Prokaryotic mRNAs are often translated before transcription is complete

(c) Prokaryotic mRNAs contain the transcript of only one gene

(d) All of these statements are correct

119. DNA indirectly directs the production of proteins based upon

(a) The order of its nitrogenous bases

(b) The order of its phosphate groups

(c) The order of its five-carbon sugars

(d) Currently unknown reasons

120. The molecule which acts as an intermediary between the DNA and the ribosomes is

(a) Messenger RNA (b) Transfer RNA

(c) Ribosomal RNA (d) Intermediate RNA

121. If a DNA strand reads TACGGCATG (where the names of the bases are abbreviated), then the RNA molecule created by transcription would read

(a) ATGCCGTAC (b) AUGCCGUAC

(c) CGTAATGCA (d) CGUAAUGCA

122. A promoter is best described as a

(a) Substance which facilitates RNA transcription

(b) Sequence of bases at which RNA polymerase begins transcription

(c) Type of RNA which promotes cell division

(d) Environmental change which cause transcription to occur

123. Which of the following is NOT a necessary component of translation?

(a) Anticodon (b) mRNA

(c) Ligase (d) Amino acid

124. Given the following DNA strand, which of the following is its complementary mRNA? G G A C T G A T T

(a) C C T G A C T A A (b) C C U G A C U A A

(c) G G A C T G A T T (d) T T A G T C A G G

125. Amino acids are joined together into a protein chain by which of the following?

(a) Transfer RNA (b) DNA polymerase

(c) Hydrogen bonds (d) Messenger RNA

126. The base pair rules states that:

(a) Replication is semiconservative

(b) A pairs with T, G pairs with C

(c) DNA is a double helix held together by hydrogen bonds

(d) A pairs with G, T pairs with C

127. Once transcription has been completed, which of the following is NOT necessary for protein synethesis to occur?

(a) tRNA (b) Ribosomes

(c) mRNA (d) DNA

128. Which site of the tRNA molecule binds to the mRNA molecule?

(a) Anticodon (b) Codon

(c) Amino acid (d) 5 prime end

129. How many different codons are possible?

(a) 3 (b) 20

(c) 64 (d) An infinite number

130. The building blocks of proteins are:

(a) Nucleotides (b) Amino acids

(c) Glucoseglycerol + fatty (d) Acidsvitamins + minerals

131. The building blocks of DNA and RNA are:

(a) Nucleotides (b) Amino acids

(c) Glucoseglycerol + fatty (d) Acidsvitamins + minerals

132. All building blocks of proteins contain these 2 chemical groups:

(a) Acid group + alcohol group

(b) Ester group + aldehyde group

(c) Ketone group + acid group

(d) amino group + acid group

133. Which of the following gives the correct base pairings in DNA?

(a) Thymine and uracil (b) Guanine and adenine

(c) Cytosine and thymine (d) Adenine and thymine

134. Which of the following are NOT part of a DNA nucleotide?

(a) Phosphate group (b) Amino group

(c) Nitrogenous basedeoxyribose (d) Sugar

135. Which of the following is a type of protein?

(a) Steroid (b) Phospholipid

(c) Enzyme (d) Starchtriglyceride

136. Which term best describes the DNA molecule?

(a) Single helix (b) Double helix

(c) Triple helix (d) Alpha helix

137. The genetic code carried by DNA lies in the order of the:

(a) Amino acids (b) Deoxyribose

(c) Sugars,phosphate groups, (d) Nitrogenous bases

138. The genetic code actually specifies the order of:

(a) Nitrogenous bases in a nucleic acid

(b) Deoxyribose molecules in a nucleic acid

(c) Glucose molecules in a carbohydrate

(d) Amino acids in a proteinnone of the above

139. The nucleotide containing guanine, phosphate, and deoxyribose is

(a) G (b) dGMP

(c) dG (d) GMP

140. The nitrogen-containing bases in RNA are

(a) A, G, T (b) A, G, C, U

(c) A, G, C, T (d) A, G, U

141. Thymine is found

(a) In RNA, but not DNA

(b) In both DNA and RNA

(c) In DNA, but not RNA

(d) In neither DNA nor RNA

142. A DNA molecule contains 44% cytosine. The percentage of guanine in the same molecule is

(a) 44% (b) 56%

(c) 28% (d) 100%

143. The complementary base sequence in the second strand of DNA for the base sequence CCGATT would be

(a) AATCGG (b) GGCTAA

(c) GGCUAA (d) CCGATT

144. A RNA molecule differs from a DNA molcule because

(a) RNA is a single strand of nucleotides

(b) RNA contains uracil

(c) RNA contains ribose

(d) All of the above

145. The mRNA in the cell

(a) Carries the genetic code from DNA to the ribosomes

(b) Carries amino acids to the ribosome for protein systhesis

(c) Combines with protein to form the ribosomes

(d) Is the smallest of the RNAs

146. What is the amino acid for the codons GUA-UGC-AAA?

(a) Val-Cys-Lys (b) et-Cys-Phe

(c) Ala-Arg-Lys (d) Val-Ser-Arg

147. An anticodon

(a) Is a sequence of three nucleotides on a tRNA

(b) Complements a mRNA codon for amino acids

(c) Is different for each of the amino acids

(d) All of the above

148. What amino acid would be carried by the tRNA with anticodon ACG?

(a) Cysteine (b) Threonine

(c) Arginine (d) Tryptophan

149. Recombinant DNA

(a) Contains a section of foreign DNA

(b) Is made from molecules of DNA called plasmids

(c) Can make many copies of the foreign DNA

(d) All of the above

150. When a mutation occurs

(a) A change occurs in the base sequence of DNA

(b) A change occurs in the sequence of amino acids in protein synthesis

(c) A genetic disease may result

(d) All of the above

151. In the operon model, control of protein synthesis is at the level of

(a) Translation (b) Translocation

(c) Replication (d) Transcription

152. In the enzyme induction model of regulation of protein systhesis,

(a) A substrate induces the synthesis of the enzymes for its utilization in the cell

(b) An end product turns off the synthesis of the enzymes needed for its use in the cell

(c) The mRNA that is produced cannot attached to the ribosome

(d) An active repressor turns off the operator

153. Viruses

(a) Contain DNA or RNA

(b) Needs a live host cell in order to reproduce

(c) Use viral RNA to construct DNA in a reverse transcription process.

(d) All of the above

154. The DNA that actually codes for a protein's primary structure in eukaryotes is:

(a) The entire gene

(b) Portions of the gene called introns

(c) Portions of the gene called exons

(d) The promoter

155. The synthesis of mRNA based on a DNA template is called:

(a) DNA replication (b) Transcription

(c) Translation (d) DNA restriction

156. Stem and loop structures are:

(a) Proteins that help partially denatured enzymes to recover their native configuration

(b) Structures in DNA caused by inverted repeats

(c) Structures at the ends of linear eukaryotic DNA molecules

(d) The bonds between adjacent DNA nucleotides in the same strand

157. During DNA replication in prokaryotes, synthesis begins on the circular chromosome:

(a) Always at the same place

(b) At any stretch of DNA that is high in AT pairs

(c) Randomly on the chromosome

(d) At the promoter

158. The sigma factor:

(a) Is responsible for recognizing the termination sequence

(b) Provides the catalytic function of cleaving off pyrophosphates during mRNA synthesis

(c) Recognizes the promotor

(d) Accomplishes DNA restriction.

159. Negative regulation of protein synthesis is accomplished by:

(a) Allosteric inhibition;

(b) The binding of RNA polymerase to the promoter

(c) The binding of a repressor to the DNA

(d) The binding of a repressor to the RNA polymerase

160. According to Chargaff's rule, the following proportion exists in DNA

(a) C=G (b) C»T

(c) C»G (d) C=T

161. The actual synthesis of DNA in *E. coli* is the function of

(a) Polymerase I (b) Primase

(c) Polymerase III (d) DNA ligase

162. The first nuclear transplant from an animal to an egg that produced a normal adult was performed on a

(a) Frog (b) Sheep

(c) Cat (d) Dog

163. All of the following were outcomes of Avery's experiment on "the active principle" except

(a) The array of the elements of purified principle agreed closely with DNA and the principle centrifuged to the same level as DNA

(b) Protein digesting enzymes did not affect the sample

(c) The extraction of lipids and proteins from the principle only slightly reduced its activity

(d) DNA-digesting enzyme, DNase, destroyed all transcription activity.

164. Polymerase III is actually a complex containing

(a) Catalytic subunits (b) Proofreading subunits

(c) "Sliding clamp" subunits (d) All of the above

165. The fact that some viruses use DNA to direct their heredity was demonstrated by

(a) Finding radioactive sulfur from a bacteriophage in a bacterium

(b) Finding radioactive phosphorus from a bacterium in a bacteriophage

(c) Finding that radioactive phosphorus from a bacteriophage had mutated in bacterium

(d) Finding radioactive phosphorus from a bacteriophage in a bacterium

166. Okazaki fragments are used to elongate

(a) The leading strand toward the replication fork

(b) The lagging strand toward the replication fork

(c) Both strands in both directions

(d) The lagging strand away from the replication fork

167. A single enzyme is specified by a single

(a) Chromosome (b) Gene

(c) Nucleotide (d) Nucleosome

168. Hammerling chose Acetabularia as his model organism because it was

(a) Large and differentiated

(b) Easier to grow than fruit flies

(c) Large and reproduced only asexually

(d) Grown abundantly in the Mediterranean area where he lived

169. In DNA guanine always pairs with

(a) Adenine (b) Cytosine

(c) Guanine (d) Thymine

170. With few exceptions, all nuclei of eukaryotes contain

(a) Genes to specify the portion of the organism in which they are found

(b) All the information needed for growing the whole organism

(c) All of the chromosomes except sex chromosomes which are restricted to sex organs

(d) Single stranded DNA

171. In DNA, thymine always pairs with

(a) Adenine (b) Cytosine

(c) Guanine (d) Thymine

172. Each individual zone of a chromosome replicates as a discrete section called

(a) Telomere (b) Euchromatin

(c) Okazaki fragment (d) Replication unit

173. Each unit of a nucleic acid consisting of a sugar, attached phosphate group, and base is a

(a) Nucleolus (b) Nucleotide

(c) Nucleosome (d) Histone

174. In replication of DNA, the helix is opened and untwisted by

(a) Ribase (b) Ligase

(c) Deoxase (d) Helicase

175. Although the work produced important results, Franklin's work was hindered by the lack of

(a) Electricity (b) Strong x-rays

(c) True crystals of DNA (d) Thin layer chromatography

176. The identification of the transforming principle proved that

(a) Dead bacteria are as lethal as live ones

(b) Streptococcus pneumoniae evolved from Pneumococcus

(c) Genetic information is contained in DNA

(d) There is no protein in chromosomes

177. For double-stranded DNA, which of the following is NOT necessarily true?

(a) A + G = T + C (b) A:T = 1

(c) A + T = C + G (d) C:G = 1

178. Of the following strands of DNA, in combination with their complement in the form of a double helix, three undergo a hyperchromic (DNA denaturing) shift with a Tm (Melting temperature) of about 42°C) One has a Tm of 52°C) Which one?

(a) TCATGCGATC (b) CTTAAATTTG

(c) TAACTACGAA (d) AAATTTGGGA

179. What were the final two pieces of data that were crucial to Watson and Crick's proposed structure of DNA?

(a) Base composition analysis and hydrogen bonding

(b) X-ray diffraction and the nucleoside structure

(c) X-ray diffraction and base composition analysis

(d) The discovery of the four nucleotides and nucleoside structure

180. Nucleotides are composed of:

(a) A nitrogenous base and a pentose sugar

(b) A nitrogenous base and a phosphate group

(c) A pentose sugar and two nitrogenous bases

(d) A nitrogenous base, a pentose sugar, and a phosphate group

181. The nitrogenous base is attached to which carbon of the pentose sugar?

(a) 5' (b) 3'

(c) 1' (d) 4'

182. Adenosine, cytidine, guanosine, and uridine are examples of:

(a) Deoxyribonucleotides (b) Phosphate groups

(c) Ribonucleotides (d) Ribonucleosides

183. The difference between a ribose and a deoxyribose sugar is:

(a) The hydroxyl group at the 3' carbon of the sugar

(b) The hydroxyl group at the 5' carbon of the sugar

(c) The phosphate group at the 5' carbon of the sugar

(d) The hydroxyl group at the 2' carbon of the sugar

184. Which of the following DNA sequences (only one strand is provided, you need to make the second strand) will be more difficult to separate into single stranded molecules?

(a) 5' GGCGGTACGCGTAGG 3'

(b) 5' AGCAAATATCGGTTCC 3'

(c) 5' AATCTAGGCTTCATG 3'

(d) 5' TTATATACGGCGTAG 3'

185. What is the complementary strand of the DNA molecule 5' AACGGGTTTATGCGT 3' ?

(a) 5' TTGCCCAAATACGCA 3'

(b) 3' AACGGGTTTATGCGT 5'

(c) 5' ACGCATAAACCCGTT 3'

(d) Both C and D are complementary strands

186. What is the complementary strand of the following RNA strand: 5' AUGCGUUUACCGA 3' ?

(a) 3' UACGCAAUGGCU 5'

(b) 3' AUGCGUUUACCGA 3'

(c) 3' AGCCAUUUGCGUA 5'

(d) 3' TACGCAAATGGCT 5

187. The genetic material must show variation. Which of the following is NOT an example of genetic variation?

(a) New mutations can occur during meiosis

(b) Different species have different numbers of chromosomes

(c) Different individuals within a species have different phenotypes

(d) Individuals of different species may have similar phenotypes

188. Why might heat-killed bacteria be useful as a vaccine?

(a) It can cause a lethal infection

(b) Heat degradation of proteins changes their shape

(c) Molecules from the cell surface are still intact and can provoke an immune response

(d) DNA molecules can transform other strains of bacteria

189. What is the key difference between 'rough' and 'smooth' S. pneumoniae?

(a) Rough bacteria can evade the immune system of a mouse

(b) Rough bacteria have more genes than smooth bacteria

(c) Smooth bacteria have more genes than rough bacteria

(d) Smooth bacteria are able to synthesize a polysaccharide capsule

190. What occurred in the transformation observed by Griffith?

(a) A rough strain underwent mutation to convert it to a smooth strain

(b) A smooth strain underwent mutation to convert it to a rough strain

(c) A rough strain passed genetic information to a smooth strain

(d) A smooth strain passed genetic information to a rough strain

191. Why was the DNase treatment used by Avery, MacLeod, and McCarty an important step?

(a) This allowed them to isolate pure DNA samples

(b) This allowed them to isolate pure protein samples

(c) This allowed them to demonstrate that removing the DNA prevents transformation

(d) This allowed them to demonstrate that mixing rough cells with DNA prevents transformation

192. Which of the following statements about Griffith's type IIR strain of bacteria is incorrect?

(a) It cannot make a polysaccharide capsule

(b) It must be competent to pick up DNA from the environment

(c) It contains a mutation in a gene that would allow synthesis of a type III capsule

(d) All of these are correct statements

193. What key characteristic of T2 bacteriophage allowed Hershey and Chase to use it in their studies of the genetic material?

(a) Its genes encode proteins that assemble to produce the viral coat

(b) It injects its genetic material into a bacterial cell

(c) It can undergo either the lytic or lysogenic life cycle

(d) It enters the bacterial cell to cause infection

194. Which of the following is a key component of a DNA nucleotide?

(a) Nitrogenous base (b) Deoxyribose sugar

(c) Phosphate group (d) All of these

195. What is the difference between purines and pyrimidines?

(a) Purines are found in RNA while pyrimidines are found in DNA

(b) Pyrimidines are found in RNA while purines are found in DNA

(c) Purines contain a double ring backbone, while pyrimidines contain a single ring backbone

(d) Pyrimidines contain a double ring backbone while purines contain a single ring backbone

196. What type of bond connects nucleotides?

(a) Phosphodiester (b) Hydrogen

(c) Ionic (d) A and B

197. Which of the following statements about the structure of DNA is incorrect?

(a) One complete turn requires 3.4nm and 10 base pairs

(b) The backbones of each strand run in opposite directions relative to each other

(c) Each pair of nucleotides is held together by three hydrogen bonds

(d) The width of the molecule is a constant 2nm

198. Which of the following is a line of evidence that indicates triplex DNA may exist *in vivo*?

(a) Triplex DNA can be synthesized in the lab

(b) Triplex DNA follows certain base pairing rules

(c) Triplex DNA is recognized by certain proteins present in cells

(d) A and B

199. What functional role is the three dimensional structure of DNA likely to accomplish?

(a) Very long molecules of DNA can be packed into a small space

(b) Information within DNA is made available to drive cellular functions

(c) Information within DNA is modified from generation to generation

(d) All of these

200. In studying a virus, you find the following proportions of nitrogenous bases present: adenine 23%, guanine 31%, cytosine 23%, unexpected base 17%. What statement(s) can you make about this virus?

(a) It probably uses RNA as its genetic material

(b) The genetic material of this virus is probably single stranded

(c) Base pairing rules in this virus probably include adenine:cytosine.

(d) A and B

201. Which statement is false?

(a) RNA and DNA are both composed of nucleotides joined by phosphodiester bonds

(b) RNA and DNA both have secondary and tertiary structure

(c) RNA is always single stranded while DNA is always double stranded

(d) A and B are both false

202. UV light damages DNA by causing

(a) Covalent bonds to form between two cytosines next to each other on the same DNA strand

(b) Covalent bonds to form between thymine nucleotides next to each other on the same DNA strand

(c) Covalent bonds to form between two thymine nucleotides on opposing DNA strands

(d) Covalent bonds to form between a thymine and an adenine on the same DNA strand

203. When DNA is damaged by UV light and is not repaired

(a) DNA replication can proceed as normal but transcription is prevented

(b) Transcription can proceed as normal but replication is prevented

(c) Neither DNA replication nor transcription can occur and the organism will probably die

(d) The damaged part will be excised and allow DNA replication to proceed

204. In light repair

(a) The covalent bonds between the thymine dimers are broken

(b) A single thymine is excised

(c) The damaged segment of DNA is excised by DNA polymerase

(d) The damaged segment of DNA is excised by DNA ligase

205. Consider the following DNA sequence: TACGGCCGTACGGC) Which of the sequences below would be produced by a nonsense mutation?

(a) TACGCCGTACGGC (b) TACGGTCCGTACGGC

(c) TACCATGCCGGGGC (d) TACGGCCGTTCGGC

206. Consider the following DNA sequence: TACGGCCCATTATC) Which of the sequences below would be produced by an inversion?

(a) TACGCCATTATC (b) TACGGGCCCATTATC

(c) TACACCCGGTTATC (d) TACTGCCCATTATC

207. Consider the following DNA sequence: TACGGCCCATTATC) Which of the sequences below would be produced by an insertion?

(a) TACGCCATTATC (b) TACGGGCCCATTATC

(c) TACACCCGGTTATC (d) TACTGCCCATTATC

208. Consider the following DNA sequence: TACGGCCCATTATC) Which of the sequences below would be produced by a transition?

(a) TACGCCATTATC (b) TACGGGCCCATTATC

(c) TACACCCGGTTATC (d) TACAGCCCATTATC

209. A(n) _____ test is a test of allelic functionality.

(a) Suppression (b) Complementation

(c) Ames (d) Functionality

210. Seymour Benzer coined the word ___ for the minimal mutable site.

(a) Cistron (b) Muton

(c) Recon (d) Nucleotide

211. Traits a and b are recessive leg mutations in a hypothetical organism. When the two mutants are mated, each offspring has legs with wild-type characteristics, not mutant. We can say that these two mutations

(a) Complement and are therefore allelic

(b) Do not complement and are therefore allelic

(c) Complement and are therefore not allelic

(d) Do not complement and are therefore not allelic

212. Which statement about the elongation of the lagging strand during DNA replication is correct?

(a) It is synthesized in a 3' —> 5' direction

(b) It progresses (grows) toward the replication fork

(c) It requires a short RNA primer to proceed

(d) It is synthesized by DNA ligase.

213. An actively dividing bacterial culture is grown in a medium containing radioactive adenine (A*). After all the adenine is labeled, the bacteria are transferred to a medium containing nonradioactive adenine (A). Following one round of DNA replication in the nonradioactive medium, the DNA is analyzeD) Which of the following sequences could represent this DNA?

(a) A* A* T T G A* T C
T T A A C T A G

(b) A* A T T G A* T C
T T A* A* C T A G

(c) A A T T G A T C
T T A A C T A G

(d) A* A* T T G A* T C
T T A* A* C T A* G

214. DNA is made up of a phosphate group, an organic base, and:

(a) A protein
(b) A sugar
(c) A molecule of ATP
(d) A fat

215. In the DNA molecule:

(a) Adenine pairs with thymine
(b) Guanine pairs with thymine
(c) Cytosine pairs with thymine
(d) Adenine pairs with cytosine

216. If one side of a DNA molecule contains the following sequence of nucleotides, AGTCCG, the complementary sequence on the other side would be: TCAGGC

(a) GCCTGA
(b) AGTCCG
(c) TCAGGC
(d) CTGAAT

217. During your summer job at Virotech, you isolate a previously unknown virus. Analysis of its genome reveals that it is composed of a double stranded DNA molecule containing 14% T (thymine). Based on this information, what would you predict the %C (cytosine) to be?

(a) 14%
(b) 28%
(c) 36%
(d) 72%

218. Wave lengths in the electromagnetic spectrum that are mutagenic are absorbed by

(a) DNA
(b) Protein
(c) Both DNA and protein
(d) Neither DNA nor protein

219. Which of the following DNA base sequences is complementary to this base sequence: ACGGATTAG

(a) TGCCTAATC
(b) GATTGCCGT
(c) UGCCUAAUC
(d) GAUUGCCGT

220. Which of the following would represent a mutation?

(a) A change in the base sequence of DNA
(b) Crossing over in meiosis

(c) Both (a) and (b)

(d) Neither (a) nor (b)

221. A change in which of the following would result in a mutation?

(a) Order of bases in DNA (b) Gene

(c) Chromosome (d) All of the above

222. A sequence of bases is found in some type of nucleic acid The sequence is AUUCCG. The nucleic acid in question must be

(a) DNA

(b) mRNA

(c) tRNA

(d) It could be either mRNA or tRNA

223. Which of the following does NOT occur in DNA replication? The original double helix

(a) Unwinds

(b) Bonds between base pairs are broken

(c) The unpaired strands split into codons

(d) All occur

224. Some point mutations are called silent mutations, i.e., the mutation does not change the amino acid after the altered mRNA is translated. This is possible because the genetic code is

(a) Universal

(b) Degenerate

(c) Both (a) and (b) are possible explanations

(d) Neither (a) nor (b) are possible explanations

225. If the arrangement of bases on DNA is CATTAG, then a corresponding strand of mRNA will be

(a) GTAATC (b) TGCCGA

(c) GUAAUC (d) UGCCGA

226. The chemical instructions in DNA are due to

(a) The way the bases are paired

(b) The order of the bases in DNA

(c) The arrangement of the sugar and phosphate molecules
(d) The codons in mRNA

227. The role of mRNA is to
(a) Carry the genetic instructions to the cytoplasm
(b) Identify and transport amino acids
(c) Provide nucleotides for RNA production
(d) Produce amino acids for protein

228. Transcription refers to
(a) DNA replication
(b) Protein synthesis using mRNA and tRNA
(c) mRNA synthesis from DNA
(d) Converting an inactive protein into its active form

229. Which of the following molecules has an amino acid attached to it?
(a) DNA (b) tRNA
(c) mRNA (d) All of the above

230. Codons may
(a) Act as start codons (b) Act as stop codons
(c) Code for a specific amino acid (d) Do all of the above

231. When a DNA molecule replicates, the resulting DNA molecules are correctly described by which of the following?
(a) Entirely new strands of DNA are made
(b) Each strand of the DNA molecule contains some old DNA and some new DNA
(c) Each new DNA molecule contains one old strand and one new strand
(d) One DNA molecule contains two old strands; the other has two new strands

232. How many amino acids are coded for in this mRNA sequence: GGUACCUUUACU
(a) 12 (b) 6
(c) 4 (d) 1

233. Thirty percent of the bases in DNA extracted from a prokaryotic cell is adenine. What percentage of cytosine is present in this DNA?

(a) 10 (b) 20

(c) 30 (d) 40

234. All cells of a species (except for the sex cells) contain the same amount of DNA) However, the cells of multicellular organisms vary in both structure and function from one tissue type to another. This specialization is evidence that most of the DNA in a cell is

(a) Inactive

(b) Removed as the cell matures

(c) Converted into protein

(d) Converted into glycoprotein

235. Which of the following is the role of the structural genes in an inducible enzyme system?

(a) To code for proteins that metabolize an inducer substance

(b) To produce a repressor that binds to the operator

(c) To stop the transcription of the operator

(d) To recognize and form a complex with the inducible enzyme

236. Substances that cause cells to increase production of an enzyme (or group of enzymes) are know as

(a) Introns (b) Repressors

(c) Inducers (d) Promoters

237. An operon is most closely associated with

(a) DNA replication (b) Point mutations

(c) Translation (d) Transcription

238. It is true that:

(a) Transcription begins at the first AUG following a ribosome binding site or Shine d'Algarno sequence

(b) Transcription begins at a stem-loop structure followed by poly-U

(c) Translation begins at a promoter

(d) Translation ends at a termination codon

239. In yeast, a eukaryotic microorganism

(a) Transcription and translation are coupled

(b) Attenuation regulates the synthesis of a number of amino acids

(c) The 5′Cap in mRNAs is recognized by the translation apparatus

(d) A poly-A tail is added to the 3' end of mRNAs in the cytoplasm

240. You wish to know whether during infection, the expression of the genes of a pathogenic dsDNA bacteriophage is accomplished by a virally encoded enzyme. It would be profitable to treat the infected cells with:

(a) The β-subunit inhibitor streptolydigin

(b) Chain terminating dideoxynucleotides

(c) The chain terminator azidothymidine

(d) (b) and (c) only

241. The nucleic acid of a virus has a 2' -OH, and its number of purines equals the number of pyrimidines. Which of these nucleic acids is the virus most likely to have as its genome?

(a) Plus-strand RNA (b) Minus-strand RNA

(c) Double-stranded RNA (d) Single-stranded DNA

242. Restriction enzymes

(a) Were discovered during study of bacterial resistance to some phage infections

(b) May leave "sticky" or "blunt" ends in DNA

(c) Are used to map DNA

(d) All of the above

243. Which is the following is true of the production of lytic bacteriophages?

(a) They can be quantitated by the plaque assay

(b) Large numbers of virions are recovered if the infected cells are opened during the eclipse phase

(c) The host decides the levels of viral mRNAs that are made throughout the infection

(d) Assembly of the viral genome and coat proteins is assisted by the host's proteins

244. An important observation initially made during study of bacteriophages is

(a) The production of large viral polyproteins

(b) The use of t-RNA as a primer for DNA replication

(c) That the entry of viral nucleic acid into a host is sufficient for the production of complete virions

(d) That the virions may contain the enzymes required to add a 5'-Cap to their transcripts

245. Which one of these statements concerning transposition of DNA is false?

(a) It produces short direct repeats in the DNA of the host

(b) It occurs only in bacteriophages

(c) It can mutate genes in the host genome

(d) It was discovered several decades ago during study of maize genetics

246. Some molecules discussed in class for controlling HIV replication are

(a) Nucleic acid chain terminators such as azidothymidine (AZT)

(b) Protease inhibitors

(c) Antisense oligonucleotides

(d) (a), (b), and (c)

247. Which of the experimental strategies below constitutes a screen?

(a) Penicillin enrichment of *E. coli* auxotrophs

(b) Growth of *E. coli* harboring the plasmid pBR322 on plates containing the antibiotic tetracycline

(c) Growth of *E. coli* lysogenic for lambda in the presence of large excess of lambda virions

(d) Detection of *lac*+ (blue) colonies of *E. coli* on plates containing the chromogenic substrate X-gal and the inducer IPTG

248. A complementation experiment requires

(a) A screen or selection

(b) A vector with an origin of replication

(c) Exchange of the defective host genes with the wild-type DNA being introduced

(d) (a) and (b)

249. The structural shape of DNA is, of course, the famous double helix. What is a "double helix"?

(a) Two Y-shaped strand

(b) Two loops, like a figure eight

(c) Two spirals, like a twisty ladder

(d) Two X-shaped strands

250. DNA is constructed of repeating units. What are these units called?

(a) Amino acids (b) Nucleotides

(c) Monosaccharides (d) Fatty acids

251. In DNA, to what base does adenine bond?

(a) Other adenines (b) Guanine

(c) Thymine (d) Cytosine

252. What bases are called purines?

(a) Thymine & cytosine (b) Adenine & thymine

(c) Adenine & guanine (d) Guanine & cytosine

253. To whom is the discovery of the structure of DNA credited?

(a) Gregor Mendel (b) Charles Darwin

(c) Watson & Crick (d) Alexander Fleming

254. What part of DNA is found at the centre of the molecule?

(a) Amino Acids (b) Sugars

(c) Bases (d) Phosphates

255. Weak hydrogen bonds are found between which molecules?

(a) Between amino acids

(b) Between bases

(c) Between sugars and phosphates

(d) Between purines

256. What process is used to copy DNA to produce another identical strand?

(a) Replication (b) Translation

(c) Transcription (d) Photocopying

257. To what molecule is the nitrogen base attached in a nucleotide?

(a) Sugar (b) Helix

(c) Phosphate (d) Lipid

258. What kind of sugar is found in DNA?

(a) Ribose (b) Deoxyribose

(c) Galactose (d) Glucose

259. The B-DNA structure found in solution is a

(a) Right-handed double helix of antiparallel chains (11 bp/turn)

(b) Left-handed double helix of antiparallel chains (~10 bp/turn)

(c) Right-handed double helix of antiparallel chains (~10 bp/turn)

(d) Left-handed zig-zag helix of antiparallel chains (12 bp/turn)

260. DNA differs from RNA in the following features

(a) DNA is resistant to base catalyzed hydrolysis; RNA is hydrolyzed by OH-

(b) DNA residues are linked by 3' —>5' phosphodiester bonds; RNA is 2' —>5' linked

(c) DNA has deoxyribose residues; RNA has ribose residues

(d) All but the second choice are correct differences

261. Because DNA is a highly charged polyanion, its stability to heat denaturation ("melting"):

(a) Does not depend on hydrophobic interactions

(b) Increases with increasing [salt]

(c) Decreases with increasing [salt]

(d) Is independent of G + C content

262. The total contour length of DNA in a human cell is about

(a) 1 mm. (b) 1 mm.

(c) 10 cm (d) 1 meter

263. Polyacrylamide and agarose gel electrophoresis separate nucleic acids based primarily on their

(a) Length (b) Ratio of mass/charge

(c) (G+C)/(A+T) content (d) Organismal origin

264. The number of supercoils in a covalently-closed, circular DNA can only be changed if

(a) At least one of the phosphodiester chains is cleaved

(b) Both of the phosphodiester chains are cleaved

(c) Histones are bound to the DNA

(d) The salt concentration is increased >1.0 M

265. Thymidine

(a) Is equal to the adenosine concentration in double-stranded DNA

(b) Is replaced by Uracil in RNA

(c) Normally forms 2 hydrogen bonds with adenosine.

(d) All of the above are correct

266. Replication in *E. coli* is initiated by the generation of short RNA primers using

(a) RNA polymerase (b) primase

(c) DNA polymerase II (d) Reverse transcriptase

267. The RNA primers used to initiate replication in *E. coli*

(a) Result in Okazaki fragments on the leading strand

(b) Are removed by Pol I "nick translation"

(c) Are joined together by DNA ligase

(d) Are removed by helicase + ATP

268. The exonuclease activities of DNA polymerase I

(a) Occur at two different active sites

(b) Are coupled during the formation of "gapped DNA"

(c) Degrade DNA in a 5' –> 3' direction

(d) Degrade DNA in a 3' –> 5' direction

269. The *E. coli* chromosome has 5x10^6 bp; a replication fork progresses at about 1000 nucleotides/sec. Therefore, the minimum time required to complete replication is about

(a) 10 min. (b) 20 min.

(c) 40 min. (d) 80 min.

270. Which feature of DNA replication in eukaryotes is different from that in prokaryotes?

(a) Not inhibited by Actinomycine

(b) More than two DNA polymerases are found

(c) RNA primers are not required

(d) Multiple origins of replication are used

271. An operon is a transcriptional unit in bacteria that contains

(a) A promoter site, an operator site, and one or more regulatory genes

(b) A promoter site, an operator site, and two or more structural genes

(c) cis-acting elements adjacent to trans-acting factors

(d) RNA polymerase loading zones

272. Consider a "radio analogy" for *E. coli* lac operon transcription; then

(a) The promoter site and startsite are the volume control

(b) CAP-cAMP is the on/off switch; the repressor is the volume control

(c) RNA polymerase is both the on/off switch and volume control

(d) The repressor is the on/off switch; CAP-cAMP is the volume control

273. The RNA polymerases that transcribe bacterial DNA or eukaryotic nuclear DNA are

(a) Multisubunit enzymes and partially homologous

(b) Monomeric and very large

(c) Multimeric and interchangeable

(d) Only active inside the cell

274. RNA chain elongation can be inhibited by

(a) Repressors

(b) Rifampicin (a Rifamycin derivative)

(c) Actinomycin

(d) RNases

275. DNA binding by proteins with the helix-turn-helix (HTH) motif does not involve

(a) Interactions with base pairs in the major groove of DNA

(b) Interactions with the sugar-phosphate backbone of DNA

(c) Hydrogen bonds, salt bridges, and vander Waals contacts

(d) Melting of the DNA at the center of symmetry

276. A promoter is

(a) A manager for a sports team

(b) A specific sequence of DNA to which RNA polymerase binds

(c) A specific sequence of DNA to which a catabolic repressor binds

(d) A specific DNA sequence to which a restriction endonuclease binds

277. The binding of lac repressor to DNA is analogous to

(a) Competitive inhibition of an enzyme

(b) Allosteric effects in enzyme regulation

(c) Uncompetitive inhibition of an enzyme

(d) Mixed-type inhibition of an enzyme

278. The enhancer sites of eukaryotic promoters

(a) Are binding sites for activating transcription factors

(b) An act at distances of <1000 base pairs

(c) Are found in eukaryotic viruses

(d) All choices are correct

279. mRNA in eukaryotes is similar to that in prokaryotes as follows

(a) The 5' ends of the primary transcript are triphosphorylated

(b) The 5' ends become capped with 7-Methyl G

(c) A polyA tail is added to the 3' end

(d) The RNAs are spliced to form mature mRNA

280. Which of the following base analogs or chemicals does not cause transition mutations?

(a) 2-Aminopurine (2AP) (b) 5-Bromouracil (5BU)

(c) Ethidium bromide (EtBr) (d) Nitrous acid (HNO2)

281. Degenerate codons differ mostly in

(a) the identities of their first base

(b) Choices three and four are both correct

(c) The identities of their third base

(d) The wobble position

282. The tertiary structure of yeast tRNAPhe,

(a) Is maintained mostly by non-Watson-Crick base pairing

(b) Involves extensive base stacking interactions

(c) Resembles the 3-dimensional structure of other tRNAs

(d) All of the choices are correct

283. The junction between the two helical portions of tRNA is stabilized by many non-standard base-base interactions between residues in the

(a) Helical stems of the D and Anticodon stems

(b) Anticodon and CCA ends

(c) Loops of the Anticodon and Acceptor stem

(d) Loops of the TYC and D arms

284. The accurate charging of tRNAIle by IleRS with Ile and not Val (50,000:1)

(a) Requires a two-step editing process

(b) Requires the binding of Ile to be 50,000-fold tighter than Val

(c) Requires the hydrolysis of 2 ATPs for each Ile charged

(d) Requires the presence of a small protein, fidelity-factor

285. Hydrolytic editing of incorrectly charged tRNAs

(a) Involves attack of PPi on the aminoacyl-AMP intermediate

(b) Occurs at a second active site on the synthetase

(c) Is a waste of ATP, that could be used for other cellular functions

(d) Is a necessary step for all of the aminoacyl-tRNA synthetases

286. Most of the identity elements for correct aminoacylation of tRNAs are found

(a) In the anticodon stem and the 3' CCA end

(b) In the T Loop and the D stem and loop

(c) In the anticodon and the acceptor stem

(d) Solely in the anticodon

287. Ribosomes select the correct tRNAs

(a) And then charge them using the appropriate synthetase

(b) Then bind to the appropriate mRNA

(c) Solely on the basis of their anticodons

(d) With the least abundant anticodons

288. The pathway of a tRNA during polypeptide elongation on the ribosome is

(a) A site —> P site —> E site

(b) P site —> A site —> E site

(c) A site —> P site —> entry site

(d) P site —> entry site —> exit site

289. Which of the following antibiotics is also an inhibitor of peptidyl transferase?

(a) Rifampicin (b) Tetracycline

(c) Ampicilin (d) Chloramphenicol

290. The peptidyl transferase reaction is catalyzed

(a) Solely by nucleotides in 23S RNA

(b) Solely by nucleotides in 16S RNA

(c) Only by 16S and 23S RNA working together

(d) By amino acid side chains of proteins, L1 and L11

291. This is an image of a _____.

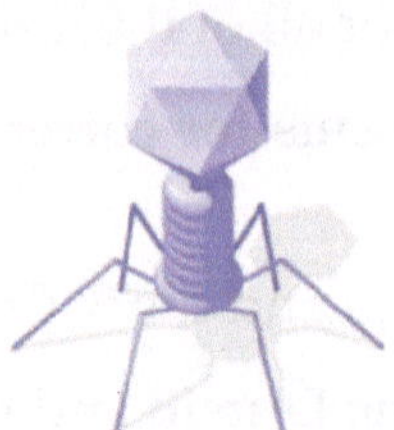

(a) Bacterium (b) Red blood cell

(c) Phage (d) Moneran

292. Who demonstrated that DNA is the genetic material of the T2 phage?

(a) Franklin (b) Hershey and Chase

(c) Meselson and Stahl (d) Watson and Crick

293. In the accompanying image, a nucleotide is indicated by the letter _____.

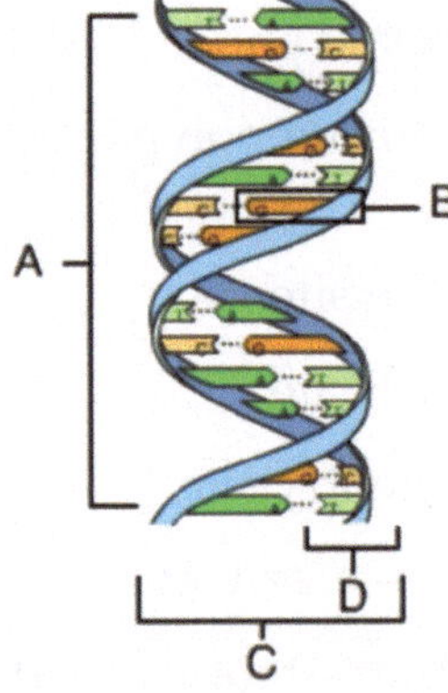

(a) A (b) B

(c) C (d) D

294. Which of these is a difference between a DNA and an RNA molecule?

(a) DNA contains uracil, whereas RNA contains thymine

(b) DNA is a polymer composed of nucleotides, whereas RNA is a polymer composed of nucleic acids

(c) DNA is double-stranded, whereas RNA is single-stranded

(d) DNA contains five-carbon sugars, whereas RNA contains six-carbon sugars

295. This is an image of a(n) _____.

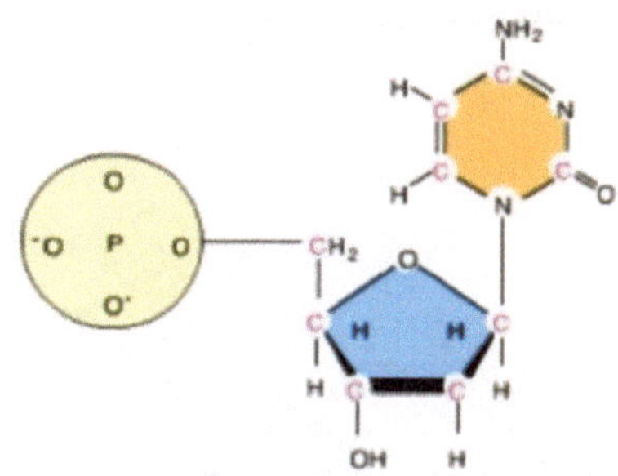

(a) Amino acid (b) Nucleic acid

(c) Thiol (d) Nucleotide

296. The letter A indicates a _____.

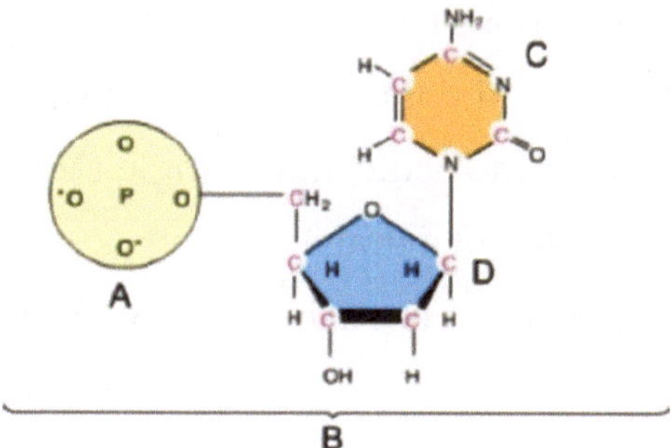

(a) Phosphate group (b) Nucleotide

(c) Nitrogenous base (d) Sugar

297. A nitrogenous base is indicated by the letter _____.

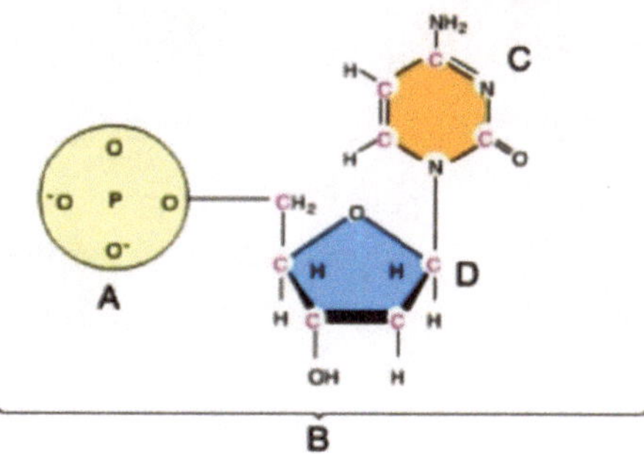

(a) A (b) B

(c) C (d) D

298. You can tell that this is an image of a DNA nucleotide and not an RNA nucleotide because you see a _____.

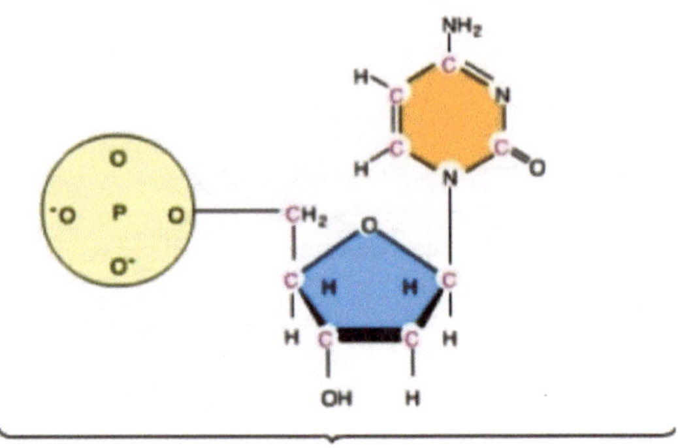

(a) Thymine nitrogenous base, not a uracil nitrogenous base

(b) Sugar with two, and not three, oxygen atoms

(c) Double-stranded molecule, not a single-stranded molecule

(d) Phosphate group, not a uracil

299. Which of these nitrogenous bases is found in DNA but not in RNA?

(a) Adenine (b) Cytosine

(c) Guanine (d) Thymine

300. Which of these is(are) pyrimidines?

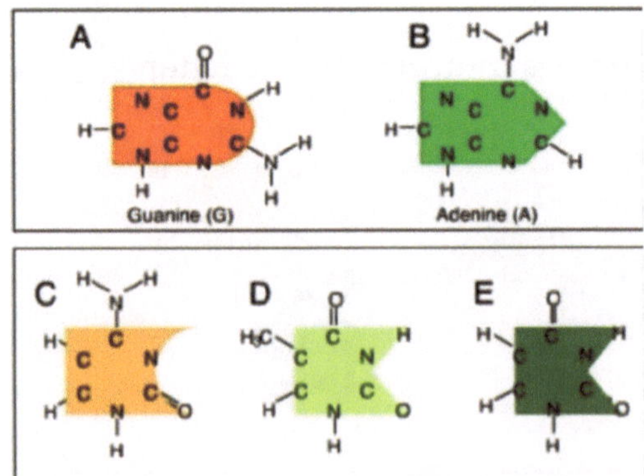

(a) C, D, and E (b) A, B, and C

(c) B and C (d) B, C, and D

301. Why is the new DNA strand complementary to the 3' to 5' strands assembled in short segments?

(a) The replication forks block the formation of longer strands

(b) DNA polymerase can assemble DNA only in the 3' to 5' direction

(c) DNA polymerase can assemble DNA only in the 5' to 3' direction

(d) It is more efficient than assembling complete new strands

302. The synthesis of a new strand begins with the synthesis of a(n) _____.

(a) Single-strand binding protein

(b) Okazaki fragment

(c) Poly(A) tail

(d) RNA primer complementary to a preexisting DNA strand

303. Which of these is responsible for catalyzing the formation of an RNA primer?

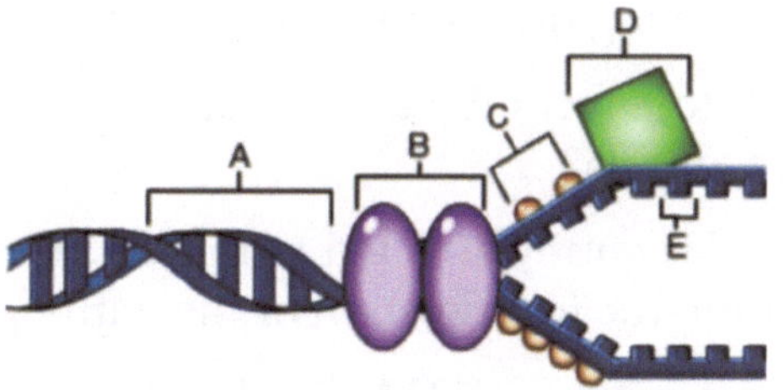

(a) A (b) B

(c) C (d) D

304. Which of these is an example of cytogenetic mapping?

(a) Hybridization studies in Drosophila determined the location of the yellow gene, which influences body color, to be near the tip of the X chromosome

(b) Karyotype analysis of humans with familial Down syndrome has identified a relatively small region of chromosome 21 containing genes that cause the Down phenotype

(c) Analysis of the chromosomes of a transgenic strain of mice indicates that the transgene has been inserted into the short arm of chromosome 15

(d) All of these

305. Cytologists can use which of the following to describe locations of a gene at a specific place on the chromosome?

(a) DNA sequence of a chromosomal region

(b) Protein expression from a chromosomal region

(c) Banding pattern of stained chromosomes

(d) Comparison to markers located within a few thousand bp

306. Cytogenetic mapping:

(a) Requires that the gene have been cloned if in situ hybridization is to be used

(b) Is a high resolution method of gene localization

(c) Can be used to determine relative order of genes located very close to each other

(d) All of these

307. Which of the following questions could be easily answered by FISH?

(a) How often does recombination occur between the ebony and sepia loci in Drosophila?

(b) In a particular family affected with Becker muscular dystrophy, is the mutation due to a deletion of the entire gene?

(c) How large a piece of DNA would you need to clone in order to clone the group of human globin genes from chromosome 11?

(d) All of these.

308. Which of these is a key characteristic of a molecular marker?

(a) It is a known gene

(b) It is located at a known site on the chromosome

(c) It is only useful for linkage and physical mapping studies

(d) None of these

309. A molecular marker which is amplified by PCR and is polymorphic by length is a(n):

(a) Restriction fragment length polymorphism (RFLP)

(b) Variable number of tandem repeats site (VNTR)

(c) Amplified fragment length polymorphism (AFLP)

(d) Single nucleotide polymorphism (SNP)

310. A polymorphism is:

(a) Any change in the DNA sequence

(b) The most common variation of a gene or marker sequence

(c) The least common variation of a gene or marker sequence

(d) A variation of gene or marker sequence present in >1% of the population

311. Which of these statements regarding RFLP analysis is correct?

(a) RFLP analysis requires Southern blotting for detection of fragments

(b) RFLPs can identify single base pair changes at any site in the chromosome

(c) An RFLP typically produces several different alleles

(d) All of these are correct

312. A monomorphic DNA segment is:

(a) A segment of DNA that exists in many forms in the population

(b) A segment of DNA that controls a single gene function

(c) A segment of DNA inherited in a dominant fashion

(d) A segment of DNA shared by over 99% of the population

313. Linkage mapping can determine the distance between which of the following pairs of DNA sequences?

(a) AFLPs and RFLPs (b) Two AFLPs

(c) Two known genes (d) All of these

314. What was the arrangement of the markers in the original parental strains?

(a) A: 5,000bp, B: 4,300bp and A: 3,350bp, B: 2,300/2000bp

(b) A: 5,000bp, B: 2,300/2,000bp and A: 3,350bp, B: 4,300bp

(c) A: 3,350bp, B: 4,300bp and A: 5,000bp, B: 4,300bp

(d) A: 3,350bp, B: 2,300/2000bp and A: 3,350bp, B: 2,300/2000bp

315. What is the distance between these two markers?

(a) 0.167mu (b) 0.20mu

(c) 16.7mu (d) 20mu

316. Which of the following would be a reasonable use of an RFLP map?

(a) Identification of the exact location of an unknown gene along the chromosome

(b) Description of the size of DNA fragment required to clone the region containing an unknown gene

(c) Identification of the region in which an unknown gene is located

(d) All of these

317. Which of these genetic markers is most likely to be highly polymorphic (have many different alleles)?

(a) An RFLP

(b) A microsatellite

(c) An SNP

(d) All of these are equally polymorphic

(e) None of these are likely to be polymorphic

318. Why might use of microsatellites in genetic mapping studies be an advantage over RFLPs?

(a) Microsatellites are easier to detect

(b) Microsatellites are more abundant than RFLPs

(c) Microsatellites have more potential alleles than RFLPs

(d) All of these

319. How are individual chromosomes identified in chromosome sorting techniques?

(a) Relative level of fluorescence when stained with a dye mixture

(b) Level of charge relative to size

(c) Size of the molecule

(d) All of these

320. Which of these describes a contig?

(a) A complete genomic library including overlapping clones

(b) A complete mRNA library

(c) A chromosome-specific library of overlapping clones

(d) An ordered genomic library

321. What vector would be best suited for creating a contig of bovine (cattle) chromosome 10?

(a) λ? (b) A plasmid

(c) A YAC) (d) Any of these

322. Which of the following would not be a critical characteristic of a YAC vector?

(a) Telomeric sequences

(b) A gene encoding a required structural protein

(c) An origin of replication

(d) A centromere

323. After cytogenetic and linkage mapping, you have identified a relatively small area of a chromosome where your gene of interest is likely to reside. You decide to attempt to use a YAC contig to map the gene by positional cloning. How might you identify the clone that may contain the gene?

(a) The clone will also contain the nearest marker mapped by linkage analysis

(b) The DNA sequence of the clone will contain an ORF (open reading frame)

(c) Cytological hybridization of the clone produces a different pattern than that of the nearest mapped marker

(d) All of these

324. Which of these is a reasonable use for the contigs and clones generated in the process of obtaining a physical map of the human genome?

(a) Correlation of physical and linkage map distances to determine recombination frequency

(b) Identification of all the known alleles of a single disease-causing gene

(c) Insertion into cells affected with genetic disease in the process of gene therapy

(d) All of these

325. Sequencing of genomes other than humans is potentially valuable because:

(a) Disorders of cellular function can be studied in relatively simple model systems

(b) Comparison of gene sequences between different species can allow prediction of disease-causing mutations

(c) Genes may have a similar function in other species, giving us a place to start with functional analysis of similar genes

(d) None of these

326. The transforming agent was identified as DNA by

(a) Griffith. (b) Avery

(c) Hershey and Chase (d) Watson and Crick

327. If a DNA molecule contains 20% A, approximately what percentage of G is present?

(a) 20% (b) 40%

(c) 30% (d) 60%

328. DNA replication is

(a) Semiconservative (b) Conservative

(c) Dispersive (d) Semidispersive

329. In the Meselson and Stahl experiment, what percentage of the DNA is labeled with light nitrogen after three generations of replication.

(a) 25% (b) 50%

(c) 75% (d) 100%

330. Enzymes called ____ catalyze the formation of covalent bonds that hold together the sugar-phosphate backbone of the DNA molecule.

(a) Helicases (b) DNA polymerases

(c) RNA polymerases (d) Ligases

331. Which strand below would make the sequence 5' AAACGCTT 3' a double stranded DNA molecule?

(a) 5' TTTGCGAA 3'
(b) 5' UUUGCGUU 3'
(c) 5' AAGCGUUU 3'
(d) 5' AAGCGTTT 3'

332. How many DNA bases make up the human genome?

(a) 3 thousand
(b) 3 million
(c) 3 billion
(d) 30 billion

333. Considering DNA replication along the 5' to 3' template strand, what event takes place first?

(a) DNA polymerase begins adding RNA nucleotides
(b) DNA polymerase begins adding DNA nucleotides
(c) Primase adds an RNA primer
(d) Primase removes the 5' phosphate

334. If Hershey and Chase found S35 in progeny phages rather than P32, their experiment would have demonstrated that

(a) Proteins contain phosphorus
(b) DNA contains sulfur
(c) Phage DNA enters the host cell
(d) Phage protein enters the host cell

335. DNA structure was proposed by

(a) Griffith
(b) Chargaff
(c) Avery
(d) Watson and Crick

336. Energy is required to break the hydrogen bonds holding the bases together. Which pair will be the most difficult to separate?

(a) A - T
(b) C - G
(c) A - C
(d) G - T

337. If a nucleic acid is found to contain 20% A and 20% T, the molecule is probably

(a) Single-stranded DNA
(b) Double-stranded DNA
(c) Single-stranded RNA
(d) Double-stranded RNA

338. Is it possible to use PCR to produce many copies of all DNA of one chromosome?

(a) Yes, if the correct primers are available

(b) Yes, if the telomeres are present

(c) No, PCR cannot copy DNA, only RNA

(d) No, PCR copies short sequences of DNA only

339. If DNA polymerase could add bases in the 3' to 5' direction, there would be no need for

(a) DNA ligase (b) Okazaki fragments

(c) helicase (d) PCR

340. Which of the following is not a component of a chromosome?

(a) Histone proteins

(b) DNA polymerase

(c) RNA

(d) Chromosomes contain all of the above

341. Place in order the following steps involved in PCR:(1) newly synthesized strands act as templates(2) temperature lowered; DNA primers and polymerase added(3) heat separates strands of target DNA(4) complementary base pairing between primers and template(5) DNA nucleotide bases added; new strand synthesized

(a) 1 - 2 - 3 - 4 (b) 3 - 5 - 4 - 2 - 3 - 1

(c) 3 - 2 - 4 - 5 - 3 - 1 (d) 2 - 3 - 4 - 5 - 2 - 1

342. Clones are identified by hybridizing them with

(a) A vector (b) An antibody

(c) A virus (d) A probe

343. The first step in the polymerase chain reaction (PCR) is

(a) Denaturation (b) Primer extension

(c) Annealing (d) Cooling

344. With the completion of the human genome project, the next frontier is

(a) Nucleonics (b) Proteomics

(c) Cytomics (d) Agrinomics

345. By using the "reference sequence" developed in the human genome project, individual differences are now being detected using

(a) NPs (b) PSNs

(c) SNPs (d) "Snaps"

346. Viruses containing RNA rather than DNA are called

(a) Riboviruses (b) Immunoviruses

(c) Bacteriophages (d) Retroviruses

347. In a polymerase chain reaction, a synthetic sequence of nucleotides are involved in

(a) Denaturing (b) Heating

(c) Priming (d) All of the above

348. Bacteria protect themselves from viruses by fragmenting viral DNA upon entry with

(a) Ligases (b) Endonucleases

(c) Methylases (d) Probes

349. In preliminary screening of clones, it is common to use

(a) Restriction enzymes (b) Dyes

(c) Antibiotics (d) Milipore filters

350. To identify an individual by DNA analysis of their blood, investigators look for

(a) Primers (b) DNA fingerprints

(c) Probes (d) Transgenic fragments

351. All fragments cut by most restriction endonucleases have

(a) Complementary double-stranded ends

(b) Supplementary single-stranded ends

(c) Double-stranded "sticky" ends

(d) Complementary single-stranded ends

352. In 1980, interferon was produced by splicing a human gene into the genome of

(a) Bacteria (b) Yeast

(c) Viruses (d) Mice

353. The second step in most genetic engineering experiments is

(a) Screening

(b) Production of recombinant DNA

(c) Cleavage of DNA

(d) Cloning

354. When "sticky ends" are paired, they can be joined by

(a) Restriction enzymes (b) pSC101

(c) Methylase (d) DNA ligase

355. A successful vector in genetic engineering has been the

(a) TMV plasmid (b) HLF virus

(c) Vaccinia virus (d) Ti plasmid

356. Humans are not affected by glyphosphate in plant crops because

(a) While it is toxic to plants it is a normal metabolite of humans

(b) Humans do not produce aromatic amino acids

(c) It is used only on those crops that convert it to a harmless substance

(d) Humans lack the enzymes to interact with it

357. In the screening process, clones that metabolize X-gal turn

(a) Yellow (b) Orange

(c) Red (d) Blue

358. A powerful way to identify an individual using a particular gene as a marker is the analysis of

(a) RFLP's (b) X-gal reaction

(c) PCR's (d) EcoRI's

359. In attempts to confer special characteristics upon plants, genetic engineers find Agrobacterium tumefaciens to be an effective vector for use with

(a) Corn (b) Rice

(c) Wheat (d) Soy beans

360. A library of DNA fragments results from the use of

(a) Restriction endonucleases (b) Virus

(c) Plasmids (d) Recombinant DNA

361. One of the most useful methods for identifying a specific gene is

(a) Thin layer chromatography (b) The Eastern blot

(c) The Western blot (d) The Southern blot

362. Plant cells receiving Ti plasmids with a TMV viral gene encoding the viral coat protein become immune to the TMV virus because

(a) The introduction of the viral gene triggered an immune response in the plant's own immune system

(b) The gene causes the cell walls to become impermeable to the virus

(c) TMV does not infect cells already infected with TMV

(d) In plants, the gene encodes an antiviral substance

363. A procedure called PCR is used to

(a) Cleave DNA (b) Produce recombinant DNA

(c) Copy gene sequences (d) Clone cells

364. Wilt-proof flowers are being engineered by making them insensitive to

(a) Methylene (b) Propylene

(c) Butylene (d) Ethylene

365. The ingestion of BST in milk is no threat to humans because

(a) It is destroyed in the stomach

(b) It cannot be digested and will pass through as waste

(c) They do not have the hormones to utilize it

(d) It is destroyed in the pasteurization process

366. Genetic engineering has successfully transferred genes from eukaryotic cells into:

(a) Bacteria (b) Plants

(c) Animals (d) All of the above

367. DNA molecules can be cut into sections by using:

(a) ATP (b) Gel electrophoresis

(c) Restriction endonucleases (d) Plasmids

368. In genetic engineering, DNA ligase is used as:

(a) A probe
(b) A sealing enzyme
(c) A restriction enzyme
(d) A mutagen
(e) None of the above

369. In Cohen and Boyer's transfer of toad genes to a bacterial cell, restriction endonucleases were used to:

(a) Isolate sections of cloned bacterial plasmids
(b) Isolate sections of toad DNA
(c) Cleave bacterial plasmids
(d) Both b and c are correct

370. A probe is used in which stage of the gene transfer process?

(a) Cleaving DNA
(b) Recombining DNA
(c) Cloning
(d) Screening

371. The Polymerase Chain Reaction is used to:

(a) Amplify a small amount of DNA
(b) Cleave bacterial plasmids
(c) Seal "sticky ends"
(d) Identify target plasmids

372. Which of the following statements is true about developing cDNA?

(a) Mature mRNA directs the formation of the DNA
(b) Mature mRNA does not contain introns
(c) DNA taken from the nucleus is used to produce the cDNA
(d) Both a and b are true

373. Genetic engineering has been used to do all of the following except:

(a) Make plants more resistant to frost
(b) Make plants more resistant to disease
(c) Make plants more resistant to herbicides
(d) All of the above are correct

374. Engineering plants that are resistant to glyphosate was an important advancement because:

(a) Glyphosate promotes frost damage

(b) Glyphosate encourages the production of fruit that is lower in protein

(c) Glyphosate is the active ingredient in herbicide

(d) Glyphosate prevents the transfer of genes into the plants

375. The use of bovine growth hormone:

(a) Has raised controversy with groups opposed to genetic engineering

(b) Increased milk production in cows

(c) Is being tested on increasing the weight gain of cows

(d) All of the above

376. Genetically identical organisms derived from a single genetic source are called

(a) Populations (b) Varieties

(c) Sibling species (d) Clones

377. Which of the following is an advantage of single-primer PCR?

(a) Less synthetic DNA is incorporated into the product

(b) Unknown sequence adjacent to a known sequence can be amplified with a single primer

(c) No primer-dimer artifacts

378. Which end of a PCR primer can be extended by polymerase?

(a) The 1' end (b) The 2' end

(c) The 3' end (d) The 4' end

379. Which nuclease activity does Taq polymerase NOT have?

(a) 5' to 3' (b) 3' to 5'

380. Which of the following significantly inhibits PCR?

(a) 12% sucrose

(b) Wooden toothpicks made of birch

(c) 0.2mM cresol red

381. What is the minimum time required for denaturation in rapid-cycle PCR?

(a) One minute (b) 30 seconds

(c) Less than one second

382. The primary importance of bacteria to mankind is

(a) Production of secondary metabolites

(b) Production of enzymes

(c) Production of chemicals

(d) For food

383. DNA (Deoxyribonucleic acid) are

(a) Blocks of protein

(b) Large organic macromolecule

(c) Synthetic acid

(d) Answer not above

384. DNA performs the following functions except

(a) Transmission of information from one cell generation to another

(b) Provide information for protein synthesis

(c) Help in digestion of protein

(d) Answer not above

385. The structure of a DNA is

(a) Sphere (b) Double helix

(c) Rod (d) Circle

386. The function of the messenger RNA is to

(a) Enable protein synthesis

(b) Provide information for repairing damages cells

(c) Enable removal of used cells

(d) Answer not above

387. The messenger RNA is produced in the

(a) Genes (b) Blood

(c) DNA (d) Bone marrows

388. How many enzymes have been identified in the body

(a) 100 (b) 15,000

(c) 3,000 (d) 6,000,000

389. Proteases are used for the following except

(a) Chillproofing of beer

(b) Biomedical applications

(c) Monitoring serum cholesterol levels

(d) Detergents

390. Hemicellulases are used for the following except

(a) Baking (b) Chillproofing of beer

(c) Brewing (d) Animal feedstuffs

391. The major threat to Africa's agricultural productivity growth

(a) Land degradation and poverty

(b) Illiteracy

(c) Riots and war

(d) Fiscal policies

392. Biotechnology started 600 years back with

(a) Use of tractors

(b) Fermentation of beer

(c) Importation of foreign materials

(d) It did not start 600 years ago

393. Pollen grain from Biotechnologically corn affects

(a) Earthworm (b) Frog

(c) Monarch butterfly caterpillars (d) Snails

394. Genetically modified potatoes as being noticed to affect which of the following animals

(a) Lizards (b) Poultry

(c) Rat (d) Pigs

395. Genes are exchanged between plants via

(a) Air (b) Pollen tube

(c) Anther (d) Filaments

396. Bonds holding the protein molecules together are

(a) Polypeptide bonds (b) Peptide bonds

(c) Chemical bonds (d) Answer not above

397. How many building blocks is the amino acid?

(a) 80 (b) 40

(c) 10 (d) 20

398. Thermal stability of lysosome has being increased with the concept of

(a) DNA science (b) Protein engineering

(c) Modern gene technology (d) Answer not above

399. Protein engineering helps in which of the following areas

(a) Improved binding of small molecules to protein receptors on cell surfaces

(b) Increasing drought resistance in plants

(c) Improving fertility of soils

(d) Answer not above

400. Proteins performs which of the following functions

(a) Energy for daily activities

(b) To refurbish dead cells/tissues

(c) Clearer vision

(d) Stronger bones

401. Who discovered that genes was located on chromosomes

(a) Thomas Hunt Morgan (b) Louis Pasteur

(c) Gregor Mendel (d) James Beal

402. The first experimental hybrid corn was developed in

(a) 1879 (b) 1820

(c) 1967 (d) 1904

403. Pasteurization was developed in

(a) 1861 (b) 1890

(c) 1209 (d) 2004

404. In 1500AD Aztecs made cakes from

(a) Spirulina algae (b) Bacteria

(c) Yeast (d) Answer not above

405. Banting, Best, Collip and MacLeod discovered what?

(a) DNA (b) Protein

(c) Genes (d) Insulin

406. In 1943 the organism used to show that DNA carries the cell's genetic information was a

(a) Fungi (b) Algea

(c) Bacteria (d) Human being

407. The hematopoietic stem cell was discovered in

(a) 1961 (b) 1903

(c) 1814 (d) 4000AD

408. The First release of genetically engineered microbes in field experiments was in

(a) 1987 (b) 1904

(c) 40 AD (d) 2004

409. Dairy farming developed as far back as

(a) 1803 (b) 4000AD

(c) 1143 (d) 2004

410. Please select a statement that is FALSE about DNA.

(a) DNA stands for deoxyribonucleic acid

(b) DNA from a mother is identical to the DNA of her child

(c) DNA of a person would remain the same with age

(d) DNA is a blueprint of life that exists in every organism

411. What is the difference between DNA and DNA profile?

(a) They two terms are the same: both are the structure that carries the information that makes up who we are

(b) DNA is a structure in our body while DNA profile is consisted of many structures of DNA

(c) DNA is a structure in our body while DNA profile is a report resulted from DNA testing that carries information that distinguishes who we are.

(d) I don't know

412. PCR is a technique used to extract DNA from blood samples.

(a) The statement above is true

(b) The statement above is false

(c) I don't know

(d) I don't know and I have never heard of PCR before

413. Which of the following statements is true about STR?

(a) STR stands for short thymine replication

(b) It is used with RFLP to extract DNA from samples

(c) STR is a test to determine whether the sample contains DNA

(d) STR is a repetition of short fragments of DNA

414. DNA fingerprinting can be applied in which of the following fields?

A. Field 1: Criminology B. Field 2: Anthropology

C. Field 3: Biochemistry

(a) Field 1 (b) Both Field 1 and 2 .

(c) Both Field 1 and 3. (d) All Fields

415. Which of the three models of DNA replication proposed that the original double helix remained entirely intact throughout the process?

(a) Conservative model (b) Semiconservative model

(c) Dispersive model (d) Destructive model

416. Which statement about bacterial DNA replication is correct?

(a) DNA replication begins at several places along the chromosome

(b) DNA replication begins at the origin and travels in both directions

(c) DNA replication begins at the origin and travels around the chromosome back to the origin

(d) DNA replication begins at a GC rich region of the chromosome

417. Why is an AT rich region part of the *E. coli* oric?

(a) DnaA proteins bind to AT rich sites

(b) DnaC proteins bind to AT rich sites

(c) AT base pairs are more easily separated than GC base pairs

(d) AT base pairs contain more hydrogen bonds than GC base pairs

418. Which enzyme produces the first piece of nucleic acid synthesized in *E. coli* DNA replication?

(a) DNA primase (b) DNA polymerase (polIII)

(c) DNA polymerase (polI) (d) DNA ligase

419. What is a key difference between DNA polIII and DNA ligase?

(a) Only DNA polIII synthesizes phosphoester bonds

(b) Only DNA ligase synthesizes phosphoester bonds

(c) DNA polIII can synthesize DNA from 3'-5'

(d) DNA ligase can use energy from ATP rather than nucleotides

420. What is the major difference between the lagging and leading strands?

(a) On the leading strand, DNA synthesis occurs from 5' to 3', while DNA synthesis occurs from 3' to 5' on the lagging strand

(b) DNA polymerase is able to continuously add new nucleotides on the leading strand while it must keep 'starting over' on the lagging strand

(c) The lagging strand requires only a single primer while the leading strand requires many

(d) Helicase opens the leading strand at a faster rate than the lagging strand

421. What explanation for Okazaki fragments is the most accurate?

(a) DNA polymerase is only able to synthesize 1000-2000 bonds before it falls off the strand

(b) DNA polymerase must stop to check its work every 1000-2000 nucleotides

(c) DNA polymerase requires a free 3' OH group to attach new nucleotides; primers are produced on the lagging strand every 1000-2000 nucleotides

(d) All of these are accurate statements

(e) None of these are accurate statements

422. What DNA sequence is required for termination of replication in *E. coli*?

(a) oric (b) ter

(c) Tus (d) Fork

423. What is the major difference between T1 and T2 ter sequences?

(a) Tus binds to T1 but not T2

(b) Tus binds to T2 but not T1

(c) T1 stops only counterclockwise-moving forks while T2 stops only clockwise-moving forks

(d) T2 stops only counterclockwise-moving forks while T1 stops only clockwise-moving forks

424. What enzyme performs decatenation?

(a) Polymerase (b) Topoisomerase

(c) Telomerase (d) Decatenase

425. Which is a correct description of the replisome?

(a) The replisome is a complex of several proteins

(b) The replisome includes DNA polymerase, DNA helicase, and primase

(c) The replisome can be found at the replication fork during DNA synthesis

(d) All of these are correct descriptions

426. Which of the following is a description of the proofreading function of DNA polymerase?

(a) Endonuclease cleavage

(b) Exonuclease cleavage

(c) Methylation

(d) An induced-fit phenomenon

427. Which of the following is not a mechanism by which *E. coli* can coordinate DNA replication with cell division?

(a) DnaA protein does not efficiently bind to hemimethylated binding sites in the oric

(b) DNA polymerase does not efficiently bind to mismatched base pairs

(c) DnaA protein is rapidly degraded after the initiation of DNA replication

(d) DnaA proteins must bind to a minimum number of binding sites in the oric

428. Which of the following is a key difference between eukaryotic and prokaryotic DNA replication?

(a) Eukaryotic origins of replication contain a high % of AT base pairs and require protein binding for correct function

(b) In eukaryotes, DNA replication proceeds bidirectionally from the origin of replication

(c) Eukaryotes contain several different DNA polymerases

(d) Telomerase is required to replicate the ends of eukaryotic chromosomes

429. What aspect of DNA polymerase function necessitates use of telomerase in eukaryotes?

(a) 3'-5' exonuclease function

(b) 5'-3' processivity

(c) Induced fit impact on incorporation of only correct bases

(d) All of these

430. Telomerase activity in humans is generally restricted to the germ line, cancerous cells, and a few specific adult cell types. How would this be expected to impact humans?

(a) Cancer cells can divide indefinitely

(b) Because there is a difference in number of divisions, sperm would be expected to have shorter telomeres than oocytes

(c) Rapidly dividing populations of cells have telomeres equal in size to slowly dividing populations of cells

(d) All of these

431. DNA replication begins at a specific site called the

(a) Initiator (b) Start codon

(c) Origin (d) Promoter

432. In DNA replication

(a) Both strands replicate in the same direction

(b) Each strand replicates in a different direction

(c) Only one strand of DNA is used as a template

(d) A single strand of DNA is copied to make two single strands of DNA

433. Semiconservative replication of DNA means

(a) Only one strand is used as a template

(b) A double-stranded DNA is split into two single-stranded DNAs

(c) Only half the genes are copied into the new cells

(d) Each DNA made contains one old strand and one new strand of DNA

434. DNA is unwound by

(a) DNA polymerase III (b) RNA primase

(c) DNA polymerase I (d) Helicase

435. DNA polymerase can add nucleotides to

(a) A free 3' OH end of a growing DNA chain

(b) A free 5' OH end of a growing DNA chain

(c) Either a 3' or a 5' end of a DNA chain

(d) A 5' end of an RNA primer

436. The leading strand of DNA is synthesized

(a) Discontinuously in a 5' to 3' direction

(b) Continuously in a 5' to 3' direction

(c) Discontinuously in a 3' to 5' direction

(d) Continuously in a 3' to 5' direction

437. Why is an RNA primer considered essential during DNA synthesis by DNA polymerase III?

(a) The enzyme requires a free 3'-PO4 group

(b) The enzyme requires a free 5'-PO4 group

(c) The enzyme requires a free 5'-OH group

(d) The enzyme requires a free 3'-OH group

438. Which molecule serves to destabilize the DNA helix in order to open it up, creating a replication fork?

(a) DNA helicase (b) DNA ligase

(c) DNA polymerase (d) SSBPs

439. In the Meselson-Stahl experiment, which mode of replication can be eliminated based on data derived after one generation of replication?

(a) Dispersive (b) Semiconservative

(c) Conservative (d) All three modes

440. What is the complementary strand of the following RNA strand: 5' GCACGUUUACCGA 3' ?

(a) 3' CGUGCAAUGGCU 5' (b) 3' AUGCGUUUACCGA 3'

(c) 3' TACGCAAATGGCT 5' (d) 3' AGCCAUUUGCGUA 5'

441. *In vivo* synthesis of DNA is

(a) 5' to 3'. (b) 3' to 5'.

(c) 5' to 2'. (d) both a and b

442. The discovery of Okazaki fragments suggested that DNA synthesis is sometimes

(a) Continuous (b) Semiconservative

(c) Discontinuous (d) 3' to 5'

443. What activity of DNA polymerase I (pol I of Kornberg's enzyme) is responsible for the removal and replacement of the RNA primer?

(a) 5' to 3' polymerase (b) Primase

(c) 5' to 3' exonuclease (d) 3' to 5' polymerase

444. Which is not a requirement for DNA replication?

(a) Primer (b) Template

(c) Ready supply of ATP (d) None of the above

445. Which one of the following does not explain how human cells can apparently replicate DNA much faster than a bacterium?

(a) The eukaryotic DNA polymerase moves much more quickly

(b) DNA polymerase III is responsible for DNA synthesis

(c) DNA synthesis occurs simultaneously at multiple points along the chromosome in eukaryotes

(d) There is more DNA polymerase in eukaryotes

446. The primer required for DNA synthesis is made by the following enzyme:

(a) DNA polymerase II (b) DNA polymerase I

(c) Replicase (d) Primase

447. Okazaki fragments

(a) Are formed in the leading strand

(b) Add nucleotides to the elongating DNA

(c) Remove the primer

(d) Are formed in the lagging strand

448. In double stranded DNA how do the four bases (C,A,T,G) pair up with each other.

(a) A always pairs with T and C always pairs with G

(b) C always pairs with T and A and G always pair with themselves

(c) C always pairs with T and A always pairs with G

(d) A always pairs with T and C and G always pair with themselves

449. During DNA replication what is the first process to occur?

(a) Sealing of the nicks between short DNA sections

(b) Synthesis of the lagging strand

(c) Unwinding of parental DNA

(d) Synthesis of the leading strand

450. DNA replicates through what process?

(a) Conservative replication

(b) Disparative replication

(c) Dispersive replication

(d) Semi-conservative replication

451. Which of the following enzymes is not as involved in DNA replication as the others?

(a) Phosphatase (b) DNA Ligase

(c) Helicase (d) DNA Polymerase

452. What is the name given to the short stretches of DNA formed on the lagging strand?

(a) Hiorshimi fragments (b) Okazaki fragments

(c) Japanese fragments (d) Okanuri fragments

453. The fragments mentioned in question 5 are approximately how many nucleotides in length?

(a) 500 (b) 1000

(c) 100 (d) 5000

454. Which of the following is NOT a function of DNA polymerase I?

(a) Polymerase (b) 3' to 5' exonuclease

(c) 5' to 3' exonuclease (d) Helicase

455. DNA helicase is found where?

(a) Replication Bubble

(b) Replication Complex

(c) Replication Fork

(d) All of these

456. What is the function of DNA polymerase III?

(a) It adds single nucleotides to the growing DNA strand

(b) It unwinds the DNA

(c) It adds nucleotide pairs to the growing DNA strand

(d) It attaches the RNA primers to the DNA strand

457. Which three people were awarded the Nobel Prize for the discovery of the structure of DNA: the double helix?

(a) James Watson, Rosalind Franklin and Maurice Wilkins

(b) Francis Crick, James Watson and Rosalind Franklin

(c) James Watson, Francis Crick and Maurice Wilkins

(d) Maurice Wilkins, Rosalind Franklin and Francis Crick

458. What is the name of the process in which DNA is copied into messenger RNA?

(a) Translocation (b) Reverse transcriptase

(c) Transcription (d) Translation

459. What does the 'r' in rRNA stand for?

(a) Ribosomal (b) Ribosome

(c) Ribose (d) None

460. Which is NOT a nitrogen base of RNA?

(a) Adenine (b) Thymine

(c) Cytosine (d) Guanine

461. What does Cytosine pair up with?

(a) Adenine (b) Guanine

(c) Thymine (d) Leucine

462. Who proposed the double-helical model of the DNA?

(a) Watson, Crick and Franklin

(b) Hershey and Chase

(c) Griffith

(d) Avery, MacLeod and McCarty

463. Nucleotides are monomers of DNA. What does each nucleotide consist of?

(a) A nitrogenous base and a pentose sugar

(b) A nitrogenous base, a pentose sugar and a phosphate group

(c) A nitrogenous base and a phosphate group

(d) A pentose sugar, and a phosphate group

464. What are Okazaki fragments?

(a) Short fragments of the promoter sequence

(b) Short fragments of newly synthesized RNA

(c) Short fragments of newly synthesized DNA

(d) Short fragments of the DNA sense strand

465. In eukaryotes, transcription of mRNA is 1)catalyzed by what type of enzyme 2)initiated by binding of transcription factors to which important promoter sequence?

(a) RNA polymerase IV ; TATA box

(b) RNA polymerase I ; Goldberg-Hogness box

(c) RNA polymerase II ; TATA box

(d) RNA polymerase III ; Goldberg-Hogness box

466. There are 20 essential amino acids in the human body. Using the triplet codon, theoretically, how many amino acids can be coded for?

(a) 20 (b) 64

(c) 36 (d) 61

467. Why does a single base substitution in DNA base sequence not necessary result in a malfunctional polypeptide?

(a) All of the above

(b) New amino acid is chemically equivalent

(c) Degeneracy of the triplet code

(d) The substitution takes place at introns

468. Translation requires the activation of amino acid into

(a) Aminoacyl adenylate

(b) Aminoacyl rRNA

(c) Aminoacyl guanylate

(d) Aminoacyl tRNA

469. Which molecule is not required during translation?

(a) Water (b) Peptidyl transferase

(c) GTP (d) DNA

470. What are involved in removal of introns from the pre-mRNA?

(a) Ribosomes (b) Spliceosomes

(c) Lac operon (d) Lysosomes

471. What is the start codon in mRNA for protein synthesis?

(a) UGA (b) UAG

(c) AUG (d) UAA

472. The outer skin of a cell is called a what?

(a) Nucleus (b) Mitochondria

(c) Cytoplasm (d) Membrane

473. The control center of a cell is what?

(a) Cytoplasm (b) Mitochondria

(c) Nucleus (d) DNA

474. What are the building blocks of protein molecules called?

(a) Fatty Acids (b) Hemoglobin

(c) Enzymes (d) Amino Acids

475. The structure of DNA was discovered by whom?

(a) James Watson and Francis Crick

(b) Carl David Anderson

(c) Paul Dirac

(d) John Napier

476. Who constructed the first advanced microscope in the late 17th and early 18th century?

(a) Wilhelm Roentgen

(b) Anton van Leeuwenhoek

(c) James Prescott Joule

(d) Emil von Behring

477. Which of the following organelles have a double membrane with pores?

(a) Mitochondria (b) Nucleus

(c) Lysosome (d) Golgi apparatus

478. Where is ribosomal RNA produced?

(a) Nucleolus (b) Endoplasmic reticulum

(c) Ribosomes (d) Golgi apparatus

479. This is (are) the site of protein synthesis in the cytoplasm.

(a) Ribosomes (b) Mitochondria

(c) Smooth endoplasmic reticulum (d) Golgi apparatus

480. Which type of organelle forms a membranous system of tubular canals, which is continuous with the nuclear envelope and branches throughout the cytoplasm?

(a) Smooth endoplasmic reticulum

(b) Rough endoplasmic reticulum

(c) Golgi apparatus

(d) Mitochondria

481. Which organelle will produce steroid hormones in the testes and adrenal cortex while detoxifying alcohol in the liver?

(a) Rough endoplasmic reticulum

(b) Mitochondria

(c) Golgi apparatus

(d) Smooth endoplasmic reticulum

482. Which organelle looks like a stack of hollow pancakes?

(a) Smooth endoplasmic reticulum

(b) Mitochondria

(c) Golgi apparatus

(d) Rough endoplasmic reticulum

483. Which organelle contains hydrolytic enzymes that can digest macromolecules

(a) Cytoskeleton

(b) Centrioles

(c) Lysosomes

(d) Endoplasmic reticulum

484. Where can cristae be found?

(a) Nucleus (b) Mitochondria

(c) Lysosome (d) Centriole

485. Which of the following is found in plant cells but not animal cells?

(a) Cytoplasm (b) Mitochondria

(c) Cell Wall (d) Ribosomes

486. Which organelle is a type of vesicle?

(a) Nucleus (b) Ribosome

(c) Lysosome (d) Endoplasmic Reticulum

487. What organelle processes and packages proteins before sending them through the cell?

(a) Ribosome (b) Endoplasmic Reticulum

(c) Golgi Apparatus (d) Vesicle

488. Where are ribosomes produced?

(a) Nucleus (b) Nucleolus

(c) Nucleoplasm (d) Nuclear Envelope

489. What is the "indent" or "groove" called where the substrate binds to the enzyme?

(a) Allosteric Site (b) Fitting Site

(c) Indented Site (d) Active Site

490. Which of the following is NOT a characteristic of an enzyme?

(a) It speeds up the rate of a reaction

(b) It raises the energy of activation (EA)

(c) It isn't changed at all during the reaction

(d) It only fits with one substrate

491. Which suffix do enzymes commonly end in?

(a) ese (b) ose

(c) ase (d) ise

492. What is the enzyme that is in stomach acid called?

(a) Amylase

(b) There aren't enzymes in stomach acid

(c) Maltase

(d) Pepsin

493. What group of macromolecules do enzymes belong to?

(a) Lipids
(b) Proteins
(c) Carbohydrates
(d) Nucleic Acids

494. What is ATP composed of?

(a) Adenosine, Ribose, 3 phosphate groups

(b) Adenine, Ribose, 3 phosphate groups

(c) Adenine, Thymine, 2 phosphate groups

(d) Amino Group, Thymine, Phosphate group

495. What is the process called when ATP gives up the third phosphate group?

(a) Photosynthesis
(b) Dehydration Synthesis
(c) Digestion
(d) Phosphorylation

496. How many oxygen atoms is a phosphate group made up of?

(a) 5
(b) 4
(c) 3
(d) 0

497. According to Chargaff's rule, the following proportion exists in DNA:

(a) C=G
(b) C>T
(c) C>G
(d) C=T

498. Which of the following sequences does NOT follow Chargaff's rules for DNA base pairing?

(a) A-T
(b) C-G
(c) A-G
(d) G-C

499. Each unit of a nucleic acid, consisting of a sugar, attached phosphate group, and a base is called a:

(a) Nucleolus
(b) Nucleotide
(c) Nucleosome
(d) Histone

500. If one side of a DNA molecule contains the following sequence of nucleotides, AGTCCG, the complementary sequence on the other side would be:

(a) GCCTGA (b) AGTCCG

(c) TCAGGC (d) CTGAAT

(e) None of the above

501. Okazaki fragments are used to elongate:

(a) The leading strand toward the replication fork

(b) The lagging strand toward the replication fork

(c) Both strands in both directions

(d) The lagging strand away from the replication fork

502. In Griffin's experiment with mice, he demonstrated that

(a) Heat-killed pathogenic forms of *S. pneumoniae* killed mice

(b) Live pathogenic forms of *S. pneumoniae* didn't kill mice

(c) Heat-killed pathogenic forms of *S. pneumoniae* transformed non-pathogenic strains, making them pathogenic and killing the mice.

(d) None

503. The fact that some viruses use DNA to direct their heredity was demonstrated by:

(a) Finding radioactive sulfur from a bacteriophage in a bacterium

(b) Finding radioactive phosphorus from a bacterium in a bacteriophage

(c) Finding radioactive phosphorus from a bacteriophage in a bacterium

(d) Finding radioactive sulfur from a bacterium in a bacteriophage

504. Which statement about the elongation of the lagging strand during DNA replication is correct?

(a) It is synthesized in a 3' —> 5' direction

(b) It progresses (grows) toward the replication fork

(c) It requires a short RNA primer to proceed

(d) It is synthesized continuously

505. Which enzyme joins DNA fragments to the lagging strand?

(a) Telomerase (b) Helicase

(c) DNA ligase (d) DNA polymerase

506. The process of DNA replication is said to be:

(a) Disjointed (b) Semiconservative

(c) Dispersive (d) Conservative

507. Transformation involves:

(a) The uptake of genetic material, often from one bacterial strain to another

(b) The creation of a strand of RNA from a DNA template

(c) Replication of DNA along the lagging strand

(d) None of the above

508. The DNA of an organism has guanine as 20% of its bases. What percentage of its bases would be cytosine?

(a) 20% (b) 30%

(c) 40% (d) 60%

509. The DNA of an organism has guanine as 20% of its bases. What percentage of its bases would be thymine?

(a) 20% (b) 30%

(c) 40% (d) 60%

510. Which of the following enzymes involved in DNA replications is NOT properly paired with its function?

(a) Helicase - unwinds the DNA helix

(b) DNA polymerase - Builds a short RNA polymer

(c) DNA ligase - joins Okazaki fragments

(d) None

511. Which research group's results supports the "one gene, one enzyme" hypothesis?

(a) Watson and Crick (b) Beadle and Tatum

(c) Meselson and Stahl (d) Hershey and Chase

512. Which scientist(s) described the double helix structure of DNA?

(a) Rosalind Franklin

(b) James Watson and Francis Crick

(c) George Beadle and Edward Tatum

(d) Matthew Meselson and Frank Stahl

513. How does the DNA molecule copy itself?

(a) Each strand serves as a template for making a new strand

(b) Of the two strands, only one makes one copy of itself

(c) Of the two strands, only one makes two copies of itself

(d) RNA servers as a template for making new DNA

514. Which of the following is characteristic of DNA replication?

(a) Doubles the amount of DNA

(b) Produces a copy of the original DNA molecule

(c) Involves the enzyme DNA polymerase

(d) All of the above are true

515. Which autosomal recessive genetic disorder is due to the change in one amino acid within the hemoglobin protein?

(a) Down's syndrome (b) Klinefelter syndrome

(c) Sickle cell anemia (d) Huntington's disease

516. In the Hershey-Chase experiment, 32P integrated into the bacteria and was released as infected cells died 35S did not get incorporated into the bacteria. What does this indicate about the genetic material?

(a) Amino acids, not DNA, contain genetic material

(b) DNA, not amino acids, contain genetic material

(c) DNA contains sulfur groups

(d) Amino acids contain phosphate groups

517. Transcription + translation is called:

(a) Gene rearrangement (b) Gene recombination

(c) Gene expression (d) DNA replication

518. According to the Central Dogma, genetic information flows in which order?

(a) Protein –>DNA –>mRNA (b) DNA –>protein –>mRNA

(c) mRNA –>DNA –>protein (d) DNA –>mRNA –>protein

519. The function of mRNA is to

(a) Provide a site for polypeptide synthesis

(b) Transport amino acids to the ribosome

(c) Travel to the ribosome to direct the assembly of polypeptides

(d) All of the above

520. Which of the following is true of eukaryotic translation?

(a) Occurs at a ribosome

(b) Occurs in the cytoplasm

(c) Involves mRNA, rRNA and tRNA

(d) All of the above

521. If a protein has 30 amino acids, how many nucleotides are needed to code for it?

(a) 30 (b) 60

(c) 90 (d) 3

522. mRNA's start codon is:

(a) ATG (b) TAC

(c) AUG (d) GUA

523. The anticodon of ACG is:

(a) TGC (b) UGC

(c) AGC (d) UCG

524. The type of RNA that carries an amino acid at one end and an anticodon at the other is:

(a) mRNA

(b) tRNA

(c) rRNA

525. Which enzyme is involved in transcription?

(a) DNA polymerase (b) RNA polymerase

(c) Peptidyl transferase (d) All of the above

526. What occurs during the elongation phase of transcription?

(a) Amino acids are added to the polypeptide in a 5' —>3' direction

(b) Nucleotides are added to the mRNA chain in the 5' —>3' direction

(c) Amino acids are added to the polypeptide in a 3' —>5' direction

(d) Nucleotides are added to the mRNA chain in the 3' —>5' direction

527. The termination of protein synthesis normally occurs:

(a) When the cell runs out of ATP

(b) Upon recognition of a stop codon

(c) Upon recognition of an exon

(d) When more than 10 ribosomes attach to the mRNA

528. In translation:

(a) A strand of mRNA is formed with nucleotides that are complementary to DNA

(b) Nucleotide sequences of tRNA are established

(c) rRNA is formed with nucleotides complementary to tRNA

(d) A protein is formed as directed by the nucleotide sequence in mRNA

529. In transcription, what is the template strand (or antisense strand)?

(a) Strand of DNA that is used to form complementary strand of mRNA

(b) Strand of DNA that has exact same sequence as mRNA

(c) Strand of DNA that has the exact same sequence as mRNA, except that DNA has thymine and RNA has uracil

(d) None

530. How does the presence of introns affect the number of proteins an organism, such as a human, can express?

(a) Increases the number of proteins an organism can express

(b) Decreses the number of proteins an organism can express

(c) Does not affect the number of proteins an organism can express

(d) None

531. Which enzyme is incorrectly matched to its function?

(a) RNA polymerase - makes mRNA from a DNA template

(b) Ligase – joins Okazaki fragments

(c) DNA polymerase - replicates DNA

(d) All of the enzymes are correctly matched to their functions

532. Adding a 5' cap and 3' poly-A tails to the mRNA are post-transcriptional modifications. What is the purpose of these modifications in eukaryotes?

(a) Prevents the mRNA from being degraded as it moves from the nucleus into the cytoplasm

(b) Prevents the DNA from being degraded as it moves from the nucleus into the cytoplasm

(c) Prevents RNA splicing, allowing the introns and exons to be translated

(d) None

533. Given the DNA strand TACCCTACG, what is the corresponding mRNA?

(a) TACCTACG (b) AUGGGAUGC

(c) GCAUCCAU (d) ATGGGATGC

534. Given the mRNA strand AUGGUCAUA, what is the corresponding amino acid sequence?

(a) Methionine - leucine – tyrosine

(b) Tyrosine - glycine – tyrosine

(c) Methionine - valine - isoleucin

(d) None

535. According to Chargaff's rule, the following proportion exists in DNA)

(a) C=G (b) C»T

(c) C»G (d) C=T

536. The actual synthesis of DNA in *E. coli* is the function of

(a) Polymerase I (b) Primase

(c) Polymerase III (d) DNA ligase

537. The first nuclear transplant from an animal to an egg that produced a normal adult was performed on a

(a) Frog (b) Sheep

(c) Cat (d) Dog

538. All of the following were outcomes of Avery's experiment on "the active principle" except

(a) The array of the elements of purified principle agreed closely with DNA and the principle centrifuged to the same level as DNA

(b) Protein digesting enzymes did not affect the sample

(c) The extraction of lipids and proteins from the principle only slightly reduced its activity

(d) DNA-digesting enzyme, DNase, destroyed all transcription activity

539. Polymerase III is actually a complex containing

(a) Catalytic subunits

(b) Proofreading subunits

(c) "Sliding clamp" subunits

(d) All of the above

540. The fact that some viruses use DNA to direct their heredity was demonstrated by

(a) Finding radioactive sulfur from a bacteriophage in a bacterium

(b) Finding radioactive phosphorus from a bacterium in a bacteriophage

(c) Finding that radioactive phosphorus from a bacteriophage had mutated in bacterium

(d) Finding radioactive phosphorus from a bacteriophage in a bacterium

541. Okazaki fragments are used to elongate

(a) The leading strand toward the replication fork

(b) The lagging strand toward the replication fork

(c) Both strands in both directions

(d) The lagging strand away from the replication fork

542. A single enzyme is specified by a single

(a) Chromosome (b) Gene

(c) Nucleotide (d) Nucleosome

543. Hammerling chose Acetabularia as his model organism because it was

(a) Large and differentiated

(b) Easier to grow than fruit flies

(c) Large and reproduced only asexually

(d) Grown abundantly in the Mediterranean area where he lived

544. In DNA guanine always pairs with

(a) Adenine (b) Cytosine

(c) Guanine (d) Thymine

545. With few exceptions, all nuclei of eukaryotes contain

(a) Genes to specify the portion of the organism in which they are found

(b) All the information needed for growing the whole organism

(c) All of the chromosomes except sex chromosomes which are restricted to sex organs

(d) Single stranded DNA

546. In DNA, thymine always pairs with

(a) Adenine (b) Cytosine

(c) Guanine (d) Thymine

547. Each individual zone of a chromosome replicates as a discrete section called

(a) Telomere (b) Euchromatin

(c) Okazaki fragment (d) Replication unit

548. Each unit of a nucleic acid consisting of a sugar, attached phosphate group, and base is a

(a) Nucleolus (b) Nucleotide

(c) Nucleosome (d) Histone

549. In replication of DNA, the helix is opened and untwisted by

(a) Ribase (b) Ligase

(c) Deoxase (d) Helicase

550. In a nucleic acid, the bases are always attached to the ____________ carbon of the sugar.

(a) 5' (b) 4'

(c) 3' (d) 1'

551. Although the work produced important results, Franklin's work was hindered by the lack of

(a) Electricity (b) Strong x-rays

(c) True crystals of DNA (d) Thin layer chromatography

552. The identification of the transforming principle proved that

(a) Dead bacteria are as lethal as live ones

(b) Streptococcus pneumoniae evolved from Pneumococcus

(c) Genetic information is contained in DNA

(d) there is no protein in chromosomes

553. Which statement about the elongation of the lagging strand during DNA replication is correct?

(a) It is synthesized in a 3' —> 5' direction

(b) It progresses (grows) toward the replication fork

(c) It requires a short RNA primer to proceed

(d) It is synthesized by DNA ligase

554. DNA is made up of a phosphate group, an organic base, and:

(a) A protein (b) A sugar

(c) A molecule of ATP (d) A fat

555. In the DNA molecule:

(a) Adenine pairs with thymine (b) Guanine pairs with thymine

(c) Cytosine pairs with thymine (d) Adenine pairs with cytosine

556. 34 If one side of a DNA molecule contains the following sequence of nucleotides, AGTCCG, the complementary sequence on the other side would be:

(a) GCCTGA (b) AGTCCG

(c) TCAGGC (d) CTGAAT

557. 35 During your summer job at Virotech, you isolate a previously unknown virus. Analysis of its genome reveals that it is composed of a double stranded DNA molecule containing 14% T (thymine). Based on this information, what would you predict the %C (cytosine) to be?

(a) 14% (b) 28%

(c) 36% (d) 72%

558. After a tautomeric shift in adenine

(a) Adenine bonds with thymine

(b) Adenine bonds with urasil

(c) Adenine bonds with cytosine

(d) Adenine is unable to bond with any molecule

559. A mutation that causes a change in a single nucleotide in DNA

(a) Will have no effect on the resulting protein

(b) Changes the corresponding nucleotide in mRNA, resulting in a different codon

(c) Causes the codon to be correct, but the anticodon to be incorrect

(d) Causes protein synthesis to stop

560. In a tautomeric shift

(a) It is always adenine that is changed

(b) Final bonding of nucleotides remains unchanged

(c) Adenine is changed so it can no longer form base pairs

(d) Hydrogen atoms move to form a base with altered hydrogen properties

561. The nucleic acid sequence in mRNA is determined by

(a) The order of amino acids in the protein

(b) Nucleotide sequence in DNA

(c) Nucleotide sequence in t-RNA

(d) All of the above

562. If a frameshift mutation causes a stop codon to be inserted into the DNA sequence

(a) The resulting protein will not be affected

(b) The phenotype will change but not the genotype

(c) The resulting protein will be too short and non-functional

(d) The resulting protein will be too long and non-functional

563. A nucleotide deletion in DNA replication

(a) Causes one amino acid of the protein to be incorrect

(b) Causes all of the amino acids of the protein to be incorrect

(c) Causes the amino acids inserted after the deletion to be incorrect

(d) Causes the amino acids inserted before the deletion to be incorrect

564. What did Frederick Griffith's 1928 experiments with a pneumonia-causing bacterium and mice show?

(a) A biochemical in a killer strain of bacteria can make a non-killing strain deadly

(b) A "transforming principle" was protein

(c) Mice can become immune to pneumonia

(d) All of these are correct

565. Which scientist found that DNA contains equal amounts of A and T and equal amounts of C and G?

(a) Martha Chase (b) Erwin Chargaff

(c) Rosalind Franklin (d) James Watson

566. What portion of the DNA molecule actually encodes the genetic information?

(a) DNA base sequences

(b) Sugar-phosphate backbone

(c) Number of hydrogen bonds

(d) Three-dimensional structure

567. What is a nucleosome?

(a) A mutation in a base pair

(b) A combination of enzymes that regulates gene activity

(c) DNA base pairs wrapped around histones

(d) An unusual sequence of bases that act as a start position for gene activity

568. Replication occurs in 5′ to 3′ direction. What is the name of the short pieces of the discontinuous strand?

(a) RNA primers (b) DNA primers

(c) Okazaki fragments (d) Nucleotides

569. Polymerase chain reaction can be used in which of the following fields?

(a) Genetic engineering (b) Forensics

(c) Ecology (d) All of these are correct.

570. Which of the following nucleotides is particularly sensitive to ultraviolet radiation?

(a) Adenine (b) Thymine

(c) Guanine (d) Cytosine

571. Which mode of DNA repair corrects small loops where complementary strands of DNA are not precisely aligned?

(a) Photoreactivation (b) Excision repair

(c) Proofreading (d) Mismatch repair

Fill in the Blanks

1. In the Hershey Chase Experiment, DNA was labeled with ____, and bacteriophage protein was labeled with _____.

(a) Radioactive phosphorous, radioactive sulfar

(b) Radioactive sulfur, radioactive phosphorous

(c) Codons, anticodons

(d) DNA Polymerase, RNA polymerase

2. In RNA, _____ codon(s) translate to ______ amino acid(s)

(a) 1, 1 (b) 3, 1

(c) 3, 3 (d) 1, 20

3. The sugar in RNA is _____, the sugar in DNA is _______

(a) Deoxyribose, ribose (b) Ribose, deoxyribose

(c) Ribose, phosphate (d) Ribose, uracil

4. RNA is synthesized on a DNA template in a process called ______, which utilizes the enzyme ______
   (a) Translation, RNA polymerase
   (b) Transcription, DNA polymerase
   (c) Transcription, RNA polymerase
   (d) Replication, DNA polymerase
5. Proteins contain ____ different amino acids, whereas DNA and RNA are composed of ___ different nucleotides
   (a) 20, 64 (b) 3, 20
   (c) 4, 20 (d) 20, 4
6. Okazaki fragments occur on the ___ and are bonded together by ______
   (a) Leading strand, polymerase (b) mRNA, anticodons
   (c) Lagging strand, ligase (d) tRNA, polymerase
7. ______________ are organic molecules that have the function of starting and regulating chemical reactions.
   (a) Carbohydrates (b) Lipids
   (c) Enzymes (d) Endoplasmic reticula
8. There are ________ different kinds of amino acids.
   (a) 20 (b) 46
   (c) 100 (d) 20,000-25,000
9. The enzyme ________ unzips and unwinds the DNA molecule.
   (a) DNA polymerase (b) Helicase
   (c) Primase (d) DNA ligase
10. Translation is terminated when a stop codon is presented at the ________ site.
   (a) A (b) P
   (c) E (d) Either A or B
11. The enzyme ________ unzips and unwinds the DNA molecule.
   (a) DNA polymerase (b) Helicase
   (c) Primase (d) DNA ligase

12. The three-nucleotide codon system can be arranged into ____________ combinations.

(a) 16 (b) 20

(c) 64 (d) 128

13. The three-nucleotide codon system can be arranged into ____________ combinations.

(a) 16 (b) 20

(c) 64 (d) 128

14. RNA is synthesized on a DNA template in a process called ______, which utilizes the enzyme _______

(a) Translation, RNA polymerase

(b) Transcription, DNA polymerase

(c) Transcription, RNA polymerase

(d) Replication, DNA polymerase

15. Proteins contain ____ different amino acids, whereas DNA and RNA are composed of ___ different nucleotides

(a) 20, 64 (b) 3, 20

(c) 4, 20 (d) 20, 4

16. Okazaki fragments occur on the ___ and are bonded together by ______

(a) Leading strand, polymerase

(b) mRNA, anticodons

(c) Lagging strand, ligase

(d) tRNA, polymerase

17. Amino acids are held together by __________ bonds to make a protein.

(a) Hydrogenionic (b) Peptide

(c) Complementary (d) Magnetic

18. By convention, the sequence of bases in a nucleic acid is usually expressed in the _________ direction.

(a) 3' to 1' (b) 3' to 5'

(c) 1' to 3' (d) 5' to 3'

19. Since the first nucleotides cannot be linked in a newly synthesized strand in DNA replication, __________ is required)

(a) A DNA primer (B) DNA polymerase

(c) Ligase (d) An RNA primer

20. In nucleic acids, the free hydroxyl group is attached to the ______________ carbon of the sugar.

(a) 5' (b) 4'

(c) 3' (d) 2'

21. DNA replication is called semiconservative because ______________ of the original duplex appears in the duplex formed in replication.

(a) None (b) Most

(c) Half (d) Hardly any

22. In a nucleic acid, the bases are always attached to the ______________ carbon of the sugar.

(a) 5' (b) 4'

(c) 3' (d) 1'

23. In nucleic acids, the phosphate group is attached to the ______________ carbon of the sugar.

(a) 5' (b) 4'

(c) 3' (d) 1'

24. ______________ join DNA fragments to the lagging strand.

(a) Telomeres (b) Centromeres

(c) Helicases (d) Ligases

25. RNA is sometimes the genetic material in __________.

(a) Yeast (b) Animals

(c) Plants (d) Viruses

26. If Hershey and Chase found only 35S in the bacterial cells, they would conclude that __________.

(a) Protein is the genetic material

(b) DNA and protein together serve as the genetic material

(c) DNA is the genetic material

(d) no conclusion is possible

27. Techniques used to identify the location of genes on a chromosome include ________.

(a) FISH (b) Reassociation kinetics

(c) Repetitive DNA studies (d) Electrophoresis

28. RNA differs from DNA in all of the following ways except ________.

(a) Thymine

(b) A sugar molecule

(c) The number of different functions of RNA vs. DNA

(d) The 5' and 3' orientation of the polynucleotide strand

29. The monomers that nucleic acids are built from are _____ and they contain not only atoms of C, H, O, and N but also ____.

(a) Amino acids; sulfur (b) Fatty acids; sulfur

(c) Bases; phosphorus (d) Nucleotides; phosphorus

30. It is possible to insert the DNA from one virus (virus A) into the protein coat of a different virus (virus B). If such a composite virus infected a cell, the resultant viruses produced in the host cell would have DNA like virus ____ and protein like ______.

(a) A;B (b) A;A

(c) B;B (d) B;A

31. In eukaryotic cells, DNA replication takes place in the ______ and protein synthesis takes place in the ________.

(a) Nucleus; cytoplasm

(b) Cytoplasm; nucleus

(c) Both take place in the nucleus

(d) Both take place in the cytoplasm

32. The frequency of incorrect nucleotide incorporation into RNA is about one error for every ___ nucleotides polymerized.

(a) 1,000 (b) 10,000

(c) 100,000 (d) 1,000,000

33. On the ribosome, mRNA binds ____; the peptidyl transferase reaction occurs ____.

(a) Between the subunits; on the large subunit

(b) Between the subunits; on the small subunit

(c) To the large subunit; on the small subunit

(d) To the small subunit; on the large subunit

34. In prokaryotes, protein synthesis rates are limited by the rate of mRNA synthesis. RNA synthesis occurs at about 50 nucleotides/sec Therefore, protein synthesis occurs at about ___ amino acids/sec

(a) 10 (b) 17

(c) 25

35. The radioactive isotope 32P labels the T2 phage's _____. (Activity 16A

(a) DNA (b) Tail

(c) Base plate (d) Protein coat

36. Hershey and Chase used _____ to radioactively label the T2 phage's proteins. (Activity 16A)

(a) ^{35}S (b) 14C

(c) 222Ra (d) 32P

37. The production of a protein from mRNA is called ______.

(a) Replication (b) Transformation

(c) Transcription (d) Translation

38. After allowing phages grown with bacteria in a medium that contained 32P and 35S, Hershey and Chase used a centrifuge to separate the phage ghosts from the infected cell. They then examined the infected cells and found that they contained _____, which demonstrated that _____ is the phage's genetic material.

(a) Labeled protein ... DNA

(b) Labeled protein .... protein

(c) Labeled DNA ... labeled protein

(d) Labeled DNA ... DNA

39. In a nucleotide, the nitrogenous base is attached to the sugar's _____ carbon and the phosphate group is attached to the sugar's _____ carbon.

(a) 2' ... 1' (b) 1' ... 5'

(c) 2' ... 3' (d) 1' ... 3'

40. Nucleic acids are assembled in the _____ direction.

(a) 1' to 5' (b) 2' to 3'

(c) 5' to 3' (d) 4' to 5'

41. In a DNA double helix an adenine of one strand always pairs with a(n) _____ of the complementary strand, and a guanine of one strand always pairs with a(n) _____ of the complementary strand

(a) Guanine ... adenine (b) Cytosine ... uracil

(c) Cytosine ... thymine (d) Thymine ... cytosine

42. Short segments of newly synthesized DNA are joined into a continuous strand by _____.

(a) Helicase (b) DNA polymerase

(c) Ligase (d) Primase

43. After DNA replication is completed, _____.

(a) Each new DNA double helix consists of one old DNA strand and one new DNA strand

(b) Each new DNA double helix consists of two new strands

(c) One DNA double helix consists of two old strands and one DNA double helix consists of two new strands

(d) Each of the four DNA strands consists of some old strand parts and some new strand parts

44. The first step in the replication of DNA is catalyzed by _____.

(a) Helicase (b) DNA polymerase

(c) Ligase (d) Primase

45. The action of helicase creates _____.

(a) Primers and DNA fragments

(b) Primers and replication bubbles

(c) DNA fragments and replication forks
(d) Replication forks and replication bubbles

46. An old DNA strand is used as a _____ for the assembly of a new DNA strand

(a) Complement (b) Primer
(c) Template (d) Source of nucleotides

47. Replication proceeds in a ____ to _____ direction.

(a) 3' to 3' (b) 3' to 5'
(c) 5' to 3' (d) 5' to 5'

48. A scientist isolates a cell line characterized by unusual cell division. In these cells, only small fragments of DNA are found. A defective ____ enzyme may be responsible.

(a) DNA polymerase (b) Ligase
(c) Primase (d) Exonuclease

49. The number of amplified pieces of DNA equals ____ after five cycles of PCR.

(a) 5 (b) 10
(c) 25 (d) 32

50. DNA replication is extremely accurate. Only about one in ____ bases is incorporated incorrectly.

(a) 100,000 (b) 1 million
(c) 10 million (d) 100 million

51. Ultraviolet radiation damages DNA by causing an extra ____ bond to form between adjacent ____.

(a) Hydrogen, pyrmidines (b) Covalent, purines
(c) Covalent, pyrmidines (d) Hydrogen, purines

52. In genome sequencing, it was found that ____% of the Drosophila genes matched other species.

(a) 45 (b) 72
(c) 98 (d) 83

53. In the discovery of introns, a DNA molecule called ______________ was formed that had the same nucleotide sequence as the gene that produced the mRNA.

(a) mDNA (b) rDNA

(c) sDNA (d) cDNA

54. Bacterial DNA is not cleaved by their own restriction enzymes because bacteria add ______________ to their own DNA.

(a) Nucleotides (b) Peptides

(c) Methyl groups (d) Glyphosphate

55. In vivo synthesis of DNA is _________.

(a) 3' to 5' (b) 5' to 3'

(c) Both 3' to 5' and 5' to 3' (d) Neither 3' to 5' nor 5' to 3'

56. The discovery of Okazaki fragments suggested that DNA synthesis is ____________.

(a) Discontinuous (b) Continuous

(c) 3' to 5' (d) Semiconservative

57. Assume that in the following cross, d+ x e+, the only gametes formed consist of de alleles. This may be due to a phenomenon called _________.

(a) Gene conversion (b) Reciprocal recombination

(c) Nonreciprocal recombination (d) Genetic exchange

58. Telomeres have all of the following properties or characteristics except _________.

(a) The telomerase enzyme

(b) Forming "hairpin" loops

(c) A link to the aging process in eukaryotic cells

(d) Being found in both eukaryotic and prokaryotic chromosomes

59. All of the following are differences between eukaryotic and prokaryotic DNA replication except _________.

(a) The type and number of polymerases involved in DNA synthesis

(b) Multiple vs. single replication origins

(c) The rate of DNA synthesis

(d) The ability to form a replication fork

60. The fundamental unit of life is the ______.

(a) Organism (b) Cell

(c) Organ system (d) Tissue

61. A threadlike material called chromatin is located within the ______.

(a) Golgi apparatus (b) Nucleus

(c) Endoplasmic reticulum (d) Lysosome

62. ______________ catalyzes the elongation of a DNA strand

(a) Telomerase (b) Helicase

(c) DNA ligase (d) DNA polymerase

63. Elongation of a DNA strand occurs in the __________ direction.

(a) 5'–>3' (b) 3'–>5'

(c) 1' –>4' (d) 4'–>1'

64. DNA is composed of two ____________ strands. One strand of DNA runs from the 5' –>3' direction, and one strand runs from the 3' –> 5' direction.

(a) Leading (b) Lagging

(c) Antiparallel (d) RNA

65. When DNA coils, it wraps around histone proteins to form _________.

(a) Nucleotides (b) Nucleosomes

(c) Histosides

66. ___________ bonds join the two strands of DNA together to form the double helix.

(a) Hydrogen (b) Covalent

(c) ionic (d) Polar covalent

67. All of the following are transcribed directly from DNA except _________.

(a) Protein (b) Exons

(c) mRNA (d) Introns

68. After mRNA is spliced, the ________ are removed, and the ________ are joined.

(a) Exons, introns (b) Spliceosomes, introns

(c) Exons, spliceosomes (d) Introns, exons

69. In a process called ______________, the initial tRNA is ejected from the ribosome.

(a) Elongation (b) Translocation

(c) Initiation (d) Transcription

70. By convention, the sequence of bases in a nucleic acid is usually expressed in the ________ direction.

(a) 3' to 1' (b) 3' to 5'

(c) 1' to 3' (d) 5' to 3'

71. Since the first nucleotides cannot be linked in a newly synthesized strand in DNA replication, __________ is required.

(a) A DNA primer (b) DNA polymerase

(c) Ligase (d) An RNA primer

72. In nucleic acids, the free hydroxyl group is attached to the ____________ carbon of the sugar.

(a) 5' (b) 4'

(c) 3' (d) 2'

73. DNA replication is called semiconservative because ____________ of the original duplex appears in the duplex formed in replication.

(a) None (b) Most

(c) Half (d) Hardly any

74. In nucleic acids, the phosphate group is attached to the ____________ carbon of the sugar.

(a) 5' (b) 4'

(c) 3' (d) 2'

75. ______________ join DNA fragments to the lagging strand.

(a) Telomeres (b) Centromeres

(c) Helicases (d) Ligases

76. English physician Archibald Garrod was the first to link inheritance and ________.

(a) DNA (b) Nucleic acids
(c) RNA (d) Protein

77. Avery, MacLeod, and McCarty confirmed the hypothesis that DNA was the genetic material by inoculating mice with __________.

(a) Heat-killed type R bacterial DNA and live type S bacteria –mice died
(b) Heat-killed type S bacterial DNA and live type R bacteria –mice lived
(c) Heat-killed type R bacterial DNA and live type S bacteria –mice lived
(d) Heat-killed type S bacterial DNA and live type R bacteria –mice died

78. The scientist(s) that confirmed that DNA was the hereditary material was (were) __________.

(a) Avery, MacLeod, and McCarty
(b) Hammerling
(c) Griffith
(d) Hershey and Chase

79. In the experiments of Hershey and Chase, DNA was shown to be the material transmitted from the virus to the bacterium because ____________.

(a) Protein contains sulfur but DNA does not
(b) DNA contains phosphorus but protein does not
(c) Radioactivity came from the phosphorus
(d) All of these are correct

80. James Watson and Francis Crick are credited with determining __________ in 1953.

(a) The nature of base-pairing in DNA
(b) The chemical composition of DNA

(c) The three-dimensional structure of DNA

(d) DNA is located in the nucleus

81. DNA contains ____________.

(a) Ribose, uracil, adenine, cytosine, guanine, phosphate

(b) Deoxyribose, uracil, adenine, cytosine, guanine, phosphate

(c) Ribose, thymine, adenine, cytosine, guanine, phosphate

(d) Deoxyribose, thymine, adenine, cytosine, guanine, phosphate

82. The two strands of DNA run opposite to each other. This is termed ______.

(a) Complementarity (b) Antiparallelism

(c) Semi-conservative (d) Conservative

83. The hypothesis of DNA replication that proposed one strand serves as a template for the replication of a new strand was called ____________.

(a) Dispersive (b) Conservative

(c) Semi-conservative (d) Complementarity

84. Enzymes that unwind and hold apart replicating DNA are __________.

(a) Polymerases (b) Ligases

(c) Helicases (d) Kinases

85. Gene amplification is a term describing the technique called ___________.

(a) Polymerase chain reaction

(b) Assisted reproductive technologies

(c) Recombinant DNA technology

(d) Genetic engineering

True/False

1. A replicating chromosome has only one replication fork at a time.

(a) True (b) False

2. Histones are unique to eukaryotes.

(a) True (b) False

3. The process of acquiring a gene from another organism is called transformation.

(a) True (b) False

4. The Hershey-Chase experiment showed that the virus protein is responsible for directing the production of new viruses.

(a) True (b) False

5. The two sides of the double helix structure of DNA are held together with hydrogen bonds.

(a) True (b) False

6. DNA replication is called semiconservative because half of each of the two new strands of DNA is "old" DNA from the original DNA molecule and half is "new" DNA)

(a) True (b) False

7. During semiconservative replication, the original double helix remains intact and a new double helix forms.

(a) True (b) False

8. DNA polymerase uses a nucleoside triphosphate, bearing phosphates on the 5' carbon to form a phosphodiester linkage to the free 3' hydroxyl group on the end of the DNA strand it is synthesizing.

(a) True (b) False

9. The elongation of Okazaki fragments during DNA synthesis progresses away from the replication fork.

(a) True (b) False

10. DNA polymerase catalyzes the elongation of a DNA strand in the 5'—>3' direction.

(a) True (b) False

11. A DNA molecule has the same amount of adenine and thymine.

(a) True (b) False

12. Many of the important events for cell division occur during interphase.

(a) True (b) False

13. In a cell with a diploid chromosome number (2n) of 12, there are 12 chromosomes per set.

(a) True (b) False

14. At the end of mitosis, the cell has divided into two daughter cells.

(a) True (b) False

15. The best way to identify an X-linked trait in fruit flies is that the trait affects one sex more often than the other.

(a) True (b) False

16. Genes on the X or Y chromosomes in humans are not always inherited in a different pattern than are genes on the autosomes.

(a) True (b) False

17. DNA can be used as forensic evidence because the DNA from every individual is unique

(a) True (b) False

18. The Ti plasmid provides a means to transfer genes into cereal plants, such as corn and rice.

(a) True (b) False

19. Genetic engineering has provided a method for the production of insulin to fight diabetes, but it remains too expensive to be practical.

(a) True (b) False

20. Piggyback vaccines work by triggering the body's defense system with a harmless virus carrying the surface proteins of a disease organism.

(a) True (b) False

21. The second stage in genetic engineering experiments is cloning the plasmid

(a) True (b) False

22. When using reverse transcriptase, DNA is made from mRNA

(a) True (b) False

23. PCR requires knowledge about parts of a target DNA sequence to be amplified True or false?

(a) True (b) False

24. One of the goals of the Human Genome Project is to address ethical issues arising from use of the information gathered

(a) True (b) False

25. Isolation of individual chromosomes can be used to map genes.
    (a) True (b) False
26. The reduced numbers of progeny with the pattern of (5,000bp, 4,300bp, 2,300bp, and 2,000bp bands) or (5,000bp, 4,300bp, and 3,350bp bands) indicates a selective disadvantage for these genotypes.
    (a) True (b) False
27. RFLPs are inherited in a simple Mendelian fashion and display codominance.
    (a) True (b) False
28. Excision repair takes place only in the presence of light.
    (a) True (b) False
29. In dark repair, the final phosphodiester bond is formed by DNA ligase.
    (a) True (b) False
30. The advent of molecular genetics has greatly improved our ability to perform research in other areas of genetics.
    (a) True (b) False
31. The genetic material must be able to contain an infinite amount of information.
    (a) True (b) False
32. Avery, MacLeod, and McCarty were able to demonstrate that their biochemical extracts were perfectly pure.
    (a) True (b) False
33. Hershey and Chase demonstrated that DNA is the genetic material for eukaryotic organisms.
    (a) True (b) False
34. T2 bacteriophage rely on the lytic life cycle to generate an ongoing infection.
    (a) True (b) False
35. DNA is the genetic material for all organisms.
    (a) True (b) False

36. A nucleoside consists of a nitrogenous base, a sugar, and up to three phosphate groups.

(a) True (b) False

37. Deoxycytidine refers to the nucleoside which includes the base cytosine and deoxyribose sugar.

(a) True (b) False

38. Hydrogen bonding is the primary chemical bond holding a strand of DNA together.

(a) True (b) False

39. Rosalind Franklin was given proper recognition for her critical role in the discovery of the structure of DNA)

(a) True (b) False

40. The backbones of the two DNA strands in a double helix run in opposite directions (one 5'-3' left to right, the other 5'-3' right to left).

(a) True (b) False

41. DNA is the only biologically relevant form of DNA.

(a) True (b) False

42. Eukaryotic DNA is packaged around proteins to form compact loops.

(a) True (b) False

43. RNA molecules follow the same base pairing rules as DNA.

(a) True (b) False

44. The addition of a single nucleotide to the DNA sequence causes a frameshift mutation.

(a) True (b) False

45. In a frameshift mutation all of the amino acids before the shift are changed.

(a) True (b) False

46. A mutation that changes a single nucleotide can result in a different amino acid being added into a protein.

(a) True (b) False

47. DNA mutations are passed on to a cell's progeny.

(a) True (b) False

48. An ATP molecule can lose 2 phosphate groups to become another molecule.

(a) True (b) False

49. Chargaff's rules states that the ratio of the purines to pyrimidines in the double-stranded DNA is always constant at 1.

(a) True (b) False

50. Is RNA polymerase required for *in vivo* DNA replication?

(a) True (b) False

51. A noncompetitive inhibitor doesn't bind to the enzyme.

(a) True (b) False

52. Bam HI restriction site is within tetracycline resistant gene. A foreign gene is inserted into a plasmid using Bam HI. When bacteria with the recombinant plasmid is grown in tetracylcine, will it die?

(a) True (b) False

53. After DNA replication two identical DNA molecules are formeD)

(a) True (b) False

54. One DNA strand is used as the template to make two new strands of DNA)

(a) True (b) False

55. The lagging strand of DNA is synthesized continuously.

(a) True (b) False

56. DNA ligase forms a bond between the 3'OH of the growing strand and the 5' phosphate in front of it, joining the DNA fragments together.

(a) True (b) False

57. Base pairing rules are critical for the ability of each DNA strand to act as a template.

(a) True (b) False

58. Meselson and Stahl's experiment was able to demonstrate in one round of division that DNA replication is a semiconservative process.

(a) True (b) False

59. DNA replication requires an origin, where the two strands can be separateD)

(a) True (b) False

60. DnaA protein is unable to bind hemimethylated DNA because its binding site includes a GATC methylation site.

(a) True (b) False

61. DNA replication proceeds only in one direction.

(a) True (b) False

62. DNA polymerase recognizes correct base pairs and incorporates new nucleotides which are correctly base paired to the template.

(a) True (b) False

63. The fidelity of DNA replication in *E. coli* is about one mistake per 100 million incorporated nucleotides.

(a) True (b) False

64. DNA polymerase moves toward the replication fork on both the lagging and leading strands.

(a) True (b) False

65. Map units used in linkage mapping correlate with a specific number of base pairs.

(a) True (b) False

66. DNA replication results in two identical daughter molecules each consisting of one old (original) strand and one newly-synthesized stranD)

(a) True (b) False

67. The point where separation of the DNA occurs is called the replication fork.

(a) True (b) False

68. The first amino acid enters through the A site.

(a) True (b) False

69. The codon of tRNA is complementary to the anticodon of mRNA)

(a) True (b) False

70. DNA replication results in two identical daughter molecules each consisting of one old (original) strand and one newly-synthesized stranD)

(a) True (b) False

71. The point where separation of the DNA occurs is called the replication fork.

(a) True (b) False

72. The synthesis of messenger RNA is called translation.

(a) True (b) False

73. Synthesis of a new strand of messenger RNA is in the 5' to 3' direction.

(a) True (b) False

74. When RNA polymerase reaches a specific sequence of nucleotides on the DNA called the transcription terminator, a hairpin loop structure forms on the DNA strand causing the RNA polymerase and mRNA to dissociate from the DNA)

(a) True (b) False

75. RNA polymerase copies the template strand of DNA making single-stranded mRNA complementary to the DNA template.

(a) True (b) False

76. DNA polymerase catalyzes the elongation of a DNA strand in the 5'—>3' direction.

(a) True (b) False

77. During semiconservative replication, the original double helix remains intact and a new double helix forms.

(a) True (b) False

78. DNA polymerase uses a nucleoside triphosphate, bearing phosphates on the 5' carbon to form a phosphodiester linkage to the free 3' hydroxyl group on the end of the DNA strand it is synthesizing.

(a) True (b) False

79. The elongation of Okazaki fragments during DNA synthesis progresses away from the replication fork.

(a) True (b) False

80. The process of acquiring a gene from another organism is called transformation.

(a) True (b) False

81. The Hershey-Chase experiment showed that the virus protein is responsible for directing the production of new viruses.

(a) True (b) False

82. The two sides of the double helix structure of DNA are held together with hydrogen bonds.

(a) True (b) False

83. DNA replication is called semiconservative because half of each of the two new strands of DNA is "old" DNA from the original DNA molecule and half is "new" DNA)

(a) Truc (b) False

84. A DNA molecule has the same amount of adenine and thymine.

(a) True (b) False

85. In situ hybridization is done on fragments of specific chromosomes.

(a) True (b) False

86. Changes in restriction enzyme sites can be used as polymorphic markers.

(a) True (b) False

Answers

Multiple Choice Question

1	2	3	4	5	6	7	8	9	10	11	12	13	14	15
C	D	B	D	B	D	B	C	A	C	B	A	B	C	B
16	17	18	19	20	21	22	23	24	25	26	27	28	29	30
C	C	A	B	C	D	D	C	A	B	C	B	A	C	B
31	32	33	34	35	36	37	38	39	40	41	42	43	44	45
D	B	B	C	C	D	A	C	B	A	B	D	A	C	D
46	47	48	49	50	51	52	53	54	55	56	57	58	59	60
D	D	B	C	D	D	C	B	A	C	C	D	B	C	B

61	62	63	64	65	66	67	68	69	70	71	72	73	74	75
A	A	A	A	B	C	C	B	A	D	C	B	C	A	B
76	77	78	79	80	81	82	83	84	85	86	87	88	89	90
C	A	B	C	D	A	C	D	B	A	D	C	A	D	B
91	92	93	94	95	96	97	98	99	100	101	102	103	104	105
A	C	D	D	E	A	D	D	D	A	A	D	C	C	B
106	107	108	109	110	111	112	113	114	115	116	117	118	119	120
C	C	A	D	B	A	D	A	A	B	C	B	B	A	A
121	122	123	124	125	126	127	128	129	130	131	132	133	134	135
B	B	C	B	A	B	D	A	C	B	A	D	D	B	C
136	137	138	139	140	141	142	143	144	145	146	147	148	149	150
B	D	D	B	B	C	A	B	D	A	A	D	A	D	D
151	152	153	154	155	156	157	158	159	160	161	162	163	164	165
D	A	D	C	B	B	A	C	C	A	C	A	C	D	D
166	167	168	169	170	171	172	173	174	175	176	177	178	179	180
D	B	A	B	B	A	D	B	D	C	C	C	A	C	D
181	182	183	184	185	186	187	188	189	190	191	192	193	194	195
C	D	D	A	D	A	D	C	D	D	C	C	B	D	C
196	197	198	199	200	201	202	203	204	205	206	207	208	209	210
D	C	C	A	D	C	B	C	A	D	C	B	D	B	B
211	212	213	214	215	216	217	218	219	220	221	222	223	224	225
C	C	A	B	A	C	C	A	A	A	D	D	C	B	C
226	227	228	229	230	231	232	233	234	235	236	237	238	239	240
B	A	C	B	D	C	C	B	A	A	C	D	D	C	A
241	242	243	244	245	246	247	248	249	250	251	252	253	254	255
C	D	A	C	B	D	D	D	C	B	C	C	C	C	B
256	257	258	259	260	261	262	263	264	265	266	267	268	269	270
A	A	B	C	D	B	D	A	A	D	D	B	A	C	D
271	272	273	274	275	276	277	278	279	280	281	282	283	284	285
B	D	A	C	D	B	A	D	A	C	B	D	D	A	B

286	287	288	289	290	291	292	293	294	295	296	297	298	299	300
C	C	A	D	A	C	B	B	C	D	A	C	B	D	A
301	302	303	304	305	306	307	308	309	310	311	312	313	314	315
C	D	D	D	C	A	B	B	C	D	A	D	D	A	C
316	317	318	319	320	321	322	323	324	325	326	327	328	329	330
C	B	D	A	C	C	B	B	A	C	B	C	A	C	D
331	332	333	334	335	336	337	338	339	340	341	342	343	344	345
D	C	C	D	D	A	B	D	B	B	C	D	A	B	C
346	347	348	349	350	351	352	353	354	355	356	357	358	359	360
D	D	B	C	B	D	A	B	D	D	B	D	A	D	A
361	362	363	364	365	366	367	368	369	370	371	372	373	374	375
D	C	C	D	A	D	C	B	D	D	A	D	D	C	D
376	377	378	379	380	381	382	383	384	385	386	387	388	389	390
D	C	C	B	B	C	B	B	B	B	C	C	C	C	B
391	392	393	394	395	396	397	398	399	400	401	402	403	404	405
A	B	C	C	B	A	B	B	C	B	A	A	A	A	D
406	407	408	409	410	411	412	413	414	415	416	417	418	419	420
C	A	A	B	B	C	B	D	D	A	B	C	A	D	B
421	422	423	424	425	426	427	428	429	430	431	432	433	434	435
C	B	C	B	D	B	B	D	B	A	C	B	D	D	A
436	437	438	439	440	441	442	443	444	445	446	447	448	449	450
B	D	A	C	A	A	C	C	D	A	D	D	A	C	D
451	452	453	454	455	456	457	458	459	460	461	462	463	464	465
A	B	B	D	D	A	A	C	A	B	B	A	B	C	C
466	467	468	469	470	471	472	473	474	475	476	477	478	479	480
B	A	D	D	B	C	D	C	D	A	B	B	A	A	B
481	482	483	484	485	486	487	488	489	490	491	492	493	494	495
D	C	C	B	C	C	C	B	D	B	C	D	B	B	D
496	497	498	499	500	501	502	503	504	505	506	507	508	509	510
B	A	C	B	C	D	C	C	C	C	B	A	A	B	B

511	512	513	514	515	516	517	518	519	520	521	522	523	524	525
B	B	A	D	C	B	C	D	C	D	C	C	B	B	B
526	527	528	529	530	531	532	533	534	535	536	537	538	539	540
B	B	D	A	A	D	A	B	C	A	C	A	C	D	D
541	542	543	544	545	546	547	548	549	550	551	552	553	554	555
D	B	A	B	B	A	D	B	D	D	C	C	C	B	A
556	557	558	559	560	561	562	563	564	565	566	567	568	569	570
C	C	C	B	D	B	C	C	A	B	A	C	C	D	B
571														
D														

Fill in the Blanks

1	2	3	4	5	6	7	8	9	10	11	12	13	14	15
A	A	B	C	D	C	C	A	B	B	B	C	C	C	D
16	17	18	19	20	21	22	23	24	25	26	27	28	29	30
A	B	D	D	C	C	D	A	D	D	A	A	D	D	B
31	32	33	34	35	36	37	38	39	40	41	42	43	44	45
A	B	D	B	A	A	D	D	B	C	D	C	A	A	D
46	47	48	49	50	51	52	53	54	55	56	57	58	59	60
C	C	A	D	D	C	D	D	C	B	A	A	D	D	B
61	62	63	64	65	66	67	68	69	70	71	72	73	74	75
B	D	A	C	B	A	A	D	B	D	D	C	C	A	D
76	77	78	79	80	81	82	83	84	85					
D	D	D	D	C	D	B	C	C	A					

True/False

1	2	3	4	5	6	7	8	9	10	11	12	13	14	15
B	A	A	B	A	A	B	A	A	A	A	A	B	B	B
16	17	18	19	20	21	22	23	24	25	26	27	28	29	30
A	A	B	B	A	B	A	A	A	A	B	A	B	A	A
31	32	33	34	35	36	37	38	39	40	41	42	43	44	45
A	B	B	A	B	B	A	B	B	A	B	A	B	A	B
46	47	48	49	50	51	52	53	54	55	56	57	58	59	60
A	A	A	A	A	B	A	A	B	B	A	A	B	A	B
61	62	63	64	65	66	67	68	69	70	71	72	73	74	75
B	A	A	A	B	A	A	B	B	A	A	B	A	B	A
76	77	78	79	80	81	82	83	84	85	86				
A	B	A	A	A	B	A	A	A	B	A				

Chapter – 12

Gene Transcription and RNA Modification

Overview of Transcription

- Gene expression requires base sequences that perform different functional roles
- The three stages of transcription are initiation, elongation, and termination
- RNA transcripts have different functions

Transcription in Bacteria

- A promoter is a short sequence of DNA that is necessary to initiate transcription
- Bacterial transcription is initiated when RNA polymerase holoenzyme binds at a promoter sequence
- The RNA transcript is synthesized during the elongation stage
- Transcription is terminated by either an RNA-binding protein or an intrinsic terminator

Transcription in Eukaryotes

- Eukaryotes have multiple RNA polymerases that are structurally similar to the bacterial enzyme

- Eukaryotic structural genes have a core promoter and regulatory elements
- Transcription of eukaryotic structural genes is initiated when RNA polymerase II and general transcription factors bind to a promoter sequence
- Transcription termination of RNA polymerase II occurs after the 3′end of the transcript is cleaved near the polyA signal sequence

RNA Modification

- Some large RNA transcripts are processed into smaller functional transcripts by enzymatic cleavage
- Introns were experimentally identified via microscopy
- Different splicing mechanisms can remove introns
- Pre-mRNA splicing occurs by the action of a spliceosome
- The ends of eukaryotic pre-mRNAs have a 5' cap and a 3' tail
- The nucleotide sequence of RNA can be modified by RNA editing

Multiple Choice Question

1. Which of the following is not required for accurate transcription?
   (a) Regulatory information within the DNA
   (b) Proteins which bind to the DNA
   (c) Proteins which copy the information from the DNA
   (d) All of these are required
   (e) None of these are required
2. The closed promoter complex consists of all of the following except:
   (a) Promoter region of the DNA (b) RNA polymerase
   (c) Transcription factors (c) Helicase
3. Which type(s) of RNA molecule has(have) the primary function of modifying other RNA molecules?
   (a) RNA of RNaseP (b) snRNa
   (c) scRNa (d) a and b
   (e) All of the above

4. If you have a strain of bacteria that produce nonfunctional sigma factor, which step of transcription would you expect will be affected?

(a) These bacteria will be unable to assemble RNA polymerase onto the DNA strand

(b) These bacteria will be unable to identify and tightly bind promoter elements

(c) These bacteria will be unable to convert a closed complex to an open complex

(d) These bacteria will be unable to synthesize RNA once an open complex is formed

5. Which of the following mutations would you expect to have the most significant effect on DNA binding by the transcription factor depicted in figure 12.6?

(a) A conservative substitution in α helix 1

(b) A conservative substitution in α helix 2

(c) A nonconservative substitution in α helix 3

(d) A nonconservative substitution in the turn region

6. What is the primary factor involved in termination of RNA synthesis?

(a) Temporary stabilization of the RNA:DNA hybrid region

(b) Synthesis of a specific termination signal in the RNA

(c) Destabilization of hydrogen bonding between DNA and RNA within the open complex

(d) Recognition of a termination signal on the DNA by termination proteins

7. Which type of transcriptional termination requires formation of a stem loop secondary structure in the RNA?

(a) ρ dependent (b) ρ independent

(c) Polymerase independent (d) Both a and b

(e) None of the above

8. Intrinsic terminators are:

(a) RNA sequences that assist in termination of transcription

(b) DNA sequences that assist in termination of transcription

(c) Proteins that assist in termination of transcription
(d) All of these
(e) None of these

9. Which statement could NOT describe both eukaryotic and prokaryotic transcription?
(a) Promoter elements in the DNA are required for polymerase binding
(b) Transcription factors control the binding of RNA polymerase
(c) RNA polymerase opens a double stranded DNA to expose the template strand
(d) RNA polymerase catalyzes RNA synthesis from 3' to 5' of the new strand
(e) None of these (all apply to both eukaryotic and prokaryotic transcription)

10. What would you expect to be the effect of a mutation of RNA polymerase that removed the rudder region?
(a) DNA would not be opened to expose the template strand
(b) Nucleotides would not base pair correctly with the template strand
(c) Nucleotides would not be connected into an RNA strand
(d) RNA would not be released from the template DNA

11. Which of the following is an example of a *cis*-acting element in eukaryotic transcription?
(a) RNA polymerase II (b) Transcription factors
(c) Enhancers (d) All of these
(e) None of these

12. Which of the following RNA modifications could best be described as "removal of intervening sequences"?
(a) Trimming (b) 5' capping
(c) Poly-A tailing (d) Splicing
(e) RNA editing or base modification

13. When studying eukaryotic enzymes in relation to their prokaryotic counterparts, you discover that some eukaryotic enzymes seem to consist of several regions, each of which performs the function of a separate enzyme in bacteria. Which of the following explanations would you be most likely to accept?

(a) Random mutation has generated these multi-function eukaryotic enzymes

(b) A polycistronic transcript in bacteria became a polycistronic transcript in eukaryotic cells

(c) Several genes located near each other in bacteria became transcribed together and the intervening sequences spliced out

(d) Similar regulatory elements for each of the prokaryotic proteins allowed them to be expressed together

14. Which of the following would be expected to demonstrate that mRNA contains introns?

(a) A comparison of the DNA and mRNA sequences

(b) A comparison of the genomic DNA and cDNA sequences

(c) A hybridization between DNA and mRNA molecules

(d) All of these

(e) None of these

15. Which of the following statements about RNA splicing is NOT correct?

(a) Some introns can splice themselves out

(b) Splicing is required for some tRNAs

(c) The final step of splicing is formation of a phosphodiester linkage

(d) Introns often have a specific function after they are removed

(e) None of these; all of the above statements are correct

16. Which of the following would you expect to find if you took apart a spliceosome?

(a) RNA only (b) Protein only

(c) Both RNA and protein (d) Neither RNA nor protein

17. Which of the following is a critical function of the 5' cap of eukaryotic mRNAs?

(a) Protection from inappropriate splicing

(b) Synthesis of the polyA tail

(c) Transport of mRNAs into the nucleus
(d) Recognition of the 5' end of the transcript by a ribosome

18. Which of the following pairs is not correctly described?
(a) 5' splice site: begins with GU, marks 5' end of intron
(b) Branch site: contains A that binds U2; branch site of lariat
(c) 3' splice site: begins with AG, marks 3' end of intron
(d) Trimer: consists of U4/U6 and U5, brings 5' splice site to branch site
(e) All of these are correctly matched

True/False

1. From the promoter of the gene, you would expect the coding strand to run 5' to 3'.
(a) True (b) False

2. Regulatory sequences are DNA sequences which encode transcription factors.
(a) True (b) False

3. During transcription, the RNA polymerase would be expected to spend the majority of its time in the open complex rather than the closed complex.
(a) True (b) False

4. The product of a structural gene is mRNA and then a polypeptide.
(a) True (b) False

5. A site designated +15 would be located within the promoter of the gene.
(a) True (b) False

6. Bacterial RNA polymerase holoenzyme consists of four subunits.
(a) True (b) False

7. Intrinsic termination requires the rho protein.
(a) True (b) False

8. Both types of transcriptional termination rely on stem-loop structures in the newly transcribed RNA.
(a) True (b) False

9. Regulation of genes in both eukaryotes and prokaryotes requires cis and trans acting factors.

(a) True (b) False

10. Only proteins can be biological catalysts

(a) True (b) False

Answers

Multiple Choice Question

1	2	3	4	5	6	7	8	9	10	11	12	13	14	15
D	D	D	B	C	C	D	A	D	D	C	D	C	D	D
16	17	18												
C	D	C												

True/False

1	2	3	4	5	6	7	8	9	10
A	B	A	A	B	B	B	A	A	B

❑❑❑

Chapter – 13

Translation of mRNA

The Genetic Basis for Protein Synthesis

- Archibald Garrod proposed that some genes code for the production of a single enzyme
- Beadle and Tatum's experiments with Neurospora led them to propose the one gene-one enzyme hypothesis
- During translation, the genetic code within mRNA is used to make a polypeptide with a specific amino acid sequence
- Exceptions to the genetic code are known to occur, which include the incorporation of selenocysteine and pyrrolysine into polypeptides
- Synthetic RNA helped to decipher the genetic code
- The use of RNA copolymers and the triplet binding assay also helped to crack the genetic code
- A polypeptide chain has directionality from its amino terminal to its carboxyl terminal end
- The amino acid sequences of polypeptides determine the structure and function of proteins
- Cellular proteins are primarily responsible for the characteristics of living cells and an organism's traits

Structure and Function of tRNA

- The function of a tRNA depends on the specificity between the amino acid it carries and its anticodon

- tRNA functions as the adaptor molecule involved in codon recognition
- Common structural features are shared by all tRNAs
- Aminoacyl-tRNA synthetases charge tRNAs by attaching the appropriate amino acid
- Mismatches that follow the wobble rule can occur at the third position in codon-anticodon pairing

Ribosome Structure and Assembly

- Bacterial and eukaryotic ribosomes are assembled from rRNA and proteins
- Components of ribosomal subunits form functional sites for translation

Stages of Translation

- The initiation stage involves the binding of mRNA and the initiator tRNA to the ribosomal subunits
- Polypeptide synthesis occurs during the elongation stage
- Termination occurs when a stop codon is reached in the mRNA
- Bacterial translation can begin before transcription is completed
- Bacterial and eukaryotic translation show similarities and differences

Multiple Choice Question

1. Which of the following was important in Garrod's proposal regarding inheritance of a defective enzyme?
   (a) Identification of a defective gene in individuals with alkaptonuria
   (b) Identification of a defective gene in parents of alkaptonuria patients
   (c) Identification of a defective enzyme in individuals with alkaptonuria
   (d) Identification of the inheritance pattern of alkaptonuria as autosomal recessive

2. Which of the following correctly restates the one gene - one enzyme hypothesis?
   (a) Every gene encodes a protein
   (b) Each structural gene encodes an essential metabolic enzyme
   (c) Each structural protein is encoded by a single gene
   (d) All of these
   (e) None of these
3. Which statement about the genetic code is not correct?
   (a) Most amino acids are specified by more than one codon
   (b) Many codons have more than one function
   (c) Some organisms use codons differently than other organisms
   (d) In some cases, the third base of a codon is nonessential
   (e) None of these (all these statements are correct)
4. What characteristic of proflavin allowed Crick to use it in his determination that the genetic code is a triplet code?
   (a) It causes mutations in DNA
   (b) It can cause large deletions in DNA
   (c) It causes single insertions or deletions in DNA sequences
   (d) It can be lethal
5. Which of the following facts is the most important evidence that the genetic code is a triplet code?
   (a) Insertions or deletions caused significant defects in protein structure
   (b) The effect of single insertions could be remedied by single deletions nearby
   (c) Two nearby insertions caused significant defects in protein structure
   (d) Three insertions near each other often led to minor defects in protein structure

6. What primarily determines the specific chemical character of a particular amino acid?
   (a) The amino (NH_3^+) group
   (b) The carboxyl (COO^-) group
   (c) The R group
   (d) All of these
   (e) None of these
7. Which of the following represents the primary structure of a protein?
   (a) Folding which allows amino acids of a polypeptide to be arranged in regions of similar characteristics (hydrophobic residues together, for ex.)
   (b) Local folding which is stabilized by hydrogen bonds
   (c) The sequence of amino acids in the polypeptide
   (d) Interactions between two or more polypeptides
8. Using Figure 13.7, which amino acid would be most likely to interact with serine in formation of secondary or tertiary structure?
   (a) Valine
   (b) Tryptophan
   (c) Asparagine
   (d) Any of these with equal affinity
   (e) None of these
9. Which of the following could not potentially explain the albino phenotype in humans?
   (a) Lack of an enzyme that synthesizes the pigment molecules
   (b) Lack of a protein that transports the pigment molecules to the appropriate location
   (c) Lack of a receptor protein that allows the pigment to be taken up into appropriate cells
   (d) All of these
   (e) None of these
10. A tRNA molecule must:
   (a) Have a base pair sequence identical to the codon in the mRNA
   (b) Be attached to a particular amino acid

(c) Bind to the protein to determine primary sequence

(d) All of these

(e) None of these

11. Which of these would be LEAST likely to lead to incorporation of the wrong amino acid into a polypeptide?

(a) A change in the wobble base of the codon

(b) A mistake in amino acid attachment to the tRNA

(c) A change in the third base of the anticodon

(d) All of these would be expected to lead to incorporation of the wrong amino acid

(e) None of these would be expected to lead to incorporation of the wrong amino acid

12. Which portion of the tRNA molecule is specifically recognized by aminoacyl-tRNA synthetase?

(a) The acceptor stem

(b) Bases in the stem loop regions

(c) The anticodon region

(d) All of these

(e) None of these

13. Which of these is a structural difference between prokaryotic and eukaryotic cytosolic ribosomes?

(a) Prokaryotic ribosomes consist of two subunits

(b) Eukaryotic ribosomes consist of both protein and RNa

(c) Eukaryotic ribosomes are smaller than prokaryotic ribosomes

(d) All of these

(e) None of these

14. Which of these represents a difference between eukaryotic and prokaryotic transcription?

(a) Initiation factors are required for forming a eukaryotic initiation complex

(b) A Kozak consensus sequence identifies the start codon in eukaryotes

(c) Binding of the initiator tRNA is required for assembly of the initiation complex in eukaryotes

(d) All of these are differences between eukaryotic and prokaryotic transcription

(e) None of these are differences between eukaryotic and prokaryotic transcription

15. Which of these is an enzymatic function of the ribosome?

(a) Polynucleotide kinase (b) Base pair recognition

(c) Methyl transferase (d) Peptidyl transferase

16. Termination codons differ from other codons in that:

(a) They only recognize a single tRNA

(b) They are capable of binding to many different tRNAs

(c) They bind to release factors instead of tRNAs

(d) They only specify a single amino acid

(e) Termination codons do not differ from other codons

17. Which of these DNA elements does NOT encode information that is recognized as RNA or protein?

(a) Shine-Dalgarno sequence

(b) Kozak consensus sequence

(c) Sorting signals

(d) All of these are recognized only as DNA

(e) None of these are recognized only as DNA

True/False

1. Translation converts information from the language of nucleic acids to the language of proteins.

(a) True (b) False

2. The genetic code is universal, meaning that any given codon specifies the same amino acid, regardless of the type of cell.

(a) True (b) False

3. A polypeptide is the same thing as a protein.
   (a) True (b) False
4. Codons may have different functions than specification of amino acids.
   (a) True (b) False
5. The organization of a polypeptide into a functional protein is typically a spontaneous process.
   (a) True (b) False
6. A particular polypeptide may either form ? helices or ? sheets.
   (a) True (b) False
7. Ribozyme activity is required for both transcription and translation.
   (a) True (b) False
8. In a polyribosome, the longest polypeptide chain would be expected to be attached to the ribosome which is farthest away from the RNA polymerase.
   (a) True (b) False

Answers

Multiple Choice Question

1	2	3	4	5	6	7	8	9	10	11	12	13	14	15
D	B	B	C	D	C	C	C	E	B	A	D	E	B	D
16	17													
B	E													

True/False

1	2	3	4	5	6	7	8
A	B	B	A	A	B	B	B

❑❑❑

Chapter – 14

Chemistry of Gene

The Structure of Proteins

- Proteins are composed of building blocks called amino acids. A string of amino acids is called a polypeptide chain. Once such a chain has folded into its working three-dimensional shape, it is a protein. Though there are tens of thousands of different proteins, all of them are put together from a starting set of 20 amino acids. It is the order in which the amino acids are linked in a polypeptide chain that determines which protein will be produced.

The Genetic Basis for Protein Synthesis

- Archibald Garrod proposed that some genes code for the production of a single enzyme
- Beadle and Tatum's experiments with Neurospora led them to propose the one gene-one enzyme theory
- During translation, the genetic code within mRNA is used to make a polypeptide with a specific amino acid sequence
- Synthetic RNA helped to decipher the genetic code
- The use of RNA copolymers and the triplet binding assay also helped to crack the genetic code
- The amino acid sequences of polypeptides determine the structure and function of proteins

- Cellular proteins are primarily responsible for the characteristics of living cells and an organism's traits

Transcription

- In a cell, transcription is the process that copies the information stored in DNA into RNA.
- Translation uses the RNA's information to construct amino acid chains that comprise proteins.
- The directional flow of genetic information from DNA to RNA to protein is known as the central dogma of molecular biology.
- Gene regulation controls the pattern of gene expression and cellular differentiation.
- The information in DNA is transferred to messenger RNA through complementary base pairing. Each "C" nucleotide in a segment of DNA being transcribed results in a "G" nucleotide being added to a segment of mRNA, and so forth. The length of mRNA that results from this base pairing is called a transcript. The enzyme RNA polymerase unwinds the DNA sequence to be transcribed and then strings together the chain of RNA nucleotides that is complementary to it, thus producing the transcript.
- Within eukaryotic gene sequences, lengths of DNA that code for amino acids are interspersed with lengths that do not code for amino acids. Both types of sequences are copied onto an initial chain of RNA nucleotides called a primary transcript. This chain then undergoes a form of editing in which non-coding sequences are removed and coding sequences are spliced back together. The product of this editing is a messenger RNA molecule. A region of a DNA sequence or a primary RNA transcript that contributes to messenger RNA is called an exon. A region of a DNA sequence or a primary RNA transcript that does not contribute to messenger RNA is called an intron. The function of introns is being investigated.
- Each three coding bases of DNA pairs with three mRNA bases, but each group of three mRNA bases then codes for a single amino acid. Each triplet of mRNA bases that codes for an amino acid is called a codon. The inventory of linkages between base triplets and the amino acids they code for is called the genetic code.

Transcription Factors

- Bacterial genes are organized into operons and coordinately regulated.
- Operon organization is rare in Eucaryotes, but has been found in the roundworm Caenorhabditis elegans.
- Most Eucaryotic genes are controlled by a complex set of transcription factors.
- A few human diseases are due to defects in transcriptional factors.

RNA Structure and Types

- RNA is copied from one strand of the double helix called the template strand.
- RNA differs from DNA in that it is single-stranded, has uracil instead of thymine, and has ribose instead of deoxyribose.
- Messenger RNA (mRNA) carries the information that specifies a particular amino acid sequence of a protein product. Each three mRNA bases in a row form a codon.
- Ribosomal RNA joins certain proteins to form ribosomes. Ribosomes physically support the other structures involved in protein synthesis, and some rRNA catalyzes formation of peptide bonds.
- Transfer RNA is cloverleaf-shaped and connects an mRNA codon to an amino acid.

RNA Processing

- In prokaryotes, RNA is translated as soon as it is transcribed.
- In eukaryotes, RNA is often altered (or modified) before it is active.
- Messenger RNA gains a modified nucleotide cap and a poly A tail.
- Many genes have intervening sequences called introns, which are transcribed and cut from the mRNA. The protein encoding sequences in mRNA, exons, are then reattached.
- The function of introns is still a mystery.
- Ribozymes are small RNAs with catalytic activity that can splice introns. They join proteins to form snurps, which associate to form spliceosomes.
- RNA editing changes the size of the protein product in different cell types.

- After being processed, the RNA must be exported from the nucleus before it is translated.

Translating a Protein

- Transfer RNA serves as a bridging molecule in protein synthesis, thanks to its ability to bond with both amino acids and nucleic acids (in the form of mRNA). A given tRNA molecule bonds with a specific amino acid in the cell's cytoplasm and then transfers that amino acid to a ribosome in which an mRNA transcript is being "read." There, another portion of the tRNA molecule, called an anticodon, bonds with the appropriate codon in the messenger RNA chain.
- Ribosomes are complex "workbenches" of protein synthesis composed of proteins and ribosomal RNA (rRNA). Each generally exists as two subunits in the cytoplasm that may come together only with the initiation of protein translation. Each bears P, A, and E binding sites to which tRNA molecules bond, thus facilitating the synthesis of an amino acid chain.
- Translation works by means of a succession of tRNA molecules arriving at a ribosome, bound to their appropriate amino acids, and then bonding to their appropriate codon in the mRNA transcript. As this process takes place, the succession of amino acids is linked together into a polypeptide chain.
- In addition to messenger RNA, transfer RNA and ribosomal RNA are transcribed from sequences of DNA. As such, it is appropriate to think of a gene as an entity that does more than code for proteins. A gene can be defined as a segment of DNA that brings about the transcription of a segment of RNA.

Structure and Function of tRNA

- The function of a tRNA depends on the specificity between the amino acid it carries and its anticodon
- Translation depends on the recognition between the codon in mRNA and the anticodon in tRNA
- tRNAs share common structural features
- Aminoacyl-tRNA synthetases charge tRNAs by attaching an appropriate amino acid
- Mismatches that follow the wobble rule can occur at the third position in codon-anticodon pairing

Deciphering the Genetic Code

- Crick and coworkers confirmed the triplet nature of the genetic code.
- The reading frame of a gene is the particular sequence of amino acids in a protein that is encoded from a certain point in a gene. Adding or subtracting 1 or 2 DNA bases to a gene disrupts the reading frame. Adding or deleting 3 contiguous bases, adds or deletes one amino acid to the protein product but does not disrupt the reading frame.
- The genetic code is nonoverlapping, continuous, virtually universal, and degenerate.
- Crick hypothesized the existence of an adaptor molecule necessary for translation.
- This was later discovered and called transfer RNA (tRNA).

Building a Protein

- As translation begins, mRNA, tRNA with bound amino acids, ribosomes, energy molecules, and protein factors assemble.
- To initiate translation, the mRNA leader sequence binds to rRNA in the small subunit of a ribosome, and the first codon attracts a tRNA to form the initiation complex.
- In elongation, the large ribosomal subunit attaches to the initiation complex. Peptide bonds form between the amino acids attached to the aligned tRNAs, building a polypeptide.
- Protein synthesis halts when a stop codon is reached.
- Translation is efficient and economical, as RNA, ribosomes, enzymes, and key proteins are recycled.

Protein Folding

- The protein folds as translation proceeds, with enzymes and chaperone proteins assisting the amino acid chain in assuming its final functional (three-dimensional) shape.
- In addition to folding properly certain proteins must be modified before they are functional.
- Errors in protein folding may be the basis of some diseases.

Reconciling Gene and Protein Number

- Differential intron splicing and exon shuffling explain how one gene can produce more than one protein.
- Intron sequences of one gene may contain the coding sequences for another protein.
- Trans-splicing involves the splicing together of two different RNAs (from separate genes) to form a translatable message.

The Human Genome Sequence Reveals Unexpected Complexity

- The genomes of an increasing number of organisms have been and are being sequenced.
- Proteomics, based on genome analysis, is the study of the collection of proteins produced by a particular cell.
- The production of proteins can be facilitated or impeded at any point in the protein synthesis pathway. This process is called genetic regulation. A critical point in genetic regulation comes during transcription, when various factors can either facilitate or impede the binding of RNA polymerase to a DNA promoter sequence. One example of transcription regulation is the bacterial lac operon system, which regulates the metabolism of lactose. The lac operon provides examples of DNA sequences that are purely regulatory. It also provides an example of feedback in genetics, in which proteins that result from transcription go on to bond with DNA sequences, thereby controlling the production of other proteins.
- The human genome contains an estimated 30,000 genes – only about 11,000 more genes than exist in the primitive roundworm C. elegans. Human complexity is in part attributable to the fact that a majority of the human genes appear to be able to code for more than one protein through the process known as alternative splicing of mRNA transcripts.

What Does the Other 98.5 Percent of the Human Genome Do?

- Ninety eight percent of the human genome does not code for protein. What is its function?
- About 1/3 of the human genome produces non-coding RNAs (i.e. rRNA, tRNA, snRNA, snoRNA, etc.).

- Part of the human genome is composed of pseudogenes. These may or may not be transcribe, and are never translated into protein.
- A large part of the human genome is composed of repeat sequences such as SINEs and LINEs.

Multiple Choice Question

1. If a Meselson-Stahl type experiment is performed, how much DNA is expected to be of intermediate density after three rounds of replication?

   (a) 0% (b) 25%

   (c) 50% (d) 75%

   (e) 100%

2. If a nucleic acid is found to contain 20% A and 20% T, the molecule is probably

   (a) ss DNA (b) ds DNA

   (c) ss RNA (d) ds RNA

   (e) DNA-RNA hybrid

3. If double-stranded DNA is found to contain 20% A and 20% T, what can you conclude about the percentage of G and C bases present?

   (a) 20% C and 2% G

   (b) G+C=A+T therefore G+C=40%

   (c) G+C=1-(A+T) therefore G+C=60%

   (d) G+C=80%

   (e) Cannot be determined

4. Considering DNA replication along the 5'—>3' template strand, what event takes place first?

   (a) DNA polymerase III begins adding RNA nucleotides

   (b) DNA polymerase III begins adding DNA nucleotides

   (c) Primase adds an RNA primer

   (d) Primase removes the 5'-phosphate

   (e) Ligase adds a free 3'- OH group

5. If Hershey and Chase found S35 in progeny phages rather than P32, their experiment would have demonstrated that
   (a) Proteins contain phosphorus
   (b) DNA contains sulfur
   (c) Phage DNA enters the host cell
   (d) Phage protein enters the host cell
   (e) Phage can kill the host cell
6. DNA replication along the 5'—>3' template strand
   (a) Is continuous (b) Is discontinuous
   (c) Cannot occur (d) Forms a leading strand
   (e) Occurs in both directions
7. DNA polymerase III
   (a) Joins together Okazaki fragments
   (b) Creates RNA primers
   (c) Removes RNA primers and fills gaps
   (d) Unwinds the DNA helix
   (e) Is the primary replicating enzyme
8. Which of the DNA molecules would have the highest melting temperature?
   (a) A+T=10% (b) A+T=30%
   (c) A+T=50% (d) G+C=30%
   (e) G+C=10%
9. The DNA replication model functioning during E.coli conjugation is called
   (a) Theta (b) Rolling circle
   (c) D-loop (d) Linear
   (e) No replication occurs
10. A nucleoside is composed of a
   (a) Purine+pyrimidine (b) Sugar+sugar
   (c) Base+sugar (d) Base+sugar+phosphate
   (e) Base+sugar+three phosphate

11. In his work with tobacco mosaic virus and Holmes rib grass virus Fraenkel-Conrat determined

(a) DNA is the genetic material used by all cellular organisms and viruses

(b) That some viruses use a complex of RNA and protein rather than DNA as their genetic information

(c) Some viruses use RNA rather than DNA as their genetic information

(d) Some viruses infect numerous species of plants and animals

(e) Proteins are the genetic information used by some viruses

12. In his experiments Fraenkel-Conrat

(a) Dissociated the RNA from the protein coats of the viruses they worked with then created hybrid viruses

(b) Complexed RNA and DNA to make functional hybrid viruses

(c) Produced infectious proteins which caused lesions to develop on tobacco plants

(d) Determined the retrovirus HIV was the cause of HIV

(e) Both a and d

13. During translation the mRNA molecule

(a) Sits between the the tRNA molecule and the large subunit of the ribosome

(b) Moves between the two subunits of the ribosome

(c) Brings amino acids from the cytoplasm to the tRNA molecule bound to the ribosome

(d) Binds amino acids together in the growing polypeptide chain

(e) Is not involved

14. tRNA molecules with bound amino acids deliver them to the

(a) P site on the ribosome

(b) 3' end of the polypeptide chain

(c) 5' end of the polypeptide chain

(d) Nucleus

(e) A site on the ribosome

15. The material that transformed harmless bacteria into virulent bacteria in Avery's experiment was

(a) DNA (b) mRNA

(c) tRNA (d) Protein

(e) RNA polymerase

16. What is the significance of Chargaff's rule in the DNA molecule?

(a) The pairing of purines and pyrimidines in DNA results in an excess of purine in some molecules

(b) Chargaff's rule says that a double helix should not form except at pH 7

(c) The rules of base pairing proposed by Chargaff result in a molecule with a constant thickness

(d) Both a and b

(e) None of the above

17. The replication of DNA molecules in a eukaryotic cell takes place

(a) Just prior to metaphase in mitosis

(b) Prior to cell division

(c) Continuously

(d) Only during meiosis

(e) Shortly after telophase

18. The production of mRNA is called

(a) Transduction (b) Transformation

(c) Transcription (d) Translation

(e) Promotion

19. Which of the following is the most accurate description of RNA polymerase?

(a) RNA polymerase is used to initiate the production of tRNA from an mRNA template

(b) RNA polymerase binds to a strand of DNA and moves along the strand to make a strand of mRNA

(c) RNA polymerase is responsible for providing a termination signal to the DNA strand at the end of rRNA synthesis

(d) DNA ligase and RNA polymerase work together to bind the nucleotides in a newly forming DNA strand

(e) RNA polymerase joins the two DNA after translation has occurred

20. During transcription

(a) Both DNA strands serve as a template for the synthesis of mRNA

(b) The two DNA strands alternate as the template for the synthesis of tRNA

(c) One of the DNA strands serves as a template for the synthesis of mRNA

(d) mRNA serves as a template for the synthesis of rRNA

(e) rRNA is assembled into complexes with proteins to make new ribosomes

21. Each amino acid is specified by

(a) Several genes

(b) An operon and a promoter

(c) A codon

(d) Markers of the surface of the cell requiring it for growth

(e) An enhancer

22. In ribosomes the rRNA is located

(a) On the surface of the large subunit

(b) Complexed with several proteins in the large subunit

(c) In both the small and the large subunits

(d) In the small subunit

(e) None of the above

23. Why can ribosomes bind to a sequence occurring at the beginning of a gene?

(a) Because of the matching, exposed sequence of rRNA nucleotides found on the small subunit of the ribosome

(b) Because of the effects of certain hormones that are associated with DNA synthesis

(c) Because the conformation of the ribosome contributes to binding with DNA

(d) Both a and b

(e) None of the above

24. What part of the tRNA molecule binds to mRNA?

(a) The single-stranded nucleotide "tail" of the tRNA molecule binds to a specific codon on the mRNA molecule

(b) The anticodon loop on the tRNA molecule binds to a specific codon on the mRNA molecule

(c) The tRNA codon binds to the anticodon on the mRNA molecule

(d) The small subunit of the tRNA binds to the promoter on the mRNA molecule

(e) The large subunit of the tRNA molecule binds to the anticodon on the mRNA molecule

25. What happens when a ribosome reaches a codon on the mRNA molecule that does not specify any of the tRNA anticodons?

(a) The ribosome begins to read the mRNA molecule in the opposite direction

(b) The ribosome releases the newly formed protein molecule

(c) The ribosome subunits separate

(d) The newly formed protein begins to break up into smaller molecules

(e) Both b and c

26. Which of the following mechanisms would function to prevent transcription?

(a) The addition of excess substrate to the environment

(b) Binding of a repressor to the operator site on the gene

(c) Binding of an enhancer to the repressor site on the gene

(d) Changing the pH of the environment

(e) Both a and d

27. What effect would the addition of lactose have on a repressed lac operon?

(a) The operator site on the operon would move

(b) It would reinforce the repression of that gene

(c) It would result in the repression of other genes in that area

(d) The lac operon would be transcribed

(e) It would have no effect whatsoever

28. Nucleosomes inhibit

(a) Activators

(b) RNA polymerase

(c) Translation

(d) Assembly of transcription factors

(e) None of the above

29. Unlike prokaryotes, the control of transcription by eukaryotes is designed to react to change by

(a) Changing (b) Ignoring change

(c) Remaining constant (d) Changing the environment

30. A form of binding motif containing a nearly identical sequence of 60 amino acids in many eukaryotes is the

(a) Homeodomain motif (d) Zinc finger motif

(c) Leucine zipper motif (d) Universal motif

(e) None of the above

31. Which of the following does not occur in the function of the catabolite activator protein (CAP) of *E. coli*?

(a) Cyclic-AMP binds to the CAP protein

(b) The protein changes shape

(c) Space is increased by the binding of tryptophan

(d) Helix-turn-helix motifs are enabled to bind to the DNA

(e) None of the above

32. All of the following can be found in a human transcription complex except

(a) Activator (b) RNA

(c) Enhancer (d) Silencer

(e) None of the above

33. Enhancers are
   (a) Proteins located adjacent to promoters
   (b) Distant sites where regulatory proteins bind
   (c) Expediters of RNA polymerase capture
   (e) Proteins that bind with repressors, deactivating them
   (e) A bacterial form of promoters

34. When tryptophan is present in the medium, the transcription of tryptophan producing genes in *E. coli* is stopped by a helix-turn-helix regulator binding to the
   (a) Trp repressor (b) Trp operon
   (c) Trp promoter (d) Trp operator
   (e) Trp polymerase

35. When a homeodomain binds to DNA, the actual binding portion of the homeodomain is
   (a) A leucine zipper (b) The operon
   (c) Zinc fingers (d) The histine
   (e) A helix-turn-helix motif

36. When tryptophan is present in the environment of *E. coli*, the tryptophan binds to the
   (a) Trp operon
   (b) Trp promoter
   (c) Trp operator
   (d) Trp repressor
   (e) Trp polymerase

37. Transcription factors appear to be unable to bind to a nucleosome because
   (a) Activators are inhibited by the configuration
   (b) Of inhibition of RNA polymerase
   (c) Of histones positioned over promoters
   (d) Nucleosomes are especially vulnerable to repressors
   (e) Operators are placed in an inaccessible position

38. In the zinc fingers motif, the spacing of the helical segments is performed by

(a) Beta sheets
(b) Helical clusters
(c) Zinc atoms
(d) Gamma helices
(e) An alpha helix

39. Translation repressor proteins may shut down translation of processed mRNA transcripts by

(a) Binding with the poly-A tail
(b) Resetting the reading frame
(c) Reinserting introns into the transcript
(d) Excising a short sequence of nucleotides
(e) Excising a short sequence of nucleotides

40. In many animals, the genes that regulate the development of stem cells are activated

(a) Once
(b) Only twice
(c) Up to 10 times
(d) Over a hundred times
(e) Not at all

41. Regulatory domains of most activators interact with

(a) The transcription factor complex
(b) RNA polymerase
(c) Repressors
(d) The regulatory factor complex
(e) The DNA binding domain

42. In order for a gene to be transcribed, RNA polymerase must have access to the DNA helix and be able to bind to the genes

(a) Activator
(b) Regulator
(c) Promoter
(d) Operator
(e) Repressor

43. In the function of the lac operon in *E. coli*, the lac genes are transcribed in the presence of lactose because

(a) RNA polymerase binds to the operator

(b) The repressor cannot bind to the promoter

(c) An isomer of lactose binds to the repressor

(d) CAP does not bind to the operator

(e) Of the absence of cAMP

44. The role of methylation of DNA is now viewed as

(a) Interfering with DNA transcription by blocking base pairing between cytosine and guanine

(b) Complexing with enhancers to prevent transcription

(c) Prevention of mutation

(d) Insuring that genes that are turned off, stay off

(e) Irrelevant to gene transcription

45. The most common form of gene expression regulation in both bacteria and eukaryotes is

(a) Translational control

(b) Transcriptional control

(c) Post-transcriptional control

(d) Post-translational control

(e) Control of passage from the nucleus

46. In eukaryotes, many genes may have to interact with each other, requiring more interacting elements than can fit around a single promoter. This physical limitation is overcome by

(a) Alternating promoters and operators

(b) Placing promoters on both sides of each gene

(c) The use of very long promoters

(d) Distant sites in a chromosome controlling transcription of a gene

(e) Having factors on one chromosome control genes on another gene

47. *E. coli* is able to use foods other than glucose in the absence of available glucose, because falling levels of glucose cause an increase of

(a) cAMP (b) CAP
(c) Lactase (d) Glu operons
(e) tRNA

48. In the absence of glucose, *E. coli* can import lactose to change into glucose and galactose because CAP binds to the

(a) cAMP (b) DNA
(c) Lac operon (d) Operator
(e) Repressor

49. Which is not part of the lac operon?

(a) Repressor (b) Activator protein
(c) Operator (d) Promotor
(e) Structural gene

50. Proteins that block the passage of RNA polymerase are called:

(a) Operons (b) Activators
(c) Repressors (d) Enhancers
(e) Promoters

51. Which of the following is part of an operon?

(a) Structural genes (b) An operator
(c) A promoter (d) A CAP binding site
(e) All of the above

52. Which of the following are not matched correctly?

(a) Exon splicing-occurs in nucleus
(b) Post-translational modifications-phosphorylation
(c) snRNA-splice out exons from transcript
(d) Activated enhancers-trigger transcription
(e) All are matched correctly

53. Which statement is not true about gene cloning?

(a) Scientists can locate a gene by visibly looking for it

(b) Gene libraries are an important part of the gene cloning process

(c) Genetic engineers use markers to identify colonies of cells that contain a specific DNA sequence

(d) All of the above are not true about gene cloning

(e) None of the above

54. What is the purpose of gene cloning in agriculutral biotechnology?

(a) To make multiple copies of a desired gene so a characteristic of that gene can be altered

(b) To make multiple colonies of cells that contain extracted DNA

(c) To create multiple copies of a gene to grow DNA

(d) To identify a gene of interest and make multiple copies of plasmids that contain the gene of interest

(e) None of the above

55. What is a gene library?

(a) A collection of living proteins that are used in the process of gene cloning

(b) A collection of books that define all the genes known by genetic engineers

(c) A collection of living bacteria colonies that have been transformed with different pieces of DNA from the organism that is the source of the gene of interest

(d) A collection of plasmids that genetic engineers screen to insure that the genes they are going to clone are not dead

(e) None of the above

56. What is the purpose of a "gene library"?

(a) To help a genetic engineer identify which gene sequences contain the DNA of interest

(b) To identify which cells encode a particular protein

(c) To help a genetic engineer identify which bacteria colonies contain the gene of interest

(d) To give the genetic engineer a quick "book-shelf" reference of written topic on genes

(e) None of the above

57. What is a restriction enzyme?

(a) A specific enzyme that reads the nucleotide sequence of the DNA, recognizes specific sequences of the DNA and then cuts the DNA sequence by breaking the bonds between nucleotides in a DNA strand

(b) A specific enzyme that restricts the growth of the DNA, so that the DNA can not be cloned

(c) An enzymes that encode the sequence of the DNA, the enzyme then produces a DNA sequence by breaking the bonds between nucleotides in a DNA strand

(d) A specific enzyme that restricts the growth of the DNA limiting the number of genes that can be cloned

(e) None of the above

58. How can genetic engineers identify a desired gene in a gene colony?

(a) By screening for a specific DNA sequence of the cloned gene

(b) By screening and detecting a specific protein that the gene encodes

(c) By using a linked DNA maker

(d) All of the above

(e) None of the above

59. How was the work of Mendel unique compared with that of others studying plant genetics at the time?

(a) Unlike the others, Mendel applied statistics to his results

(b) Mendel crossed hybrids only

(c) Mendel used artificial pollination

(d) All of these are correct

(e) None of the above

60. Mendel's law of segregation states that

(a) Gametes segregate into traits during fertilization

(b) Traits segregate into alleles during fertilization

(c) Traits segregate when gametes form

(d) Gametes segregate when traits form

(e) None of the above

61. When the genotype of an individual is unknown, what kind of cross should be made?

(a) Monohybrid cross (b) Dihybrid cross

(c) Test cross (d) Reciprocal cross

(e) None of the above

62. An organism has an unknown genotype and a geneticist is trying to determine whether it is homozygous dominant (red-skinned) or heterozygous. If the result of a test cross yielded a population of all red-skinned organisms, the organism must have been

(a) Homozygous dominant

(b) Heterozygous

(c) Homozygous recessive

(d) The results are inconclusive using the information provided

(e) None of the above

63. Which human disorder is inherited as an autosomal dominant?

(a) Phenylketonuria (b) Huntington disease

(c) Cystic fibrosis (d) Sickle cell disease

(e) None of the above

64. Using the product rule, what is the likelihood of obtaining a RRYY plant from dihybrid (RrYy) parents?

(a) 1/4 (b) 1/12

(c) 1/16 (d) 1/64

(e) None of the above

65. Lethal alleles are usually

(a) Homozygous dominant alleles

(b) Heterozygous alleles

(c) Homozygous recessive alleles

(d) Carried on the Y chromosome

(e) None of the above

66. ABO blood type is an example of which type of dominance?

(a) Codominance

(b) Incomplete dominance

(c) Simple dominant/recessive traits

(d) None of these is correct

(e) None of the above

67. In his work with tobacco mosaic virus and Holmes rib grass virus Fraenkel-Conrat determined

(a) DNA is the genetic material used by all cellular organisms and viruses

(b) That some viruses use a complex of RNA and protein rather than DNA as their genetic information

(c) Some viruses use RNA rather than DNA as their genetic information

(d) Some viruses infect numerous species of plants and animals

(e) Proteins are the genetic information used by some viruses

68. In his experiments Fraenkel-Conrat

(a) Dissociated the RNA from the protein coats of the viruses they worked with then created hybrid viruses

(b) Complexed RNA and DNA to make functional hybrid viruses

(c) Produced infectious proteins which caused lesions to develop on tobacco plants

(d) Determined the retrovirus HIV was the cause of HIV

(e) Both a and d

69. During translation the mRNA molecule

(a) Sits between the the tRNA molecule and the large subunit of the ribosome

(b) Moves between the two subunits of the ribosome

(c) Brings amino acids from the cytoplasm to the tRNA molecule bound to the ribosome

(d) Binds amino acids together in the growing polypeptide chain

(e) Is not involved

70. tRNA molecules with bound amino acids deliver them to the

(a) P site on the ribosome

(b) 3' end of the polypeptide chain

(c) 5' end of the polypeptide chain

(d) Nucleus

(e) A site on the ribosome

71. The material that transformed harmless bacteria into virulent bacteria in Avery's experiment was

(a) DNA (b) mRNA

(c) tRNA (d) Protein

(e) RNA polymerase

72. What is the significance of Chargaff's rule in the DNA molecule?

(a) The pairing of purines and pyrimidines in DNA results in an excess of purine in some molecules

(b) Chargaff's rule says that a double helix should not form except at pH 7

(c) The rules of base pairing proposed by Chargaff result in a molecule with a constant thickness

(b) Both a and b

(e) None of the above

73. The replication of DNA molecules in a eukaryotic cell takes place

(a) Just prior to metaphase in mitosis

(b) Prior to cell division

(c) Continuously

(d) Only during meiosis

(e) Shortly after telophase

74. The production of mRNA is called

(a) Transduction (b) Transformation

(c) Transcription (d) Translation

(e) Promotion

75. Which of the following is the most accurate description of RNA polymerase?

(a) RNA polymerase is used to initiate the production of tRNA from an mRNA template

(b) RNA polymerase binds to a strand of DNA and moves along the strand to make a strand of mRNA

(c) RNA polymerase is responsible for providing a termination signal to the DNA strand at the end of rRNA synthesis

(d) DNA ligase and RNA polymerase work together to bind the nucleotides in a newly forming DNA strand

(e) RNA polymerase joins the two DNA after translation has occurred

76. During transcription

(a) Both DNA strands serve as a template for the synthesis of mRNA

(b) The two DNA strands alternate as the template for the synthesis of tRNA

(c) One of the DNA strands serves as a template for the synthesis of mRNA

(d) mRNA serves as a template for the synthesis of rRNA

(e) rRNA is assembled into complexes with proteins to make new ribosomes

77. Each amino acid is specified by

(a) Several genes

(b) An operon and a promoter

(c) A codon

(d) Markers of the surface of the cell requiring it for growth

(e) An enhancer

78. In ribosomes the rRNA is located

(a) On the surface of the large subunit

(b) Complexed with several proteins in the large subunit

(c) In both the small and the large subunits

(d) In the small subunit

(e) None of the above

79. Why can ribosomes bind to a sequence occurring at the beginning of a gene?

(a) Because of the matching, exposed sequence of rRNA nucleotides found on the small subunit of the ribosome

(b) Because of the effects of certain hormones that are associated with DNA synthesis

(c) Because the conformation of the ribosome contributes to binding with DNA

(d) Both a and b

(e) None of the above

80. What part of the tRNA molecule binds to mRNA?

(a) The single-stranded nucleotide "tail" of the tRNA molecule binds to a specific codon on the mRNA molecule

(b) The anticodon loop on the tRNA molecule binds to a specific codon on the mRNA molecule

(c) The tRNA codon binds to the anticodon on the mRNA molecule

(d) The small subunit of the tRNA binds to the promoter on the mRNA molecule

(e) The large subunit of the tRNA molecule binds to the anticodon on the mRNA molecule

81. What happens when a ribosome reaches a codon on the mRNA molecule that does not specify any of the 64 tRNA anticodons?

(a) The ribosome begins to read the mRNA molecule in the opposite direction

(b) The ribosome releases the newly formed protein molecule

(c) The ribosome subunits separate

(d) The newly formed protein begins to break up into smaller molecules

(e) Both b and c

82. Which of the following mechanisms would function to prevent transcription?

(a) The addition of excess substrate to the environment

(b) Binding of a repressor to the operator site on the gene

(c) Binding of an enhancer to the repressor site on the gene

(d) Changing the pH of the environment

(e) Both a and d

83. What effect would the addition of lactose have on a repressed lac operon?

(a) The operator site on the operon would move

(b) It would reinforce the repression of that gene

(c) It would result in the repression of other genes in that area

(d) The lac operon would be transcribed

(e) It would have no effect whatsoever

84. The Fraenkel Conrat experiment showed

(a) Viruses are made up of RNA

(b) Viruses are made up of DNA

(c) The virulent part of the tobacco mosaic virus is the protein coat

(d) The virulent part of the tobacco mosaic virus is RNA

(e) None of the above

85. DNA is made up of a phosphate group, an organic base, and

(a) A protein (b) A sugar

(c) A molecule of ATP (d) A fat

(e) None of the above

86. In the DNA molecule

(a) Adenine pairs with thymine

(b) Guanine pairs with thymine

(c) Cytosine pairs with thymine

(d) Guanine pairs with cytosine

(e) None of the above

87. If one side of a DNA molecule contains the following sequence of nucleotide, AGTCCG, the complementary sequence on the other side would be

(a) GCCTGA (b) AGTCCG

(c) TCAGGC (d) CTGAAT

(e) None of the above

88. Which one of the following is not a kind of RNA?

(a) tRNA (transfer RNA) (b) mRNA (messenger RNA)

(c) rRNA (ribosomal RNA) (d) nRNA (nuclear RNA)

(e) None of the above

89. The direct result of transcription is

(a) A duplicate DNA molecule (b) nRNA

(c) A protein (d) mRNA

(e) None of the above

90. The direct result of translation is

(a) A duplicate DNA molecule (b) nRNA

(c) A protein (d) mRNA

(e) None of the above

91. Which one of the following is not associated with transcription?

(a) rRNA (b) mRNA

(c) RNA polymerase (d) DNA

(e) None of the above

92. Which one of the following is not associated with translation?

(a) rRNA (b) mRNA

(c) tRNA (d) DNA

(e) None of the above

93. In the process of translation, the amino acids are arranged to form specific proteins because they are attached to

(a) rRNA (b) tRNA

(c) mRNA (d) Promoter

(e) None of the above

94. Proteins that block the passage of RNA polymerase are called

(a) Operons. (b) Activators

(c) Repressors (d) Enhancers

(e) None of the above

95. Which of the following is part of an operon?

(a) Protein-encoding genes (b) An operator

(c) A promoter (d) a, b, and c

(e) None of the above

96. Using genetic engineering genes can be

(a) Moved from one organism to potentially any other organism

(b) Easily converted to produce any product we desire

(c) Turned off or on at our whim

(d) Used to cure all known diseases

(e) Create plants resistant to all insect pests

97. Genetic engineering has produced salmon

(a) With shorter reproductive cycles

(b) That produce four times more offspring

(c) Resistant to viral infection

(d) That are heavier than non-engineered salmon

(e) Both a and d

98. Sticky ends are the result of

(a) Treatment of a nucleotide sequence with DNA ligase

(b) Exposure of eukaryotic DNA to a prokaryotic plasmid

(c) Infection of a cell with a bacteriophage

(d) DNA breaking down in the presence of reverse transcriptase

(e) Cutting by restriction enzymes "off center" in a specific nucleotide sequence

99. In order to produce a DNA fragment that can be inserted into the DNA of a second organism one would need to

(a) Create a compatible DNA segment from an mRNA template

(b) Find a virus that could serve as a vector for transfer of the donor DNA

(c) Make sure both organisms have compatible DNA

(d) Be sure both donor and recipient DNA were prokaryotic or eukaryotic

(e) Cut the DNA of the donor and recipient cells with the same restriction enzyme

100. Interleukins:

(a) Stimulate white cell production

(b) Activate white blood cells. Used to treat HIV and cancer

(c) Halt the production of viruses

(d) Used to treat individuals with kidney disorders

(e) None of the above

101. Erythropoietin:

(a) Stimulates white cell production

(b) Activates white blood cells. Used to treat HIV and cancer

(c) Halts the production of viruses

(d) Used to treat individuals with kidney disorders

(e) None of the above

102. Interferons:

(a) Stimulate white cell production

(b) Activate white blood cells. Used to treat HIV and cancer

(c) Halt the production of viruses

(d) Used to treat individuals with kidney disorders

(e) None of the above

103. Colony-stimulating factors:

(a) Stimulate white cell production

(b) Activate white blood cells. Used to treat HIV and cancer

(c) Halt the production of viruses

(d) Used to treat individuals with kidney disorders

(e) None of the above

104. Genes can be inserted into cells by

(a) Shooting them into the cell with a DNA particle gun

(b) Using a plasmid as a gene vector

(c) Causing them to be produced in the recipient cell

(d) All of the above

(e) Only a and b

105. Glyphosate prevents plants from growing by

(a) Cleaving critical DNA in the plant and halting cell division

(b) Preventing the production of gametes

(c) Preventing the formation of spindle fibers and halting mitosis

(d) Stopping them from synthesizing proteins

(e) Both a and d

106. You are conducting research on eight species of Tribolium flour beetles and you want to compare their proteins. Which of the following techniques might you use?

(a) Genetic engineering (b) Gene amplification

(c) Gel electrophoresis (d) Polymerase chain reaction

(e) Gene therapy

107. The key to isolating a particular gene for use in recombinant DNA research is to

(a) Find a restriction enzyme to cut out the gene

(b) Find a compatible species to receive the gene

(c) Identify copies of the gene on several chromosomes

(d) Both a and b

(e) None of the above

108. Cuts in DNA are sealed with

(a) Restriction enzymes

(b) Ligases

(c) Reverse transcriptase

(d) Polymerase

(e) Plasmids

109. DNA is made from RNA with the help of

(a) Restriction enzymes (b) Ligases

(c) Reverse transcriptase (d) Polymerase

(e) Plasmids

110. Sticky ends are created by

(a) Adding ligase to the end of the DNA fragment

(b) Adding polymerase to the end of the DNA fragment

(c) Cutting DNA so that there are short, single strands at each end

(d) Cutting the DNA fragment in several places

(e) Both a and d

111. When investigating a crime scene detectives find a single hair on the floor. The sample is not large enough to perform many tests. Which of the following methods could be used to make sufficient DNA to allow for comparison to the DNA of the prime suspect?

(a) Reverse transcriptase could be used to make DNA from RNA in the hair sample

(b) The hair root could be tissue cultured to make additional copies of the DNA

(c) The gene could be amplified by use of electrophoresis

(d) Polymerase chain reaction could be used to make many copies of the DNA

(e) The gene of interest could be spliced into a bacterium that would make a new copy every time it replicated

112. cDNA is significant when

(a) You need several thousand copies of a gene

(b) Cloning eukaryotic DNA in prokaryotic cells

(c) It is necessary to copy introns in prokaryotic cells

(d) A primary transcript is not available

(e) Processed mRNA is not available

113. In order for plants to resist the effects of the herbicide Roundup, they must be

(a) Monocots

(b) Dicots

(c) Annuals

(d) Able to synthesize glyphosate

(e) Resistant to glyphosate

114. Although the Ti plasmid has revolutionized plant genetic engineering, one limitation of its use is that it

(a) Is only about 50% effective

(b) Can only be used on fruit-bearing plants

(c) Is only effective on plants that are resistant to Roundup

(d) Does not infect cereal plants

(e) Cannot transmit prokaryotic genes

115. Genetic engineering has successfully transferred genes from eukaryotic cells into

(a) Bacteria (b) Plants

(c) Animals (d) All of the above

(e) None of the above

116. DNA molecules can be cut into sections by using

(a) ATP (b) Gel electrophoresis

(c) Restriction enzymes (d) Plasmids

(e) None of the above

117. In genetic engineering, ligase is used as

(a) A probe
(b) A sealing enzyme
(c) A restriction enzyme
(d) A mutagen
(e) None of the above

118. A restriction enzyme is used to

(a) Isolate sections of cloned bacterial plasmids
(b) Isolate sections of DNA
(c) Cleave the source DNA into fragments
(d) Both b and c are correct
(e) None of the above

119. A probe is used in which stage of the gene transfer process?

(a) Cleaving DNA
(b) Recombining DNA
(c) Cloning
(d) Screening
(e) None of the above

120. Polymerase chain reactions are used to

(a) Create DNA "fingerprints"
(b) Cleave bacterial plasmids
(c) Seal "sticky ends"
(d) Identify target plasmids
(e) None of the above

121. Which of the following statements is(are) true about processed mRNA?

(a) Processed mRNA directs the formation of the primary mRNA transcript
(b) Introns have been removed from processed mRNA
(c) Processed mRNA directs the formation of cDNA
(d) Both b and c are true
(e) None of the above

122. Genetic engineering has been used to do all of the following except

(a) Make plants more resistant to frost
(b) Make plants more resistant to disease
(c) Make plants more resistant to herbicides

(d) All of the above are correct

(e) None of the above

123. Engineering plants that are resistant to glyphosate was an important advancement because

(a) Glyphosate promotes frost damage

(b) Glyphosate encourages the production of fruit that is lower in protein

(c) Glyphosate is the active ingredient in herbicide

(d) Glyphosate prevents the transfer of genes into the plants

(e) None of the above

124. The use of bovine growth hormone has

(a) Raised controversy with groups opposed to genetic engineering

(b) Increased milk production

(c) Increased the weight gain of cows

(d) Resulted in a, b, and c

(e) None of the above

125. Which of the following is not required for accurate transcription?

(a) Regulatory information within the DNA

(b) Proteins which bind to the DNA

(c) Proteins which copy the information from the DNA

(d) All of these are required

(e) None of these are required

126. The closed promoter complex consists of all of the following except:

(a) Promoter region of the DNA

(b) RNA polymerase

(c) Transcription factors

(d) Helicase

(e) None of the above

127. Which type(s) of RNA molecule has(have) the primary function of modifying other RNA molecules?

(a) RNA of RNaseP
(b) snRNA
(c) scRNA
(d) A and B
(e) All of the above

128. If you have a strain of bacteria that produce nonfunctional sigma factor, which step of transcription would you expect will be affected?

(a) These bacteria will be unable to assemble RNA polymerase onto the DNA strand

(b) These bacteria will be unable to identify and tightly bind promoter elements

(c) These bacteria will be unable to convert a closed complex to an open complex

(d) These bacteria will be unable to synthesize RNA once an open complex is formed

(e) None of the above

129. Which of the following mutations would you expect to have the most significant effect on DNA binding by the transcription factor depicted in figure 12.5?

(a) A conservative substitution in α helix 1

(b) A conservative substitution in α helix 2

(c) A nonconservative substitution in α helix 3

(d) A nonconservative substitution in the turn region

(e) None of the above

130. What is the primary factor involved in termination of RNA synthesis?

(a) Temporary stabilization of the RNA:DNA hybrid region

(b) Synthesis of a specific termination signal in the RNA

(c) Destabilization of hydrogen bonding between DNA and RNA within the open complex

(d) Recognition of a termination signal on the DNA by termination proteins

(e) None of the above

131. Which type of transcriptional termination requires formation of a stem loop secondary structure in the RNA?

(a) ρ dependent

(b) ρ independent

(c) Polymerase independent

(d) Both A and B

(e) None of the above

132. Intrinsic terminators are:

(a) RNA sequences that assist in termination of transcription

(b) DNA sequences that assist in termination of transcription

(c) Proteins that assist in termination of transcription

(d) All of these

(e) None of these

133. Which statement could NOT describe both eukaryotic and prokaryotic transcription?

(a) Promoter elements in the DNA are required for polymerase binding

(b) Transcription factors control the binding of RNA polymerase

(c) RNA polymerase opens a double stranded DNA to expose the template strand

(d) RNA polymerase catalyzes RNA synthesis from 3' to 5' of the new strand

(e) None of these (all apply to both eukaryotic and prokaryotic transcription)

134. What would you expect to be the effect of a mutation of RNA polymerase that removed the rudder region?

(a) DNA would not be opened to expose the template strand

(b) Nucleotides would not base pair correctly with the template strand

(c) Nucleotides would not be connected into an RNA strand

(d) RNA would not be released from the template DNA

(e) None of the above

135. Which of the following is an example of a cis acting factor in eukaryotic transcription?

(a) RNA polymerase II
(b) Transcription factors
(c) Enhancers
(d) All of these
(e) None of these

136. Which of the following RNA modifications could best be described as "removal of intervening sequences"?

(a) Trimming
(b) 5' capping
(c) Poly-A tailing
(d) Splicing
(e) RNA editing or base modification

137. When studying eukaryotic enzymes in relation to their prokaryotic counterparts, you discover that some eukaryotic enzymes seem to consist of several regions, each of which performs the function of a separate enzyme in bacteria. Which of the following explanations would you be most likely to accept?

(a) Random mutation has generated these multi-function eukaryotic enzymes
(b) A polycistronic transcript in bacteria became a polycistronic transcript in eukaryotic cells
(c) Several genes located near each other in bacteria became transcribed together and the intervening sequences spliced out
(d) Similar regulatory elements for each of the prokaryotic proteins allowed them to be expressed together
(e) None of the above

138. Which of the following would be expected to demonstrate that mRNA contains introns?

(a) A comparison of the DNA and mRNA sequences
(b) A comparison of the genomic DNA and cDNA sequences
(c) A hybridization between DNA and mRNA molecules
(d) All of these
(e) None of these

139. Which of the following statements about RNA splicing is NOT correct?

(a) Some introns can splice themselves out

(b) Splicing is required for some tRNAs

(c) The final step of splicing is formation of a phosphodiester linkage

(d) Introns often have a specific function after they are removed

(e) None of these; all of the above statements are correct

140. Which of the following would you expect to find if you took apart a spliceosome?

(a) RNA only (b) Protein only

(c) Both RNA and protein (d) Neither RNA nor protein

(e) None of the above

141. Which of the following is a critical function of the 5' cap of eukaryotic mRNAs?

(a) Protection from inappropriate splicing

(b) Synthesis of the polyA tail

(c) Transport of mRNAs into the nucleus

(d) Recognition of the 5' end of the transcript by a ribosome

(e) None of the above

142. Which of the following pairs is not correctly described?

(a) 5' splice site: begins with GU, marks 5' end of intron

(b) Branch site: contains A that binds U2; branch site of lariat

(c) 3' splice site: begins with AG, marks 3' end of intron

(d) Trimer: consists of U4/U6 and U5, brings 5' splice site to branch site

(e) All of these are correctly matched

143. Which of the following experiments or discoveries did NOT help to identify the "triplet nature" of the DNA code?

(a) Frameshift mutations (b) Anticodons

(c) Start and stop codons (d) RNA homopolymers

(e) Universality of the genetic code

144. Which of the following does NOT involve introns?

(a) Alternative splicing (b) Ribozymes

(c) Intervening sequences (d) Heteroduplexes

(e) Antisense oligonucleotides

145. Arrange the flow of genetic information in the cell in order: 1. mRNA, 2. DNA, 3. protein.

(a) 3, 1, 2 (b) 2, 1, 3

(c) 3, 2, 1 (d) 1, 2, 3

(e) 2, 3, 1

146. Where in the cell is the DNA transcribed into mRNA?

(a) Vacuole (b) Cell cytoskeleton

(c) Nucleus (d) Cytoplasm

(e) Golgi

147. A DNA template strand has the following sequence: TACGGGATT. What is the mRNA sequence transcribed from the DNA?

(a) AUGCCCUAA (b) AAUCAGUTA

(c) TACGGGATT (d) ATGCCCTAA

(e) UACGGGAUU

148. Once a strand of mRNA starts being translated:

(a) Many proteins are made simultaneously by the ribosome

(b) DNA transcription shuts down

(c) Another mRNA is made simultaneously from the DNA

(d) It is degraded so that only one polypeptide is made from this mRNA

(e) Only a single protein will be transcribed

149. Insertion of a single base in a DNA sequence results in what kind of mutation?

(a) Frameshift mutation (b) Chromosome breakage

(c) Transition (d) Point mutation

(e) Deletion

150. In a repeating copolymer assay involving A and U, what would you expect to happen? (Use the codon chart in your book to answer this question.)

(a) Isoleucine and tyrosine will both be incorporated into the protein

(b) Arginine will be incorporated into the protein

(c) Tyrosine (tyr) will be incorporated into the protein

(d) Isoleucine (ile) will be incorporated into the protein

(e) There will be no protein incorporation

151. Occasionally, a mutation is discovered that suppresses termination and allows a chain terminating codon to be read as an amino acid code without changing the codon. What do you think these mutations will do? The mutations:

(a) Cause a release factor to allow the polypeptide to be released

(b) Alter a terminating tRNA so that it carries an amino acid

(c) Change the anticodon sequence for an amino-acid carrying tRNA

(d) Change the ribosome binding capacity

(e) Cause a mutation in a release factor

152. Transcription in prokaryotes ends at

(a) A region beyond the end of a single gene

(b) A region beyond the end of one or more tandem genes

(c) A gene's stop codon

(d) A region called the "TATA" box

(e) A region at the head of every gene

153. You set up an experiment in which a piece of eukaryotic DNA is hybridized to its messenger RNA. You examine the product with an electron microscope. You will see:

(a) Loops appear along the DNA strand

(b) Loops appear along the RNA strand

(c) Both strands pair exactly and are of the same length

(d) Both strands pair exactly but are of differing lengths

154. Which of the following does NOT happen during hnRNA processing?

(a) A poly A tail is added

(b) Exons are spliced together

(c) A 7-methylguanosine cap is added to the 5' end of the RNA

(d) Ribosomes bind and begin translation

(e) Introns are spliced out

155. How does a zygote's cytoplasm affect cell determination?

(a) It has no effect

(b) It slows down determination

(c) The cytoplasm initiates differential transcription

(d) It speeds up determination

(e) It controls determination completely

156. The sequence of the coding strand of DNA is the same as the

(a) Template strand (b) mRNA

(c) Protein (d) Two of these

(e) None of these

157. A DNA strand with the sequence 5'-GCAATGCAA-3' is transcribed. What is the sequence of the mRNA molecule synthesized?

(a) 5'-CGTTACGTT-3' (b) 5'-CGUUACGUU-3'

(c) 5'-UUGCAUUGC-3' (d) 5'-TTGCAUUGC-3'

(e) 5'-AACGUAAGC-3'

158. In prokaryotes, the element found 10 bases upstream of the first base transcribed is the

(a) Pribnow box (b) TATA box

(c) CAAT box (d) Rho region

(e) Enhancer region

159. Posttranscriptional modification of prokaryotic mRNAs include

(a) a 5' cap

(b) a 3' cap

(c) a 3' poly- A tail

(d) a 3' poly- A tail and a 5'- cap

(e) Nothing, prokaryotic mRNA is unmodified

160. Noncoding regions of DNA interspersed among coding regions are known as

(a) Promoters (b) Sigma factors

(c) Introns (d) Exons

(e) Enhancers

161. The transcribing enzyme is primarily

(a) RNA polymerase (b) RNA primase

(c) DNA polymerase III (d) DNA polymerase II

(e) Reverse transcriptase

162. The sigma factor recognizes the

(a) Promoter (b) Intron

(c) Exon (d) Enhancer

(e) Rho protein

163. DNA can act as a messenger at the ribosome only

(a) In prions (b) *In vitro*

(c) With reverse transcriptase (d) In certain DNA phages

(e) Never

164. In eukaryotes, the enzyme that primarily transcribes the nucleolar organizer is

(a) RNA polmerase I (b) RNA polmerase II

(c) RNA polmerase III (d) Primase

(e) Reverse transcriptase

165. Polymerase Chain Reaction (PCR) is used to:

(a) Amplify, or make more copies of a small amount of DNA

(b) Cleave bacterial plasmids

(c) Seal "sticky ends"

(d) Identify target plasmids

(E) None of the above

166. Which is the correct order for the 3 steps of polymerase chain reaction (PCR)?

(A) Denaturation of DNA, annealing of primers, primer extension

(B) Primer extension, annealing of primers, denaturation of DNA

(c) Annealing of primers, primer extension, denaturation of DNA

(d) All the above

(e) None of the above

167. In each cycle of PCR, the quantity of the DNA is:

(a) Tripled (b) Doubled

(c) Quadrupled (d) Increased by 50%

(e) None of the above

168. What is the purpose of the Southern blot?

(a) Amplify DNA

(b) Identify a specific gene in a DNA sequence

(c) Seal the sticky ends of DNA cut with restriction enzymes

(d) Separate DNA fragments based upon charge

(e) None of the above

169. In preliminary screening of a clone library, it is common to use:

(a) Restriction enzymes (b) Dyes

(c) Antibiotics (d) Radiation

(e) Milipore filters

170. Plasmids usually contain genes for antibiotic resistance. Which statement is true in regards to screening a clone library?

(a) All bacterial colonies that contain plasmids will die in the presence of antibiotics

(b) All bacterial colonies that do not contains plasmids will die in the presence of antibiotics

(c) Only plasmids without genes for antibiotic resistance contain genes of interest

(d) All the above

(e) None of the above

171. Bacteria protect themselves from viruses by fragmenting viral DNA with:

(a) Restriction ligases

(b) Restriction enzymes

(c) Restriction methylases

(d) Vectors

(e) Probes

172. What is the purpose of the Southern blot?

(a) Amplify DNA

(b) Identify a specific gene in a DNA sequence

(c) Seal the sticky ends of DNA cut with restriction enzymes

(d) Separate DNA gragments based upon charge

(e) None of the above

173. In PCR, how are the DNA segments sorted?

(a) Gel electrophoresis (b) Centrifugation

(c) Chromatography (d) Any of the above

(e) None of the above

174. When "sticky ends" are paired, they can be joined by:

(a) Restriction enzymes (b) RNA polymerase

(c) Methylase (d) DNA ligase

(e) None of the above

175. Which statement(s) is/are true regarding single nucleotide polymorphisms (SNPs)?

(a) SNP vary among individuals

(b) SNPs that differ from a reference sequence may indicate the likelihood of disease

(c) a and b

(d) All the above

(e) None of the above

176. Which is an example of a widely used medical application of biotechnology?

(a) Production of insulin

(b) Production of "piggyback" vaccine

(c) Gene therapy - replace defective genes with a function copy of the gene

(d) None of these are widely used applications of biotechnology

(e) None of the above

177. Gel electrophoresis separates molecules based upon:

(a) Size (b) Shape

(c) Charge (d) All of the above

(e) None of the above

178. Genetically identical organisms derived from a single genetic source are called:

(a) Populations (b) Varieties

(c) Sibling species (d) Clones

(e) None of the above

179. When tryptophan is present in the environment of *E. coli*, the tryptophan binds to the:

(a) Trp operon (b) Trp promoter

(c) Trp operator (d) Trp repressor

(e) None of the above

180. Which statement(s) is/are true if tryptophan is present in the environment of *E. coli*?

(a) The repressor isn't bound to the operator

(b) RNA polymerase binds to the promoter

(c) Transcription occurs

(d) The repressor is bound to the operator

(e) A, B and C are true

181. Which statement(s) is/are true if tryptophan is NOT present in the environment of *E. coli*?

(a) The repressor isn't bound to the operator

(b) RNA polymerase binds to the promoter

(c) Transcription occurs

(d) The repressor is bound to the operator

(e) A, B and C are true

182. Recall that the bacteria *E. coli* prefers to use glucose as a food source. However, it can also use lactose as a food source. How is the transcription of lac genes controlled when glucose is present in the environment?

(a) Genes to metabolize lactose are not transcribed

(b) Genes to metabolize lactose are transcribed

(c) Genes to metabolize lactose are translated, but not transcribed

(d) All the above

(e) None of the above

183. When lactose is present, but glucose is absent, how is transcription of lac genes controlled in *E. coli*?

(a) Genes to metabolize lactose are not transcribed

(b) Genes to metabolize lactose are transcribed

(c) Genes to metabolize lactose are translated, but not transcribed

(d) All the above

(e) None of the above

184. When glucose, but not lactose, is present in *E. coli*'s environment, which statement is true?

(a) The repressor remains bound to the operator

(b) RNA polymerase binds the promoter

(c) Transcription occurs

(d) B and C are true

(e) None of the above

185. When lactose, but not glucose, is present in *E. coli*'s environment, which statement is true?

   (a) The repressor remains bound to the operator
   (b) RNA polymerase binds the promoter
   (c) Transcription occurs
   (d) B and C are true
   (e) None of the above

186. Which of the following statements describes the catabolite activator protein (CAP) of *E. coli*?

   (a) CAP protein binds cyclic AMP (cAMP)
   (b) CAP (with cAMP bound to it) attaches to the promoter sequence of DNA
   (c) CAP (with cAMP bound to it) acts as an activator
   (d) All of the above
   (e) None of the above

187. When glucose, but not lactose is present in *E. coli*'s environment,

   (a) cAMP levels are high and the lac operon is not activated
   (b) cAMP levels are high and the lac operon is activated
   (c) cAMP levels are low, and the lac operon is not activated
   (d) cAMP levels are low and the lac operon is activated
   (e) None of the above

188. Following transcription, gene expression can be controlled by:

   (a) RNA processing
   (b) Degradation of mRNA
   (c) Translation repressor proteins
   (d) All of the above
   (e) None of the above

189. Experiments with Dolly, the sheep cloned from the mammary cell of another sheep, demonstrated that:

   (a) Only stem cells can be used for cloning
   (b) Determination is irreversible
   (c) Determined cells can be reversed under specific circumstances

(d) All the above

(e) None of the above

190. This enzyme increases the length of the telomeres:

(a) Telomerase
(b) Ligase
(c) RNA polymerase
(d) Amylase
(e) None of the above

191. Transcription is initiated when RNA polymerase binds to

(a) A promoter
(b) An initiator
(c) A transcriptor
(d) A codon
(e) None of the above

192. The promoter sequence in eukaryotes is

(a) TAATAA
(b) TATAAA
(c) TTGACA
(d) GTTAAA
(e) None of the above

193. Each time a nucleotide is added as the transcription bubble passes down the DNA, the RNA-DNA complex

(a) Elongates
(b) Rotates
(c) Shrinks
(d) Disassembles
(e) None of the above

194. Eukaryotic mRNA transcripts are protected from modification by

(a) 5' caps
(b) 5' poly-A caps
(c) 3' caps
(d) 5'-3' poly tails
(e) None of the above

195. The order in which nucleotides are moved along the ribosomes binding sites is

(a) APE
(b) PEA
(c) EPA
(d) EAP
(e) None of the above

196. In eukaryotes, the "start" codon also specifies the amino acid,

(a) Phenylalanine (b) Valine

(c) Aspartate (d) Methionine

(e) Methionine

197. Together with proteins, rRNA

(a) Provides a site for polypeptide synthesis

(b) Transports amino acids to the ribosome

(c) Travels to the ribosome to direct the assembly of polypeptides

(d) Transcribes DNA

(e) Translates DNA

198. The genetic code consists of groups of three nucleotides called

(a) Codons (b) Introns

(c) Anticodons (d) Reading frames

(e) Triplets

199. The function of tRNA is to

(a) Provide a site for polypeptide synthesis

(b) Transport amino acids to the ribosome

(c) Travel to the ribosome to direct the assembly of polypeptides

(d) Transcribe DNA

(e) Translate DNA

200. Enzymes called amino acyl-tRNA synthetases

(a) Synthesizes tRNA

(b) Attaches amino acids to tRNA

(c) Strips tRNA from its amino acid in the process of translation

(d) Destroys excess tRNA molecules

(e) Helps tRNA synthesize amino acids

201. In mRNA the "start" sequence is

(a) UAA (b) UAG

(c) UGA (d) AUG

(e) GUU

202. The function of mRNA is to

(a) Provide a site for polypeptide synthesis

(b) Transport amino acids to the ribosome

(c) Travel to the ribosome to direct the assembly of polypeptides

(d) Transcribe DNA

(e) Translate DNA

203. In the process of translation,

(a) A strand of mRNA is formed with nucleotide sequences complementary to those of DNA

(b) Nucleotide sequences of tRNA are established

(c) A polypeptide is formed in response to the rRNA nucleotide sequence

(d) rRNA is synthesized with sequences complementary to those of tRNA

(e) A polypeptide is formed as dictated by the nucleotide sequence in mRNA

204. In mRNA, the series of nucleotides CCC specifies

(a) Serine (b) Proline

(c) Alanine (d) Arginine

(e) Stop

205. A molecule of tRNA with the anticodon AAA will transport the amino acid

(a) Phenylalanine (b) Lysine

(c) Proline (d) Glycine

(e) Arginine

206. In the process of transcription

(a) The base sequence of DNA is copied into tRNA

(b) A polypeptide is formed as specified by the genes in a chromosome

(c) rRNA is specified by exons in DNA

(d) A strand of mRNA is formed with base sequences complementary to those of DNA

(e) mRNA is formed as coded by introns

207. In messenger RNA, the nucleotide series UAG specifies

(a) Arginine (b) Serine

(c) Stop (d) Proline

(e) Aspartate

208. Which enzyme is incorrectly matched to its function?

(a) RNA polymerase – transcription

(b) Aminoacyl-tRNA synthetase – attaches an amino acid to a tRNA

(c) Peptidyl transferase – translation

(d) DNA ligase – joins Okazaki fragments

(e) All of the above are paired correctly

209. Identify the polypeptide that would be produced as a result of transcribing and translating the following DNA sequence. DNA: ...3' A C C A A G T C T 5'...

(a) arg - phe - trp (b) arg - leu - gly

(c) thr - lys - ser (d) trp - phe - arg

(e) gly - leu - arg

210. The strand of DNA used in the previous problem undergoes a mutation, such that the second C from the 3' end is changed to a T. The mutant DNA is shown below. What would change in the polypeptide?

211. Mutant DNA: ...3' A C T A A G T C T 5'...

(a) No change

(b) Polypeptide would be 1 amino acid shorter

(c) Polypeptide would be 1 amino acid longer

(d) 1 amino acid would substitute for another

(e) Translation would terminate within this stretch

212. Which one of the following is not a kind of RNA?

(a) tRNA (transfer RNA)

(b) mRNA (messenger RNA)

(c) rRNA (ribosomal RNA)

(d) nRNA (nuclear RNA)

(e) All of the above are kinds of RNA

213. The direct result of transcription is:

(a) A duplicate DNA molecule (b) nRNA

(c) A protein (d) mRNA

(e) None of the above

214. The direct result of translation is:

(a) A duplicate DNA molecule (b) nRNA

(c) A protein (d) mRNA

(e) All of the above

215. Which one of the following is not associated with transcription?

(a) tRNA (b) mRNA

(c) RNA polymerase (d) DNA

(e) A promoter

216. Which one of the following is not associated with translation?

(a) rRNA (b) mRNA

(c) tRNA (d) Anticodon

(e) All are involved with translation

217. In the process of translation, the amino acids are arranged to form specific proteins because they are attached to:

(a) rRNA (b) tRNA

(c) mRNA (d) Promoter

(e) None of the above

218. Who first described the "central dogma" that presents the relationship between nucleic acids and proteins?

(a) Francis Jacob (b) Jacques Monod

(c) Watson and Crick (d) Martha Chase

(e) None of the above

219. How does RNA differ from DNA?

(a) RNA is single-stranded (b) RNA has uracil

(c) RNA can be catalytic (d) All of these are correct.

(e) None of the above

220. How does the lac operon turn genes on or off in bacteria?

(a) The lac operon negatively controls transcription by removing a block

(b) The lac operon positively controls transcription by adding a block

(c) The lac operon initiates enzymatic activity only at certain times

(d) The lac operon activates the promoter

(e) None of the above

221. If a DNA template strand has the sequence CTACCGTAC, what will be the transcribed mRNA sequence?

(a) GAUGGCAUG (b) GATGGCATG

(c) CGTATCCGT (d) AAGTTGCCA

(e) None of the above

222. To facilitate movement of mRNA from the nucleus into the cytoplasm, what is added to the 3′ end of mRNA?

(a) A cap of modified nucleotides (b) Modified tRNA molecules

(c) Ribosomes (d) A poly A tail

(e) None of the above

223. How many RNA bases encode one amino acid?

(a) 2 (b) 3

(c) 6 (d) 9

(e) None of the above

224. Consider the following sequence of mRNA: AUGAAUCUUCACGGWhat is the corresponding sequence of amino acids, using Table 13.4 in your text as a guide?

(a) arg-his-pho-stop-val (b) met-asn-leu-his-gly

(c) stop-ilu-phe-thr-ala (d) glu-ser-ser-arg-ala

(e) None of the above

225. What is required for elongation to occur in translation?

(a) Completion of the initiation complex

(b) Attachment of fMET

(c) Attachment of the next amino acid

(d) Attachment of the large ribosomal subunit

(e) None of the above

226. A change from UAC (tyrosine) to UAA ("stop") is an example of what type of mutation?

(a) Transition (b) Missense

(c) Nonsense (d) Spontaneous

(e) None of the above

227. The flow of genetic material in microbial cells usually proceeds from

(a) Proteins through RNA to DNA

(b) RNA through DNA to proteins

(c) DNA through RNA to proteins

(d) All the above

(d) None of the above

228. A new copy of a DNA molecule is precisely synthesized during a process called

(a) Translation (b) Transcription

(c) Replication (d) All the above

(e) None of the above

229. The final step in gene expression is protein synthesis, or

(a) Translation (b) Transcription

(c) Replication (d) All the above

(e) None of the above

230. The chromosome of almost all bacteria is in the shape of

(a) An open circle (b) A closed circle

(c) A linear chromosome (d) All the above

(e) None of the above

231. The enzyme that catalyzes the synthesis of DNA is called

(a) Helicase (b) DNA gyrase

(c) DNA polymerase (d) All the above

(e) None of the above

232. DNA polymerase catalyzes the synthesis of DNA in the

(a) 3' to 5' direction (b) 5' to 3' direction

(c) All the above (d) None of the above

(e) Can't say

233. The bases of one strand of DNA match-up with the bases of the second strand according to base pairing rules, therefore, the two strands are said to be

(a) Oppositely charged (b) Complementary

(c) Identical (d) All the above

(e) None of the above

234. Eukaryotes are able to replicate their chromosomes in a relatively short period of time because

(a) Each chromosome contains multiple replicons

(b) The eucaryotic replicative machinery is 1000 times faster than the procaryotic replicative machinery

(c) Eucaryotic DNA is always single stranded

(d) All the above

(e) None of the above

235. The enzymes that unwind short stretches of DNA helix immediately ahead of a replication fork are

(a) Single-stranded binding proteins

(b) Topoisomerases

(c) Helicases

(d) All the above

(e) None of the above

236. The enzyme that repairs nicked DNA by forming a phosphodiester bond between adjacent nucleotides is

(a) DNA gyrase (b) Helicase

(c) DNA ligase (d) All the above

(e) None of the above

237. The approximate rate of DNA replications in procaryotes is

(a) 10 bases per second (b) 1,000 bases per second

(c) 100,000 bases per second (d) All the above

(e) None of the above

238. Another name for the three stop codons which aid in termination of translation is

(a) Nonsense codons (b) Missense codons

(c) Pause codons (d) All the above

(e) None of the above

239. The enzyme that catalyzes the synthesis of RNA from a DNA template is

(a) DNA gyrase (b) DNA ligase

(c) RNA polymerase (d) All the above

(e) None of the above

240. Another name for the RNA polymerase recognition/binding site upstream of a gene is the

(a) Promoter (b) Terminator

(c) Facilitator (d) All the above

(e) None of the above

241. The Shine-Dalgarno sequence is

(a) A short sequence that acts as a ribosomal binding site

(b) Another name for a stop codon

(c) The reading frame of a gene

(d) All the above

(e) None of the above

242. Most prokaryotic proteins begin with this modified amino acid

(a) N-formylleucine
(b) N-formylmethionine
(c) N-formylserine
(d) All the above
(e) None of the above

243. Mutations may be characterized by

(a) Changes in genotype
(b) Changes in phenotype
(c) Changes in either genotype or phenotype
(d) All the above
(e) None of the above

244. Mutations that arise without exposure to external agents are called

(a) Induced mutations
(b) Analogous mutations
(c) Spontaneous mutations
(d) All the above
(e) None of the above

245. A mutagen is

(a) A chemical or physical agent that induces mutations
(b) An enzyme that repairs mutations
(c) A molecule which stabilizes DNA thus prevents mutations from occurring
(d) All the above
(e) None of the above

246. A mutation that affects only a single nucleotide is called

(a) A regional mutation
(b) A site mutation
(c) A point mutation
(d) All the above
(e) None of the above

247. A nonsense mutation results in

(a) An abnormal elongation of a polypeptide
(b) A premature termination of the synthesis of a polypeptide
(c) A large deletion within the reading frame of a gene

(d) All the above

(e) None of the above

248. The process by which thymine dimers are directly repaired with the help of visible light is called

(a) Photosynthesis

(b) Excision repair

(c) Photoreactivation

(d) All the above

(e) None of the above

249. DNA repair mechanisms are able to distinguish newly synthesized DNA strands from older strands because

(a) New strands do not contain cytosine bases

(b) Old strands are methylated while new strands are not

(c) New strands are methylated while old strands are not

(d) All the above

(e) None of the above

250. The ability of the anticodon of tRNA to interact with multiple codons of mRNA is called

(a) Promiscuity (b) Wobble

(c) Stagger (d) All the above

(e) None of the above

251. Which of the following was important in Garrod's proposal regarding inheritance of a defective enzyme?

(a) Identification of a defective gene in individuals with alkaptonuria

(b) Identification of a defective gene in parents of alkaptonuria patients

(c) Identification of a defective enzyme in individuals with alkaptonuria

(d) Identification of the inheritance pattern of alkaptonuria as autosomal recessive

(e) None of the above

252. Which of the following correctly restates the one gene – one enzyme theory?

(a) Every gene encodes a protein

(b) Each structural gene encodes an essential metabolic enzyme

(c) Each structural protein is encoded by a single gene

(d) All of these

(e) None of these

253. Which statement about the genetic code is not correct?

(a) Most amino acids are specified by more than one codon

(b) Many codons have more than one function

(c) Some organisms use codons differently than other organisms

(d) In some cases, the third base of a codon is nonessential

(e) None of these (all these statements are correct)

254. What characteristic of proflavin allowed Crick to use it in his determination that the genetic code is a triplet code?

(a) It causes mutations in DNA

(b) It can cause large deletions in DNA

(c) It causes single insertions or deletions in DNA sequences

(d) It can be lethal

(e) None of the above

255. Which of the following facts is the most important evidence that the genetic code is a triplet code?

(a) Insertions or deletions caused significant defects in protein structure

(b) The effect of single insertions could be remedied by single deletions nearby

(c) Two nearby insertions caused significant defects in protein structure

(d) Three insertions near each other often led to minor defects in protein structure

(e) None of the above

256. What primarily determines the specific chemical character of a particular amino acid?

(a) The amino (NH_3^+) group.
(b) The carboxyl (COO^-) group.
(c) The R group
(d) All of these
(e) None of these

257. Which of the following represents the primary structure of a protein?

(a) Folding which allows amino acids of a polypeptide to be arranged in regions of similar characteristics (hydrophobic residues together, for ex.)
(b) Local folding which is stabilized by hydrogen bonds
(c) The sequence of amino acids in the polypeptide
(d) Interactions between two or more polypeptides
(e) None of the above

258. Which amino acid would be most likely to interact with serine in formation of secondary or tertiary structure?

(a) Valine
(b) Tryptophan
(c) Asparagine
(d) Any of these with equal affinity
(e) None of these

259. Which of the following could not potentially explain the albino phenotype in humans?

(a) Lack of an enzyme that synthesizes the pigment molecules
(b) Lack of a protein that transports the pigment molecules to the appropriate location
(c) Lack of a receptor protein that allows the pigment to be taken up into appropriate cells
(d) All of these
(e) None of these

260. A tRNA molecule must:

(a) Have a base pair sequence identical to the codon in the mRNA

(b) Be attached to a particular amino acid

(c) Bind to the protein to determine primary sequence

(d) All of these

(e) None of these

261. Which of these would be LEAST likely to lead to incorporation of the wrong amino acid into a polypeptide?

(a) A change in the wobble base of the codon

(b) A mistake in amino acid attachment to the tRNA

(c) A change in the third base of the anticodon

(d) All of these would be expected to lead to incorporation of the wrong amino acid

(e) None of these would be expected to lead to incorporation of the wrong amino acid

262. Which portion of the tRNA molecule is specifically recognized by aminoacyl-tRNA synthetase?

(a) The acceptor stem

(b) Bases in the stem loop regions

(c) The anticodon region

(d) All of these

(e) None of these

263. Which of these is a structural difference between prokaryotic and eukaryotic cytosolic ribosomes?

(a) Prokaryotic ribosomes consist of two subunits

(b) Eukaryotic ribosomes consist of both protein and RNA

(c) Eukaryotic ribosomes are smaller than prokaryotic ribosomes

(d) All of these

(e) None of these

264. Which of these represents a difference between eukaryotic and prokaryotic transcription?

(a) Initiation factors are required for forming a eukaryotic initiation complex

(b) A Koxak consensus sequence identifies the start codon in eukaryotes

(c) Binding of the initiator tRNA is required for assembly of the initiation complex in eukaryotes

(d) All of these are differences between eukaryotic and prokaryotic transcription

(e) None of these are differences between eukaryotic and prokaryotic transcription

265. Which of these is an enzymatic function of the ribosome?

(a) Polynucleotide kinase

(b) Base pair recognition

(c) Methyl transferase

(d) Peptidyl transferase

(e) None of the above

266. Termination codons differ from other codons in that:

(a) They only recognize a single tRNA

(b) They are capable of binding to many different tRNAs

(c) They bind to release factors instead of tRNAs

(d) They only specify a single amino aciD. E. Termination codons do not differ from other codons

(e) None of the above

267. Which of these DNA elements does NOT encode information that is recognized as RNA or protein?

(a) Shine-Dalgarno sequence

(b) Kozak consensus sequence

(c) Sorting signals

(d) All of these are recognized only as DNA

(e) None of these are recognized only as DNA

268. Synthesis of mRNA is

(a) In the 5' to 3' direction with new nucleotides being added to the 5' end of the mRNA molecule

(b) In the 3' to 5' direction with new nucleotides being added to the 5' end of the mRNA molecule

(c) In the 5' to 3' direction with new nucleotides being added to the 3' end of the mRNA molecule

(d) In the 3' to 5' direction with new nucleotides being added to the 3' end of the mRNA molecule

(e) Able to proceed either from the 5' to 3' or 3' to 5' direction

269. Sigma factor is

(a) A subunit of mRNA

(b) A subunit of RNA polymerase

(c) A subunit of DNA

(d) An enzyme needed for transcription to proceed

(e) The place on RNA where RNA polymerase attaches

270. Transcription begins when RNA polymerase binds to the

(a) Terminator on DNA (b) Initiator on DNA

(c) Promoter on DNA (d) Promoter on RNA

(e) None of the above

271. New amino acids (other than the initial f-Met) enter at which site?

(a) E (b) P

(c) A (d) Any of the above

(e) None of the above

272. At the E site

(a) Transfer RNA is released

(b) Anticodons match with codons

(c) Peptide bonds are formed between amino acids

(d) Transcription occurs

(e) None of the above

273. After the ribosome moves to the next codon on the mRNA, the growing peptide chain is found in which position(s)?

(a) A (b) P

(c) E (d) A, P and E

(e) None of the above

274. Which of the following is NOT an outcome of posttranslational modifications of a protein?

(a) Removal of N-terminus amino acid

(b) Addition of phosphate groups

(c) Addition of metals to create tertiary or quaternary structures

(d) Removal of introns

(e) None of the above

275. What can we gain from an understanding of a protein's function that cannot be deduced by the DNA sequence of the gene (that encodes the protein) alone?

(a) The protein's function determines the DNA sequence

(b) The possible amino acids coded for

(c) An understanding of biochemical pathways involving the protein

(d) A point mutation

(e) None of the above

276. In Neurospora, a mutation was induced that causes the loss of function of an enzyme essential to the biosynthesis of the vitamin pyridoxine. On which medium will such organisms NOT grow?

(a) Minimal

(b) Minimal + all vitamins

(c) Neither minimal nor minimal + pyridoxine

(d) Complete

(e) Minimal + pyridoxine

277. Which of the following is NOT true about tRNA molecules?

(a) There are four single-stranded loops in each tRNA

(b) There is about as much double-stranded as single-stranded structure in a tRNA

(c) Amino acids are carried on the 5′ end

(d) The anticodon area is at the tip of the molecule

(e) Some of the nucleotide bases in tRNAs are unusual and are not found in any other molecule in the cell

278. The process of using ATP to drive the loading of a tRNA with its appropriate amino acid is known as .

(a) Charging

(b) Burdening

(c) Uploading

(d) RNA processing

(e) Filling

279. An isoaccepting tRNA is one that:

(a) Can pair its anticodon with only one codon

(b) Can be loaded with any amino acid by any of the amino acyl synthetases

(c) Can only be loaded with the amino acid isoleucine

(d) Can fit in any amino acyl synthetase in the case of cellular emergencies

(e) Belongs to a subgroup of tRNAs that can be successfully loaded with a particular amino acid by a particular synthetase

280. Which of the following is TRUE about tRNAs?

(a) tRNAs act as scaffolds to help polypeptides fold correctly

(b) tRNAs are only found in the cytoplasm of eukaryotic cells

(c) tRNAs form part of the structure and function of ribosomes

(d) There are more prokaryotic tRNAs than in eukaryotes

(e) Prokaryotic and eukaryotic tRNAs are virtually identical

281. In a polysome, which ribosomes will have synthesized the longest peptide chains at any one time?

(a) The ones nearest the 3' end of the mRNA molecule

(b) The ones nearest the AUG codon in the mRNA

(c) The ones nearest the 5' end of the mRNA molecule

(d) The ones nearest the middle of the mRNA molecule

(e) The ones nearest both the 3' and 5' ends of the mRNA molecule

282. Based on the schematic animation that you have just seen, which of the following would you say is true?

(a) The large subunit acts primarily to decode mRNA

(b) The large subunit acts primarily to form peptide bonds

(c) The small subunit acts primarily to form peptide bonds

(d) The small subunit acts primarily to decode tRNA

(e) Both answers B and D are correct

283. Inborn disorders of metabolism result from:

(a) Viruses

(b) Neutral mutations

(c) Ingestion of a toxin from the environment

(d) Mutations in protein coding genes which result in enzymes with altered catalytic properties

(e) None of the above

284. Most inborn disorders of metabolism associated with the phenylalanine - tyrosine pathway have deleterious effects because:

(a) The cells cannot obtain enough energy to function normally

(b) Intermediate compounds that are not catalyzed at the blocks accumulate to toxic levels

(c) The specific disorder causes large numbers of cells to die

(d) Necessary metabolic end products are not synthesized

(e) The defective enzyme responsible for the block accumulates to toxic levels

285. Albinism is a consequence of:

(a) An enzyme that converts phenylalanine to tyrosine

(b) A gene mutation that results in the conversion of melanin to a lighter colored pigment

(c) A mutation that blocks the pathway to melanin production

(d) A mutation that prevents the transcription of several other genes

(e) A gene mutation that results in excessive melanin production

286. The experiments of Beadle and Tatum:

(a) Neither supported nor refuted the one gene - one enzyme hypothesis

(b) Were not related to the one gene- one enzyme hypothesis

(c) Supported the one gene - one enzyme hypothesis

(d) Did not support the one gene - one enzyme hypothesis

(e) Clearly disproved the one gene - one enzyme hypothesis

287. Neurospora crassa is an excellent organism for investigating the effects of mutations on metabolic pathways because:

(a) All the gametes from a single meiosis are packaged in a single sac

(b) It is easy to grow

(c) It can reproduce asexually

(d) Mutations can be easily induced and isolated

(e) All of the above

288. Neurospora crassa is an excellent organism for investigating the effects of mutations on nutrient synthesis because:

(a) Spores with mutations affecting pathways of nutrient biosynthesis can grow on minimal medium

(b) The wild type can only grow on minimal medium

(c) The wild type can only grow on complete medium

(d) Spores with mutations affecting pathways of nutrient biosynthesis will not grow on minimal medium

(e) Spores that are produced are all mutated the same way

289. What would be the advantages of using asexual spores of Neurospora to study metabolic pathways?

(a) Investigate the effects of genetic recombination

(b) Avoid the confounding effects of recombination in their experiments

(c) Maintain clonal populations of mutants

(d) Both A. and C. above are correct

(e) Both B. and C. above are correct

290. The one gene - one enzyme hypothesis is the idea that:

(a) Metabolic pathways are not controlled by genes

(b) Each step in a metabolic pathway is catalyzed by a specific enzyme coded for by a specific gene

(c) All steps in a metabolic pathway are catalyzed by one enzyme which is the product of one gene

(d) None of the above

291. Mutation of one of several genes can result in the same nutritional requirement in different organisms

(a) If the pathway of nutrient biosynthesis involves only one step

(b) Because the one gene - one enzyme hypothesis is incorrect

(c) Because each step in a metabolic pathway involves multiple enzymes

(d) Because some genes can keep other genes from being transcribed

(e) If the pathway of nutrient biosynthesis involves several steps

292. What sequence on the template strand of DNA corresponds to the first amino acid inserted into a protein?

(a) AUG (b) ATG

(c) UAC (d) TAC

(e) ATG

293. During translation, codons are read from

(a) DNA (b) mRNA

(c) tRNA (d) rRNA

(e) Protein

294. Transcription and translation of a gene composed of 30 nucleotides would form a protein containing no more than how many amino acids?

(a) 10 (b) 15

(c) 30 (d) 60

(e) 90

295. The enzyme peptidyl transferase

(a) Charges tRNA

(b) Synthesizes long, artificial mRNA's

(c) Recognizes the prokaryotic ribosomes

(d) Creates peptide bonds between amino acids

(e) Is not an enzyme involved in translation

296. A charged tRNA enters the protein-synthesizing machinery at the ribosomes by recognition of its

(a) Codon (b) Anticodon

(c) Second genetic code (d) Wobble

(e) Amino acid

297. What amino acid corresponds to the mRNA sequence GGG?

(a) pro (b) trp

(c) ser (d) lys

(e) gly

298. What amino acid corresponds to the DNA template sequence GGG?

(a) pro (b) trp

(c) ser (d) lys

(e) gly

299. What is the consequence of changing the codon UGA to UGG?

(a) The peptide will be shorter than normal

(b) The peptide will be longer than normal

(c) The peptide will have one altered amino acid

(d) The protein will not be made

(e) The protein will be normal

300. Which of the following is an example of the degeneracy of the genetic code?

(a) A given amino acid has more than one codon

(b) Each codon specifies more than one amino acid

(c) The first two bases specify the amino acid

(d) A cell has only 50 tRNAs

(e) The genetic code is not degenerate

301. Which of the following statements is false?

(a) Penicillin is an antibiotic that interferes with translation at the ribosome)

(b) Most organisms use the AUG codon as a start codon

(c) The genetic code is non-overlapping

(d) The genetic code is a triplet

(e) A frameshift can result from an incorrect translational starting point

302. Which is true of transcription?

(a) It transcribes the entire DNA in a cell into mRNA

(b) It produces a mature mRNA transcript that contains both introns and exons that will be translated into a protein

(c) It produces an exact copy of one of the strands of DNA

(d) It occurs in the nucleus of eukaryotic cells

(e) None of the above

303. Which is true of translation of the mRNA transcript?

(a) It utilizes tRNA that acts as the "translator" between the molecular languages of amino acids and nucleic acids

(b) It utilizes the genetic information encoded in sequential mRNA codons to determine the sequence of monosaccharides in a polysaccharide chain

(c) It occurs in the nucleus of eukaryotic cells

(d) It always produces a fully mature protein once the ribosome detaches from the mRNA at the end of translation

(e) None of the above

304. Which of the following is a function of RNA polymerase?

(a) To catalyze the reaction that links amino acids together in the growing polypeptide chain

(b) To unwind DNA during transcription

(c) To remove introns from mRNA transcripts

(d) To "read" the mRNA sequence during production of polypeptides

(d) None of the above

305. Which of the following is NOT a component of a ribonucleotide?

(a) Phosphate group (b) 5-carbon sugar

(c) Thymine (d) Guanine

(e) None of the above

306. How does RNA differ from DNA?

(a) RNA is found only in the cytoplasm; DNA is found only in the nucleus

(b) RNA is double stranded, while DNA is single stranded

(c) In living cells, DNA replicates itself while RNA does not

(d) DNA has four different nitrogenous bases, while RNA has five

307. Which of the following might cause a shift in the entire "reading frame" of the mRNA transcript during translation?

(a) A substitution of one nucleotide for another occurs

(b) A substitution of amino acids in a protein occurs

(c) A substitution of an entire codon occurs

(d) A deletion of a single nucleotide base pair occurs

(e) None of the above

308. An accidental change in the monomer sequence in which of the following molecules could affect the function of a protein?

(a) DNA (b) The protein itself

(c) mRNA (d) All of the above

(e) None of the above

309. Which of the following would have the greatest impact on the ability of a mammal to transport oxygen in the blood?

(a) An accidental alteration in the sequence of amino acids during construction of the hemoglobin protein

(b) An accidental alteration in the ribonucleotide sequence of an mRNA transcript transcribed from the hemoglobin gene

(c) An accidental alteration in the sequence of nucleotides in the hemoglobin gene during

(d) An inherited mutation in DNA that altered the sequence of nucleotides in the gene for hemoglobin

(e) None of the above

310. The DNA base sequence TTGCGATCG would code for which mRNA transcript?

(a) AACGCUAGC (b) TTGCGATCG

(c) AACGCTAGC (d) TTGCGUTCG

(e) None of the above

311. The process of translation is completed when

(a) A stop codon moves into the A site

(b) The cell runs out of amino acids

(c) The cell runs out of ribonucleotides

(d) The translation machinery reaches the end of the DNA molecule

(e) None of the above

312. Which of the following is a correct statement concerning translation?

(a) The mRNA will load into the large ribosomal subunit to begin the translation process

(b) New codons that need to be "read" move into the P-site of the ribosome

(c) The large ribosomal subunit does not join the translation process until after the first tRNA has arrived

(d) Amino acids attach to their corresponding tRNA molecules after the tRNA molecule attaches to the ribosome

(e) None of the above

313. Which of the following is a correct statement concerning mRNA?

(a) An mRNA is always translated in the nucleus

(b) An mRNA will always have an equal amount of adenine and uracil

(c) During transcription, two mRNAs are always produced together; one from each side of the separated DNA

(d) A single mRNA molecule may be translated again and again by multiple ribosomes

(e) None of the above

314. A protein that is made up of 50 amino acids would have been translated from an mRNA that had at least... (level 1) [Hint]

(a) 500 nucleotides
(b) 300 nucleotides
(c) 150 nucleotides
(d) 1000 nucleotides
(e) None of the above

315. Which of the following polymers would have the greatest number of monomers?

(a) The mRNA that was transcribed from the 1-micrometer long piece of DNA

(b) The protein that was translated from the mRNA transcript described in answer A

(c) A section of DNA that was exactly 1 micrometer long

(d) All the above

(e) None of the above

316. Which of the following is a correct statement concerning the codons found in the genetic code? (level 1) [Hint]

(a) Some codons do not code for any amino acids

(b) The codon AUG is usually the start codon for every mRNA

(c) Some amino acids are coded for by 4 different codons

(d) All of the above are correct statements

(e) None of the above

317. When an RNA molecule is first produced it often contains erroneous information in it that must be eliminated before translation can begin. This erroneous information is referred to as

(a) Enterons
(b) Exitrons
(c) Introns
(d) Exons
(e) None of the above

318. Which of the following is a correct statement concerning a gene?

(a) A gene is a section of the DNA that is transcribed to form rRNA that becomes part of a ribosome

(b) A gene is a section of the DNA that is transcribed into an mRNA that is used to build a protein

(c) A gene is a section of the DNA that is transcribed to form tRNA that is used in the translation process

(d) All of the above are correct definitions of a gene

(e) None of the above

319. Some of the DNA in your genome is not used to code for proteins. Approximately how much of this non-coding DNA is found in the human genome?

(a) < 10% (b) > 95%

(c) < 5% (d) < 50%

(e) None of the above

320. Monod and Jacob discovered that a system dubbed an "operon" was responsible for regulating the gene activity of certain *E. coli* genes. According to this system, whenever there was an increase in the concentration of the sugar lactose you would also expect to see.

(a) A decrease in the number of promoter sites

(b) An increase in the production of b-galactosidase

(c) An increase in the repressor protein concentration.

(d) A decrease in the repressor protein concentration.

(e) None of the above

Fill in the Blanks

1. Adenine and guanine have nitrogen-containing bases with __________ ring(s) while thymine and cytosine have nitrogen-containing bases with ________ ring(s).

(a) Two carbon, one carbon (b) One carbon, one carbon

(c) Two carbon, two carbon (d) One carbon, two carbon

(e) Three carbon, two carbon

2. X-ray diffraction experiments conducted by _________ led to the determination that DNA was a helical molecule.
   (a) Francis Crick
   (b) James Watson
   (c) Heinz Fraenkel-Conrat
   (d) Rosalind Franklin
   (e) Martha Chase

3. Mutations that throw the reading of the gene out of register are __________ mutations.
   (a) Point
   (b) Transpositions
   (c) Frame-shift
   (d) Chromosomal rearrangements
   (e) Insignificant

4. Mutations involving the alteration of one or a few base pairs in the coding sequence are __________ mutations.
   (a) Point
   (b) Transpositions
   (c) Frame-shift
   (d) Chromosomal rearrangements
   (e) Insignificant

5. Diesel exhaust, pesticides, and cigarette tar have been implicated in ______ cancer.
   (a) Breast
   (b) Skin
   (c) Nose
   (d) Lung
   (e) Bladder

6. After transcription, the mRNA molecule must _________ before it is translated into a protein molecule.
   (a) Assume the correct three-dimensional sequence
   (b) Adjust to its new environment
   (c) Have its introns removed
   (d) Have its exons removed
   (e) Complex with a tRNA molecule

7. Chemical or physical damage to DNA can cause ________ mutations involving one or a few nucleotides.

(a) Single (b) Simple
(c) Insignificant (d) Point
(e) Patterned

8. The assembly of transcription factors on a promoter begins some 25 nucleotides upstream where it binds to a start ____________ sequence.

(a) ATAT (b) AATT
(c) TTAA (d) TAAT
(e) TATA

9. The leucine zipper motif involves the cooperation of two ____________ subunits.

(a) Leucine (b) Protein
(c) RNA (d) Polymerase
(e) Histone

10. The operon that controls tryptophan producing genes in *E. coli* consists of __________.

(a) Only one gene (b) Two genes
(c) Three genes (d) Four genes
(e) Five genes

11. In order for the helix-turn-helix motif to bind to DNA, the ____________ must fit into the major groove of the DNA.

(a) Homeotic switches
(b) Zinc fingers
(c) Operator
(d) Recognition helix
(e) Protein link

12. In an operon the location of the regulatory region occurs _______ the structural genes.

(a) After (b) Within
(c) Before (d) All the above
(e) None of the above

13. In Mendel's crosses with single traits in pea plants, the trait that masked the other was said to be ________________.

(a) Recessive (b) Dominant
(c) Elementen (d) True-breeding
(e) None of the above

14. The basic unit of inheritance is a(n) __________.

(a) Allele (b) Trait
(c) Gene (d) Chromosome
(e) None of the above

15. An individual with two identical alleles for a gene is called ____________.

(a) Homologous (b) Homozygous
(c) Heterozygous (d) Dominant
(e) None of the above

16. If an organism has been designated Tt, it is ______________.

(a) Homozygous dominant (b) Heterozygous
(c) Homozygous recessive (d) Homologous
(e) None of the above

17. Mendel's pea plants had either yellow or green unripe pods. These characteristics represent an organism's ____________.

(a) Adaptedness (b) Genotype
(c) Phenotype (d) Chromosomes
(e) None of the above

18. Organisms that vary from the wildtype are called ____________.

(a) Mutant (b) Recessive
(c) Independent (d) Dominant
(e) None of the above

19. The typical phenotypic ratio expected from a monohybrid cross involving simple dominance is __________.

(a) 1:1 (b) 4:0
(c) 3:1 (d) 2:2
(e) None of the above

20. In a typical pedigree chart, a half-filled circle indicates a ____________.

(a) Recessive parent
(b) Non-affected male
(c) Affected female
(d) Carrier female
(e) None of the above

21. If 60 out of 100 cats that inherit the dominant allele for polydactyly actually have extra digits, the allele is said to be ____________.

(a) Expressed
(b) Completely penetrant
(c) Codominant
(d) 60% penetrant
(e) None of the above

22. Marfan syndrome and porphyria were historically difficult to trace through families or determine their mode of inheritance because of the variety of symptoms of each disease. Disorders or traits with many symptoms are __________.

(a) Pleiotropic
(b) Epistatic
(c) Incompletely dominant
(d) Codominant
(e) None of the above

23. A cross of a red-flowered snapdragon plant with a white-flowered snapdragon produces a plant with pink flowers. This is an example of ____________.

(a) Pleiotropy
(b) Epistasis
(c) Incomplete dominance
(d) Codominance
(e) None of the above

24. A ____________ appears to be an inherited trait but is instead caused by an environmental factor.

(a) Multifactorial trait
(b) Phenocopy
(c) Variably expressive trait
(d) Epistatic gene
(e) None of the above

25. A complex trait determined by one or more genes plus the environment is ____________.

(a) Polygenic
(b) Multifactorial
(c) Quantitative
(d) Genetically heterogenetic
(e) None of the above

26. Identical twins result when a single fertilized egg divides in two. Another term for identical twin is ___________.

(a) Fraternal (b) Maternal

(c) Monozygotic (d) Dizygotic

(e) None of the above

27. A cross of a red-flowered snapdragon plant with a white-flowered snapdragon produces a plant with pink flowers. This is an example of ___________.

(a) Pleiotropy (b) Epistasis

(c) Incomplete dominance (d) Codominance

(e) None of the above

28. Adenine and guanine have nitrogen-containing bases with __________ ring(s) while thymine and cytosine have nitrogen-containing bases with ________ ring(s).

(a) Two carbon, one carbon

(b) One carbon, one carbon

(c) Two carbon, two carbon

(d) One carbon, two carbon

(e) Three carbon, two carbon

29. X-ray diffraction experiments conducted by _________ led to the determination that DNA was a helical molecule.

(a) Francis Crick (b) James Watson

(c) Heinz Fraenkel-Conrat (d) Rosalind Franklin

(e) Martha Chase

30. Mutations that throw the reading of the gene out of register are _________ mutations.

(a) Point

(b) Transpositions

(c) Frame-shift

(d) Chromosomal rearrangements

(e) Insignificant

31. Mutations involving the alteration of one or a few base pairs in the coding sequence are __________ mutations.

(a) Point
(b) Transpositions
(c) Frame-shift
(d) Chromosomal rearrangements
(e) Insignificant

32. Diesel exhaust, pesticides, and cigarette tar have been implicated in ______ cancer.

(a) Breast
(b) Skin
(c) Nose
(d) Lung
(e) Bladder

33. After transcription, the mRNA molecule must __________ before it is translated into a protein molecule.

(a) Assume the correct three-dimensional sequence
(b) Adjust to its new environment
(c) Have its introns removed
(d) Have its exons removed
(e) Complex with a tRNA molecule

34. Chemical or physical damage to DNA can cause __________ mutations involving one or a few nucleotides.

(a) Single
(b) Simple
(c) Insignificant
(d) Point
(e) Patterned

35. The wobble hypothesis involves __________.

(a) tRNA
(b) mRNA
(c) rRNA
(d) Both mRNA and rRNA
(e) Both mRNA and tRNA

36. RNA synthesis on a DNA template is called __________.

(a) Transformation
(b) Translation
(c) Transduction
(d) Transcription
(e) Transfection

37. Areas of a eukaryotic gene that are represented in the final protein product are termed _________.

(a) Introns (b) Mutons

(c) Recons (d) Exons

(e) Cistrons

38. Enhancer elements are found _________.

(a) In none of the stated regions; enhancers are environmental agents

(b) In any of these regions

(c) Within coding sequences

(d) Downstream from the start site

(e) Upstream from the start site

39. Regarding enhancers and transcription factors (TFs) and their effects on gene expression, _________.

(a) Both are trans-activating

(b) Enhancers are cis-activating, while TFs are trans-activating

(c) Enhancers are trans-activating, while TFs are cis-activating

(d) Both are cis-activating

(e) Neither are cis- or trans-activating; they are repressors

40. The genetic code is considered "degenerate" because _________.

(a) The code originated from bacterial genomes

(b) DNA is unstable

(c) Degenerate is the name of the person who deciphered the code

(d) More than one nucleotide triplet codes for the same amino acid

(e) None of the above

41. The wobble hypothesis states that ___.

(a) Several mRNA codons may pair with a single transfer RNA

(b) An mRNA codon may pair with more than one transfer RNA

(c) There are too many tRNAs present to account for the number of amino acids

(d) An mRNA codon can pair with only a single transfer RNA

(e) None of the above

42. RNA polymerase specificity for specific genes is governed by ___.

(a) The delta subunit (b) The sigma subunit

(c) The beta subunit (d) The gamma subunit

(e) The alpha subunit

43. In prokaryotic transcription, ________ may bind to the mRNA strand as it is being transcribed.

(a) Ribosomes

(b) DNA polymerase

(c) tRNAs

(d) Single strand binding protein

(e) Sigma factors

44. The progressive restriction of future developmental capacity is ________.

(a) Determination (b) Maternal effect

(c) Differentiation (d) Transdetermination

(e) Predestination

45. Homeotic mutants in Drosophila are those in which ________.

(a) Recombination has taken place

(b) A position-effect is evident

(c) One structure is replaced with another

(d) Translocations have occurred

(e) A structure is missing

46. Differentiation of cells is controlled at both the ________ and the ________ levels.

(a) Transcriptional/translational

(b) Transcriptional/DNA repair

(c) Cell cycle control/translational

(d) Mitotic/meiotic

(e) DNA replication/translational

47. In multicellular organisms, cell-to-cell interactions help regulate development by ________.

(a) Exchanging DNA

(b) Signal transduction between cells

(c) Crowding each other

(d) Competing for nutrients

(e) Cell-to-cell interactions have no role in development

48. Viruses use_____ to make a double stranded DNA molecule that is inserted into the host DNA

(a) RNA polymerase
(b) RNA primase
(c) DNA polymerase III
(d) DNA polymerase II
(e) Reverse transcriptase

49. Researchers use a radioactively labeled ______ to search a clone library for a gene of interest.

(a) Vector
(b) Antibody
(c) Virus
(d) Probe
(e) None of the above

50. After DNA is cut with restriction enzymes, ________________ that are generated have complementary bases.

(a) Smooth ends
(b) Sticky ends
(c) Probes
(d) Plasmids
(e) None of the above

51. A strand that contains both elephant DNA and ostrich DNA is called ________ DNA.

(a) Recombinant
(b) Restriction
(c) Transformed
(d) Paired
(e) None of the above

52. Diseases caused by non-functional enzymes may in the future be treated through ______________, where a functional copy of the gene is inserted into cells.

(a) Gene therapy

(b) Gene rearrangement

(c) PCR

(d) Plasmid translocation

(e) None of the above

53. The maintenance of a stable internal environment is called ____________.

(a) Gene expression (b) Operon

(c) Activation (d) Homeostasis

(e) None of the above

54. In order for a gene to be transcribed, RNA polymerase must be able to bind to the gene's ____________.

(a) Activator (b) Regulator

(c) Promoter (d) Operator

(e) Repressor

55. A(n) ______________ is a sequence of DNA within the promoter.

(a) Repressor (b) Activator

(c) Operator (d) All the above

(e) None of the above

56. If a(n) ___________ binds to the operator, it prevents transcription.

(a) Repressor (b) Activator

(c) Promoter (d) Operator

(e) None of the above

57. When the amino acid tryptophan is present in the environment of *E. coli*, transcription of genes to produce tryptophan in *E. coli* is ____________.

(a) Stopped (b) Started

(c) Unchanged (d) All the above

(e) None of the above

58. Bacteria and eukaryotes most commonly regulate gene expression regulation at the __________ level.

(a) Translational (b) Transcriptional

(c) Post-transcriptional (d) Post-translational

(e) None of the above

59. __________ enzymes are produced only in the presence of the substances they act upon, such as lactose.

(a) Inducible (b) Repressible

(c) All the above (d) None of the above

60. ____________ genes determine the fate of each body segment - (develops into a wing, leg, etc.)

(a) Totipotent (b) Stem cell

(c) Homeobox-containing (d) Lac operon

(e) None of the above

61. When cells are no longer able to divide, they have reached ____________ and die by a process called ____________.

(a) Determination, apoptosis

(B) Apoptosis, senescence

(c) Senescence, apoptosis

(d) Senility, senescence

(e) None of the above

62. In eukaryotes, there are ______________ codons that specify amino acids.

(a) 21 (b) 24

(c) 61 (d) 64

(e) 60

63. In mitochondrial genomes, ______________ is a "stop" codon.

(a) UGA (b) UUU

(c) AUA (d) UAA

(e) AGA

64. In transcription, the nucleotide sequence CAT in DNA would specify _______________ in mRNA.

(a) TAC (b) GAU

(c) CAT (d) GTU

(e) GUA

65. In the formation of an initiation complex, a _______________ is positioned first.

(a) Met-tRNA (b) ser-tRNA

(c) tyr-rRNA (d) mval-tRNA

(e) cyst-tRNA

66. In a process called _______________, the initial tRNA is ejected from the ribosome.

(a) Elongation (b) Translocation

(c) Initiation (d) Transcription

(e) Elimination

67. The process of _______________ cuts introns from the primary transcript and the final "processed" mRNA is produced.

(a) RNA cleaving (b) RNA translocation

(c) RNA elongation (d) RNA splicing

(e) RNA releasing

68. As polypeptides are formed at the ribosome, elongation continues until _______________ is exposed.

(a) A release factor (b) An intron

(c) A nonsense codon (d) An exon

(e) Polypeptidase

69. The process of synthesizing RNA from a DNA template is called ________.

(a) Replication (b) Transcription

(c) Translation (d) Translocation

(e) None of the above

70. The process of synthesizing proteins using a mRNA molecule is called ___________.

(a) Replication (b) Transcription
(c) Translation (d) Translocation
(e) None of the above

71. The group of proteins that aid in the initiation of transcription are called ___________.

(a) RNA polymerase (b) Transcription factors
(c) Nuclear matrix proteins (d) Promoters
(e) None of the above

72. Like DNA replication, the first major step in transcription is ___________.

(a) Binding of the transcription factors
(b) Elongation
(c) Unwinding of the double helix
(d) Recognition of the promoter sites
(e) None of the above

73. The triplet of nucleotides in tRNA that is complementary to a triplet of nucleotides in mRNA is called a(n) ___________.

(a) Codon (b) Anticodon
(c) Antiparallel triplet (d) Repressor
(r) None of the above

74. Parts of genes called ___________ are transcribed into mRNA but are later removed.

(a) Exons (b) Prions
(c) Introns (d) Caps
(e) None of the above

75. _______________ help(s) the protein fold correctly after elongation halts.

(a) Chaperone proteins (b) The initiation complex
(c) The "stop" codon (d) The large ribosomal subunit
(e) None of the above

76. If there is an error in the sequence of a gene, the error is called a(n) ________.

(a) Mutation
(b) Missense
(c) Transversion
(d) Transition
(e) None of the above

77. Natural mutagens include ____________.

(a) Ultraviolet light
(b) Natural radiation from rocks
(c) Cosmic rays from outer space
(d) All of these are correct.
(e) None of the above

78. The addition of one or two nucleotides in a DNA sequence, for example ATGGCT to ATGCCGCT, would cause a ___________ mutation.

(a) Point
(b) Frameshift
(c) Missense
(d) Nonsense
(e) None of the above

79. The actual process of protein synthesis takes place on ribosomes that serve as workbenches, with ____________ acting as the blueprint or template.

(a) DNA
(b) mRNA
(c) tRNA
(d) All the above
(e) None of the above

80. The ____________ is nontranslated sequence that is located between the transcription start site and the translation start site

(a) Reading frame
(b) Leader sequence
(c) Trailer sequence
(d) All the above
(e) None of the above

81. Ribosomal RNAs in eukaryotes and prokaryotes are different in all of the following ways except __________.

(a) Location in cell
(b) Size of subunits

(c) Number of proteins associated with ribosomes

(d) Function

(e) None of the above

82. The quaternary structure of proteins is represented by the _________.

(a) Amino acid sequence

(b) Alpha helix

(c) Arrangement of subunits

(d) Beta-pleated sheet

(e) Three-dimensional conformation

83. Alkaptonuria is a metabolic disorder that results in the accumulation of _________.

(a) Phenylalanine

(b) Homogentisic acid

(c) Tyrosine

(d) Phenylalanine hydroxylase

(e) None of the above

84. The process of tRNA "charging" does NOT involve _________.

(a) rRNA

(b) tRNA

(c) Aminoacyl synthetase

(d) ATP

(e) An amino acid

85. Polyribosomes consist of _________.

(a) Several ribosomes

(b) One mRNA

(c) Several tRNAs

(d) One mRNA, several tRNAs, and several ribosomes

(e) One mRNA and several ribosomes

86. The increase of a polypeptide chain by one amino acid is called __.

(a) Translation

(b) Transcription

(c) Elongation

(d) Initiation

(e) Hydrolysis

87. The formation of the peptide bond is catalyzed by __.

(a) EF-Tu
(b) Tryptophan
(c) Peptidyl transferase
(d) Aminoacyl tRNA synthetase
(e) GTP

88. The charged initiator tRNA binds initially to the __ in protein synthesis.

(a) E site
(b) 3' end of mRNA
(c) A site
(d) UGA base sequence of mRNA
(e) P site

89. In prokaryotes, the initiation codon for protein synthesis is normally __.

(a) UAC (b) UGA
(c) UUC (d) AUG
(e) AAG

90. The shift of mRNA by 3 nucleotides in order to bring another codon into position in the ribosome is called __.

(a) Translation (b) Transcription
(c) Trans-isomerism (d) Transport
(e) Translocation

91. The final location in the ribosome that the uncharged tRNA is thought to move to before exiting the ribosome is called the __.

(a) Small subunit (b) Termination codon
(c) E site (d) P site
(e) A site

92. IF2 and GTP are examples of __.

(a) Elongation factors (b) tRNA factors
(c) Ranslation factors (d) RNA factors
(e) Initiation factors

93. The growing polypeptide chain in the ribosome during translation is attached to the __.

(a) 5' end of a tRNA
(b) 3' end of the mRNA
(c) Small subunit
(d) 5' end of the mRNA
(e) 3' end of a tRNA

94. EF-Tu is an example of __.

(a) An initiation factor
(b) An anticodon
(c) An enzyme catalyzing peptide bond formation
(d) An elongation factor
(e) A termination factor

95. Translation of mRNA in a ribosome proceeds __.

(a) From either the 5' or 3' end, depending on the enzymes present
(b) From the 5' end to the 3' end
(c) From the 3' end to the 5' end
(d) From the middle simultaneously toward the 3' and 5' ends
(e) From the middle to the 3' end and then from the middle to the 5' end

96. Multiple sites translating simultaneously the code sequence on different parts of a single mRNA molecule constitute a __.

(a) Polymorphism
(b) Polymerase chain
(c) Polypeptide
(d) Polynucleotide
(e) Polyribosome

97. A mutation in a gene coding for a protein can result in ___.

(a) A change in three dimensional shape of a protein
(b) A change in the catalytic properties of the protein
(c) A change in the higher order structure of the protein
(d) A change in the amino acid sequence of the protein
(e) All of the above

True/False

1. From the promoter of the gene, you would expect the coding strand to run 5' to 3'.

   (a) True (b) False

2. Regulatory sequences are DNA sequences which encode transcription factors.

   (a) True (b) False

3. During transcription, the RNA polymerase would be expected to spend the majority of its time in the open complex rather than the closed complex.

   (a) True (b) False

4. The product of a structural gene is mRNA and then a polypeptide.

   (a) True (b) False

5. A site designated +15 would be located within the promoter of the gene.

   (a) True (b) False

6. Bacterial RNA polymerase holoenzyme consists of four subunits.

   (a) True (b) False

7. Intrinsic termination requires the rho protein.

   (a) True (b) False

8. Both types of transcriptional termination rely on stem-loop structures in the newly transcribed RNA.

   (a) True (b) False

9. Regulation of genes in both eukaryotes and prokaryotes requires cis and trans acting factors.

   (a) True (b) False

10. Only proteins can be biological catalysts.

    (a) True (b) False

11. Transcription takes place in the nucleus using the DNA in the nucleus as a template for the formation of proteins.

    (a) True (b) False

12. Genes are composed of segments of "extra," nonessential materials called exons.

(a) True (b) False

13. Aminoacyl-tRNA synthetase directs the synthesis of tRNA molecules from a DNA template.

(a) True (b) False

14. Just as one amino acid may be specified by more than one codon, one codon may specify more than one amino acid.

(a) True (b) False

15. For a given gene, RNA is transcribed from both strands of the double helix simultaneously.

(a) True (b) False

16. The process of transcription is only loosely regulated.

(a) True (b) False

17. Bacteria are the only organisms that employ operons to regulate transcription.

(a) True (b) False

18. mRNA is this the product of DNA transcription.

(a) True (b) False

19. RNA polymerase binds to one strand of a DNA molecule at a site called the promoter.

(a) True (b) False

20. Activators are regulatory genes that allow enhancers to bind to the promoter site on a DNA molecule.

(a) True (b) False

21. DNA loops around so the enhancer is positioned near the promoter it regulates.

(a) True (b) False

22. In eukaryotic gene regulation, the location of the promoter is always before and the location of the enhancer always after the gene being regulated.

(a) True (b) False

23. A single gene may use a regulatory site to control the expression of that gene, but genes rarely have multiple regulatory sites.

(a) True (b) False

24. The law of independent assortment states that inheritance of one trait does not depend on the transmission of a gene for another trait.

(a) True (b) False

25. The Hershey-Chase team of researchers used viruses to prove DNA was the genetic material causing the transformation observed in Griffith's and Avery's experiments.

(a) True (b) False

26. Hershey and Chase found that the hereditary material in bacteriophages was a protein/DNA complex.

(a) True (b) False

27. mRNA is this the product of DNA transcription.

(a) True (b) False

28. RNA polymerase binds to one strand of a DNA molecule at a site called the promoter.

(a) True (b) False

29. Activators are regulatory genes that allow enhancers to bind to the promoter site on a DNA molecule.

(a) True (b) False

30. DNA loops around so the enhancer is positioned near the promoter it regulates.

(a) True (b) False

31. Changes that alter the sequence of DNA nucleotides are called point mutations.

(a) True (b) False

32. Genes that can move about from one location on a chromosome to another are called transposons.

(a) True (b) False

33. A single gene may use a regulatory site to control the expression of that gene, but is unable to turn a gene off and on.

(a) True (b) False

34. The process of acquiring a gene from another organism is called transformation.

(a) True (b) False

35. The Hershey-Chase experiment showed that the virus protein is responsible for directing the production of new viruses.

(a) True (b) False

36. Chargaff's rule suggests that the purines and pyrimidines in the DNA molecule combine in a nonrandom pattern.

(a) True (b) False

37. The two sides of the double helix structure of DNA are held together with hydrogen bonds.

(a) True (b) False

38. In DNA replication, DNA polymerase assembles a complementary new strand on each old one.

(a) True (b) False

39. Transcription takes place in the nucleus using the DNA in the nucleus as a template for the formation of proteins.

(a) True (b) False

40. Each three-nucleotide block in a gene corresponds to a particular amino acid.

(a) True (b) False

41. The vast variety of characteristics results from the fact that an almost unlimited number of codons can be formed.

(a) True (b) False

42. Most genes in your cells exist in clusters of almost identical sequences.

(a) True (b) False

43. Genes are composed of segments of "extra," nonessential materials called exons.

(a) True (b) False

44. Sanger earned his second Nobel Prize for development of a technique called DNA fingerprinting.

(a) True (b) False

45. Cloned human genes are localized by using radioactive probes to detect binding between the probe and the complementary sequence on the chromosome.

(a) True (b) False

46. One possible risk of genetic engineering is the accidental production of dangerously resistant microorganisms.

(a) True (b) False

47. Although risks are possible, gene engineering appears to offer more of a contribution to human welfare than threat.

(a) True (b) False

48. The Ti plasmid provides a means to transfer genes into cereal plants, such as corn and rice.

(a) Truc (b) False

49. Genetic engineering has provided a method for the production of insulin to fight diabetes, but it remains too expensive to be practical.

(a) True (b) False

50. Piggyback vaccines work by triggering the body's defense system with a harmless virus carrying the surface proteins of a disease organism.

(a) True (b) False

51. Interferons are used to disrupt the reproduction of viruses.

(a) True (b) False

52. The human genome refers to the entire collection of genes found within the human cell.

(a) True (b) False

53. cDNA can be used as a probe to screen bacterial colonies for the presence of a desired gene

(a) True (b) False

54. TNF should prove effective in relieving the symptoms of cystic fibrosis.

(a) True (b) False

55. The second stage in gene transfer is cloning the plasmid.

(a) True (b) False

56. Factor VIII is a protein found in anticoagulants used to treat heart attack patients.

(a) True (b) False

57. Genetic engineering can make plants toxic to insect pests that feed on the plant, but leave the plant harmless to humans

(a) True (b) False

58. Translation converts information from the language of nucleic acids to the language of proteins.

(a) True (b) False

59. The genetic code is universal, meaning that any given codon specifies the same amino acid, regardless of the type of cell.

(a) True (b) False

60. A polypeptide is the same thing as a protein.

(a) True (b) False

61. Codons may have different functions than specification of amino acids.

(a) True (b) False

62. The organization of a polypeptide into a functional protein is typically a spontaneous process.

(a) True (b) False

63. A particular polypeptide may either form á helices or â sheets.

(a) True (b) False

64. Ribozyme activity is required for both transcription and translation.

(a) True (b) False

65. In a polyribosome, the longest polypeptide chain would be expected to be attached to the ribosome which is farthest away from the RNA polymerase.

(a) True (b) False

66. When RNA polymerase reaches a specific sequence of nucleotides on the DNA called the transcription terminator, a hairpin loop structure forms on the DNA strand causing the RNA polymerase and mRNA to dissociate from the DNA.

(a) True (b) False

67. RNA polymerase copies the template strand of DNA making single-stranded mRNA complementary to the DNA template.

(a) True (b) False

68. The codon on tRNA matches up with the complementary anticodon on mRNA.

(a) True (b) False

69. A stop codon codes for an amino acid as well as the signal to stop.

(a) True (b) False

70. The Hershey-Chase team of researchers used viruses to prove DNA was the genetic material causing the transformation observed in Griffith's and Avery's experiments.

(a) True (b) False

71. Hershey and Chase found that the hereditary material in bacteriophages was a protein/DNA complex.

(a) True (b) False

Answers

Multiple Choice Question

1	2	3	4	5	6	7	8	9	10	11	12	13	14	15
B	B	C	C	D	B	E	A	B	C	C	A	B	E	A
16	17	18	19	20	21	22	23	24	25	26	27	28	29	30
C	B	C	B	C	C	D	A	B	E	B	D	D	C	A
31	32	33	34	35	36	37	38	39	40	41	42	43	44	45
C	B	B	C	E	D	C	A	E	A	A	C	C	D	B
46	47	48	49	50	51	52	53	54	55	56	57	58	59	60
D	A	B	B	C	E	C	A	D	C	A	C	D	A	C
61	62	63	64	65	66	67	68	69	70	71	72	73	74	75
C	A	B	A	C	A	C	A	B	E	A	C	B	C	B
76	77	78	79	80	81	82	83	84	85	86	87	88	89	90
C	C	D	A	B	E	B	D	D	B	A	C	D	D	C

91	92	93	94	95	96	97	98	99	100	101	102	103	104	105
A	D	B	C	D	A	E	E	E	B	D	C	A	E	D
106	107	108	109	110	111	112	113	114	115	116	117	118	119	120
C	A	B	C	C	D	B	E	D	D	C	B	D	D	A
121	122	123	124	125	126	127	128	129	130	131	132	133	134	135
D	D	C	D	D	D	D	B	C	C	D	A	D	D	C
136	137	138	139	140	141	142	143	144	145	146	147	148	149	150
D	C	D	D	C	D	C	E	E	B	C	A	C	A	A
151	152	153	154	155	156	157	158	159	160	161	162	163	164	165
C	B	A	D	C	B	C	A	E	C	A	A	E	A	A
166	167	168	169	170	171	172	173	174	175	176	177	178	179	180
A	A	B	D	C	B	B	B	A	A	C	A	D	D	D
181	182	183	184	185	186	187	188	189	190	191	192	193	194	195
E	A	B	A	D	D	C	D	C	B	A	B	B	A	A
196	197	198	199	200	201	202	203	204	205	206	207	208	209	210
D	A	A	B	B	D	C	E	B	A	D	C	E	D	E
211	212	213	214	215	216	217	218	219	220	221	222	223	224	225
D	D	D	C	A	E	B	C	A	A	A	D	B	B	D
226	227	228	229	230	231	232	233	234	235	236	237	238	239	240
C	C	C	A	B	C	B	B	A	C	C	B	A	C	A
241	242	243	244	245	246	247	248	249	250	251	252	253	254	255
A	B	C	C	A	C	B	C	B	B	D	B	B	C	D
256	257	258	259	260	261	262	263	264	265	266	267	268	269	270
C	C	C	E	B	A	D	E	B	D	C	E	C	B	C
271	272	273	274	275	276	277	278	279	280	281	282	283	284	285
C	A	B	D	C	A	C	A	E	E	A	B	D	B	C
286	287	288	289	290	291	292	293	294	295	296	297	298	299	300
C	E	D	E	B	E	D	B	A	D	B	E	A	B	A
301	302	303	304	305	306	307	308	309	310	311	312	313	314	315
A	D	A	B	C	C	D	D	D	A	A	C	D	C	C

316	317	318	319	320										
D	C	D	B	B										

Fill in the Blanks

1	2	3	4	5	6	7	8	9	10	11	12	13	14	15
A	D	C	A	D	C	D	E	B	E	D	C	B	C	D
16	17	18	19	20	21	22	23	24	25	26	27	28	29	30
B	C	A	C	D	D	A	C	B	B	C	C	A	D	C
31	32	33	34	35	36	37	38	39	40	41	42	43	44	45
A	D	D	D	E	D	D	B	B	D	A	B	A	A	C
46	47	48	49	50	51	52	53	54	55	56	57	58	59	60
A	B	B	B	B	D	D	D	D	C	A	A	B	A	A
61	62	63	64	65	66	67	68	69	70	71	72	73	74	75
B	C	E	E	A	B	D	C	B	C	B	C	B	C	A
76	77	78	79	80	81	82	83	84	85	86	87	88	89	90
A	D	B	B	B	D	C	B	A	D	C	C	E	D	E
91	92	93	94	95	96	97								
C	E	E	D	B	E	E								

True/False

1	2	3	4	5	6	7	8	9	10	11	12	13	14	15
A	B	A	A	B	B	B	A	A	B	B	B	B	B	B
16	17	18	19	20	21	22	23	24	25	26	27	28	29	30
B	B	A	A	B	A	B	B	A	A	B	A	A	B	A
31	32	33	34	35	36	37	38	39	40	41	42	43	44	45
A	A	B	A	B	A	A	A	B	A	B	A	B	B	A
46	47	48	49	50	51	52	53	54	55	56	57	58	59	60
A	A	B	B	A	A	A	A	B	B	B	A	A	B	B
61	62	63	64	65	66	67	68	69	70	71				
A	A	B	B	B	B	A	B	B	A	B				

❑❑❑

Chapter – 15

Gene Mutation and DNA Repair Technology

General Effect of Mutation

- Gene mutations are molecular changes in the DNA sequence of a gene
- Gene mutations can alter the coding sequence within a gene
- Gene mutations are also given names that describe how they affect the wild-type genotype and phenotype
- Gene mutations can occur outside of the coding sequence and still influence gene expression
- DNA sequences known as trinucleotide repeats may cause mutation
- Changes in chromosome structure can affect the expression of a gene
- Mutations can occur in germ line or somatic cells

Effect of Mutation on Proteins

- A mutation is a change in a gene's DNA (genotype), and a mutant is the corresponding expression (phenotype).
- Mutations include single base changes, deletions, additions, or moved sequences in genes that encode proteins or in regulatory genes.
- Learning how a mutation alters a protein can explain how a disease arises.

- Germline mutations can be inherited, where as somatic mutations cannot.

The Beta Globin Gene

- Sickle cell disease is a recessive disorder associated with a mutation caused by a single nucleotide substitution in the beta globin gene on chromosome 11.
- In sickle cell disease, a mutation causes hemoglobin to crystallize in a low-oxygen environment, bending red blood cells into sickle shapes that block circulation, causing diverse symptoms.
- In beta thalassemia, beta globin is absent or scarce, causing too few complete hemoglobin molecules and buildup of free alpha globin chains and iron.

Disorders of Orderly Collagen

- Collagen is a highly symmetrical protein that is found in a variety of tissues, including bone, cartilage, skin, ligaments, and tendons.
- Mutations in collagen genes often disrupt the protein's precise organization.

A Mutation that Causes Early-Onset Alzheimer Disease

- In one form of Alzheimer disease, a mutation in a receptor protein may lead to beta-amyloid buildup.

Causes of Mutation

Spontaneous Mutation

- Mutations occur spontaneously when rare tautomers of bases are incorporated into replicating DNA, causing a base mismatch.
- Genes spontaneously mutate at different rates.
- Because bacteria and viruses reproduce frequently, they have higher spontaneous mutation rates.
- Spontaneous mutations are more likely in or near repetitive or symmetrical DNA sequences.

Induced Mutations

- Researchers use mutagens (chemicals or radiation) to cause mutations in genes, which they study to reveal normal gene function.
- Site-directed mutagenesis is PCR based technique using primers with intentional mismatches to engineer and amplify specific mutations.
- Accidental exposure to mutagens may come from nuclear accidents, weapons, medical treatments, cosmic rays, and radioactive isotopes in rocks.

Natural Exposure to Mutagens

- Natural environmental mutagens include exposure to ionizing radiation and chemical mutagens.
- The effects of radiation damage to DNA depend on the function of mutated genes.
- The risk that exposure to a chemical will cause a mutation can be difficult to predict since people vary in their susceptibility to certain chemicals.

Types of Mutations

Point Mutations

- A point mutation is a change in a single base. In a transition, a purine replaces a purine or a pyrimidine replaces a pyrimidine. In a transversion, a purine replaces a pyrimidine or vice versa.
- A missense mutation is a point mutation that changes one amino acid to a different one.
- A nonsense mutation is a point mutation that changes an amino acid encoding codon to a stop codon, which halts translation. A stop codon that changes to an amino-acid coding codon, lengthens the protein.

Deletions and Insertions Can Cause Frameshifts

- Inserting or deleting bases in DNA alters the gene's reading frame, causing a frameshift mutation.
- A tandem duplication repeat a section of a gene.

Pseudogenes and Transposons Revisited

- Pseudogenes are nonfunctional sequences that are very similar to a nearby functional gene.
- Although pseudogenes are not expressed, they can disrupt meiotic pairing. Gaucher disease is an example of a disorder that can result from a crossover between the working gene and its pseudogene.
- A transposable element is DNA that moves. It can insert into and disrupt transcription of a gene.

Expanding Repeats Lead to Protein Misfolding

- Myotonic dystrophy, Fragile X syndrome, and Huntington disease are examples of diseases that are caused by expanding triplet repeats.
- Expanded triplet repeat disorders are described as "dominant toxic gain of function" and often disturb brain function.
- The extra amino acids in an expanded gene may impart a new function to the expanded protein.
- Some triple repeat mutations affect the export or translation of a mRNA.

The Importance of a Mutation's Position in the Gene

- Whether a mutation alters the phenotype and how it does so, depends upon where in the protein the change occurs.

Globin Variants

- Mutations in the globin genes are well studied and diverse.
- Globin mutations occur in either the alpha or beta globin genes and they can cause anemia, affect oxygen binding, or be clinically silent.

Inherited Susceptibility to Prion Disorders

- Certain mutations in the prion gene predispose individuals to at least two inherited prion disorders.
- The disorders fatal familial insomnia and Creutzfeldt-Jakob syndrome both involve mutations in two key parts of the prion protein (amino acids 129 and 178).

Factors that Lessen the Effects of Mutation

- Many mutations in the third codon-position do not alter the specified amino acid.
- Changes in the second codon position often replace and amino acid with a structurally similar one.
- The phenotypes of conditional mutations are expressed only in certain environments.

DNA Repair

- Damaged bases can be directly repaired
- Base excision repair removes a damaged base
- Nucleotide excision repair systems remove segments of damaged DNA
- Mismatch repair systems recognize and correct a base pair mismatch
- Damaged DNA can be repaired by recombination
- Actively transcribed DNA is repaired more efficiently than nontranscribed DNA
- Damaged DNA may be replicated by translesion DNA polymerases

Three Types of DNA Repair

- Many genes encode enzymes that search for errors in DNA and correct them.
- UV light dimerizes adjacent pyrimidines, kinking the DNA and disrupting replication.
- Pyrimidine dimers and other types of DNA damage can be corrected by photoreactivation, excision repair, or mismatch repair.

DNA Repair Disorders

- p53 is one of a variety of proteins that can slow down the cell cycle and permit damaged cells to repair their DNA.
- If damage is to severe, p53 and other proteins can induce cells to undergo apoptosis (cell death).
- Abnormal repair genes cause disorders such as HNPCC (deficient mismatch repair), and XP (deficient excision repair).

- A defect in a cell cycle gene affects DNA repair in the disorder Ataxia Telangiectasis (AT), which is an autosomal recessive disorder. Heterozygotes with a defective AT gene (0.4-1.5% of the population) have an increased risk for breast and other cancers.
- People with DNA repair disorders often have broken chromosomes and increased cancer risks for certain types of cancer.

Multiple Choice Question

1. Mutations:

(a) Are permanent changes in the DNA sequence or structure

(b) Produce allelic variation

(c) Are more likely to be harmful than beneficial

(d) All of the above

(e) None of the above

2. Mutations that cause loss of a chromosome would be termed:

(a) Structural mutations

(b) Chromosome mutations

(c) Genome mutations

(d) Single-gene mutations

(e) None of these

3. A frameshift mutation could be caused by:

(a) A transition (b) A transversion

(c) A deletion of 3 bp (d) All of these

(e) None of these

4. Which of the following point mutations would be most likely to affect protein function?

(a) TAA to TGA

(b) CAA to TAA

(c) AGG to AGA

(d) CTT to CTC

(e) All of these are likely to affect protein function

5. A wild type allele:
   (a) Is the allele that is the most common in the population
   (b) Is the allele with the highest level of functional protein
   (c) Is the most favorable allele in the population
   (d) All of the above
   (e) None of the above

6. The mutation which causes sickle cell anemia in humans:
   (a) Is a base substitution (b) Is a deleterious mutation
   (c) Is a missense mutation (d) All of the above
   (e) None of the above

7. The gene that is mutated in cases of cystic fibrosis encodes an ion transporter protein normally expressed in the respiratory and digestive tracts along with other locations. Individuals homozygous for the most common mutation in this gene (ΔF508) died at very early ages prior to the advent of advanced medical intervention. Despite this, the mutation exists at relatively high frequency in populations of European descent. One potential explanation is that heterozygotes for such mutations may be relatively immune to the lethal effects of diseases such as typhoid fever. Which statement below is most accurate?
   (a) The ΔF508 cystic fibrosis mutation is a beneficial mutation
   (b) The ΔF508 cystic fibrosis mutation is a lethal mutation
   (c) The ΔF508 cystic fibrosis mutation is a conditional mutation
   (d) The ΔF508 cystic fibrosis mutation is a neutral mutation
   (e) The ΔF508 cystic fibrosis mutation is a suppressor mutation

8. In Himalayan rabbits, the ears, feet, and muzzle are pigmented while the rest of the body is white. This is because they express an enzyme variant that is ineffective at body temperature but functions at the lower temperatures experienced by cells of the extremities. This is an example of:
   (a) A wild type allele (b) A lethal mutation
   (c) A suppressor mutation (d) A conditional mutation
   (e) None of the above

9. A woman with mild symptoms of myotonic muscular dystrophy has a daughter with a very severe form of the disease. Which of the following statements is most correct?
   (a) The daughter probably has a mutation with the same number of repeats as her mother
   (b) The mother probably has a mutation with a larger number of repeats than the daughter
   (c) This must be an X-linked dominant condition
   (d) This is an example of anticipation
   (e) None of the above
10. Which form of RNA acts as a blueprint for polypeptide biosynthesis by the ribosome?
   (a) mRNA (b) tRNA
   (c) rRNA (d) All the above
   (e) None of the above
11. RNA is synthesized under the direction of DNA by this enzyme
   (a) DNA polymerase (b) DNA ligase
   (c) RNA polymerase (d) All the above
   (e) None of the above
12. Which component of RNA polymerase facilitates the recognition of promoter sequences?
   (a) A subunit (b) B subunit
   (c) S subunit (d) All the above
   (e) None of the above
13. The first hexameric sequence typically found in procaryotic promoters is located
   (a) At the transcription start site
   (b) Approximately 35 bases upstream of the transcription start site
   (c) Approximately 10 bases upstream of the transcription start site
   (d) All the above
   (e) None of the above

14. What is the usual sequence of a Pribnow box?

(a) TTGACA (b) TATAAT

(c) UUUUUU (d) All the above

(e) None of the above

15. Following binding to a promoter, RNA polymerase next

(a) Unwinds a short segment of DNA

(b) Nicks one strand of DNA

(c) Begins to replicate DNA

(d) All the above

(e) None of the above

16. Which of the following components is found in all procaryotic transcription terminators?

(a) rho factor (b) a poly-U region

(c) a hairpin structure (d) All the above

(e) None of the above

17. Posttranscriptional modification of RNA is more common in

(a) Procaryotes (b) Eucaryotes

(c) All the above (d) None of the above

18. Which structure is found on eucaryotic mRNA but not procaryotic mRNA?

(a) 3' poly-A tail

(b) 5' cap

(c) All the above

(d) None of the above

(e) 5' poly-A tail

19. Which regions of interrupted genes code for RNA that will be present in the final RNA product?

(a) Introns (b) Exons

(c) Promoter (d) All the above

(e) None of the above

20. What is a ribozyme?
(a) A protein that synthesizes RNA
(b) An RNA molecule that catalyzes a self-splicing reaction
(c) A DNA molecule that acts as a template for RNA synthesis
(d) All the above
(e) None of the above

21. What happens during amino acid activation?
(a) A methyl group is attached to an amino acid
(b) An amino acid is bound to mRNA
(c) An amino acid is bound to tRNA
(d) All the above
(e) None of the above

22. Which portion of a tRNA molecule is complementary to the mRNA triplet encoding a particular amino acid?
(a) D arm
(b) Anticodon triplet
(c) Amino acid stem
(d) All the above
(e) None of the above

23. Most procaryotic proteins begin with this modified amino acid
(a) N-formylleucine
(b) N-formylmethionine
(c) N-formylserine
(d) All the above
(e) None of the above

24. The most common initiator codon is
(a) GUA
(b) UAA
(c) AUG
(d) All the above
(e) None of the above

25. Which part of translation involves the addition of amino acids to an existing polypeptide chain?
(a) Initiation
(b) Elongation
(c) Termination
(d) All the above
(e) None of the above

26. What molecule catalyzes the transpeptidation reaction?
(a) Peptidyl transferase
(b) RNA polymerase
(c) DNA ligase
(d) All the above
(e) None of the above

27. Aminoacyl-tRNA binds to which site of the ribosome during elongation?
(a) P site
(b) A site
(c) E site
(d) All the above
(e) None of the above

28. Which of the following is NOT a nonsense codon?
(a) AUG
(b) UAA
(c) UGA
(d) All
(e) None

29. What is the role of molecular chaperones?
(a) To facilitate binding of ribosomes to mRNA
(b) To degrade newly synthesized polypeptides that contain inaccurate sequences
(c) To aid a newly synthesized polypeptide in folding to its proper shape
(d) All the above
(e) None of the above

30. An enzyme whose amount is reduced by the presence of an end product are:
(a) Inducible enzymes
(b) Repressible enzymes
(c) Corepressors
(d) All the above
(e) None of the above

31. A small molecule that causes the increase in levels of an inducible enzyme
(a) Repressor
(b) Corepressor
(c) Inducer
(d) All the above
(e) None of the above

32. The operator is

(a) Another name for the terminator of a gene

(b) A DNA site bound by repressor proteins

(c) A protein that coordinates gene expression

(d) All the above

(e) None of the above

33. Procaryotes have developed global regulatory systems

(a) To aid in DNA replication

B) To regulate the regulation of many genes under varying environmental conditions

(c) To determine the exact location of the cell on the globe

(d) All the above

(e) None of the above

34. A molecule of RNA that is complementary to a segment of another RNA molecule to which it will bind is called

(a) Antisense RNA (b) Missense RNA

(c) Nonsense RNA (d) All the above

(e) None of the above

35. Which is not an essential part of a two-component phosphorelay system?

(a) Sensor kinase (b) Response regulator

(c) Autoinducer (d) All the above

(e) None of the above

36. Which of these is NOT a mechanism by which a chromosome rearrangement could affect gene function?

(a) The breakpoint is in the middle of the coding region

(b) The breakpoint is between the coding region and the regulatory elements

(c) The new arrangement puts a gene near new regulatory elements

(d) A gene is moved near a locus control region

(e) None (all of these may be caused by rearrangements and affect gene function)

37. Which of the following would be an example of a germline mutation?

(a) Exposure to excessive UV radiation causes changes in the DNA of a skin cell, leading to basal cell carcinoma (skin cancer)

(b) In a very early human embryo, a mistake in mitosis causes loss of a Y chromosome in one daughter cell. The resulting child is a fertile male who is chromosomally partially male (XY) and partially female with Turner's syndrome (XO).

(c) A man with normal chromosomes has a hip X-ray without a protective shield. Ten months later, his wife gives birth to a child with a chromosomal deletion

(d) An embryo missing one copy of the third chromosome is miscarried very early in pregnancy

(e) None of the above

38. Mitochondrial DNA polymerase does not have an error checking mechanism such as that of nuclear DNA polymerase. This would be expected to lead to a higher rate of which type of mutation in mitochondrial DNA?

(a) Spontaneous mutations (b) Induced mutations

(c) Deletions (d) All of these

(e) None of these

39. The main difference between the directed mutation theory and the random mutation theory is:

(a) The chemical nature of the mutagens

(b) The effect of the mutation on the phenotype

(c) The cause of the mutation

(d) The heritability of the trait

(e) None of the above

40. In a haploid organism, lethal alleles would be expected to:

(a) Have a high mutation frequency

(b) Have a high mutation rate

(c) Be caused only by induced mutations

(d) All of the above

(e) None of the above

41. In which situation would you expect the mutation frequency to increase over time?

(a) The deleterious effect of the mutation is balanced with the mutation rate

(b) A lethal allele is created at a high mutation rate

(c) A neutral allele is created at low mutation rate

(d) All of the above

(e) None of the above

42. A bacterial cell experiences a mutation as a result of exposure to nitrogen mustard and then divides several times to produce a total of eight cells. How many of the resulting cells would you expect to contain the mutation?

(a) One (b) Two

(c) Four (d) Eight

(e) There is no way to tell

43. Which of the following forms of DNA repair does NOT require DNA polymerase?

(a) Direct DNA repair

(b) Base excision repair

(c) Nucleotide excision repair

(d) Mismatch repair

(e) Recombinational repair

44. Why is alkyltransferase only able to be used once?

(a) The protein is hydrolyzed in the process of repairing the DNA

(b) The protein is chemically changed by the addition of a methyl group

(c) The protein becomes attached to the DNA strand

(d) All of these

(e) None of these

45. How does recombinational repair differ from nucleotide excision repair (NER)?

(a) Unlike NER, recombinational repair is unable to repair damage caused by thymine dimers

(b) NER replaces the thymine dimmer while recombinational repair leaves it in place

(c) NER requires DNA polymerase and ligase function, while recombinational repair does not

(d) NER involves nick translation, which recombinational repair does not

(e) None of the above

46. A new chemical pesticide developed for agricultural use was subjected to the Ames test. The rate of mutation seen in samples treated with the pesticide and liver extract was approximately half that of the samples treated with the liver extract alone; in samples treated with only pesticide, the mutation rate was about the same as that of untreated samples. Which of the following statements could you make about this pesticide?

(a) The pesticide is modified by liver enzymes

(b) The pesticide is unlikely to be a mutagen

(c) The pesticide may be protective against spontaneous mutation

(d) All of the above

(e) None of the above

47. You want to determine if DNA replication has any effect on mutations resulting from UV exposure. You mutagenize group 1 of your bacteria and hold them in the dark for 24 hours then return them to daylight. You repeat this with group 2, but you also inhibit DNA replication chemically during this time. Where will you find the most mutations?

(a) Neither group 1 nor group 2

(b) Group 1

(c) Group 2

(d) Both group 1 and group 2

(e) None of the above

48. The excision repair process may begin by recognition of a damaged nucleotide or by recognition of a damaged base. Damaged bases, often a consequence of chemical mutagens, are recognized by a variety of cell enzymes called

(a) Polymerases (b) Ligases

(c) Photoreactivating enzymes (d) Peroxidases

(e) Glycosylases

49. Studies involving fusion of cells from individuals with xeroderma pigmentosum have identified about 7 different genes that can cause the disease. A patient comes to you with an apparent case of xeroderma pigmentosum. You decide to fuse cells from this patient with existing cell lines derived from other patients. You discover that when you fuse cells from your patient with the cells from patient X, the hybrid fusion product does not repair UV damage. If patient X is deficient in the protein that binds to thymine dimers, XPA, you conclude

(a) Your patient has suffered at least 5 mutations

(b) Your patient is also deficient in this gene product

(c) Your patient is deficient in a different aspect of the repair process

(d) Your patient has a new gene

(e) Your patient has acquired a mutagenic virus

50. The bacteria used for the Ames test is

(a) Shigella dysenteriae (b) Thermus aquaticus

(c) Salmonella typhimurium (d) Neisseria flavescens

(e) Escherichia coli

51. Liver extract is used in the Ames test because

(a) The liver enzymes activate the bacterial enzymes

(b) The liver does not produce enzymes that interfere with the test

(c) A liver extract is necessary for the bacteria to produce histidine revertants

(d) The bacteria requires the nutrients present in the liver extract for growth

(e) Liver enzymes may activate some innocuous compounds, making them mutagenic

52. The strains of Salmonella typhimurium used for the Ames test are

(a) Prototrophs his-
(b) Auxotrophs his-
(c) Auxotrophs his+
(d) Prototrophs leu+
(e) Auxotrophs leu+

53. The strains of S. typhimurium used in the Ames test are sensitive to mutation because they

(a) Have an increased cell wall permeabilit
(b) Don't have a lipopolysaccharide barrier to protect the bacterial surface
(c) Have a mutation in the excision repair of DNA
(d) All of the above
(e) None of the above

54. When an incorrect base is not repaired during DNA synthesis, it may be repaired after synthesis is completed. In this case, cell processes can identify the original non-mutated strand because

(a) There are missing nucleotides within a gene
(b) There are extra nucleotides within a gene
(c) The sugar-phosphate backbone is not distorted
(d) It is methylated
(e) It contains a gap in the sugar-phosphate backbone

55. Repair of DNA errors involving recA, lexA, and other genes in bacteria is called

(a) Proofreading repair
(b) The SOS response
(c) Replication repair
(d) UV response
(e) Mismatch repair

56. Chromosome breaks can lead to

(a) Mutations
(b) Chromosome rearrangements
(c) Healthier cells
(d) Both a. and b. are correct
(e) None of the above

57. What enzyme is responsible for proofreading repair after replication?

(a) RNA polymerase
(b) Glycolysase
(c) Ligase
(d) Primase
(e) DNA polymerase

58. Using genetic engineering genes can be

(a) Moved from one organism to potentially any other organism
(b) Easily converted to produce any product we desire
(c) Turned off or on at our whim
(d) Used to cure all known diseases
(e) Create plants resistant to all insect pests

59. Genetic engineering has produced salmon

(a) With shorter reproductive cycles
(b) That produce four times more offspring
(c) Resistant to viral infection
(d) That are heavier than non-engineered salmon
(e) Both a and d

60. Sticky ends are the result of

(a) Treatment of a nucleotide sequence with DNA ligase
(B) exposure of eukaryotic DNA to a prokaryotic plasmid
(c) Infection of a cell with a bacteriophage
(d) DNA breaking down in the presence of reverse transcriptase
(e) Cutting by restriction enzymes "off center" in a specific nucleotide sequence

61. In order to produce a DNA fragment that can be inserted into the DNA of a second organism one would need to

(a) Create a compatible DNA segment from an mRNA template
(b) Find a virus that could serve as a vector for transfer of the donor DNA
(c) Make sure both organisms have compatible DNA

(d) Be sure both donor and recipient DNA were prokaryotic or eukaryotic

(e) Cut the DNA of the donor and recipient cells with the same restriction enzyme

62. Interleukins:

(a) Stimulate white cell production

(b) Activate white blood cells. Used to treat HIV and cancer

(c) Halt the production of viruses

(d) Used to treat individuals with kidney disorders

(e) None of the above

63. Erythropoietin:

(a) Stimulates white cell production

(b) Activates white blood cells. Used to treat HIV and cancer

(c) Halts the production of viruses

(d) Used to treat individuals with kidney disorders

(e) None of the above

64. Interferons:

(a) Stimulate white cell production

(b) Activate white blood cells. Used to treat HIV and cancer

(c) Halt the production of viruses

(d) Used to treat individuals with kidney disorders

(e) None of the above

65. Colony-stimulating factors:

(a) Stimulate white cell production

(b) Activate white blood cells. Used to treat HIV and cancer

(c) Halt the production of viruses

(d) Used to treat individuals with kidney disorders

(e) None of the above

66. Genes can be inserted into cells by

(a) Shooting them into the cell with a DNA particle gun

(b) Using a plasmid as a gene vector

(c) Causing them to be produced in the recipient cell

(d) All of the above

(e) Only a and b

67. Glyphosate prevents plants from growing by

(a) Cleaving critical DNA in the plant and halting cell division

(b) Preventing the production of gametes

(c) Preventing the formation of spindle fibers and halting mitosis

(d) Stopping them from synthesizing proteins

(e) Both a and d

68. You are conducting research on eight species of Tribolium flour beetles and you want to compare their proteins. Which of the following techniques might you use?

(a) Genetic engineering (b) Gene amplification

(c) Gel electrophoresis (d) Polymerase chain reaction

(e) Gene therapy

69. The key to isolating a particular gene for use in recombinant DNA research is to

(a) Find a restriction enzyme to cut out the gene

(b) Find a compatible species to receive the gene

(c) Identify copies of the gene on several chromosomes

(d) Both a and b

(e) None of the above

70. Cuts in DNA are sealed with

(a) Restriction enzymes

(b) Ligases

(c) Reverse transcriptase

(d) Polymerase

(e) Plasmids

71. DNA is made from RNA with the help of

(a) Restriction enzymes (b) Ligases

C) Reverse transcriptase (d) Polymerase

(e) Plasmids

72. Sticky ends are created by

(a) Adding ligase to the end of the DNA fragment

(b) Adding polymerase to the end of the DNA fragment

(c) Cutting DNA so that there are short, single strands at each end

(d) Cutting the DNA fragment in several places

(e) Both a and d

73. When investigating a crime scene detectives find a single hair on the floor. The sample is not large enough to perform many tests. Which of the following methods could be used to make sufficient DNA to allow for comparison to the DNA of the prime suspect?

(a) Reverse transcriptase could be used to make DNA from RNA in the hair sample

(b) The hair root could be tissue cultured to make additional copies of the DNA

(c) The gene could be amplified by use of electrophoresis

(d) Polymerase chain reaction could be used to make many copies of the DNA

(e) The gene of interest could be spliced into a bacterium that would make a new copy every time it replicated

74. cDNA is significant when

(a) You need several thousand copies of a gene

(b) Cloning eukaryotic DNA in prokaryotic cells

(c) It is necessary to copy introns in prokaryotic cells

(d) A primary transcript is not available

(e) Processed mRNA is not available

75. In order for plants to resist the effects of the herbicide Roundup, they must be

(a) Monocots

(b) Dicots

(c) Annuals

(d) Able to synthesize glyphosate

(e) Resistant to glyphosate

76. Although the Ti plasmid has revolutionized plant genetic engineering, one limitation of its use is that it

(a) Is only about 50% effective

(b) Can only be used on fruit-bearing plants

(c) Is only effective on plants that are resistant to Roundup

(d) Does not infect cereal plants

(e) Cannot transmit prokaryotic genes

77. Clones are identified by hybridizing them with

(a) A vector (b) An antibody

(c) A virus (d) A probe

(e) None of the above

78. The first step in the polymerase chain reaction (PCR) is

(a) Denaturation (b) Primer extension

(c) Annealing (d) Cooling

(e) None of the above

79. Viruses containing RNA rather than DNA are called

(a) Riboviruses (b) Immunoviruses

(c) Bacteriophages (d) Rotaviruses

(e) Retroviruses

80. In a polymerase chain reaction, a synthetic sequence of nucleotides are involved in

(a) Denaturing (b) Heating

(c) Priming (d) Copying

(e) All of the above

81. Bacteria protect themselves from viruses by fragmenting viral DNA upon entry with

(a) Ligases
(b) Endonucleases
(c) Methylases
(d) Vectors
(e) Probes

82. In preliminary screening of clones, it is common to use

(a) Restriction enzymes
(b) Dyes
(c) Antibiotics
(d) Radiation
(e) Milipore filters

83. To identify an individual by DNA analysis of their blood, investigators look for

(a) Primers
(b) DNA fingerprints
(c) Probes
(d) Nucleosomes
(e) Transgenic fragments

84. All fragments cut by most restriction endonucleases have

(a) Complementary double-stranded ends
(b) Supplementary single-stranded ends
(c) Double-stranded "sticky" ends
(d) Complementary single-stranded ends
(e) Double-stranded supplementary ends

85. In 1980, interferon was produced by splicing a human gene into the genome of

(a) Bacteria
(b) Yeast
(c) Viruses
(d) Mice
(e) Rabbits

86. The second step in most genetic engineering experiments is

(a) Screening
(b) Production of recombinant DNA
(c) Cleavage of DNA

(d) Cloning

(e) Testing

87. When "sticky ends" are paired, they can be joined by

(a) Restriction enzymes (B) pSC101

(c) Methylase (d) X-gal

(e) DNA ligase

88. A successful vector in genetic engineering has been the

(a) TMV plasmid (b) HLF virus

(c) Vaccinia virus (d) Ti plasmid

(e) Retrovirus

89. Humans are not affected by glyphosphate in plant crops because

(a) While it is toxic to plants it is a normal metabolite of humans

(b) Humans do not produce aromatic amino acids

(c) It is used only on those crops that convert it to a harmless substance

(d) Humans lack the enzymes to interact with it

(e) Humans have EPSP synthetase

90. In the screening process, clones that metabolize X-gal turn

(a) Yellow (b) Orange

(c) Red (d) Blue

(e) Colorless

91. A powerful way to identify an individual using a particular gene as a marker is the analysis of

(a) RFLP's (b) X-gal reaction

(c) PCR's (d) EcoRI's

(e) BST's

92. In attempts to confer special characteristics upon plants, genetic engineers find Agrobacterium tumefaciens to be an effective vector for use with

(a) Corn (b) Rice

(c) Wheat (d) Soy beans

(e) Barley

93. A library of DNA fragments results from the use of

(a) Restriction endonucleases (b) Virus

(c) Plasmids (d) Recombinant DNA

(e) DNA ligase

94. One of the most useful methods for identifying a specific gene is

(a) Thin layer chromatography (b) The Eastern blot

(c) The Western blot (d) Magnetic resonance imaging

(e) The Southern blot

95. Plant cells receiving Ti plasmids with a TMV viral gene encoding the viral coat protein become immune to the TMV virus because

(a) The introduction of the viral gene triggered an immune response in the plant's own immune system

(b) The gene causes the cell walls to become impermeable to the virus

(c) TMV does not infect cells already infected with TMV

(d) in plants, the gene encodes an antiviral substance

(e) None of the above

96. A probe is used in which stage of the gene transfer process?

(a) Cleaving DNA (b) Recombining DNA

(c) Cloning (d) Screening

(e) Both a and b

97. The Polymerase Chain Reaction is used to:

(a) Amplify a small amount of DNA

(b) Cleave bacterial plasmids

(c) Seal "sticky ends"

(d) Identify target plasmids

(e) None of the above

98. Which of the following statements is true about developing cDNA?

(a) Mature mRNA directs the formation of the DNA

(b) Mature mRNA does not contain introns

(c) DNA taken from the nucleus is used to produce the cDNA
(d) Both a and b are true
(d) None of the above are true

99. Genetic engineering has been used to do all of the following except:
(a) Make plants more resistant to frost
(b) Make plants more resistant to disease
(c) Make plants more resistant to herbicides
(d) Improve the nutritional balance of plants
(d) All of the above are correct.

100. Engineering plants that are resistant to glyphosate was an important advancement because:
(a) Glyphosate promotes frost damage
(b) Glyphosate encourages the production of fruit that is lower in protein
(c) Glyphosate is the active ingredient in herbicide
(d) Glyphosate prevents the transfer of genes into the plants
(e) None of the above

Fill in the Blanks

1. Many chaperones often are called ____________________ because of their expression under high temperatures.
(a) Heat-shock proteins (b) Cold-shock proteins
(c) Thermostable protein (d) All the above
(e) None of the above

2. The tryptophan operon in *E. coli* demonstrates a level of regulation called ____________ by the terminatation of transcription if tryptophan is abundant in cells.
(a) Activation (b) Corepression
(c) Attenuation (d) All the above
(e) None of the above

3. A collection of genes or operons that is controlled by a common regulatory proteins is called ______

(a) A gene cluster (b) A genome

(c) A regulon (d) All the above

(e) None of the above

4. Based on the physiological adaptation theory, bacterial populations that had not been exposed to T1 infection ___________

(a) Should have a high level of resistance to T1 phage

(b) Should be completely susceptible to infection by T1 phage

(c) Should replicate more rapidly in the presence of T1 phage

(d) All of the above

(e) None of the above

5. Examples of environmental mutagens include _______________

(a) Alkylating agents, which chemically modify bases

(b) Nucleotide base analogues

(c) Ionizing radiation

(d) All of these

(e) None of these

6. In the Ames test _______________

(a) Mutagens cause lethal mutations, reducing the number of colonies

(b) Mutagens cause mutations that disrupt the ability of the cell to produce histidine

(c) Mutagens will cause an increase in the number of revertants

(d) Only mutagens that cause transitions can be identified

(e) None of the above

7. Individuals with Cockayne syndrome are more likely to ___________

(a) Have an increased rate of somatic mutation

(b) Have an increased rate of germ line mutation

(c) Have a decreased rate of spontaneous mutation

(d) Have a decreased rate of induced mutation

(e) None of the above

8. A human disorder thought to be due to a mutation eliminating or reducing DNA repair activity is ________.

(a) Lesch-Nyhan syndrome

(b) Xeroderma pigmentosum

(c) Tay-Sachs disease

(d) Albinism

(e) Down syndrome

9. The role of tautomerism in causing mutations relates to the fact that the process ultimately affects the ________.

(a) Covalent bonding affinities of the nitrogenous base

(b) Hydrogen bonding affinities of the nitrogenous base

(c) Deoxyribose sugar

(d) Ability of DNA to replicate at all

(e) Phosphate group

10. When comparing ultraviolet light and X rays, ________.

(a) X rays have longer wavelengths and are thus less mutagenic

(b) X rays and UV light are of the same wavelength and are equally mutagenic

(c) X rays have a shorter wavelength than does UV light

(d) X rays have longer wavelengths and are thus more mutagenic

(e) X rays have shorter wavelengths and are thus less mutagenic

11. Myotonic dystrophy patients exhibit ________.

(a) No dystrophin and a single nucleotide change

(b) A triplet sequence repeated many times

(c) Trisomy

(d) A single nucleotide change

(e) No dystrophin

12. The Ames test assays whether or not a compound causes ________.

(a) Chromosome mutations (b) Gene mutations

(c) DNA repair (d) Crossing over

(e) Cancer

13. A regulatory mutation has been observed randomly throughout a population of one million. Families with members carrying this mutation are not related genetically. This mutation is probably ________.

(a) A spontaneous mutation

(b) A gametic mutation

(c) A somatic mutation

(d) A spontaneous, somatic mutation

(e) An induced gametic mutation

14. A frameshift mutation is one of the most severe types of mutations because ________.

(a) They occur only in gametes

(b) More than one gene is affected

(c) They cannot be reversed

(d) More than one amino acid or entire proteins are affected

(e) Translation is stopped

15. The discovery of transposons showed that genetic information ________.

(a) Is not fixed in the genome

(b) Can be altered

(c) Is passed from one generation to another

(d) Can be transferred between species

(e) None of the above

16. Base substitutions (point mutations) in a gene may not result in a change in the gene product because__________

(a) Of the redundancy of the genetic code

(b) Most DNA is non-coding

(c) Single bases are largely unimportant; it is the overall structure of the gene that matters

(d) None of the above

(e) None of the above

17. Frameshift mutations result from ______________

(a) Base substitutions

(b) Cancer

(c) Incorporation of base analogues

(d) Tautomeric shifts

(e) Base deletions or additions

18. Tautomers of nitrogenous bases are isomers that differ from each other in __________

(a) The number of hydrogen molecules in the molecule

(b) The number of carbon atoms in the molecule

(c) The bonding location of one carbon atom

(d) The bonding location of the nitrogen atoms

(e) The bonding location of one hydrogen atom

19. The biologically important tautomers of nucleic acid bases __________

(a) Are found in approximately equal proportions of DNA

(b) Can form hydrogen bonds to themselves

(c) Are chemically indistinguishable

(d) Have identical bonding properties because they have the same elemental composition

(e) Have different bonding properties

20. Molecules that can substitute for nitrogenous bases in nucleic acid molecules are called ________

(a) Base analogues

(b) Base stand-ins

(c) Substitutes.

(d) Tautomers

(e) Isomers

21. 5-bromouracil (keto) is a base analogue of thymine. A tautomeric shift to a 5-bromouracil (enol) will result in bonding affinity for ____________

(a) Any purine (b) Any pyrimidine

(c) Cytosine (d) Guanine

(e) Adenine

22. Chemical mutagens are compounds that ____________

(a) Can, because of their size, intercalate ("wedge") between base pairs in a DNA molecule

(b) Can chemically modify a nitrogenous base, thereby altering its bonding properties

(c) Break apart DNA into small fragments

(d) Different chemical mutagens can do A. and B., above

(e) None of the above

23. Sulfur containing molecules in the mustard gas family of compounds are mutagenic through their action as ____________

(a) Intercalating agents (b) Alkylating agents

(c) Base deletion agents (d) Decarboxylating agents

(e) Deaminating agents

24. Deamination of cytosine converts it to uracil, a base analogue of ________ in a DNA molecule ____________.

(a) Guanine (b) Uracil

(c) Adenine (d) Thymine

(e) Cytosine

25. One possible mechanism for the generation of frameshift mutations by chemical mutagens is ____________

(a) Chemical modification of nitrogenous bases

(b) Intercalation of very large molecules, thus distorting the shape of the DNA double helix

(c) Functioning as polymerases to make the DNA longer

(d) Functioning as enzymes, catalyzing the insertion of bases into the DNA molecule

(e) Intercalation of molecules approximately the same size as base pairs

26. Light-mediated cell processes are often examined by determining an action spectrum (i.e., by asking in what area of the light spectrum does one get an effect). An action spectrum for PRE would show ________ mutations in blue light as compared with red, yellow, or green light.

(a) More (b) Fewer

(c) The same (d) All the above

(e) None of the above

27. UV radiation is part of the natural spectrum of radiation from the sun. Some protection comes from the Earth's ozone layer which is becoming thinner and developing holes, especially in the southern hemisphere, as a consequence of human activities. If you lived in Australia, you should

________________.

(a) Be very concerned (repair systems aren't perfect)

(b) Not be concerned (UV damage is repaired by two different repair systems)

(c) All the above

(d) None of the above

28. In genome sequencing, it was found that ____% of the Drosophila genes matched other species.

(a) 45 (b) 72

(c) 98 (d) 83

(e) None of the above

29. With the completion of the human genome project, the next frontier is

____

(a) Nucleonics (b) Proteomics

(c) Cytomics (d) Agrinomics

(e) None of the above

30. By using the "reference sequence" developed in the human genome project, individual differences are now being detected using ____________

(a) NPs (b) PSNs

(c) SNPs (d) "Snaps"

(e) None of the above

31. In the discovery of introns, a DNA molecule called ______________ was formed that had the same nucleotide sequence as the gene that produced the mRNA.

(a) mDNA (b) rDNA

(c) sDNA (d) gDNA

(e) cDNA

32. Bacterial DNA is not cleaved by their own restriction enzymes because bacteria add ______________ to their own DNA.

(a) Nucleotides (b) Peptides

(c) Methyl groups (d) Glyphosphate

(e) Somototropin

33. A procedure called PCR is used to ____________

(a) Cleave DNA (b) Produce recombinant DNA

(c) Copy gene sequences (d) Clone cells

(e) Screen clones

34. Wilt-proof flowers are being engineered by making them insensitive to _________

(a) Methylene (b) Propylene

(c) Butylene (d) Acetylene

(e) Ethylene

35. The ingestion of BST in milk is no threat to humans because __________

(a) It is destroyed in the stomach

(b) It cannot be digested and will pass through as waste

(c) They do not have the hormones to utilize it

(d) It is destroyed in the pasteurization process

(e) The FDA has ruled that such milk cannot be sold

36. Genetic engineering has successfully transferred genes from eukaryotic cells into___

(a) Bacteria (b) Plants

(c) Animals (d) All of the above

(e) None of the above

37. DNA molecules can be cut into sections by using ____________

(a) ATP

(b) Gel electrophoresis

(c) Restriction endonucleases

(d) Plasmids

(e) A probe

38. In genetic engineering, DNA ligase is used as ____________

(a) A probe

(b) A sealing enzyme

(c) A restriction enzyme

(d) A mutagen

(e) None of the above

39. In Cohen and Boyer's transfer of toad genes to a bacterial cell, restriction endonucleases were used to ____________

(a) Isolate sections of cloned bacterial plasmids

(b) Isolate sections of toad DNA

(c) Cleave bacterial plasmids

(d) Both a and b are correct

(e) Both b and c are correct

40. The use of bovine growth hormone ____________

(a) Has raised controversy with groups opposed to genetic engineering

(b) Increased milk production in cows

(c) Is being tested on increasing the weight gain of cows

(d) Is being tested as a treatment for dwarfism in humans

(e) All of the above

41. Genetically identical organisms derived from a single genetic source are called ____________

(a) Populations (b) Varieties

(c) Sibling species (d) Ecotypes

(e) Clones

True/False

1. Because they only affect a single amino acid, missense mutations do not have a significant effect on protein function.

   (a) True (b) False

2. One possible risk of genetic engineering is the accidental production of dangerously resistant microorganisms.

   (a) True (b) False

3. Although risks are possible, gene engineering appears to offer more of a contribution to human welfare than threat.

   (a) True (b) False

4. Sanger earned his second Nobel Prize for development of a technique called DNA fingerprinting.

   (a) True (b) False

5. Cloned human genes are localized by using radioactive probes to detect binding between the probe and the complementary sequence on the chromosome.

   (a) True (b) False

6. The SOS response system is error prone.

   (a) True (b) False

7. All compounds that are carcinogenic are also mutagenic.

   (a) True (b) False

8. A splice site mutation leading to loss of an exon can affect production of regions of the protein encoded in downstream exons even if they are spliced correctly.

   (a) True (b) False

9. Provided the number of base pairs is a multiple of three, insertion of additional DNA in the coding region of a protein is unlikely to have a dramatic effect on protein function.

   (a) True (b) False

10. Increasing environmental levels of mutagen is likely to lead to increasing frequency of deleterious alleles.

    (a) True (b) False

11. The Luria-Delbruck fluctuation test was designed to differentiate between appearance of a trait due to random, spontaneous mutations vs. appearance of the trait due to adaptation to environmental conditions.

(a) True (b) False

12. Chromosomal rearrangements are unlikely to produce an obvious phenotype, since they do not remove any DNA.

(a) True (b) False

13. DNA repair systems typically work in three major steps: detection of the error, removal of the abnormality, and replacement of the removed DNA.

(a) True (b) False

14. Mutational hot spots are sites in the DNA where a mutation is most likely to cause a change in protein structure.

(a) True (b) False

15. The Ti plasmid provides a means to transfer genes into cereal plants, such as corn and rice.

(a) True (b) False

16. Genetic engineering has provided a method for the production of insulin to fight diabetes, but it remains too expensive to be practical.

(a) True (b) False

17. Piggyback vaccines work by triggering the body's defense system with a harmless virus carrying the surface proteins of a disease organism.

(a) True (b) False

18. The second stage in genetic engineering experiments is cloning the plasmid.

(a) True (b) False

19. When using reverse transcriptase, DNA is made from mRNA.

(a) True (b) False

Answers

Multiple Choice Question

1	2	3	4	5	6	7	8	9	10	11	12	13	14	15
D	C	E	B	A	D	B	D	D	A	C	C	B	B	A
16	17	18	19	20	21	22	23	24	25	26	27	28	29	30
C	B	B	B	B	C	B	B	C	B	A	B	A	C	B
31	32	33	34	35	36	37	38	39	40	41	42	43	44	45
C	B	B	A	C	E	C	A	C	E	C	C	A	B	B
46	47	48	49	50	51	52	53	54	55	56	57	58	59	60
D	B	E	B	C	E	B	D	D	B	C	E	A	E	E
61	62	63	64	65	66	67	68	69	70	71	72	73	74	75
E	B	D	C	A	E	D	C	A	B	C	C	D	B	E
76	77	78	79	80	81	82	83	84	85	86	87	88	89	90
D	D	A	E	E	B	C	B	D	A	B	E	D	B	D
91	92	93	94	95	96	97	98	99	100					
A	D	A	E	C	D	A	D	E	C					

Fill in the blanks

1	2	3	4	5	6	7	8	9	10	11	12	13	14	15
A	C	C	E	D	C	A	B	B	C	B	B	D	D	A
16	17	18	19	20	21	22	23	24	25	26	27	28	29	30
A	E	E	E	A	D	D	B	D	E	B	A	D	B	C
31	32	33	34	35	36	37	38	39	40	41				
E	C	C	E	A	D	C	B	E	E	E				

True/False

1	2	3	4	5	6	7	8	9	10	11	12	13	14	15
B	A	A	B	A	A	B	A	B	A	A	B	A	B	B
16	17	18	19											
B	A	B	A											

❑❑❑

Chapter – 16

Recombination and Transposition at the Molecular Level

Sister Chromatid Exchange and Homologous Recombination

- The staining of harlequin chromosomes can reveal recombination between sister chromatids
- The holliday model describes a molecular mechanism for the recombination process
- More recent models have refined the molecular steps of homologous recombination
- Various proteins are necessary to facilitate homologous recombination
- Gene conversion may result from DNA gap repair synthesis or DNA mismatch repair

Site-Specific Recombination

- The integration of viral genomes can occur by site-specific recombination
- Antibody diversity in the immune system is produced by site-specific recombination

Transposition

- McClintock found that chromosomes of corn plants contain loci that can move

- Transposable elements and retroelements move via one of three transposition pathways
- Each type of transposable element has a characteristic pattern of DNA sequences
- Transposase catalyzes the excision and insertion of transposable elements
- Replicative transposition requires both transposase and resolvase
- Retroelements utilize reverse transcriptase and integrase for retrotransposition
- Transposons have become important tools in molecular biology

Multiple Choice Question

1. Which statement regarding recombination is false?

(a) Recombination creates new combinations of alleles on a chromosome

(b) Recombination commonly occurs between homologous sequences

(d) Sister chromatid exchange is a form of recombination

(d) Gene conversion may occur during recombination

(e) None of the above

2. Which of these describes a Holliday junction?

(a) A section of DNA where base pairing is not exact

(b) A strand of DNA containing genetic material from two different chromosomes

(c) An interaction of two strands of DNA from homologous chromosomes

(d) A three stranded DNA structure where single stranded DNA has invaded a double helix

(e) None of the above

3. Considering the steps listed below, identify the correct sequence of events proposed by the double-stranded break model of recombination.

I. Strand invasion and D loop formation

II. DNA gap repair synthesis

III. Generation of single stranded DNA segments

IV. Resolution

(a) I, III, II, IV (b) III, I, II, IV

(c) III, II, I, IV (d) IV, III, I, II

4. What is gene conversion?

(a) A process by which a gene is rendered nonfunctional

(b) A process that changes an allele to the allele existing on the homologous chromosome

(c) A process that restores the wild type function of a gene

(d) A process that creates a new allele of a gene

(e) None of these

5. If double stranded break recombination and DNA gap repair synthesis result in gene conversion, how is the original allele lost?

(a) It isn't, it is merely rearranged

(b) It is eliminated by DNA repair enzymes

(c) It is prevented from replication due to selective pressure

(d) It is digested away after a break within the gene

(e) None of the above

6. By what process is the λ viral DNA inserted into the *E. coli* chromosome during lysogeny?

(a) Site specific recombination

(b) Homologous recombination

(c) Integral ligation

(d) Strand invasion

(e) None of these

7. Which of these mechanisms is NOT critical to generation of antibody variability?

(a) Site specific recombination

(b) Presence of many V and J regions in a single gene

(c) Imprecise joining at recombination signal sequences

(d) All of the above

(e) None of the above

8. What λ DNA protein would you expect to be similar in function to RAG1 and RAG2?

(a) Recombinase
(b) Splicease
(c) Excisionase
(d) Integrase
(e) None of these

9. Barbara McClintock noted that, in a particular strain of corn, chromosome 9 tended to break at the same site at a relatively high rate. What caused this breakage?

(a) Random double stranded breaks
(b) Inaccurate excision of a transposable element
(c) Insertion of a transposable element
(d) Holliday junctions
(e) None of the above

10. After identification of the first "mutable site", McClintock was able to demonstrate that the element that caused the mutable site was able to move. What was her evidence?

(a) In some offspring from the original strain, the breakpoint was in a different location
(b) In some offspring from the original strain, the phenotype was no longer mutable
(c) In some offspring from the original strain, the Ac element had been lost
(d) Some strains had breakpoints in different locations than other strains
(e) None of the above

11. Which type of transposable element is only found in eukaryotes (not prokaryotes)?

(a) Transposable elements that move by simple transposition
(b) Transposable elements that move by replicative transposition
(c) Retrotransposons, which move via an RNA intermediate

(d) All of these

(e) None of these

12. Simple transposons do not replicate themselves during transposition but they are able to proliferate. How?

(a) They are replicated after excision from the DNA strand, creating extra copies to be inserted at new sites

(b) Excision is stimulated by DNA repair processes

(c) These transposons separate into single stranded elements after excision. Insertion of two ingle strands followed by DNA gap repair effectively duplicates the element

(d) All of these

(e) None of these

13. Composite transposons can be recognized by which of the following characteristics?

(a) Inverted repeats

(b) Direct repeats

(c) Two pairs of inverted repeats flanking an intervening gene

(d) Two pairs of direct repeats flanking a transposase gene

(e) None of the above

14. What would you predict would happen if the inverted repeats at one end of a composite transposon were involved in simple transposition, rather than the two repeats at the ends?

(a) The rest of the transposon would be eliminated

(b) The intervening sequence would be "stranded" and no longer able to move

(c) The second pair of inverted repeats would be highly likely to jump out

(d) All of these

(e) None of these

15. Which of the following elements would be considered to be a complete transposable element?

(a) An insertion sequence that includes a transposase gene between the inverted repeats

(b) A composite element that carries an antibiotic resistance gene

(c) A replicative transposon that includes a reverse transcriptase gene

(d) A retrotransposon containing long terminal repeats

(e) None of the above

16. Which of the following describes the key difference between simple transposition and replicative transposition?

(a) In simple transposition, transposase catalyzes the removal and reintegration of the transposon, while transposase is not involved in replicative transposition

(b) In simple transposition, a double stranded piece of DNA is moved from one site to another, while in replicative transposition a single strand is moved, leaving a single strand behind

(c) Simple transposition requires DNA synthesis while replicative transposition does not

(d) Replicative transposition requires DNA synthesis while simple transposition does not

(e) None of the above

17. In exon shuffling, a new exon may be inserted into a gene via transposition, leading to a protein with a modified function. Which of these statements is most likely to be true?

(a) The new exon is probably inserted into an existing exon

(b) The new exon changes the reading frame of the original protein

(c) The new exon is probably inserted into an intron

(d) All of the above are likely to be true

(e) None of the above are likely to be true

18. Why is transposon tagging easier for X linked genes?

(a) These genes are more likely to have an interesting phenotype

(b) These genes are hemizygous in males

(c) The transposon is more likely to insert into sites on the X chromosome

(d) The X chromosome does not undergo recombination

19. Which statement regarding recombination is false?
   (a) Recombination creates new combinations of alleles on a chromosome
   (b) Recombination commonly occurs between homologous sequences
   (c) Sister chromatid exchange is a form of recombination
   (d) Gene conversion may occur during recombination

20. Harlequin chromosomes:
   (a) Have been used to demonstrate sister chromatid exchange
   (b) Have sister chromatids that stain differently
   (c) Can only exist in prophase or metaphase
   (d) All of the above
   (e) None of the above

21. Which of these describes a Holliday junction?
   (a) A section of DNA where base pairing is not exact
   (b) A strand of DNA containing genetic material from two different chromosomes
   (c) An interaction of two strands of DNA from homologous chromosomes
   (d) A three stranded DNA structure where single stranded DNA has invaded a double helix

22. After isomerization and resolution, the resulting chromosomes are:
   (a) Unchanged from the original chromosome
   (b) Nonrecombinant chromosomes with a heteroduplex region
   (c) Recombinant chromosomes with a heteroduplex region
   (d) Recombinant chromosomes without a heteroduplex region
   (e) Harlequin chromosomes

23. Considering the steps listed below, identify the correct sequence of events proposed by the double-strand break model of recombination.
   I. Strand invasion and D loop formation
   II. DNA gap repair synthesis

III. Generation of single stranded DNA segments

IV. Resolution

(a) I, III, II, IV (b) III, I, II, IV

(c) III, II, I, IV (d) IV, III, I, II

24. What is gene conversion?

(a) A process by which a gene is rendered nonfunctional

(b) A process that changes an allele to the allele existing on the homologous chromosome

(c) A process that restores the wild type function of a gene

(d) A process that creates a new allele of a gene

(e) None of these

25. If double stranded break recombination and DNA gap repair synthesis result in gene conversion, how is the original allele lost?

(a) It isn't, it is merely rearranged

(b) It is eliminated by DNA repair enzymes

(c) It is prevented from replication due to selective pressure

(d) It is digested away after a break within the gene

26. By what process is the ? viral DNA inserted into the *E. coli* chromosome during lysogeny?

(a) Site specific recombination

(b) Homologous recombination

(c) Integral ligation

(d) Strand invasion

(e) None of these

27. Antibodies are:

(a) Proteins produced by the immune system

(b) Able to bind specifically to a single antigen

(c) An example of proteins with quaternary structure

(d) All of the above

(e) None of the above

28. Which of these mechanisms is NOT critical to generation of antibody variability?

(a) Site specific recombination

(b) Presence of many V and J regions in a single gene

(c) Imprecise joining at recombination signal sequences

(d) All of the above

(e) None of the above

29. What ? DNA protein would you expect to be similar in function to RAG1 and RAG2?

(a) Recombinase (b) Splicease

(c) Excisionase (d) Integrase

(e) None of these

30. Barbara McClintock noted that, in a particular strain of corn, chromosome 9 tended to break at the same site at a relatively high rate. What caused this breakage?

(a) Random double stranded breaks

(b) Inaccurate excision of a transposable element

(c) Insertion of a transposable element

(d) Holliday junctions

31. After identification of the first "mutable site", McClintock was able to demonstrate that the element that caused the mutable site was able to move. What was her evidence?

(a) In some offspring from the original strain, the breakpoint was in a different location

(b) In some offspring from the original strain, the phenotype was no longer mutable

(c) In some offspring from the original strain, the Ac element had been lost

(d) Some strains had breakpoints in different locations than other strains

32. Which type of transposable element is only found in eukaryotes (not prokaryotes)?

(a) Transposable elements that move by simple transposition

(b) Transposable elements that move by replicative transposition

(c) Retrotransposons, which move via an RNA intermediate

(d) All of these

(e) None of these

33. Simple transposons do not replicate themselves during transposition but they are able to proliferate. How?

(a) They are replicated after excision from the DNA strand, creating extra copies to be inserted at new sites

(b) Excision is stimulated by DNA repair processes

(c) These transposons separate into single stranded elements after excision. Insertion of two single strands followed by DNA gap repair effectively duplicates the element

(d) All of these

(e) None of these

34. Composite transposons can be recognized by which of the following characteristics?

(a) Inverted repeats

(b) Direct repeats

(c) Two pairs of inverted repeats flanking an intervening gene

(d) Two pairs of direct repeats flanking a transposase gene

35. What would you predict would happen if the inverted repeats at one end of a composite transposon were involved in simple transposition, rather than the two repeats at the ends?

(a) The rest of the transposon would be eliminated

(b) The intervening sequence would be "stranded" and no longer able to move

(c) The second pair of inverted repeats would be highly likely to jump out

(d) All of these

(e) None of these

36. Which of the following elements would be considered to be a complete transposable element?

(a) An insertion sequence that includes a transposase gene between the inverted repeats

(b) A composite element that carries an antibiotic resistance gene

(c) A replicative transposon that includes a reverse transcriptase gene

(d) A retrotransposon containing long terminal repeats

37. Which of the following describes the key difference between simple transposition and replicative transposition?

(a) In simple transposition, transposase catalyzes the removal and reintegration of the transposon, while transposase is not involved in replicative transposition

(b) In simple transposition, a double stranded piece of DNA is moved from one site to another, while in replicative transposition a single strand is moved, leaving a single strand behind

(c) Simple transposition requires DNA synthesis while replicative transposition does not

(d) Replicative transposition requires DNA synthesis while simple transposition does not

38. In exon shuffling, a new exon may be inserted into a gene via transposition, leading to a protein with a modified function. Which of these statements is most likely to be true?

(a) The new exon is probably inserted into an existing exon

(b) The new exon changes the reading frame of the original protein

(c) The new exon is probably inserted into an intron

(d) All of the above are likely to be true

(e) None of the above are likely to be true

39. In transposon tagging:

(a) Insertion of a transposon randomly inactivates a gene of interest

(b) Transposable elements are directed to inactivate a specific gene

(c) Presence of a transposon in a known gene allows the cloning of the transposon

(d) All of these

(e) None of these

40. Why is transposon tagging easier for X linked genes?

(a) These genes are more likely to have an interesting phenotype

(b) These genes are hemizygous in males

(c) The transposon is more likely to insert into sites on the X chromosome

(d) The X chromosome does not undergo recombination

41. In a modification of transposon tagging known as enhancer trapping, reporter gene coding regions with a minimal promoter but no regulatory elements are placed inside a composite transposon, which is then "jumped" into the chromosomes. Activation of the reporter gene in a specific tissue:

(a) Is unlikely to occur due to the flanking elements of the transposon

(b) Indicates that the transposon has disrupted a gene whose function is critical for correct function of that tissue

(c) Indicates that the transposon has integrated near tissue specific regulatory elements

(d) Is likely to be recessive and thus difficult to identify

Fill in the blanks

1. Harlequin chromosomes __________

(a) Have been used to demonstrate sister chromatid exchange

(b) Have sister chromatids that stain differently

(c) Can only exist in prophase or metaphase

(d) All of the above

(e) None of the above

2. After isomerization and resolution, the resulting chromosomes are ______

(a) Unchanged from the original chromosome

(b) Nonrecombinant chromosomes with a heteroduplex region

(c) Recombinant chromosomes with a heteroduplex region

(d) Recombinant chromosomes without a heteroduplex region

(e) Harlequin chromosomes

3. Antibodies are ____________

(a) Proteins produced by the immune system

(b) Able to bind specifically to a single antigen

(c) An example of proteins with quaternary structure

(d) All of the above

(e) None of the above

4. In transposon tagging ____________

(a) Insertion of a transposon randomly inactivates a gene of interest

(b) Transposable elements are directed to inactivate a specific gene

(c) Presence of a transposon in a known gene allows the cloning of the transposon

(d) All of these

(e) None of these

5. In a modification of transposon tagging known as enhancer trapping, reporter gene coding regions with a minimal promoter but no regulatory elements are placed inside a composite transposon, which is then "jumped" into the chromosomes. Activation of the reporter gene in a specific tissue ____________

(a) Is unlikely to occur due to the flanking elements of the transposon

(b) Indicates that the transposon has disrupted a gene whose function is critical for correct function of that tissue

(c) Indicates that the transposon has integrated near tissue specific regulatory elements

(d) Is likely to be recessive and thus difficult to identify

(e) None of the above

True/False

1. The existence of single Holliday junctions is strong support for the Holliday model of recombination.

(a) True (b) False

2. Retrotransposons replicate themselves during transposition.

   (a) True (b) False

3. Crossing over occurs in both mitosis and meiosis.

   (a) True (b) False

Answers

Multiple Choice Question

1	2	3	4	5	6	7	8	9	10	11	12	13	14	15
C	C	B	B	D	A	E	D	B	A	C	E	C	A	A
16	17	18	19	20	21	22	23	24	25	26	27	28	29	30
B	C	B	C	D	C	C	B	B	D	A	D	E	D	B
31	32	33	34	35	36	37	38	39	40	41				
A	C	E	C	B	A	B	C	B	B	C				

Fill in the blanks

1	2	3	4	5
B	C	D	A	C

True/False

1	2	3
B	A	A

❑❑❑

Chapter – 17

Recombinant DNA Technology

Recombinant DNA Technology

- A transgenic organism contains foreign DNA.
- Biotechnology is the industry dedicated to using altered cells or molecules for various applications.
- Recombinant DNA technology (i.e. gene cloning) was the first of the modern biotechnologies.
- Gene targeting uses precision techniques to "knockout" a gene or substitute a gene for one on a chromosome.
- Bioethics questions, gene patenting, and potential uses and risks of biotechnology pose treat questions for society.

Constructing Recombinant DNA Molecules

- Recombinant DNA molecules are made by covalently linking foreign DNA with the DNA of a transport molecule known as a cloning vector (i.e. plasmid DNA).
- Recombinant DNA is amplified by introducing it into a bacterium or other single cell host.
- Guidelines for using recombinant DNA technology were first drawn up at a 1975 meeting at Asilomar, CA.

Selecting Recombinant DNA Molecules

- Phenotype selection can be used to identify host cells with recombinant DNA.
- A popular technique identifies host cells with recombinant DNA using antibiotics and a color test.

Isolating the Gene of Interest

- Bacteria and other single cell organisms are also used as hosts of DNA libraries.
- DNA libraries can be constructed from the DNA fragments of chromosomes (genomic libraries) or can contain only processed coding sequences when mRNA is used as the source of the genetic information (cDNA libraries).
- DNA probes are used to identify and isolate specific DNA fragments of interest from DNA libraries.

Gene Cloning

- Cloning experiments usually involve two kinds of DNA molecules: chromosomal DNA and vector DNA
- Enzymes are used to cut DNA into pieces and join the pieces together
- Gene cloning involves the insertion of DNA fragments into vectors, which are then propagated within host cells
- In the first gene cloning experiment, Cohen, Chang, Boyer, and Helling inserted a kanamycinR gene into a plasmid vector
- cDNA can be made from mRNA via reverse transcriptase
- Restriction mapping is used to locate the restriction sites within a vector
- Polymerase chain reaction (PCR) can also be used to make many copies of DNA

Detection of Genes and Gene Products

- A DNA library is a collection of many cloned fragments of DNA
- Blotting methods can be used to detect DNA sequences
- Gene products can also be detected by blotting methods

- Techniques can be used to detect the binding of proteins to DNA sequences

Analysis and Alteration of DNA Sequences

- The dideoxy method of DNA sequencing is based on our knowledge of DNA replication
- In vitro site-directed mutagenesis is a technique to alter DNA sequences

Applications of Recombinant DNA Technology

- The products of recombinant DNA technology are used in health care, food technology, agriculture, and forensics.

Transgenic Organisms

Delivering DNA

- Introducing foreign DNA into a single germline cell creates a transgenic organism.
- Entry of foreign DNA into cells requires the use of chemical agents (calcium phosphate, liposomes), physical methods (electroporation, microinjection, particle bombardment), and biological vehicles (viruses or Agrobacterium).
- The foreign DNA is then transmitted to each cell as the organism develops.

Transgenic Pharming from Milk and Semen

- Animals are transgenically altered to "pharm" useful human proteins in their milk or germline cells (eggs or sperm).
- Transgenic animals are also valuable research tools as models of human disease.

Gene Targeting

- In contrast to transgenic organisms, where DNA is randomly inserted into chromosomes, gene targeting in mice involves homologous recombination and the controlled insertion of a gene at a specific chromosomal locus an embryonic stem (ES) cell.

- The ES cell is then incorporated into a developing embryo from another individual, which is implanted into a surrogate.

Gene-Targeted Mice as Models

- Animals with phenotypes indicating that they harbor cells with the targeted gene are bred to each other to yield homozygotes for the targeted gene.
- Swapping an inactivated allele for a gene of interest produces a knockout mouse, and replacing a gene with another that has an altered function creates a knockin mouse.

When Knockouts Are Normal

- Knockout mice can often serve as useful animal models for human diseases.
- Sometimes animals with supposedly vital genes knocked out are perfectly healthy and the importance of the gene product must then be reconsidered at the level of the entire organism.

Multiple Choice Question

1. Which of the following is an example of a DNA vector?

(a) A prophage (b) An Hfr element

(c) An F factor (d) All of these

(e) None of these

2. Which of the following would NOT be a useful selectable marker?

(a) A gene encoding a protein that degrades the antibiotic ampicillin

(b) A gene encoding a protein that allows the cell to synthesize histidine

(c) A gene encoding a protein that is an essential structural component of the cell

(d) All of these are useful selectable markers

(e) None of these are useful selectable markers

3. Which statement regarding restriction endonucleases is NOT correct?

(a) They recognize a specific base sequence in the DNa

(b) They are produced by bacterial cells as a primitive immune system

(c) They digest DNA by removing nucleotides from a free 3' end

(d) They often generate short single stranded sequences

(e) None of the above

4. Which of these palindromes is most similar to a restriction endonuclease recognition site?

(a) MADAM, I'M ADAM

(b) A MAN, A PLAN, A CANAL: PANAMA!

(c) BOB

(d) None: Restriction endonuclease recognition sites are not palindromic

(e) None of the above

5. Formation of what type of chemical bond is catalyzed by DNA ligase?

(a) Glycosidic (b) Hydrogen

(c) Phosphoester (d) Ester

(e) None of the above

6. Which of the enzymes whose recognition sites are shown here would produce sticky ends compatible with BstAUI (5'-G | GTACC-3')? (Hint: Only one strand is shown. The enzyme cuts at the | on this strand, and at an equivalent position in the palindrome on the opposite strand

(a) BclI (5'-T | GATCA-3') (b) SunI (5'-C | GTACG-3')

(c) HindIII (5'-A | AGCTT-3') (d) KpnI (5'-GGTAC | C-3')

(e) None of the above

7. If you are using a vector with lacZ as a screenable marker, which of the following bacterial strains would you want to use as host cells?

(a) lacO- (b) lacI-

(c) lacZ- (d) Any of these would work

(e) None of these

8. In the first gene cloning experiment:

(a) Researchers successfully identified a human gene responsible for disease

(b) Researchers successfully inserted a gene for kanamycin resistance into a plasmid vector

(c) Researchers demonstrated that many different DNA fragments could insert into a plasmid vector

(d) Researchers produced a strain of bacteriophage with increased ability to infect *E. coli*

(e) None of the above

9. You discover a novel eukaryotic organism that glows in the dark. You believe this trait is due to a single gene, and you wish to clone the gene. Which of the following strategies is most likely to be successful?

(a) Isolate the genomic DNA from the organism, digest with a restriction endonuclease, insert into a plasmid vector and transform into bacteria. Screen colonies for the ability to glow in the dark

(b) Isolate the genomic DNA from the organism, digest with a restriction endonuclease, insert into a plasmid vector and transform into eukaryotic cells such as yeast. Screen colonies for the ability to glow in the dark

(c) Isolate mRNA from the organism, reverse transcribe and generate cDNA, insert into a plasmid vector and transform into bacteria. Screen colonies for the ability to glow in the dark

(d) Isolate mRNA from the organism, reverse transcribe and generate cDNA, insert into a plasmid vector and transform into eukaryotic cells such as yeast. Screen colonies for the ability to glow in the dark

(e) None of the above

10. Which of these would be an example of subcloning?

(a) A recircularized vector is digested with a second restriction enzyme

(b) Part of the insert from a hybrid vector is inserted into a different vector

(c) Several different restriction enzymes are used to cut a hybrid vector, individually and in combinations

(d) All of these

(e) None of these

11. You restriction map a vector using BamHI, EcoRI, and HaeIII and obtained fragments of the following sizes. Draw a restriction map of the vector and determine which of the following statements is correct.

Results:

BamHI alone: 4,000bp;

EcoRI alone: 4,000bp;

HaeIII alone: 3,000bp and 1,000bp;

BamHI and EcoRI: 3,500bp, 500bp;

BamHI and HaeIII: 2,300bp, 1,000bp, 700bp;

EcoRI and HaeIII: 2,800bp, 1,000bp, 200bp;

BamHI, EcoRI, and HaeIII: 2,300bp, 1,000bp, 500bp, 200bp

(a) There are three HaeIII sites in this vector

(b) BamHI is 1,000bp away from EcoRI

(c) BamHI cuts the smaller fragment from the HaeIII digestion

(d) All of the above are correct

(e) None of the above are correct

12. Which of the following would NOT be an advantage to using DNA sequencing rather than restriction mapping?

(a) DNA sequencing allows the identification of all known restriction enzyme recognition sites in one experiment

(b) Restriction mapping is limited to identification of restriction sites that are relatively few in number

(c) Restriction sites that are very close together are difficult to restriction map

(d) None of these (all are advantages)

(e) None of the above

13. What key feature of Taq polymerase allows PCR to be conveniently performed?

(a) Taq polymerase does not require primers

(b) Taq polymerase does not require a template

(c) Taq polymerase is not damaged by heating

(d) Taq polymerase can work at very low temperatures

(e) None of the above

14. Which of the following would you expect to be a drawback to PCR techniques?

(a) PCR has the potential to create positive results from very low levels of contamination

(b) PCR can be used with very small samples such as single sperm cells

(c) PCR is very rapid; in some cases results can be available in a few hours

(d) All of these

(e) None of these

15. In choosing a vector, viral vectors typically can carry much larger pieces of DNA than plasmid vectors, but plasmid vectors are much easier to work with since they do not kill their host cells. In which of the following situations would you anticipate preferring a viral vector?

(a) You need to clone an entire gene, including regulatory sequences

(b) You are generating a genomic library from an organism with a large genome

(c) You are creating an mRNA library to identify the gene encoding a very large protein

(d) All of these

(e) None of these

16. You hypothesize that the Himalayan phenotype in rabbits is caused by expression of the tyrosinase gene only at low temperatures, rather than an effect of temperature on protein function. The tyrosinase gene has already been cloned and sequenced in humans. What technique is best suited for you to test your hypothesis?

(a) Creation of a genomic library from Himalayan rabbits, then use of the human tyrosinase sequence as a probe in colony screening

(b) Digestion of genomic DNA from Himalayan and wild type rabbits, then Southern blotting using the human tyrosinase gene as a probe

(c) Isolation of mRNA from black and white regions of Himalayan rabbits, then Northern blotting using the human tyrosinase gene as a probe

(d) Any of these are equally suited for your question

(e) None of these are suited to this question

17. What type of probe would you use for a Western blot experiment?

(a) A known DNA sequence (b) An RNA molecule

(c) A purified protein (d) An antibody

(e) None of the above

18. A gel retardation assay would be useful for identification of which of the following?

(a) The parts of a gene sequence that are introns

(b) The parts of a gene sequence that encode protein

(c) The parts of a gene that function as enhancers

(d) All of these

(e) None of these

19. Site-directed mutagenesis:

(a) Is a technique to produce specific mutants

(b) Can be used to alter gene function in specific ways

(c) Can create mutant genes to be studied in living organisms

(d) All of these

(d) None of these

20. DNA transformation involves the transfer of DNA via

(a) Pili (b) Naked DNA in solution

(c) A restriction endonuclease (d) A bacteriophage

(e) None of the above

21. The mismatch repair system will repair

(a) Only mismatching donor sequences

(b) Only mismatching recipient sequences

(c) Both sides of mismatching sequences

(d) Either the donor or the recipient strand
(e) None of the above

22. When DNA from an ampicillin resistant organism is brought into an ampicillin sensitive organism by transformation and the product is grown on agar containing ampicillin
(a) Only the transformed cells will grow
(b) Only donor cells will grow
(c) All cells grow
(d) No cells grow
(e) Only ampicillin sensitive cells grow

23. Which is the most specific recombinant DNA library?
(a) Genomic
(b) Protein
(c) cDNA
(d) Chromosomal
(e) cRNA

24. Why are Type I and Type III restriction endonucleases so infrequently used in recombinant DNA technology?
(a) They are much more difficult to purify and obtain than Type II nucleases
(b) They don't cut the DNA in at specific recognition sequences
(c) They require that the DNA be in short fragments in order to cut it up
(d) They are more expensive to use than Type II nucleases
(e) They only work on virus DNA, not bacterial or eukaryotic DNA

25. Restriction enzymes that produce staggered cuts known as "sticky ends" are useful because:
(a) DNA fragments cut with the same enzyme can anneal to each other
(b) Sticky ends stick to almost anything
(c) Blunt cuts are never desirable in DNA engineering
(d) Staggered cuts won't join with other DNA fragments regardless of how they were cut
(e) Sticky ends allow fragment ends to reanneal, which is usually desirable

26. The most common recognition sequence among Type II restriction nucleases used in recombinant DNA technology is:
    (a) 6 base pairs long
    (b) 4 base pairs long
    (c) Of varying length
    (d) 8 base pairs long
    (e) 2 base pairs long

27. Which of the following vectors can carry the largest insert?
    (a) Phages
    (b) Macroplasmids
    (c) Plasmids
    (d) Cosmids
    (e) YACs and BACs

28. The types of vectors based on the fertility plasmid (F factor) are:
    (a) Ti plasmids
    (b) Bacterial artificial chromosomes (BACs)
    (c) Yeast and artificial chromosomes
    (d) Cosmids
    (e) None of the above

29. Which of the following is an application of PCR technology?
    (a) Forensics
    (b) *In vitro* mutagenesis
    (c) Gene mapping
    (d) Epidemiology
    (e) All of the above

30. Which of the following is not required for a PCR reaction?
    (a) RNA transcriptase
    (b) A target sequence
    (c) Taq polymerase
    (d) dNTPs
    (e) A primer

31. In a PCR reaction, which of the following statements is not true?
    (a) Restriction digestion of the DNA is necessary
    (b) Polymerization occurs from 5' to 3'
    (c) PCR fragments can be run in a gel and isolated for cloning
    (d) A thermostable enzyme is necessary
    (e) A primer is necessary

32. Which end of the primer is extended by the Taq polymerase?

(a) 1' (b) 3'

(c) 5' (d) 2'

(e) 4'

33. Under what circumstances should a researcher consider using the chromosome walking technique?

(a) When you are investigating complex eukaryotic genes

(b) When there are no time restrictions

(c) When bacterial conjugation fails

(d) When the approximate location of the gene is already known

(e) Under no circumstances should this technique be used

34. Choose the type of library that should be used in order to study a selection of specific genes and chromosome organization:

(a) Chromosome-specific libraries

(b) cDNA libraries

(c) Genomic libraries

(d) Expression libraries

(e) Ti plasmid library

35. A library representing the genes active in a specific cell type at a specific time (genes being expressed) is called a:

(a) Phage library

(b) Chromosome-specific library

(c) Genomic library

(d) Transposon library

(e) cDNA library

36. A method for analyzing long stretches of DNA, in which the end of a cloned segment is used as a probe to identify other clones that overlap the first clone, is called:

(a) Library screening (b) Nucleic acid blotting

(c) Southern blotting (d) Filter hybridization

(e) Chromosome walking

37. cDNA can be synthesized from mRNA by using the following enzyme

(a) S1 nuclease (b) Reverse transcriptase

(c) DNA polymerase (d) Lysozyme

(e) RNA polymerase

38. Which type of library would you screen in order to identify the promoter of the hemoglobin gene?

(a) Expression library (b) Genomic library

(c) cDNA library (d) None of the above

(e) None of the above

39. A library that contains at least one copy of all the sequences represented in the genome is:

(a) Genomic library (b) Expression libraries

(c) Chromosome-specific library (d) cDNA library

(e) None of the above

40. In the production of cDNA from mRNA, the enzyme used to cleave the hairpin formed in the single stranded cDNA is:

(a) DNA polymerase (b) EcoRI

(c) RNAseH (d) S1 nuclease

(e) Reverse transcriptase

41. A characteristic of almost all eukaryotic mRNA molecules (but not prokaryotic mRNA) that allows the construction of cDNA libraries is:

(a) The presence of a poly A tail at the 3' end of the mRNA molecule

(b) The presence of a poly A tail at the 5' end of the mRNA molecule

(c) The presence of a poly T tail at the 5' end of the mRNA

(d) The presence of a poly T tail at the 3' end of the mRNA

(e) The presence of a promoter sequence

42. The method used to screen a library for a recombinant that contains a specific gene is called:

(a) Hybridization analysis (b) Southern blotting

(c) Vector specific analysis (d) Conjugation

(e) DNA complementation analysis

43. In the Southern blot, the following component is labeled to make possible the visualization of the DNA hybrids formed:
   (a) The probe (single stranded DNA)
   (b) The hybridization bag
   (c) The restriction enzymes
   (d) The gel used for DNA separation
   (e) None of the above

44. You want to clone a fragment of DNA into a plasmid vector such as pBR322. The plasmid is 4.36 kb long and has a single restriction site for the restriction enzyme BamHI. If you digest pBR322 with BamHI, how many linear fragments will result?
   (a) One (b) Four
   (c) Two (d) Three
   (e) None, the plasmid will remain circular

45. You have made a mouse DNA library by cloning mouse DNA fragments into the vector pBR322 at the BamHI site and then transforming the plasmid into *E. coli*. You select one bacterial colony (derived from a single individual) and extract plasmid DNA. You digest the plasmid with BamHI. You get a single 4.36 kb fragment (pBR322 is 4.36 kb in size). Does this plasmid contain a cloned gene?
   (a) No (b) Yes

46. You select another bacterial colony and extract plasmid DNA. Assume that you have a 2 kb cloned fragment in this plasmid. If you digest the plasmid with BamHI, how many fragments will you get?
   (a) Two, a 2 kb and a 4.36 kb fragment
   (b) One 6.36 kb fragment
   (c) Two, a 3 kb and a 6 kb fragment
   (d) One 2 kb fragment
   (e) One 4.36 kb fragment (plasmid size)

47. A 700 bp fragment of DNA is digested with EcoRI. Two fragments result, a 300 bp and a 400 bp fragment. When the same DNA is digested with HaeIII, two fragments result, a 200 bp and a 500 bp. A double digestion

produces three fragments, 100 bp, 200 bp, and 400 bp. Map the fragment. Is the HaeIII restriction site within the EcoRI 400 bp fragment?

(a) Yes (b) No

48. pBR322 (a plasmid vector) has a single BamHI site and a single BspMI site. A double digestion produces two fragments of approximately 690 bp and 3670 bp. Map the two sites on the circular plasmid. If the BamHI site is at the beginning of the plasmid tetracycline resistance gene and the gene is about 1200 bp long, is the BspMI site also in the gene?

(a) Yes

(b) You can't determine the answer from the data given.

(c) No

(d) None of the above

49. When DNA is electrophoresed in a gel, DNAs that are small (20 base pairs or less) will –

(a) Be found near the top of the gel after the DNA has been electrophoresed

(b) Not migrate in the gel

(c) Remain stationary

(d) Move through the gel more quickly than DNAs that are 100 bp long

(e) Move through the gel more slowly than DNAs that are 100 bp long

50. The DNA sequence obtained by both the manual and automatic sequencing corresponds to:

(a) The messenger RNA

(b) The complementary DNA strand

(c) The template DNA strand

(d) The restriction sequences for certain restriction enzymes

(e) The tRNA

51. What sequence on the template strand of DNA corresponds to the first amino acid inserted into a protein?

(a) AUG (b) ATG

(c) UAC (d) TAC

(e) ATG

52. During translation, codons are read from

(a) DNA (b) mRNA
(c) tRNA (d) rRNA
(e) Protein

53. Transcription and translation of a gene composed of 30 nucleotides would form a protein containing no more than how many amino acids?

(a) 10 (b) 15
(c) 30 (d) 60
(e) 90

54. The enzyme peptidyl transferase

(a) Charges tRNA
(b) Synthesizes long, artificial mRNA's
(c) Recognizes the prokaryotic ribosomes
(d) Creates peptide bonds between amino acids
(e) Is not an enzyme involved in translation

55. A charged tRNA enters the protein-synthesizing machinery at the ribosomes by recognition of its

(a) Codon (b) Anticodon
(c) Second genetic code (d) Wobble
(e) Amino acid

56. What amino acid corresponds to the mRNA sequence GGG?

(a) pro (b) trp
(c) ser (d) lys
(e) gly

57. What amino acid corresponds to the DNA template sequence GGG?

(a) pro (b) trp
(c) ser (d) lys
(e) gly

58. What is the consequence of changing the codon UGA to UGG?

(a) T he peptide will be shorter than normal

(b) The peptide will be longer than normal

(c) The peptide will have one altered amino acid

(d) The protein will not be made

(e) The protein will be normal

59. Which of the following is an example of the degeneracy of the genetic code?

(a) Given amino acid has more than one codon

(b) Each codon specifies more than one amino acid

(c) The first two bases specify the amino acid

(d) A cell has only 50 tRNAs

(e) The genetic code is not degenerate

60. Which of the following statements is false?

(a) Penicillin is an antibiotic that interferes with translation at the ribosome

(b) Most organisms use the AUG codon as a start codon

(c) The genetic code is non-overlapping

(d) The genetic code is a triplet

(e) A frameshift can result from an incorrect translational starting point

61. The deliberate modifications of an organism's genetic information by directly changing its nucleic acid content.

(a) Genetic engineering (b) Population genetics

(c) Microbiology (d) All the above

(e) None of the above

62. Enzymes that recognize and cleave specific 4 to 8 base pair sequences

(a) DNA ligase (b) Helicases

(c) Restriction endonucleases (d) All the above

(e) None of the above

63. What is the normal role of these enzymes in bacterial cells?

(a) To degrade the bacterial chromosome into small pieces during replication

(b) To degrade invading phage DNA

(c) To produce RNA primers for replication

64. Which type of restriction endonuclease cuts the DNA within the recognition site?

(a) Type I (b) Type II

(c) Type III (d) All the above

(e) None of the above

65. Bacterial cells protect their own DNA from degradation by restriction endonucleases

(a) By methylating the DNA at the sites that the enzyme recognizes

(b) By deleting all recognition sites from the genome

(c) By not producing any restriction endonucleases

(d) All the above

(e) None of the above

66. Vectors are:

(a) Molecules that degrade nucleic acids

(b) Molecules that are able to covalently bond to and carry foreign DNA into cells

(c) Molecules that protect host cells from invasion by foreign DNA

(d) All the above

(e) None of the above

67. Separation of charged molecules are separated based on varying rates of migration through a solid matrix when subjected to an electric field

(a) Photoreactivation (b) Gel electrophoresis

(c) Autoradiography (d) All the above

(e) None of the above

68. The Southern blotting technique depends on

(a) Similarities between the sequences of probe DNA and experimental DNA

(b) The molecular mass of proteins

(c) The amino acid sequence of a protein

(d) All the above

(e) None of the above

69. A short (2-20 nucleotide) molecule

(a) Plasmid (b) Vector

(c) Oligonucleotide (d) All the above

(e) None of the above

70. A molecular technique in which the DNA sequences between two oligonucleotide primers can be amplified

(a) Southern blotting (b) Northern blotting

(c) Polymerase Chain Reaction (d) All the above

(e) None of the above

71. The advantage to using DNA polymerases from thermophilic organisms in PCR is:

(a) The DNA polymerases of these bacteria are much faster than those from other organisms

(b) The DNA polymerases of these bacteria can withstand the high temperatures needed to denature the DNA strands

(c) The DNA polymerases of these bacteria never make mistakes while replicating DNA

(d) All the above

(e) None of the above

72. The enzyme used to covalently bond foreign DNA to a vector plasmid

(a) DNA polymerase (b) Restriction endonuclease

(c) DNA ligase (d) All the above

(e) None of the above

73. A genomic library is:

(a) A database where the sequence of an organisms genome is stored

(b) A collection of many clones possessing different DNA fragments from the same organisms bound to vectors

(c) A book that describes how to isolate DNA from a particular organism

(d) All the above

(e) None of the above

74. DNA that is obtained using processed mRNA molecules as a template

(a) rDNA (b) mDNA

(c) cDNA (d) All the above

(e) None of the above

75. For gene probes to be useful they must:

(a) Be large enough to contain gene-specific sequences

(b) Be labeled in some manner to allow detection

(c) Both of the above

(d) All the above

(e) None of the above

76. Which of the following is not commonly used as vector?

(a) Artificial chromosome (b) Cosmid

(c) Fungi (d) All the above

(e) None of the above

77. In genetic engineering, a chimera is:

(a) An enzyme that links DNA molecules

(b) A plasmid that contains foreign DNA

(c) A virus that infects bacteria

(d) All the above

(e) None of the above

78. Which of the following vector can maintain the largest fragment of foreign DNA?

(a) YAC (b) Cosmid

(c) Plasmid (d) All the above

(e) None of the above

79. An animal that has gained new genetic information from the acquisition of foreign DNA

(a) A chimera (b) A transgenic animal

(c) A vector (d) All the above

(e) None of the above

80. Electroporation is:

(a) The process of separating charged molecules through a gel maintained in an electric field

(b) The process of combining foreign DNA to an electrically charged vector molecule

(c) The process of introducing DNA into cells by the application of high voltage pulses

(d) All the above

(e) None of the above

81. The piece of equipment that introduces DNA into cells by via DNA-coated microprojectiles

(a) Laser (b) DNA probe

(c) Gene gun (d) All the above

(e) None of the above

82. Problems in obtaining large amounts of proteins encoded by recombinant genes can often be overcome by the use of these special cloning vectors

(a) BACs

(b) Expression vectors

(c) YACs

(d) All the above

(e) None of the above

83. In which ways are recombinant DNA techniques are playing an important role in human health

(a) Helping to determine the molecular causes of disease

(b) Producing effective means to express proteins such as insulin and growth hormone

(c) Both of the above

(d) All the above

(e) None of the above

84. Agrobacterium tumefaciens is:

(a) A disease in humans that causes loss of sight

(b) A bacterium that can be used to introduce DNA into plants

(c) A fungi that is used to produce antibiotics in large amounts

(d) All the above

(e) None of the above

85. Which of the following is an example of a DNA vector?

(a) A prophage

(b) An Hfr element

(c) An F factor

(d) All of these

(e) None of these

86. Which of the following would NOT be a useful selectable marker?

(a) A gene encoding a protein that degrades the antibiotic ampicillin

(b) A gene encoding a protein that allows the cell to synthesize histidine

(c) A gene encoding a protein that is an essential structural component of the cell

(d) All of these are useful selectable markers

(e) None of these are useful selectable markers

87. Which statement regarding restriction endonucleases is NOT correct?

(a) They recognize a specific base sequence in the DNA

(b) They are produced by bacterial cells as a primitive immune system

(c) They digest DNA by removing nucleotides from a free 3' end

(d) They often generate short single stranded sequences

88. Which of these palindromes is most similar to a restriction endonuclease recognition site?

(a) MADAM, I'M ADAM

(b) A MAN, A PLAN, A CANAL: PANAMA!

(c) BOB

(d) None: Restriction endonuclease recognition sites are not palindromic

89. Formation of what type of chemical bond is catalyzed by DNA ligase?

(a) Glycosidic (b) Hydrogen

(c) Phosphoester (d) Ester

90. Which of the enzymes whose recognition sites are shown here would produce sticky ends compatible with BstAUI (5'-G | GTACC-3')? (Hint: Only one strand is shown. The enzyme cuts at the | on this strand, and at an equivalent position in the palindrome on the opposite stranD.)

(a) BclI(5'-T | GATCA-3') (b) SunI(5'-C | GTACG-3')

(c) HindIII(5'-A | AGCTT-3') (d) KpnI(5'-GGTAC | C-3')

91. A DNA vector and genomic DNA are both cut with the same restriction enzyme. The genomic DNA is shaded. Which of these diagrams depicts a recombinant vector, which can be used to clone the DNA?

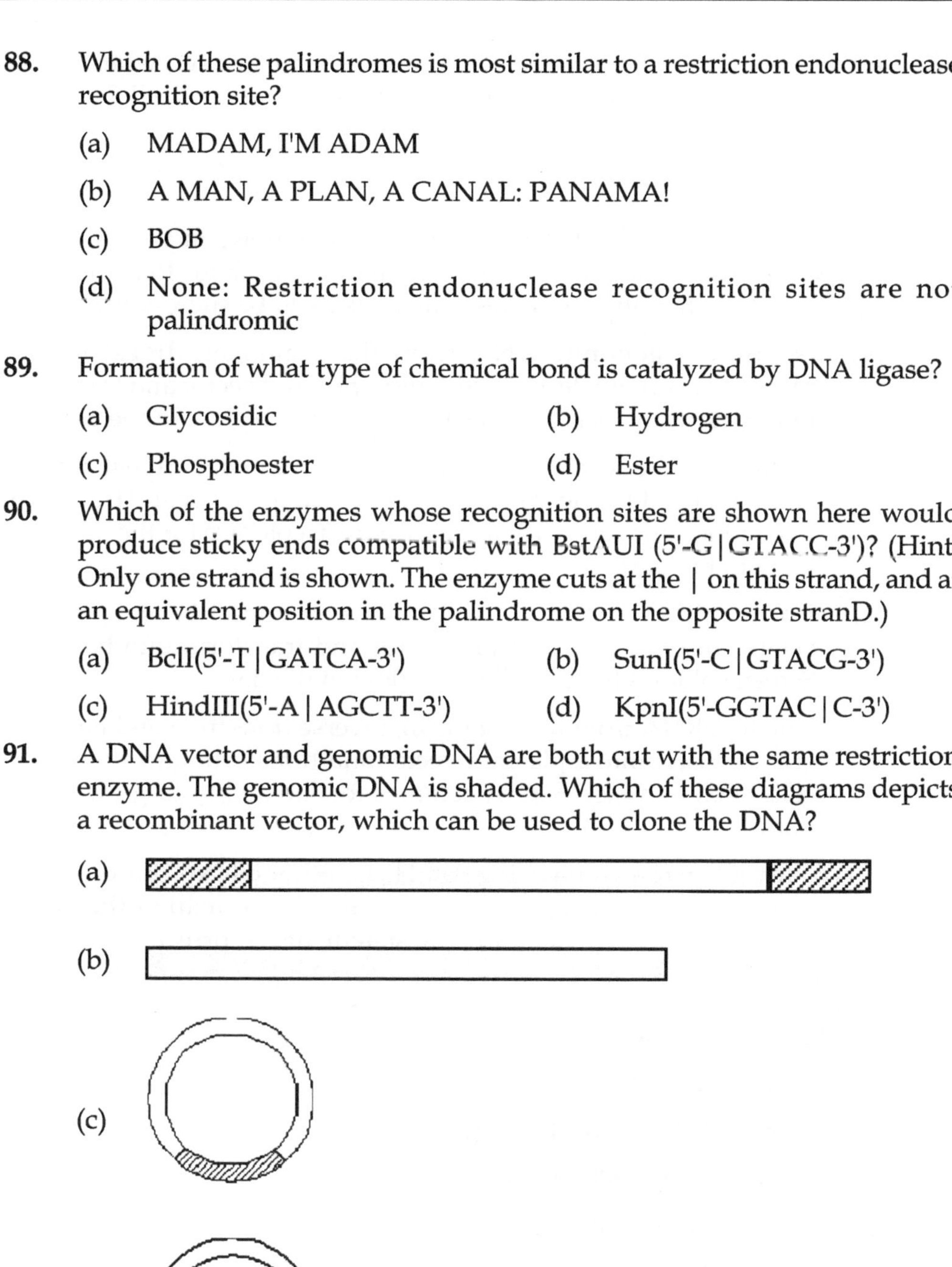

92. If you are using a vector with lacZ as a screenable marker, which of the following bacterial strains would you want to use as host cells?

(a) *lacO-* (b) *lacI-*

(c) *lacZ-*

93. You discover a novel eukaryotic organism that glows in the dark. You believe this trait is due to a single gene, and you wish to clone the gene. Which of the following strategies is most likely to be successful?

(a) Isolate the genomic DNA from the organism, digest with a restriction endonuclease, insert into a plasmid vector and transform into bacteria. Screen colonies for the ability to glow in the dark

(b) Isolate the genomic DNA from the organism, digest with a restriction endonuclease, insert into a plasmid vector and transform into eukaryotic cells such as yeast. Screen colonies for the ability to glow in the dark

(c) Isolate mRNA from the organism, reverse transcribe and generate cDNA, insert into a plasmid vector and transform into bacteriA. Screen colonies for the ability to glow in the dark

(d) Isolate mRNA from the organism, reverse transcribe and generate cDNA, insert into a plasmid vector and transform into eukaryotic cells such as yeast. Screen colonies for the ability to glow in the dark

94. You restriction map a vector using BamHI, EcoRI, and HaeIII and obtained fragments of the following sizes. Draw a restriction map of the vector and determine which of the following statements is correct.

Results:

BamHI alone: 4,000bp;

EcoRI alone: 4,000bp;

HaeIII alone: 3,000bp and 1,000bp;

BamHI and EcoRI: 3,500bp, 500bp;

BamHI and HaeIII: 2,300bp, 1,000bp, 700bp;

EcoRI and HaeIII: 2,800bp, 1,000bp, 200bp;

BamHI, EcoRI, and HaeIII: 2,300bp, 1,000bp, 500bp, 200bp

(a) There are three HaeIII sites in this vector

(b) BamHI is 1,000bp away from EcoRI

(c) BamHI cuts the smaller fragment from the HaeIII digestion

(d) All of the above are correct

(e) None of the above are correct

95. Which of the following would NOT be an advantage to using DNA sequencing rather than restriction mapping?

(a) DNA sequencing allows the identification of all known restriction enzyme recognition sites in one experiment

(b) Restriction mapping is limited to identification of restriction sites that are relatively few in number

(c) Restriction sites that are very close together are difficult to restriction map

(d) None of these (all are advantages)

96. What key feature of Taq polymerase allows PCR to be conveniently performed?

(a) Taq polymerase does not require primers

(b) Taq polymerase does not require a template

(c) Taq polymerase is not damaged by heating

(d) Taq polymerase can work at very low temperatures

97. Which of the following would you expect to be a drawback to PCR techniques?

(a) PCR has the potential to create positive results from very low levels of contamination

(b) PCR can be used with very small samples such as single sperm cells

(c) PCR is very rapid; in some cases results can be available in a few hours

(d) All of these

(e) None of these

98. In choosing a vector, viral vectors typically can carry much larger pieces of DNA than plasmid vectors, but plasmid vectors are much easier to work with since they do not kill their host cells. In which of the following situations would you anticipate preferring a viral vector?

(a) You need to clone an entire gene, including regulatory sequences

(b) You are generating a genomic library from an organism with a large genome

(c) You are creating an mRNA library to identify the gene encoding a very large protein

(d) All of these

(e) None of these

99. You hypothesize that the Himalayan phenotype in rabbits is caused by expression of the tyrosinase gene only at low temperatures, rather than an effect of temperature on protein function. The tyrosinase gene has already been cloned and sequenced in humans. What technique is best suited for you to test your hypothesis?

(a) Creation of a genomic library from Himalayan rabbits, then use of the human tyrosinase sequence as a probe in colony screening

(b) Digestion of genomic DNA from Himalayan and wild type rabbits, then Southern blotting using the human tyrosinase gene as a probe

(c) Isolation of mRNA from black and white regions of Himalayan rabbits, then Northern blotting using the human tyrosinase gene as a probe

(d) Any of these are equally suited for your question

(e) None of these are suited to this question

100. What type of probe would you use for a Western blot experiment?

(a) A known DNA sequence (b) An RNA molecule

(c) A purified protein (d) An antibody

101. A gel retardation assay would be useful for identification of which of the following?

(a) The parts of a gene sequence that are introns

(b) The parts of a gene sequence that encode protein

(c) The parts of a gene that function as enhancers

(d) All of these

(e) None of these

102. Site-directed mutagenesis:

(a) Is a technique to produce specific mutants

(b) Can be used to alter gene function in specific ways

(c) Can create mutant genes to be studied in living organisms

(d) All of these

(e) None of these

Fill in the blanks

1. Northern blots probe _________.

(a) DNA and RNA (b) DNA

(c) RNA and proteins (d) Proteins

(e) RNA

2. Recombinant DNA research is dependent on _________.

(a) Cloning (b) Hosts

(c) Vectors (d) Restriction endonucleases

(e) All of the above are correct.

3. The polymerase chain reaction is a technique that _________.

(a) Selectively replicates RNA

(b) Probes and selectively replicates DNA

(c) Selectively replicates DNA

(d) Probes DNA

(e) Probes RNA

4. A "YAC" is a useful _________.

(a) Vector (b) Screen

(c) Host (d) Probe

(e) Library

5. Restriction enzymes are involved in all of the following genetic engineering techniques except _________.

(a) Cloning DNA into vectors

(b) Mapping studies

(c) Identification of genetic markers

(d) PCR

(e) None of the above

6. Chromosome walking is a technique used to _________.

(a) Move chromosomes around the nucleus

(b) Move a fragment of chromosomal DNA from one area of a chromosome to another. This is a casual term for the work that genetic engineers do

(c) Recombination between chromosomal DNA of two different species

(d) A method used to locate a gene using a set of clones from a DNA library

(e) None of the above

7. A "probe" in molecular biology is _________.

(a) An instrument used to manipulate cells in culture

(b) A type of vector system

(c) A DNA or an RNA molecule used in hybridization reactions

(d) Probes are not used in molecular biology

(e) None of the above

8. Type II restriction endonucleases cut DNA at ___.

(a) Random DNA sequences (b) Specific DNA sequences

(c) Single-stranded sequences (d) A and C

(e) None of the above

9. The tools used to cut genes in recombinant DNA technology are___.

(a) DNA ligases (b) Restriction enzymes

(c) Helicases (d) Reverse transcriptases

(e) Vector enzymes

10. Insertion of a plasmid into a bacterium is done by using ___.

(a) Heat shock

(b) Incubation of bacteria with the plasmid

(c) Injecting the bacteria into the plasmid

(d) Mixing in vitro bacterial chromosomal DNA and the plasmid

(e) Both A and B

11. Bacteria protect their own genome from degradation by endonucleases by ___.

(a) Phosphorylating their palindromic sequences

(b) Methylating their own DNA

(c) Cutting only phage or plasmid DNA

(d) Not having palindromic sequences in their chromosomal DNA

(e) Making double-stranded cuts

12. _______ seals a nick in a strand of DNA by creating a sugar phosphate bond between the adjacent, disjointed nucleotides.

(a) DNAse (b) DNA ligase

(c) Restriction endonuclease (d) EcoRI

(e) A and C

13. The process by which plasmid vectors are introduced into bacteria is called _________.

(a) Bacterial induction (b) Transformation

(c) Plasmolysis (d) Conjugation

(e) Polymerization

14. A piece of DNA is introduced into the ampicillin resistant gene of the plasmid, which also has a tetracycline resistant gene. Bacteria transformed with this plasmid will be ___.

(a) Sensitive to both antibiotics (b) Resistant to ampicillin only

(c) Resistant to both antibiotics (d) Resistant to tetracycline only

(e) Sensitive to tetracycline only

15. Taq polymerase is a commonly used enzyme in PCR because this enzyme is ___.

(a) A primer replicase (b) A faster polymerase

(c) Thermostable (d) Not prone to errors

(e) Thermolabile

16. At the beginning of each cycle the temperature of the PCR reaction is raised in order to ___.

(a) Attach the primer
(b) Denature the double DNA strands
(c) Elongate the primer
(d) Polymerize the DNA
(e) Renature the double DNA strands

17. The three main steps in a PCR cycle are ___.

(a) Denaturation, annealing, elongation
(b) Polymerization, chain elongation, recombination
(c) Denaturation, recombination, primer synthesis
(d) Annealing, primer synthesis, recombination
(e) Polymerization, primer synthesis, DNA synthesis

18. A primer is ___.

(a) A regulator region a short distance from the 5' end of a gene
(b) A short DNA or RNA molecule that acts as starting point for 3' chain growth
(c) An autonomously replicating piece of DNA that serves as a template
(d) A long DNA or RNA sequence that elongates the template
(e) None of the above

19. Taq polymerase is a thermostable enzyme isolated from ___.

(a) Haemophilus influenza
(b) Agrobacterium tumefaciens
(c) Escherichia coli
(d) Bacteriophage T4
(d) Thermus aquaticus

20. For PCR, DNA can be isolated from the following source: ___.

(a) Fossil specimens (b) Frozen tissue
(c) Spermatic cells (d) Preserved tissue
(e) All of the above

21. In manual Sanger sequencing, _____ sequencing reactions are performed.

(a) Two (b) Five

(c) Three (d) One

(e) Four

22. Dideoxy nucleoside triphosphate _____.

(a) Will terminate DNA synthesis when incorporated into a growing DNA strand

(b) Cannot bind to a growing DNA chain

(c) Has an oxygen at the 3' carbon

(d) Has an oxygen at the 2' carbon

(d) Can form a sugar phosphate bond with a new nucleoside triphosphate during DNA synthesis

23. Manual DNA sequencing gels are read from _____.

(a) From the left side (when viewed from the loading wells)

(b) From backside using infrared light

(c) The bottom up

(d) The top down

(e) From the right side (when viewed from the loading wells)

True/False

1. Recombinant DNA technology is a modification of natural recombination.

(a) True (b) False

2. Regulatory elements of a gene could be cloned by recombinant DNA technology.

(a) True (b) False

3. The function of a host cell is to prevent environmental influences from causing mutations in the cloned gene.

(a) True (b) False

4. In transformation, both strands of donor DNA enter the recipient cell.

(a) True (b) False

5. Donor DNA is combined with recipient DNA by homologous recombination.
   (a) True (b) False
6. A cDNA library can be used to clone any piece of DNA from the organism.
   (a) True (b) False
7. Reverse transcriptase can be used with PCR to clone genes from their mRNA.
   (a) True (b) False
8. Competent cells are created under unique laboratory conditions.
   (a) True (b) False
9. Restriction mapping can be used to identify the location of a gene on a chromosome.
   (a) True (b) False
10. Western blotting would be a useful technique to determine expression of a gene in a particular tissue.
   (a) True (b) False
11. A dideoxynucleotide will stop DNA synthesis due to lack of a 2' OH group.
   (a) True (b) False
12. Recombinant DNA technology is a modification of natural recombination.
   (a) True (b) False
13. A dideoxynucleotide will stop DNA synthesis due to lack of a 2' OH group.
   (a) True (b) False
14. Western blotting would be a useful technique to determine expression of a gene in a particular tissue.
   (a) True (b) False
15. A cDNA library can be used to clone any piece of DNA from the organism.
   (a) True (b) False
16. Reverse transcriptase can be used with PCR to clone genes from their mRNA.
   (a) True (b) False

17. Restriction mapping can be used to identify the location of a gene on a chromosome.

(a) True (b) False

18. Regulatory elements of a gene could be cloned by recombinant DNA technology.

(a) True (b) False

19. Competent cells are created under unique laboratory conditions.

(a) True (b) False

20. The function of a host cell is to prevent environmental influences from causing mutations in the cloned gene.

(a) True (b) False

Answers

Multiple Choice Question

1	2	3	4	5	6	7	8	9	10	11	12	13	14	15
C	C	C	C	C	B	C	B	D	B	E	D	C	A	D
16	17	18	19	20	21	22	23	24	25	26	27	28	29	30
C	D	C	D	D	D	A	C	B	A	A	E	B	E	A
31	32	33	34	35	36	37	38	39	40	41	42	43	44	45
A	B	D	A	E	E	B	B	A	D	A	A	A	A	A
46	47	48	49	50	51	52	53	54	55	56	57	58	59	60
A	B	B	D	B	D	B	A	D	B	E	A	B	A	A
61	62	63	64	65	66	67	68	69	70	71	72	73	74	75
A	C	B	B	A	B	B	A	C	C	B	C	D	C	C
76	77	78	79	80	81	82	83	84	85	86	87	88	89	90
C	B	A	B	C	C	B	C	B	C	C	C	C	C	B
91	92	93	94	95	96	97	98	99	100	101	102			
C	C	D	E	D	C	A	D	C	D	C	C			

Fill in the blanks

1	2	3	4	5	6	7	8	9	10	11	12	13	14	15
E	E	C	A	D	D	C	B	B	E	B	B	B	D	C
16	17	18	19	20	21	22	23							
B	A	B	B	E	E	A	C							

True/False

1	2	3	4	5	6	7	8	9	10	11	12	13	14	15
A	A	B	B	A	A	A	B	B	A	B	A	B	A	A
16	17	18	19	20										
A	B	A	B	B										

❑❑❑

Chapter – 18

Genome Analysis and Its Uses

Uses of Microorganisms in Biotechnology

- Somatostatin was the first human peptide hormone produced by recombinant bacteria
- Many important medicines are produced by recombinant microorganisms
- Bacterial species can be used as biological control agents
- The release of recombinant microorganisms into the environment is sometimes controversial
- Microorganisms can reduce environmental pollutants

New Methods for Genetically Manipulating Plants and Animals

- The integration of a cloned gene into a chromosome can result in gene addition or gene replacement
- Molecular biologists can produce mice that contain gene replacements
- Agrobacterium tumefaciens can be used to make transgenic plants
- Researchers have succeeded in cloning mammals from somatic cells
- Stem cells have the ability to divide and differentiate into different cell types

Cytogenetic and Linkage Mapping

- Cytogenetic mapping provides a way to determine the locations of genes by microscopically examining the chromosomes
- In situ hybridization can localize genes along particular chromosomes
- Linkage mapping can use molecular markers
- The distance between two linked RFLPs can be determined
- An RFLP map describes the locations of many different RFLPs throughout the genome
- RFLP analysis can be used to follow the inheritance of disease-causing alleles
- Genetic mapping also uses molecular markers called microsatellites

Physical Mapping

- Specific chromosomes can be isolated by pulsed-field gel electrophoresis or chromosome sorting
- A physical map of a chromosome is constructed by creating a contiguous series of clones from a chromosome-specific library
- YAC cloning vectors are used to make contigs of eukaryotic chromosomes
- Positional cloning can be achieved using chromosome walking
- The Human Genome Project has stimulated genomic research
- Many genome sequences have been determined

Applications of Transgenic Plants and Animals

- Gene replacements in mice can be used to understand gene function and human disease
- Biotechnology holds promise in producing transgenic livestock
- Transgenic plants can be given characteristics that are agriculturally useful

DNA Fingerprinting

- DNA fingerprinting provides a unique banding pattern for every individual

- DNA fingerprinting is used for identification and relationship testing

Human Gene Therapy

- Gene therapy involves the introduction of cloned genes into human cells
- Adenosine deaminase deficiency was the first inherited disease treated with gene therapy
- Aerosol sprays may be used to treat cystic fibrosis

Multiple Choice Question

1. Human genes are not always readily produced by bacteria. Which of the following modifications can be useful?

(a) Inserting a cDNA sequence rather than the entire gene

(b) Inserting a synthetic DNA sequence if the coding region is very short

(c) Production of a hybrid or fusion protein

(d) All of these

(e) None of these

2. Prior to production of recombinant human insulin, diabetic patients could use insulin isolated from cattle. Which of the listed items would not be a drawback to this procedure?

(a) Insulin from cattle is slightly different from human insulin, potentially leading to allergic reactions

(b) Members of some religions consider cattle sacred and thus are unable to use insulin isolated from cattle

(c) Infectious diseases may be transferred in the isolated insulin

(d) Despite the relatively stable supply of cattle pancreases from slaughter houses, insulin isolated from cattle is still relatively expensive and in somewhat limited supply

(e) None of these; all are drawbacks to use of cattle insulin

3. Which of these is an example of biodegradation?

(a) A strain of bacteria chemically captures heavy metal ions into chemical compounds

(b) A population of microorganisms metabolized toluene from a massive leak of jet fuel in Hanahan, SC into carbon dioxide and other small molecules, reducing toluene levels from over 5,000 parts per billion to undetectable levels

(c) Naturally occurring microorganisms in Picatinny, NJ and other sites are able to dechlorinate common industrially produced contaminants such as trichloroethane, generating a nontoxic compound

(d) All of these are biodegradation

(e) None of these are biodegradation

4. To generate a gene knockout, which of the following is required?

(a) A method for screening to verify insertion of the cloned gene

(b) Regions of homology between the cloned gene and the target gene

(c) A method to verify that the ends of the cloned gene were removed due to homologous recombination rather than nonhomologous recombination

(d) All of these

(e) None of these

5. Some chimeric animals may not produce offspring carrying the gene replacement, while other chimeric animals with the same gene replacement can easily produce offspring carrying the gene replacement. Why?

(a) Mosaic animals are unable to produce offspring

(b) The gene replacement prevents correct gamete formation

(c) Some chimeras have a germline entirely composed of cells from the blastocyst rather than the modified cells

(d) Some chimeras have a germline entirely composed of modified cells rather than cells from the blastocyst

(e) None of the above

6. Which of the following explains why production of transgenic plants is easier than production of transgenic animals?

(a) Plant cells can grow in cell culture

(b) Plant cells have a lower number of potentially lethal genes

(c) Plant cells are totipotent

(d) Production of mutant plants poses less ethical dilemmas than production of mutant animals

(e) None of these

7. Which of the following groups of individuals could be considered to be clones?

(a) Fraternal twins

(b) Several strawberry plants that grew from runners from the same parent

(c) Littermates from an inbred strain of mice

(d) All of these

(e) None of these

8. What process causes shortening of telomeres during cell division?

(a) DNA synthesis

(b) Exonuclease removal of primers

(c) Ligation of DNA fragments

(d) Telomerase activity

(e) None of the above

9. Which of these is an example of an adult stem cell?

(a) A red blood cell

(b) A neuron

(c) A cell from a blastocyst

(d) Skin cells that produce new epithelial layers

(e) None of the above

10. Which of these uses of transgenic technology is not directly relevant to agricultural applications?

(a) Production of a pharmaceutical protein in goat's milk

(b) Production of salmon that grow all year round instead of just in warm weather

(c) Production of soybean plants that are resistant to herbicides

(d) Production of "super mice" that carry the human growth hormone gene

(e) None of the above

11. How many different bands would you expect an individual to have in a VNTR test?

(a) Two
(b) Between 20 and 40
(c) Up to 100
(d) Thousands
(e) None of the above

12. How many different bands would expect an individual to have for a particular STR?

(a) One
(b) Two
(c) Up to 20
(d) 20 to 100
(e) None of the above

13. Which of these is an accurate description of gene therapy?

(a) A form of site directed mutagenesis
(b) Introduction of cloned genes into living cells to cure disease
(c) Cloning of a disease gene in order to understand the disease
(d) All of these
(e) None of these

14. Which of the following best describes the process of alternative splicing?

(a) Various ways of splicing out introns in eukaryotic pre-mRNAs resulting in one gene producing several different mRNAs and protein products

(b) The succession of codons determined by reading nucleotides in groups of three from a specific initiation codon

(c) The alternation of gametophyte and sporophyte stages in the life cycle of a plant

(d) The reaction that removes introns and joins together exons in eukaryotic nuclear primary RNA transcripts

(e) None of the above

15. Which of the following best explains why a genome project is not finished when the sequence has been completed?

(a) Genomes change too fast and must constantly be updated

(b) Individuals within a species are so unique that having a single sequence is insufficient to characterize a species' genome

(c) Without knowing the number, the function, and the location of genes within a genome, the sequence is not very useful

(d) Many sections of a genome are too difficult to sequence and have not actually been included in 'complete' genomes

(e) None of the above

16. Which of the following best defines gene annotation?

(a) Shorthand method of documenting a genes function

(b) Describing different regions of the code and identifying which regions can be called genes

(c) Determining the sequence of nucleotide bases for a gene

(d) Identifying which regions of cDNA are separate exons

(e) None of the above

17. Which of these is an example of cytogenetic mapping?

(a) Hybridization studies in Drosophila determined the location of the yellow gene, which influences body color, to be near the tip of the X chromosome

(b) Karyotype analysis of humans with familial Down syndrome has identified a relatively small region of chromosome 21 containing genes that cause the Down phenotype

(c) Analysis of the chromosomes of a transgenic strain of mice indicates that the transgene has been inserted into the short arm of chromosome 15

(d) All of these

(e) None of these

18. Cytologists can use which of the following to describe locations of a gene at a specific place on the chromosome?

(a) DNA sequence of a chromosomal region

(b) Protein expression from a chromosomal region

(c) Banding pattern of stained chromosomes

(d) Comparison to markers located within a few thousand bp

(e) None of the above

19. Which of the following questions could be easily answered by FISH?

(a) How often does recombination occur between the ebony and sepia loci in Drosophila?

(b) In a particular family affected with Becker muscular dystrophy, is the mutation due to a deletion of the entire gene?

(c) How large a piece of DNA would you need to clone in order to clone the group of human globin genes from chromosome 11?

(d) All of these

(e) None of these

20. Which of these is a key characteristic of a molecular marker?

(a) It is a known gene

(b) It is located at a known site on the chromosome

(c) It is only useful for linkage and physical mapping studies

(d) None of these

(e) None of the above

21. Which of these statements regarding RFLP analysis is correct?

(a) RFLP analysis requires Southern blotting for detection of fragments

(b) RFLPs can identify single base pair changes at any site in the chromosome

(c) An RFLP typically produces several different alleles

(d) All of these are correct

(e) None of these are correct

22. A monomorphic DNA segment is:

(a) A segment of DNA that exists in many forms in the population

(b) A segment of DNA that controls a single gene function

(c) A segment of DNA inherited in a dominant fashion

(d) A segment of DNA shared by over 99% of the population

(e) All of these

23. Linkage mapping can determine the distance between which of the following pairs of DNA sequences?

(a) AFLPs and RFLPs

(b) Two AFLPs

(c) Two known genes

(d) A known gene and any type of molecular marker

(e) All of these

The next set of questions all refer to the following situation:

You have set up the following experiment to linkage map two RFLPs in pigs. For RFLP A, you can detect either a 5,000bp band or a 3,350bp band; for RFLP B you can detect either a 4,300bp band or a pair of bands of 2,300 and 2,000bp. Beginning with strains homozygous for different forms of each RFLP, you generate compound heterozygotes, which you then cross to pigs that are homozygous for the smaller forms of each RFLP. By analysis of several litters you get the following results: 52 individuals with 5,000bp and 4,300bp bands; 48 individuals with 5,000bp, 4,300bp, 3,350bp, 2,300bp, and 2,000bp bands; 11 individuals with 5,000bp, 4,300bp, 2,300bp, and 2,000bp bands; and 9 individuals with 5,000bp, 4,300bp, and 3,350bp bands.

24. What was the arrangement of the markers in the original parental strains?

(a) A: 5,000bp, B: 4,300bp and A: 3,350bp, B: 2,300/2000bp

(b) A: 5,000bp, B: 2,300/2,000bp and A: 3,350bp, B: 4,300bp

(c) A: 3,350bp, B: 4,300bp and A: 5,000bp, B: 4,300bp

(d) A: 3,350bp, B: 2,300/2000bp and A: 3,350bp, B: 2,300/2000bp

(e) None of the above

25. What is the distance between these two markers?

(a) 0.167mu

(b) 0.20mu

(c) 16.7mu

(d) 20mu

(e) None of these

26. Which of the following would be a reasonable use of an RFLP map?

(a) Identification of the exact location of an unknown gene along the chromosome

(b) Description of the size of DNA fragment required to clone the region containing an unknown gene

(c) Identification of the region in which an unknown gene is located

(d) All of these

(e) None of these

27. Which of these genetic markers is most likely to be highly polymorphic (have many different alleles)?

(a) An RFLP

(b) A microsatellite

(c) An SNP

(d) All of these are equally polymorphic

(e) None of these are likely to be polymorphic

28. Why might use of microsatellites in genetic mapping studies be an advantage over RFLPs?

(a) Microsatellites are easier to detect

(b) Microsatellites are more abundant than RFLPs

(c) Microsatellites have more potential alleles than RFLPs

(d) All of these

(e) None of these

29. How are individual chromosomes identified in chromosome sorting techniques?

(a) Relative level of fluorescence when stained with a dye mixture

(b) Level of charge relative to size

(c) Size of the molecule

(d) All of these

(e) None of these

30. Which of these describes a contig?

(a) A complete genomic library including overlapping clones

(b) A complete mRNA library

(c) A chromosome-specific library of overlapping clones

(d) An ordered genomic library

(e) None of the above

31. What vector would be best suited for creating a contig of bovine (cattle) chromosome 10?

(a) λ

(b) A plasmid

(c) YAC

(d) Any of these

(e) None of these

32. Which of the following would not be a critical characteristic of a YAC vector?

(a) Telomeric sequences

(b) A gene encoding a required structural protein

(c) An origin of replication

(d) A centromere

(e) All of these are critical characteristics of a YAC

33. After cytogenetic and linkage mapping, you have identified a relatively small area of a chromosome where your gene of interest is likely to reside. You decide to attempt to use a YAC contig to map the gene by positional cloning. How might you identify the clone that may contain the gene?

(a) The clone will also contain the nearest marker mapped by linkage analysis

(b) The DNA sequence of the clone will contain an ORF (open reading frame)

(c) Cytological hybridization of the clone produces a different pattern than that of the nearest mapped marker

(d) All of these

(e) None of these

34. Which of these is a reasonable use for the contigs and clones generated in the process of obtaining a physical map of the human genome?

(a) Correlation of physical and linkage map distances to determine recombination frequency

(b) Identification of all the known alleles of a single disease-causing gene

(c) Insertion into cells affected with genetic disease in the process of gene therapy

(d) All of these

(e) None of these

35. Which of these is an example of cytogenetic mapping?

(a) Hybridization studies in Drosophila determined the location of the yellow gene, which influences body color, to be near the tip of the X chromosome

(b) Karyotype analysis of humans with familial Down syndrome has identified a relatively small region of chromosome 21 containing genes that cause the Down phenotype

(c) Analysis of the chromosomes of a transgenic strain of mice indicates that the transgene has been inserted into the short arm of chromosome 15

(d) All of these

(e) None of these

36. Cytologists can use which of the following to describe locations of a gene at a specific place on the chromosome?

(a) DNA sequence of a chromosomal region

(b) Protein expression from a chromosomal region

(c) Banding pattern of stained chromosomes

(d) Comparison to markers located within a few thousand bp

37. Cytogenetic mapping:

(a) Requires that the gene have been cloned if in situ hybridization is to be used

(b) Is a high resolution method of gene localization

(c) Can be used to determine relative order of genes located very close to each other

(d) All of these

(e) None of these

38. Which of the following questions could be easily answered by FISH?

(a) How often does recombination occur between the *ebony* and *sepia* loci in *Drosophila*?

(b) In a particular family affected with Becker muscular dystrophy, is the mutation due to a deletion of the entire gene?

(c) How large a piece of DNA would you need to clone in order to clone the group of human globin genes from chromosome 11?

(d) All of these

(e) None of these

39. *In situ* hybridization is done on fragments of specific chromosomes.

(a) True (b) False

40. Which of these is a key characteristic of a molecular marker?

(a) It is a known gene

(b) It is located at a known site on the chromosome

(c) It is only useful for linkage and physical mapping studies

(d) None of these

41. A molecular marker which is amplified by PCR and is polymorphic by length is a(n):

(a) Restriction fragment length polymorphism (RFLP)

(b) Variable number of tandem repeats site (VNTR)

(c) Amplified fragment length polymorphism (AFLP)

(d) Single nucleotide polymorphism (SNP)

42. A polymorphism is:

(a) Any change in the DNA sequence

(b) The most common variation of a gene or marker sequence

(c) The least common variation of a gene or marker sequence

(d) A variation of gene or marker sequence present in >1% of the population

(e) None of these

43. Which of these statements regarding RFLP analysis is correct?

(a) RFLP analysis requires Southern blotting for detection of fragments

(b) RFLPs can identify single base pair changes at any site in the chromosome

(c) An RFLP typically produces several different alleles

(d) All of these are correct

(e) None of these are correct

44. A monomorphic DNA segment is:

(a) A segment of DNA that exists in many forms in the population

(b) A segment of DNA that controls a single gene function

(c) A segment of DNA inherited in a dominant fashion

(d) A segment of DNA shared by over 99% of the population

(e) All of these

45. Linkage mapping can determine the distance between which of the following pairs of DNA sequences?

(a) AFLPs and RFLPs

(b) Two AFLPs

(c) Two known genes

(d) A known gene and any type of molecular marker

(e) All of these

The next set of questions all refer to the following situation:

You have set up the following experiment to linkage map two RFLPs in pigs. For RFLP A, you can detect either a 5,000bp band or a 3,350bp band; for RFLP B you can detect either a 4,300bp band or a pair of bands of 2,300 and 2,000bp. Beginning with strains homozygous for different forms of each RFLP, you generate compound heterozygotes, which you then cross to pigs that are homozygous for the smaller forms of each RFLP. By analysis of several litters you get the following results: 52 individuals with 5,000bp and 4,300bp bands; 48 individuals with 5,000bp, 4,300bp,

3,350bp, 2,300bp, and 2,000bp bands; 11 individuals with 5,000bp, 4,300bp, 2,300bp, and 2,000bp bands; and 9 individuals with 5,000bp, 4,300bp, and 3,350bp bands.

46. What was the arrangement of the markers in the original parental strains?

(a) A: 5,000bp, B: 4,300bp and A: 3,350bp, B: 2,300/2000bp

(b) A: 5,000bp, B: 2,300/2,000bp and A: 3,350bp, B: 4,300bp

(c) A: 3,350bp, B: 4,300bp and A: 5,000bp, B: 4,300bp

(d) A: 3,350bp, B: 2,300/2000bp and A: 3,350bp, B: 2,300/2000bp

47. What is the distance between these two markers?

(a) 0.167mu

(b) 0.20mu

(c) 16.7mu

(d) 20mu

(e) None of these

48. Which of the following would be a reasonable use of an RFLP map?

(a) Identification of the exact location of an unknown gene along the chromosome

(b) Description of the size of DNA fragment required to clone the region containing an unknown gene

(c) Identification of the region in which an unknown gene is located

(d) All of these

(e) None of these

49. Which of these genetic markers is most likely to be highly polymorphic (have many different alleles)?

(a) An RFLP

(b) A microsatellite

(c) An SNP

(d) All of these are equally polymorphic

(e) None of these are likely to be polymorphic

50. Why might use of microsatellites in genetic mapping studies be an advantage over RFLPs?

(a) Microsatellites are easier to detect

(b) Microsatellites are more abundant than RFLPs

(c) Microsatellites have more potential alleles than RFLPs

(d) All of these

(e) None of these

51. Which of these describes a contig?

(a) A complete genomic library including overlapping clones

(b) A complete mRNA library

(c) A chromosome-specific library of overlapping clones

(d) An ordered genomic library

52. What vector would be best suited for creating a contig of bovine (cattle) chromosome 10?

(a) λ

(b) A plasmid

(c) A YAC

(d) Any of these

(e) None of these

53. Which of the following would not be a critical characteristic of a YAC vector?

(a) Telomeric sequences

(b) A gene encoding a required structural protein

(c) An origin of replication

(d) A centromere

(e) All of these are critical characteristics of a YAC

54. After cytogenetic and linkage mapping, you have identified a relatively small area of a chromosome where your gene of interest is likely to reside. You decide to attempt to use a YAC contig to map the gene by positional cloning. How might you identify the clone that may contain the gene?

(a) The clone will also contain the nearest marker mapped by linkage analysis

(b) The DNA sequence of the clone will contain an ORF (open reading frame)

(c) Cytological hybridization of the clone produces a different pattern than that of the nearest mapped marker

(d) All of these

(e) None of these

55. Which of these is a reasonable use for the contigs and clones generated in the process of obtaining a physical map of the human genome?

(a) Correlation of physical and linkage map distances to determine recombination frequency

(b) Identification of all the known alleles of a single disease-causing gene

(c) Insertion into cells affected with genetic disease in the process of gene therapy

(d) All of these

(e) None of these

56. Sequencing of genomes other than humans is potentially valuable because:

(a) Disorders of cellular function can be studied in relatively simple model systems

(b) Comparison of gene sequences between different species can allow prediction of disease-causing mutations

(c) Genes may have a similar function in other species, giving us a place to start with functional analysis of similar genes

(d) None of these

Fill in the blanks

1. Bacteria that are unable to synthesize chemicals that allow formation of ice crystals have been engineered. Commercial fruit farmers can spray this bacterial strain on their fruit plants, with the idea that they will compete with unmodified bacteria and reduce frost damage. This is an example of __________

(a) Bioremediation (b) Biological control

(c) Biodegradation (d) Bioeradication

(e) None of the above

2. In gene knockout technology ______________

(a) A gene is inserted by nonhomologous recombination

(b) A mutant gene is replaced by a functional copy

(c) A functional gene is replaced by a mutant copy
(d) A functional gene is inserted in addition to the mutant copy
(e) None of the above

3. Identify the best description of a stem cell ______________
(a) A cell that has differentiated into a specialized cell type
(b) A cell that is capable of division
(c) A cell that remains undifferentiated
(d) A cell that can grow in cell culture
(e) None of the above

4. Diseases amenable to treatment by gene therapy ______________
(a) Must be relatively mild
(b) Must be caused by a defect in cells that are clinically accessible
(c) Must have other types of treatments available in case gene therapy fails
(d) All of these
(e) None of these

5. Series of codons from a single strand of DNA sequence which can be "read" in three different ways, depending on whether one starts at the first nucleotide position, the second or third ______________
(a) Reading Frame (RF) (b) Alternative Splicing
(c) Open Reading Frame (ORF) (d) mRNA Processing
(e) None of the above

6. Modifications made after transcription has been completed (in eukaryote cells), including the addition of a 5'cap and 3' polyA tail and the removal of noncoding regions________
(a) Alternative Splicing
(b) Annotation
(c) mRNA Processing
(d) Ab initio gene prediction
(e) None of the above

7. Transcribed as part of the normal gene primary transcript, but these sequences are not found in the functional mRNA outside of the nucleus____________

(a) Exon (b) mRNA

(c) Intron (d) Reading Frame

(e) None of the above

8. A section of a sequence that begins with an initiation (methionine, ATG) codon and ends with a nonsense (stop) codon ____________

(a) Untranslated Region (UTR) (b) Open Reading Frame (ORF)

(c) Exon (d) Intron

(e) None of the above

9. The portion of an mRNA from the 5' end to the position of the first codon used in translation, and also the portion of an mRNA from the 3' end of the mRNA to the position of the last codon ____________

(a) Exon (b) Intron

(c) Open Reading Frame (d) Untranslated Region (UTR)

(e) None of the above

10. A region of a gene that is present in the final functional mRNA transcript; these segments of the mRNA are spliced together and translated into protein ____________

(a) Alternative Splicing (b) Exon

(c) Intron (d) Untranslated Region (UTR)

(e) None of the above

11. Cytogenetic mapping ____________

(a) Requires that the gene have been cloned if in situ hybridization is to be used

(b) Is a high resolution method of gene localization

(c) Can be used to determine relative order of genes located very close to each other

(d) All of these

(e) None of these

12. A molecular marker which is amplified by PCR and is polymorphic by length is a(n) ___

(a) Restriction fragment length polymorphism (RFLP)

(b) Variable number of tandem repeats site (VNTR)

(c) Amplified fragment length polymorphism (AFLP)

(d) Single nucleotide polymorphism (SNP)

(d) None of the above

13. A polymorphism is ______________

(a) Any change in the DNA sequence

(b) The most common variation of a gene or marker sequence

(c) The least common variation of a gene or marker sequence

(d) A variation of gene or marker sequence present in >1% of the population

(e) None of these

14. Sequencing of genomes other than humans is potentially valuable because ___________

(a) Disorders of cellular function can be studied in relatively simple model systems

(b) Comparison of gene sequences between different species can allow prediction of disease-causing mutations

(c) Genes may have a similar function in other species, giving us a place to start with functional analysis of similar genes

(d) None of these

(e) None of the above

True/False

1. Recombinant microorganisms are fundamentally different from nonrecombinant strains.

(a) True (b) False

2. Transgenic organisms contain recombinant DNA integrated into their genome.

(a) True (b) False

3. The ability to grow embryonic stem cells in the laboratory has made gene replacement technology in mammals feasible.

(a) True (b) False

4. A pluripotent cell is capable of dividing and differentiating into every type of cell in the adult, and thus can form a new individual.

(a) True (b) False

5. DNA fingerprinting provides information about predisposition for genetic disease.

(a) True (b) False

6. Knockout mice can serve as model organisms for study of human disease.

(a) True (b) False

7. Gene therapy by gene addition is best able to treat diseases in which a gene function is missing.

(a) True (b) False

8. An AFLP is an example of a sequence tagged site.

(a) True (b) False

9. Changes in restriction enzyme sites can be used as polymorphic markers.

(a) True (b) False

10. In situ hybridization is done on fragments of specific chromosomes.

(a) True (b) False

11. Map units used in linkage mapping correlate with a specific number of base pairs.

(a) True (b) False

12. RFLPs are inherited in a simple Mendelian fashion and display codominance.

(a) True (b) False

13. One of the goals of the Human Genome Project is to address ethical issues arising from use of the information gathered.

(a) True (b) False

14. Isolation of individual chromosomes can be used to map genes.

(a) True (b) False

15. The reduced numbers of progeny with the pattern of (5,000bp, 4,300bp, 2,300bp, and 2,000bp bands) or (5,000bp, 4,300bp, and 3,350bp bands) indicates a elective disadvantage for these genotypes.

    (a) True (b) False

16. An AFLP is an example of a sequence tagged site.

    (a) True (b) False

17. Changes in restriction enzyme sites can be used as polymorphic markers.

    (a) True (b) False

18. Map units used in linkage mapping correlate with a specific number of base pairs.

    (a) True (b) False

19. One of the goals of the Human Genome Project is to address ethical issues arising from use of the information gathered.

    (a) True (b) False

20. Isolation of individual chromosomes can be used to map genes.

    (a) True (b) False

21. The reduced numbers of progeny with the pattern of (5,000bp, 4,300bp, 2,300bp, and 2,000bp bands) or (5,000bp, 4,300bp, and 3,350bp bands) indicates a selective disadvantage for these genotypes.

    (a) True (b) False

22. RFLPs are inherited in a simple Mendelian fashion and display codominance.

    (a) True (b) False

Answers

Multiple Choice Question

1	2	3	4	5	6	7	8	9	10	11	12	13	14	15
D	C	B	D	C	C	B	B	D	D	B	B	B	A	C
16	17	18	19	20	21	22	23	24	25	26	27	28	29	30
B	D	C	B	B	A	D	E	A	C	C	B	D	A	C

31	32	33	34	35	36	37	38	39	40	41	42	43	44	45
C	B	B	A	D	C	A	B	B	B	C	D	A	D	E
46	47	48	49	50	51	52	53	54	55	56				
A	C	C	B	D	C	C	B	B	A	C				

Fill in the blanks

1	2	3	4	5	6	7	8	9	10	11	12	13	14	
B	C	C	B	A	C	C	B	D	B	A	C	D	C	

True/False

1	2	3	4	5	6	7	8	9	10	11	12	13	14	15
B	A	A	B	B	A	A	A	A	B	B	A	A	A	B
16	17	18	19	20	21	22								
A	A	B	A	A	B	B								

❑❑❑

Chapter – 19

Functional Genomics, Proteomics and Bioinformatics

Functional Genomics

- Expressed genes can be identified in a cDNA library
- A microarray can identify genes that are transcribed
- The coordinate regulation of many genes is revealed by a DNA microarray analysis

Proteomics

- The proteome is much larger than the genome
- 2D gel electrophoresis is used to separate a mixture of cellular proteins
- Mass spectrometry is used to identify proteins
- Protein microarrays can be used to study protein expression and function

Bioinformatics

- Sequence files are analyzed by computer programs
- The scientific community has collected sequence files and stored them in large computer databases
- Different computational strategies can identify functional genetic sequences

- Several computer-based approaches can identify structural genes within a nucleotide sequence
- Computer programs can identify homologous sequences
- A simple dot matrix can compare the degree of similarity between two sequences
- A database can be searched to identify homologous sequences
- Genetic sequences can be used to predict the structure of RNA and proteins

Multiple Choice Question

1. What would be a likely explanation for the existence of pseudogenes?
   (a) Gene duplication
   (b) Gene duplication and mutation events
   (c) Mutation events
   (d) Unequal crossing over
   (e) Evolutionary pressure
2. The collection of proteins that can be produced by a given species is:
   (a) Considered that species' genetic complement
   (b) Correlated with the size of the organism
   (c) Called the proteome
   (d) All of these
   (e) None of these
3. Which of these questions could be answered using subtractive hybridization?
   (a) What genes are required for basic cellular functions?
   (b) What site is bound by a kidney specific transcription factor?
   (c) What genes are expressed in cardiac but not skeletal muscle?
   (d) All of these
   (e) None of these

4. Which of these might be an advantage to genetic testing of individuals via microarrays?

(a) Many different potential mutations in a single gene could be tested at once

(b) Expression patterns of many different genes can be analyzed simultaneously

(c) Microarray analysis can provide information on relative levels of expression of particular genes

(d) All of these

(e) None of these

5. Which of these would not be a valid reason that use of microarray technology to differentiate between closely related bacterial species and subspecies is important?

(a) Certain strains of bacteria are more pathogenic than other related strains

(b) Some strains of bacteria are more active in bioremediation than other related strains

(c) Infection by different strains of bacteria may require different therapeutic approaches

(d) In many cases, critical information about characteristics of a bacterium causing an infection needs to be immediately available

(e) All of the above are valid reasons

6. Which of the following statements regarding the proteome is the most correct?

(a) A large proportion of the proteome is expressed by each cell of a species

(b) Levels of gene expression at the mRNA level generally correlate highly with levels of functional protein

(c) The proteins produced by a specific cell depend on cell type and environmental conditions

(d) All of the above

(e) None of the above

7. Which of these conclusions might be drawn from the results of a 2D gel electrophoresis experiment?
   (a) Levels of mRNA expression for two different genes are lower under one set of conditions than another
   (b) In a mutant cell, the lack of protein expression is due to production of unstable mRNA, which is rapidly degraded
   (c) A mutation prevents proper posttranslational modification of a protein
   (d) All of these are reasonable conclusions
   (e) None of these are reasonable conclusions
8. If computers were not able to access the entire genomic sequence of an organism, which of the following techniques might allow determination of the gene sequence that encodes a particular protein?
   (a) PCR amplification of related gene sequences
   (b) Hybridization of a genomic library with a degenerate probe
   (c) Production of synthetic peptides
   (d) All of these
   (e) None of these
9. Which of these concerns would apply to functional protein microarrays but not antibody microarrays?
   (a) Antibodies recognize relatively short peptide sequences from a protein, and presence of the same peptide sequence in several proteins is possible
   (b) Three dimensional structure of a protein is essential for correct function, and the process of microarray preparation may interfere with this structure
   (c) Certain functions, such as structural support of a cell, may not be amenable to analysis on a microarray
   (d) All of these apply to functional protein microarrays but not antibody microarrays
   (e) All of these apply equally to both types of microarrays

10. Which of the following pieces of information would not be required for a computer program to determine exon/intron structure of a gene?
    (a) A table correlating codons to amino acids
    (b) The nucleotide sequence of the mRNA
    (c) Splice site signals for the species under study
    (d) All of these pieces of information would be required. E. None of these pieces of information would be required. This would be a required piece of information; at least the start and stop codons must be recognized by the computer program to identify potential open reading frames
    (e) None of the above
11. How many potential open reading frames are present in a DNA sequence?
    (a) One (b) Three
    (c) Six (d) More than six
    (e) None of these
12. You have cloned a new gene from your favorite species of experimental organism, which has had very few genes cloned so far. Which of the following techniques would you expect to provide information about the function of this gene?
    (a) Comparison of regulatory sequences to other DNA sequences in this organism
    (b) Comparison of regulatory sequences to DNA sequences from other organisms
    (c) Comparison of the gene sequence to other DNA sequences in this organism
    (d) Comparison of the gene sequence to DNA sequences from other organisms
    (e) None of these
13. Why might you want to search a database for a protein motif?
    (a) A specific motif may impart a specific function to the molecule. You could then identify groups of proteins that may have similar functions
    (b) Presence of a specific motif in several proteins indicates that they are likely to be all from the same species

(c) Absence of a specific motif from one of a pair of otherwise similar proteins indicates that they are produced by alternative splicing of the same gene

(d) All of the above

(e) None of the above

14. Which of these would not be an example of sequence element?

(a) A recognition site for binding of a particular transcription factor

(b) A restriction endonuclease cut site

(c) An open reading frame

(d) A stop codon

(e) None of these

15. In an analysis of eukaryotic gene, you identify several nonoverlapping open reading frames, but they are not all in the same frame. Which explanation makes the most sense?

(a) By random chance, a second reading frame within the gene also has an open reading frame

(b) This gene includes introns which are not multiples of three

(c) This is a mutant allele that has had several small insertions

(d) All of these

(e) None of these

16. You clone a novel gene from your favorite experimental organism, and identify the sequence of the gene itself and the protein. You would like to know if this gene is regulated by any known transcription factor. Which technique would be best suited to provide you evidence to address this question?

(a) BLAST analysis of the protein sequence

(b) BLAST analysis of the DNA from the coding sequence

(c) BLAST analysis of the DNA from the upstream regions

(d) None of these

(e) All the above

17. The collection of proteins that can be produced by a given species is:

(a) Considered that species' genetic complement

(b) Correlated with the size of the organism

(c) Called the proteome

(d) All of these

(e) None of these

18. Which of these might be an advantage to genetic testing of individuals via microarrays?

(a) Many different potential mutations in a single gene could be tested at once

(b) Expression patterns of many different genes can be analyzed simultaneously

(c) Microarray analysis can provide information on relative levels of expression of particular genes

(d) All of these

(e) None of these

19. Which of these would not be a valid reason that use of microarray technology to differentiate between closely related bacterial species and subspecies is important?

(a) Certain strains of bacteria are more pathogenic than other related strains

(b) Some strains of bacteria are more active in bioremediation than other related strains

(c) Infection by different strains of bacteria may require different therapeutic approaches

(d) In many cases, critical information about characteristics of a bacterium causing an infection needs to be immediately available

(e) All of the above are valid reasons

20. The two most common processes that lead to production of multiple functional proteins from the same DNA sequence are:

(a) RNA editing and alternative splicing

(b) Protein folding and posttranslational covalent modifications

(c) Alternative splicing and posttranslational covalent modifications

(d) Posttranslational covalent modification and transcriptional regulation

(e) RNA editing and base modification

21. Which of the following statements regarding the proteome is the most correct?

(a) A large proportion of the proteome is expressed by each cell of a species

(b) Levels of gene expression at the mRNA level generally correlate highly with levels of functional protein

(c) The proteins produced by a specific cell depend on cell type and environmental conditions

(d) All of the above

(e) None of the above

22. In two-dimensional gel electrophoresis:

(a) Different forms of the same protein will tend to migrate at the same position

(b) Up to about a hundred different proteins can be distinguished from each other

(c) Proteins with similar functions will be located near each other

(d) All of these

(e) None of these

23. Which of these conclusions might be drawn from the results of a 2D gel electrophoresis experiment?

(a) Levels of mRNA expression for two different genes are lower under one set of conditions than another

(b) In a mutant cell, the lack of protein expression is due to production of unstable mRNA, which is rapidly degraded

(c) A mutation prevents proper posttranslational modification of a protein

(d) All of these are reasonable conclusions

(e) None of these are reasonable conclusions

24. If computers were not able to access the entire genomic sequence of an organism, which of the following techniques might allow determination of the gene sequence that encodes a particular protein?

(a) PCR amplification of related gene sequences

(b) Hybridization of a genomic library with a degenerate probe

(c) Production of synthetic peptides

(d) All of these

(e) None of these

25. Which of these concerns would apply to functional protein microarrays but not antibody microarrays?

(a) Antibodies recognize relatively short peptide sequences from a protein, and presence of the same peptide sequence in several proteins is possible

(b) Three dimensional structure of a protein is essential for correct function, and the process of microarray preparation may interfere with this structure

(c) Certain functions, such as structural support of a cell, may not be amenable to analysis on a microarray

(d) All of these apply to functional protein microarrays but not antibody microarrays

(e) All of these apply equally to both types of microarrays

26. Which of the following pieces of information would not be required for a computer program to determine exon/intron structure of a gene?

(a) A table correlating codons to amino acids

(b) The nucleotide sequence of the mRNA

(c) Splice site signals for the species under study

(d) All of these pieces of information would be required

(e) None of these pieces of information would be required

27. How many potential open reading frames are present in a DNA sequence?

(a) One

(b) Three

(c) Six

(d) More than six

(e) None of these

28. You have cloned a new gene from your favorite species of experimental organism, which has had very few genes cloned so far. Which of the following techniques would you expect to provide information about the function of this gene?

(a) Comparison of regulatory sequences to other DNA sequences in this organism

(b) Comparison of regulatory sequences to DNA sequences from other organisms

(c) Comparison of the gene sequence to other DNA sequences in this organism

(d) Comparison of the gene sequence to DNA sequences from other organisms

(e) None of these

29. Why might you want to search a database for a protein motif?

(a) A specific motif may impart a specific function to the molecule. You could then identify groups of proteins that may have similar functions

(b) Presence of a specific motif in several proteins indicates that they are likely to be all from the same species

(c) Absence of a specific motif from one of a pair of otherwise similar proteins indicates that they are produced by alternative splicing of the same gene

(d) All of the above

(e) None of the above

30. Which of these would not be an example of sequence element?

(a) A recognition site for binding of a particular transcription factor

(b) A restriction endonuclease cut site

(c) An open reading frame

(d) A stop codon

(e) None of these

31. In an analysis of eukaryotic gene, you identify several nonoverlapping open reading frames, but they are not all in the same frame. Which explanation makes the most sense?

(a) By random chance, a second reading frame within the gene also has an open reading frame

(b) This gene includes introns which are not multiples of three

(c) This is a mutant allele that has had several small insertions

(d) All of these

(e) None of these

32. Homologous genes:

(a) Would be expected to have very similar sequences in related organisms

(b) Would be expected to be more similar in distantly related organisms than in organisms that are closely related

(c) May have become similar to each other by random mutation

(d) All of these

(e) None of these

33. You clone a novel gene from your favorite experimental organism, and identify the sequence of the gene itself and the protein. You would like to know if this gene is regulated by any known transcription factors. Which technique would be best suited to provide you evidence to address this question?

(a) BLAST analysis of the protein sequence

(b) BLAST analysis of the DNA from the coding sequence

(c) BLAST analysis of the DNA from the upstream regions

(d) None of these

34. Secondary structure of RNA molecules:

(a) Depends on complementary base pairing

(b) Is generated by covalent bonding between sections of the RNA molecule

(c) Can be described as interactions between portions of the backbone of the molecule

(d) Does not have an impact on function of the molecule

Fill In the blanks

1. Two-dimensional PAGE are used to _________.
   (a) Separate DNA fragments
   (b) Separate RNA fragments
   (c) Separate different proteins
   (d) Observe a protein in two dimensions
   (e) Separate DNA from RNA
2. A cDNA library:
   (a) Can also be called an expressed sequence tag (EST) library
   (b) Consists of coding sequences from genes that are expressed
   (c) Is specific to the set of conditions under which the original mRNA was generated
   (d) All of these
   (e) None of these
3. The two most common processes that lead to production of multiple functional proteins from the same DNA sequence are:
   (a) RNA editing and alternative splicing
   (b) Protein folding and posttranslational covalent modifications
   (c) Alternative splicing and posttranslational covalent modifications
   (d) Posttranslational covalent modification and transcriptional regulation
   (e) RNA editing and base modification
4. In two dimensional gel electrophoresis:
   (a) Different forms of the same protein will tend to migrate at the same position
   (b) Up to about a hundred different proteins can be distinguished from each other
   (c) Proteins with similar functions will be located near each other
   (d) All of these
   (e) None of these

5. Homologous genes:
   (a) Would be expected to have very similar sequences in related organisms
   (b) Would be expected to be more similar in distantly related organisms than in organisms that are closely related
   (c) May have become similar to each other by random mutation
   (d) All of these
   (e) None of these
6. Secondary structure of RNA molecules:
   (a) Depends on complementary base pairing
   (b) Is generated by covalent bonding between sections of the RNA molecule
   (c) Can be described as interactions between portions of the backbone of the molecule
   (d) Does not have an impact on function of the molecule

True/False

1. Western blotting of 2D gel electrophoresis would be a high resolution method of demonstrating a change in a particular protein as a result of mutation.
   (a) True (b) False
2. A cluster analysis is the study of groups of genes that seem to be regulated together.
   (a) True (b) False
3. A single microarray may have a surface area of less than three square inches, yet may contain unique spots of tens of thousands of gene sequences.
   (a) True (b) False
4. The technique of subtractive hybridization allows identification of genes that are selectively activated under a certain set of conditions.
   (a) True (b) False

5. Examples of irreversible posttranslational covalent modification include acetylation and phosphorylation.

   (a) True (b) False

6. A computer program must be written in a specific language in order for the computer to understand what needs to be done.

   (a) True (b) False

7. Functional protein microarrays can identify protein functions only if their sequence is known.

   (a) True (b) False

8. Tandem mass spectrometry can be used to determine the specific amino acid sequence of a protein.

   (a) True (b) False

9. Identification of a gene that does not fit the typical patterns for eukaryotic gene structure would not have a dramatic effect on bioinformatics.

   (a) True (b) False

10. A multiple sequence alignment of related genes can identify amino acids required for protein function.

    (a) True (b) False

11. A single microarray may have a surface area of less than three square inches, yet may contain unique spots of tens of thousands of gene sequences.

    (a) True (b) False

12. A multiple sequence alignment of related genes can identify amino acids required for protein function.

    (a) True (b) False

13. Identification of a gene that does not fit the typical patterns for eukaryotic gene structure would not have a dramatic effect on bioinformatics.

    (a) True (b) False

14. A computer program must be written in a specific language in order for the computer to understand what needs to be done.

    (a) True (b) False

15. Functional protein microarrays can identify protein functions only if their sequence is known.

(a) True (b) False

16. Tandem mass spectrometry can be used to determine the specific amino acid sequence of a protein.

(a) True (b) False

17. Western blotting of 2D gel electrophoresis would be a high resolution method of demonstrating a change in a particular protein as a result of mutation.

(a) True (b) False

18. Examples of irreversible posttranslational covalent modification include acetylation and phosphorylation.

(a) True (b) False

19. A cluster analysis is the study of groups of genes that seem to be regulated together.

(a) True (b) False

Answers

Multiple Choice Question

1	2	3	4	5	6	7	8	9	10	11	12	13	14	15
B	C	C	D	D	C	E	B	C	B	C	D	A	C	B
16	17	18	19	20	21	22	23	24	25	26	27	28	29	30
D	C	D	B	C	C	E	E	B	C	B	C	D	A	C
31	32	33	34											
B	A	D	A											

Fill in the blanks

1	2	3	4	5	6
C	D	C	E	A	A

True/False

1	2	3	4	5	6	7	8	9	10	11	12	13	14	15
A	A	A	A	B	A	B	B	B	A	B	A	B	A	B
16	17	18	19											
B	A	B	A											

❑❑❑

Chapter – 20

Genomics and the Future

Genome Sequencing: A Continuation of Genetics

- The human genome project involves the assembly of a DNA sequence map for the 3 billion base pairs present in all 23 human chromosomes.
- Two groups, the International Consortium and Celera Genomics, used two opposite strategies to obtain a draft sequence of the human genome.
- The use of bacterial artificial chromosomes (BACs), shotgun clones, and positional cloning has yielded a detailed look at the human genome and is facilitating the identification of disease causing genes.

The Sanger Method of DNA Sequencing

- The Sanger method of DNA sequencing involves an enzymatic extension reaction to generate complementary copies of an unknown DNA sequence.
- Incorporation of "dideoxy" base causes chain termination and separation of extension products on polyacrylamide gels reveals the sequence bases in the unknown DNA.
- DNA sequencing occurs on a vast scale today through the use of automated DNA sequencing machines.

Technology Drives the Sequencing Effort

- Expressed sequenced tags (EST) technology enabled researchers to focus on protein encoding genes.
- The speed of the project depended upon the technological developments of automated DNA sequencing machines and DNA sequence alignment computer software.

Future Work

- Mycoplasm genitalium has the smallest known genome and represents a "near-minimal set of genes for independent life."
- Large percentages of the known genomes have an unknown function.
- Comparing genomes of different species indicate shared ancestry.
- Homologies between human genes and genes of other organisms may allow the other organisms to serve as test-beds for human therapies.

Multiple Choice Question

1. Shotgun cloning differs from the clone-by-clone method in which of the following ways?

(a) The location of the clone being sequenced is known relative to other clones within the genomic library in shotgun cloning

(b) Genetic markers are used to identify clones in shotgun cloning

(c) Computer software assembles the clones in the clone-by-clone method

(d) The entire genome is sequenced in the clone-by-clone method, but not in shotgun sequencing

(e) No genetic or physical maps of the genome are needed to begin shotgun cloning

2. Gene duplication has been found to be one of the major reasons for genome expansion in eukaryotes. In general, what would be the selective advantage of gene duplication?

(a) If one gene copy is nonfunctional, a backup is available

(b) Larger genomes are more resistant to spontaneous mutations

(c) Duplicated genes will make more of the protein product

(d) Gene duplication will lead to new species evolution

(e) None of the above

3. How are so many different antibodies produced from fewer than 300 major genes?

(a) Gene duplication

(b) Alternative splicing mechanisms

(c) The formation of polyproteins

(d) The formation of nonspecific B cells

(e) Recombination, deletions, and random assortment of DNA segments

4. Why was the first genome sequence considered a milestone?

(a) It proved that viruses contained nucleic acids

(b) It proved that DNA could be isolated from a virus

(c) It showed the coding sequences for all the proteins produced by a virus

(d) It proved that the coding sequences within a genome could be located and identified

(e) None of the above

5. How do we recognize where a gene begins within a long sequence of DNA nucleotides?

(a) Most genes begin with the nucleotide adenine

(b) The process of trial and error eventually leads investigators to locate where a gene begins

(c) A gene starts with a three-nucleotide "start" codon

(d) Genes begin at either end of a DNA sequence, never in the middle

(e) None of the above

6. When the first free-living organism, the bacterium Haemophilus influenzae, was sequenced in the mid-1990s, about what percentage of its genes were shown to have a known function?

(a) About half

(b) Around 10%

(c) Nearly all

(d) Less than 1%

(e) None of the above

7. Why did the first organism that had their genomes sequenced have comparatively small genome sizes?

   (a) Organisms that could be kept easily in the lab were the focus of early research

   (b) Until the advent of automated DNA sequencers, researchers had to use organisms with small genomes because the procedure was very time-consuming

   (c) Only organisms that reproduced rapidly were selected and these had small genomes

   (d) It was only coincidental that the first organisms to be sequenced had small genome sizes

   (e) None of the above

8. What researcher proposed "shotgun sequencing"?

   (a) Todd Golub
   (b) Frederick Sanger
   (c) Craig Venter
   (d) Stephen Fodor
   (e) None of the above

9. In the Sanger method of sequencing DNA, how does the researcher differentiate between the four nucleotide bases?

   (a) Each nucleotide base has a unique radioactive label and the results are read on X-ray film

   (b) Each nucleotide base is fluorescently colored and a spectrophotometer reads the results

   (c) Computers read the results of the gel electrophoresis

   (d) Time-consuming chemical assays distinguish one nucleotide base from another

   (e) None of the above

10. How many genes constitute the human genome?

    (a) More than 1 million
    (b) 4 Mb
    (c) About 30,000
    (d) 46
    (e) None of the above

11. What interesting or unexpected outcome(s) resulted from the Human Genome Project?

(a) Genes are not distributed evenly over the genome

(b) A human gene is fragmented

(c) There are four kinds of mRNA as there are genes

(d) All of these are correct

(e) None of the above

12. Which of these is not a type of noncoding human DNA?

(a) Structural DNA

(b) Repeated sequences

(c) Constitutive heterochromatin

(d) None of these. All are types of noncoding DNA

(e) None of the above

13. What is true about variation within the human genome?

(a) There is more variation among regional groups than between individuals

(b) There is tremendous variation among individual human genomes

(c) There is comparatively little variation between individuals

(d) Nothing is yet known of the variation within the human genome

(e) None of the above

14. Genome duplication has led to polyploidy in plants. What event(s) led to the genome in domestic wheat?

(a) Transposons from bacteria combined with genes from a grass species

(b) Two allopolyploid events occurred in the grass species Triticum.

(c) Two separate grass species exchanged DNA through sexual reproduction

(d) None of the above are correct

(e) None of the above

15. What occurred when the development-regulating Pax6 mouse gene was inserted into a fruit fly genome?
    (a) The Pax6 gene was expressed and a fruit fly eye formed on the leg of the fly
    (b) The Pax6 gene was evidently not expressed and nothing happened
    (c) The normal fly eyes began to degenerate
    (d) A mouse eye formed in the fruit flies
    (e) None of the above

Fill in the blanks

1. CpG islands and codon bias are tools used in eukaryotic genomics to __________.
    (a) Identify open reading frames
    (b) Differentiate between eukaryotic and prokaryotic DNA sequences
    (c) Find regulatory sequences
    (d) Look for DNA-binding domains
    (e) Identify a gene's function
2. As the complexity of an organism increases, all of the following characteristics emerge except __________.
    (a) The gene density decreases
    (b) The number of introns increases
    (c) The gene size increases
    (d) An increase in the number of chromosomes
    (e) Repetitive sequences are present
3. All of the genes and other DNA of an organism is called its ________
    (a) Open reading frame
    (b) Genome
    (c) Intron
    (d) Multigene family
    (e) None of the above

4. Groups of distinctly different genes that often occur together in a cluster are called ___

(a) Single-copy genes
(b) Segmental duplications
(c) Multigene families
(d) Tandem clusters
(e) None of the above

5. Whole blocks of genes that have been copied from one chromosome to another are referred to as ___________

(a) Single-copy genes
(b) Segmental duplications
(c) Multigene families
(d) Tandem clusters
(e) None of the above

6. Bits of DNA that literally jump from one chromosome to another are ________

(a) Introns
(b) Transposable elements
(c) Pseudogenes
(d) Tandem clusters
(e) None of the above

7. Comparisons of the genomes of different organisms suggests that ______

(a) The larger the organism, the less noncoding DNA is found in the genome
(b) Genes can amplify within chromosomes of a given organism
(c) There is no movement of genes between organisms
(d) Genes have moved laterally between organisms
(e) None of the above

8. The conservation of blocks of genes that have moved to different chromosomes is called ___________

(a) Hitchhiking DNA
(b) Lateral gene transfer
(c) Conserved synteny
(d) Gene expression
(e) None of the above

9. A discrete collection of gene fragments on a stamp-sized chip is called a ________

(a) Reference sequence (b) SNP profile

(c) Gene microarray (d) Semiconductor

(e) None of the above

10. A fast-growing new field of science that seeks to predict the structure of a protein from its nucleotide sequence is called ________

(a) Genomics (b) Proteomics

(c) Bioinformatics (d) None of these is correct

(e) None of the above

Answers

Multiple Choice Question

1	2	3	4	5	6	7	8	9	10	11	12	13	14	15
E	A	E	D	C	A	B	C	A	C	D	D	B	B	A

Fill in the blanks

1	2	3	4	5	6	7	8	9	10
A	D	B	C	B	B	D	C	C	C

❑❑❑

Chapter – 21

Genetics of Cancer

Genetic Analysis of Diseases

- A genetic basis for a human diseases may be suggested from a variety of observations
- Inheritance patterns of human diseases may be determined via pedigree analysis
- Many genetic disorders are heterogeneous
- Genetic testing can identify many inherited human diseases
- Prions are infectious particles that alter protein function posttranslationally

Genetic Basis of Cancer

- Certain viruses can cause cancer by carrying viral oncogenes into the cell
- DNA isolated from malignant mouse cells can transform normal mouse cells into malignant cells
- Many oncogenes have abnormalities that affect proteins involved in cell division pathways
- Genetic changes in cellular proto-oncogenes convert them to cellular oncogenes
- Tumor suppressor genes play a role in preventing the proliferation of cancer cells

- The vertebrate p53 gene is a master tumor-suppressor gene that senses DNA damage
- Many types of tumor suppressor genes can promote cancer when their function is lost
- Most forms of cancer involve multiple genetic changes leading to malignancy
- Inherited forms of cancers may be caused by defects in tumor-suppressor genes and DNA repair genes

Cancer as a Genetic Disorder

Cancer in Families

- The search for cancer genes in families has given way to a genomic approach to identifying cancer genes.
- Carcinogens, most of which are mutagens, are cancer producing agents.
- More than 100 oncogenes and 30 tumor suppressor genes have been identified.
- DNA microarrays can be used to identify and distinguish different types of tumors.

A Loss of Cell Cycle Control

- Cancer is caused by a loss of cell division control.
- Genes involved in the onset and progression of cancer include those encoding growth factors, transcription factors, DNA repair genes, and telomerase.

Inherited Versus Sporadic Cancer

- Several mutations may contribute to the development of cancer.
- Most cancers are sporadic and therefore, the causative mutation occurs in cells of the affected tissue.
- Cancer may develop when an environmental trigger causes mutations in a somatic cell or when a somatic mutation compounds an inherited susceptibility.

Characteristics of Cancer Cells

- Cancer cells do not respond to normal cell cycle control signals and as a result they grow vigorously and continuously.
- The cancer phenotype is heritable and transmitted to daughter cells.
- A cancer cell differs from normal cells in that it is transplantable (can be removed grow in a healthy animal), dedifferentiated (less specialized), lacks contact inhibition (will pile up on one another), and has surface structures that allow it to invade healthy tissue and metastasize (spread).

Genes that Cause Cancer

Oncogenes

- Many genes that normally control the cell cycle are proto-oncogenes.
- Proto-oncogenes are transformed into oncogenes by mutations that increase the expression or alter the biological activity of their protein product.

Tumor Suppressors

- Mutations in tumor-suppressor genes that cause cancer are deletions causing loss of gene function.
- Loss of tumor-suppressor genes enables cells to ignore extracellular constraints on cell division.
- The RB, p53, and BRCA1 genes encode tumor suppressors.

A Series of Genetic Changes Causes Some Cancers

A Rapidly Growing Brain Tumor

- Astrocytoma and FAP are two cancers that require several mutations to develop.
- A series of genetic alterations recorded for astrocytoma involve loss of function for both p53 alleles, loss of interferon and tumor suppressor genes on chromosome 9, activation of an oncogene on chromosome 7, and loss of chromosome 10.

Colon Cancer

- FAP begins in early childhood with precancerous growths called polyps.
- An autosomal dominant pattern of inheritance is associated with FAP and it occurs with a frequency of one in 5000 people in the United States.
- A dominant mutation in the APC gene on chromosome 5 is the first step in the multi-step model for colon cancer.
- Subsequent mutations in chromosomes 12p (Ras), 18q (DCC) and 17p (p53) cause the transformation of the polyp into a tumor.

Cancer Prevention and Treatment

Diet-Cancer Associations

- Epidemiological studies are used to identify risk factors in a population.
- Avoiding fats along with eating more fruits, vegetables, and whole grain cereals is associated with a lower cancer risk.

Diagnosing and Treating Cancer

- Testing can identify family members at risk of inherited cancers.
- Treatments for cancer target the characteristics of cancer cells.
- Surgery removes tumors.
- Chemotherapy and radiation nonselectively destroy rapidly dividing cells.
- Newer treatments target receptors on cancer cells, block telomerase, trigger redifferentiation, or attack a tumor's blood supply.

Multiple Choice Question

1. Which of these factors does not provide evidence that the disorder may have a genetic cause?
   (a) An increased frequency of the disorder in genetic relatives of an affected individual than in the general population
   (b) A higher concordance in identical twins than in fraternal twins

(c) Individuals that live with an affected individual are more likely to also exhibit the disorder

(d) All of these indicate a genetic cause

(e) None of these indicate a genetic cause

2. Which of these statements would indicate that the trait in question has a genetic component?

(a) Several adopted individuals living in the same family all have exceptionally strong fingernails

(b) In the last two weeks, every child at your daughter's daycare has had a cold

(c) Your brother, two cousins, and a great uncle all have been diagnosed with Duchenne muscular dystrophy, a disorder that affects about 1/7000 individuals

(d) All of these indicate a genetic component

(e) None of these indicate a genetic component

3. Which of these characteristics of Tay-Sachs indicates that it is likely to be genetic?

(a) The frequency of the disorder is approximately 1/3,600 in the Ashenazi Jewish population, but much less common in other Caucasian populations

(b) The disorder leads to death at a very young age

(c) Tay-Sachs often appears in a pedigree with no known family history of the disorder

(d) All of these

(e) None of these

4. Which of the following observation is most conclusive that Gaucher disease is inherited in an autosomal recessive pattern? (All these observations are of a single family, who all live together).

(a) Brothers David and Joshua are both affected with Gaucher disease

(b) Mom has no family history of Gaucher disease

(c) Mom and Dad are both unaffected with Gaucher disease

(d) The boys experience different severities of the disease

(e) None of these indicate an autosomal recessive pattern

5. Which of the following would be most likely to be inherited in an autosomal recessive pattern in humans?
   (a) A mutant gene has a new function that creates a phenotype
   (b) A deletion on the X chromosome inactivates a gene
   (c) A point mutation in a regulatory element reduces the level of gene expression of a gene on chromosome 12
   (e) None of these
   (e) None of the above
6. Why are males more likely to exhibit an X linked recessive trait than females?
   (a) Most genes on the X chromosome are not shared with the Y
   (b) Male sex hormones affect expression of genes from the X chromosome
   (c) X linked traits tend to control male secondary sexual development
   (d) All of these
   (e) None of these
7. A couple comes to your genetic counseling clinic concerned that they might have a second child affected with the same X-linked recessive disease displayed by their son. Neither of the parents displays the trait. What do you tell them?
   (a) They have a ¼ chance of having another affected child. Since it's a recessive trait, they must both be carriers
   (b) They have a ¼ chance of having another affected child but only if the child is a boy. They could try to select "girl" sperm to avoid the disease
   (c) They have a ½ chance of having another affected child, but only if the child is a girl. They could try to select "boy" sperm to avoid the disease
   (d) They are not at risk of having another affected child
   (e) None of the above
8. Which of these could be considered a genetic test?
   (a) A midwife licks the newborn's forehead to see if it tastes salty. If it does, she suggests to the family that the child may have cystic fibrosis

(b) A man of Ashkenazi Jewish descent submits a tissue sample to determine if his level of protein produced by the gene affected in Gaucher disease is normal or reduced

(c) An African American woman submits a DNA sample to find out if she is a carrier for the allele that causes sickle cell anemia

(d) All of these

(e) None of these

9. Which definition of cancer is the most accurate?

(a) A group of diseases characterized by uncontrolled cell division

(b) A disorder in which cells produce a localized tumor

(c) A disorder in which tissues of the body are replaced by nonfunctional cells

(d) All of these

(e) None of these

10. Which of these would be likely to lead to oncogenesis?

(a) Reduced levels of function of a repressor of cell division

(b) A mutation that causes the growth hormone receptor to be constitutively active

(c) A mutation that prevents activated Ras from being deactivated

(d) All of these

(e) None of these

11. Genetic changes that can convert proto-oncogenes to oncogenes include all of the following EXCEPT:

(a) Point mutations

(b) Increase in the copy number of the proto-oncogene

(c) A change in the relationship between a coding region and regulatory elements

(d) Deletion of the proto-oncogene

(e) All of these can convert proto-oncogenes to oncogenes

12. Which of these factors might influence the creation of translocations known to cause chronic myelogenous leukemia (the Philadelphia chromosome) and Burkitt's lymphoma?
    (a) The sites of translocation show a lower frequency of breakage than other chromosomal sites
    (b) Sites on non-homologous chromosomes may have enough similarity of sequence to stimulate "semi-homologous" recombination
    (c) Sister chromatid exchange can occur during mitosis
    (d) All of these
    (e) None of these
13. Which of these are likely as a result of the two hit model?
    (a) Predisposition to cancer is inherited as a dominant trait
    (b) Tumor formation is inherited as a dominant trait
    (c) Individuals who have inherited a mutant copy of a tumor suppressor are more likely to develop tumors, but they usually occur at later ages of onset
    (d) All of these
    (e) None of these
14. What is the normal interaction between the Rb protein and the E2F transcription factor?
    (a) Rb inhibits E2F activity, effectively preventing cell division
    (b) Rb enhances E2F activity, effectively preventing cell division
    (c) Rb inhibits E2F activity, increasing the rate of cell division
    (d) Rb enhances E2F activity, increasing the rate of cell division
    (e) None of the above
15. Which of the following is a common type of tumor suppressor genes?
    (a) Genes that increase the rate of the cell cycle or stimulate cell division
    (b) Genes involved in performing DNA synthesis
    (c) Genes involved in maintenance of cell cycle checkpoints
    (d) All of these
    (e) None of these

16. Why are cancer cells likely to accumulate mutations, including changes in chromosome number, over time?

(a) They are less likely to attach chromosomes correctly to the mitotic spindle

(b) They are more likely to make mistakes in DNA synthesis

(c) They are allowed to divide despite potential errors in DNA synthesis or spindle assembly

(d) All of these

(e) None of these

17. Why is familial retinoblastoma known to be inherited as an autosomal dominant trait whe the responsible gene is recessive?

(a) This recessive gene needs only one copy to express this phenotype

(b) People with this mutation are always homozygous for this allele

(c) If you are born with one allele mutated in every cell, the chance that the other allele will be mutated in one of the retinal cells is extremely high

(d) Retinoblastoma was mischaracterized originally when it was called an autosomal dominant trait

(e) None of the above

Fill in the blanks

1. Alteration of gene sequences in individuals affected by genetic disease ________

(a) Demonstrates the importance of environmental factors affecting disease expression

(b) Can be demonstrated by gene cloning and DNA sequencing

(c) Are the first line of evidence pursued to determine if the disease is in fact genetic

(d) All of these

(e) None of these

2. Jeremy's grandfather, uncle, and sister all have been diagnosed with the same genetic disorder. The most likely pattern of inheritance for this disorder is ________

   (a) Autosomal dominant, since it is present in all three generations

   (b) Autosomal recessive, since Jeremy's parents are unaffected

   (c) X-linked recessive, since it affects more males than females

   (d) You cannot determine the pattern of inheritance from this information

   (e) None of the above

3. Genetic heterogeneity could be described as ____________

   (a) A population in which there are several mutant alleles for a specific gene

   (b) A situation where different alleles of a gene cause different severity of disease

   (c) A situation where people with the same genotype might have different phenotypes

   (d) A situation where people with the same phenotype have different genotypes

   (e) None of the above

4. Genes which promote cancer are called _____________

   (a) Tumor suppressors
   (b) Oncogenes
   (c) Growth factors
   (d) Malignancy enhancers
   (e) None of the above

5. The first oncogene to be identified __________________

   (a) Was c-src

   (b) Causes sarcomas in chickens

   (c) Is normally expressed at high levels in eukaryotic cells

   (d) All of these

   (e) None of these

6. When a cell is stimulated by a growth hormone ___________
   (a) The presence of the hormone directly activates gene expression
   (b) A cascade of protein interactions leads to changes in gene expression
   (c) Cell surface receptors are deactivated
   (d) All of these
   (e) None of these
7. The role of a tumor suppressor gene _______________
   (a) Is to encode a protein that stimulates cell division
   (b) Is to prevent mistakes in DNA replication
   (c) Is to allow cell division only under appropriate conditions
   (d) All of these
   (e) None of these
8. Oncogenes _________.
   (a) Are both derived from proto-oncogenes and induce uncontrolled cell proliferation
   (b) Are tumor suppressor genes
   (c) Are derived from proto-oncogenes
   (d) Induce uncontrolled cell proliferation
   (e) Are viral genes
9. The function of the ras gene family involves _________.
   (a) Signal transduction
   (b) Tumor suppression
   (c) Chromosome segregation
   (d) Cell cycle control
   (e) Chromatin condensation
10. Cancer is best described as _________.
   (a) A genetic disorder at the cellular level
   (b) A genetic disease at the somatic cell level
   (c) A genetic disease at the gametic cell level
   (d) An environmental disease at the gametic cell level
   (e) A bacterial disease at the somatic cell level

11. Apoptosis represents genetically programmed _________.
    (a) Determination
    (b) Tumor suppression
    (c) Cell death
    (d) Proto-oncogene conversion
    (e) Differentiation
12. Ionizing radiation, viruses, and chemicals are all examples of _________.
    (a) Carcinogens
    (b) Precursors to oncogenes
    (c) Tumor suppressors
    (d) Environmental hazards
    (e) None of the above
13. _________ and _________ are two classes of proteins known to regulate the cell cycle checkpoints.
    (a) Protein kinases/cyclins
    (b) Cyclin-dependent kinases/cyclins
    (c) p53/polymerases
    (d) Ligase/polymerases
    (e) Protein kinases/ATP synthase

True/False

1. The breakpoints in the bcr and abl genes that lead to translocation and creation of the Philadelphia chromosome demonstrate an unusually high level of chromosome breakage.
    (a) True (b) False
2. Activation of a cell signaling pathway in absence of a growth factor signal can lead to uncontrolled cell division.
    (a) True (b) False

3. Rous sarcoma virus (RSV) causes cancer by integrating into genes that normally control cell division and eliminating their function.

(a) True (b) False

4. Prenatal genetic testing requires collection of fetal cells, usually by amniocentesis or chorionic villus sampling.

(a) True (b) False

5. p53 will stimulate a badly damaged cell to undergo apoptosis rather than proceeding through the cell cycle.

(a) True (b) False

6. The difference between benign tumors and malignant cancer is the type of mutation.

(a) True (b) False

7. Most human traits are influenced by both genetic and environmental factors.

(a) True (b) False

8. Identification of a genetic disorder in a model species allows analysis of that disorder that can then be applied to humans.

(a) True (b) False

9. For most dominant genetic diseases, homozygous dominant individuals are more severely affected than those that are heterozygous.

(a) True (b) False

10. You are certain to know someone who has undergone genetic testing.

(a) True (b) False

Answers

Multiple Choice Question

1	2	3	4	5	6	7	8	9	10	11	12	13	14	15
C	C	A	C	C	A	B	D	A	D	D	B	A	A	C
16	17													
C	C													

Fill in the blanks

1	2	3	4	5	6	7	8	9	10	11	12	13
B	B	D	B	B	B	C	A	A	A	C	A	B

True/False

1	2	3	4	5	6	7	8	9	10
A	A	B	A	A	B	A	A	A	A

❑❑❑

Chapter – 22

Developmental Genetics

Invertebrate Development

- The early stages of embryonic development determine the pattern of structures in the adult organism
- The study of Drosophila mutants with disrupted development patterns has identified genes that control development
- The generation of a body pattern depends on the positional information that each cell receives during development
- The gene products of maternal effect genes are deposited asymmetrically into the oocyte and establish the anteroposterior and dorsoventral axes at a very early stage of development
- Gap, pair-rule, and segment-polarity genes act sequentially to divide the Drosophila embryo into segments
- The expression of homeotic genes controls the phenotypic characteristics of segments
- The developmental fate of each cell in the nematode Caenorhabditis elegans is known
- Heterochronic mutations disrupt the timing of developmental changes in C. elegans

Vertebrate Development

- Researchers have identified homeotic genes in vertebrates

- Genes that encode transcription factors also play a key role in cell differentiation

Plant Development

- Plant growth occurs from meristems that are formed during embryonic development
- Plant homeotic genes control flower development

Sex Determination in Animals and Plants

- In Drosophila, sex determination involves a regulatory cascade that includes alternative splicing
- In C. elegans the ratio of X chromosomes to autosomes initiates a regulatory cascade that determines sex
- In mammals, the Sry gene on the Y chromosome determines maleness
- In sexually dimorphic plants, the male plant is usually heteromorphic

Multiple Choice Question

1. Which of these would NOT be an example of positional information?

(a) A specific cell in the embryo knows that it needs to divide into daughter cells

(b) Cells at the anterior end of the embryo develop into structures of the head

(c) Cells between the digits of the human hand die by apoptosis

(d) Cells that form the germline migrate from their initial position to colonize the gonads

(e) None of these (all are examples of positional information)

2. A mutation eliminating the function of which of these genes is most likely to create an embryo with two abdominal regions?

(a) Bicoid

(b) Nanos

(c) Torso

(d) Toll

(e) Abdominal a

3. Activation of hunchback by Bicoid demonstrates which of the following concept?
   (a) Cell:cell recognition
   (b) Homeotic mutation
   (c) Critical threshold concentration
   (d) Segmentation
   (e) None of these
4. Which of these would be an example of a homeotic phenotype?
   (a) The wings are shrunken and useless
   (b) The eyes of the fly are brown instead of the normal red
   (c) The first abdominal segment has legs
   (d) All of these
   (e) None of these
5. If college students were cells, which of these would be determined but not differentiated?
   (a) A freshman student who is undecided as to his major
   (b) A freshman student that has declared a major
   (c) A senior student that is nearing graduation
   (d) An alumnus that has begun her career
   (e) None of the above
6. In addition to potentially similar effects on pattern formation, Hox genes in vertebrates and HOM-C (homeotic) genes in Drosophila have which of the following similarities?
   (a) Both sets of genes are located in a single chromosomal cluster in their respective species
   (b) Each gene is known to contain a leucine zipper motif, leading to DNA binding of dimers of the encoded protein
   (c) The linear order of the genes along the chromosome is reflected in the anterior limits of expression of the genes along the AP axis of the animal
   (d) All of these
   (e) None of these

7. Which statement comparing plant and animal development is the most correct?
   (a) Organization of both plants and animals occurs around specific axes
   (b) Cell migration is a major feature of pattern formation in both groups
   (c) Plants cells have relatively limited potential, compared to animal cells, which are totipotent
   (d) There are no critical differences between plant and animal development
   (e) There are no important similarities between plant and animal development
8. What function do homeotic genes in plants and animals have in common?
   (a) They activate the same group of target genes in both groups
   (b) They encode identical proteins in each group
   (c) They are transcriptional activators that regulate other genes
   (d) They prevent cell division when they are active
   (e) None of the above
9. Which of the following events is the primary mechanism used for sex determination in Drosophila?
   (a) Male-determining genes on the Y chromosome function as transcriptional activators of genes that control male development
   (b) A cascade of transcription factor activation specifies development in the male pathway; lack of activation of this cascade will result in a female
   (c) Alternative splicing of specific gene products leads to development of a female if two X chromosomes are present
   (d) All of these
   (e) None of these
10. A loss of function mutation in which of these genes would be most likely to lead to an XX male phenotype?
   (a) SXL (b) FRU
   (c) MSL-2 (d) All of these
   (e) None of these

11. How is sex determination in C. elegans similar to sex determination in Drosophila?
    (a) In both cases, the end result of the process is either a male or a hermaphrodite
    (b) In both cases, alternative splicing is a major mechanism of control of gene activity
    (c) In both cases, expression of genes on the X chromosome is reduced in the process of dosage compensation
    (d) In both cases, the ratio of X chromosomes to autosomes determines which pathway the individual will follow
    (e) None of the above

12. In which of this species is an X chromosome required to produce a normal male?

    i. Drosophila   ii. C. elegans
    iii. Mus musculus (mouse)

    (a) III only   (c) I only
    (d) I and II   (e) I, II, and III
    (e) II and III

13. Which of the following statement regarding mammalian sex determination is most likely to be correct?
    (a) An XX male is more likely to be fertile than an XY female
    (b) An XY female is more likely to be fertile than an XX male
    (c) Loss of function mutations in several different genes are required to produce an XY female
    (d) All of the above are correct
    (e) None of the above are correct

14. Which of these mechanisms is common to sex determination in Drosophila, C. elegans, mammals, and plants?
    (a) Transcriptional regulation
    (b) Alternative splicing
    (c) Environmental influences

(d) Y linked genes that determine maleness
(e) None of the above

15. The most common pattern of sex phenotypes in plants is:
(a) Monomorphic
(b) Hermaphroditic
(c) Dioecious
(d) Sexually dimorphic
(e) None of the above

16. In regard to development, the term pattern refers to:
(a) The proteins produced by an individual cell
(b) The stages of development through which an organism proceeds
(c) The arrangement of the regions of the body of an organism
(d) The process of differentiation of specific cell types

17. In Drosophila, the pattern is established:
(a) During production of the oocyte
(b) As a result of environmental interactions with the oocyte
(c) During gastrulation
(d) After embryogenesis has completed
(e) None of these

18. Segmentation genes:
(a) Can be grouped into three major classes
(b) Control the development of the segments of the Drosophila embryo
(c) Often produce embryonic lethal phenotypes when mutant
(d) All of these
(e) None of these

19. Which of these would NOT be an example of positional information?
(a) A specific cell in the embryo knows that it needs to divide into daughter cells
(b) Cells at the anterior end of the embryo develop into structures of the head
(c) Cells between the digits of the human hand die by apoptosis

(d) Cells that form the germline migrate from their initial position to colonize the gonads

(e) None of these (all are examples of positional information)

20. A morphogen:

(a) Conveys positional information to cells within the embryo

(b) Is typically present in similar amounts in all cells

(c) Functions only at very low concentrations

(d) All of these

(e) None of these

21. A mutation eliminating the function of which of these gene is most likely to create an embryo with two abdominal regions?

(a) Bicoid (b) Nanos

(c) Torso (d) Toll

(e) Abdominal A

22. Activation of hunchback by bicoid demonstrates which of the following concept?

(a) Cell:cell recognition

(b) Homeotic mutation

(c) Critical threshold concentration

(d) Segmentation

(e) None of these

23. In Drosophila development, an embryonic segment:

(a) Is the same thing as a parasegment

(b) Will contribute to two different segments in the adult fly

(c) Is a region of spatial control of gene expression

(d) All of the above

(e) None of the above

24. Which of these would be an example of a homeotic phenotype?

(a) The wings are shrunken and useless

(b) The eyes of the fly are brown instead of the normal red

(c) The first abdominal segment has legs

(d) All of these

(e) None of these

25. Homologous genes:

(a) Are genes that have evolved to have similar sequences

(b) Are genes that have similar sequences due to common descent

(c) Are genes that have similar functions

(d) Are present only in vertebrates

26. If college students were cells, which of these would be determined but not differentiated?

(a) A freshman student who is undecided as to his major

(b) A freshman student that has declared a major

(c) A senior student that is nearing graduation

(d) An alumnus that has begun her career

27. In addition to potentially similar effects on pattern formation, Hox genes in vertebrates and HOM-C (homeotic) genes in Drosophila have which of the following similarities?

(a) Both sets of genes are located in a single chromosomal cluster in their respective species

(b) Each gene is known to contain a leucine zipper motif, leading to DNA binding of dimers of the encoded protein

(c) The linear order of the genes along the chromosome is reflected in the anterior limits of expression of the genes along the AP axis of the animal

(d) All of these

(e) None of these

28. Myogenic basic helix-loop-helix (bHLH) proteins:

(a) Control differentiation of cells into muscle cell types

(b) Have been shown to exist in all vertebrates, Drosophila, and C. elegans

(c) Function as transcriptional regulators

(d) All of the above

(e) None of the above

29. Which statement comparing plant and animal development is the most correct?

(a) Organization of both plants and animals occurs around specific axes

(b) Cell migration is a major feature of pattern formation in both groups

(c) Plants cells have relatively limited potential, compared to animal cells, which are totipotent

(d) There are no critical differences between plant and animal development

(e) There are no important similarities between plant and animal development

30. In plant development, an organized group of actively dividing stem cells is:

(a) Unusual

(b) A meristem

(c) A signaling center

(d) Only present at the apical tip of each shoot

(e) None of these

31. What function do homeotic genes in plants and animals have in common?

(a) They activate the same group of target genes in both groups

(b) They encode identical proteins in each group

(c) They are transcriptional activators that regulate other genes

(d) They prevent cell division when they are active

32. Which of the following event is the primary mechanism used for sex determination in Drosophila?

(a) Male-determining genes on the Y chromosome function as transcriptional activators of genes that control male development

(b) A cascade of transcription factor activation specifies development in the male pathway; lack of activation of this cascade will result in a female

(c) Alternative splicing of specific gene products leads to development of a female if two X chromosomes are present

(d) All of these

(e) None of these

33. A loss of function mutation in which of these genes would be most likely to lead to an XX male phenotype?

(a) SXL (b) FRU

(c) MSL-2 (d) All of these

(e) None of these

34. How is sex determination in C. elegans similar to sex determination in Drosophila?

(a) In both cases, the end result of the process is either a male or a hermaphrodite

(b) In both cases, alternative splicing is a major mechanism of control of gene activity

(c) In both cases, expression of genes on the X chromosome is reduced in the process of dosage compensation

(d) In both cases, the ratio of X chromosomes to autosomes determines which pathway the individual will follow

(e) None of the above

35. In which of the species is an X chromosome required to produce a normal male?

I. Drosophila II. C. elegans

III. Mus musculus (mouse)

(a) I only (b) I and II

(c) I, II, and III (d) II and III

(e) III only

36. A translocation of the Sry gene to an autosome would be likely to cause which of the following:

(a) An XY female

(b) An XX male

(c) Infertility in an XY male

(d) Infertility in an XX female

(e) No impact on sexual development

37. Which of the following statements regarding mammalian sex determination is most likely to be correct?

(a) An XX male is more likely to be fertile than an XY female

(b) An XY female is more likely to be fertile than an XX male

(c) Loss of function mutations in several different genes are required to produce an XY female

(d) All of the above are correct

(e) None of the above are correct

38. Which of these mechanisms is common to sex determination in Drosophila, C. elegans, mammals, and plants?

(a) Transcriptional regulation

(b) Alternative splicing

(c) Environmental influences

(d) Y linked genes that determine maleness

39. The most common pattern of sex phenotypes in plants is:

(a) Monomorphic (b) Hermaphroditic

(c) Dioecious (d) Sexually dimorphic

Fill in the blanks

1. In regard to development, the term pattern refers to ______________

(a) The proteins produced by an individual cell

(b) The stages of development through which an organism proceeds

(c) The arrangement of the regions of the body of an organism

(d) The process of differentiation of specific cell types

(e) None of the above

2. In Drosophila, the pattern is established ______________

(a) During production of the oocyte

(b) As a result of environmental interactions with the oocyte

(c) During gastrulation

(d) After embryogenesis has completed

(e) None of these

3. Segmentation genes ________________

(a) Can be grouped into three major classes

(b) Control the development of the segments of the Drosophila embryo

(c) Often produce embryonic lethal phenotypes when mutant

(d) All of these

(e) None of these

4. A morphogen ______________

(a) Conveys positional information to cells within the embryo

(b) Is typically present in similar amounts in all cells

(c) Functions only at very low concentrations

(d) All of these

(e) None of these

5. In Drosophila development, an embryonic segment __________

(a) Is the same thing as a parasegment

(b) Will contribute to two different segments in the adult fly

(c) Is a region of spatial control of gene expression

(d) All of the above

(e) None of the above

6. Homologous genes _______________

(a) Are genes that have evolved to have similar sequences

(b) Are genes that have similar sequences due to common descent

(c) Are genes that have similar functions

(d) Are present only in vertebrates

(e) None of the above

7. Myogenic basic helix-loop-helix (bHLH) proteins ____________
   (a) Control differentiation of cells into muscle cell types
   (b) Have been shown to exist in all vertebrates, Drosophila, and C. elegans
   (c) Function as transcriptional regulators
   (d) All of the above
   (e) None of the above
8. In plant development, an organized group of actively dividing stem cells is ________
   (a) Unusual
   (b) A meristem
   (c) A signaling center
   (d) Only present at the apical tip of each shoot
   (e) None of these
9. A translocation of the Sry gene to an autosome would be likely to cause which of the following____________
   (a) An XY female
   (b) An XX male
   (c) Infertility in an XY male
   (d) Infertility in an XX female
   (e) No impact on sexual development

True/False

1. Gap genes directly regulate pair-rule genes.
   (a) True (b) False
2. A morphogen is a molecule that controls positional information by controlling cell adhesion.
   (a) True (b) False
3. During the early stages of Drosophila development, mitosis is not accompanied by cytokinesis.
   (a) True (b) False

4. In C. elegans, the father and the mother of an individual worm could be the same individual.

(a) True (b) False

5. Cell differentiation is the process by which cells specialize into their adult form and function.

(a) True (b) False

6. Change in expression of a single gene can lead to major changes in determination or differentiation of a cell.

(a) True (b) False

7. Knockouts of Hox genes may not cause a homeotic phenotype.

(a) True (b) False

8. Homeotic mutations have only been seen in invertebrate species.

(a) True (b) False

9. During the early stages of Drosophila development, mitosis is not accompanied by cytokinesis.

(a) True (b) False

10. In C. elegans, the father and the mother of an individual worm could be the same individual.

(a) True (b) False

11. Cell differentiation is the process by which cells specialize into their adult form and function.

(a) True (b) False

12. Change in expression of a single gene can lead to major changes in determination or differentiation of a cell.

(a) True (b) False

13. A morphogen is a molecule that controls positional information by controlling cell adhesion.

(a) True (b) False

14. Gap genes directly regulate pair-rule genes.

(a) True (b) False

15. Knockouts of Hox genes may not cause a homeotic phenotype.

(a) True (b) False

16. Homeotic mutations have only been seen in invertebrate species.

(a) True (b) False

Answers

Multiple Choice Question

1	2	3	4	5	6	7	8	9	10	11	12	13	14	15
E	A	C	C	D	C	A	C	C	A	D	B	B	A	A
16	17	18	19	20	21	22	23	24	25	26	27	28	29	30
C	A	D	E	A	A	C	E	C	B	B	C	D	A	B
31	32	33	34	35	36	37	38	39						
C	C	A	D	E	B	B	A	A						

Fill in the blanks

1	2	3	4	5	6	7	8	9
C	A	D	A	E	B	D	B	B

True/False

1	2	3	4	5	6	7	8	9	10	11	12	13	14	15
A	B	A	A	A	A	A	B	A	A	A	A	B	A	A
16														
B														

❑❑❑

Chapter – 23

Quantitative Genetics

Quantitative Traits

- Quantitative traits exhibit a continuum of phenotypic variation that may follow a normal distribution
- Statistical methods are used to evaluate a frequency distribution quantitatively
- Some statistical methods compare two variables with each other

Polygenic Inheritance

- Polygenic inheritance and environmental factors create overlaps between genotypes and phenotypes
- Polygenic inheritance explains DDT-resistance in Drosophila
- Quantitative trait loci (QTLs) are now mapped by linkage to molecular markers

Heritability

- Phenotypic variance is due to the additive effects of genetic variance and environmental variance
- Heritability is the relative amount of phenotypic variation that is due to genetic variation
- Heritability of dermal ridge count in human fingerprints is very high

- Heritability values are only relevant to particular groups raised in a particular environment
- Selective breeding of species can alter quantitative traits dramatically
- Heterosis may be explained by dominance or overdominance

Multiple Choice Question

1. Which of these is an example of a quantitative trait?
   (a) A study compares rats with and without pigment in the irises of their eyes
   (b) Chinese strains of pigs have larger numbers of piglets per litter than American strains
   (c) Caterpillars exposed to juvenile hormone mimics are unable to metamorphose into adults
   (d) All of these are quantitative traits
   (e) None of these are quantitative traits
2. You want to design a study to gauge the extent to which coat color in Irish Setters is similar between parents and offspring. What statistical measure would be most appropriate?
   (a) Variance (b) Covariance
   (c) Correlation coefficient (d) Regression analysis
   (e) None of the above
3. In a study of rate of weight gain in pigs, scientists crossed two very different breeds of pigs and interbred the offspring for several generations. When they analyzed the F4 offspring, even though all the pigs were raised on the same farm and given the same feed, there were dramatic differences in the rate of weight gain, measured both as pounds of gain per day and number of days to reach market weight. When graphing either statistic for a large number of pigs, a continuous distribution was found. Which of the following statements is the most correct?
   (a) Rate of weight gain in pigs is primarily controlled by environmental factors
   (b) There are probably only a few genetic factors controlling this trait
   (c) This trait is controlled by a large number of genetic factors

(d) This trait is not affected by environmental factors

(d) None of the above statements are accurate

4. Regarding the Nilsson-Ehle wheat experiment, which of the following statements is true?

(a) Wheat are heterozygous for three hull color loci

(b) Wheat plants can have a range of intensity of hull color

(c) Hull color in wheat plants is controlled by a single dominant allele

(d) All of the above are true

(e) None of the above are true

5. What color would you expect for the hulls of a wheat plant heterozygous for all three hull color loci?

(a) Light red

(b) Intermediate red

(c) Medium red

(d) Dark red

(e) Extra dark red

6. How would you expect an increase in molecular marker density to affect QTL mapping studies?

(a) Genes affecting quantitative traits can be directly identified

(b) Segments of the genome which control quantitative traits can be more quickly identified

(d) QTLs will be smaller than those identified in previous studies

(d) All of these

(e) None of these

7. Which of these definitions of heritability is most correct?

(a) The total amount of phenotypic variation in a population

(b) The proportion of phenotypic variation in a population that is due to genetic factors

(d) The total amount of genetic variation in a population

(d) The proportion of phenotypic variation in a population that is due to environmental factors

(e) None of these are correct definitions of heritability

8. High heritability is best demonstrated in which of the following situations?

   (a) Within a wild population of clover plants, there is a very low amount of variation in the number of leaves per stem

   (b) When studying obesity in a human population, there tends to be less variation within families than across the entire population

   (c) In a study of adopted children, the children's height is much more similar to that of their biological parents than their adoptive parents

   (d) The DNR maintains records of deer shot each hunting season. In drought years, the average weight is much lower than in years of sufficient rain

   (e) None of the above

9. Which of these is the best example of the dominance hypothesis?

   (a) Two inbreed wheat strains are homozygous for deleterious recessive alleles for different genes, which mildly decrease yield. Hybrids of these two inbred strains are heterozygous for each locus, and display a dramatic increase in yield over either parent

   (b) Two inbreed wheat strains are homozygous for deleterious dominant alleles for different genes, which mildly decrease yield. Hybrids of these two inbred strains are heterozygous for each locus, and display a dramatic increase in yield over either parent

   (c) Two inbreed wheat strains are homozygous for different codominant alleles at a single locus. Hybrids of these two inbred strains are heterozygous for this locus and display an increase in yield over either parent due to an interaction between the gene products of both alleles

   (d) All of the above illustrate the dominance hypothesis

   (e) None of the above illustrate the dominance hypothesis

10. What is the main difference between the dominance and overdominance hypotheses?

   (a) Overdominance requires the presence of recessive alleles for the genes involved

   (b) Overdominance requires interactions between alleles of different genes

(c) Overdominance requires interactions between alleles of the same gene

(d) All of these are correct

(e) None of these are correct

11. In a cross involving polygenic inheritance, only 2/125 of the offspring (F2) were as extreme as one of the P1 parents. How many gene pairs are involved?

(a) 3 (b) 1

(c) 4 (d) 5

(e) 2

12. In a cross involving polygenic inheritance where three gene pairs control plant height, the shortest and tallest plants are 6 cm and 24 cm, respectively. What height should all F1s display if homozygous 12-cm and 24-cm plants are crossed, assuming their environments are the same?

(a) 15 cm (b) 21 cm

(c) 9 cm (d) 12 cm

(e) 18 cm

13. Inheritance of complex traits differs from inheritance of quantitative traits in which of the following ways?

(a) Traits run in families

(b) No clear Mendelian pattern of inheritance is observed

(c) More than one gene is involved

(d) There is no difference

(e) None of the above

14. Which of the following term is defined as "a measure of the variation in a phenotype (for a polygenic trait in a given population) due to genetic factors"?

(a) Broad-sense heritability

(b) Narrow-sense heritability

(c) Variance

(d) Biometry

(e) None of the above

15. Which of the following are mismatched?

(a) QTLs = RFLPs

(b) RFLPs = DNA markers

(c) QTLs = more than one loci

(d) Discontinuous variation = QTLs

(e) Cosegregation = genetic mapping

16. Four additive loci control ear length in Easter bunnies. Two extreme strains are crossed; the F1 are interbred to produce the F2, which cover the complete range of ear lengths. How many of the F2 are expected to have ears as extreme as one parental (P1) bunny or the other?

(a) 1/4 (b) 1/16

(c) 1/32 (d) 1/64

(e) 1/256

17. If the P1 bunnies from question #3 had ears measuring 10mm and 30mm, how long were the ears, on average, of F1 bunnies?

(a) 10mm (b) 15mm

(c) 20mm (d) 25mm

(e) 30mm

18. The correlation coefficient between identical twins for body weight was calculated to be 0.42. What is the heritability of body weight?

(a) 0 (b) 0.21

(c) 0.42 (d) 0.84

(e) Over 1.0

19. Consider the polygenic trait, tail length, in hypothetical zeebogs. Which of the following genotypes would have tails of intermediate length?

(a) AABBCCDD

(b) AABbCcDd

(c) AABBccdd

(d) aaBbCCDD

(e) aabbccdd

20. Consider the following body weights (g) for the Madagascar hissing cockroach: 7,9,6,13,10,8,8,9. Approximately what is the mean of these values?

(a) 7 (b) 8

(c) 9 (d) 10

(e) 13

21. Consider the following body weights (g) for the Madagascar hissing cockroach: 7,9,6,13,10, 8,8,9. Approximately what is the sample variance of these values?

(a) 1 (b) 3

(c) 5 (d) 8

(e) 10

22. Consider the following body weights (g) for the Madagascar hissing cockroach: 7,9,6,13,10,8,8,9. Approximately what is the standard deviation of these values?

(a) 2 (b) 3

(c) 4 (d) 5

(e) 6

23. A corn population created by selective breeding over many years shows a very low VG for number of kernels per ear. Ear size in this strain of corn can probably be increased by further selective breeding.

(a) A mutation in the wobble base of a lysine codon of the gene for an essential metabolic protein

(b) A nonsense mutation in a gene required for correct assembly of the cytoskeleton

(c) A mutation in an intron splice recognition site in a unique aminoacyltransferase

(d) All of these might be considered neutral mutations

(e) None of these could be considered neutral mutations

24. Random genetic drift in small isolated subpopulations of desert toads:

(a) Could create subspecies monomorphic for different coat color alleles

(b) Could lead to a loss of genetic diversity in one subpopulation

(c) Could interact with new mutations to produce one subpopulation that is very different than the others

(d) All of these

(e) None of these

25. Which of these factors probably most influenced the frequency of hemophilia in the royal families of Europe?

(a) Restriction of marriage partners of royalty to other royals

(b) Differential access to medical care

(c) Differential mutation rates

(d) All of the above

(e) None of the above

26. Which of these illustrates genetic load in a population?

(a) Individuals homozygous for a recessive lethal allele do not survive to reproductive age

(b) A native species of tree from Florida is unlikely to survive after being transplanted to Wisconsin

(c) Allele frequencies are expected to remain constant in the absence of selection

(d) All of these

(e) None of these

Fill in the blanks

1. In a normal distribution ________________.

(a) Equal numbers of individuals are found above and below the mean

(b) The mean is also the most frequent value

(c) 68% of individuals are within one standard deviation of the mean

(d) All of these

(e) None of these

2. Most behavioral traits are affected by many different genetic influences as well as environment. The best descriptor for such traits is ____________.
   (a) Mendelian
   (b) Polygenic
   (c) Highly heritable
   (d) None of these descriptors are accurate
   (e) None of the above
3. Environmental factors most affect the expression of __________.
   (a) Codominant alleles
   (b) Additive alleles
   (c) Recessive alleles
   (d) Dominant alleles
   (e) Multiple alleles
4. Polygenic inheritance centers around __________.
   (a) Epistasis and is qualitative
   (b) Additive alleles and is quantitative
   (c) Multiple alleles and is quantitative
   (d) Additive alleles and is qualitative
   (e) Multiple alleles and is qualitative
5. Monozygotic twins are used in human heritability studies because ______.
   (a) They are genetically identical
   (b) They are raised together in the same environment
   (c) Environmental effects are not important to the expression of their genotypes
   (d) Monozygotic twins are genetically similar in the same way siblings are; however, they shared the same womb
   (e) They are not used in human heritability studies
6. Traits exhibiting quantitative inheritance are called ________________.
   (a) Continuous
   (b) Metrical
   (c) Quantitative
   (d) Additive
   (e) All of the choices

7. In an additive model of polygenic inheritance, (A+a)2n represents ______.
   (a) The number of different types of gametes produced by an F1 multihybrid
   (b) The number of different F2 genotypes produced by the mating of F1 multihybrids
   (c) The number of F2 phenotypes produced by the mating of F1 multihybrids that are as extreme as one or the other of the original homozygous parentals (P1)
   (d) The distribution pattern of F2 phenotypes produced by the mating of F1 multihybrids
8. A quantitative trait locus (QTL) ___________.
   (a) Contains a known gene which affects a polygenic trait
   (b) Contains a single gene which affects a polygenic trait
   (c) Controls all of the variation for a polygenic trait
   (d) All of the above
   (e) None of the above

True/False

1. A corn population created by selective breeding over many years shows a very low Vg for number of kernels per ear. Ear size in this strain of corn can probably be increased by further selective breeding.
   (a) True (b) False
2. For quantitative traits, two individuals with the same genotype may have different phenotypes.
   (a) True (b) False
3. Polygenic traits are traits that are only affected by genetic influences.
   (a) True (b) False
4. Quantitative traits are probably controlled by single genes.
   (a) True (b) False
5. You do a study of heritability of insecticide resistance in European corn borers in Iowa corn crops, and calculate a value of 0.78. This tells you

that the trait of insecticide resistance in primarily (78%) influenced by genetic factors, and environment only has a minor impact (22%).

(a) True (b) False

6. Dominance can be easily confused with overdominance if QTLs are involved.

(a) True (b) False

Answers

Multiple Choice Question

1	2	3	4	5	6	7	8	9	10	11	12	13	14	15
B	C	C	B	C	C	B	C	A	C	A	E	B	A	D
16	17	18	19	20	21	22	23	24	25	26				
E	C	C	C	C	C	A	A	D	B	E				

Fill in the blanks

1	2	3	4	5	6	7	8
D	B	B	B	A	E	D	E

True/False

1	2	3	4	5	6
B	A	B	B	B	A

❑❑❑

Chapter – 24

Population Genetics

Genes in Populations

- A population is a group of interbreeding individuals who share a gene pool
- Some genes are monomorphic, and others are polymorphic
- Population genetics is concerned with allele and genotype frequencies

The Hardy-Weinberg Equilibrium

- The Hardy-Weinberg equation can be used to calculate genotype frequencies based on allele frequencies
- Nonrandom mating may occur in natural and human populations

Factors that Change Allele Frequencies in Populations

- Mutations provide the source of genetic variation
- In small populations, allele frequencies can be altered by random genetic drift
- Migrations between two populations can alter allele frequencies
- Natural selection favors the survival of the fittest
- Balanced polymorphisms may exist due to heterozygote superiority or heterogeneous environments
- Industrial melanism in the moth *Biston betularia* is a modern example of natural selection
- Genetic Load Is the Negative Consequence Of Genetic Variation

Multiple Choice Question

1. Which of these is the best example of two individuals from the same population?
   (a) A Siberian hawk owl and a mouse that occupy the same area of the Siberian steppes
   (b) Two large-mouth bass fish that occupy the same lake in Minnesota, USA
   (c) Two dingos living several hundred miles apart in the Outback, Australia
   (d) A ginko tree in its native China and a ginko tree in downtown Ames, Iowa, USA
   (e) None of these pairs of individuals are from the same population
2. Which of these genes would be polymorphic?
   (a) In Caucasians of European descent, the mutant allele that causes cystic fibrosis is present in about 1/20 individuals
   (b) Alleles used for tissue typing in the African cheetah are identical in all members of the population
   (c) In a population of sea urchins in the Atlantic Ocean, isocitrate dehydrogenase has two allozymes, one of which is present at 0.5% frequency
   (d) None of these are examples of polymorphic genes
   (e) All of these are examples of polymorphic genes
3. In a population of Drosophila, 200 individuals have bright red eyes and are homozygous for the wild type allele of a specific gene, 28 individuals have bright red eyes and are heterozygous at that locus, and 13 individuals have dull brown eyes and are homozygous for a mutant version of the gene. What is the homozygous recessive genotype frequency?
   (a) 0.05 (b) 0.11
   (c) 0.17 (d) 0.83
   (e) 0.95
4. Which of these is not one of the requirements for presence of an equilibrium in allele frequencies in a population?
   (a) Large population size

(b) No new mutations in the gene of interest

(c) No natural selection for the gene of interest

(d) Mate selection based on phenotype of interest

(e) None of these (all are requirements)

5. Albinism in humans is controlled by a single gene. You observe a population of humans where 82% are albino and the rest are normally pigmented. Which trait is recessive?

(a) Normal pigmentation

(b) Albino pigmentation

(c) Neither; these alleles are codominant

(d) There is not enough information to determine which trait is recessive

(e) None of the above

6. Albinism in humans is controlled by a single gene, and is recessive. You observe a population of humans where 82% are albino and the rest are normally pigmented. What is the allelic frequency of the dominant allele?

(a) 0.094 (b) 0.088

(c) 0.170 (d) 0.180

(e) 0.820

7. What effect will inbreeding tend to have on genotypic frequencies in a population?

(a) It will increase the frequency of genotypes that produce a dominant phenotype

(b) It will decrease the frequency of genotypes that produce a dominant phenotype

(c) It will increase the frequency of homozygous genotypes

(d) It will increase the frequency of heterozygous genotypes

(e) It will have no effect on genotypic frequencies

8. Which of these might be considered a neutral mutation?

(a) A mutation in the wobble base of a lysine codon of the gene for an essential metabolic protein

(b) A nonsense mutation in a gene required for correct assembly of the cytoskeleton

(c) A mutation in an intron splice recognition site in a unique aminoacyltransferase

(d) All of these might be considered neutral mutations

(e) None of these could be considered neutral mutations

9. Which of these (rather bizarre) scenarios would illustrate the founder effect?

(a) An outbreak of SARS decimates the population of China, leaving behind only those few resistant to infection or lucky enough to be caught in the first wave while hospital care was still available

(b) A spaceship lands on Earth, collects several families from rural Kentucky, and transports them to a new planet to start a colony

(c) Society's obsession with ultra-thin women has progressed to the point where most successful women remain thin by hiring surrogate females to gestate a zygote created from their oocyte and donor sperm. However, surrogate females with broad "child bearing" hips are in short supply, and the frequency of difficulty in childbirth is increasing

(d) All of these illustrate the founder effect

(e) None of these illustrate the founder effect

10. Which of these factors probably most influenced the frequency of hemophilia in the royal families of Europe?

(a) Restriction of marriage partners of royalty to other royals

(b) Differential access to medical care

(c) Differential mutation rates

(d) All of the above

(e) None of the above

11. Which of these is not a mode of natural selection?

(a) Stabilizing selection (b) Disruptive selection

(c) Operational selection D) Directional selection

(e) None of the above

12. Which of these individuals would show a higher Darwinian fitness?

(a) A female sparrow that attracts a large number of mates

(b) A female sparrow that lays a large number of eggs

(c) A female sparrow that hatches a large number of offspring

(d) A female sparrow that lives an unusually long life

(e) None of the above

13. Which of these illustrates genetic load in a population?

(a) Individuals homozygous for a recessive lethal allele do not survive to reproductive age

(b) A native species of tree from Florida is unlikely to survive after being transplanted to Wisconsin

(c) Allele frequencies are expected to remain constant in the absence of selection

(d) All of these

(e) None of these

14. All members of an isolated village were genotyped for a particular RFLP. Of the 1000 individuals, 200 were homozygous for the presence of the restriction site, 700 were heterozygous, and 100 were homozygous for its absence. What is the frequency of the "restriction site present" allele, p?

(a) 0.55 (b) 0.45

(c) 0.9 (d) 0.2

(e) None of the above

15. Which of the following is NOT a condition of the Hardy-Weinberg law?

(a) Mating within the population occurs at random with respect to the loci under consideration

(b) The population is infinitely large or, in practical terms, is large enough that sampling errors and random effects are negligible

(c) No more than one other factor—mutation, migration, or random genetic drift—can be acting at one time during the period under study

(d) There is no selective advantage for any of the genotypes under consideration; all genotypes are equally viable and fertile

(e) All of the above are conditions of the Hardy-Weinberg law

16. A population is in Hardy-Weinberg equilibrium for a gene locus that has four alleles at the following allelic frequencies: a1 (0.1), a2 (0.2), a3 (0.4), and a4 (0.3). What is the expected frequency of the a2a3 genotype?

(a) 0.2 (b) 0.08

(c) 0.16 (d) 0.4

(e) None of the above

17. Which of the following is true about mutations in population genetics?

(a) Mutation frequencies can be calculated from the frequency of homozygous recessive genotypic class alone

(b) Except in situations of Hardy-Weinberg equilibrium, mutation plays a significant role in changing allelic frequencies

(c) Mutation rate, mu, is equal to the allelic frequency of the recessive allele divided by the initial frequency of the recessive allele.

(d) Mutation is a major force in generating genetic variability, but by itself plays a relatively insignificant role in changing allelic frequencies

(e) None of the above are correct

18. Which of the following is NOT one of the classes of natural selection?

(a) Disruptive selection (b) Neutral selection

(c) Stabilizing selection (d) Directional selection

(e) None of the above

19. For a locus where there are two alleles, B and b, and for which bb is homozygous lethal and BB and Bb have identical fitness, what is the expected allele frequency of b after 10 generations if the frequency of b is 0.1 at generation zero?

(a) 0.1 (b) 0.05

(c) 0.0 (d) 0.5

(e) None of the above

20. Which of the following statements does NOT provide an explanation for hybrid vigor?

(a) Under certain circumstances, heterozygotes are superior to either possible homozygote

(b) Disease-causing, homozygous recessive phenotypes from either parent are masked in the hybrids

(c) Offspring from a hybrid cross usually possess the best of two desirable parents

(d) Inherently, hybrids have no deleterious mutations

(e) Hybrids often have more than one form of many of the genes present, giving them an edge against environmental pressures

21. In a population of 100 people, 75 are homozygous dominant, 20 are heterozygotes, and 5 are homozygous recessive. What is the frequency of the dominant allele?

(a) 0.5 (b) 0.87

(c) 0.15 (d) 0.35

(e) 0.75

22. A rare metabolic disease is found in 18 out of 100,000 individuals studied. What is the frequency of carriers for this disease?

(a) 0.97 (b) 0.03

(c) 0.026 (d) 0.00018

(e) 0.013

23. In a population that is in genetic equilibrium for a single locus trait with two alleles, the frequency of one homozygous genotype is 0.09. What is the frequency of heterozygotes?

(a) 0.91 (b) 0.49

(c) 0.7 (d) 0.18

(e) 0.42

24. For which phenotypes in the ABO blood group system is it possible to determine genotype?

(a) Phenotypes A, B, and AB, but not O

(b) Only phenotype O

(c) Only phenotypes AB and O

(d) Only phenotypes A and B

(e) Phenotypes A, B, and O, but not AB

25. If 42% of a population shows the recessive phenotype, approximately what is the frequency of the recessive allele?

(a) 0.35 (b) 0.42

(c) 0.58 (d) 0.65

(e) Cannot be calculated

26. If 42% of a population shows the recessive phenotype, what is the expected frequency of heterozygotes?

(a) 0.18 (b) 0.23

(c) 0.43 (d) 0.46

(e) Cannot be determined

27. In a sample of 100 persons from Nashville, 80 had Rh + blood and 20 had Rh- blood (recessive condition). Is the population in Hardy-Weinberg proportions?

(a) Yes, chi-square < critical chi-square value

(b) Yes, chi-square > critical chi-square value

(c) No, chi-square< critical chi-square value

(d) No, chi-square> critical chi-square value

(e) Cannot test this population for equilibrium

28. In a sample of 100 individuals from Nashville, 10 had type A blood, 30 had type B blood, 10 had type AB blood, and 50 had type O blood. Approximately what is the frequency of the B allele?

(a) 0.10 (b) 0.18

(c) 0.55 (d) 0.63

(e) 0.71

29. In a population where the O allele is missing, 45 individuals were of blood type A, 20 were of type AB, and 35 were of type B. What is the approximate frequency of the B allele?

(a) 0.28 (b) 0.45

(c) 0.59 (d) 0.74

(e) Cannot be determined

30. Is the population in question #7 in equilibrium?

(a) Yes, calculated chi-square > critical chi-square value

(b) Yes, calculated chi-square < critical chi-square value

(c) No, calculated chi-square > critical chi-square value

(d) No, calculated chi-square < critical chi-square value

(e) Cannot test this population for equilibrium

31. In a mixed-sex population in equilibrium, the frequency of sex-linked colorblind males is 20%. What percentage of the females in this population are normal but carry the colorblind allele?

(a) 0.20 (b) 0.32

(c) 0.45 (d) 0.55

(e) Cannot be determined

32. In a mixed-sex population in equilibrium, the frequency of sex-linked colorblind females is 20%. What percentage of the females in this population are normal but carry the colorblind allele?

(a) 0.04 (b) 0.20

(c) 0.32 (d) 0.50

(e) Cannot be determined

33. At a particular locus, there are two alleles, G and g. The mutation rate of G to g is 2.0 x 10-4 whereas the mutation rate of g to G is 5.0 x 10-8. What is the equilibrium frequency of the g allele assuming no other factor is operating in this population to disturb Hardy-Weinberg equilibrium?

(a) 1 x 10-11 (b) 0.0002

(c) 0.0003 (d) 0.9998

(e) Cannot be calculated

34. Given a population of about 1,000 individuals and a frequency of the g allele of 0.80, what is the probability of the loss of the g allele by random genetic drift?

(a) 0.16

(b) 0.20

(c) 0.50

(d) 0.80

(e) Cannot be predicted from information given

35. In a hypothetical population, the fitnesses of the genotypes of locus G are determined. Three genotypes, GG, Gg, and gg, have fitnesses of 0.75, 1.0, and 0.4 respectively. What is the equilibrium frequency of the g allele?

(a) 0.286 (b) 0.294

(c) 0.706 (d) 0.714

(c) Cannot be determined

36. In which model of selection are both alleles maintained in a population?

(a) Selection against one allele

(b) Selection against homozygotes

(c) Selection against recessive homozygotes

(d) Selection against heterozygotes

(e) None of the above

37. Examination of segregation of alleles at the t-allele (tailless) locus in the house mouse reveals that as many as 95% of the gametes of heterozygotes carry the recessive allele t. This phenomenon is referred to as

(a) Zygotic selection (b) Gametic selection

(c) Fecundity selection (d) Meiotic drive

(e) Two of these

38. In a circumstance in which a population was subjected to migration from an outside group, the B allele of the ABO blood system was determined to be 0.3 in migrants and 0.6 in the conglomerate population. The migration rate was 0.2. What was the frequency of the allele in the original natives?

(a) 0.225 (b) 0.325

(c) 0.675 (d) 0.775

(e) Can not be calculated

39. Two populations of equal size, each in Hardy-Weinberg proportions, merge. In population one there are 16 AA, 48 Aa, and 36 aa individuals. In population two there are 64 AA, 32 Aa, and 4 aa individuals. After one

generation of random mating, what proportion of the population is expected to be heterozygous?

(a) 0.40 (b) 0.48

(c) 0.52 (d) 0.60

(e) None of the above

40. Which form of selection will be the most effective in the short run?

(a) Mutation-selection equilibrium

(b) Selection against heterozygotes

(c) Selection against homozygotes

(d) Selection against a recessive phenotype

(e) Selection against a dominant phenotype

41. Which of the following statements is false?

(a) The apparent decrease in heterozygotes in conglomerate populations is due to the Wahlund effect

(b) A founder effect occurs when natural selection removes some alleles in the population

(c) Stabilizing selection acts equally well on individuals at both ends of the distribution

(d) All of the above statements are true

(e) None of the above

42. Which of these is an example of a quantitative trait?

(a) A study compares rats with and without pigment in the irises of their eyes

(b) Chinese strains of pigs have larger numbers of piglets per litter than American strains

(c) Caterpillars exposed to juvenile hormone mimics are unable to metamorphose into adults

(d) All of these are quantitative traits

(e) None of these are quantitative traits

43. In a normal distribution:

(a) Equal numbers of individuals are found above and below the mean

(b) The mean is also the most frequent value

(c) 68% of individuals are within one standard deviation of the mean
(d) All of these
(e) None of these

44. You want to design a study to gauge the extent to which coat color in Irish Setters is similar between parents and offspring. What statistical measure would be most appropriate?

(a) Variance (b) Covariance
(c) Correlation coefficient (d) Regression analysis

45. Most behavioral traits are affected by many different genetic influences as well as environment. The best descriptor for such traits is:

(a) Mendelian
(b) Polygenic
(c) Highly heritable
(d) None of these descriptors are accurate

46. In a study of rate of weight gain in pigs, scientists crossed two very different breeds of pigs and interbred the offspring for several generations. When they analyzed the F4 offspring, even though all the pigs were raised on the same farm and given the same feed, there were dramatic differences in the rate of weight gain, measured both as pounds of gain per day and number of days to reach market weight. When graphing either statistic for a large number of pigs, a continuous distribution was found. Which of the following statements is the most correct?

(a) Rate of weight gain in pigs is primarily controlled by environmental factors
(b) There are probably only a few genetic factors controlling this trait
(c) This trait is controlled by a large number of genetic factors
(d) This trait is not affected by environmental factors
(e) None of the above statements are accurate

47. Which of these definitions of heritability is most correct?

(a) The total amount of phenotypic variation in a population
(b) The proportion of phenotypic variation in a population that is due to genetic factors

(c) The total amount of genetic variation in a population

(d) The proportion of phenotypic variation in a population that is due to environmental factors

(e) None of these are correct definitions of heritability

48. High heritability is best demonstrated in which of the following situations?

(a) Within a wild population of clover plants, there is a very low amount of variation in the number of leaves per stem

(b) When studying obesity in a human population, there tends to be less variation within families than across the entire population

(c) In a study of adopted children, the children's height is much more similar to that of their biological parents than their adoptive parents

(d) The DNR maintains records of deer shot each hunting season. In drought years, the average weight is much lower than in years of sufficient rain

Fill in the blanks

1. Subpopulations (also known as local populations) are ____________.

(a) Subsets of a population whose individuals are more likely to mate with each other than with other members of the general population

(b) Often separated by a geographic barrier

(c) Able to mate with other members of the same general population

(d) All of the above

(e) None of the above

2. You have an acquaintance that is extremely self centered and stubborn. She is currently separated from her fourth husband. All of her husbands have also been self centered and stubborn. This would be an extreme example of ___________.

(A) Assortative mating

(b) Disassortative mating

(c) Inbreeding

(d) All of the above

(e) None of the above

3. A coefficient of inbreeding ____________.

(a) Measures the frequency with which inbreeding occurs in a particular population

(b) Indicates the difference between expected genotypic frequencies and those observed in an inbreeding population

(c) Indicates the frequency with which an inbred individual will be homozygous due to inheritance from a common ancestor

(d) All of the above

(e) None of the above

4. Random genetic drift in small isolated subpopulations of desert toads ________.

(a) Could create subspecies monomorphic for different coat color alleles

(b) Could lead to a loss of genetic diversity in one subpopulation

(c) Could interact with new mutations to produce one subpopulation that is very different than the others

(d) All of these

(e) None of these

5. A critical fact underlying both the bottleneck and founder effects is ______.

(a) Large populations tend to have less genetic diversity than small populations

(b) Small populations tend to have less genetic diversity than large populations

(c) Alleles in small populations are less likely to become fixed than those in large populations

(d) New mutations are more common in small populations than large populations

(e) None of the above

6. A selection coefficient ____________.

(a) Relates to the effect of an allele on an individual

(b) Measures the selection against a particular genotype

(c) Can be calculated from the variation from the expected Hardy-Weinberg equilibrium

(d) All of these

(e) None of these

7. In a balanced polymorphism ______________.

(a) Selection against a homozygous recessive condition may not lead to a reduction in allele frequency

(b) Migration between members of a diverse population counteracts directional selection

(c) Increased reproductive success of heterozygous individuals may not change allele frequencies

(d) All of these

(e) None of these

8. All of the following can have a significant effect on allele frequencies except ________.

(a) Migration

(b) Natural selection

(c) Genetic drift

(d) Heterozygote superiority

(e) Nonrandom mating

9. A population is defined as _________.

(a) Members of the same species, living in the same area, that have the potential to interbreed

(b) Members of the same genus, living in the same area, that have the potential to interbreed

(c) Interbreeding members of the same species

(d) Members of the same species

(e) Members of the same genus

10. A population is in genetic equilibrium when ___________.

(a) The mutation rate is high

(b) Allele and genotype frequencies do not change from generation to generation

(c) Mating is not at random

(d) Selection is occurring

(e) Allele frequencies change across generations but genotype frequencies do not

11. In the MN blood group system it is possible to know the genotype from the phenotype because ______________.

(a) The alleles are codominant

(b) The 'N' allele is recessive

(c) There are only two genotypes possible

(d) There are only two alleles

(e) None of the above; the statement is not correct: it is not possible to determine genotype from phenotype

12. Directional selection ____________.

(a) Shifts the phenotypic distribution toward one end of the distribution

(b) Is only a theoretical concept; it does not occur in nature

(c) Occurs when phenotypes near the center of a phenotypic distribution have the lowest fitness

(d) Involves selection against phenotypes near both ends of a phenotypic distribution

(e) Occurs when phenotypes near the center of a phenotypic distribution have the highest fitness

13. Stabilizing selection ___________.

(a) Is only a theoretical concept; it does not occur in nature

(b) Occurs when phenotypes near the center of a phenotypic distribution have the highest fitness

(c) Occurs when phenotypes near the center of a phenotypic distribution have the lowest fitness

(d) Shifts the phenotypic distribution toward one end of the distribution

(e) Involves selection against phenotypes near both ends of a phenotypic distribution

14. Disruptive selection _____________.

(a) Involves selection against phenotypes near both ends of a phenotypic distribution

(b) Occurs when phenotypes near the center of a phenotypic distribution have the lowest fitness

(c) Is only a theoretical concept; it does not occur in nature

(d) Shifts the phenotypic distribution toward one end of the distribution

(e) Occurs when phenotypes near the center of a phenotypic distribution have the highest fitness

15. Spread of pesticide resistance through a population of insects is an example of ______.

(a) Stabilizing selection (b) Directional selection

(c) Mutation (d) Genetic equilibrium

(e) Disruptive selection

16. In a population in Hardy-Weinberg equilibrium, the proportion of homozygous dominant genotypes is __________

(a) p2 (b) 2pq

(c) q2 (d) p2+2pq

(e) q2+2pq

17. In a population in Hardy-Weinburg equilibrium, the proportion of individuals with a dominant phenotype is ____________

(a) p2 (b) 2pq

(c) q2 (d) p2+2pq

(e) q2+2pq

18. Consider genetic drift continually occurring in 100 separate populations each with 100 individuals and allele frequencies of 0.5. After approximately 2N generations, populations are fixating for one or the other allele at a rate of _______ populations per generation.

(a) 1/N (b) 1/2N

(c) N (d) 2N

(e) None of the above

True/False

1. Directional selection will lead to fixation of beneficial alleles.

   (a) True (b) False

2. Mutation rates are generally too low to have a significant effect on allele frequencies.

   (a) True (b) False

3. It is estimated that humans, on average, carry 7 deleterious recessive alleles in the heterozygous state. Since any individual deleterious recessive allele is very rare, diseases caused by these mutations are also very rare except in cases of inbreeding. In the case of inbreeding, you would expect the frequency of such a disease to be equal to the coefficient of inbreeding.

   (a) True (b) False

4. Inbreeding is extremely rare in humans.

   (a) True (b) False

5. In most natural populations, many genes are polymorphic.

   (a) True (b) False

6. The gene pool is the variety of alleles of a particular gene present in a given species.

   (a) True (b) False

7. A corn population created by selective breeding over many years shows a very low Vg for number of kernels per ear. Ear size in this strain of corn can probably be increased by further selective breeding.

   (a) True (b) False

8. You do a study of heritability of insecticide resistance in European corn borers in Iowa corn crops, and calculate a value of 0.78. This tells you that the trait of insecticide resistance in primarily (78%) influenced by genetic factors, and environment only has a minor impact (22%).

   (a) True (b) False

9. For quantitative traits, two individuals with the same genotype may have different phenotypes.

   (a) True (b) False

10. Quantitative traits are probably controlled by single genes.

(a) True (b) False

Answers

Multiple Choice Question

1	2	3	4	5	6	7	8	9	10	11	12	13	14	15
B	A	A	D	D	A	C	A	B	B	C	C	E	A	C
16	17	18	19	20	21	22	23	24	25	26	27	28	29	30
C	D	B	B	D	B	C	E	C	D	D	E	B	B	C
31	32	33	34	35	36	37	38	39	40	41	42	43	44	45
B	D	D	B	B	B	E	C	B	E	B	B	D	C	B
46	47	48												
C	B	C												

Fill in the blanks

1	2	3	4	5	6	7	8	9	10	11	12	13	14	15
D	A	C	D	B	C	D	E	A	B	A	A	B	B	B
16	17	18												
A	D	B												

True/False

1	2	3	4	5	6	7	8	9	10
B	A	B	B	A	A	B	B	A	B

❑❑❑

Chapter – 25

Evolutionary Genetics

Origin of Species

- A biological species is a group of reproductively isolated individuals
- Speciation usually occurs via a branching process called cladogenesis
- Divergent evolution can be allopatric, parapatric, or sympatric
- Müntzing was able to re-create an allotetraploid species
- Evolution can proceed gradually or be punctuated by periods of rapid change

Evolution at the Molecular Level

- Homologous genes are derived from a common ancestral gene
- Variation in gene sequences can be used to construct phylogenetic trees
- Horizontal gene transfer also contributes to the evolution of species
- Scientists can examine the relationships between living and extinct flightless birds by analyzing ancient DNA and then comparing DNA sequences
- Genetic variation at the molecular level is associated with neutral changes in gene sequences
- Speciation is associated with changes in chromosome structure and number

Multiple Choice Question

1. Which of these would be the best example of macroevolution?
   - (a) Genetic drift in two populations of salamander eventually prevents production of viable offspring by interbreeding between them
   - (b) The founder effect leads to very different frequencies of alleles for a gene that affects beak shape in two populations of ground finch in the Galapagos Islands
   - (c) A group of humans travels from Texas to northern Alaska. At first they are uncomfortable but eventually they are quite at home
   - (d) All of these
   - (e) None of these
2. Which of these examples would illustrate two individuals that are members of the same species?
   - (a) Walleye and sauger are both fish that naturally occur in many lakes and reservoirs in the Midwestern United States. They may mate to create saugeye offspring, which can mate with either parental species but such matings only show about 10% reproductive success
   - (b) Two plants flower at different times of the year, so the pollen from one plant is not available to pollenate the flowers of the other
   - (c) You are trying to breed Springer spaniels. However, your neighbor has a randy Jack Russell terrier that managed to not only get over the fence but also inseminated your female before you realized she was coming into heat. The puppies that resulted looked a little odd but were healthy and two have since had healthy puppies of their own
   - (d) All of the above
   - (e) None of the above
3. One argument occasionally made in an attempt to refute evolution runs along the lines of "If humans evolved from chimps, why are there still chimps?" What fundamental error underlies this argument?
   - (a) This argument assumes that all species were created by cladogenesis
   - (b) This argument assumes that all species were created by anagenesis

(c) This argument assumes that all species were derived from the same common ancestor

(d) This argument assumes that all extant species are unlikely to respond to further selection

(e) None of the above

4. Which of these events may lead to sympatric speciation?

(a) Construction of a highway through the center of a population range of a sedentary species

(b) A dominant mutation changes the pattern of markings on male finches

(c) A lethal recessive allele arises in the population

(d) All of these

(e) None of these

5. Which of these is required for parapatric speciation?

(a) Migration of individuals

(b) Geographic separation between the developing species

(c) Selection against hybrid offspring

(d) All of these

(e) None of these

6. Which of these illustrates a punctuated equilibrium?

(a) A couple has a stable lifestyle with minor changes until their first baby arrives

(b) A population of fish in a pet store aquarium is constantly changing

(c) A group of seagulls lives on the same Pacific island for many generations

(d) All of these

(e) None of these

7. Which of these would not be expected to contribute to sudden change in an established species?

(a) Migration of individuals of a species to a new location

(b) A new mutation that increases reproductive success occurs in one individual of the population

(c) A new species of plant enters the environment of the species

(d) One individual in the population experiences a large inversion of one chromosome

(e) None of these

8. Why was the ability to create an "artificial" G. tetrahit important?

(a) It implies that direct intervention was required to create the naturally occurring species

(b) This experiment demonstrated that fusion of two diploid cells could generate tetraploid individuals

(c) The ability to mimic natural processes provides insight into how they might have occurred

(d) All of these

(e) None of these

9. Which of these techniques would give you the most detailed information regarding molecular differences between two species?

(a) DNA sequencing (b) Hybridization

(c) Restriction mapping (d) cytogenetic analysis

10. Which of these is an example of a pair of paralogous genes?

(a) A gene encoding myoglobin is present in both Drosophila and Xenopus

(b) Genes encoding α-globin exist in all mammals

(c) Some genes also exist in nonfunctional copies known as pseudogenes

(d) All of these

(e) None of these

11. Which piece of data is most compellingly in support of the neutral theory of evolution?

(a) The mutation rate can vary due to differences in environmental conditions

(b) Beneficial mutations are very rare compared to neutral and deleterious mutations

(c) Fixation of neutral mutations in a population is completely random

(d) All of these

(e) None of these

12. Which of the following is not an example of a prezygotic reproductive isolating mechanism?

(a) Populations live in the same region but occupy different habitats

(b) Populations exist in the same region but are sexually mature at different times

(c) Populations are isolated by incompatible premating behavior

(d) Cross-fertilization is prevented by incompatible differences in reproductive structures

(e) All of the above are examples of prezygotic isolation

13. The amino acids making up a protein in dogs and chickens are determined. Of the eight amino acids in the protein, three differ between the two species. What is the average number of amino acid substitutions, per site, between dog and chicken?

(a) 10-9 (b) 0.375

(c) 0.47 (d) 0.625

(e) Cannot be determined

14. Which of the following is a classic example of heterozygote advantage?

(a) Albinism (b) Achondroplasia

(c) Sickle-cell anemia (d) Huntington disease

(e) All of the above

15. Periods of stasis followed by rapid evolutionary change occur specifically in which model?

(a) Phyletic gradualism

(b) Punctuated equilibrium

(c) Cladogenesis

(d) Anagenesis

(e) Two of the above

16. Regions in which previously isolated populations come into contact and produce hybrids are called

(a) Hybrid zones
(b) Allopatric zones
(c) Parapatric zones
(d) Polymorphic areas
(e) None of the above

17. Microevolution is:

(a) A process by which new species of microorganism arise
(b) The process by which any new species arises
(c) The change of allele frequencies in a population
(d) All of these
(e) None of these

18. Which of these would be the best example of macroevolution?

(a) Genetic drift in two populations of salamander eventually prevents production of viable offspring by interbreeding between them
(b) The founder effect leads to very different frequencies of alleles for a gene that affects beak shape in two populations of ground finch in the Galapagos Islands
(c) A group of humans travels from Texas to northern Alaska. At first they are uncomfortable but eventually they are quite at home
(d) All of these
(e) None of these

19. The process of natural selection:

(a) Requires genetic variation among individuals of a population
(b) Results in "survival of the fittest"
(c) May lead to fixation of beneficial alleles in a population
(d) All of these
(e) None of these

20. Which of these examples would illustrate two individuals that are members of the same species?

(a) Walleye and sauger are both fish that naturally occur in many lakes and reservoirs in the Midwestern United States. They may mate to create saugeye offspring, which can mate with either parental species but such matings only show about 10% reproductive success

(b) Two plants flower at different times of the year, so the pollen from one plant is not available to pollenate the flowers of the other

(c) You are trying to breed Springer spaniels. However, your neighbor has a randy Jack Russell terrier that managed to not only get over the fence but also inseminated your female before you realized she was coming into heat. The puppies that resulted looked a little odd but were healthy and two have since had healthy puppies of their own

(d) All of the above

(e) None of the above

21. Two closely related species of orchid share the same geological distribution but do not interbreed. The petals of one species are specialized to deposit pollen on honeybees as they suck nectar; the pollen is subsequently transferred to another orchid as the honeybee moves on. The other species has larger flowers, and cannot use honeybees as a pollination vector. This is an example of:

(a) Habitat isolation

(b) Temporal isolation

(c) Sexual isolation

(d) Gametic isolation

(e) Hybrid inviability

22. One argument occasionally made in an attempt to refute evolution runs along the lines of "If humans evolved from chimps, why are there still chimps?" What fundamental error underlies this argument?

(a) This argument assumes that all species were created by cladogenesis

(b) This argument assumes that all species were created by anagenesis

(c) This argument assumes that all species were derived from the same common ancestor

(d) This argument assumes that all extant species are unlikely to respond to further selection

(e) None of the above

23. The process of evolution creates species that are ideally fitted to their niche.

(a) True (b) False

24. A large population of salamanders lives at the south end of a large lake. As this population expands to the north, subpopulations on opposite sides of the lake become so different that eventually the northernmost subpopulations are unable to interbreed. This is an example of:

(a) Allopatric speciation (b) Sympatric speciation

(c) Parapatric speciation (d) Hypermorphic speciation

(e) None of the above

25. Which of these events may lead to sympatric speciation?

(a) Construction of a highway through the center of a population range of a sedentary species

(b) A dominant mutation changes the pattern of markings on male finches

(c) A lethal recessive allele arises in the population

(d) All of these

(e) None of these

26. Which of these is required for parapatric speciation?

(a) Migration of individuals

(b) Geographic separation between the developing species

(c) Selection against hybrid offspring

(d) All of these

(e) None of these

27. The rate of evolution is relatively stable and dependent on the mutation rate.

(a) True (b) False

28. Which of these illustrates a punctuated equilibrium?

(a) A couple has a stable lifestyle with minor changes until their first baby arrives

(b) A population of fish in a pet store aquarium is constantly changing

(c) A group of seagulls lives on the same Pacific island for many generations

(d) All of these

(e) None of these

29. Which of these would not be expected to contribute to sudden change in an established species?

(a) Migration of individuals of a species to a new location

(b) A new mutation that increases reproductive success occurs in one individual of the population

(c) A new species of plant enters the environment of the species

(d) One individual in the population experiences a large inversion of one chromosome

(e) None of these

30. Why was the ability to create an "artificial" G. tetrahit important?

(a) It implies that direct intervention was required to create the naturally occurring species

(b) This experiment demonstrated that fusion of two diploid cells could generate tetraploid individuals

(c) The ability to mimic natural processes provides insight into how they might have occurred

(d) All of these

(e) None of these

31. Which of these techniques would give you the most detailed information regarding molecular differences between two species?

(a) DNA sequencing (b) Hybridization

(c) Restriction mapping (d) Cytogenetic analysis

32. Which of these is an example of a pair of paralogous genes?

(a) A gene encoding myoglobin is present in both Drosophila and Xenopus

(b) Genes encoding α-globin exist in all mammals

(c) Some genes also exist in nonfunctional copies known as pseudogenes

(d) All of these

(e) None of these

33. A diagram that describes the evolutionary relationships between different species is:

(a) A morphological branching diagram

(b) A distributed analysis

(c) A phylogenetic tree

(d) A blasphemous suggestion

(e) None of these

34. The transfer of genetic material from parents to offspring over many generations is:

(a) Vertical gene transfer

(b) Horizontal gene transfer

(c) The only way genetic material is transferred

(d) All of these

(e) None of these

35. Which piece of data is most compellingly in support of the neutral theory of evolution?

(a) The mutation rate can vary due to differences in environmental conditions

(b) Beneficial mutations are very rare compared to neutral and deleterious mutations

(c) Fixation of neutral mutations in a population is completely random

(d) All of these

(e) None of these

36. A synteny group is:

(a) A group of genes that share a common DNA sequence

(b) A group of genes that are linked in several species

(c) A group of genes that are located on the sex chromosomes

(d) All of these

(e) None of these

Fill in the blanks

1. Microevolution is:
   (a) A process by which new species of microorganism arise ________
   (b) The process by which any new species arises
   (c) The change of allele frequencies in a population
   (d) All of these
   (e) None of these
2. The process of natural selection ____________
   (a) Requires genetic variation among individuals of a population
   (b) Results in "survival of the fittest"
   (c) May lead to fixation of beneficial alleles in a population
   (d) All of these
   (e) None of these
3. Two closely related species of orchid share the same geological distribution but do not interbreed. The petals of one species are specialized to deposit pollen on honeybees as they suck nectar; the pollen is subsequently transferred to another orchid as the honeybee moves on. The other species has larger flowers, and cannot use honeybees as a pollination vector. This is an example of _______________
   (a) Habitat isolation (b) Temporal isolation
   (c) Sexual isolation (d) Gametic isolation
   (e) Hybrid inviability
4. A large population of salamanders lives at the south end of a large lake. As this population expands to the north, subpopulations on opposite sides of the lake become so different that eventually the northernmost subpopulations are unable to interbreed. This is an example of _________
   (a) Allopatric speciation (b) Sympatric speciation
   (c) Parapatric speciation (d) Hypermorphic speciation
   (e) None of the above

5. A diagram that describes the evolutionary relationships between different species is ____

   (A) A morphological branching diagram
   (B) A distributed analysis
   (C) A phylogenetic tree
   (D) A blasphemous suggestion
   (E) None of these

6. The transfer of genetic material from parents to offspring over many generations is ____

   (a) Vertical gene transfer
   (b) Horizontal gene transfer
   (c) The only way genetic material is transferred
   (d) All of these
   (e) None of these

7. A synteny group is ______________

   (a) A group of genes that share a common DNA sequence
   (b) A group of genes that are linked in several species
   (c) A group of genes that are located on the sex chromosomes
   (d) All of these
   (e) None of these

8. Until recently, evolutionary biologists believed that __________ speciation was the general rule.

   (a) Allopatric (b) Parapatric
   (c) Sympatric (d) a and b
   (e) b and c

9. Haplodiploidy in the eusocial hymenoptera produces females with a genetic similarity of ________% with their sisters while only a _______% similarity with their own offspring.

   (a) 100, 50 (b) 50, 50
   (c) 25, 75 (d) 75, 50
   (e) 50, 100

10. A squirrel is likely to have all its alleles still represented in the next generation if it sacrifices itself for

    (a) Two or more siblings (b) Four or more uncles

    (c) Four of more nieces (d) Eight cousins

    (e) Any of the above

11. In ________ mimicry, a vulnerable organism gains a selective advantage by looking like a dangerous or distasteful organism.

    (a) Mullerian (b) Batesian

    (c) Selective (d) Two of these

    (e) One of these

12. The low fitness of sickle-cell anemia homozygotes involves ________ selection.

    (a) Soft (b) Hard

    (c) Frequency dependent (d) Density dependent

    (e) Several of these

True/False

1. Transformation can lead to horizontal gene transfer in both prokaryotes and eukaryotes.

    (a) True (b) False

2. The rate of evolution is relatively stable and dependent on the mutation rate.

    (a) True (b) False

3. The process of evolution creates species that are ideally fitted to their niche.

    (a) True (b) False

4. Sexual selection tends to operate on female traits more than male traits.

    (a) True (b) False

5. Darwin's theories about evolutionary processes grew naturally out of influences from scientists/writers of his time and his own observations of the natural world.

    (a) True (b) False

Answers

Multiple Choice Question

1	2	3	4	5	6	7	8	9	10	11	12	13	14	15
A	C	B	B	C	A	E	C	A	B	B	E	C	C	B
16	17	18	19	20	21	22	23	24	25	26	27	28	29	30
A	C	A	D	C	C	B	B	A	B	C	B	A	E	C
31	32	33	34	35	36									
A	E	C	A	B	B									

Fill in the blanks

1	2	3	4	5	6	7	8	9	10	11	12
C	D	C	A	C	A	B	A	D	E	B	B

True/False

1	2	3	4	5
A	B	A	B	A

❑❑❑

Chapter – 26

Relative Reading

When Allele Frequencies Stay Constant

The Importance of Knowing Allele Frequencies

- Population genetics concerns the study of allele frequencies in a population.
- Genes in a population comprise the gene pool.
- Gene flow is movement of alleles between populations.
- Microevolution reflects changes in gene frequencies in populations, which can be traced using the Hardy-Weinberg equation.
- Allele frequencies are affected by mutation, migration, genetic drift, selection, and non-random mating.

When Allele Frequencies Stay Constant-Hardy Weinberg Equilibrium

- In Hardy-Weinberg equilibrium, gene frequencies stay relatively constant.
- Hardy-Weinberg equilibrium occurs in large populations where mating is random and there is no migration, mutation, or natural selection.
- For a population in Hardy-Weinberg equilibrium where frequencies are known for two alleles, the proportion of homozygous dominants equals the square of the frequency of the dominant allele (p^2); the proportion of homozygous recessives equals the square of the

frequency of the recessive allele (q^2); proportion of heterozygotes equals 2pq.

- Allele frequencies can be calculated from the proportions of genotypes (either p or q). The frequency of the recessive allele (q) equals the proportion of homozygous recessives plus 1/2 that of the carriers. The frequency of the dominant allele equals the proportion of homozygous dominants plus 1/2 that of the carriers.

Practical Applications of Hardy-Weinberg Equilibrium

- In real life situations, the value of q^2 can be derived by knowing the frequency of affected individuals in a population. The Hardy-Weinberg equation can then be used to predict the likelihood of someone being a carrier.
- For X-linked recessive traits, the frequency of the recessive allele in males is q and in females it is q^2.
- For very rare inherited disorders, p approaches 1, so the carrier frequency is 2q (approximately twice the frequency of the disease-causing allele).

DNA Fingerprinting-A Practical Test of Hardy-Weinberg Assumptions

- Polymorphisms are DNA sequences that vary among individuals with a frequency of at least 0.01.
- Polymorphisms that alter restriction enzyme cutting sites (RFLPs) can be used to distinguish among individuals.

DNA Patterns Distinguish Individuals

- DNA fingerprints are based on differences in the numbers of DNA repeats.
- Polyacrylamide gel electrophoresis is used to separate pieces of DNA, and distinguish among individuals on the basis of the sizes of certain DNA regions.

Population Statistics are Used to Interpret DNA Fingerprints

- Population statistics derived from databases are applied to DNA fingerprinting results to determine the probability that the same pattern is in two individuals. A limitation of the method is that databases may not adequately represent real human populations.

- Developing narrower ethnic databases and considering historical and social factors may help develop more realistic population statistics.

DNA Fingerprinting to Identify World Trade Center Victims

- PCR analysis of short tandem repeats (STRs) can be used to obtain a unique fingerprint pattern.
- STR analysis was used to identify victims in the World Trade Center attack on Sept. 11, 2001.

Changing Allele Frequencies

Nonrandom Mating

- Human mating is nonrandom. People choose their mates for many reasons and frequently they choose people with traits similar to themselves.
- At the population level, human mating is nonrandom because individuals do not contribute equally to the next generation (ie, some people have more children than others).
- For some traits, such as blood types, mating can be considered random and allele frequencies are often in Hardy-Weinberg equilibrium.
- Populations that practice marriage between related members (consanguinity) experience increases in the proportion of homozygotes at the expense of heterozygotes.

Migration

Historical Clues

- Migration disrupts gene frequencies by introducing new alleles into a population, although it may take many generations for different ethnic groups to intermarry.
- We can make inferences about the directions in which ancient people traveled by looking at the distribution of allele frequencies in modern populations.

Geographical and Linguistic Clues

- Clines are changes in allele frequencies between neighboring populations.
- Geographical barriers and language clues often result in clines with gradual or abrupt changes in allele frequencies.

Genetic Drift

- Genetic drift occurs when a subset of a population has different gene frequencies, often reflecting a physical separation.

The Founder Effect

- The founder effect occurs when a few individuals leave a community to start a new settlement.
- The resulting population may, by chance, lack some alleles present in the original population, or have high frequencies of others.

Population Bottlenecks

- In a population bottleneck, many members of a population die, and only a few individuals contribute genetically to the next generation. A population bottleneck may also occur if a small number of individuals colonize an island.

Mutation

- Mutation alters gene frequencies by introducing new alleles.
- Heterozygotes and mutations maintain the frequencies of deleterious alleles in populations, even if homozygotes die.
- Genetic load is the collection of deleterious alleles in a population.

Natural Selection

Tuberculosis Ups and Downs and Ups

- Differential survival based on phenotype is called natural selection.
- The bacteria causing tuberculosis (TB) was first identified by Robert Koch in 1882.

- As a result of natural selection operating on bacterial and human populations the effect of TB on human health evolved from outbreaks of an acute systemic infection to an increasingly rare chronic lung infection.
- Starting in the late 1980s s there has been a resurgence of TB as a result of natural selection and mutation. One in seven TB cases is resistant to several drugs and 5% of these patients die.

Evolving HIV

- HIV is a retrovirus and viral mutations accumulate rapidly in its RNA genome.
- During the course of an HIV infections environmental conditions select for viral variants that can multiply rapidly and evade the immune system.

Balanced Polymorphism

- Balanced polymorphisms exist in populations when heterozygotes carrying a disease-causing allele enjoy a health advantage.
- The recessive disease causing alleles for sickle cell disease, G6PD deficiency, PKU, Tay-Sachs disease, and Cystic fibrosis represent balanced polymorphisms since they are associated with conferring resistance to infectious diseases in the heterozygous condition.
- A changing environment may make previously selected alleles a disadvantage, such as in diabetes mellitus and P-glycoprotein.

Gene Genealogy

- Frequencies of different mutations in different populations provide information on the natural history of alleles.

PKU Revisited

- Mutational analysis in various populations indicates that PKU originated more than once, and that genetic drift, balanced polymorphism, and perhaps mutation have influenced its prevalence.

CF Revisited

- Studies of allele prevalence place the origin of CF farther back than previously thought.

Human Origins and Evolution

Human Origins

- Propliopithecus and *Aegyptopithecus* were monkeylike species that lived about 30 to 40 million years ago. Both are possible ancestors to gibbons, apes, and humans.
- Hominoids were ancestral to apes and humans.
- Dryopithecus lived 22 to 32 million years ago and may have walked onto grasslands.
- Hominids were ancestral to humans only.

The Australopithecines-And Others

- Hominids (human ancestors) appeared about 4 million years ago.
- Four million years ago bipedalism opened up new habitats for australopithecines who walked upright and used tools.
- There were several types of australopithecines, and one, *A. garhi*, may have been a direct ancestor of *Homo*.
- A newly described species, *Keynanthropus platyops*, may have been yet another australopithecine or a contemporary.

Homo

- By 2 million years ago, *Australopithecus* coexisted with the more humanlike *Homo habilis*.
- Later *H. habilis* coexisted with *H. erectus*, who used tools in more complex societies.
- *H. sapiens* either coexisted with or arose from *H. erectus* 300,000-500,000 years ago.
- The Neanderthals were contemporaries of *Homo erectus* and were displaced by the Cro-Magnons. By about 28,000 years ago the Neanderthals no longer existed while the Cro-Magnons gave rise to modern humans.

Modern Humans

- Evidence from the discovery of cave art demonstrates that by 14,000 years ago our ancestors had the ability to use symbols and had developed fine hand coordination.
- A preserved man from 5,300 years ago is genetically like us.

Molecular Evolution

- Evolutionary distances between living and extinct organisms can be estimated by studies that examine similarities and differences among chromosomes and sequences of proteins and nucleic acids.
- The more alike two organisms are on a molecular level, the more likely it is that they share a common descent.

Comparing Genomes

- Whole genome comparisons indicate that humans differ from chimps by only 0.5 percent of their protein-encoding genes.
- Small differences at the genome level can translate to large differences in phenotype. This may result from changes in developmental control genes where a mutation in a single gene can exert great effects on phenotype.
- The similarity of all vertebrate genomes is seen in the comparison of the human and pufferfish genomes.
- Only 40% of the human odorant receptor genes are functional. During the later stages of human evolution, a sense of smell may not have been the survival factor that it was in our ancestors.

Comparing Chromosomes

- Chromosome banding patterns is highly conserved in higher primates.
- All mammals have identically banded X chromosomes.
- Synteny is the correspondence of gene order in chromosomes in different species.
- Identifying regions of synteny can reveal information about species relatedness and the evolutionary history of chromosomes.

Comparing Protein Sequences

ü Many proteins are very similar in amino acid sequence in different species.

ü Homeobox proteins are transcriptional factors that control developmental processes in many organisms.

ü Homeobox genes are highly conserved throughout evolution.

ü Mutations in homeobox genes can cause disease in humans or bizarre morphological anomalies in flies and mice.

ü Evidence for functional relationships between conserved genes comes from experiments where genes encoding proteins of one species are transplanted into a distantly related species.

Comparing DNA Sequences

- The degree of evolutionary relatedness between two species can be estimated using DNA hybridization and observing how quickly the DNA forms hybrid double helices.
- Comparison of DNA sequences between organisms reveals evolutionary relationships and is an important step in determining the function of newly isolated genes.
- DNA obtained from preserved extinct organisms can be amplified and compared to DNA sequences in modern species.

Molecular Clocks

- Molecular clocks apply mutation rates to time scales in order to estimate when two individuals or types of organisms most recently shared ancestors.
- Parsimony analysis selects likely evolutionary trees from DNA data.

Neanderthals Revisited

- Molecular clocks have been used to examine the relationship of Neanderthals to modern humans.
- Analysis of available DNA sequence data suggests that humans and Neanderthals did not ever interbreed.
- Mitochondrial DNA analysis suggests that Neanderthals and modern humans last shared a common ancestor 550,000 to 700,000 years ago.

Choosing Clues

- Genes change (mutate) at different rates. Sometimes different conclusions arise from comparing different sequences.
- Mitochondrial DNA clocks trace maternal lineages, and Y chromosome sequences trace paternal lineages.
- Analyses of mitochondrial DNA and Y chromosome information has been used to examine the origin of modern humans in Africa or in many regions and the migration pattern of native Americans from Asia.
- Mitochondrial DNA and Y chromosomal DNA support a "out of Africa" origin of Homo sapiens about 200,000 years ago.
- An "out of Mongolia" model for the origin of Native Americans is supported by mitochondrial DNA and Y chromosomal DNA. Contributions from other Siberian populations are evident.

Eugenics

- Eugenics is the control of individual reproduction for societal goals.
- In the early twentieth century several different eugenic policies were promoted and implemented.
- Positive eugenic policies aimed to maximize the genetic contribution of those deemed acceptable or superior (positive eugenics).
- Negative eugenic polices were designed to minimize the contribution of those considered inferior.
- Some aspects of genetic technology also affect reproductive choices and has been compared to eugenics.
- The goal of genetic screening is to alleviate human suffering rather than to change society.
- Laws have been proposed and passed in many nations around the world to prevent genetic discrimination.

Genetics of Immunity

The Importance of Cell Surfaces

- Foreign antigens (molecules) elicit an immune response from a host.
- Antibodies and cytokines produced by the immune system attack foreign antigens.

- As many as 20,000 genes in the human genome may be, directly or indirectly, involved in an immune response.

Blood Groups

- Results of early transfusions were inconsistent.
- With the discovery of human blood groups systems, properly matched donor and recipient types resulted in successful transfusions.
- Blood types, including the ABO groups, Rh factor, and others, result from self antigen patterns on red blood cells.
- Rh incompatibility may put a fetus at risk.
- RhoGAM can be used to prevent Rh incompatibility.

The Human Leukocyte Antigens

- In addition to blood group antigens, a large number of cell surface molecules are recognized by the immune system.
- Many of the cell surface proteins that help to establish immunity are encoded by the approximately seventy genes of the major histocompatibility complex (MHC) located on chromosome 6 in humans.
- Class I and II genes of the MHC encode human leukocyte antigens (HLA).
- Class III genes produce plasma proteins involved in the immune response.
- HLA antigens on leukocytes are involved in processing of foreign antigens.
- HLA proteins are each encoded by several genes with many alleles and as a result only 2 in every 20,000 unrelated people match for the six major HLA genes by chance.
- Individuals with certain HLA combinations have an increased risk of developing certain HLA-linked diseases.

The Immune System

- At the cellular level the immune system consists of various types of lymphocytes and macrophages.

- The immune response consists of an immediate, generalized, innate immunity and a slower, more specific adaptive immunity.

Physical Barriers and the Innate Immune Response

- Skin, mucous membranes, tears, and saliva are examples of physical and chemical barriers that keep pathogens from entering the body.
- Pathogens that breach this barrier encounter an innate immune response consisting of inflammation, phagocytosis, complement, collectins, and cytokines.

The Adaptive (Acquired) Immune Response

- The innate response is rapid (minutes) while the aquired immune response must be stimulated to action and takes days to respond.
- The acquired immune response is divided into the humoral and cellular immune responses. Both of these responses differ from the innate immune response in that they are more complex, highly specific and each has a cellular component with memory.
- The humoral response involves B cells that secrete antibodies in order to neutralize, clump, and stimulate the destruction of pathogens by recognizing and binding specific foreign antigens.
- Antibodies are made of Y-shaped polypeptides consisting of constant and variable regions.
- The astounding diversity of antibody binding activities is due to a shuffling of gene pieces (exons) encoding antibody polypeptide products in B cells.
- In the cellular immune response, helper T cells stimulate B cells to manufacture antibodies and cytotoxic T cells to secrete cytokines.
- Some T cells bind to nonself cells and virus-covered cells and burst them. Other T cells function to coordinate the immune response.

Abnormal Immunity

- Inherited Immune Deficiencies
- Inherited immune deficiencies represent defects in the genes that encode proteins involved in immunity.

Acquired Immune Deficiency Syndrome

- Acquired immune deficiency syndrome is caused by HIV.
- HIV replicates very rapidly, and T cell production matches it until the immune system is overwhelmed and AIDS begins.
- HIV is a retrovirus that injects its RNA into host cells by binding receptors.
- Reverse transcriptase then copies viral RNA into DNA.
- HIV uses the cell's protein synthesis machinery to mass produce itself; then the cell bursts, releasing virus.
- Reverse transcriptase and protease inhibitors have been effective in slowing down HIV.
- A new fusion inhibitor was introduced in 2002.
- HIV continually mutates, and may become resistant to drugs.

Autoimmunity

- In autoimmune disorders, autoantibodies attack healthy tissue.
- These conditions may be caused by a virus that borrows a self antigen, Tcells that never learn to recognize self, or healthy cells bearing antigens that resemble nonself antigens.
- Some conditions thought to be autoimmune may actually reflect an immune system response to retained fetal cells.
- Mutations in some genes may present the symptoms of an autoimmune disease.

Allergies

- An overly sensitive immune system causes allergies.
- In an allergic reaction, allergens bind to IgE antibodies on mast cells, which release allergy mediators.
- A subset of helper T cells secrete cytokines that contribute to allergy symptoms.

Altering Immune Function

- Vaccines
- Vaccines are disabled pathogens or their parts that elicit an immune response, protecting against infection by the active pathogen.

- Immunotherapy
- Immunotherapy uses immune system components to fight disease.
- Hybridomas are artificial cells that consist of a B cell fused with a cancer cell and produce monoclonal antibodies (Mabs) that can target specific antigens.
- Cytokines boost immune function and destroy cancer cells.

Transplantation

- Autografts transfer tissue from one part of a person's body to another.
- Isografts are tissue transfers between identical twins.
- Allografts involve tissue transfers between members of the same species. These transplants can cause tissue rejection reactions.
- A Xenograft is a cross species transplant. A danger of these transplants is that they can set off a hyperacute rejection.
- Graft-versus-host disease involves a rejection of recipient tissues by transplanted bone marrow.
- The success rate of transplants is improved by the use of immunosuppressive drugs, by stripping antigens from donor tissue, and by matching donor to the recipient.

A Genomic View of Immunity-The Pathogen's Perspective

- Analyzing the genomes of pathogens may reveal the molecular basis of pathogenesis.

Crowd Diseases

- Crowd diseases readily spread through populations and can cause epidemics.
- Native populations with no immunity can be devastated by introduction of new diseases.

Bioweapons

- A variety of pathogens have been adapted to military use including anthrax.

Chapter – 27

Glossary

A : Adenine residue, in either DNA or RNA.

A priori : Deduced from first principles; without prior knowledge.

A site : The binding site for the aminoacyl-tRNA on the ribosome.

Ab : See antibody.

Abaxial : Directed away from the stem of a plant; pertaining to the lower surface of a leaf (see adaxial).

Abiotic : Pertaining to the absence of life, as diseases not caused by living organisms.

Abiotic stress : The effect of non-living factors which can harm living organisms. These non-living factors include drought, extreme temperatures, pollutants, etc.

Ablation experiment : An experiment designed to produce an animal deficient in one or a few cell types, in order to study cell lineage or cell function. The idea is to make a transgenic mouse with a toxin gene (often diphtheria toxin) under control of a specialized promoter which activates only in the target cell type. When embryo development progresses to the point where it starts to form the target tissue, the toxin gene is activated, and that specific tissue dies. Other tissues are unaffected.

Abortive transduction : An event where the fragment of DNA introduced by transduction fails to be recombined into the recipient chromosome and because the DNA fragment lacks an origin of replication it is only inherited by one of the daughter cells at each cell division.

Abscise : To separate from a plant, as leaves, flowers, and fruits do when they fall.

Abscisic acid : A plant growth regulator involved in abscission, dormancy, stomatal opening/closure, and inhibition of seed germination. It also affects the regulation of somatic cell embryogenesis in some plant species.

Absciss; abscissa : The horizontal axis of a graph. cf ordinate.

Abscission : The shedding of leaves or other plant parts as the result of physical weakness in a specialized layer of cells (the abscission layer) that develops at the base of the structure.

Absorb (l ab, away + sorbere, to suck in): To suck up, or to take in. In the cell, materials are taken in (absorbed) from a solution. cf adsorb.

Absorption: In general: the process of absorbing; taking up of water and nutrients by assimilation or imbibition. The taking up by capillary, osmotic, chemical or solvent action, such as the taking up of a gas by a solid or liquid, or taking up of a liquid by a solid. cf adsorption.

In biology: the movement of a fluid or a dissolved substance across a cell membrane.

In plants: water and mineral salts are absorbed from the soil by roots.

In animals: solubulized food material is absorbed into the circulatory system through cells lining the alimentary canal.

Abzyme : See catalytic antibody.

Acaricide : A pesticide used to kill or control mites or ticks.

Acceptable Daily Intake (ADI) : A measure of the level of a chemical, i.e. a pesticide residue, that is believed to be able to be consumed on a daily basis over a lifetime without harm; used by the U.S. government to establish safe levels of potentially hazardous substances in food.

Accessory bud : Lateral bud occurring at the base of a terminal bud or at the side of an axillary bud.

Acclimatization : The adaptation of a living organism (plant, animal or micro-organism) to a changed environment that subjects it to physiological stress. Acclimatization should not be confused with adaptation (q.v.). cf acquired.

Acellular : Describing tissues or organisms that are not made up of separate cells but often have more than one nucleus. cf syncytium.

Acentric chromosome : Chromosome fragment lacking a centromere.

Acervulus (pl acervuli) : Erumpent, cushionlike fruiting body bearing conidiophores, conidia, and sometimes setae.

Acetanilide : The active ingredient in a class of selective herbicides used predominantly in corn. Acetanilides are pre-emergent herbicides.

Acetyl co-Enzyme A; acetyl coa : A compound formed in the mitochondria when an acetyl group (CH_3CO-) - derived from breakdown of fats, proteins, or carbohydrates - combines with the thiol group (-SH) of co-enzyme A.

Achondroplasia : The most common and well known form of short limbed dwarfism characterized by a normal trunk size with disproportionally short arms and legs, and a disproportionally large head; autosomal dominant condition.

Acid : Having a pH of less than 7.

Aacid precipitation : Precipitation (fog, rain, snow) with a low pH, due to the presence of nitric and sulfuric acid formed by the reaction of air pollutants (NO_x and SO_2) with water.

Acid rain : Precipitation with a low pH, due to the presence of nitric and sulfuric acid formed by the combination of air pollutants (NO_x and SO2) with water.

Acquired : Developed in response to the environment, not inherited, such as a character trait (acquired characteristic) resulting from environmental effect(s). cf acclimatization.

Acquired genetic mutation : See: somatic cell genetic mutation.

Acridine dyes : A class of positively charged polycyclic molecules that intercalate into DNA and induce frameshift mutations.

Acrocentric : A chromosome that has its centromere near the end.

Acropetal : Upward from the base to the apex of a shoot of a plant; in fungi, the production of spores in succession in the direction of the apex so that the apical spore is the youngest (see basipetal).

Acropetal : 1. Developing or blooming in succession towards the apex, such as leaves or flowers developing acropetally. 2. Transport or movement of substances towards the apex, such as the movement of water through the plant. The opposite tendency is termed basipetal.

Acrosome : An apical organelle in the head of a spermatozoon, q.v.

Acrylamide : Used to make polyacrylamide gels for separation of macromolecules by electrophoresis. Polyacrylamide gels are produced by polymerization of acrylamide into linear chains and cross-linking the acrylamide chains with bis-acrylamide (N,N'-methylene-bis-acrylamide). Polymerization is initiated by adding ammonium persulfate and the reaction is accelerated by TEMED (N,N,N',N'-tetramethylethylenediamine) which catalyzes the formation of free radicals from ammonium persulfate. Acrylamide gels are generally poured between glass plates because oxygen inhibits the polymerization because oxygen radicals formed can interact with acrylamide and terminate chain elongation.

Acrylamide gels : A polymer gel used for electrophoresis of DNA or protein to measure their sizes (in daltons for proteins, or in base pairs for DNA). Acrylamide gels are especially useful for high-resolution separations of DNA in the range of tens to hundreds of nucleotides in length.

Actin : One of the two contractile proteins in muscle (the other being myosin). Actin is also found in the microfilaments that form part of the cytoskeleton of all cells.

Actinomycete : A group of microorganisms similar to bacteria that produce long filaments.

Activated charcoal; activated carbon : Charcoal which has been treated to remove hydrocarbons and to increase its adsorptive properties. It acts by condensing and holding a gas or solute onto its surface; thus inhibitory substances in nutrient medium may be adsorbed to charcoal included in the medium. Rooting factors such as phenolamines present as contaminants in charcoal may stimulate growth *in vitro*. Its addition to rooting medium may stimulate root initiation in some plant species. Activated charcoal may differ in origin and in composition. cf charcoal; phenolic oxidation.

Activator : A gene product (usually protein) that positively regulates transcription. Activators may either increase binding of RNA polymerase to the promoter (closed complex formation) or stimulate RNA polymerase to begin transcription (open complex formation).

Active collection : In PGR: Defined in the International Undertaking on Plant Genetic Resources (FAO, 1983) as a collection which complements a base collection (q.v.) and is a collection from which seed samples are drawn for distribution, exchange and other purposes such as multiplication and evaluation.

Active site : 1. A site on the surface of a catalyst at which activity occurs 2. The site on the surface of an enzyme molecule that binds the substrate molecule.

Acute : Pertaining to symptoms that develop suddenly(see chronic).

Acute toxicity : Ability of a single dose of a compound to poison (see chronic toxicity).

Adaptation : The process by which individuals (or parts of individuals), populations, or species change form or function in such a way to better survive under given environmental conditions. Also the result of this process.

Adaptation traits : In AnGR: The complex of traits related to reproduction and survival of the individual in a particular production environment. Adaptation traits contribute to individual fitness; they are the traits subjected to selection during the evolution of animal genetic resources. By definition, these traits are also important to the ability of the animal genetic resource to be sustained in the production environment.

Adapter : A short chemically synthesized DNA double strand which can be used to link the ends of two DNA molecules.

Adaption : The ability to physiologically adjust to a new environment while exposed to that particular environment. Typically cells de-adapt when transferred to different growth conditions.

Adaptive radiation : The evolution of new forms, sub-species or species from one species of plant or animal in order to exploit new habitats or food sources. a.k.a. divergent evolution.

Adaptor : 1. A synthetic double-stranded oligonucleotide that has a blunt end, while the other end has a nucleotide extension that can base pair with a cohesive end created by cleavage of a DNA molecule with a specific type II restriction endonuclease. The blunt end of the adaptor can be ligated to the ends of a target DNA molecule and the construct can be cloned into a vector by using the cohesive ends of the adaptor. 2. A synthetic single-stranded oligonucleotide that, after self-hybridization, produces a molecule with cohesive ends and an internal restriction endonuclease site. When the adaptor is inserted into a cloning vector by means of the cohesive ends, the internal sequence provides a new restriction endonuclease site.

Adaxial : Directed toward the stem of a plant; pertaining to the upper surface of a leaf (see abaxial).

Addendum : In formulation of tissue culture media: an item or a constituent substance to be added.

Additive allelic effects : Effects of alleles at a locus, where the heterozygote is exactly intermediate between the two homozygotes.

Additive gene effects : Additive allelic effects summed across all the loci that contribute to genetic variation in a quantitative trait.

Additive genes : A form of allelic interaction in which there is no dominance. The heterozygote is intermediate in phenotype between homozygotes for the alternative alleles. For multiple-gene traits, approximately equal contributions made by many loci.

Additive genetic effects : When the combined effects of alleles at different loci are equal to the sum of their individual effects. See also: anticipation, complex trait.

Adenilate cyclase : The enzyme that catalyses the formation of cyclic AMP.

Adenine : One of the four bases in DNA that make up the letters ATGC, adenine is the "A". The others are "G" for guanine, "C" for cytosine, and "T" for thymine. Adenine always pairs with thymine. Cytosine always pairs with guanine. These letters are used as shorthand for the sequences of fragments of DNA e.g. CCAAGTAC. These sequences are the code for genetic information.

Adenosine disphosphate : (ADP) See ATP.

Adenosine monophosphate : (AMP) See ATP.

Adenosine triphosphate (ATP) : A compound with high energy phosphate bonds that provide the energy for many cellular processes.

Adenovirus : A group of DNA viruses which cause diseases in animals. In man, they produce acute respiratory tract infections with symptoms resembling common cold. They are used in gene cloning, as vectors for expressing large amounts of recombinant proteins in animal cells. They are also used to make live-virus vaccines against more dangerous pathogens. See viral vaccines.

ADEPT : A way to target a drug to a specific tissue. The drug is administered as an inactive pro-drug, and then converted into an active drug by an enzyme administered with a second injection. The enzyme is coupled to an antibody that concentrates it in the target tissue. When the enzyme arrives at the target tissue, the pro-drug is activated to form the active drug, while elsewhere it remains inactive. See drug delivery; targeted drug delivery.

Adhesion : The attraction of dissimilar molecules for each other. A sticking together of unlike substances, such as soil and water.

A-DNA : A right-handed DNA double helix that has 11 base pairs per turn. DNA exists in this form when partially dehydrated.

ADP : (adenosine diphosphate) See ATP.

Adpressed (syn appressed) : Closely flattened down or pressed against a surface.

Adsorb : See adsorption.

Adsorbent : Noun: A substance to which compounds adhere. In tissue culture, an adsorbent is added to the culture medium to adsorb compounds released by cultured cells or tissues, thus minimizing any adverse effect on the subsequent growth in culture. A common adsorbent in tissue culture is activated charcoal, q.v.

Adsorption : The formation of a layer of gas, liquid or solid on the surface of a solid. cf absorption.

Adult cloning : The creation of identical copies of an adult animal by nuclear transfer (q.v.) from differentiated adult tissue. See also cloning; Dolly.

Advanced : Applied to an organism or a part thereof, implying considerable development from the ancestral stage or from the explant stage.

Advanced maternal age : Women over age 34 (age 35 at delivery) at increased risk for nondisjunction trisomy in fetus.

Adventitious : A structure arising at sites other than the usual ones, e.g., shoots from roots or leaves, and embryos from any cell other than a zygote.

Aeciospore : Dikaryotic spore of a rust fungus produced in an aecium; in heteroecious rusts, a spore stage that infects the alternate host.

Aecium (pl aecia; adj aecial) : The fruiting body of a rust fungus in which the first dikaryotic spores (aeciospores) are produced.

Aerate : To supply with or mix with air or gas. The process is aeration.

Aerial : Occurring in the air.

Aerification : The act of infusing or forcing air into, for example, soil.

Aerobe : A micro-organism that grows in the presence of oxygen. Opposite: anaerobe.

Aerobic : The presence of oxygen. Often used to describe an organism that requires oxygen for growth or growth conditions with oxygen.

Aerobic bacteria : Bacteria that can live in the presence of oxygen.

Aerobic respiration : A type of respiration in which foodstuffs are completely oxidized to carbon dioxide and water, with the release of chemical energy, in a process requiring atmospheric oxygen.

Affected relative pair : Individuals related by blood, each of whom is affected with the same trait. Examples are affected sibling, cousin, and avuncular pairs. See also: avuncular relationship.

Affinity : The strength of an interaction between two molecules, such as the binding of an antigen to an antibody.

Affinity chromatography : A method for separating molecules by exploiting their ability to bind specifically to other molecules. There are several types of biological affinity chromatography. A biological molecule can be immobilized and a smaller molecule (ligand, q.v.,) to which it is to bind can be stuck to it, or the smaller ligand can be immobilized and the macromolecule stuck to it. A variant is to use an antibody as the immobilized molecule and use it to "capture" its antigen: this is often called immuno-affinity chromatography. A variation is pseudo-affinity chromatography, in which a compound which is like a biological ligand is immobilized on a solid material, and enzymes or other proteins are bound to it. Other techniques include metal affinity chromatography, where a metal ion is immobilized on a solid support: metal ions bind tightly and specifically to many biomolecules. The metal ion is bound to a chelator or chelating group, a chemical group that binds specifically and extremely tightly to that metal.

Affinity tag; purification tag : An amino acid sequence that has been engineered into a protein to make its purification easier. These can work in a number of ways. The tag could be another protein, which binds to some other material very tightly and thus allowing the protein to be purified by affinity chromatography (q.v.). The tag could be a short amino acid sequence, which is recognized by an antibody. The antibody would then bind to the protein whereas it would not have done so before. One such short peptide, called FLAG, has been designed so that it is particularly easy to make antibodies against it. The tag could be a few amino acids, which are then used as a chemical tag on the protein. For example, a string of positively charged amino acids will bind very strongly to a negatively charged filter: this could be used as the basis of a separation system. Some amino acids bind metals very strongly, especially in pairs: this chemical property can be exploited by using a filter with metal atoms chemically linked onto it to pull a protein out of a mixture of proteins. cf affinity chromatography.

Aflatoxin : Toxic compounds, produced by moulds (fungi) of the Aspergillus flavus group, that bind to DNA and prevent replication and transcription. Aflatoxins can cause acute liver damage and cancer. Animals may be poisoned by eating stored food or feed contaminated with the mould.

AFLP : See amplified fragment length polymorphism.

Ag : See antigen.

Agar : (Malay, agar-agar) A polysaccharide solidifying agent used in nutrient media preparations and obtained from certain types of red algae (Rhodophyta). Both the type of agar and its concentration can affect the growth and appearance of cultured explants.

Agarose : An uncharged polysaccharide purified from agar. Agarose melts when heated to 100 C and resolidifies when cooled below about 50°C. When it solidifies agarose forms a matrix. The size of the pores in the gel matrix can be varied by using different concentrations of agarose: the higher the concentration of agarose, the smaller the pore size.

Agarose gel electrophoresis : A method to analyze the size of DNA (or RNA) fragments. In the presence of an electric field, larger fragments of DNA move through a gel slower than smaller ones, producing different migrating "bands". Usually, these are visualized by soaking the gel in a dye (ethidium bromide) which makes the DNA fluoresce under UV light. This is the gel of choice for DNA or RNA in the range of thousands of bases in length, or even up to 1 megabase if you are using pulsed field gel electrophoresis. (See also electrophoresis).

Aggregate : 1. A clump or mass formed by gathering or collecting units. 2. A body of loosely associated cells, such as a friable callus or cell suspension.3. Coarse inert material, such as gravel, that is mixed with soil to increase its porosity. 4. A serological reaction (aggregation) in which the antibody and antigen react and precipitate out of solution.

Aggregation technique : A technique used in model organism studies in which embryos at the 8-cell stage of development are pushed together to yield a single embryo (used as an alternative to microinjection). See also: model organisms.

Aggressiveness : Relative ability of a plant pathogen to colonize and cause damage to plants (see virulence).

Agonist : A drug, hormone or transmitter substance that forms a complex with a receptor site that is capable of triggering an active response from a cell.

Agricultural biological diversity : See agrobiodiversity.

Agrobacterium : A genus of bacteria that includes several plant pathogenic species, causing tumour-like symptoms. See Agrobacterium tumefaciens; crown gall; hairy root culture; Ri plasmid; Ti plasmid.

Agrobacterium tumefaciens : A bacterium that causes crown gall disease in some plants. The bacterium infects a wound, and injects a short stretch of DNA into some of the cells around the wound. The DNA comes from a large plasmid - the Ti (tumour induction) plasmid - a short region of which (called T-DNA, = transferred DNA) is transferred to the plant cell, where it causes the cell to grow into a tumour-like structure. The T-DNA contains genes which inter alia allows the infected plant cells to make two unusual compounds, nopaline and octopine, that are characteristic of transformed cells. The cells form a gall, which hosts the bacterium. This DNA-transfer mechanism is exploited in the genetic engineering of plants. The Ti plasmid is modified so that a foreign gene is transferred into the plant cell along with or instead of the nopaline synthesis genes. When the bacterium is cultured with isolated plant cells or with wounded plant tissues, the "new" gene is injected into the cells and ends up integrated into the chromosomes of the plant.

Agrobacterium tumefaciens-mediated transformation : A naturally occurring process of DNA transfer from the bacterium A. tumefaciens to plants.

Agrobiodiversity; agricultural biological diversity : That component of biodiversity that is relevant to food and agriculture production. The term agrobiodiversity Encompasses within-species, species and ecosystem diversity. (Based on FAO, 1999).

AI : See artificial insemination.

AIDS (acquired immunodeficiency syndrome) : The usually fatal human disease in which the immune system is destroyed by a retrovirus (Human Immunodeficiency Virus, HIV). The virus infects and destroys helper T-cells, which are essential for combating infections.

Air pollution : Any air contaminant that causes undesirable effects on living organisms or materials.

Airlift fermenter : A cylindrical fermentation vessel in which the cells are mixed by air introduced at the base of the vessel and that rises through the column of culture medium. The cell suspension circulates around the column as a consequence of the gradient of air bubbles in different parts of the reactor.

Alate : winged (see apterous).

Albinism : Hereditary absence of pigment in an organism. Albino animals have no colour in their skin, hair and eyes. The term is also used for absence of chlorophyll in plants.

Albino : 1. An organism lacking pigmentation, due to genetic factors. The condition is albinism, q.v. 2. A conspicuous plastome (plastid) mutant involving loss of chlorophyll.

Alcoholism : A chronic and progressive condition characterized by the inability to control the consumption of alcohol.

Aleurone : The outermost layer of the endosperm in a seed.

Algal biomass : Single-celled plants, such as Chlorella spp. and Spirulina spp., are grown commercially in ponds to make feed materials. Chlorella is grown commercially to make into fish food: it is fed to zooplankton, and these in turn are harvested as feed for fish farms. This is a means of converting sunlight into food in a way more convenient and controllable than normal farming.

Alginate : Alkylating agents Chemicals that transfer alkyl (methyl, ethyl, etc.) groups to the bases in DNA.

Alkaline : Having basic (nonacidic) properties; pH greater than 7.

Alkaline phosphatase : An enzyme which catalyzes the hydrolysis of phosphomonoesters of the 5' nucleotides. Used to dephosphorylate (remove phosphate groups from) the 5' ends of DNA or RNA molecules, to facilitate 5' end-labeling with 32P added back by T4 polynucleotide kinase; or to dephosphorylate the 5' ends of DNA molecules to prevent unwanted ligation reactions during cloning.

Alkaloids : Nitrogen-containing ring compounds produced by plants that cause physiological effects in animals.

Alkylating agent : An organic compound able to transfer an alkyl group to nucleotides.

Allele : (Gr. allelon, of one another, mutually each other); allelomorph (adj: allelic, allelomorphic). One of a pair, or series, of variant forms of a gene that occur at a given locus in a chromosome. Alleles are symbolized with the same basic symbol (e.g., B for dominant and b for recessive); B1, B2, ..., Bn for n additive alleles at a locus). In a normal diploid cell there are two alleles of any one gene (one from each parent), which occupy the same relative position (locus) on homologous chromosomes. Within a population there may be more than two alleles of a gene. See multiple alleles.

Allele frequency : The number of copies of an allele in a population, expressed as a proportion of the total number of copies of all alleles at a locus in a population.

Allele specific suppressor : A second-site mutation that repairs the mutant phenotype but only in strains with certain, specific mutations at the first-site.

Allele-specific amplification (ASA) : The use of polymerase chain reaction (PCR) at a sufficiently high stringency that only a primer with exactly the same sequence as the target DNA will be amplified. A powerful means of genotyping for single-locus disorders that have been characterized at the molecular level.

Allele-specific ligation : Technique permitting discrimination of two allele at locus by providing two short synthetic oligonucleotides that would bind ajjacent to teach other on amplified DNA fragment, and would be ligated in presence of DNA ligase; if one of alleles containing mutation overlapped by 3'-end of one oligonucleotide, its ligation to oligonucleotide bound 3' to it would be prevented; to deduce identity of unknown allele, differentially labelled oligonucleotide pairs may be designed for each allele and their ligation efficiency compared in presence of unknown allele.

Allele-specific PCR (AS-PCR) : Refers to amplification of specific alleles, or DNA sequence varients at same locus; specificity is achieved by designing one or both PCR primers so that they partially overlap site of sequence difference between amplified alleles.

Allelic exclusion : A phenomenon whereby only one functional allele of an antibody gene can be assembled in a given B lymphocyte. The "allele" on the other homologous chromosome in a diploid mammalian cell cannot undergo a functional re-arrangement, which would result in the production of two different antibodies by a single plasma cell.

Allelomorph : See allele.

Allelopathy : The phenomenon by which the secretion of chemicals, such as phenolic and terpenoid compounds, by a plant inhibits the growth or reproduction of other plant species with which it is competing.

Allergen : An antigen that provokes an immune response.

Allogamy : Cross fertilization in plants. See fertilization.

Allogeneic : Variation in alleles among members of the same species.

Allometric : When the growth rate of one part of an organism differs from that of another part or of the rest of the body.

Allopatric speciation : Speciation occurring at least in part because of geographic isolation.

Allopolyploid : A polyploid organism (usually a plant) having multiple sets of chromosomes derived from different species. Hybrids are usually sterile, because they do not have sets of homologous chromosomes and therefore pairing cannot take place. However, if doubling of the chromosome number occurs in a hybrid derived from two diploid (2n) species, the resulting tetraploid (4n) is a fertile plant, since it contains two sets of homologous chromosome and pairing may occur; this tetraploid is an allotetraploid.

Allosteric : The stereospecific modification of a protein by an effector to influence the activity of another site of the protein.

Allosteric control : See allosteric regulation.

Allosteric enzyme : An enzyme that has two structurally distinct forms, one of which is active and the other inactive. Active forms of allosteric enzymes tend to catalyse the initial step in a pathway leading to the synthesis of molecules. The end product of this synthesis can act as a feedback inhibitor, converting the enzyme to the inactive form, thus controlling the amount of product synthesized.

Allosteric regulation : A catalysis-regulating process in which the binding of a small effector molecule to one site on an enzyme affects the activity at another site.

Allosteric transition : A reversible interaction of a small molecule with a protein molecule, resulting in a change in the shape of the protein and a consequent alteration of the interaction of that protein with a third molecule.

Allotetraploid : An organism with four genomes derived from hybridization of different species. Usually, in forms that become established, two of the four genomes are from one species and two are from another species. See allopolyploid.

Allozygote : A diploid individual that is homozygous at a locus in which the two genes are not identical by descent from a common ancestor.

Allozyme : See allosteric enzyme.

Alpha-complementation : The ability of a short N-terminal fragment of beta-galactosidase to interact with the C-terminal portion of beta-galactosidase to form a functional enzyme.

Alpha-fetoprotein (AFP) : A protein excreted by the fetus into the amniotic fluid and from there into the mother's bloodstream through the placenta.

Alphalactalbumin : Protein component of milk.

Alternate host : One of two kinds of plant on which a parasitic fungus (e.g. rust) must develop to complete its life cycle.

Alternation of generations : A reproductive cycle in which a haploid phase alternates with a diploid phase.

Alternative host : A plant other than the main host that a parasite can colonize; alternative hosts are not required for completion of the developmental cycle of the parasite.

Alternative mRNA splicing : The inclusion or exclusion of different exons to form different mRNA transcripts. See RNA.

Alternative splicing : Refers to fact that certain genes retain or omit particular exons in the final spliced transcript.

Alu repetitive sequence : The most common dispersed repeated DNA sequence in the human genome accounting for 5% of human DNA. The name is derived from the fact that these sequences are cleaved by the restriction endonuclease Alu.

Alu sequence : A family of sequence-related elements about 300 bp in length approximately 500 000 copies of which are scattered along the human genome.

Amber codon : The UAG nonsense codon. This is one of the three normal chain termination or stop codons that does not code for any amino add in the universal code.

Amber mutation : A mutation that creats an amber codon (UAG) from a different codon.

Amber suppressor : A mutant tRNA gene that recognizes the UAG codon but is aminoacylated and inserts the amino acid at the corresponding position in the growing polypeptide.

Ambient temperature : Air temperature at a given time and place; not radiant temperature.

Ames test : A genetic test for the identification of carcinogens based upon their mutagenic activity initially developed by Bruce Ames. The test relies upon the ability of a chemical or physical agent to promote the reversion of different classes of mutations that cause histidine auxotrophy.

AMH : See Atlantic Modal Haplotype.

Amino acid : Any of a class of 20 molecules that are combined to form proteins in living things. Consisting of the basic formula NH2-CHR-COOH, where "R" is the side chain which defines the amino acid:The sequence of amino acids in a protein and hence protein function are determined by the genetic code.

Amino acid analog : See analog.

Amino acid sequence : The linear order of the amino acids in a protein or peptide.

Amino Terminus : Refers to the NH2 end of a peptide chain (by custom drawn at the left of a protein sequence).

Aminoacyl site; A-site : One of two sites on ribosomes to which the incoming aminoacyl tRNA binds.

Aminoacyl tRNA synthetase : Enzyme that attaches each amino acid to its specific tRNA molecule.

Amitosis : Cell division (cytokinesis), including nuclear division through constriction of the nucleus, without chromosome differentiation as in mitosis. The maintenance of genetic integrity and diploidy during amitosis is uncertain. This process occurs in the endosperm of flowering plants.

Amniocentesis : A procedure for obtaining amniotic fluid from a pregnant mammal for the diagnosis of some diseases in the unborn foetus. Cells are cultured, and metaphase chromosomes are examined for irregularities (e.g., Down syndrome, spina bifida, etc., in humans).

Amniocyte : Cells obtained by amniocentesis.

Amnion : The thin membrane that lines the fluid-filled sac in which the embryo develops in higher vertebrates, reptiles and birds.

Amniotic fluid : Liquid contents of the amniotic sac of higher vertebrates, containing cells of the embryo (not of the mother). Both fluid and cells are used for diagnosis of genetic abnormalities in the embryo or foetus.

Amorph; null mutation : A mutation that obliterates gene function.

AMP (adenosine monophosphate) : See ATP.

Amp resistance : See "Antibiotic resistance".

Amphid : A chemosensory organ located in the anterior region of a nematode.

Amphidiploid : A species or type of plant derived from doubling the chromosomes in the F1 hybrid of two species; an allopolyploid. In an amphidiploid the two species are known, whereas in other allopolyploids they may not be known.

Amphigynous : Having an antheridium through which the oogonium grows, as in many Phytophthora species (see paragynous).

Amphimixis : Copulation of two unrelated cells and nuclei, e.g., egg and sperm; reproduction by a sexual process.

Amphimobile : Able to move both upward and downward in phloem, as do some pesticides.

Ampicillin (Amp) : An antibiotic that inhibits crosslinking of peptidoglycan chains in the cell wall of eubacteria. Cells growing in the presence of ampicillin synthesize weak cell walls, causing them to burst due to the high internal osmotic pressure. Amp resistance encoded by most transposons and plasmids is due to a periplasmic Beta-lactamase that breaks the Beta-lactam ring of ampicillin. Ampicillin is a derivative of penicillin.

Amplification : 1. Treatment (e.g., use of chloramphenicol) designed to increase the proportion of plasmid DNA relative to that of bacterial (host) DNA. 2. Replication of a gene library in bulk.3. Duplication of gene(s) within a chromosomal segment. 4. Creation of many copies of a segment of DNA by the polymerase chain reaction (PCR).

Amplified fragment length polymorphism (AFLP) : A type of DNA marker, generated by digestion of genomic DNA with two restriction enzymes to create many DNA fragments, ligation of specific sequences of DNA (called adaptors) to the ends of these fragments, amplification of the fragments via PCR (using a set of primers with sequences corresponding to the adapters, plus various random combinations of three additional bases at the end), and visualization of fragments via gel electrophoresis. The PCR will amplify any fragment whose sequence happens to start with any of the three-base sequences in the set of primers. AFLPs have the important advantage that many markers can be generated with relatively little effort. They are a very useful means of quantifying the extent of genetic diversity within and between populations. Their major disadvantage is that they are not specific to a particular locus and, because they are scored as the presence or absence of a band, heterozygotes cannot be distinguished from homozygotes, i.e., they are inherited in a dominant fashion.

Amplify : To increase the number of copies of a DNA sequence, either *in vivo* by inserting into a cloning vector that replicates within a host cell, or *in vitro* by polymerase chain reaction (PCR).

Ampometric : See enzyme electrode.

Amylase : A group of enzymes that degrade starch, glycogen and other polysaccharides, producing a mixture of glucose and maltose. Plants have both α- and β-amylase; animal have only α-amylase.

Amylolytic : The capability of breaking down starch into sugars.

Amylopectin : A polysaccharide comprising highly branched chains of glucose molecules. The water-insoluble portion of starch.

Amylose : A polysaccharide consisting of linear chains of 100 to 1000 glucose molecules. The water-soluble portion of starch.

Anabolic pathway : A pathway by which a metabolite is synthesized; a biosynthetic pathway.

Anaerobe : An organism that can grow in the absence of oxygen. Opposite: aerobe.

Anaerobic : The absence of oxygen. Often used to describe an organism that is sensitive to oxygen or growth conditions without oxygen.

Anaerobic digestion : Digestion of materials in the absence of oxygen. See anaerobic respiration.

Anaerobic respiration : Respiration in which foodstuffs are partially oxidized, with the release of chemical energy, in a process not involving atmospheric oxygen, such as alcoholic fermentation, in which one of the end products is ethanol.

Analog (or the British spelling Analogue) : A chemical that is similar, but not identical, to another chemical. For example, thioproline is an analog of the amino acid proline, and azaguanine is an analog of the nucleotide guanine. Certain analogs are toxic and can be used to select for mutants that affect the synthesis or metabolism of the corresponding compound.

Analogous : Features of organisms or molecules that are superficially or functionally similar but have evolved in a different way or contain different compounds.

Analysis of variance : A statistical analysis by which F values (variance ratios) are composed in such a manner as to determine the probability that differences among populations or treatments are too large to be due to chance.

Anamorph (adj anamorphic; syn imperfect state) : The asexual form in the life cycle of a fungus, when asexual spores (such as conidia) or no spores are produced (see holomorph and teleomorph).

Anaphase (Gr ana, up + phais, appearance) : The stage of mitosis or meiosis during which the daughter chromosomes (sister chromatids) pass from the equatorial plate to opposite poles of the cell (toward the ends of the spindle). Anaphase follows metaphase and precedes telophase.

Anastomosis (pl anastomoses) : Fusion between branches of the same or different structures (e.g. hyphae) to make a network.

Anchor gene : A gene that has been positioned on both the physical map and the linkage map of a chromosome.

Anchor sequence : A hydrophobic amino acid sequence which fixes a segment of a newly synthesized, translocating protein within the lipid bilayer membrane of the endoplasmic reticulum.

Androgen : Any hormone that stimulates the development of male secondary sexual characteristics, and contributes to the control of sexual activity in vertebrate animals. Usually synthesized by the testes.

Androgenesis : Male parthenogenesis, i.e., the development of a haploid embryo from a male nucleus. The maternal nucleus is eliminated or inactivated subsequent to fertilization of the ovum, and the haploid individual (referred to as androgenetic) contains in its cells the genome of the male gamete only. Androgenesis is detected by cytological staining. See anther culture; gynogenesis; parthenogenesis.

Aneuploid (Gr aneu, without + ploid) : An organism or cell having a chromosome number that is not an exact multiple of the monoploid (x) with one chromosome being present in greater (e.g., trisomic 2n + 1) or lesser (e.g., monosomic 2n - 1) number than the normal diploid number.

Angelman syndrome : A rare genetically inherited form of mental retardation due to the deletion or inactivation of specific genes on chromosome 15. The inheritance of this syndrome is subject to genome imprinting. Children with Angleman syndrome typically also have small heads, experience seizures, have pronounced speech impairment, are hyperactive, and have balance disorders.

Angstrom : A unit of measurement that was widely used until recently to describe molecular dimensions, but the unit nanometer (nm) is now more commonly used. One angstrom equals 10 nm.

Animal breeding : The application of selecting beneficial traits that are economically important to meat quality, health, reproduction and production for improvement in each successive generation of livestock. This process involves the application of genetics to livestock production through various tools, such as direct and correlated response to selection, genetic evaluation and prediction, marker assisted breeding, genomics, and artificial insemination.

Animal cell immobilization : Entrapment of animal cells in some solid material in order to produce some natural product or genetically engineered protein. Animal cells have the advantage that they already produce many proteins of pharmacological interest, and that genetically engineered proteins are produced by them with the post-translation modifications normal to animals. However, because animal cells are much more fragile than bacterial ones, they cannot tolerate a commercial fermentation process. Typical materials are hollow fibre membrane bioreactors, or porous carriers made of polysaccharide, protein, plastic or ceramic materials with microscopic holes inside which the cells grow.

Animal cloning : See cloning.

Animal genetic resources databank : A databank that contains inventories of farm animal genetic resources and their immediate wild relatives, including any information that helps to characterize these resources. (Source: FAO, 1999).

Animal genome (gene) bank : A planned and managed repository containing animal genetic resources. Repositories include the environment in which the genetic resource has developed, or is now normally found (in situ) or facilities elsewhere (ex situ - *in vivo* or *in vitro*). For *in vitro*, ex situ genome bank facilities, germplasm is stored in the form of one or more of the following: semen, ova, embryos and tissue samples. (Source: FAO, 1999).

Animal model : See: model organisms.

Anion : A negatively charged ion; opposite: cation.

Anneal : The act of two nucleic acid sequences hydrogen bonding through complementarity of the bases determining their sequences. Generally synonymous with "hybridize".

Annealing : Pairing of a single-stranded nucleic acid with the complementary strand to form a duplex.

Annotation : Adding pertinent information such as gene coded for, amino acid sequence, or other commentary to the database entry of raw sequence of DNA bases. See also: bioinformatics.

Annual : a plant that completes its life cycle and dies within one year (see biennial, perennial).

Annual ring : Single-year growth of xylem in a woody stem.

Annual (L annualis, within a year) : 1. (adj:) Taking one year, or occurring at intervals of one year. 2. Noun: In botany, a plant that completes its life cycle within one year. During this time the plant germinates, grows, flowers, produces seeds, and dies. See biennial, perennial.

Annulations : grooved bands in the cuticle of some nematodes.

Anonymous DNA marker : A DNA marker (q.v.) detectable by virtue of variation in its sequence, irrespective of whether or not it actually occurs in or near a coding sequence. Microsatellites are typical anonymous DNA markers.

Antagonism : A general term for interference between organisms that may include antibiosis or competition for nutrients or space; action of two or more pesticides that reduces the effectiveness of one or all (see synergism).

Antagonist : A compound that inhibits the effect of an agonist in such a way that the combined biological effect of the two becomes smaller than the sum of their individual effects.

Anterior : Toward the front or head (see posterior).

Anther : Microsporangium bearing microspores which develop into pollen (microgametophytes). The upper part of a stamen, containing pollen sacs within which are numerous pollen grains.

Anther culture : The aseptic culture of anthers for the production of haploid plants from microspores. See androgenesis; gynogenesis; parthenogenesis.

Antheridium (pl antheridia) : Male sexual organ (male gametangium) found in some fungi.

Anthesis : The flowering period or efflorescence. Anthesis is the time of full bloom, which lasts till fruit set.

Anthocyanin : Water-soluble blue, purple and red flavonoid pigments found in vacuoles of cells.

Anthracnose : disease caused by acervuli-forming fungi (order Melanconiales) and characterized by sunken lesions and necrosis.

Antiauxin : A chemical that interferes with the auxin response. Antiauxin may or may not involve prevention of auxin transport or movement in plants. Some antiauxins are said to promote morphogenesis *in vitro*, such as 2,3,5-tri-iodobenzoate (TIBA; f.w. 499.81), or 2,4,5-trichlorophenoxyacetate (2,4,5-T; m.w. 255.49), which stimulate the growth of some cultures.

Antibiosis : An association between organisms, or between an organism and a metabolic product of another organism, that is harmful to one of them.

Antibiotic : A class of natural and synthetic compounds that inhibit the growth of or kill some micro-organisms. Antibiotics such as penicillin are often used to control (to some extent kill) contaminating organisms. However, resistance to particular antibiotics can be acquired through mutations. Some contaminating organisms are only suppressed or their metabolism slowed to an insignificant level. See antibiotic resistance; bactericide; bacteriostat.

Antibiotic resistance : Resistance conferred to the host the ability to survive a given antibiotic by plasmids containing resistance genes. If such a plasmid is present in a host, that host will not be killed by (moderate levels of) ampicillin or tetracycline. By transforming plasmids containing antibiotic resistance genes, one can get rid of all the contaminating bacteria which does not have such plasmid, thus ensuring that the plasmid will be propagated as the surviving cells divide.

Antibiotic resistance : The ability of a micro-organism to produce a protein that disables an antibiotic or prevents transport of the antibiotic into the cell.

Antibody : An immunoglobulin protein produced by B-lymphocytes of the immune system that binds to a specific antigen molecule. (See monoclonal antibodies, polyclonal antibodies).

Antibody class : The class to which an antibody belongs, depending on the type of heavy chain present. In mammals, there are five classes of antibodies: IgA, IgD, IgE, IgG, and IgM.

Antibody structure : Antibodies have a well-defined structure. Each antibody has two identical "light" chains and two identical "heavy" chains. Each chain comprises a constant region, i.e., a region that is the same between antibodies of the same class and sub-class, and a variable region that differs between. The antigen-binding region or binding site - complementarity determining region - is in the variable region. The antibody can be cut by proteases into several fragments, known as Fab, Fab', and Fc.

Antibody-mediated (humoral) immune response : The synthesis of antibodies by B cells in response to an encounter of the cells of the immune system with a foreign immunogen.

Anticipation : Each generation of offspring has increased severity of a genetic disorder; e.g., a grandchild may have earlier onset and more severe symptoms than the parent, who had earlier onset than the grandparent. See also: additive genetic effects, complex trait.

Anticlinal : The plane of cell wall orientation or cell division perpendicular (at right angles) to the surface of an organ. See tunica; periclinal.

Anticoding strand : The strand of the DNA double helix that is actually transcribed. Also known as the antisense or template strand.

Anticodon : The 3-base sequence of a tRNA that base-pairs with the mRNA codon to effect translation of the mRNA into polypeptide sequence. (See transfer RNA and messanger RNA).

Antigen : Any foreign substance, such as a virus, bacterium, or protein that elicits an immune response by stimulating the production of antibodies.

Antigenic determinant : A surface feature of a micro-organism or macromolecule, such as a glycoprotein, that elicits an immune response. See epitope.

Antigenic drift : The accumulation of mutations that modify a molecule on the cell surface of a pathogen and thereby alters recognition by the host immune system. These changes prevent preexisting antibodies from recognizing the invading pathogen. This process is most often used to describe viruses. (See Antigenic variation below).

Antigenic switching : The altering of a micro-organism's surface antigens through genetic re-arrangement, to elude detection by the host's immune system.

Antigenic variation : Mechanism to ensure rapid sequence variation of the gene(s) encoding homologues of an individual protein antigen; usually involving multiple, related gene copies.

Antihaemophilic globulin : Blood globulin that reduces the clotting time of haemophilic blood.

Anti-idiotype antibodies : Antibodies which recognize the binding sites of other antibodies. Their binding sites are complementary to the binding sites of another immunoglobulin. When an animal becomes immune to something, it not only acquires antibodies against that something, it also acquires antibodies against those antibodies. This forms a network of antibodies which can all bind to each other to various degrees, helping to regulate the immune response. Some allergic responses are in part due to the breakdown of this sort of regulation.

Antimicrobial agent : Any chemical or biological agent that harms the growth of micro-organisms.

Antimutator : A mutation that decreases the overall mutation rate, often by modifying DNA polymerase.

Antinutrients : Compounds that inhibit normal uptake of nutrients.

Anti-oncogene : A gene whose product prevents the normal growth of tissue. cf recessive oncogene.

Antioxidant : Compound that slows the rate of oxidation reactions.

Antioxidant solution : Pre-treatment solution (e.g., Vitamin C; citric acid) that retards senescence and browning of tissue. It is employed to incubate explants prior to surface sterilization.

Antiparallel orientation : The normal arrangement of the two strands of a DNA molecule, and of other nucleic-acid duplexes (DNA-RNA, RNA-RNA), in which the two strands are oriented in opposite directions so that the 5′-phosphate end of one strand is aligned with the 3′-hydroxyl end of the complementary strand.

Antisense : Nucleic acid that has a sequence exactly opposite to an mRNA molecule made by the body; binds to the mRNA molecule to prevent a protein from being made. See also: transcription.

Antisense construct : Plasmid that contains coding region of gene placed adjacent to promoter in such orientation as to cause gene to be transcribed from DNA strand complementary to that from which RNA is normally transcribed.

Antisense DNA : 1. The strand of chromosomal DNA that is transcribed. 2. A DNA sequence that is complementary to all or part of an mRNA molecule.

Antisense gene : A gene that produces a transcript (mRNA) that is complementary to the pre-mRNA or mRNA of a normal gene (usually constructed by inverting the coding region relative to the promoter).

Antisense RNA : An RNA molecule that can hybridize to an mRNA molecule. Hybridization of an antisense RNA molecule to the translation start sites on an mRNA prevents the initiation of translation. For example, an antisense RNA is involved in the regulation of transposase expression from Tn10.

Anti-sense strand : See Sense strand.

Antisense therapy : The *in vivo* treatment of a genetic disease by blocking translation of a protein with a DNA or an RNA sequence that is complementary to a specific mRNA.

Antiseptic : Any substance that kills or inhibits the growth of disease-causing micro-organism (a micro-organism capable of causing sepsis), but is essentially non-toxic to cells of the body.

Antiserum : The fluid portion of the blood of an animal (after coagulation of the blood), containing antibodies.

Anti-terminator : A type of protein which enables RNA polymerase to ignore certain transcriptional stop or termination signals and read through them to produce longer mRNA transcripts.

Antitranspirant : A compound designed to reduce transpiration when sprayed or painted on leaves of newly transplanted trees, shrubs or vines, or used as a dip for cuttings in lieu of misting; may interfere with photosynthesis and respiration if the coating is too thick or unbroken.

AP endonuclease : An endonuclease which recognizes an AP site and cuts the defective strand on the 5' side of the missing base.

AP site : A molecule of single-stranded or double-stranded DNA missing a purine or pyrimidine base.

AP-1 site : The binding site on DNA at which the transcription "factor" AP-1 binds, thereby altering the rate of transcription for the adjacent gene. AP-1 is actually a complex between c-fos protein and c-jun protein, or sometimes is just c-jun dimers. The AP-1 site consensus sequence is (C/G) TGACT(C/A) A. Also known as the TPA-response element (TRE). (TPA is a phorbol ester, tetradecanoyl phorbol acetate, which is a chemical tumor promoter).

Apert syndrome : a condition caused by the premature closure of the sutures of the skull bones, resulting in an altered head shape, with webbed fingers and toes. Autosomal dominant.

Apex (L apex, a tip, point, or extremity; pl apices) : The tip, point or angular summit. The tip of a leaf; the portion of a root or shoot containing apical and primary meristems. Usually used to designate the apical tip of the meristem.

Aphid : small, sucking insect of the family Aphididae (order Homoptera) that produces honeydew and injures plants when in large populations.

Apical cell : A meristematic initial in the apical meristem of shoots or roots of plants. As this cell divides, new tissues are formed.

Apical dominance : The phenomenon of inhibition of growth of lateral (axillary) buds in a plant by the presence of the terminal (apical) bud on the branch, due to auxins produced by the apical bud.

Apical meristem : A region of the tip of each shoot and root of a plant in which cell division is continually occurring to produce new stem and root tissue, respectively. Two regions are visible in the apical meristem: (i)

An outer 1-4-cell layered region (called the tunica), where cell divisions are anticlinal, i.e., perpendicular to the surface; and below the tunica, (ii) the corpus, where the cells divide in all directions, giving them an increase in volume.

Apoenzyme : Inactive enzyme that has to be associated with a specific organic molecule called a co-enzyme in order to function. The apoenzyme/co-enzyme complex is called a holoenzyme.

Apomixis(Gr apo, away from + mixis, a mingling; adj apomictic) : The asexual production of diploid offspring without the fusion of gametes. The embryo develops by mitotic division of the maternal or paternal gamete, or, in the case of plants, by mitotic division of a diploid cell of the ovule. cf androgenesis; gynogenesis; panmixis; parthenogenesis.

Apoptosis : The process of cell death, which occurs naturally as a part of normal development, maintenance and renewal of tissue in an organism. Apoptosis differs from necrosis, in which cell death is caused by a toxic substance.

Aporepressor : A protein that requires binding of a small molecule co-repressor to negatively regulate gene expression.

Apothecium (pl apothecia) : Open, cuplike or saucerlike, ascus-bearing fungal fruiting body (ascocarp), often supported on a stalk.

Appressed (syn adpressed) : Closely flattened down or pressed against a surface.

Appressorium (pl appressoria) : Swollen, flattened portion of a fungal filament that adheres to the surface of a higher plant, providing anchorage for invasion by a fungus.

Apterous : Wingless (see alate).

Apurinic or apyrimidinic (AP) site : A molecule of single-stranded or double-stranded DNA missing a purine or pyrimidine base.

Aquaculture : Growing of water plants and animals, rather than harvesting them from wherever they happen to grow in rivers or seas. Usually aquaculture uses fresh water; when it uses sea water it can be called mariculture. It is considered to be a part of biotechnology (although peripheral) because it is a new commercial development, and because it often involves growing organisms in large volumes of water, which has similarities to growing large volumes of yeast or bacteria. Biotechnology also provides clean, well-aerated water for the animals to grow in; food, such as krill or powdered synthetic food; and food additives, such as astaxanthins to ensure that fish and prawns have the right colour. Aquaculture has also been used to mass-produce macro- and micro-algae for chemicals, vitamins and pigments. For both animals and plants, biotechnologists have been using genetic methods to produce triploid and tetraploid organisms, and hybrid algae through plant cell fusions. Triploid trout, for example, are sterile, and can be used for biocontrol of weeds without the threat of their being able to breed themselves.

Arabidopsis : A genus of flowering plants in the Cruciferae. A. thaliana is used in research as a model plant because it has a small genome (5 pairs of chromosomes; 2n = 10) and can be cultured easily, with a generation time of two months.

Arable : Able to be cultivated for agriculture.

Arbuscular mycorrhiza (abbr AM; syn endomycorrhiza) : Symbiotic association between a nonpathogenic or weakly pathogenic fungus and the roots of plants in which fungal hyphae invade cortical cells of the root and produce vesicles and arbuscles (see ectomycorrhiza).

Archae : One of the three domains of living organisms: Archae, Bacteria, and Eukarya. Although they share a basic morphology with bacteria and they are also prokaryotes (i.e. they lack a nuclear membrane), in many molecular details they resemble eukaryotes more than bacteria. Previously called Archaebacteria. First described by Carl Woese at the University of Illinois.

Arm animal genetic resources utilizationf : See utilization of farm animal genetic resources.

Arrayed library : Individual primary recombinant clones (hosted in phage, cosmid, YAC, or other vector) that are placed in two- dimensional arrays in microtiter dishes. Each primary clone can be identified by the identity of the plate and the clone location (row and column) on that plate. Arrayed libraries of clones can be used for many applications, including screening for a specific gene or genomic region of interest as well as for physical mapping. Information gathered on individual clones from various genetic linkage and physical map analyses is entered into a relational database and used to construct physical and genetic linkage maps simultaneously; clone identifiers serve to interrelate the multilevel maps. Compare library, genomic library.

ARS (autonomous replicating sequence) : Any eukaryotic DNA sequence that initiates and supports chromosomal replication; they have been isolated in yeast cells. Also called autonomous(ly) replicating segment.

Arthropod : Member of the phylum Arthropoda, which consists of animals with articulated bodies and limbs and which includes insects, arachnids, and crustaceans.

Artificial inembryonation : Non-surgical transfer of embryo(s) to a recipient female. As *in vitro* embryo technology develops, artificial inembryonation will gradually replace artificial insemination.

Artificial insemination : the placement of sperm into a female reproductive tract or the mixing of male and female gametes by other than natural means.

Artificial medium : See culture medium.

Artificial seed : Encapsulated or coated somatic embryos (embryoids) that are planted and treated like seed.

Artificial selection : The practice of choosing individuals from a population for reproduction, usually because these individuals possess one or more desirable traits.

ASA : See allele-specific amplification.

Ascocarp (syn ascoma) : sexual fruiting body of an ascomycetous fungus that produces asci and ascospores; e.g. apothecium, ascostroma, cleistothecium, perithecium, pseudothecium.

Ascogenous : pertaining to ascus-producing hyphae.

Ascogonium (pl ascogonia) : A specialized cell that gives rise to the hyphae that produce asci.

Ascoma (pl ascomata; syn ascocarp) : Sexual fruiting body of an ascomycetous fungus that produces asci and ascospores; e.g. Apothecium, ascostroma, cleistothecium, perithecium, pseudothecium.

Ascomycetes (adj ascomycetous) : A group of fungi, also called the ascomycota, that produces sexual spores (ascospores) within a saclike structure called an ascus.

Ascorbic acid; vitamin C ($C_6H_8O_6$; fw 17612) : A water-soluble vitamin present naturally in some plants, and also synthetically produced. Aside from its role as a vitamin, it is used as an antioxidant in plant tissue culture; and included in disinfection solutions.

Ascospore : One of the spores contained in the ascus (q.v.) of certain fungi.

Ascostroma (pl ascostromata) : A fruiting body containing bitunicate (double-walled) asci in locules (cavities); usually dark with multiple locules, but sometimes single (see pseudothecium).

Ascus (pl asci) : Reproductive sac in the sexual stage of a type of fungi (Ascomycetes) in which ascospores are produced.

Aseptic : Asepsis or sterile. The state of being free of contaminating organisms (bacteria, fungi, algae and all micro-organisms except viruses) but not necessarily free of internal symbionts. See axenic.

Asexual (Gr a, without + L sexualis, sexual) : Any type of reproduction not involving meiosis or the union of gametes.

Asexual embryogenesis : The sequence of events whereby embryos develop from somatic cells. a.k.a. somatic cell embryogenesis.

Asexual propagation : Vegetative, somatic, non-sexual reproduction of a plant without fertilization. cf apomixis.

Asexual reproduction : Reproduction that does not involve the formation and union of gametes from the different sexes or mating types. It occurs mainly in lower animals, micro-organisms and plants. In plants, asexual reproduction is by vegetative propagation (e.g., bulbs, tubers, corms) and by formation of spores.

A-site : See aminoacyl site.

Asn : See asparagine.

Asparagine (abbr Asn; $C_4H_8N_2O_3$; fw 13212) : One of the 20 essential amino acids. It is occasionally included in plant tissue culture media as a source of reduced nitrogen.

Aspartic acid (abbr Asp; C4H7NO4; fw 13212) : An amino acid necessary for nucleotide synthesis and occasionally included in plant tissue culture media.

Assay : 1. To test or evaluate. 2. The procedure for measuring the quantity of a given substance in a sample (chemically or by other means). 3. The substance to be analysed.

Assembly : Putting sequenced fragments of DNA into their correct chromosomal positions.

Assortative mating : A mating scheme that relies on the pairing of unrelated individuals with similar phenotypes to obtain consistency of type and reinforce desirable traits.

Assortment : See segregation.

Asynapsis : The failure or partial failure in the pairing of homologous chromosomes during the meiotic prophase.

ATG or AUG : The codon for methionine; the translation initiation codon. Usually, protein translation can only start at a methionine codon (although this codon may be found elsewhere within the protein sequence as well). In eukaryotic DNA, the sequence is ATG; in RNA it is AUG. Usually, the first AUG in the mRNA is the point at which translation starts, and an open reading frame follows - i.e. the nucleotides taken three at a time will code for the amino acids of the protein, and a stop codon will be found only when the protein coding region is complete.

Atlantic Modal Haplotype (AMH) : The descriptive term used by James F. Wilson to characterize the most common haplotype in parts of Europe. The markers and most common repeat values for the AMH are;

DYS19 = 14

DYS388 = 12

DYS390 = 24

DYS391 = 11

DYS392 = 13

DYS393 = 13

Atoll : A region of DNA that is absent on the chromosome of an organism but absent from closely related organisms (i.e., the opposite of a genetic island). The loss of such genetic material may confer a distinctive phenotype. For *E. coli* enhances virulence.

ATP (adenosine triphosphate) : A nucleotide of fundamental importance as a carrier of chemical energy in all living organisms. It consists of adenosine with three phosphate groups, linked together linearly. The phosphates are attached to adenosine through its ribose (sugar) portion. Upon hydrolysis, bonds yield either one molecule of ADP (adenosine diphosphate) and an inorganic phosphate, or one molecule of AMP (adenosine monophosphate) and pyrophosphate; in both cases releasing energy that is used to power biological processes. ATP is regenerated by rephosphorilation of AMP and ADP, using chemical energy derived from the oxidation of food.

ATP-ase : An enzyme that brings about the hydrolysis of ATP, by the cleavage of either one phosphate group with the formation of ADP and inorganic phosphate, or of two phosphate groups, with the formation of AMP and pyrophosphate.

Att site : See attachment site.

Attachment site (att) : The specific sequences on phage (attP) and bacterial (attB) chromosomes between which site-specific recombination occurs in order to integrate the phage genome into the bacterial chromosome.

Attenuated vaccine : A virulent organism that has been modified to produce a less virulent form, but nevertheless retains the ability to elicit antibodies against the virulent form.

Attenuation : A mechanism of regulating the level of transcription by interfering with mRNA elongation. Slowed translation through a regulatory region allows formation of a RNA secondary structure that promotes transcription termination. Requires coupled transcription and translation and thus is restricted to prokaryotes. Two examples of genes regulated by attenuation include the trp and his operons in *E. coli.*

Attenuator : A region of leader mRNA that can form alternative secondary structures that determine whether transcription is terminated or proceeds into downstream genes. See Attenuation.

AUDPC (abbr for Area Under Disease Progress Curve) : A measure of the total amount of disease over a period of time, determined from graphs of disease vs. time, which can be used to compare epidemics quantitatively.

Authentic protein : A recombinant protein that has all the properties - including any post-translational modifications - of its naturally occurring counterpart.

Autocatalysis : Catalysis in which one of the products of the reaction is a catalyst for the reaction. Usually the catalysis starts slowly and increases as the quantity of the catalyst increases, falling off as the product is used up.

Autocatalytic reaction : See autocatalysis.

Autoclave : 1. An enclosed chamber in which substances are heated under pressure to sterilize utensils, liquids, glassware, etc., using steam. The routine method uses steam pressure of 103.4×103 Pa at 121°C for 15 minutes, or longer to allow large volumes to reach the critical temperature. 2. A pressure cooker used to sterilize growth medium and instruments for tissue culture work.

Autoecious : In reference to rust fungi, producing all spore forms on one species of host plant (see heteroecious).

Autogamy : Self-fertilization.

Autogenous regulation (autoregulation) : The gene product that controls its own expression. See autogenous repressor.

Autogenous repressor : The product of a regulator gene which binds to its own operator and prevents transcription of the associated structural genes.

Auto-immune disease : Disorder in which the immune systems of affected individuals produce antibodies against molecules that are normally produced by those individuals (called self antigens).

Auto-immunity : A disorder in the body's defence mechanism in which an immune response is elicited against its own (self) tissues.

Autologous cells : Cells that are taken from an individual, cultured (or stored), and, possibly, genetically manipulated before being infused back into the original donor.

Autolysis : The process of self destruction of a cell, cell organelle, or tissue. It occurs by the action of lysosomic enzymes.

Autonomous : A term applied to any biological unit that can function on its own, i.e., without the help of another unit, such as a transposable element that encodes an enzyme for its own transposition.

Autopolyploid : A polyploid that has multiple and identical or nearly identical sets of chromosomes (genomes) all derived from the same species. A polyploid species with genomes derived from the same original species.

Autoradiography : A technique that captures the image formed in a photographic emulsion as a result of the emission of either light or radioactivity from a labelled component that is placed next to unexposed film. The technique is used for detecting the location of an isotope in a tissue, cell or molecule. The sample is placed in contact with a photographic emulsion, usually an X-ray film. The emission of ?-particles from the sample activates the silver halide grains in the emulsion and allows them to reduce to metallic silver when the film is developed. In genetic engineering, autoradiography is most commonly used to detect the hybridization of a radioactive DNA (probe) molecule to denatured DNA in either the Southern transfer or colony hybridization procedures.

Autosomal dominant : A gene on one of the non-sex chromosomes that is always expressed, even if only one copy is present. The chance of passing the gene to offspring is 50% for each pregnancy. See also: autosome, dominant, gene.

Autosome : A chromosome not involved in sex determination. The diploid human genome consists of 46 chromosomes, 22 pairs of autosomes, and 1 pair of sex chromosomes (the X and Y chromosomes).

Autosome : A chromosome not involved in sex determination. The diploid human genome consists of 46 chromosomes, 22 pairs of autosomes, and 1 pair of sex chromosomes (the X and Y chromosomes).

Autotroph : An organism which does not require any organic carbon for its energy source or for its growth.

Autotrophic : Self-nourishing organisms capable of utilizing carbon dioxide or carbonates as the sole source of carbon and obtaining energy for life processes from radiant energy or from the oxidation of inorganic elements, or compounds such as iron, sulphur, hydrogen, ammonium and nitrites. See heterotrophic.

Auxin (Gr auxein, to increase) : A group of plant growth regulators (natural or synthetic) which stimulate cell division, enlargement, apical dominance, root initiation, and flowering. One naturally produced auxin is indole-acetic acid (IAA).

Auxin-cytokinin ratio : The relative proportion of auxin to cytokinin present in plant-tissue-culture media. Varying the relative amounts of these two hormone groups in tissue culture formulae affects the proportional growth of shoots and roots *in vitro*. As the ratio is increased (increased auxin or decreased cytokinin), roots are more likely to be produced, and as it is decreased root growth declines and shoot initiation and growth are promoted. This relationship was first recognized by C.O. Miller and F. Skoog in the 1950s.

Auxotrophic requirement : A nutritional supplement required for growth of an auxotrophic strain.

Availability : A reflection of the form and location of nutritional elements and their suitability for absorption. In tissue culture media this is related to the abundance of each nutritional element, the osmotic concentration and pH of the medium, the stability and solubility of the item in question, the presence of absorbing agents in the media, and other factors.

Avidin : A protein that binds biotin with a very high affinity. Used for the detection of biotinylated probes.

Avirulence (avr) gene : Gene in a pathogen that usually causes a hypersensitive reaction, is associated with active plant defense reactions in a resistant plant, and causes disease in a susceptible plant.

Avirulent : Inability of a pathogen to produce a disease on its host.

Avoidance : Principle of plant disease control in which plants are grown at times or locations where the pathogen is inactive or not present.

Avuncular relationship : The genetic relationship between nieces and nephews and their aunts and uncles.

Awn : Bristlelike structure at the apex of the outer bract of some cereal and grass flowers.

Axenic : Culture in the absence of living bacteria or other organisms; pure culture.

Axenic culture : Free of external contaminants and internal symbionts; generally not possible with surface sterilization alone, and incorrectly used to indicate aseptic culture, q.v.

Axil : The angle formed by the leaf petiole and the stem.

Axillary : Pertaining to or placed within an axil.

Axillary bud : A bud found at the axil of a leaf (synonymous with lateral bud).

B cells : An important class of white blood cells that mature in bone marrow and produce antibodies. They are largely responsible for the antibody-mediated or humoral immune response; they give rise to the antibody-producing plasma cells and some other cells of the immune system. See B lymphocytes.

BABS (biosynthetic antibody binding sites) : See dabs.

BAC (bacterial artificial chromosome) : A cloning vector constructed from bacterial fertility (F) factors; like YAC vectors, they accept large inserts of size 200 to 500 kb. See cloning vector; YAC.

Bacilliform : Shaped like short rods with rounded ends.

Bacillus : A rod-shaped bacterium.

Bacillus thuringiensis (Bt) : A naturally occurring bacterium present in soil and used successfully by home gardeners and organic farmers to control certain insects for more than 40 years. When ingested by a target insect, the protein produced by Bt controls the insect by disturbing the digestive system. The Bt protein is harmless to other insects, people and animals.

Back Mutation : Reverse the effect of a point or frame-shift mutation that had altered a gene; thus it restores the wild-type phenotype (see Revertant).

Backcross : Crossing a hybrid back to one of the original parents, called the recurrent parent. In agronomy, the recurrent parent is usually a single genotype, often a named variety. In forestry, the recurrent parent is usually a different genotype from the same species or population. (This is done to prevent inbreeding depression.) Example: collecting pollen from a pitch pine x loblolly pine F1 hybrid and pollinating a pure loblolly pine. The seed produced would be called first backcross and the loblolly pine the recurrent parent.

Backcross Breeding : A system of breeding whereby recurrent backcrosses are made to one of the parents of a hybrid, accompanied by selection for a specific character or characters.

Backward selection : Selection of parent trees based on results from a progeny test. (See Forward selection).

Bacteria : Plural of bacterium, q.v.

Bacterial artificial chromosome (BAC) : A chromosome-like structure, constructed by genetic engineering. BAC is a cloning vector capable of carrying between 100 and 300 kilobases of target sequence. They are propagated as a mini-chromosome in a bacterial host. The size of the typical BAC is ideal for use as an intermediate in large-scale genome sequencing projects. Entire genomes can be cloned into BAC libraries, and entire BAC clones can be shotgun-sequenced fairly rapidly.

Bacterial streaming : Large populations of bacteria that exude from the cut surface of infected plant tissue when observed with a microscope.

Bacterial toxin : A toxin produced by a bacterium, such as Bt toxin by Bacillus thuringiensis.

Bactericide : A chemical or physical agent that kills bacteria.

Bacteriocin : A protein antibiotic, one or more types of which can be produced and excreted by certain strains of bacteria.

Bacteriophage : A virus that infects bacteria. Also called simply phage. Altered forms are used in DNA cloning work, where they are convenient vectors. The bacteriophages most used are derived from two "wild" phages, called M13 and lambda (?). Lambda phages are used to clone segments of DNA in the range of around 10-20 kb. They are lytic phages, i.e., they replicate by lysing their host cell and releasing more phages. On a bacteriological plate, this results in a small clear zone - a plaque. Some lambda vectors have also been developed which are expression vectors (q.v.). The M13 system can grow inside a bacterium, so that it does not destroy the cell it infects but causes it to make new phages continuously. It is a single-stranded DNA phage, and is used for the Sanger di-deoxy DNA sequencing method (see DNA sequencing). Both of these phages grow on Escherichia coli as a host bacterium.

Bacteriostat : A substance that inhibits or slows down growth and reproduction of bacteria.

Bacteriostatic : A condition which prevents the growth of bacteria without killing them.

Bacterium (Gr bakterion, a stick; pl bacteria) : Common name for the class Schizomycetes: minute (0.5-5 □m), unicellular organisms, without a distinct nucleus. Bacteria are prokaryotes, and most of them are identified by means of Gram staining (q.v.). They are classified on the basis of their oxygen requirement (aerobic vs anaerobic) and shape (spherical = coccus; rodlike = bacillus; spiral = spirillum; comma-shaped = vibrio; corkscrew-shaped = spirochaete; filamentous). Bacteria usually reproduce asexually, by simple cell division, although a few undergo a form of sexual reproduction, termed conjugation. A few bacteria can photosynthesize (including green-blue cyanobacteria), some are saprophytes and others are parasites and can cause diseases. They are major agents of fermentation, putrefaction and decay, and frequently a source of contamination in tissue culture. In plant pathology, strains of bacteria causing disease in specific plant cultivars are called pathovars (q.v.).

Baculovirus : Baculoviruses are a class of insect virus which have been used to make DNA cloning vectors for gene expression in eukaryotic cells. Baculoviruses have a gene which is expressed at very high levels late in their infection cycle, filling the nucleus of the cell with many-sided bodies full of a protein which is not needed to produce more viruses, but is necessary for the virus's spread in the wild. In a vector cloning system, this gene is replaced by one that the biotechnologist wants expressed. Production of the protein can be up to 50% of the cells' protein content, and several proteins can be made simultaneously, so that multi-sub-unit enzymes can be made by this system. Being an animal expression system, baculoviruses produce proteins that are glycosylated (addition of carbohydrates) like the proteins in animals, making it an attractive option for the production of biopharmaceuticals. In addition, baculoviruses are non-infective and non-pathogenic to vertebrates.

Balance : The condition in which genetic components are adjusted in proportions that give satisfactory development. Balance applies to individuals and populations.

Balanced lethal system : A system for maintaining a recessive lethal allele at each of two loci on the same pair of chromosomes. In a closed population with no crossing-over between the loci, only the double heterozygotes for the lethal mutations survive.

Balanced polymorphism : Two or more types of individuals maintained in the same breeding population.

Band : A pattern of light and dark regions by Giemsa staining that can serve as landmarks on chromosomes.

Band shift assay : See Gel shift assay.

Bank : See gene bank.

Bar : A unit used for pressure of fluid. 1 bar = 105 Pa. 1 bar is approximately equivalent to 1 atmosphere.

Bark : All tissues outside the vascular cambium of a woody plant, generally including the cork layers, cork cambium (phellogen), and phloem.

Barr body : The condensed single X-chromosome seen in the nuclei of somatic cells of female mammals. base pair a pair of hydrogen-bonded nitrogenous bases (one purine and one pyrimidine) that join the component strands of the DNA double helix.

Basal : 1. Located at the base of a plant or a plant organ. 2. A fundamental formulation of a tissue culture medium.

Basal body : Small granule to which a cilium or flagellum is attached. cf kinetosome.

Basal knob (syn stylet knob) : Structure at the base of a nematode stylet.

Base : Having a pH of greater than 7.

Base : A cyclic, nitrogen-containing compound that is one of the essential components of nucleic acids. Exists in five main forms (adenine, A; guanine, G; thymine, T; cytosine, C; uracil, U). A and G have a similar structure and are called purines; T, C and U have a similar structure and are called pyrimidines. A base joined to a ribose sugar joined to a phosphate group is a nucleotide - the building block of nucleic acids.

Base analog : A purine or pyrimidine base that differs slightly in structure from the normal bases found in DNA or RNA. Some analogs may be incorporated into nucleic acids in place of the normal base, often resulting in a base substitution mutation. Some examples include aminopurine, azaguanine, azauracil, and 5-bromodeoxyuridine.

Base collection : In PGR: Defined in the International Undertaking on Plant Genetic Resources (FAO, 1983) as a collection of seed stock or vegetative propagating material (ranging from tissue cultures to whole plants) held for long-term security in order to preserve the genetic variation for scientific purposes and as a basis for plant breeding as multiplication and evaluation. cf active collection.

Base pair (bp) : Two nitrogenous bases (adenine and thymine or guanine and cytosine) held together by weak bonds. Two strands of DNA are held together in the shape of a double helix by the bonds between base pairs. A set of two bonded nucleotides on opposite strands of DNA. There are two possible base pairs: C-G and A-T. These letters are used as shorthand for the sequences of fragments of DNA e.g. CCAAGTAC. These sequences are the code for genetic information. Strung together in chains each base reaches across and forms a pair with its complementary base on the opposite strand; like the rungs of a ladder. Base pairing ensures that the genetic information, the sequence of bases in the DNA, is passed securely from generation to generation in a process called DNA replication.

Base sequence : A partnership of organic bases found in DNA and RNA; adenine forms a base pair with thymine (or uracil) and guanine with cytosine in a double-stranded nucleic acid molecule.

Base sequence analysis : A method, sometimes automated, for determining the base sequence.

Base substitution : Replacement of one base by another in a DNA molecule. See transition; transversion.

Base substitution mutation : A mutation resulting in the replacement of one base for a different base.

Basic number : The number of chromosomes in ancestral diploid ancestors of polyploids, represented by x.

Basidiocarp (syn basidioma) : Sexual fruiting body of a basidiomycetous fungus.

Basidiomycetes (adj basidiomycetous) : A fungus group, also called the Basidiomycota, characterized by the formation of external basidiospores on basidia.

Basidiospore : haploid (1N) sexual spore produced on a basidium.

Basidium (pl basidia; adj basidial) : Specialized cell or organ, often club-shaped, in which karyogamy and meiosis occur, followed by production of externally-borne basidiospores (generally four) that are haploid. There are several types of basidia.

Basipetal : Downward from the apex toward the base of a shoot; referring to development in the direction of the base so that the apical part is oldest (see acropetal).

Basophil : A type of white blood cell (leucocyte), produced by stem cells in the red bone marrow.

Batch culture : Growth of bacteria in a fixed volume of liquid medium in a closed vessel, with no additions or removals made during the period of incubation.

Batch fermentation : A process in which cells or micro-organisms are grown for a limited time. At the beginning of the fermentation, an inoculum is introduced into fresh medium, with no addition or removal of medium for the duration of the process.

Baysian analysis : A mathematical method to further refine recurrence risk taking into account other known factors.

B-DNA : Conformation of the Watson-Crick double helix in which the two strands form a right-handed double helix.

Becker muscular dystrophy : X-linked condition characterized by progressive muscle weakness and wasting; manifests later in life with progression less severe than Duchenne muscular dystrophy.

Beetle : Any insect of the order Coleoptera characterized by elytra (thickened outer wings), chewing mouth parts, and complete metamorphosis.

Behavioral genetics : The study of genes that may influence behavior.

Bench-scale process : A small- or laboratory-scale process; commonly used in connection with fermentation.

Best linear prediction (BLP) : A statistical method which utilities matrix algebra to predict breeding values for any trait or selection index; in BLP fixed effects are assumed to be known. BLP is especially suited for analyses of messy or unbalanced data. (See unbalanced).

Best linear unbiased prediction (BLUP) : A statistical method which predicts breeding values for any trait or selection index; unlike BLP, in BLUP the fixed effects are estimated. Like BLP, BLUP is well suited for analyses of messy or unbalanced data. (See unbalanced).

Beta-DNA : See β-DNA.

Beta-galactosidase : See β-galactosidase.

Beta-lactamase : See β-lactamase.

Bicellular : Two-celled.

Bidirectional replication : Two replication forks proceed in opposite directions from the same origin of replication. Sometimes called omega-replication.

Biennial (L biennium, a period of two years) : In botany, a plant which completes its life cycle within two years and then dies. For most biennial plants, the two growing seasons have to be separated by a period of cold temperature sufficient to induce flowering and fruit formation.

Biflagellate : Having two flagella.

Bifunctional vector : See shuttle vector.

Binary fission : A type of asexual reproduction in which two cells, usually of similar size and shape, are formed by the growth and division of one cell.

Binary vector system : A two-plasmid system in Agrobacterium tumefaciens for transferring into plant cells a segment of T-DNA that carries cloned genes. One plasmid contains the virulence gene (responsible for transfer of the T-DNA), and another plasmid contains the T-DNA borders, the selectable marker and the DNA to be transferred. See also cDNA; carrier DNA; plasmid; vector.

Binding : The ability of molecules to stick to each other because of the exact shape and chemical nature of parts of their surfaces. Many biological molecules bind extremely tightly and specifically to other molecules: enzymes to their substrates; antibodies to their antigens; DNA strands to their complementary strands; and so on. Binding can be characterized by a binding constant or association constant (Ka), or its inverse, the dissociation constant (Kd).

Binding site : A place on cellular DNA to which a protein (such as a transcription factor) can bind. Typically, binding sites might be found in the vicinity of genes, and would be involved in activating transcription of that gene (promoter elements), in enhancing the transcription of that gene (enhancer elements), or in reducing the transcription of that gene (silencers).

Note : that whether the protein in fact performs these functions may depend on some condition, such as the presence of a hormone, or the tissue in which the gene is being examined. Binding sites could also be involved in the regulation of chromosome structure or of DNA replication.

Binomial expansion : The probability that an event will occur 0, 1, 2, ..., n times out of n is given by the successive terms of the expression (p + q)n, where p is the probability of the event occurring, and q = 1 - p.

Binomial nomenclature : In biology, each species is generally identified by two terms: the first is the genus to which it belongs, and the second is the specific epithet that distinguishes it from others in that genus (e.g., Quercus suber, cork oak). The genus name always has an initial capital; the specific epithet is never capitalized, even though it may be derived from a proper name (e.g., keranda nut, Elaeocarpus bancroftii). Both terms in the binomial are italicized. Based on the system of classification developed by Carolus Linnaeus.

Binomial, Latin : The scientific name of an organism, composed of two names, the first designating the genus and second the specific epithet, together making the species name

Binucleate : Having two nuclei

Bio- : A prefix derived from bios and used in scientific words to associate the concept of "living organisms." Usually written with a hyphen before vowels, for emphasis or in neologisms; otherwise usually without a hyphen.

Bio-accumulation : In an organism, concentration of materials which are not components critical for that organism's survival. Usually it refers to the accumulation of metals or other compounds (e.g., DDT). Many organisms - plants, fungi, protists, bacteria, etc. - accumulate metals when grown in a solution of them, either as part of their defence mechanism against the poisonous effect of those compounds, or as a side-effect of the chemistry of their cell walls. Bio-accumulation is important as part of the microbial mining cycle (q.v.), removing toxic metals from wastewater, as a purification (bioremediation) process, etc. See also biosorption; microbial mining.

Bio-assay : A procedure for the assessment of a substance by measuring its effect in living cells or on organisms. Animals have been used extensively in drug research in bio-assays for the pharmacological activity of drugs. However, bio-assays are now usually developed using bacteria or animal or plant cells, as these are usually much easier to handle than whole animals or plants, are cheaper to make and keep, and avoid the ethical problems associated with testing of animals. Sometimes used to detect minute amounts of substances that influence or are essential to growth.

Bio-augmentation : Increasing the activity of bacteria that decompose pollutants; a technique used in bioremediation.

Biobased products : Fuels, chemicals, building materials, or electric power or heat produced from biological material(s). The term may include any energy, commercial or industrial products, other than food or feed, that utilizes biological products or renewable domestic agricultural (plant, animal and marine), or forestry materials.

Biocatalysis; biocatalyst : Use of enzymes to catalyse chemical reactions. See biotransformation.

Biocide : A compound toxic to all forms of life.

Biocontrol : The control of living organisms (especially pests) by biological means. Any process using deliberately introduced living organisms to restrain the growth and development of other, very often pathogenic, organisms, such as the use of spider mites to control cassava mealy bug, or the introduction of myxomatosis into Australia to control rabbits. The term also applies to use of disease-resistant crop cultivars. Biotechnology approaches biocontrol in various ways, such as using fungi, viruses or bacteria which are known to attack an insect or weed pest.

Bioconversion : Conversion of one chemical into another by living organisms, as opposed to their conversion by enzymes (which is biotransformation) or by chemical processes. The usefulness of bioconversion is much the same as that of biotransformation - in particular its extreme specificity and ability to work in moderate conditions. However, bioconversion has several other properties, including the possibility of having several chemical steps. A major commercial application is in the manufacture of steroids. The "basic" steroid molecule, often isolated from plants, is itself a very complicated molecule, and not one that is easy to modify by normal chemical means to produce the very specific molecules needed for drug use. However, a particular type of bioconversion that attacks only specific bits of the molecule can be used. Bioconversion is particularly useful for introducing chemical changes at specific points in large, complex molecules.

Biodegradation : The breakdown by living organisms of a compound to its chemical constituents. Materials that can be easily biodegraded are colloquially termed biodegradable.

Biodiversity : The word - a contraction of 'biological diversity' - refers to the number, variety and variability of living organisms. It encompasses three basic levels of organization in living systems: the genetic, species and ecosystem levels. Plant and animal species are the most commonly recognized units of biodiversity.

Bio-energetics : The study of the flow and the transformations of energy that occur in living organisms.

Bio-engineering : The use of artificial tissues, organs and organ components to replace parts of the body that are damaged, lost or malfunctioning.

Bio-enrichment : Adding nutrients or oxygen to increase microbial breakdown of pollutants.

Bio-ethics : The branch of ethics that deals with the life sciences and their potential impact on society. At one extreme, it can be enormously useful in focusing attention on problems that need to be confronted; at the other extreme, it can become a name-calling argument between the "pro-biotechnology" and "anti-biotechnology" schools of thought, which, as it reduces discussion to epithets and clichés, can make better sound bites.

Biofuel : A gaseous, liquid or solid fuel that contains energy derived from a biological source. For example, rapeseed oil or fish liver oil can be used in place of diesel fuel in modified engines. A commercial application is the use of modified rapeseed oil, which - as rapeseed methyl ester (RME) - can be used in modified diesel engines, and is sometimes named bio-diesel. cf biogas.

Biogas : A mixture of methane and carbon dioxide resulting from the anaerobic decomposition of waste such as domestic, industrial and agricultural sewage. a.k.a. gobar.

Biogenesis : The principle that a living organism can only arise from other living organisms similar to itself and can never originate from non-living material.

Bioinformatics : The management and analysis of data (especially DNA sequence data) using advanced computing techniques. Bioinformatics is an important field of genomics research, because of the complexity of searching and comparing the large number of DNA sequences generated.

Biolistic transformation (see gene gun) : A method used for genetic engineering in which plant cells are bombarded with metal particles coated with foreign genes using a "gene gun".

Biolistics : A means of introducing DNA into cells that involves bombardment with high-velocity microprojectiles coated with DNA.

Biological ageing : See senescence.

Biological boundaries : A concept that differentiates one organism from another and suggests that organisms cannot or should not exchange genetic material. An alternative concept is that genes are defined not by the organism from which they came, but by their function. As scientists have identified genes in seemingly non-related organisms such as plants and humans, they have found identical genes in each.

Biological containment : One of the precautionary measures taken to prevent the replication of recombinant DNA molecules in microorganisms in the natural environment. Biological containment involves the use of vectors and host organisms that have been modified so that they will not survive outside of the laboratory.

Biological control (syn biocontrol) : Exploitation by humans of the natural competition, parasitism and/or antagonism of organisms for management of pests and pathogens.

Biological diversity : See biodiversity. (Source: FAO, 1999).

Biomass : 1. The cell mass produced by a population of living organisms. 2. The organic mass that can be used either as a source of energy or for its chemical components. 3. All the organic matter that derives from the photosynthetic conversion of solar energy.

Biomass concentration : The amount of biological material in a specific volume.

Biome : A major ecological community or complex of communities, extending over a large geographical area and characterized by a dominant type of vegetation.

Biometry : The application of statistical methods to the analysis of biological problems.

Biopesticide : A compound that kills organisms by virtue of specific biological effects rather than as a broader chemical poison. Specific types include bio-insecticides and bio-fungicides. Bio-pesticides differ from biocontrol agents in that bio-pesticides are passive agents, whereas biocontrol agents are active, seeking out the pest to be destroyed. There are some extremely attractive anti-pest materials, such as Bacillus thuringiensis (Bt) toxin, which specifically interferes with the absorption of food from the guts of some insects but is harmless to mammals. The rationale behind developing bio-pesticides is that they are more likely to be biodegradable and are targeted at specific elements of the pest's metabolism.

Biopolymer : Any large polymeric molecule (protein, nucleic acid, polysaccharide, lipid) produced by a living organism.

Bioprocess : Any process that uses complete living cells or their components (e.g., enzymes, chloroplasts) to effect desired physical or chemical changes.

Bioreactor : A tank in which cells, cell extracts or enzymes carry out a biological reaction. Often refers to a growth chamber (fermenter, fermentation vessel) for cells or micro-organisms.

Bioremediation : The use of biological organisms such as plants or microbes to aid in removing hazardous substances from an area.

Biosensor : A device that uses an immobilized agent (such as an enzyme, antibiotic, organelle or whole cell) to detect or measure a chemical compound. A reaction between the immobilized agent and the molecule being analysed is transduced into an electric signal.

Biosphere : The part of the earth and its atmosphere that is inhabited by living organisms.

Biosynthesis : Synthesis of compounds by living cells, which is the essential feature of anabolism.

Biotechnology : 1. The use of biological processes or organisms for the production of materials and services of benefit to humankind. Biotechnology includes the use of techniques for the improvement of the characteristics of economically important plants and animals and for the development of micro-organisms to act on the environment. 2. The scientific manipulation of living organisms, especially at the molecular genetic level, to produce new products, such as hormones, vaccines or monoclonal antibodies.

Biotic : Relating to life, as disease caused by living organisms.

Biotic factor : Other living organisms that are a factor of an organism's environment, and form the biotic environment, affecting the organism in many ways.

Biotic stress : Stress resulting from living organisms which can harm plants, such as viruses, fungi, bacteria, parasitic weeds and harmful insects. cf abiotic stress.

Biotin : A vitamin of the B complex. It is a co-enzyme for various enzymes that catalyse the incorporation of carbon dioxide into various compounds. It is essential for the metabolism of fats. Biotin is attached to pyruvate carboxylase by a long, flexible chain like that of lipoamide in the pyruvate dehydrogenase complex. Adequate amounts are normally produced by the intestinal bacteria in animals. a.k.a. vitamin H (in USA).

Biotin labelling : 1. The attachment of biotin to another molecule. 2. The incorporation of a biotin-containing nucleotide into a DNA molecule.

Biotinylated-DNA : A DNA molecule labelled with biotin by incorporation of biotinylated-dUTP into a DNA molecule. It is used as a non-radioactive probe in hybridization experiments, such as Southern transfer. The detection of the labelled DNA is achieved by complexing it with streptavidin (an antibiotic with a high affinity for biotin) to which is attached a colour-generating agent such as horseradish peroxidase that gives a fluorescent green colour upon reaction with various organic reagents.

Biotope : A small habitat in a large community.

Biotoxin : A naturally produced toxic compound which shows pronounced biological activity and presumably has some adaptive significance to the organism which produces it.

Biotransformation : The conversion of one chemical or material into another using a biological catalyst: a near synonym is biocatalysis, and hence the catalyst used is called a biocatalyst. Usually the catalyst is an enzyme, or a whole, dead micro-organism that contains an enzyme or several enzymes.

Biotroph (syn obligate parasite) : An organism that can live and multiply only on another living organism (see necrotroph).

Biotype : A group of individuals with the same genotype. Biotypes may be homozygous or heterozygous.

Birth defect : Any harmful trait, physical or biochemical, present at birth, whether a result of a genetic mutation or some other nongenetic factor. See also: congenital, gene, mutation, syndrome.

Bitunicate : having a double ascus wall (see unitunicate).

Bivalent : A pair of synapsed or associated homologous chromosomes (one of maternal origin; the other of paternal origin) that have each undergone duplication. Each duplicated chromosome comprises two chromatids. Thus a bivalent comprises four chromatids.

BLAST : A computer program that identifies homologous (similar) genes in different organisms, such as human, fruit fly, or nematode.

Blasting : A symptom of plant disease characterized by shedding of unopened buds; classically, the failure to produce fruit or seed.

Blastocyst (also blastocist) : A mammalian embryo (fertilized ovum) in the early stages of development, approximately up to the time of implantation. It consists of a hollow ball of cells.

Blastomere : Any one of the cells formed from the first few cleavages in animal embryology. The embryo usually divides into two, then four, then eight blastomeres, and so on.

Blastula : In animals, an early embryo form that follows the morula stage; typically, a single-layered sheet (blastoderm) or ball of cells (blastocyst).

Bleach: A fluid, powder or other whitening (bleaching) or cleaning agent, usually with free chlorine ions. Commercial bleach contains calcium hypochlorite or sodium hypochlorite, and is a common disinfectant used for cleaning working surfaces, tools and plant materials in plant tissue culture and grafting.

Bleeding : Used to describe the occasional purplish-black coloration of media due to phenolic products given off by (usually fresh) transfers.

Blending theory : An incorrect 19th century theory about the inheritance of characteristics. It proposed that inherited traits blend from generation to generation. Through his plant cross-breeding experiments, Gregor Mendel proved that this was wrong.

Blight : Sudden, severe, and extensive spotting, discoloration, wilting, or destruction of leaves, flowers, stems, or entire plants.

Block : In a field study arranged in a randomized complete-block design, a block is a unit of land upon which at least one plot of all of the entries (families, clones provenances) are planted. (In this case a replication and a block are the same).

Blot : 1. As a verb, this means to transfer DNA, RNA or protein to an immobilizing matrix. 2. As a noun, it usually refers to the autoradiograph produced during the Southern or northern blotting procedures. The variations on this theme depend on the molecules:

- **Southern blot:** the molecules transferred are DNA molecules, and the probe (q.v.) is DNA.
- **Northern blot:** the molecules transferred are RNA, and the probe is DNA.
- **Western blot:** the molecules transferred are protein, and the probe is labelled antibody.
- **Southwestern blot:** the molecules transferred are protein, and the probe is DNA.
- **dot blot:** DNA, RNA or protein are dotted directly onto the membrane support, so that they form discrete spots.
- **colony blot:** the molecules (usually DNA) are from colonies of bacteria or yeast growing on a bacteriological plate. See DNA probes.

Blotch : Necrotic area of tissue irregular in form.

Blotting : A technique for detecting one RNA within a mixture of RNAs (a Northern blot) or one type of DNA within of a mixture of DNAs (a

Southern blot). A blot can prove whether that one species of RNA or DNA is present, how much is there, and its approximate size. Basically, blotting involves gel electrophoresis, transfer to a blotting membrane (typically nitrocellulose or activated nylon), and incubating with a radioactive probe. Exposing the membrane to X-ray film produces darkening at a spot correlating with the position of the DNA or RNA of interest. The darker the spot, the more nucleic acid was present there.

Blunt end : A terminus of a duplex DNA molecule which ends precisely at a base pair, with no overhang (unpaired nucleotide) in either strand. Some but not all restriction endonucleases leave blunt ends after cleaving DNA. Blunt-ended DNA can be ligated nonspecifically to other blunt-ended DNA molecules (compare with Sticky End).

5'-->3'

NNNCCC GGGNNN SmaI cut, no overhang.

NNNGGG CCCNNN

3'<--5'

Blunt-end cut : To cleave phospho-diester bonds in the backbone of duplex DNA between the corresponding nucleotide pairs on opposite strands. This cleavage process results in both strands finishing at the same residue, i.e., there are no nucleotide extensions on either strand. a.k.a. flush-end cut.

Blunt-end ligation : Joining (ligation) of the nucleotides that are at the ends of two blunt-ended DNA duplex molecules.

Borer : Insect or insect larva that forms tunnels or cavities in the bark or within the wood of trees.

Boring platform : Sterile bottom half of a Petri dish used for preparing explants with a cork borer.

Bound water : Water held by the cell and not released if freezing occurs in the intercellular space. cf free water.

Boundary layer : The calm layer of air on surfaces below the layer of more turbulent air.

Bovine somatotropin : A natural protein produced in the pituitary glands of all cattle, bST helps adult cows produce milk. Marketed by Monsanto as Posilac bovine somatotropin.

Bovine spongiform encephalopathy (BSE) : a.k.a. mad cow disease. See proteinaceous infectious particle.

Box : Refers to a short nucleic acid consensus sequence or motif that is universal within kingdoms of organisms. Examples of DNA boxes are the Pribow box (TATAAT) for RNA polymerase, the Hogness box (TATA) that has a similar function in eukaryotic organisms, and the homeo box. RNA boxes have also been described, such as Pilipenko's Box-A motif that may be involved in ribosome binding in some viral RNAs.

Bract : A modified leaf that subtends flowers or inflorescences and may appear to be a petal.

Branch migration : A "zipper-like" pairing of two homologous DNA strands during genetic recombination.

Breaking : Disease symptom, usually caused by a virus, involving addition or loss of flower color to create a variegated pattern. (see flower break).

Breed : Noun: In AnGR, either (i) a sub-specific group of domestic livestock with definable and identifiable external characteristics that enable it to be separated by visual appraisal from other similarly defined groups within the same species or (ii) a group of domestic livestock for which geographical and/or cultural separation from phenotypically similar groups has led to acceptance of its separate identity. cf breed at risk; breed not at risk; critical breed; critical-maintained breed; endangered-maintained breed. (Source: FAO, 1999).

Breed at risk : In AnGR: Any breed that may become extinct if the factors causing its decline in numbers are not eliminated or mitigated. (Source: FAO, 1999) Breeds may he in danger of becoming extinct for a variety of reasons. Risk of extinction may result from, inter alia, low population size; direct and indirect impacts of policy at the farm, country or international levels; lack of proper breed organization; or lack of adaptation to market demands. Breeds are categorized as to their risk status on the basis of, inter alia, the actual numbers of male and/or female breeding individuals and the percentage of pure-bred females. FAO has established categories of risk status: critical, endangered, critical-maintained, endangered-maintained, and not-at-risk. (Source: FAO, 1999).

Breed not at risk : In AnGR: A breed where the total number of breeding females and males is greater than 1 000 and 20 respectively; or the population size approaches 1 000 and the percentage of pure-bred females is close to 100%, and the overall population size is increasing. cf breed at risk. (Source: FAO, 1999).

Breeder seed : Seed produced by the agency sponsoring a variety and used to produce foundation seed.

Breeding : The process of sexual reproduction and production of offspring.

Breeding line : Plant strain used in a plant breeding program and usually containing one or more desirable agronomic or breeding characteristics.

Breeding orchard (breeding seed orchard, breeding arboretum) : A planting of selected trees, usually clonally propagated, designed to ease breeding work.

Breeding population : A group of individuals selected from a wild population for use in a breeding program. Usually phenotypically selected for desirable traits. In species with large natural ranges, there are usually several to many, more-or-less separate breeding populations, each designed to provide progeny suited to a particular geographic region (See seed zone and breeding zone).

Breeding value : In quantitative genetics, the part of the deviation of an individual phenotype from the population mean that is due to the additive effects of alleles. In practical terms: if an animal is mated with a random sample of animals from a population, that animal's breeding value for a certain trait is twice the average deviation of its offspring from the population mean for that trait.

Breeding zone : An area within which a single population of improved trees can be planted without fear of mis-adaptation. (See seed zone).

Brewer's yeast : Strains of yeast, often Saccharomyces cerevisiae, that are used in the production of beer.

Brewing : The process by which beer is made. In the first stage the barley grain is soaked in water and allowed to germinate (malting), during which the natural enzymes of the grain convert the seed starch to maltose, and then to glucose. Grain is then dried, crushed, and added to water at a specific temperature (steeping) and any remaining starch is converted to sugar. The resulting liquid (wort) is the raw material to which yeast is added to convert sugar to alcohol. Hops (female flowers of Humulus lupulus) are added during this process to give a characteristic flavour.

Bridge : A filter paper or other substrate used as a wick and support structure for a plant tissue in culture when a liquid medium is used.

Broad host range : The ability to infect and reproduce in a wide variety of different organisms. For example, a phage or plasmid that can grow in many different bacterial species, or a bacterium that can produce disease in many different eukaryotic hosts.

Broadcast application : Application by spreading or scattering on the soil surface.

Broad-host-range plasmid : A plasmid that can replicate in a number of different bacterial species.

Broad-sense heritability : In quantitative genetics, the proportion of the total phenotypic variation due to genetic variation.

Brooming : Profuse branching of woody stems from single stem position.

Broth culture : Microorganisms grown in a liquid medium.

Brown rot (of wood) : Wood decay resulting from selective removal of cellulose and hemicellulose, leaving a brown amorphous residue that usually cracks into cubical blocks and consists largely of slightly modified lignin.

Browning : Discoloration due to phenolic oxidation of freshly cut surfaces of explant tissue. In later stages of culture, such discoloration may indicate a nutritional or pathogenic problem, generally leading to necrosis.

Brucellosis : Disease caused by infection with organisms of the genus Brucella.

BSE : Bovine spongiform encephalopathy. See proteinaceous infectious particle.

Bt crops : Crops that are genetically engineered to carry a gene from the soil bacterium Bacillus thuringiensis (Bt). The bacterium produces proteins that are toxic to target pests but non-toxic to humans and other mammals. Crops containing the Bt gene are able to produce this toxin, thereby providing protection for the plant. Bt corn and Bt cotton are examples of commercially available Bt crops.

Bubble column fermenter : A fermentation vessel, or bioreactor, in which the cells or micro-organisms are kept suspended in a tall cylinder by rising air, which is introduced at the base of the vessel.

Bud : A region of meristematic tissue with the potential for developing into leaves, shoots, flowers or combinations; generally protected by modified scale leaves. A terminal (or apical) bud exists at the tip of a stem or branch, while axillary (or lateral) buds develop in the axils of leaves.

Bud scale : A specialized protective leaf of a bud.

Bud scar : A scar left on a shoot when the bud or bud scales drop.

Bud sport : A somatic mutation arising in a bud and producing a genetically different shoot. Bud sports includes changes due to gene mutation, somatic reduction, chromosome deletion or polyploidy.

Bud wood : Wood consisting of strong, young shoots bearing buds suitable for use in budding.

Budding : 1. A method of asexual reproduction in which a new individual is derived from an outgrowth (bud) that becomes detached from the body of the parent. 2. Among fungi, budding is characteristic of the yeast Saccharomyces cerevisiae. 3. A form of grafting in which a single vegetative bud is taken from one plant and inserted into stem tissue of another plant so that the two will grow together. The inserted bud develops into a new shoot. See grafting.

Buffer : A solution that resists change in pH when an acid or alkali is added, or when solutions are diluted.

Bug : Any insect of the order Hemiptera characterized in part by piercing-sucking mouth parts, a triangular scutellum, two pairs of wings, and gradual metamorphosis.

Bulb : A short, flattened, usually globose or disc-shaped, underground, perennial, storage organ composed of concentric layers of overlapping fleshy scale leaves attached to a stem plate at the base; essentially a subterranean bud.

Bulk breeding : The growing of genetically diverse populations of self-pollinated crops in a bulk plot with or without mass selection, followed by single-plant selection.

Bunt ball : Smut sorus filled, with teliospores, that replaces a cereal or grass kernel but is covered by plant tissue at maturity.

Buoyant density : The density possessed by a molecule or particle when suspended in an aqueous salt or sugar solution.

Burr knot : A rough outgrowth, often present on the trunks or roots of certain trees.

Bursa : Extension or flap of cuticle at the side of the male nematode sex organ, used for orienting during mating.

Burst size : The average number of phage released from a single infected bacterium during lytic growth. The burst size depends upon both the phage itself and the growth conditions of the bacteria. The burst size for characterized phage varies from only a few progeny phage per bacterium to over a hundred progeny phage per bacterium.

Bypass suppressor : A second site mutation that activates a new pathway that eliminates the need for the original mutant pathway. For an example of bypass suppression, see the argE suppression of proline biosynthesis mutants.

C : Cytosine residue in either DNA or RNA.

CAAT box (also CAT box) : A conserved sequence found within the promoter region of the protein-encoding genes of many eukaryotic organisms. It has the consensus sequence GGCCAATCT; it occurs around 75 bases prior to the transcription initiation site; and it is one of several sites for recognition and binding of regulatory proteins called transcription factors.

Calf scours : A watery diarrhoea in calves.

Callose : Amorphous, hardened carbohydrate constituent of plant cell walls, commonly developing upon injury.

Callus culture : A technique of tissue culture; it is usually on solidified medium and initiated by inoculation of small explants or sections from established organ or other cultures (the inocula). Callus culture is used as the basis for organogenic (shoot, root) cultures, cell cultures or proliferation of embryoids. Callus cultures can be indefinitely maintained through regular sub-culturing.

Callus (L callum, thick skin; pl calluses or calli) : 1. A protective tissue, consisting of parenchyma cells, that develops over a cut or damaged plant surface. 2. Mass of unorganized, thin-walled parenchyma cells induced by hormone treatment. 3. Actively dividing non-organized masses of undifferentiated and differentiated cells often developing from injury (wounding) or in tissue culture in the presence of growth regulators.

Calorie (abbr cal) : Equivalent to the amount of heat required to raise the temperature of 1 gram of water from 14.5°C to 15.5°C (□4.19 J). cf kilocalorie.

Calyx (Gr kalyx, a husk, cup) : All the sepals of a flower considered collectively. The outermost whorl of flower parts.

Cambial zone : Region in stems and roots consisting of the cambium and its recent derivatives.

Cambium (L cambium, one of the alimentary body fluids supposed to nourish the body organs; pl cambia) : A layer, usually regarded as one or two cells thick, of persistently meristematic tissue between the xylem and phloem tissues, and which gives rise to secondary tissues, thus resulting in an increase in diameter. The two most important cambia are the vascular (fascicular) cambium and the cork cambium.

Cancer : A disease process initiated by transformation of a cell to behavior lacking in normal proliferative control, and often involving invasive behavior and localized vascularization.

Candidate gene : A gene whose function suggests that it may be involved in the genetic variation observed for a particular trait, e.g., the gene for growth hormone is a candidate gene for body weight.

Candidate tree : A tree that has been tentatively selected for inclusion in a breeding program, but has not yet been measured or compared with surrounding trees. (See plus tree and elite tree).

Candidate-gene strategy : An experimental approach in which knowledge of the biochemistry and/or physiology of a trait is used to draw up a list of genes whose protein products could be involved in the trait.

Canine : Pertaining to dogs.

Canker : A plant disease characterized (in woody plants) by the death of cambium tissue and loss and/or malformation of bark, or (in non-woody plants) by the formation of sharply delineated, dry, necrotic, localized lesions on the stem; "canker" may also be used to refer to the lesion itself, particularly in woody plants.

Canola : Any of several cultivars of oilseed rape (more fully: canola oil); the vegetable oil high in mono-unsaturated fatty acid obtained from these cultivars.

Canopy : The expanded leafy top of a plant or plants.

Cap : All eukaryotes have at the 5' end of their messages a structure called a "cap", consisting of a 7-methylguanosine in 5'-5' triphosphate linkage with the first nucleotide of the mRNA. It is added post-transcriptionally, and is not encoded in the DNA.

Cap site : Two usages: 1] In eukaryotes, the cap site is the position in the gene at which transcription starts, and really should be called the "transcription initiation site". The first nucleotide is transcribed from this site to start the nascent RNA chain. That nucleotide becomes the 5' end of the chain, and thus the nucleotide to which the cap structure is attached (see "Cap"). 2] In bacteria, the CAP site (note the capital letters) is a site on the DNA to which a protein factor (the Catabolite Activated Protein) binds.

Capacitation : The final stage in the maturation process of a spermatozoon, taking place inside the female genital tract as the sperm penetrates the ovum.

Capillary array : Gel-filled silica capillaries used to separate fragments for DNA sequencing. The small diameter of the capillaries permit the application of higher electric fields, providing high speed, high throughput separations that are significantly faster than traditional slab gels.

Capsduction : A DNA transfer system in Rhodobacter capsulatus where the donor DNA is carried in a phage-like structure known as a gene transfer agent (GTA). The donor DNA derived from the GTA-producing cells can enter a recipient cell and recombine with th e recipient genome.

Capsid (syn coat protein) : The protective layer of protein surrounding the nucleic acid core of a virus; the protein molecules which make up this layer.

Capsule : Gel-like material surrounding a bacterial cell.

Carbohydrate : An organic compound based on the general formula Cx (H2O)y, performing many vital roles in living organisms. The simplest carbohydrates are the sugars (saccharides), including glucose and sucrose. Polysaccharides are carbohydrates of much greater molecular weight and complexity; examples are starch, which serves as energy store in plant seeds and tubers; cellulose and lignin that form the cell walls and woody tissue of plants of plants; glycogen, etc.

Carbon cycle : The continuous circulation of carbon atoms from inorganic carbon dioxide (CO_2) to organic molecules and back to CO_2.

Carbon dioxide (CO_2) : A colorless, odorless, unreactive gas that is a normal part of the atmosphere. It is produced during the combustion of fossil fuels when the carbon in the fuels reacts with oxygen. It is also produced when living organisms respire. It is essential for plant nutrition. While relatively innocuous in itself, carbon dioxide and other greenhouse gases can trap the sun's heat in the atmosphere and contribute to Global Warming.

Carbon source : A nutrient which provides the carbon required for cellular biosynthesis.

Carbowax : See polyethylene glycol.

Carboxyl terminus : Refers to the COOH end of a peptide chain (by custom drawn at the right of a protein sequence).

Carboxypeptidases : Two enzymes (A and B) found in pancreatic juice. Their role is to remove the C-terminal amino-acid from a peptide; the A form removes any amino acid; the B form removes only lysine or arginine. Used when sequencing peptides.

Carcinogen : A physical or chemical agent that causes cancer.

Carcinoma : A malignant tumour derived from epithelial tissue, which forms the skin and the outer cell layers of internal organs.

Carotene (L carota, carrot) : A reddish-orange plastid pigment involved in light reactions in photosynthesis.

Carotenoid : Red to yellow pigments responsible for the characteristic colour of many plant organs or fruits, such as tomatoes, carrots, etc. Oxidation products of carotene are called xanthophylls. Carotenoids serve as light-harvesting molecules in photosynthetic assemblies and also play a role in protecting prokaryotes from the deleterious effects of light.

Carpel: Female reproductive organ of flowering plants, consisting of stigma, style and ovary. In some plants, one or more carpels unite to form the pistil.

Carrier : In genetics, typically an individual that has one recessive mutant allele for some defective condition that is "masked" by a dominant normal allele at the same locus, i.e., an individual that is heterozygous for a recessive harmful allele and a dominant normal allele; the phenotype is normal, but the individual passes the defective (recessive) allele to half of its offspring.

Carrier DNA : DNA of undefined sequence content which is added to the transforming (plasmid) DNA used in physical DNA-transfer procedures. This additional DNA increases the efficiency of transformation in electroporation and chemically mediated DNA-delivery systems. The mechanism responsible for this effect is not known. See also binary vector; plasmid; chimeric gene.

Carrier gas : The gas that carries the sample in gas chromatography.

Carrier molecule : 1. A molecule that plays a role in transporting electrons through the electron transport chain. Carrier molecules are usually proteins bound to non-protein groups and able to undergoing oxidation and reduction relatively easily, thus allowing electrons to flow. 2. A lipid-soluble molecule that can bind to lipid-insoluble molecules and transport them across membranes. Carrier molecules have specific sites that interact with the molecules they transport. Efficiency of carrier molecules may be modified by modifying the interacting sites through genetic engineering.

Cascade : A series of events that result in transmission and usually amplification of a weak signal.

Casein : A group of proteins found in milk.

Casein hydrolysate : Mixture of amino acids and peptides produced by enzymatic or acid hydrolysis of casein. cf organic complex; undefined.

Cassette : A gene fragment that can be cloned into a site to confer some property. For example, cassettes encoding antibiotic resistance are often used to disrupt a cloned gene.

Casting : Premature loss of abscised leaves or twigs.

CAT assay : An enzyme assay. CAT stands for chloramphenicol acetyl transferase, a bacterial enzyme which inactivates chloramphenicol by acetylating it. CAT assays are often performed to test the function of a promoter. The gene coding for CAT is linked onto a promoter (transcription control region) from another gene, and the construct is "transfected" into cultured cells. The amount of CAT enzyme produced is taken to indicate the transcriptional activity of the promoter (relative to other promoters which must be tested in parallel). It is easier to perform a CAT assay than it is to do a Northern blot, so CAT assays were a common method for testing the effects of sequence changes on promoter function. Largely supplanted by the reporter gene luciferase.

Catabolic pathway : A pathway by which an organic molecule is degraded in order to release energy for growth and other cellular processes; degradative pathway.

Catabolism : The metabolic breakdown of large molecules in living organism, with accompanying release of energy.

Catabolite repression : A global regulatory system resulting in decreased expression of many genes due to the addition of an efficient carbon source such as glucose. Characteristically mediated by a complex between the CRP protein and cAMP in enteric bacteria.

Catalysis : The process of changing the rate of a chemical reaction by use of a catalyst.

Catalyst (Gr katalyein, to dissolve) : A substance that promotes a chemical reaction by lowering the activation energy of a chemical reaction, without itself undergoing any permanent chemical change. The process is catalysis. cf catalytic antibody; catalytic RNA.

Catalytic antibody (= abzyme) : An antibody selected for its ability to catalyse a chemical reaction by binding to and stabilizing the transition-state intermediate.

Catalytic RNA (= ribozyme; gene shears) : A natural or synthetic RNA molecule that cuts an RNA substrate.

Cataract : An impairment of vision caused by the lenses of the eyes becoming cloudy. Cataracts are common in elderly people. They may be inherited or caused by diabetes and environmental factors.

Catenane : A circular DNA molecule with two or more interlinked rings.

Cation : A positively charged ion; opposite is *anion*.

Catkin : Conelike cluster of male or female flowers; conelike fruit of angiosperms.

Caulogenesis : Stem organogenesis; induction of shoot development from callus.

Causal agent : Organism or agent that incites and governs disease or injury.

CCAAT box (CAT box, CAAT box, and other variants) : A sequence found in the 5' flanking region of certain genes which is necessary for efficient expression. A transcription factor (CCAAT-binding protein, CBP) binds to this site.

CCC : An abbreviation for "circular, covalently closed" DNA molecules.

CD molecules; cluster of differentiation molecules : Any group of antigens that is associated with a specific sub-population of T cells. There are designations for surface molecules on various cells of the immune system, e.g., CD4 is present on the surface of helper T cells.

cDNA clone "complementary DNA"; : A piece of DNA copied from an mRNA. The term "clone" indicates that this cDNA has been spliced into a plasmid or other vector in order to propagate it. A cDNA clone may contain DNA copies of such typical mRNA regions as coding sequence, 5'-untranslated region, 3' untranslated region or poly(A) tail. No introns will be present, nor any promoter sequences (or other 5' or 3' flanking regions). A "full-length" cDNA clone is one which contains all of the mRNA sequence from nucleotide #1 through to the poly(A) tail.

cDNA cloning : A method of cloning the coding sequence of a gene, starting with its mRNA transcript. It is normally used to clone a DNA copy of a eukaryotic mRNA. The cDNA copy, being a copy of a mature messenger molecule, will not contain any intron sequences and may be readily expressed in any host organism if attached to a suitable promoter sequence within the cloning vector.

cDNA library : A collection of cDNA clones that were generated *in vitro* from the mRNA sequences isolated from an organism or a specific tissue or cell type or population of an organism. cf library.

cDNA; complementary DNA : The double-stranded DNA complement of an mRNA sequence; synthesized *in vitro* from a mature RNA template using reverse transcriptase (to create a single strand of DNA from the RNA template) and DNA polymerase (to create the double-stranded DNA). Preparation of cDNAs is often the first step in cloning DNA sequences of interest. Used as specific and sensitive probes in hybridization studies, because cDNAs usually do not include regulatory or other controlling sequences, and so they can be used to identify (probe) and isolate genes and their associated sequences from genomic DNA. See binary vector; carrier DNA.

CDR (complementarity-determining regions) : These are regions of the variable (V) regions of light and heavy antibody chains that make contact with the antigen. The primary amino acid sequences of these regions are highly variable among antibodies of the same class.

Cell : The lowest denomination of life thought to be possible. Most organisms consist of more than one cell which become specialized into particular functions to enable the whole organism to function properly. Cells contain DNA and many other elements to enable the cell to function.

Cell culture : The *in vitro* growth of cells derived from multi-cellular organisms. The cells are usually of one type.

Cell cycle : Sequence of stages that a cell passes through between one division and the next. The cell cycle oscillates between mitosis and the interphase, $_2$. In the G phase there is a high rate of biosynthesis and growth; in the S phase there is the doubling of the DNA content as a consequence of chromosome replication; in the G_2 phase the final preparations for cell division (cytokinesis) are made.

Cell differentiation : Continuous loss of physiological and cytological characters of young cells, resulting in getting the characters of adult cells. The unspecialized cells become modified and specialized for the performance of specific functions. Differentiation results from the controlled activation and de-activation of genes.

Cell division : Formation of two or more daughter cells from a single mother cell. The nucleus divides first, followed by the formation of a cell membrane between the daughter nuclei. Division of cytoplasm and nucleus into two or more parts by formation of a cell plate.

Cell extract : A preparation consisting of a large number of broken cells and their released contents.

Cell fusion : Formation of a single hybrid cell from two cells of different species, cultured *in vitro*. The cells fuse and coalesce, but their nuclei may remain separated. During subsequent cell division, a single spindle is formed so that each daughter cell has a single nucleus containing sets of chromosomes from each parental line. Subsequent divisions often result in the loss of chromosomes and therefore of genes. The cell fusion technique can be used to determine the control of specific genes and their assignment to chromosomes. cf cell hybridization.

Cell generation time : The interval between the beginning of consecutive divisions of a cell. The time that it takes for a population of single-celled organisms to double its cell number. Successive generations of cells or organisms within a population are separated by a time interval called generation time. The cell regeneration time can be determined with the aid of time-lapse microcinematography.

Cell hybridization : The fusion of two or more dissimilar cells leading to the formation of a somatic hybrid. cf cell fusion.

Cell line : A cell lineage that can be maintained in culture. A cell line arises from a primary culture. It implies that cultures from it consist of several lineages of the cells originally present in the primary culture.

Cell number : The number of cells per unit volume of a culture.

Cell plate : The precursor of the cell wall, formed as cytokinesis starts during cell division. The cell plate develops in the region of the equatorial plate and arises from membranes in the cytoplasm.

Cell sap : Water and dissolved substances, sugar, amino acids, waste substances, etc., in the plant cell vacuole.

Cell selection : The process of selecting cells within a group of genetically different cells. Select cells or cell lines are sub-cultured onto fresh medium for continued selection and often are exposed to an increased level of the selection agent. The final objective is to regenerate plants exhibiting the traits selected for at the cellular level.

Cell strain : A strain of cells having specific properties or markers derived from a primary culture of a cell line by selection or cloning. The selected properties must persist during subsequent cultivation. a.k.a. single-cell line.

Cell suspension : Cells in culture in moving or shaking liquid medium, often used to describe suspension cultures of single cells and cell aggregates. See suspension culture.

Cell wall : Protective, resistant, but permeable structure secreted externally to the cell membrane in plants, bacteria, fungi, and certain other organisms.

Cell (L cella, small room) : The smallest structural unit of living matter capable of functioning independently; a microscopic mass of protoplasm surrounded by a semi-permeable membrane, usually including one or more nuclei and various non-living products, capable - either alone or by interacting with other cells - of performing all the fundamental functions of life.

Cell-free translation system : A cell extract containing all the components required for protem synthesis (i.e. ribosomal subunits, tRNAs, amino acids, enzymes and cofactors) and able to translate added mRNA molecules.

Cell-mediated immune response : The activation of T cells of the immune system in response to the presence of a foreign antigen.

Cellular oncogene (proto-oncogene) : A normal mammalian or avian gene that when mutated or improperly expressed contributes to the development of cancer. See oncogene.

Cellulose : Carbohydrate polysaccharide composed of glucose units; major component of plant primary cell wall.

Cellulose nitrate : See nitrocellulose.

Cellulose (cell + ose, a suffix indicating a carbohydrate) : A complex carbohydrate composed of long, unbranched chains of beta-glucose ((1.4)-linked- -d-glucose) molecules, which contribute to the structural framework of plant cell walls. It comprises 40% to 55% by weight of the plant cell wall.

Cellulosome : A multi-protein aggregate that is present in some cellulolytic micro-organisms and contains multiple copies of all the enzymes required to completely break down cellulose. This complex is often found on the outer surface of cellulolytic micro-organisms.

Centimorgan (cM) : A unit of measure of the statistical probability recombination frequency between alleles. One centimorgan is equal to a 1% chance that a marker at one genetic locus will be separated from a marker at a second locus due to crossing over in a single generation. In human beings, 1 centimorgan is equivalent, on average, to 1 million base pairs.

Central dogma : A phrase that refers to the concept of information flow proceeding only from DNA to RNA to protein.

Central mother cell : A subsurface cell located in a plant apical meristem and characterized by a large vacuole.

Centres of origin : The locations in the world where particular domesticated plants originated. These areas show the highest variation, and are rich in wild alleles.

Centrifugation : Separating molecules by size or density using centrifugal forces generated by a spinning rotor. G-forces of several hundred thousand times gravity are generated in ultracentrifugation. See density gradient centrifugation.

Centrifuge : A device in which solid or liquid particles of different densities are separated by rotating them in a tube in a horizontal circle. The denser particles tend to move along the length of the tube to a greater radius of rotation, displacing the lighter particles to the other end.

Centriole : An organelle in many animal cells that appears to be involved in the formation of the spindle during mitosis. During cell division, the two centrioles move to opposite sides of the nucleus to form the ends of the spindle.

Centromere : A specialized constricted region of a chromosome to which spindle fibers attach during cell division at which two sister chromatids (the two exact copies of each chromosome that are formed after replication) are joined, and which attach to the spindle during cell division.

Cephem-type antibiotic : An antibiotic that shares the basic chemical structure of cephalosporin.

Certification : Seeds, propagative plant material, or nursery stock produced and sold under inspection to maintain genetic identity and purity, freedom from harmful pathogens, insect pests, and weed seeds. It is approved and certified by an official certifying agency.

Certified seed : Seed used for commercial crop production produced from foundation, registered, or certified seed under regulation of a legally constituted agency.

Cfu (abbr for colony forming unit) : The number of colonies formed per unit of volume or weight of a cell or spore suspension.

Chain-termination ("nonsense" or "stop") codon : A codon signalling the termination of polypeptide synthesis. The three standard stop codons are UAG (amber), UAA (ochre), and UGA (opal).

Chaperone : A protein that facilitates the folding of other proteins or assembly of multi-protein complexes.

Chaperone proteins : A series of proteins present in the endoplasmic reticulum which prevent proteins from folding prematurely and guide the proper folding of secreted proteins through a complex series of binding and release reactions.

Character : An attribute of an organism resulting from the interaction of a gene or genes with the environment.

Character (trait) : A distinctive but not necessarily invariable feature exhibited by all individuals of a group and capable of being described or measured;

e.g. color, size, performance. A character of a given individual will have a certain phenotype as determined by the individual's genotype and environment.

Characterization : 1. Of AnGR: All activities associated with the description of AnGR aimed at better knowledge of these resources and their state. (Source: FAO, 1999) Characterization by a country of its AnGR will incorporate development of necessary descriptors for use; identification of the country's sovereign AnGR; baseline and advanced surveying of these populations, including their enumeration and visual description, their comparative genetic description in one or more production environments, their valuation, and ongoing monitoring of those AnGR at risk. (Source: FAO, 1999) 2. Of PGR: Systematic recording of descriptors that are independent of environmental factors.

Charcoal : The black porous residue of partly burnt wood, bones, etc; a form of carbon. When treated to purify it and increase its adsorptive power, it is called activated charcoal (q.v.) in which form it is added to nutrient medium in order to prevent or decrease the effect of browning.

Charcot-Marie Tooth disease : A condition characterized by degeneration of the motor and sensory nerves that control movement and feeling in the arm below the elbow and the leg below the knee; transmitted in autosomal dominant, autosomal recessive and X-linked forms.

Chelate : Noun: Complex organic molecule that can combine with cations and does not ionize. Chelates can supply micronutrients to plants at slow, steady rates. Usually used to supply iron to plant cells.

Chelator : A substance that binds particular ions, removing them from solution. Some chelators are relatively nonspecific, binding to a variety of ions with similar charge, and other chelators bind much more strongly to specific ions. A few examples include the divalent ion chelators citrate and EDTA, the calcium chelator EGTA, and the iron chelator desferrioxamine.

Chemical mutagen : A chemical or product capable of causing genetic mutation in living organisms exposed to it.

Chemically-defined medium : When all of the chemical components of a plant tissue culture medium are fully known and defined. cf undefined; organic complex.

Chemiluminescence : The emission of light from a chemical reaction.

Chemostat : A continuous and open culture in which growth rate and cell density are maintained constant by a fixed rate of input of a growth-limiting nutrient. cf phytostat.

Chemotaxis (syn chemotropism) : Movement or growth of an organism in response to changing concentration of a chemical stimulus, often in relation to food or for mating.

Chemotherapy : The treatment of disease, especially infections or cancer, by means of chemicals. In treating cancers, it involves administering chemicals toxic to malignant cells.

Chi site : An 8-nucleotide DNA sequence (5' GCTGGTGG 3') which is a hot-spot for homologous recombination. When the RecBCD exonuclease encounters a chi site the RecD subunit is released and the RecBC complex then functions as a helicase, unwinding the DNA and providing single stranded which can invade a homologous region of double-stranded DNA. Chi sites typically occur about once every 5 kb on the chromosomes of enteric bacteria and most are oriented in the same direction as DNA replication.

Chiasma : A cross-over between strands of two non-sister chromatids during recombination. The junction where two homologous chromosomes appear to exchange genetic material during recombination. (Chi is the greek letter c which resembles a genetic cross-over).

Chimera (or chimaera) : From chimera, a mythological creature with the head of a lion, the body of a goat and the tail of a serpent. An organism whose cells are not all derived from the same zygote. 1. Animal. An individual exhibiting two or more genotypes in patches derived from two or more embryos. An individual derived from two embryos by experimental intervention. 2. Plant. Part of a plant with a genetically different constitution as compared with other parts of the same plant. It may result from different zygotes that grow together, or from artificial fusion (grafting); it may either be periclinal chimera, in which one tissue lies over another as a glove fits a hand; mericlinal chimera, where the outer tissue does not completely cover the inner tissue; and sectoral chimera, in which the tissues lie side by side. 3. A recombinant DNA molecule that contains sequences from different organisms.

Chimeraplasty : An experimental targeted repair process in which a desirable sequence of DNA is combined with RNA to form a chimeraplast. These molecules bind selectively to the target DNA. Once bound, the chimeraplast activates a naturally occurring gene-correcting mechanism. Does not use viral or other conventional gene-delivery vectors. See also: gene therapy, cloning vector.

Chimeric DNA : A recombinant DNA molecule containing unrelated genes.

Chimeric gene : A semi-synthetic gene, consisting of the coding sequence from one organism, fused to promoter and other sequences derived from a different gene. Most genes used in transformation are chimeric. See carrier DNA; binary vector; plasmid; transformation; vector.

Chimeric plasmid : A plasmid used in gene cloning and constructed from two or more different plasmids.

Chimeric selectable marker gene : A gene that is constructed from parts of two or more different genes and allows the host cell to survive under conditions where it would otherwise die.

Chip, DNA : See DNA chip.

Chi-squared test (χ^2 test) : A significance test used to statistically assess the goodness of fit of observed data to a prediction.

Chitin : A nitrogenous polysaccharide occurring as skeletal material in many invertebrates and fungi.

Chitinase : An enzyme which breaks down chitin.

Chlamydospore : Thick-walled or double-walled asexual resting spore formed from hyphal cells (terminal or intercalary) or by transformation of conidial cells that can function as an overwintering stage.

Chloramphenicol (Cam) : An antibiotic that inhibits protein synthesis by binding to the 50s ribosomal subunit and blocking the peptidyltransferase reaction. Chloramphenicol resistance encoded by many plasmids (e.g. pBR328) is due to a cytoplasmic chloramphenicol acyltransferase which inactivates chloramphenicol by covalently acetylating it.

Chlorenchyma (Gr chloros, green + enchyma, a suffix meaning tissue) : Tissue containing chloroplasts, including leaf mesophyll and other parenchyma cells.

Chlorophyll (Gr chloros, green + phyllon, leaf) : One of the two pigments responsible for the green colour of most plants. It is essential in the absorption of light energy for photosynthesis.

Chloroplast chromosome : Circular DNA found in the photosynthesizing organelle chloroplast of plants instead of the cell nucleus where most genetic material is located.

Chloroplast (Gr chloros, green + plastos, formed) : Specialized cytoplasmic organelle that contains chlorophyll. Lens-shaped and bounded by a double membrane, chloroplasts contain membranous structures (thylakoids) piled up into stacks, surrounded by a gel-like matrix (stroma). They are the site of solar energy transfer and important reactions of starch or sugar synthesis. Chloroplasts have their own DNA and are inherited cytoplasmically, independent of nuclear genes.

Chloroplastid : See chloroplast.

Chlorosis (Gr chloros, green + osis, diseased state) : Failure of chlorophyll development, and appearance of yellow colour in plants, because of a nutritional disturbance or because of an infection by a virus, bacteria or fungus.

Chorionic villus sampling : An invasive prenatal diagnostic procedure involving removal of villi from the human chorion to obtain chromosomes and cell products for diagnosis of disorders in the human embryo.

Chromatid : A single, continuous double stranded DNA molecule with its unique, complete grouping of genetic information, associated proteins, higher-order structures, and centromeric and telomeric regions necessary for separation and maintenance after replication.

Chromatin fibres : A basic organizational unit of eukaryotic chromosomes, consisting of DNA and associated proteins assembled into strands of 30 nm average diameter.

Chromatin (Gr chroma, colour) : Substance of which eukaryotic chromosomes are composed. It consists of primarily DNAm with some proteins (mainly histones), and small amounts of RNA. Originally named because of the readiness with which it stains with certain dyes (chromaticity).

Chromatography (Gr chroma, colour + graphein, meaning to draw or write) : 1. A method for separating and identifying the components of mixtures of molecules having similar chemical and physical properties. 2. The term originally used by Mikhail Tswett (1906) to describe the separation of a mixture of leaf pigments on a calcium carbonate column.

Chromocentre : Body produced by fusion of the heterochromatic regions of the chromosomes in the polytene tissues (e.g., the salivary glands) of certain Diptera.

Chromogenic substrate : A substrate that changes color when modified by a specific enzyme.

Chromomere : One of the serially aligned beads or granules of a eukaryotic chromosome, resulting from local coiling of a continuous DNA thread.

Chromonema (pl chromonemata) : An optically single thread forming an axial structure within each chromosome.

Chromoplast : Plastid containing pigments, such a chloroplast, or one in which carotenoids predominate.

Chromosomal aberration : Any change in chromosome structure or number. Although it can be a mechanism for enhancing genetic diversity, such alterations are usually fatal or ill-adaptive, especially in animals.

Chromosomal deletion : The loss of part of a chromosome's DNA.

Chromosomal integration site : A chromosomal location where foreign DNA can be integrated, often without impairing any essential function in the host organism.

Chromosomal inversion : Chromosome segments that have been turned 180 degrees. The gene sequence for the segment is reversed with respect to the rest of the chromosome.

Chromosomal polymorphism : The occurrence of one to several chromosomes in two or more alternative structural forms within a population; the structurally changed chromosomes are the result of chromosome mutations (i.e., any structural change involving the gain, loss or re-location of chromosome segments).

Chromosome aberration : Abnormal structure or number of chromosomes; includes deficiency, duplication, inversion, translocation, aneuploidy, polyploidy, or any other change from the normal pattern.

Chromosome banding : Staining of chromosomes in such a way that light and dark areas occur along the length of the chromosomes in repeatable patterns. Lateral comparisons identify pairs. Each chromosome can be identified by its banding pattern.

Chromosome jumping : A technique whereby one starts with a piece of DNA from one region of a chromosome, and obtains clones from nearby regions without cloning everything in between (as in chromosome walking; see below). One round of jumping yields new clones at distances of several tens of kb away from the starting point. In practice, this method is used when classical genetics proves that a known piece of DNA is located on the chromosome close to a gene you would like to clone (such as a human disease gene). By cloning fragments some distance away in both directions from the known fragment, one might obtain (1) fragments further from the desired gene (which are discarded); (2) fragments even more closely linked to the desired gene (in which case one goes for another round of jumping); or (3) fragments from within the desired gene - the optimal result.

Chromosome mutation : A change in the gross structure of a chromosome, usually causing severely deleterious effects in the organism. They are often due to an error in pairing during the crossing-over stage of meiosis. The main types of chromosome mutation are translocation, duplication, deletion and inversion.

Chromosome number : The number of chromosomes found in a typical individual of a particular species. Typically the sex cells of a species have one-half the number of chromosomes found in the vegetative portion and are said to be haploid. (See polyploid).

Chromosome painting : Attachment of certain fluorescent dyes to targeted parts of the chromosome. Used as a diagnositic for particular diseases, e.g. types of leukemia.

Chromosome region p : A designation for the short arm of a chromosome.

Chromosome region q : A designation for the long arm of a chromosome.

Chromosome theory of inheritance : The theory that chromosomes carry the genetic information and that their behaviour during meiosis provides the physical basis for segregation and independent assortment.

Chromosome walking : A technique for cloning everything in the genome around a known piece of DNA (the starting probe). You screen a genomic library for all clones hybridizing with the probe, and then figure out which one extends furthest into the surrounding DNA. The most distal piece of this most distal clone is then used as a probe, so that ever more distal regions can be cloned. This has been used to move as much as 200 kb away from a given starting point (an immense undertaking). Typically used to "walk" from a starting point towards some nearby gene in order to clone that gene. Also used to obtain the remainder of a gene when you have isolated a part of it.

Chromosome (Gr chroma, colour + soma, body) : 1. A single DNA molecule, a tightly coiled strand of DNA, condensed into a compact structure *in vivo* by complexing with accessory histones or histone-like proteins. 2. A group of nuclear bodies containing genes which are largely responsible for the differentiation and activity of a eukaryotic cell; one of the bodies into which the nucleus resolves itself at the beginning of mitosis and from which it is derived at the end of mitosis. Chromosomes contain most of the cell's DNA. Chromosomes exist in pairs in eukaryotes - one paternal (from the male parent) and one maternal (from the female parent). Each eukaryotic species has a characteristic number of chromosomes. Bacterial and viral cells contain only a single chromosome, consisting of a single or double strand of DNA or, in some viruses, RNA, without histones.

Chronic : Pertaining to slow-developing, persistent, or recurring symptoms (see acute).

Chronic disease : An illness that lasts for a long period of time or indefinitely. In contrast, an acute disease is one with a rapid onset and a short but usually severe course.

Chronic toxicity : Poisoning due to low levels of exposure to a compound over a period of time (see acute toxicity).

Chymosin : An enzyme that clots milk; it is used in the manufacture of cheese.

Chytridiomycetes : A fungus group, also called the Chytridiomycota, characterized by the production of motile zoospores and resting sporangia; the plant-pathogenic species are all biotrophs that are restricted to the cells of their host.

Cilium (pl cilia; adj ciliate) : Hairlike locomotor structure on certain cells; a locomotor structure on a ciliate protozoan.

Circadian : Of physiological activity, etc.: occurring or recurring about once a day. cf diurnal.

Circularization : A DNA fragment generated by digestion with a single restriction endonuclease will have complementary 5′ and 3′ extensions (sticky ends). If these ends are annealed and ligated, the DNA fragment will have been converted to a covalently-closed circle, or circularized.

Circularly permuted DNA : A population of linear DNA molecules produced as if by breaking open circular molecules at different points.

Circulative transmission (syn persistent transmission) : Virus transmission characterized by a long period of acquisition of the virus by a vector (typically an insect), a latent period of several hours before the vector is able to transmit the virus, and retention of the virus by the vector for a long period, usually several days; the virus circulates in the body of the vector.

Cirrhus (pl cirrhi) : A curled, tendril-like mass of exuded spores, held together by a slimy matrix.

Cis : As used in molecular biology, an interaction between two sites which are located within the same molecule. However, a cis-acting protein can either be one which acts only on the molecule of DNA from which it was expressed, or a protein which acts on itself (e.g., self-proteolysis).

Cis configuration : See coupling.

Cis heterozygote : A heterozygote that contains two mutations arranged in a cis configuration (e.g., a+ b+ / a b).

Cis-acting sequence : A nucleotide sequence that only affects the expression of genes located on the same chromosome.

Cis-dominance : A mutation that prevents the expression of downstream genes. Polar mutations in an operon are typically cis-dominant.

Cis-trans test : A genetic test to determine whether two mutations are in the same or in different genes.

Cistron : An obsolete term for a gene. Often used to refer to the number of genes encoded by a single transcript; monocistronic refers to a transcript that encodes one gene, and polycistronic refers to a transcript that encodes multiple genes.

Clade : From the Greek word klados, meaning branch. A branch of biological taxa or species that share features inherited from a common ancestor. A single phyletic group or line. Also cladus. Monophyletic group of taxa.

Cladistics : School of phylogenetic analysis emphasizing the branching patterns of monophyletic taxa relying on synapomorphies (vs. symplesiomorphies) to unite sister taxa.

Cladogram : A diagram, in the form of a stylized tree, showing inferred historical branching patterns among taxa.

Claims : The section of a patent that states, in detail, the uses and possible applications of the invention described in the patent.

Clamp connection clavate (or claviform) : A bridge- or buckle-hyphal protrusion in basidiomycetous fungi, formed at cell division and connecting the newly divided cells.

Class switching : The process during which a plasma cell stops producing antibodies of one class and begins producing antibodies of another class.

Cleared lysate : A cell extract that has been centrifuged to remove cell debris, subcellular particles, and much of the chromosomal DNA.

Cleave : To break phospho-diester bonds of double-stranded DNA, usually with a type II restriction endonuclease. a.k.a. to cut or digest.

Cleft lip/palate : Congenital condition with cleft lip alone, or with cleft palate; cause is thought to be multifactorial.

Cleistothecium (pl cleistothecia) : A spherical ascocarp that is closed at maturity.

Cline : An environmental gradient (temperature, rainfall, soil ph . . .) and a corresponding phenotypic gradient in a population of plants or animals. Where these have been evaluated by provenance tests, clines are often found to have a genetic basis.

Clonal: A population of cells derived from a single cell and thus expected to be genetically identical. Genetic differences in a "clonal" population may arise from random spontaneous mutations during growth of the cells.

Clonal propagation : Asexual propagation of many new plants (ramets) from an individual (ortet); all have the same genotype.

Clonal selection : The production of a population of plasma cells all producing the same antibody in response to the interaction between a B lymphocyte producing that specific antibody and the antigen bound by that antibody.

Clonal test : A field planting of several to many vegetatively propagated plants. Such tests furnish estimates of the relative performance of different genotypes, but do not necessarily provide information on breeding behavior.

Clone : Cells derived from a single cell and thus expected to be genetically identical.

Clone (verb) : The act of duplicating genetic material within a vector. To clone a piece of DNA, one would insert it into some type of vector (say, a plasmid) and put the resultant construct into a host (usually a bacterium) so that the plasmid and insert replicate with the host. An individual bacterium is isolated and grown and the plasmid containing the "cloned" DNA is re-isolated from the bacteria, at which point there will be many millions of copies of the DNA - essentially an unlimited supply. Actually, an investigator wishing to clone some gene or cDNA rarely has that DNA in a purified form, so practically speaking, to "clone" something involves screening a cDNA or genomic library for the desired clone. See also "Probe" for a description of how one might start a cloning project, and "Screening" for how the probe in used. One can also clone more complex organisms, with considerable difficulty. The much-publicized Scottish research that resulted in the sheep 'Dolly' exemplifies this approach.

Clone bank : See genomic library.

Cloned DNA : Any DNA fragment that passively replicates in the host organism after it has been joined to a cloning vector.

Cloned strain or line : A strain or line descended directly from a clone.

Cloning : Using specialized DNA technology to produce multiple, exact copies of a single gene or other segment of DNA to obtain enough material for further study. This process, used by researchers in the Human Genome Project, is referred to as cloning DNA. The resulting cloned (copied) collections of DNA molecules are called clone libraries. A second type of cloning exploits the natural process of cell division to make many copies of an entire cell. The genetic makeup of these cloned cells, called a cell line, is identical to the original cell. A third type of cloning produces complete, genetically identical animals such as the famous Scottish sheep, Dolly. See also: cloning vector.

Cloning vector : DNA molecule originating from a virus, a plasmid, or the cell of a higher organism into which another DNA fragment of appropriate size can be integrated without loss of the vectors capacity for self-replication; vectors introduce foreign DNA into host cells, where it can be reproduced in large quantities. Examples are plasmids, cosmids, bacterial artificial chromosomes (BAC) and yeast artificial chromosomes (YAC); vectors are often recombinant molecules containing DNA sequences from several sources.

Cloning vehicle : See cloning vector.

Closed continuous culture : A continuous culture in which inflow of fresh medium is balanced by outflow of corresponding volumes of spent medium. Cells are separated mechanically from outflowing medium and added back to the culture. cf open continuous culture; batch culture; continuous culture.

Cluster of differentiation : See CD.

Coalesce : Grow together into one body or spot.

Coat protein (syn capsid) : The protective layer of protein surrounding the nucleic acid core of a virus; the protein molecules which make up this layer.

Coccus (pl cocci) : A spherical bacterium. Cocci may occur singly, in pairs, in groups of four or more, and in cubical packets.

Coconut milk : Liquid endosperm of the coconut, often used to supply organic nutrients to cultured cells and tissues. See addendum; organic complex; undefined.

Cocoon : A protective coverage for eggs and/or larvae produced by many invertebrates, such as the silkworm moth.

Co-culture : The joint culture of two or more types of cells, such as a plant cell and a micro-organism, or two types of plant cells. Used in various dual-culture systems or in nurse-culture, q.v.

Code : See codon and genetic code.

Coding : The specification of a peptide sequence, by the code contained in DNA molecules.

Coding region : A sequence of DNA that encodes a polypeptide.

Coding sequence : The portion of a gene or an mRNA which actually codes for a protein. Introns are not coding sequences; nor are the 5' or 3' untranslated regions (or the flanking regions, for that matter - they are not even transcribed into mRNA). The coding sequence in a cDNA or mature mRNA includes everything from the AUG (or ATG) initiation codon through to the stop codon, inclusive.

Coding strand : The strand of duplex DNA which contains the same base sequence (after substituting U for T) found in the mRNA molecule resulting from transcription of that segment of DNA. a.k.a. sense strand. The mRNA molecule is transcribed from the other strand, known as the template or antisense strand. Coding strand.

5′ ATGAAAGCTTTAGTGGGCGCCCGTAT 3′ Template strand 3′ TACTTTCGAAATCACCCGCGGGCATA

5′ mRNA 5′ AUGAAAGCUUUAGUGGGCGCCCGUAU 3′

Codominance: The observation of a phenotype caused by independent expression of both alleles of a gene in a diploid.

Codominant alleles : Two alleles that have different effects that are distinguishable in a heterozygous individual (e.g. AB blood groups).

Codon : A group of three consecutive nucleotides within messenger RNA (mRNA) that encodes a message to initiate translation, to incorporate a specific amino acid into the growing polypeptide chain, or to stop translation. The sequence of codons in the mRNA unambiguously defines the primary structure of the final protein. Of course, the codons in the mRNA were also present in the genomic DNA, but the sequence may be interrupted by introns. (See genetic code also).

Codon bias : The tendency for an organism or virus to use certain codons more than others to encode a particular amino acid. An important detrminant of codon bias is the guanosine-cytosine (GC) content of the genome. An organism that has a relatively low G+C content of 30% will be less likely to have a G or C at the third position of a codon (wobble position) than a A or T to specify an amino acid that can be represented by more than one codon.

Codon optimization : An experimental strategy in which codons within a cloned gene - ones not generally used by the host cell translation system - are changed by *in vitro* mutagenesis to the preferred codons, without changing the amino acids of the synthesized protein.

Coefficient : A number expressing the amount of some change or effect under certain conditions (e.g., the coefficient of inbreeding).

Coelomycetes : A group of fungi in the deuteromycetes (Imperfect Fungi) that produce pycnidia or acervuli (see hyphomycetes).

Coenocytic : having multiple nuclei embedded in cytoplasm without cross walls; nonseptate.

Co-enzyme : An organic molecule of low molecular weight and usually non-protein, such as a vitamin, that binds to an enzyme and promotes its catalytic activity.

Co-evolution : The evolution of complementary adaptations in two species caused by the selection pressure that each exerts on the other. It is common in symbiotic associations, in insect-pollinated plants, etc.

Co-factor : An organic molecule or inorganic ion necessary for the normal catalytic activity of an enzyme.

Co-fermentation : The simultaneous growth of two micro-organisms in one bioreaction.

Co-generation : Production of both electricity and process heat (steam) in an industrial plant.

Cohesion : Holding together; a force holding a solid to a solid or a solid to a liquid, owing to attraction between like molecules.

Cohesive ends : Single-stranded complementary sequences at two ends of a DNA molecule (examples include the ends of the linear phage lambda chromosome, and the ends resulting from digestion of DNA with many restriction endonucleases).

Coincidence : The ratio of the observed frequency of double crossings-over to the expected frequency, where the expected frequency is calculated by assuming that the two crossing-over events occur independently of each other.

Coinheritance frequency : The ratio of recombinants that acquire both the selected and unselected marker vs the total number of recombinants.

Cointegrate : A circular molecule of DNA formed during replicative transposition by joining two separate circular replicons. The resulting structure has two copies of the transposon, one at each junction point, orientated in direct repeats.

Co-integrate vector system : A two-plasmid system for transferring cloned genes to plant cells. The cloning vector has a T-DNA segment that contains cloned genes. After introduction into Agrobacterium tumiefasciens, the cloning vector DNA undergoes homologous recombination with a resident disarmed Ti plasmid to form a single plasmid carrying the genetic information for transferring the genetically engineered T-DNA region to plant cells.

Coisogenic or congenic : Nearly identical strains of an organism; they vary at only a single locus.

Colchicine (L colchicum, meadow saffron, from colchis, ancient Mingrelia) : An alkaloid obtained from Colchicum autumnale, autumn crocus or meadow saffron, which inhibits spindle formation in cells during mitosis, so that chromosomes cannot separate during anaphase, thus inducing multiple sets of chromosomes. Also used to halt mitosis at metaphase - the stage when chromosomes are most visible.

Cold sensitive mutant : A mutation that results in a gene product that is active at a high temperature (e.g. 42°C) but inactive at low temperatures (e.g. 30°C). Cold sensitive mutations often affect hydrophobic interactions between protein complexes or between proteins and membrane componants.

Cold spot : Sites within a gene (or genome)Êat which mutations occur with much lower frequency than at other sites.

Coleoptile : Protective sheath covering the shoot apex of the embryo in monocotyledenous plants. It protects the plumule as it emerges through the soil.

Coleorhiza (Gr koleos, sheath + rhiza, root) : A protective sheath surrounding the radicle of monocotyledenous plants.

Colicin : A polypeptide secreted by bacteria which inhibit or prevent the growth of related bacteria which lack the corresponding colicin immunity proteins. Colicins are encoded by a group of naturally occurring plasmids first found in *E. coli* (e.g., the Col E1 plasmid). Colicins kill other bacteria by a variety of mechanisms depending on the type of colicin -- some colicins block protein synthesis, some colicins induce the degradation of chromosomal DNA, some colicins form transmembrane holes in the cytoplasmic membrane, etc.

Colinear : Markers that occur in a sequential order in a DNA or protein sequence.

Co-linearity : Refers to an exact correspondence between information encoded in DNA and the polypeptide product ultimately translated from transcripts made from the DNA. Most prokaryotic genes are co-linear with their products; eukaryotic genes often are not.

Coliphage : A phage that infects Escherichia coli.

Collar : the portion of the seedling or plant near the surface of the soil; in grafted woody plants, the scion portion of the plant near the soil surface.

Collection : In PGR: see base collection; active collection.

Collenchyma (Gr kolla, glue + enchyma, a suffix, derived from parenchyma and denoting a type of cell tissue) : A tissue of living cells, the walls being unevenly thickened with cellulose and hemicellulose, but never lignified; it functions in mechanical support in young, short-lived or non-woody organs and is thus found in midribs and leaf petioles.

Colonization : Establishment and ramification of a pathogen within a host plant.

Colonize : To infect and ramify through plant tissue with the growth of a pathogen.

Colony : A visible group of cells arising from a single cell plated on solid medium.

Colony blots : See Colony hybridization.

Colony forming unit (abbr cfu) : the number of colonies formed per unit of volume or weight of a cell or spore suspension.

Colony hybridization : Use of in situ hybridization with a labeled nucleotide probe to identify bacterial colonies that contain DNA sequences homologous to a probe.

Colony PCR : Use of a bacterial colony for direct polymerase chain reaction to amplify a specific nucleotide sequence.

Combinatorial library : During the ligation reaction with cDNAs of light and heavy antibody chains into a bacteriophage lambda (?) vector, many novel combinations consisting of one heavy and one light chain coding region are formed. The library comprises these combinations, each in a separate vector.

Combining ability : General, average performance of a strain in a series of crosses. Specific, deviation from performance predicted on the basis of the general combining ability.

Combining ability, general : Good general combining ability (GCA) is the ability of an individual to produce progeny with high genetic quality, when crossed to many other individuals in the population.

Combining ability, specific : Good specific combining ability (SCA) is when the progeny from a particular full-sib cross perform better than what would be predicted from the general combining ability of either parent.

Commensalism : The interaction of two or more dissimilar organisms where the association is advantageous to one without affecting the other(s). cf parasitism; symbiosis.

Companion cell : Living cell associated with the sieve cell of phloem tissue in vascular plants.

Comparative gene mapping : The comparison of map locations of genes between species. The results of these comparisons indicate substantial conservation of blocks of genes and even large segments of chromosomes between species. Great use can be made of this conservation of map position. For example, in the case of mammals, it means that if a gene has been mapped in one or both of the intensely-mapped species (humans and mice), then the likely location of that gene in other mammals can be predicted with considerable confidence. Conversely, if a mapped anonymous DNA marker has an effect on a quantitative trait (this being indicative of the marker being linked to a quantitative trait locus (QTL)) in, say, cattle, then knowledge of the comparative map between cattle and humans can identify genes in the homologous region of the human genome that could correspond to the QTL. Such genes are called comparative positional candidate genes (q.v.).

Comparative positional candidate gene : A gene that is likely to be located in the same region as a DNA marker that has been shown to be linked to a single-locus trait or to a quantitative trait locus (QTL), where the gene's likely location in the genome of the species in question is based on its known location in the map of another species, i.e., is based on the comparative map between the two species.

Compartmentalization : Isolation of a specific tissue area by host barrier tissues.

Compatibility : The ability of two different types of plasmid to coexist in the same cell.

Competence : The transient physiological state required for a bacterial cell to take up transforming DNA.

Competency : An ephemeral state, induced by treatment with cold cations, during which bacterial cells are capable of taking up foreign DNA.

Competent : Bacterial cells which are capable of accepting foreign extra-chromosomal DNA. There are a variety of processes by which cells may be made competent.

Complement : Definition determined by context: 1] The complementary sequence to a nucleic acid sequence under study. 2] To provide a function that is missing or altered because of a mutational event.

Complement proteins : Proteins that bind to antibody-antigen complexes and help degrade the complexes by proteolysis.

Complementarity : The relationship between the two strands of a double helix of DNA. Thymine in one strand pairs with adenine in the other strand, and cytosine in one strand pairs with guanine in the other strand.

Complementary : Two polynucleotide chains that can base-pair to form a double-stranded molecule.

Complementary DNA (cDNA) : A DNA sequence synthesized from a messenger RNA molecule, using reverse transcriptase enzyme. cDNAs can be used experimentally to determine the sequence of messenger RNAs after their introns (non-protein-coding sections) have been spliced out. The single- stranded form is often used as a probe in physical mapping.

Complementary entity : 1. One of a pair of nucleotide bases that form hydrogen bonds with each other. Adenine (A) pairs with thymine (T) [or with Uracil (U) in RNA], and guanine (G) pairs with cytosine (C). 2. One of a pair of segments or strands of nucleic acid that will hydridize (join by hydrogen bonding) with each other.

Complementary genes : Two or more interdependent genes, such that (in the case of dominant complementarity) the dominant allele from either gene can only produce an effect on the phenotype of an organism if the dominant allele from the other gene is also present; or (in the case of recessive complementarity) only double homozygous recessive show the effect.

Complementary homopolymeric tailing : The process of adding complementary nucleotide extensions to different DNA molecules, e.g., dG (deoxyguanosine) to the 3′-hydroxyl ends of one DNA molecule and dC (deoxycytidine) to the 3′-hydroxyl ends of another DNA molecule to facilitate, after mixing, the joining of the two DNA molecules by base pairing between the complementary extensions. Also called dG - dC tailing, dA - dT tailing.

Complementary nucleotides : Members of the pairs adenine-thymine, adenine-uracil, and guanine-cytosine that have the ability to hydrogen bond to one another. See nucleotide.

Complementary sequence : Nucleic acid base sequence that can form a double-stranded structure with another DNA fragment by following base-pairing rules (A pairs with T and C with G). The complementary sequence to GTAC for example, is CATG.

Complementation : The ability of a gene to produce a functional gene product which compensates for the mutant phenotype caused by a mutation in another gene. Typically, the complementing gene produces a gene product (e.g. a repressor protein or an enzyme for synthesis of a metabolite) that is diffusible and thus functions in trans.

Complementation test; trans test : Introduction of two mutant chromosomes into the same cell to determine whether the mutants are alleles of the same gene. If the mutations are non-allelic, the genotype will be m1 +/+ M2 , and the phenotype will be wild-type (normal), because each chromosome "covers" for the other. In contrast, if they are allelic, the mutant phenotype will result.

Complete digest : The treatment of a DNA preparation with an endonuclease for sufficient time for all of the potential target sites within that DNA to have been cleaved. cf partial digest.

Complete resistance : A resistant reaction in which some aspect of disease development, usually symptom expression or pathogen reproduction, is completely stopped (see partial resistance).

Complex : Containing multiple macromolecules in an organized structure. For example, the RecBCD proteins form a complex that functions as an exonuclease and this activity requires all three of the proteins to function together.

Complex trait : Trait that has a genetic component that does not follow strict Mendelian inheritance. May involve the interaction of two or more genes or gene-environment interactions. See also: Mendelian inheritance, additive genetic effects.

Composite transposon : A transposible element flanked by two copies of an IS element.

Compost : a mixture of organic residues and soil that is allowed to decompose biologically.

Compound chromosome : A chromosome formed by the union of two separate chromosomes, as in attached-X chromosomes or attached-X-Y chromosomes.

Concatemer : Tandem arrays of monomeric DNA molecules with complementary ends. Intra-molecular reassociation of such molecules leads to circularisation while inter-molecular reaction produces concatemers.

Concentric : one circle within another with a common center.

Concordance : Identity of matched pairs or groups for a given trait, such as sibs expressing the same trait.

Conditional lethal mutation : A mutation that is lethal under one set of environmental conditions - the restrictive conditions - but is viable under another set of environmental conditions - the permissive conditions, e.g., temperature-sensitive mutations.

Conditional mutant : A mutant that can grow under one set (permissive) of environmental conditions but cannot grow under different (restrictive or nonpermissive) conditions. Some conditional mutants survive under the nonpermissive conditions but others (conditional lethal mutants) die under the nonpermissive conditions.

Conditioning : 1. The effects on phenotypic characters of external agents during critical developmental stages. 2. The undefined interaction between tissues and culture medium resulting in the growth of single cells or small aggregates. Conditioning may be accomplished by immersing cells or callus contained within a porous material (such as dialysis tubing) into fresh medium for a period dependent on cell density and a volume related to the amount of fresh medium.

Conelet : An immature cone (strobilus) in the conifers.

Confidentiality : In genetics, the expectation that genetic material and the information gained from testing that material will not be available without the donor's consent.

Conformation : The 3-dimensional structure of a molecule.

Conformational epitope : Also called a "global epitope". An epitope comprised of contiguous but physically discontinuous components of the immunogenic molecule. In a proteinaceous antigen this could be sequences from different stretches within a polypeptide, or even from different polypeptide subunits.

Congenic : Strains that are identical except for a small region of the chromosome.

Congenital : Any trait present at birth, whether the result of a genetic or nongenetic factor. See also: birth defect.

Congression : The uptake of more than one transforming DNA fragment by a single competent cell (occurs most often when a high concentration of transforming DNA is available).

Conidioma (pl conidiomata) : Specialized conidia-bearing structure, e.g. acervulus, pycnidium, sporodochium, synnema.

Conidiophore: Simple or branched hypha on which conidia are produced.

Conidium (pl conidia) : An asexual, nonmotile fungal spore that develops externally or is liberated from the cell that formed it.

Conjugation : The establishment of a bridge between a donor and a recipient cell and the transfer of DNA from one cell to the other. Conjugation is mediated by certain plasmids and transposons. Conjugation requires direct contact between the donor and recipient cells.

Conjugative functions : Plasmid-based genes and their products that facilitate the transfer of a plasmid from one bacterium to another.

Conjugative plasmid : A bacterial plasmid that encodes functions required for conjugation.

Conjugative transposon : A transposon that encodes functions required for conjugation.

Conk : A shelf-like, typically hardened basidiocarp of a wood decaying fungus, usually a polypore.

Consanguinity : Genetic relationship. Consanguineous individuals have at least one common ancestor in the preceding few generations.

Consensus sequence : A 'nominal' sequence inferred from multiple, imperfect examples. Multiple lanes of shotgun sequence can me merged to show a consensus sequence. The optimal sequence of nucleotides recognized by some factor. A DNA binding site for a protein may vary substantially, but one can infer the consensus sequence for the binding site by comparing numerous examples. For example, the (fictitious) transcription factor ZQ1 usually binds to the sequences AAAGTT, AAGGTT or AAGATT. The consensus sequence for that factor is said to be AARRTT (where R is any purine, i.e. A or G). ZQ1 may also be able to weakly bind to ACAGTT (which differs by one base from the consensus).

Conservation : Identical parts of genes that are present in two distinct organisms are said to be conserved. Conservation can be detected by measuring the similarity of the two sequences at the base (RNA or DNA) or amino-acid (protein) level.

Conservation of farm animal genetic resources : In AnGR: Refers to all human activities, including strategies, plans, policies and actions, undertaken to ensure that the diversity of farm animal genetic resources is being maintained to contribute to food and agricultural production and productivity; now and in the future. (Source: FAO, 1999).

Conservation tillage : Crop production methods that de-emphasize use of the plow for weed removal and encompass a range of new farming production practices like reduced tillage, or no tillage. In general, these methods all include reduced use of the plow and increased use of crop mulch and cover on the fields.

Conservative (nonreplicative) transposition : A transposition event where the transposable element is lost from its original location and inserted at a new location.

Conservative change : An amino acid change that does not affect significantly the function of the protein.

Conservative substitution : A nucleotide mutation which alters the amino acid sequence of the protein, but which causes the substitution of one amino acid with another which has a side chain with similar charge/ polarity characteristics (see Amino Acid). The size of the side chain may also be an important consideration. Conservative mutations are generally considered unlikely to profoundly alter the structure or function of a protein, but there are many exceptions (see Nonconservative Substitution).

Conserved sequence : A nucleotide or protein sequence which is shared between different regions or organisms, typically because the sequence fulfills an important function (e.g. a DNA-binding site or protein motif) and thus is retained during evolution.

Constant domains : Regions of antibody chains that have the same amino acid sequence in different members of a particular class of antibody molecules.

Constitutive : An organism is said to be constitutive for the production of an enzyme or other protein if that protein is always produced by the cells under all physiological conditions. See inducible.

Constitutive ablation : Gene expression that results in cell death.

Constitutive enzyme : An enzyme that is synthesized continually regardless of growth conditions. cf inducible enzyme; repressible enzyme.

Constitutive expression : Constantly expressed at some appreciable level.

Constitutive gene expression : A gene or operon which is expressed at all times independant of the environmental conditions.

Constitutive promoter : An unregulated promoter that allows for continual transcription of its associated gene. See promoter.

Constitutive synthesis : Continual production of RNA or protein by an organism.

Contact fungicide (syn protectant fungicide) : A fungicide that remains on the surface where it is applied; no after-infection activity (see systemic fungicide).

Contaminant : Bacterial, fungal or algal micro-organism accidentally introduced into a culture or culture medium. Contaminant may overgrow the plant cells and consequently inhibit their growth. Working under aseptic conditions with a rigorous exclusion of potential contaminants must be practised in plant tissue culture. See disinfectation; disinfestation.

Contig: Several uses, all nouns. The term comes from a shortening of the word 'contiguous'. A 'contig' may refer to a map showing placement of a set of clones that completely, contiguously cover some segment of DNA in which you are interested. Also called the 'minimal tiling path'. More often, the term 'contig' is used to refer to the final product of a shotgun sequencing project. When individual lanes of sequence information are merged to infer the sequence of the larger DNA piece, the product consensus sequence is called a 'contig'.

Contig map : A map depicting the relative order of a linked library of small overlapping clones representing a complete chromosomal segment.

Contiguous genes : Genes physically close on a chromosome that when acting together express a phenotype.

Contiguous map; contig map : The alignment of sequence data from large, adjacent regions of the genome to produce a continuous nucleotide sequence across a chromosomal region. See mapping.

Continuous cropping : Growing the same crop in the same location repeatedly.

Continuous culture : The culture of microorganisms in liquid medium under controlled conditions, with regular additions of depleted nutrients and removal of excreted metabolites from the medium over a lengthy period of time.

Continuous fermentation : A process in which cells or micro-organisms are maintained in culture in the exponential growth phase by the continuous addition of fresh medium that is exactly balanced by the removal of cell suspension from the bioreactor. cf continuous culture.

Continuous variation : Variation not represented by distinct classes. Phenotypes grade into each other, and measurement data are required for analysis. Multiple genes are usually responsible for this type of variation. a.k.a. quantitative variation. cf discontinuous variation.

Control : 1. Noun: Unchanged (standard) protocol or treatment for comparison with the experimental treatment. The term is commonly used for untreated organisms. 2. Verb: To direct or regulate cultures with addition of plant growth regulators.

Control pollination : To purposely pollinate the female flowers of a tree with pollen from a known source, usually one specific tree. Usually the flowers are Protected from undesirable pollen by covering them with a pollen-tight cloth or paper bag before they are receptive and adding known-source pollen at receptivity. Full-sib families are produced.

Controlled environment : The environment in which parameters, such as light, temperature, relative humidity and sometimes the partial gas pressure, are fully controlled.

Controlling element : In eukaryotes, transposable elements which control the activity of standard genes. A controlling element may, in the simplest case, inhibit the activity of a gene through becoming integrated in, or close to, that gene. Occasionally, either in germinal or somatic tissue, it may be excised from this site, and due to excision the activity of the gene is more or less restored, while the element may become reintegrated elsewhere in the genome where it may affect the activity of another gene. For example, in maize, a controlling element such as Ac or Ds is capable of influencing the expression of a nearby gene. See transposable element.

Conversion : The development of a somatic embryo into a plant. See regeneration; micropropagation; organogenesis.

Coordinate gene expression : Transcription of a group of genes at the same time due to a common regulatory mechanism.

Coordinate repression : Correlated regulation of the structural genes in an operon by a molecule that interacts with the operator sequence.

Coorperativity : For DNA binding proteins, cooperativity generally refers to the increased binding of a protein to a DNA site due to the prior binding of another protein nearby. (Note that more rigorous definations of cooperativity are sometimes used in biochemistry to refer to binding of small molecules, that can result in positive cooperativity (enhanced binding) and negative cooperativity (poorer binding) of a second molecule following binding of a first molecule).

Co-polymers : Mixtures consisting of more than one monomer; for example, polymers of two kinds of organic bases, such as uracil and cytosine (poly-UC) have been combined for studies of the genetic code.

Coppice : Regrowth of woody plants by sprouts from stumps or roots.

Copy DNA : See cDNA.

Copy number: The number of molecules of a particular plasmid present in a bacterium.

Coremium (pl coremia; syn synnema) : Compact or fused, generally upright conidiophores, with branches and spores forming a headlike cluster.

Co-repressor : A small molecule that binds to an aporepressor protein, producing a conformational change that causes it to function as a transcriptional repressor.

Cork : External protective tissue of a stem or root, impermeable to water and gasses; the primary component of bark.

Cornelia de Lange syndrome : Condition involving growth deficiency, significant developmental delay, anomalies of the extremities and a characteristic facial appearance.

Corolla : Petals, collectively.

Corpus : The corpus is found below the tunica (q.v.) and is a part of the apical meristem. In the corpus, cells divide in all directions, giving them an increase in volume. See apical meristem.

Correlation : A statistical association between variables.

Cortex : Primary tissue of a stem or root, bounded externally by the epidermis and internally in the stem by the phloem and in the root by the pericycle.

Cos ends : The 12-base, single-strand, complementary extensions of bacteriophage lambda (?) DNA. a.k.a. cos sites.

Cos site : The sequence that is cut to produce the cohesive, single-stranded extensions located at the ends of the linear DNA molecules of certain phages (e.g. lambda).

Cosanguine : Related by a common ancestor.

Co-segregation : When two genetic conditions appear to be inherited together.

Cosmid : A type of artificially constructed vector used for cloning 35-45 kb of DNA. These are plasmids carrying a phage l cos site (which allows packaging into l capsids), an origin of replication and an antibiotic resistance gene. A plasmid of 40 kb is very difficult to put into bacteria, but can replicate once there. Cosmids, however, have a cos site, and thus can be packaged into l phage heads (a reaction which can be performed *in vitro*) to allow efficient introduction into bacteria.

Cot curve : When duplex DNA is heated, it dissociates into single strands. When the temperature is lowered, complementary strands tend to anneal or re-nature. The extent of re-naturation depends on the product of DNA concentration in moles of nucleotides per litre, and time in seconds. A graph showing the proportion of re-natured DNA against cot is known as a cot curve. The cot at which half the DNA has re-natured is the half-cot, a parameter indicating the degree of complexity of the DNA.

Cotransduction : Homologous recombination of two closely linked genes or two mutations within the same gene and brought into a cell on the same fragment of transducing DNA.

Co-transfection : In baculovirus expression systems, the procedure by which the baculovirus and the transfer vector are simultaneously introduced into insect cells in culture.

Cotransformation : The simultaneous uptake of two genetic markers via transformation. Cotransformation of two markers occurs at a frequency greater than 50 percent indicates that the two markers are genetically linked.

Cotyledons : Leaflike structures at the first node of the seedling stem. In some dicotyledons, they contain stored food for the young plant not yet able to photosynthesize its own food. Often referred to as seed leaves.

Counterselection : A condition that prevents growth of the donor in a genetic cross.

Coupling : Linked recessive alleles occur in one homologous chromosome and their dominant alternatives occur in the other chromosome. Opposed to repulsion in which one dominant and one recessive occur in each member of the pair of homologous chromosomes.

Covalent bond : A bond in which an electron pair is equally shared by protons in two adjacent atoms.

Covalently closed-circular (CCC) : A completely double-stranded circular DNA molecule, with no nicks or discontinuities, usually with a supercoiled conformation.

Co-variance : A measure of the statistical association between variables; the extent to which two variables vary together.

cpDNA : The DNA of plant plastids, including chloroplasts.

CpG islands : areas of multiple CG repeats in DNA.

Cri-du-chat syndrome : A chromosomal condition (monosomy 5p). Name comes from the distinctive mewing cry of affected infants; characterized by significant mental deficiency, low birthweight, failure to thrive and short stature; deletion of a small section of the short arm of chromosome 5.

Critical breed : In AnGR: A breed where the total number of breeding females is less than 100 or the total number of breeding males is less than or equal to five; or the overall population size is close to, but slightly above 100 and decreasing, and the percentage of pure-bred females is below 80%. (Source: FAO, 1999).

Critical-maintained breed; endangered-maintained breed In AnGR : Categories where critical or endangered breeds are being maintained by an active public conservation programme or within a commercial or research facility. (Source: FAO, 1999).

CRM : Cross reacting material. A molecule that reacts with an antibody. A common test for the presence of an inactive protein in cells.

Crop rotation : The successive planting of different crop species; often used to improve soil fertility or to reduce disease and pest problems.

Cross : In genetic studies, the mating of two individuals or populations. Also called mating.

Cross hybridization : The hydrogen bonding of a single-stranded DNA sequence that is partially but not entirely complementary to a single-stranded substrate. Often, this involves hybridizing a DNA probe for a specific DNA sequence to the homologous sequences of different species.

Cross pollination : Fertilization of a plant from a plant with a different genetic makeup.

Cross pollination efficiency : Efficiency of pollen from one plant reaching the stigma of another plant.

Cross reacting material : See CRM.

Cross-breeding : Mating between members of different populations (lines, breeds, races or species).

Cross-fertilization : A process in which sexual reproduction occurs as a result of the fusion of sex cells from different individuals (see self-fertilization).

Crossing groups : Any group of individuals which comprises a unique set of parents. A diallel CG is one in which controlled crosses are made between each pair of parents in the group but crosses with parents outside the group are excluded. A factorial CG is one in which a limited number of parents are used as male testers in controlled crosses with an unlimited number of female parents. An open-pollinated CG is one in which all parents in a breeding population are included in a progeny test or series of tests and which may be used for reforestation with genetically improved material.

Crossing over: The breaking during meiosis of one maternal and one paternal chromosome, the exchange of corresponding sections of DNA, and the rejoining of the chromosomes. This process can result in an exchange of alleles between chromosomes. Compare recombination.

Crossing-over unit : A measure of distance between two loci on genetic maps that is based on the average number of crossing-over events that take place in the interval between those two loci during meiosis. A map interval that is one crossing-over unit in length (a centiMorgan) describes an interval between two loci such that one in every hundred gametes recovered from meiosis is recombinant in that interval, i.e., the allele at the first locus is maternal in origin, while the allele at the other locus in that same gamete is paternal in origin.

Cross-over : The site where two homologous DNA strands originating from homologous chromosomes are resealed to form the recombinant chromosome. The reciprocal exchange of genetic material to produce genetic recombinants. Sometimes abbreviated as X-over.

Cross-pollination : the mating of two genetically different plants of the same species. Usually, the term is used in reference to the crossing of two pure breeding (homozygous) plants.

Crown: The region at the base of the stem of cereals and forage species from which tillers or branches arise. In woody plants, the root-stem junction. In forestry, the top portions of the tree.

Crown gall (AS gealla, gall) : A bulbous growth that occurs at the base of certain plants as result of infection, especially by Agrobacterium tumefaciens; a bacterial gene carried by the Ti plasmid is transferred by the bacteria into a higher plant cell, where it causes a tumour-like growth. a.k.a. crown gall tumour. See Agrobacterium; hairy root disease.

Crozier : the hook of an ascogenous hypha before ascus development; the curved apical portion of a blighted stem.

CRP : Catabolite receptor protein, also called CAP or catabolite activator protein. The interaction of CRP with cAMP modulates many aspects of catabolite repression in enteric bacteria.

Cry1A : A protein derived from the bacterium Bacillus thuringiensis that is toxic to some insects when ingested. This bacterium occurs widely in nature and has been used for decades as an insecticide although it constitutes less than 2 percent of the overall insecticides used.

Cryobiological preservation; cryopreservation; freeze preservation : The preservation of germplasm resources in a dormant state by cryogenic techniques, as currently applied to storage of plant seeds and pollen, micro-organisms, animal sperm, and tissue culture cell lines. See ex situ conservation; gene bank.

Cryogenic : At very low temperature.

Cryopreservation : See cryobiological preservation.

Cryoprotectant : Compound preventing cell damage during freezing and thawing processes. Cryoprotectants are agents with high water solubility and low toxicity. Two types of cryoprotectant agent are commonly used: permeating (glycerol and DMSO (q.v.)) and non-permeating (sugars, dextran, ethylene glycol, polyvinyl pyrolidone and hydroxyethyl starch).

Cryptic : 1. Structurally heterozygous individuals not identifiable on the basis of abnormal meiotic-chromosome pairing configurations ('cryptic structural hybrids'). 2. A form of polymorphism controlled by recessive genes ('cryptic polymorphism'). 3. Any mutation which is exposed by a sensitizing mutation and otherwise poorly detected (such mutations probably escape detection because of the plasticity of composition of the corresponding polypeptide). 4. Phenotypically very similar species (cryptic species) which do not hybridize under normal conditions. 5. Cryptic genetic variation refers to the existence of, for example, alleles conferring high performance for a trait, in a breed that has low performance for that trait.

Cryptobiosis (hidden life) : The ability to enter a state of suspended metabolic activity during unfavorable environmental conditions (drying, cold); survival strategy of some plant pathogenic nematodes.

C-terminus : The end of a polypeptide chain that has a free carboxylic acid (-COOH) group.

Culm : stem of grasses, cereals and bamboos.

Cultivar (from cultivated + variety) (abbr cv) : A category of plants that are, firstly, below the level of a sub-species taxonomically, and, secondly, found only in cultivation. It is an international term denoting certain cultivated plants that are clearly distinguishable from others by stated characteristics and that retain their distinguishing characters when reproduced under specific conditions.

Cultural practices : The manner in which plants are grown, such as: application of nutrients, irrigation practices, type of cultivation; may be used for disease management.

Culture : A population of plant or animal cells or micro-organisms that is grown under controlled conditions.

Culture alteration : A term used to indicate a persistent change in the properties of a culture's behaviour (e.g., altered morphology, chromosome constitution, virus susceptibility, nutritional requirements, proliferative capacity, etc.). The term should always be qualified by a precise description of the change which has occurred in the culture.

Culture medium : Any nutrient system for the cultivation of cells of plants, bacteria or other organisms; usually a complex mixture of organic and inorganic nutrients. cf medium.

Culture room : Room for maintaining cultures and often in a controlled environment of light, temperature and humidity. cf growth cabinet; incubator.

Curing : A treatment that promotes the loss of a resident prophage or plasmid from a cell. For example, acridine orange is sometimes used to cure bacteria of a plasmid.

Curl : Distoration, puffing, and crinkling of a leaf resulting from unequal growth.

Cut : Slang: to make a double-stranded break in DNA, usually with a type II restriction endonuclease. E.g., "The DNA was cut with EcoRI and run out on a 1% agarose gel." cf nick; cleave.

Cut and paste : Jargon for the excision of a DNA fragment from one molecule and insertion into another DNA molecule. Commonly used to describe cloning a DNA fragment by cleavage with a restriction enzyme and ligation into a vector. Also used to describe transposition mechanisms where the transposon is excised from one DNA molecule and inserted into a second DNA molecule without replication of the transposon.

Cuticle (adj cuticular) : Noncellular outer layer of an insect or a nematode; water-repellent, waxy layer of epidermal cells of plant parts, such as leaves, stems and fruit.

Cuticle (L cuticula, diminutive of cutis, the skin) : Layer of cutin or wax on the outer surface of leaves and fruits and that reduces water loss.

Cutting : Noun: A detached plant part that under appropriate cultural conditions can regenerate the complete plant without a sexual process.

cv (abbr for cultivar; syn variety) : A plant type within a species, resulting from deliberate manipulation, which has recognizable characteristics (color, shape of flowers, fruits, seeds and height or form).

Cybrid : A cytoplasmic hybrid, originating from the fusion of a cytoplast (the cytoplasm without nucleus) with a whole cell, as in nuclear transfer (although the term is not used in that context). Note that the nucleus and cytoplasm of the fused cell products are from different genetic sources.

Cyclic AMP (cAMP) : An adenosine monophosphate molecule with the phosphate covalently attached to both the 3' and 5' carbons of the ribose.

Cystein : An amino acid.

Cystic fibrosis : A genetically inherited disease in children that results in chronic fluid development in the lungs, making breathing difficult. This disease also prevents normal absorption of fats and other nutrients from food. Cystic fibrosis occurs as a result of inheriting a recessive allele for it from both parents. This is ultimately a fatal disease, but with modern medical care, about 2/3 of the people with it survive into early adulthood. About 30,000 people have cystic fibrosis in the U.S. today.

Cytogenetics : Area of biology concerned with chromosomes and their implications in genetics, cellular activity and variability.

Cytokine : In immunology, any of many soluble molecules that cells produce to control reactions between other cells.

Cytokinesis : Cytoplasmic division and other changes exclusive of nuclear division that are a part of mitosis or meiosis. See cell division.

Cytokinin : Plant growth regulators (hormones) characterized as substances that induce cell division and cell differentiation (e.g., BA, kinetin, and 2-iP). In tissue culture, these substances are associated with enhanced callus and shoot development. The compounds are derivatives of adenine.

Cytological band : An area of the chromosome that stains differently from areas around it.

Cytological map : A type of chromosome map whereby genes are located on the basis of cytological findings obtained with the aid of chromosome mutations.

Cytology : The study of the structure and function of cells.

Cytolysis : Cell disintegration.

Cytopathology : The study of changes induced by disease at the cellular level.

Cytoplasm : Living protoplasm in a cell, except the nucleus.

Cytoplasm (Gr kytos, a hollow vessel + plasma, form) : The living material of the cell, exclusive of the nucleus, consisting of a complex protein matrix or gel. The part of the cell in which essential membranes and cellular organelles (mitochondria, plastids, etc.) reside.

Cytoplasmic (uniparental) inheritance : See: cytoplasmic trait.

Cytoplasmic genes : DNA-containing bodies in the cell but external to the nucleus.

Cytoplasmic inheritance : Hereditary transmission dependent on the cytoplasm or structures in the cytoplasm rather than the nuclear genes; extrachromosomal inheritance. Thus, plastid characteristics in plants are inherited by a mechanism independent of nuclear genes.

Cytoplasmic male sterility : Genetic defect due to defective functions of mitochondria in the pollen. Fertilization will not occur. Exploited in certain plant breeding strategies, such as F1-hybrid maize cultivars.

Cytoplasmic organelles : Discrete sub-cellular structures located in the cytoplasm of cells; these allow division of labour within the cell.

Cytoplasmic trait : A genetic characteristic in which the genes are found outside the nucleus, in chloroplasts or mitochondria. Results in offspring inheriting genetic material from only one parent.

Cytosine : One of the four bases in DNA that make up the letters ATGC, cytosine is the "C". The others are "A" for adenine, "G" for guanine, and "T" for thymine. Cytosine always pairs with guanine. Adenine always pairs with thymine. These letters are used as shorthand for the sequences of fragments of DNA e.g. CCAAGTAC. These sequences are the code for genetic information.

Cytosol : The fluid portion of the cytoplasm, i.e., the cytoplasm minus its organelles.

Cytotoxic T cell : See killer T cell.

Cytotype : A maternally inherited cellular condition in Drosophila that regulates the activity of transposable P elements.

Dabs (single-domain antibodies) : Antibodies with only one (instead of two) protein chain derived from only one of the domains of the antibody structure. Dabs exploit the finding that, for some antibodies, half of the antibody molecule binds to its target antigen almost as well as the whole molecule. The potential advantages of dabs are that they can be made easily by bacteria or yeasts, and offer a way to clone antibody-like molecules into bacteria, and hence to be able to easily screen millions of antibodies. Related ideas are single-chain antigen binding technology (SCA), biosynthetic antibody binding sites (BABS), minimum recognition units (MRUs), and complementary determining regions (CDRs).

Dalton (symbol Da) : A unit of atomic mass roughly equivalent to the mass of a hydrogen atom. 1.67 × 10-24 g. Named after the famous nineteenth-century chemist, John Dalton (1766-1844)). Used in shorthand expressions of molecular weight, especially as kilo- (kDa) or megadaltons (MDa), which are equal to respectively to 1 × 103 and 1 × 106 daltons.

Damping-off : Death of a seedling before or shortly after emergence due to decomposition of the root and/or lower stem; it is common to distinguish between preemergence damping-off and postemergence damping-off.

Darwinian cloning : Selection of a clone from a large number of essentially random starting points, rather than isolating a natural gene or making a carefully designed artificial one. Molecules which are more similar to those needed are selected, mutated to generate new variants, and re-selected. The cycle proceeds until the required molecule is found. The advantage of the system is that the selection is from a vast number of possibilities.

Data warehouse : A collection of databases, data tables, and mechanisms to access the data on a single subject.

Database search : Once an open reading frame or a partial amino acid sequence has been determined, the investigator compares the sequence with others in the databases using a computer and a search algorithm. This is usually done in a protein database such as PIR or Swiss-Prot. Nucleic acid sequences are in GenBank and EMBL databases. The search algorithms most commonly used are BLAST and FASTA.

Days to harvest : Term for the prescribed minimum number of days required by U.S. government regulations between a pesticide application and the harvest of the crop.

ddNTP : See di-deoxynucleotide.

de novo (L "from the beginning, anew") : Arising, anew, afresh, once more. Also ex novo.

Death phase : The final growth phase, during which nutrients have been depleted and cell number decreases. See growth phases.

Decay : The gradual decomposition of organic matter.

Deceleration phase : The phase of declining growth rate, following the linear phase and preceding the stationary phase in most batch-suspension cultures. See growth phases.

Deciduous : Describing a tree that sheds its leaves completely at the end of its annual growth period.

Decoy crop : Crop that stimulates germination of seeds of a parasitic plant such as witchweed (Striga spp.), but is not susceptible to infection by the parasitic plant; helps reduce seed populations of the parasite in soil so a susceptible crop can be planted.

De-differentiation : The process by which cells lose their specialization and proliferate by cell division to form a mass of cells which, in response to appropriate stimuli, may later differentiate again to form either the same cell type or a different one. De-differentiation occurs in response to wounding and in tissue cultures. See re-differentiation.

Defective phage : A phage that lacks one or more functions that are required for reproduction.

Deficiency : Insufficiency or absence of one or more usable forms of enzymatic, nutritional or environmental requirements, so that development, growth or physiological functions are affected.

Defined : 1. Fixed conditions of medium, environment and protocol for growth. 2. Precisely known and stated elements of a tissue culture medium. cf undefined.

Defined medium : A bacterial growth medium in which all the components are known. Typically only includes the minimal growth requirements.

Defoliation : Loss of leaves from a plant, whether normal or premature.

Degeneracy : In molecular biology, this term refers to the fact that multiple different codons may encode the same amino acid. However, a given codon does not encode more than one amino acid within the nucleus of an organism.

Degenerate codon : A codon that specifies the same amino acid as another codon.

Degeneration : 1. Changes in cells, tissues or organs due to disease. 2. The reduction in size or complete loss of organs during evolution.

Degree-day : The departure of the average daily temperature from a defined base (e.g. the minimum recognized temperature for the growth of a plant species). The number of degree-days may be totaled to assess the accumulated warmth of a particular year's growing season.

Dehalogenation : The removal of halogen atoms (chlorine, iodine, bromine, fluorine) from molecules, usually during biodegradation.

Dehiscence : The spontaneous and often violent opening of a fruit, seed pod or anther to release and disperse the seeds or pollen.

Dehydrogenase : An enzyme that catalyses the remove of hydrogen atoms in biological reactions.

Dehydrogenation : A chemical reaction in which hydrogen is removed from a compound.

De-ionized water : Water which is free of most inorganic (not completely free, since Na is present in ample quantities) and most organic compounds.

Deletion : The loss of one or more bases or base pairs from a molecule of DNA.

Deletion map: A description of a specific chromosome that uses defined mutations --specific deleted areas in the genome-- as 'biochemical signposts,' or markers for specific areas.

Deletion mapping : The use of deletion mutations to determine the position of overlapping point mutations for fine structure mapping of a genetic locus.

Deliberate release : Putting something into the outside world; in biotechnology it means putting a genetically modified organism (GMO) into field trials.

Deme : A group of organisms in the same taxon.

Demicyclic : a rust fungus that lacks the urediniospore (repeating) stage (e.g. many species of Gymnosporangium) (see macrocyclic, microcyclic).

De-mineralize : To remove the mineral content (salts, ions) from a substance, especially water. Removal methods include distillation and electrodialysis. The process is de-mineralization.

Denaturated DNA : Duplex DNA that has been converted to single strands by breaking the hydrogen bonds of complementary nucleotide pairs. Usually achieved by heating.

Denaturation : With respect to nucleic acids, refers to the conversion from double-stranded to the single-stranded state, often achieved by heating or alkaline conditions. This is also called "melting" DNA. With respect to proteins, refers to the disruption of tertiary and secondary structure, often achieved by heat, detergents, chaotropes, and sulfhydryl-reducing agents.

Denaturation of proteins : The unfolding of a protein molecule, usually by high temperatures or ionic detergents such as sodium dodecyl sulfate (SDS).

Denaturing Gel : An agarose or acrylamide gel run under conditions which destroy secondary or tertiary protein or RNA structure. For protein, this usually means the inclusion of 2-ME (which reduces disulfide bonds between cysteine residues) and SDS and/or urea in an acrylamide gel. For RNA, this usually means the inclusion of formaldehyde or glyoxal to destroy higher ordered RNA structures. In DNA sequencing gels, urea is included to denature dsDNA to ssDNA strands. In denaturing gels, macromolecules tend to be separated on the basis of size and (to some extent) charge, while shape and oligomerization of molecules are not important. Contrast with Native Gel.

Denitrification : A chemical process in which nitrates in the soil are reduced to molecular nitrogen, which is released to the atmosphere.

Density gradient centrifugation : High-speed centrifugation in which molecules "float" at a point where their density equals that in a gradient of caesium chloride or sucrose. The density gradient may either be formed before centrifugation by mixing two solutions of different density (as in sucrose density gradients) or it can be formed by the process of centrifugation itself (as in CsCl and Cs2SO4 density gradients). See centrifugation.

Density shift labeling : The addition of either a heavy or a light isotope to the growth medium which changes the density of the newly synthesized nucleic acids. For example, N14 is the natural isotope of nitrogen, so incorporation of N15 into DNA incr eases the density of the DNA.

Density-gradient centrifugation : Separation of molecules and particles on the basis of buoyant density, by centrifugation in a concentrated sucrose or caesium chloride solution.

Deoxyribonuclease (DNase) : Any enzyme that hydrolyses DNA.

Deoxyribonucleicacid (abbr DNA) : The double-stranded, helical molecule that contains genetic code information; each repeating unit, or nucleotide, is composed of deoxyribose (a sugar), a phosphate group, and a purine (adenine or guanine) or a pyrimidine (thymine or cytosine) base.

Deoxyribonucleotide : See nucleotide.

Deoxyribose : A type of sugar that is one component of DNA (deoxyribonucleic acid).

De-repression : The process of "turning on" the expression of a gene or set of genes whose expression has been repressed (turned off). Displacement of a repressor protein from a promoter region of DNA. When attached to the DNA, the repressor protein prevents RNA polymerase from initiating transcription. The "turning on" of a gene.

Derivative : 1. Resulting from or derived from. 2. Term used to identify a variant during meristematic cell division.

Desiccant : Any compound used to remove moisture or water.

Desiccate : To dry, exhaust or deprive of water or moisture. Any chemical used for this purpose is called a desiccant. An apparatus for drying and preventing hygroscopic samples from rehydrating is a desiccator. The process is desiccation.

Desoxyribonucleic acid : Obsolete spelling of deoxyribonucleic acid (DNA).

Dessicator : Apparatus for drying or depriving of moisture.

Desulphurization (USA desulfurization) : Technology for removing sulphur from oil and coal by use of bacteria. Sulphur residues in fuels end up as sulphur dioxide when the fuel is burned, resulting in acid rain. Bacteria may oxidize sulphites (insoluble) into sulphates (soluble), which can be washed away with the bacteria.

Detergent : Substance which lowers the surface tension of a solution, improving its cleaning properties (e.g., Tween-20TM, a surfactant and wetting agent). See surfactant; wetting agent.

Determinate : Ceasing vegetative growth when the first flower or reproductive structure forms.

Determinate growth : Growth determined and limited in time, as in most floral meristems and leaves. The differentiation process is irreversibly established. Determinate growth contrasts with the usual culture growth, which is infinite and indeterminate.

Determination : Process by which undifferentiated cells in an embryo become committed to develop into specific cell types, such as neurons, fibroblasts or muscle cells.

Determined : Describing embryonic tissue at a stage when it can develop only as a certain kind of tissue.

Deuteromycetes (syn Fungi Imperfecti) : Group of fungi without a sexual stage; the asexual stage of ascomycetes and basidiomycetes.

Development : The sum total of events that contribute to the progressive elaboration of an organism. The two major aspects of development are growth and differentiation.

Deviation : 1. In statistics: the difference between an actual observation and the mean of all observations. 2. An alteration from the typical form, function or behaviour. Mutation or stress are the common reasons behind deviation.

Dextrin : An intermediate polysaccharide compound resulting from the hydrolysis of starch to maltose by amylase enzymes.

Dextrose : See glucose.

dG dC tailing : See complementary homopolymeric tailing.

Diabetes : an inherited metabolic disorder in which there are abnormally high blood sugar levels. In advanced stages, this often results in blindness from cataracts, nerve damage, gangrene in the feet and legs leading to amputation, heart disease, and kidney failure. Type 1 diabetes melitis (juvenile onset diabetes) is due to decreased production of insulin by the pancreas. Type 2 diabetes melitis is due to increased resistance of cells in the body to insulin. The gene or genes for diabetes are incompletely penetrant.

Diagnostic (n diagnosis) : A distinguishing characteristic important for the identification of a disease or other condition.

Diagnostic procedure : A test or assay used to determine the presence of an organism, substance or nucleic acid sequence alteration.

Diakinesis : A stage of meiosis just before metaphase I, in which the separation of homologous chromosomes is almost completed.

Diallel cross, complete : The crossing in all possible combinations of a series of genotypes.

Diallel, complete or full- : A mating design and subsequent progeny test resulting from the crossing of 'n' parents in all possible 'n2' combinations including selfs and reciprocals. (Because of severe inbreeding depression in the selfs, these are often skipped and the test still called a "full" diallel).

Diallel, incomplete or partial : A partial sampling - any individual family or type of family may be omitted. In either the complete or incomplete diallel, identities of both seed and pollen parents are maintained for each family. (See Pedigree).

Diapause : A period of spontaneous dormancy, independent of environmental conditions, interrupting developmental activity in an embryo, larva or pupa.

Diauxic growth : A biphasic growth curve, typically resulting from the sequential use of multiple nutrients (e.g. glucose utilization followed by lactose utilization).

Diazotroph : An organism that can fix atmospheric nitrogen.

Dicentric chromosome : A chromosome having two centromeres.

Dichotomous : branching, often successively, into two more or less equal arms.

Dicot (syn dicotyledon) : Plant with two cotyledons or seed leaves (see monocot).

Dicotyledon (Gr dis, twice + kotyledon, a cup-shaped hollow) : A plant with two cotyledons, or seed leaves. One of the two classes of plants in the Angiosperms (the other class is the monocotyledons). Colloquially called a dicot. Examples include many crop plants (potato, pea, beans), ornamentals (rose, ivy) and timber trees (oak, beech, lime).

Dideoxy sequencing : Enzymatic determination of DNA or RNA sequence by the method of Sanger and colleagues, based on the incorporation of chain terminating dideoxynucleotides in a growing nucleic acid strand copied by DNA polymerase or reverse transcriptase from a DNA or RNA template. Separate reactions include dideoxynucleotides containing A,C, G, or T bases. The reaction products represent a collection of new. labeled DNA strands of varying lengths, all terminating with a dideoxynucleotide at the 3' end (at the site of a complementary base in the template nucleic acid), and are separated in a polyacrylamide/urea gel to generate a sequence "ladder". This method is more commonly used than "Maxam-Gilbert" (chemical) sequencing.

Dideoxynucleotide : A modified nucleotide that lacks the 3' hydroxyl group, preventing further chain elongation when incorporated into a growing polynucleotide.

didN : See di-deoxynucleotide (ddNTP).

Dieback (v die back) : Progressive death of shoots, leaves, or roots, beginning at the tips.

Differential centrifugation : A method of separating sub-cellular particles according to their sedimentation coefficients, which are roughly proportional to their size. Cell extracts are subjected to a succession of centrifuge runs at progressively faster rotation speeds. Large particles, such as nuclei or mitochondria, will be precipitated at relatively slow speeds; higher G forces will be required to sediment small particles, such as ribosomes.

Differential host (syn differential cultivar) : A plant host that on the basis of disease symptoms serves to distinguish between various strains or races of a given plant pathogen.

Differential medium : Culture medium that is used to distinguish between organisms that can grow on it.

Differentially permeable : Referring to a membrane, through which different substances diffuse at different rates; some substances may be unable to diffuse through such a membrane.

Differentiation (L differre, to carry different ways) : A process in which unspecialized cells develop structures and functions characteristic of a particular type of cell. Development from one cell to many cells, accompanied by a modification of the new cells for the performance of particular functions. In tissue culture, the term is used to describe the formation of different cell types.

Diffusion (L diffusus, spread out) : The movement of molecules from a region of higher concentration to a region of lower concentration.

Digest : To cut DNA molecules with one or more restriction endonucleases. See cleave.

Digitate : Having lobes radiating from a common center.

Dihaploid : An individual which arises from a doubled haploid.

Dihybrid : An individual that is heterozygous for two pairs of alleles; the progeny of a cross between homozygous parents differing at two loci.

Dikaryon (adj dikaryotic) : Having two sexually compatible haploid nuclei per cell, that divide simultaneously; this phase is called the dikaryophase.

Dilution plating : A method to obtain pure colonies of bacteria and fungi in which infected plant material or infested soil is diluted in sterilized water. Small samples of the water are spread on the medium surface of several petri plates to find which dilution will produce pure colonies.

Dilution streaking : Repeated streaking of bacteria on the surface of a nutrient medium with a sterile metal loop to allow pure colonies to grow.

Dimer : Association of two molecules.

Dimethyl sulphoxide; dimethyl sulfoxide (C_2H_6OS; mw 7813) : A highly hygroscopic liquid and powerful solvent with little odour or colour. It is an organic co-solvent used in small quantities to dissolve neutral organic substances in tissue culture media preparation. DMSO also has uses as a cryoprotectant.

Dimorphic : Having two distinct shapes or forms.

Dimorphism : The existence of two distinctly different types of individuals within a species. An obvious example is the sexual dimorphism in certain animals.

Dioecious : Plants in which staminate and pistillate flowers occur on different individuals.

Diplochromosome : See endoreduplication.

Diploid : A full set of genetic material, consisting of paired chromosomes one chromosome from each parental set. Most animal cells except the gametes have a diploid set of chromosomes. The diploid human genome has 46 chromosomes.

Diploid cell : A cell which contains two sets of chromosomes. cf haploid cell.

Diplonema (adj diplotene) : Stage in prophase of meiosis I following the pachytene stage, but preceding diakinesis, in which one pair of sister chromatids begin to separate from the other pair, i.e., the centromeres begin to disjoin.

Diplophase : Phase with 2n chromosomes.

Diplotene : The stage of meiosis which follows pachytene and during which the four chromatids of each bivalent move apart in two pairs but remain attached in the region of the chiasmata.

Direct embryogenesis : Embryoids form directly in culture, without an intervening callus phase, on the surface of zygotic or somatic embryos or on explant tissues (leaf section, root tip, etc.).

Direct organogenesis : Formation of organs directly on the surface of cultured intact explants. The process does not involve callus formation.

Direct penetration : Penetration of plant tissues by a pathogen through barriers such as leaf cuticle by chemical and physical means (e.g. penetration peg).

Direct repeat : Two or more stretches of DNA within a single molecule which have the same nucleotide sequence in the same orientation. Direct repeats may be either adjacent to one another or far apart on the same molecule. For example TATTA...TATTA

ATAAT...ATAAT

Directed evolution : This term is used in two ways. (i) A laboratory process using mechanisms of natural selection to produce mutants adaptated to defined environmental challenges. This approach can be used to produce adaptions in single molecules, such as a RNA molecule or an enzyme. Since chemical reaction can occur very fast, multiple "generations" of molecules can be produced quickly. Microorganisms such as bacteria can also reproduce quickly, allowing multiple generations of selection in a short period of time. This approach has been used to produce novel enzymes with industrial and biomedical applications, and to produce bacteria with potential bioremediation applications. (i) The concept that under certain stressful conditions, there is a selection for those mutational changes that favor survival. This controversial concept, also called adaptive evolution, is essentially a molecular version of Lamarkian evolution.

Directed mutagenesis : The process of generation of nucleotide changes in cloned genes by any one of several procedures, including site-specific and random mutagenesis. Also called *in vitro* mutagenesis.

Directed sequencing : Successively sequencing DNA from adjacent stretches of chromosome.

Directional cloning : The technique by which DNA insert and vector molecules are digested with two different restriction enzymes to create non-complementary sticky ends at either end of each restriction fragment, so allowing the insert to be ligated to the vector in a specific orientation and preventing the vector from re-circularizing. See cloning.

Disaccharide : A carbohydrate consisting of two linked sugar units.

Disarm : To delete from a plasmid or virus genes that are cytotoxic or tumour inducing.

Discomycetes : A group of the ascomycetes in which the hymenium is exposed at maturity; one in which the fruiting body is an apothecium or discocarp.

Discontinuous variation : Phenotypic variation involving distinct classes, such as red versus white, tall versus dwarf. See continuous variation.

Discordant : Members of a pair showing different, rather than similar, characteristics.

Disease : A departure from normal metabolism and a reduction of its normal potential for.

Disease cycle : Succession of all of events and interactions among the host, parasite and environment that occur in a disease, from initial infection of the plant by a causal agent, through pathogenesis, to over-seasoning, until another infection occurs.

Disease incidence : Number of plants affected by a disease within a population.

Disease progress curve : Graph of some measure of disease (i.e. severity) over time.

Disease pyramid disease severity : A memory aid similar to the disease triangle but including, in addition, the factor of time in the development of a disease.

Disease resistance : The ability to remain healthy by resisting disease or the disease agent. Disease resistance or tolerance is a subject of intense interest in biotechnology.

Disease triangle : A memory aid that diagrams the three important components necessary for disease: susceptible plant, virulent pathogen and favorable environment.

Disease-associated genes : Alleles carrying particular DNA sequences associated with the presence of disease.

Disease-free : A plant or animal certified through specific tests as being free of specified pathogens. Disease-free should be interpreted to mean "free from any known diseases" as "new" diseases may yet be discovered to be present. cf disease indexing.

Disease-indexing : Disease-indexed organisms have been assayed for the presence of known diseases according to standard testing procedures. cf disease-free.

Disinfect : To eliminate a pathogen from infected plant tissues.

Disinfectation : Full elimination of internal micro-organisms from a culture; disinfectation is rarely obtained. cf disinfestation.

Disinfest : To kill pathogens that have not yet initiated disease, or other contaminating microoganisms, that occur in or on inanimate objects as such soil or tools, or that occur on the surface of plant parts such as seed.

Disinfestation : The elimination or inhibition of the activity of surface-adhering micro-organisms. cf disinfection.

Disjunction : Separation of homologous chromosomes during anaphase I of meiosis; separation of sister chromatids during anaphase of mitosis and anaphase II of meiosis. As soon as the sister chromatids have separated, they are each called a chromosome. See non-disjunction.

Disomy (adj disomic) : The presence of a pair of specific chromosome. This is the normal condition, and abnormal occurrences are monosomy (q.v.), trisomy (q.v.) and nullisomy (with respectively one chromosome of a pair, three or none). There are also abnormal disomic conditions, such as when both chromosomes of the pair were inherited from the same parent.

Dispense : Portion out a nutrient medium into containers, such as test tubes, jars, Erlenmeyer flasks, Petri dishes, etc.

Dispersal (syn dissemination) : Spread of infectious material (inoculum) from diseased to healthy plants.

Dissecting microscope : A microscope with a low magnifying power of about 50×, used to examine or excise small plant or animal parts.

Dissection (L dissectio, a dissecting or being dissected) : Separation of a tissue by cutting for analysis or observation.

Dissemination (syn dispersal) : Spread of infectious material (inoculum) from diseased plants to healthy plants.

Dissolve : Pass chemicals into solution.

Distal : Far from the point of attachment or origin; in a direction away from main body (see proximal).

Distillation (L distillatio, a distilling process) : The process of heating a mixture to separate the more volatile from the less volatile parts, and then cooling and condensing the resulting vapour so as to produce a more nearly pure or refined substance.

Di-sulphide bond : A chemical bond that stabilizes the three-dimensional structure of proteins, and hence the protein's normal function. They form between cysteine residues in the same or different peptide molecules. a.k.a. di-sulphide bridge.

Di-sulphide bridge : See di-sulphide bond.

Ditype : In fungi, a tetrad that contains two kinds of meiotic products (spores), e.g., 2AB and 2ab. See tetrad (1).

Diurnal : Term describing the occurrence of an event at least once every 24 hours. cf circadian.

Divergent evolution : See adaptive radiation.

Dizygotic twins : Two-egg twins, i.e., a pair of individuals that shared the same uterus at the same time, but which arose from separate and independent fertilization of two ova.

D-loop (displacement loop) : The structure formed when ssDNA pairs with a dsDNA molecule, disrupting the pairing of the original duplex DNA to form a hybrid between the ssDNA and one of the strands of the duplex DNA, and displacing the other strand from the duplex.

DMD : Duchenne muscular dystrophy.

DMSO : See dimethyl sulphoxide.

DNA (deoxyribonucleic acid) : The molecule that encodes genetic information. DNA is a doublestranded molecule held together by weak bonds between base pairs of nucleotides. The four nucleotides in DNA contain the bases: adenine (A), guanine (G), cytosine (C), and thymine (T). In nature, base pairs form only between A and T and between G and C; thus the base sequence of each single strand can be deduced from that of its partner.

DNA amplification : Multiplication of a piece of DNA in a test-tube into many thousands of millions of copies. The most commonly used process is the polymerase chain reaction (PCR) system, but other systems are being developed, including ligase chain reaction (LCR), nucleic acids sequence-dependent amplification, and the Q-□ system.

DNA bank : In AnGR: Storage of DNA, which may or may not be the complete genome, but should always be accompanied by inventory information. (Note: at the present time, animals cannot be re-established from DNA alone.) (Source: FAO, 1999).

DNA chip : See DNA micro-array.

DNA cloning : See gene cloning.

DNA construct : A DNA molecule inserted into a cloning vector, usually a plasmid.

DNA delivery system : A generic term for any procedure that transports DNA into a recipient cell.

DNA diagnosis : The use of DNA polymorphisms to detect the presence of a specific allele (often associated with a disease or syndrome) or DNA sequence.

DNA fingerprint : The unique pattern of DNA fragments identified originally by Southern hybridization (using a probe that binds to a polymorphic region of DNA) or now by polymerase chain reaction (PCR) (using primers flanking the polymorphic region). See genetic fingerprinting.

DNA fingerprint technique : A method employed to determine differences in amino acid sequences between related proteins; relies upon the presence of a simple tandem-repetitive sequences that are scattered throughout the human genome.

DNA gyrase : A topoisomerase that removes supercoils from DNA by first producing double-strand breaks and then sealing them.

DNA helicase (gyrase) : An enzyme that catalyses the unwinding of the complementary strands of a DNA double helix.

DNA hybridization : The pairing of two DNA molecules, often from different sources, by hydrogen bonding between complementary nucleotides. This technique is frequently used to detect the presence of a specific nucleotide sequence in a DNA sample.

DNA ligase : Enzymatic activity responsible for creating phosphodiester bonds between the 5' end of one strand of DNA and the 3' end of another strand. Requires the presence of a 5' phosphate on one strand, and a 3' Hydroxyl group on the second strand.

DNA marker : A gene or other fragment of DNA whose location in the genome is known.

DNA micro-array : A small glass surface to which has been fixed an array of DNA fragments, each with a defined location. A typical DNA chip would contain 10 000 discrete spots (each containing a different DNA fragment) in an area of just a few square centimetres. When a solution of fluorescently labelled DNA fragments is hybridized to the chip, spots to which hybridization occurs are visible as fluorescence. If the spots on the chip are genes (expressed sequence tags, q.v.), hybridization with cDNA from a particular tissue shows which genes are expressed in that tissue. If the spots are short, synthesized oligonucleotides (approximately 25 bases) corresponding to that part of a gene containing a single nucleotide polymorphisms (SNP)(q.v.), with a separate spot for each of the 4 possible bases at that site, hybridization with genomic DNA from an individual plant or animal enables that individual to be genotyped at as many SNP loci as are represented on the chip. The big advantage of DNA chips is the extent to which the process of genotyping can be automated, thereby enabling huge numbers of plants or animals to be genotyped for a huge number of loci.

DNA polymerase : Enzymatic activity responsible for catalyzing the polymerization of DNA. Is dependent upon an annealed primer from which to initiate polymerization, and a DNA template from which to copy.

DNA polymorphism : The existence of two or more alternative forms (alleles) of a chromosomal locus that differ in nucleotide sequence or have variable numbers of repeated nucleotide units. See allele.

DNA primase : The enzyme that normally synthesizes the RNA primers required for initiating DNA synthesis.

DNA probe : A labelled (tagged) segment of DNA that is able, after a DNA hybridization reaction, to detect a specific DNA sequence in a mixture of sequences. If the tagged sequence is complementary to any one in the mixture, the two sequences will form a double helix. This will be identified thanks to its label (either by radioactivity or fluorescence).

DNA Repair : A variety of different mechanisms that remove or modify damaged DNA.

DNA repair enzymes : Enzymes that catalyse the repair of DNA.

DNA repair genes : Genes encoding proteins that correct errors in DNA sequencing.

DNA replication : The use of existing DNA as a template for the synthesis of new DNA strands. In humans and other eukaryotes, replication occurs in the cell nucleus.

DNA sequence : The relative order of base pairs, whether in a fragment of DNA, a gene, a chromosome, or an entire genome. See base sequence analysis.

DNA sequencing : Procedures for determining the nucleotide sequence of a DNA fragment. There are two common methods for doing this: · the Maxam and Gilbert technique (chemical degradation), that uses different chemicals to break the DNA into fragments at specific bases; or · the Sanger technique (called the di-deoxy or chain-terminating method) uses DNA polymerase to make new DNA chains, with di-deoxy nucleotides (chain terminators) to stop the chain randomly as it grows. In both cases, the DNA fragments are separated according to length by polyacrylamide gel electrophoresis, enabling the sequence to be read directly from the gel.

DNA topo-isomerase : An enzyme that catalyses the introduction or removal of supercoils in DNA.

DNA transformation : See transfection; transformation.

DNA typing : The analysis of sections of DNA for purposes of identification.

DNase (deoxyribonuclease) : An enzyme that catalyses the cleavage of DNA. DNase I is a digestive enzyme secreted by the pancreas, that degrades DNA into shorter nucleotide fragments. Many other endonucleases and exonucleases are involved in DNA repair and replication. cf nuclease.

dNTP : Deoxynucleotide triphosphates used in the synthesis of DNA -- deoxy-ATP, deoxy-TTP, deoxy-GTP, or deoxy-CTP.

Dodder : parasitic seed plant (Cuscuta spp.) without leaves; a yellow filamentous vine.

DOE : Department of Energy.

Dolipore septum : Cross wall found in basidiomycetes and characterized by special swellings and membranes in association with the septal pore.

Dolly : The [name of the] first mammal to be created by cloning a cell from an adult animal. In this particular case, the cell came from the mammary tissue of an adult ewe. The creation of Dolly showed that the process of differentiation into adult tissue is not, as previously thought, irreversible. The result was achieved by nuclear transfer (q.v.). Dolly's birth was announced in 1997. Since then, cattle and mice have also been cloned from adult cells.

Domain : A discrete, independantly folded region of a protein. Different functions of a multifunctional protein are usually localized in separate domains.

Domestic animal diversity (DAD) : In AnGR: The spectrum of genetic differences within each breed, and across all breeds within each domestic animal species, together with the species differences; all of which are available for the sustainable intensification of food and agriculture production. (Source: FAO, 1999).

Dominance : In classical Mendelian genetics, the masking of the action of one allele by another. If an individual with red flowers is crossed with an individual with white flowers and all progeny have red flowers, then the allele for red pigment is fully dominant over the allele for white flowers.

Dominance, partial : The incomplete masking of the action of one allele by another. If an individual with red flowers is crossed with an individual with white flowers and all progeny have pink flowers, then the allele for red pigment is partially dominant over the allele for white flowers.

Dominant : The phenotype that is observed when two alleles of a gene are present in a cell. The gene whose phenotype is expressed is the dominant allele and the allele whose phenotype is masked is the recessive allele. If the phenotype of the diploid is inte rmediate between that of the two haploid genes, then the alleles are co-dominant.

Dominant allele : One that determines the phenotype even when there is only one copy (i.e. in a heterozygous individual).

Dominant marker selection : Selection of cells via a gene encoding a product that enables only the cells that carry the gene to grow under particular conditions. For example, plant and animal cells that express the introduced Neor gene are resistant to the compound G418, while cells that do not carry the Neor gene are killed by G418. a.k.a. positive selection; see also positive selectable marker.

Dominant selectable marker gene : A gene that allows the host cell to survive under conditions where it would otherwise die.

Dominant(-acting) oncogene : A gene that stimulates cell proliferation and contributes to oncogenesis when present in a single copy. See oncogene.

Donor : The source of DNA in a genetic cross. For example, the transducing phage brought into a bacterial cell.

Donor Parent : The parent from which one or a few genes are transferred to the recurrent parent in backcross breeding.

Dormancy (F dormir, from L dormire, to sleep) : An inactive period in the life of an animal or plant during which growth slows or completely ceases. Physiological changes associated with dormancy help the organism survive adverse environmental conditions. Annual plants survive the winter as dormant seeds, while many perennial plants survive as dormant tubers, rhizomes, or bulbs. Hibernation and aestivation in animals help them survive extremes of cold and heat, respectively. cf rest period.

Dosage compensation : A phenomenon whereby inactivation of all but one of the X chromosomes in female mammals results in males and females producing the same quantity of peptide from X-linked genes.

Dot blot : A technique for measuring the amount of one specific DNA or RNA in a complex mixture. The samples are spotted onto a hybridization membrane (such as nitrocellulose or activated nylon, etc.), fixed and hybridized with a radioactive probe. The extent of labeling (as determined by autoradiography and densitometry) is proportional to the concentration of the target molecule in the sample. Standards provide a means of calibrating the results.

Double cross : A cross between two F1 hybrids.

Double crossing-over : Two simultaneous reciprocal breakage and reunion events between the same two chromatids.

Double digestion : Cleavage of a DNA molecule with two different restriction endonucleases, either concurrently or consecutively.

Double fertilization : A process, unique to flowering plants, in which two male nuclei, which have travelled down the pollen tube, separately fuse with different female nuclei in the embryo sac. The first male nucleus fuses with the egg cell to form the zygote; the second male nucleus fuses with the two polar nuclei to form a triploid nucleus that develops into the endosperm.

Double helix : The helical shape assumed by DNA in which the two complementary strands hydrogen bond together in opposite orientations (i.e. have opposite polarities).

Double recessive : An organism homozygous for a recessive allele at each of two loci.

Double-stranded complementary DNA (dscDNA) : A double-strand DNA molecule created from a cDNA template.

doubling time : See generation time.

Down syndrome : A type of mental deficiency due to trisomy (three copies) of autosome 21, a translocation of 21 or mosaicism.

Downstream : A sequence located after a particular site relative to the direction of transcription and translation (i.e. located to the 3' side of a particular site). For example, the lacZ structural gene is located downstream of the lac promoter.

Downy mildew : A plant disease in which the fungus appears as a downy growth on the host surface; caused by a member of the oomycetes.

Draft sequence : The sequence generated by the HGP as of June 2000 that, while incomplete, offers a virtual road map to an estimated 95% of all human genes. Draft sequence data are mostly in the form of 10,000 base pair-sized fragments whose approximate chromosomal locations are known. See also: sequencing, finished DNA sequence, working draft DNA sequence.

Drift : Changes in allele frequencies over time due to chance (as opposed to selection or mutation).

Drought : Insufficient soil water for normal plant growth.

Drug : See therapeutic agent.

Drug delivery : Method by which a therapeutic agent is delivered to its site of action. For traditional therapeutic agents this is another name for formulation. However, biotechnology has allowed the development of a range of new therapeutic-agent delivery systems, such as liposomes and other encapsulation techniques, and a range of mechanisms that target a therapeutic agent to a particular cell or tissue. See therapeutic agent.

Dry weight : The moisture-free weight of tissue obtained by drying at high (oven-drying) or low (freeze-drying) temperatures for an interval sufficient to remove all water.

dscDNA : See double-stranded complementary DNA.

dsDNA : Double-stranded DNA.

DT>Structural protein : A protein which fulfill a purely structural role (i.e. not enzymatic). This includes phage capsid proteins, some ribosomal proteins, "histone-like" proteins, etc.

Dual culture : A culture made of a plant tissue and one organism (such as a nematode) or an obligate parasite/micro-organism (such as a fungus). Dual culture techniques are used for a variety of purposes, including assessing host-parasite interactions and the production of axenic cultures.

Duchenne/Becker muscular dystrophy : The most common and severe form of muscular dystrophy; transmitted as an X-linked trait. X-linked recessive. Symptoms include onset at 2-5 years with difficulty with gait and stairs, enlarged calf muscles, progression to wheelchair by adolescence, shortened life span.

Duplex : A nucleic acid molecule in which two strands are base paired with each other.

Duplex DNA : A double-stranded DNA molecule.

Duplication : A region of DNA that is present in two copies. The DNA is present as an adjacent, direct repeat in tandem duplications. It is also possible for the duplicated DNA to be present at distant sites on a chromosome.

Durable resistance : Resistance that remains effective during prolonged and widespread use in an environment favorable to disease.

dUTPas : An enzyme which degrades dUTP to prevent the incorpation of dU into DNA.

Dwarfing : Underdevelopment of a plant or plant organs, which may be caused by disease, inadequate nutrition, or unfavorable environmental conditions.

Dwarfism : Conditions of short stature with adult height under 4'10" as adult, usually with normal intelligence and lifespan. Ehlers Danlos Syndrome connective tissue condition including problems with tendons, ligaments, skin, bones, cartilage, and membranes surrounding blood vessels and nerves. Symptoms include joint laxity, elastic skin, dislocations. Many forms: autosomal dominant, autosomal recessive, X-linked forms.

DYS : D = DNA, Y = Y chromosome, S = a unique DNA segment. This nomenclature is controlled by the HUGO Gene Nomenclature Committee, with the assignment of new DYS numbers. This guideline determines each part of the symbol for naming arbitrary DNA fragments and loci.

Dysgenic : An action or process that is detrimental to the genetic qualities of a population. Usually applied to human actions, such as high grading or land clearing, which would reduce the local gene pool.

Dystonia : Neurologic condition involving repeated twisting and movement. Involves a variety of muscle groups. Intelligence not effected. Three forms: childhood - autosomal dominant, autosomal recessive, adult-acquired.

E coli : Common bacterium that has been studied intensively by geneticists because of its small genome size, normal lack of pathogenicity, and ease of growth in the laboratory.

E site : See exit site.

EBV : See estimated breeding value.

EC : See Enzyme Commission.

Ecdysone : A steroid hormone in insects that stimulates moulting and metamorphosis. It acts on specific genes, stimulating the synthesis of proteins involved in these bodily changes.

Echinulate : having small spines projecting from cell walls.

Eclipse period : The period of time after infection with a virus during which new virions are formed, but not released (contrast with latent period).

Eclosion : 1. Emergence of an adult insect from the pupal stage. 2. Beginning of germination of fungal spores.

Ecological diversity : See biodiversity.

Ecology : The study of the interactions between organisms and their natural environment, both living and non-living.

Economic threshold : The pathogen density at or above which the value of crop losses (in the absence of management efforts) would exceed the cost of management practices.

Economic trait locus (ETL) : A locus influencing a trait that contributes to income. The plural form (economic trait loci) is also abbreviated as ETL.

Ecosystem : The complex of a living community and its environment, functioning as an ecological unit in nature. cf abiotic factors; biotic factors.

Ecotype : A sub-population of a species that occurs in a particular well-defined environment, usually showing better adaptation to that environment than the species as a whole. (See cline).

Ectomycorrhiza (pl ectomycorrhizae) : symbiotic association between a nonpathogenic or weakly pathogenic fungus and the roots of plants with fungal hyphae between and external to root cells (see arbuscular mycorrhiza).

Ectoparasite : Parasite that feeds from the exterior of its host (see endoparasite).

Ectotrophic : Fungal development primarily over the root surface.

Edema (also oedema; syn intumescence) : Blisters produced on leaves and other plant parts under conditions of high moisture and restricted transpiration.

EDTA : See ethylene-diamine tetra-acetic acid.

Effective population size (Ne) : The size of a hypothetical stable, randomly-mating population that would have the same rate of gene loss or increase in inbreeding as the real population (size N). As all finite populations are inbred to some degree and generally do not choose mates at random, Ne is typically 1/10 N or less, particularly if fewer males breed than females.

Effector cells : Cells of the immune system that are responsible for cell-mediated cytotoxicity.

Effector molecule : A molecule that influences the behaviour of a regulatory molecule, such as a repressor protein, thereby influencing gene expression.

Egg : 1. The fertilized ovum (zygote) in egg-laying animals after it emerges from the body. 2. The mature female reproductive cell in animals and plants. See also ovum.

Egg mass : Group of eggs held together by a gelatinous matrix.

EGS : External guide sequence. See guide sequence.

Elastin : A fibrous protein that is the major constituent of the yellow elastic fibres of connective tissue.

Electro-blotting : The electrophoretic transfer of macromolecules (DNA, RNA or protein) from a gel, in which they have been separated, to a support matrix, such as a nitrocellulose sheet. A transfer usually used in techniques such as Southern and northern blotting.

Electrochemical sensor : Type of biosensor in which a biological process is harnessed to an electrical sensor system, such as an enzyme electrode. Other types couple a biological event to an electrical one via a range of mechanisms, such as those based on oxygen and pH. See enzyme electrode.

Electromagnetic radiation : Electromagnetic waves, including ultraviolet (UV), X-rays, and gamma radiation (□ rays). Electromagnetic radiation is used to produce mutant cells or organisms, or, in the case of UV, disinfestation and sterilization, in tissue culture.

Electromagnetic spectrum : The range of wavelengths or frequencies over which electromagnetic radiation extends.

Electron microscope : Microscope that uses a focused beam of electrons to produce a greatly enlarged image of minute objects, such as a virus, in the same way that light is used in a compound microscope.

Electrophoresis : A method of separating large molecules (such as DNA fragments or proteins) from a mixture of similar molecules. An electric current is passed through a medium containing the mixture, and each kind of molecule travels through the medium at a different rate, depending on its electrical charge and size. Separation is based on these differences. Agarose and acrylamide gels are the media commonly used for electrophoresis of proteins and nucleic acids.

Electrophoresis of DNA : DNA fragments have a constant charge/length ratio due to the net negative charge of the phosphate backbone. Therefore, DNA migrates toward the (+) electrode. The rate of migration of linear DNA mainly depends on the pore size of the gel and length of the DNA. During electrophoresis DNA molecules seem to "snake" through the pores in the gel "head first". As the pore size decreases (i.e. the agarose or acrylamide concentration increases) it is harder for longer DNA molecules to orient properly to snake through the pores: smaller DNA fragments snake through the pores easier and hence migrate faster. Thus, the rate of migration of linear double stranded DNA is inversely proportional to the log of its molecular weight. By using gels with different concentrations of agarose or acrylamide, a wide range of DNA fragment sizes can be separated. The shape of the DNA molecule also affects its ability to snake through the pores in the gel. In general features that make the DNA less flexible or less compact slow the migration in a gel. The rate of migration of different forms of plasmid DNA is usually: supercoiled > linear > nicked circles. In addition, secondary structure and bends in linear DNA may affect the rate of migration.

Electroporation : A method for introducing foreign nucleic acid into bacterial or eukaryotic cells that uses a brief, high voltage DC charge which renders the cells permeable to the nucleic acid. Also useful for introducing synthetic peptides into eucaryotic cells.

Elicitor : A molecule produced by the host (or pathogen) that induces a response by the pathogen (or host).

ELISA (acronym for Enzyme-Linked ImmunoSorbent Assay) : A serological test in which the sensitivity of the reaction is increased by attaching an enzyme that produces a colored product to one of the reactants.

Elite tree : A tree that has been shown by progeny testing to produce superior offspring (See plus tree and candidate tree).

Elongation factor : A protein(s) that associates with the ribosome cyclically to assist in loading tRNA into the the A site of the ribosome.

ELSI : Ethical, legal and social implications (of HGP).

Emasculation : Removal of the anthers from a flower.

Embryo cloning : The creation of identical copies of an embryo by embryo splitting (q.v.) or by nuclear transfer (q.v.) from undifferentiated embryonic cells.

Embryo culture : The culture of embryos on nutrient media.

Embryo multiplication and transfer (EMT) : The cloning of animal embryos and their subsequent transfer to recipients (via artificial inembryonation (q.v.)). The cloned embryos can be clones of an embryo or of an adult.

Embryo sac : A large thin-walled space within the ovule of the seed plant in which the egg and, after fertilization, the embryo develop; the mature female gametophyte in angiosperms. Generally a seven-celled structure. The seven cells are two synergids, one egg cell, three antipodal cells (each with a single haploid nucleus) and one endosperm mother cell with two haploid nuclei.

Embryo sexing : The determination of the sex of an embryo, typically by means of PCR (q.v.) involving amplification from a small sample of embryonic tissue, using primers specific for a locus on the Y chromosome.

Embryo splitting : The splitting of young embryos into several sections, each of which develops into an animal. A form of animal cloning, i.e., of producing animals that are genetically identical. In practice, the number of identicals (identical organisms) that can be produced from a single embryo is less than 10.

Embryo technology : Generic name for any modification of mammalian embryos. It encompasses embryo cloning, embryo splitting, *in vitro* fertilization, and embryo storage.

Embryo transfer (ET) : See multiple ovulation and embryo transfer.

Embryo (Gr en, in + bryein, to swell) : An immature organism in the early stages of development. In mammals, this occurs in the first months in the uterus. In plants, it is the structure that develops in the megagametophyte, as result of the fertilization of an egg cell, or without fertilization. In aseptic cultures, adventitious embryos show polarization, followed by the growth of a shoot from one end and a root from the other end.

Embryogenesis : 1. (General) Development of an embryo. 2. (In plants) *In vitro* formation of plants from plant tissues, through a pathway closely resembling normal embryogeny from the zygote; if this development in culture involves somatic cells and not the zygote, it can be indicated by using the term adventitious embryogenesis or somatic embryogenesis. The generation of embryos has two stages: initiation and maturation. Initiation needs a high level of the group of plant hormones called auxins; maturation needs a lower level. Other chemicals have to be at suitable levels. The procedure involves the explanting of a piece of plant tissue and putting it on a high-auxin medium, where the cells grow into a mass of callus. This is then transferred to a maturation medium, where the callus starts to initialize organs, ultimately growing a root and a shoot.

Embryoid : An embryo-like body developing *in vitro*. It forms a complete, self-contained platelet with no vascular connection with the callus. The term embryoid is no longer commonly used. See embryo.

Embryonic stem (ES) cells : An embryonic cell that can replicate indefinitely, transform into other types of cells, and serve as a continuous source of new cells.

Embryonic stem cells : Cells of the early embryo that can give rise to all differentiated cells, including germ line cells.

Emergence : growth of the seedling shoot through the surface of the soil.

Empirical : Relating to or based upon practical experience, trial and error, direct observation or observation alone, without benefit of scientific method, knowledge or theory.

EMT : See embryo multiplication and transfer.

Enation : Abnormal outgrowth from the surface of a stem or leaf.

Encapsidate : To cover virus nucleic acid with a protein coat.

Encapsidation : The process by which a virus' nucleic acid is enclosed in a capsid. See capsid; coat protein.

Encapsulating agents : Anything which forms a shell around an enzyme or bacterium, although the agents used are usually polysaccharides such as alginate or agar. The agents are inert and allow nutrients and oxygen to diffuse into and out of the sphere readily, and are easy to convert from gel (solid) to sol (liquid) or solution form by altering the temperature or the concentration of ions.

Encapsulation : Any method of getting something, usually an enzyme or bacterium, into a small package or capsule while it is still working or alive. It is a method for immobilizing cells for use in a bioreactor.

Encode : To specify, after decoding by transcription and translation, the sequence of amino acids in a protein.

Encyst : To form a cyst or protective covering.

End labeling : The technique of adding a radioactively labeled group to one end (5' or 3' end) of a DNA strand.

Endangered breed : In AnGR: A breed where the total number of breeding females is between 100 and 1 000 or the total number of breeding males is less than or equal to 20 and greater than five; or the overall population size is close to, but slightly above 100 and increasing, and the percentage of pure-bred females is above 80%; or the overall population size is close to, but slightly above, 1 000 and decreasing, and the percentage of pure-bred females is below 80%. (Source: FAO, 1999).

Endangered species : A plant or animal species in immediate danger of extinction because its population numbers have reached a critical level or its habitats have been drastically reduced.

Endemic : 1. Describing a plant or animal species whose distribution is restricted to one or a few localities. 2. Describing a disease or a pest that is always present in an area.

End-labelling : The introduction of a radioactive atom at the end of a DNA or RNA molecule. A commonly used method is to use T4 polynucleotide kinase to introduce a 32P atom onto the end of a DNA molecule.

Endocarp : Inner layer of a fruit wall.

Endoconidium (pl endoconidia) : A conidium produced inside a hypha or conidiophore.

Endocrine gland : Any gland in an animal that manufactures hormones and secretes them directly into the bloodstream to act at distant sites in the body, known as target organs or cells. cf exocrine gland. See gland.

Endocrine interference : Interference with the normal balance hormones.

Endocytosis : The process by which materials enter a cell without passing through the cell membrane. The membrane folds around material outside the cell, resulting in the formation of a saclike vesicle into which the material is incorporated. This vesicle is then pinched off from the cell surface so that it lies within the cell. See phagocytosis; pinocytosis.

Endoderm : The internal layer of cells of the gastrula, which will develop into the alimentary canal (gut) and digestive glands of the adult.

Endodermis : (Gr. endon, within + derma, skin) The layer of living cells, with various characteristically thickened walls and no intercellular spaces, which surrounds the vascular tissue of certain plants and occurs in nearly all roots and certain stems and leaves. The endodermis separates the cortical cells from cells of the pericycle.

Endogamy : The fusion of reproductive cells from closely related parents, i.e., inbreeding. cf exogamy.

Endogenote : The part of the bacterial chromosome that is homologous to a genome fragment (exogenote) transferred from the donor to the recipient cell in the formation of a merozygote.

Endogenous (Gr endon, within, + genos, race, kind) : Developed or added from within the cell or organism.

Endomitosis : Duplication of chromosomes without division of the nucleus, resulting in increased chromosome number within a cell. Chromosome strands separate, but the cell does not divide.

Endomycorrhiza (pl endomycorrhizae; syn arbuscular mycorrhiza) : Symbiotic association between a nonpathogenic or weakly pathogenic fungus and the roots of plants in which fungal hyphae invade cortical cells of the root (see ectomycorrhiza).

Endonuclease : An enzyme that makes breaks in a molecule of DNA by hydrolyzing internal phosphodiester bonds. An endonuclease may be specific for either single or double stranded DNA or RNA.

Endoparasite : Parasitic organism that lives and feeds from inside its host (see ectoparasite).

Endophyte : Plant developing inside another organism; also used for endoparasitic fungi found in grass species.

Endoplasmic reticulum (Gr endon, within + plasma, anything formed or moulded; L reticulum, a small net) : A cytoplasmic net of membranes, adjacent to the nucleus, made visible by the electron microscope. Any system of paired membranes within the cytoplasm. Frequently abbreviated to ER. They are sites of protein synthesis.

Endopolyploidy : The result of nuclear divisions without subsequent cytoplasmic division (cytokinesis); the polyploids so obtained are called endopolyploids. See polyploidy.

Endoprotease : An enzyme that cleaves the peptide bonds between amino acids within a protein. Cleavage is usually at one or more specific sites.

Endoreduplication : Chromosome reproduction during interphase. 4-chromatid chromosomes (diplochromosomes) are seen during this phase.

Endosperm mother ce ll : One of the seven cells of the mature embryo sac, containing the two polar nuclei and, after reception of a sperm cell, gives rise to the primary endosperm cell from which the endosperm develops.

Endosperm (Gr endon, within + sperma, seed) : Nutritive tissue that develops in the embryo sac of most angiosperms. It usually forms after the fertilization of the two fused primary endosperm nuclei of the embryo sac with one of the two male gamete nuclei. In most diploid plants, e.g., cereals, the endosperm is triploid (3n), but in some (e.g., lily) it is often consumed as the seed matures.

Endospore : A metabolically inactive, nonreplicating form of certain bacteria, including the genera Bacillus and Clostridium. Endospores tend to be highly resistant to physical and chemical damage.

Endotoxin : A component of the cell wall of Gram-negative bacteria that elicits, in mammals, an inflammatory response and fever.

End-product inhibition : The inhibition of an enzyme by a metabolite. The enzyme is sometimes the first enzyme in a biosynthetic pathway, and the metabolite is generally the product of the last step in the pathway. See feedback inhibition.

Enhancer : An enhancer is a cis-acting nucleotide sequence to which transcription factor(s) bind, and which increases the transcription of a gene. It is not part of a promoter; the basic difference being that an enhancer can be moved around anywhere in the general vicinity of the gene (within several thousand nucleotides on either side or even within an intron), and it will still function. It can even be clipped out and spliced back in backwards, and will still operate. A promoter, on the other hand, is position- and orientation-dependent. Enhancers are ususlly around 70-80 bp in length and are found, for e.g in viral DNA molecules. Some enhancers are "conditional" - in other words, they enhance transcription only under certain conditions, for example in the presence of a hormone.

Enhancer element; enhancer sequence : 1. A sequence found in eukaryotes and certain eukaryotic viruses which can increase transcription of a gene when located (in either orientation) up to several kilobases from the gene concerned. These sequences usually act as enhancers when on the 5′ side (upstream) of the gene in question. However, some enhancers are active when placed on the 3′ side (downstream) of the gene. In some cases enhancer elements can activate transcription of a gene with no (known) promoter. 2. A substance or object that increases a chemical activity or a physiological process. 3. A major or modifier gene that increases a physiological process.

Enrichment : A condition which increases the proportion of a particular organism or mutant strain from a mixed population. For example, a penicillin enrichment kills growing cells and allows mutants unable to grow under a particular condition to survive; however, for each round of penicillin enrichment typically only 1% of the growing cells are killed, thus each round of enrichment increases the mutant ratio about 100-fold. Compare with "selection".

Enterotoxin : A bacterial protein that, following release into the intestine, causes cramps, diarrhoea and nausea.

Enucleated ovum : Egg cell from which the nucleus has been removed.

Environment : The aggregate of all the external conditions and influences affecting the life and development of an organism.

Environmental Impact Quotient (EIQ) : A universal indicator that integrates the various environmental impacts of individual pesticides into a single field value per acre (hectare). EIQ incorporates the impact of the key toxicity and environmental exposure data on farm workers, consumers and ecology, providing a consistent and comprehensive measure of the non-intentional impact of different pesticides on the environment. EIQ is not universally accepted and is an indicator only; it does not take into account all environmental issues and impacts.

Enzyme bioreactor : A reactor in which a chemical conversion reaction is catalysed by an enzyme.

Enzyme commission (EC) number : Systematic name and number which identify an enzyme in technical literature. Assigned by the Enzyme Commission, the EC Number consists of four numbers separated by dots: the first classifies the enzyme into one of the six broad groups: 1. Oxidoreductase; 2. Transferases; 3. Hydrolases; 4. Lyases; 5. Isomerases; 6. Ligases. Each group is subdivided into sub-groups, each sub-group into sub-sub-groups, and the last number is specific for the enzyme, e.g., EC 3.1.21.1 is deoxyribonuclease I.

Enzyme electrode : A type of biosensor, in which an enzyme is immobilized onto the surface of an electrode. When the enzyme catalyses its reaction, electrons are transferred from the reactant to the electrode, and so a current is generated. There are two types of enzyme electrodes: · ampometric, where the electrode is kept as near zero voltage as possible. When the enzyme catalyses its reaction, electrons flow into the electrode, and so a current flows; and · potentiometric, when the electrode is held at a voltage which counteracts the voltage determined by the enzyme's tendency to push electrons into its. Usually enzymes transfer their electrons inefficiently to the electrode, so a mediator compound is coated onto the electrode to help the transfer.

Enzyme stabilization using antibodies : A method of stabilizing enzymes by binding antibodies to them. The antibodies should not block the active site of the enzyme, as otherwise the protein is stabilized but is inactive as a catalyst. Monoclonal antibodies are usually used as they bind to specific bits of the protein surface. If the enzyme tries to unfold into an inactivate structure, it must not only overcome its own binding energy but also throw off all the bound antibodies; this requires more energy, and so is a correspondingly slower process.

Enzyme (Gr en, in + zyme, yeast or leaven) : A protein produced in living cells, which, even in very low concentration, catalyses specific chemical reactions but is not used up in the reaction. Enzymes are classified into six major groups, according to the type of reaction they catalyse: 1. Oxidoreductase; 2. Transferases; 3. Hydrolases; 4. Lyases; 5. Isomerases; 6. Ligases. The names of most individual enzymes are usually derived from the substrate on which they act, with the suffix -ase. Thus lactase is the enzyme that acts to breakdown lactose; it is classified as a hydrolase. See Enzyme Commission Number.

enzyme-linked immunosorbent assay (acronym ELISA) : A serological test in which the sensitivity of the reaction is increased by attaching an enzyme that produces a colored product to one of the reactants.

EPA (acronym for Environmental Protection Agency) : The federal agency responsible for controlling the various aspects of air, water, and soil pollution, including pesticide regulations.

EPD : See expected progeny difference.

Epicotyl (Gr epi, upon + kotyledon, a cup-shaped hollow) : The upper portion of the axis of a plant embryo or seedling, above the cotyledons.

Epidemic : An increase of disease in a population; a general and serious outbreak of disease (see epiphytotic).

Epidemiology (adj epidemiologic) : The study of factors influencing the initiation, development, and spread of infectious disease; the study of disease in populations of plants.

Epidermis (Gr epi, upon + derma, skin) : 1. The outmost layer of cells of the body of an animal. In invertebrates the epidermis is normally only one cell thick and is covered by an impermeable cuticle. In vertebrates the epidermis is the thinner of the two layers of skin. 2. The outermost layer of cells covering a plant. It is overlaid by a cuticle and its functions are principally to protect the plant from injury and to reduce water loss. Some epidermal cells are modified to form guard cells or hairs of various types. In woody plants the functions of the shoot epidermis are taken over by the periderm tissues and in mature roots the epidermis is sloughed off and replaced by the hypodermis.

Epigenesis : Describes the developmental process whereby each successive stage of normal development is built up on the foundations created by the preceding stages of development; an embryo is built up from a zygote, a seedling from an embryo, and so on.

Epigenetic : The inheritance of a particular trait that is not encoded in the nucleotide sequence. For example, methylation of certain DNA sequences can influence gene expression, and the methylation pattern may be maintained following DNA synthesis by preferential methylation of hemimethylated sequences.

Epigenetic variation : Non-hereditary and reversible variation; often the result of a change in gene expression.

Epinasty : A process by which the growth of branches or petioles is abnormally pointing downward. This phenomenon is caused by the more rapid growth of the upper side. Epinasty may result from either nutritional deficiencies or irregularities at the plant growth regulator level. Not to be confused with wilting, as epinastic tissues are turgid.

Epiphyllous : On the upper surface of a leaf.

Epiphyte : A plant that grows upon another plant, but is neither parasitic on it nor rooted in the ground.

Epiphytic : Living on the surface of plants, but not as a parasite.

Epiphytotic : An unarrested spread of a plant disease.

Episome : A genetic element that can exist either as an autonomous replicating plasmid or can insert into the bacterial chromosome. This is an old term that is rarely used anymore. Such elements are simply called plasmids (for example, F-prime plasmid).

Epistasis : A condition where one gene masks the expression of another gene. For example, if there are two mutations that affect a single metabolic pathway, a mutation that disrupts the gene product which acts at an earlier step in the pathway will usually obscure the mutation that affects a later step in the pathway.

Epitope : A portion of an antigen recognized by an antibody binding site. For protein antigens the epitope is typically 5-8 amino acids. Gene fusions to epitopes can be used as a tag to recognize specific proteins.

Epizootic : A disease affecting a large number of animals simultaneously.

EPS (extracellular polysaccharides) : Sugar polymers that contribute to the slimy appearance of bacteria probably promote colonization of plant tissues and disease development in bacteria.

Equational division : Mitotic-type division that is usually the second division in the meiotic sequence; somatic mitosis and the non-reductional division of meiosis. A chromosome division in which the two chromatids of each duplicated chromosome separate longitudinally, prior to being incorporated into two daughter nuclei.

Equatorial plate : The figure formed by the chromosomes in the centre (equatorial plane) of the spindle in mitosis.

Equilibrium : A state of dynamic systems in which there is no net change.

Equilibrium density gradient centrifugation : A procedure used to separate macro-molecules based on their density (mass per unit volume).

Equimolar : Identical molar concentrations. See molarity; mole.

ER : See endoplasmic reticulum.

Eradication : Control of plant disease by eliminating the pathogen after it is established or by eliminating the plants that carry the pathogen.

ERE : Estrogen Response Element. A binding site in a promoter to which the activated estrogen receptor can bind. The estrogen receptor is essentially a transcription factor which is activated only in the presence of estrogens. The activated receptor will bind to an ERE, and transcription of the adjacent gene will be altered. See also "Response element".

Ergot : A disease of certain grasses and cereals, especially rye, caused by Claviceps spp.; a sclerotium, or resting structure, produced by Claviceps species and other closely related fungi in infected flowers of parasitized grain plants.

Ergotism : A disease caused by ingestion of grain contaminated with alkaloids of ergot fungi, often called the holy fire or St. Anthony's fire in the Middle Ages.

Erlenmeyer flask : A conical flat-bottomed laboratory flask with a narrow neck, designed by E. Erlenmeyer. Widely used for culturing micro-organisms.

Error prone repair : A mechanism for repair of DNA damage that often results in mutations (e.g. SOS repair).

Erumpent : Bursting or erupting through the substrate surface.

Erythrocytes : The hemoglobin-containing cell found in the blood of vertebrates.

ES cells : See embryonic stem cells.

Escape : Failure of inherently susceptible plants to become diseased, even though disease is prevalent.

Escherichia coli : A commensal bacterium inhabiting the colon of many species. *E. coli* is widely used in biology, both as a simple model of cell biochemical function and as a host for molecular cloning experiments. In environmental studies, it is a key indicator of water pollution due to human sewage effluent.

Essential amino acid : An amino acid that cannot be synthesized by animals and therefore has to be ingested with feed or food.

Essential element : Any of a number of elements required by living organisms to ensure normal growth, development and maintenance.

Essential nutrient : Any substance required by living organisms to ensure normal growth, development and maintenance.

Essential requirement : A nutrient is essential when it is mandatory for growth, development and reproduction. In tissue culture, it comprises inorganic salts, including all of the elements necessary for plant metabolism; organic factors (amino acids, vitamins); usually also endogenous plant growth regulators (auxins, cytokinins and often gibberellins); as well as a carbon source (sucrose or glucose).

EST : Expressed sequence tag. See sequence tagged site.

Established culture : 1. An aseptic viable explant (See micropropagation stages I & II). 2. A suspension culture subjected to several passages with a constant cell number per unit time.

Estimated breeding value (EBV) : Twice the expected progeny difference (q.v.). The difference is doubled because breeding value is a reflection of all the genes of an animal, in contrast to progeny difference, which is a reflection of a sample half of an animal's genes. The predicted performance of the offspring of the mating between any two animals is the average of their EBVs (averaged because each parent makes an equal contribution to each offspring).

Estrogen : See oestrogen.

ET : See multiple ovulation and embryo transfer.

Ethanol (ethyl alcohol) (C_2H_6O; fw 4607) : Commonly used to disinfest plant tissues, glassware utensils and working surfaces in tissue culture manipulations. The concentration used is 70% (v/v) for disinfecting and 95% (v/v) when flaming tools. Ethanol is also used to dissolve water-insoluble additions (addendums) to culture media.

Ethanol precipitation : Precipitation of nucleic acid molecules by ethanol plus salt. Commonly used to concentrate DNA from aqueous solutions.

Ethephon (2-chloroethyl) phosphonic acid ($ClC_2PO_3H_6$; fw 14450) : Through a spontaneous degradation of ethephon, ethylene is produced. Ethephon is a synthetic compound commonly used to treat cultured cells or unripened fruit with ethylene (q.v.).

Ethics : The study of fundamental principles which defines values and determines moral duty and obligation.

Ethidium bromide : A fluorescent chemical that intercalates between base pairs in a double stranded DNA molecule. Commonly used to detect DNA following gel electrophoresis. Ethidium bromide intercalates between the stacked bases of DNA and RNA. When excited by UV light between 254 nm (short wave) and 366 nm (long wave) it emits fluorescent light at 590 nm. Due to the decreased rotation possible when it is intercalated, the DNA-ethidium bromide complex produces about 50 times more fluorescence than free ethidium bromide. When ethidium bromide stained gels are photographed on a UV transilluminator a UV filter is required to screen out the background UV from the transilluminator. Ethidium bromide is a known carcinogen.

Ethyl alcohol : See ethanol.

Ethylene : A plant hormone influencing various aspects of vegetative growth, fruit ripening, abscission of plant parts, and the senescence of flowers.

Ethylenediamine tetraacetic acid (EDTA) : A chelating compound. In tissue culture it is used to keep nutrients, such as iron, bound in a form that leaves them still available to the plant but which prevents them from precipitating out.

Etiolation : An abnormal increase in stem elongation, accompanied by poor or absent leaf development. Physiological etiolation is caused by a lack of chlorophyll, and is typical of plants growing under low light intensity or in complete darkness. It can also be caused by disease.

Etiology : The study of the causes of diseases.

ETL : See economic trait locus.

Eucaryote : An organism with a nuclear membrane and membrane bound organelles (e.g. mitochondria), and a mitotic spindle. This group of organisms is more correctly called Eucarya.

Euchromatin : Genetic material that is stained less intensely by certain dyes during interphase, and that comprises many different kinds of genes. cf heterochromatin.

Eugenics : The improvement of humanity by altering its genetic composition by encouraging breeding of those presumed to have desirable genes.

Eukaryote (Gr eu, true + karyon, true nucleus) : Any organism characterized by having the nucleus enclosed by a membrane. Eukaryotic organisms include animals, plants, fungi and some algae. They also possess membrane-bound functional organelles, such as mitochondria and chloroplasts, in the cytoplasm of their cells. cf prokaryote.

Euploid : An organism or cell having a chromosome number that is an exact multiple of the monoploid (n) number. Terms used to identify different levels in an euploid series are diploid, triploid, tetraploid, and so on. cf aneuploid.

Evaluation : In AnGR: Measurement of the characteristics that are important for production and adaptation, either of individual animals or of populations, most commonly in the context of comparative evaluation of the traits of animals or of populations. (Source: FAO, 1999).

Evapotranspiration (L evaporare, e, out of, + vapor, vapour + F transpirer, to perspire) : The process of water loss in vapour form from a unit surface of land both directly and through leaf surfaces during a specific period of time.

Evolution : The process by which the present diversity of plant and animal life arose from the earliest organisms, a process believed to have been continuing for at least 3 000 million years.

Evolutionarily conserved : See conserved sequence.

Evolutionary clock : Defined by the rate at which mutations accumulate within a given gene.

ex novo **:** See de novo.

ex situ **conservation :** A conservation method which entails the actual removal of germplasm resources (seeds, pollen, sperm, individual organisms) from the original habitat or natural environment. cf gene bank; cryobiological preservation; in situ conservation.

ex situ **conservation of farm animal genetic diversity :** In AnGR: All conservation of genetic material *in vivo*, but out of the environment in which it developed, and *in vitro* including, inter alia, the cryoconservation of semen, oocytes, embryos, cells or tissues. Note that ex situ conservation and ex situ preservation are considered here to be synonymous. (Source: FAO, 1999).

ex situ **management :** The management of planted stands of trees to protect them and insure survival, growth and identity; the conservation or preservation of trees as seed, pollen, tissue culture, or excised plant parts. (See germplasm conservation, ex situ).

ex vitro **(L "from glass") :** Organisms removed from tissue culture and transplanted; generally plants to soil or potting mixture.

ex vivo **gene therapy :** The delivery of a gene or genes to the isolated cells of an individual. After culturing, the transformed cells are introduced back into the individual by transfusion, infusion or injection, to alleviate a genetic disorder.

Excinuclease : The endonuclease-containing protein complex that excises a segment of damaged DNA during excision repair.

Excision : 1. The natural or *in vitro* enzymatic release (removal) of a DNA segment from a chromosome or cloning vector. 2. Cutting out and preparing a tissue, organ, etc., for culture. 3. Removing adventitious shoots from callus tissue.

Excision repair : A repair system that removes nucleotides from a damaged strand of DNA and then replaces them with a new tract of DNA synthesised using the undamaged complementary strand as a template.

Exclusion : Control of disease by excluding the pathogen or infected plant material from crop production areas (e.g. by quarantines and embargoes).

Excrete : To transport a compound out of a cell. a.k.a. to secrete; to export.

Exit site (E site) : The ribosome binding site that contains the free tRNA prior to its release.

Exo III : See ase III.

Exocrine gland : In animals, a gland that secretes through a duct. cf endocrine gland. See gland.

Exodeoxyribonuclease III : See exonuclease III.

Exogamy : The fusion of reproductive cells from distantly related or unrelated organisms, i.e., outbreeding. cf endogamy.

Exogenote : Chromosomal fragment homologous to an endogenote and donated to a merozygote.

Exogenous (Gr exe, out, beyond + genos, race, kind) : Produced outside of, originating from, or due to external causes. Opposite of endogenous.

Exogenous DNA : DNA originating outside an organism that has been introducted into the organism.

Exon : Those portions of a genomic DNA sequence which will be represented in the final, mature mRNA ie. A contiguous segment of genomic DNA that codes for a polypeptide in a gene.The term "exon" can also be used for the equivalent segments in the final RNA. Exons may include coding sequences, the 5' untranslated region or the 3' untranslated region.

Exon amplification : A procedure that is used to amplify exons.

Exons : portion of a gene included in the transcript of a gene and survives processing of the RNA in the cell nucleus to become part of a spliced messenger of a structural RNA in the cell cytoplasm; an exon specifies the amino acid sequence of a portion of the complete polypeptide.

Exonuclease : An enzyme that digests a molecule of nucleic acid by removing successive nucleotides from the 5' or 3' end. Exonucleases may be specific for single or double stranded DNA or RNA, and may be specific for either 3' or 5' ends.

Exonuclease III : An *E. coli* enzyme that removes nucleotides from the 3′-hydroxyl ends of double stranded DNA. a.k.a. exo III; exodeoxyribonuclease III.

Exopolysaccharide : A high-molecular-weight polymer that is composed of sugar residues and is secreted by a micro-organism into the surrounding environment.

Exotic : Broad definition: A non-native population introduced into a new area. Narrow definition: a species introduced from another country.

Exotoxin : A toxin released by a bacterium into the medium in which it grows.

Expected progeny difference (EPD) : The predicted performance of the future offspring of an animal for a particular trait, calculated from measurement(s) of the animal's own performance and/or the performance of one or more of its relatives, for the trait in question and/or for one or more correlated traits. Typically, the prediction is expressed as a deviation from a well-defined base population, assuming the animal in question is mated to a sample of animals whose average genetic merit equals that of the base population. The predicted performance of the offspring of the mating between any two animals is the sum of their EPDs.

Explant : A plant part aseptically excised and prepared for culture in a nutrient medium. cf donor plant.

Explant donor : The source plant or mother plant from which is taken the explant used to initiate a culture. cf donor plant.

Explantation : The removal of cells, tissues or organs of animals and plants for observation of their growth and development in appropriate culture media.

Exponential phase : The growth stage where cells undergo their maximum rate of cell division. The exponential phase follows the lag phase, and precedes the linear growth phase. See growth phases. a.k.a logarithmic phase.

Export : To transport a compound out of a cell. a.k.a. to secrete; to excrete.

Express : To transcribe and translate a gene's message into a peptide product.

Expressed gene : See gene expression.

Expressed sequence tag (EST) : A short piece of DNA sequence corresponding to a fragment of a complementary DNA (made from a cell's messenger RNA). ESTs have been used to hunt for genes, so hundreds of thousands are present in sequence databases. (See Sequence tagged sites).

Expression : To "express" a gene is to cause it to function. A gene which encodes a protein will, when expressed, be transcribed and translated to produce that protein. A gene which encodes an RNA rather than a protein (for example, a rRNA gene) will produce that RNA when expressed.

Expression clone : This is a clone (plasmid in a bacteria, or maybe a lambda phage in bacteria) which is designed to produce a protein from the DNA insert. Mammalian genes do not function in bacteria, so to get bacterial expression from your mammalian cDNA, you would place its coding region (i.e. no introns) immediately adjacent to bacterial transcription/ translation control sequences. That artificial construct (the "expression clone") will produce a pseudo-mammalian protein if put back into bacteria. Often, that protein can be recognized by antibodies raised against the authentic mammalian protein, and vice versa.

Expression library : A population of different DNA molecules encoding peptides, that has been cloned into one kind of expression vector. See library.

Expression system : Combination of host and vector which provides a genetic context for making a cloned gene function, i.e., produce peptide, in the host cell.

Expression vector : A cloning vector designed so that a foreign gene inserted into the vector will be expressed in the host organism. The inserted DNA is often placed under the control of a strong, regulatable promoter (e.g. the lacZ or araBAD promoter) so that expression of the gene product will only occur when desired.

Expressivity : Degree of expression of a trait controlled by a gene. A particular gene may show different degrees of expression in different individuals. See variable expressivity.

Extension : Single-stranded DNA region consisting of one or more nucleotides at the end of a strand of duplex DNA. a.k.a. protruding end; sticky end; overhang; cohesive end. See extension.

External guide sequence (EGS) : See guide sequence.

Extinct breed : In AnGR: A breed where it is no longer possible to recreate the breed population. Extinction is absolute when there are no breeding males (semen), breeding females (oocytes), nor embryos remaining. (Source: FAO, 1999).

Extracellular polysaccharides (EPS) : sugar polymers that aid in movement, help prevent desiccation, and contribute to the slimy appearance of bacteria, fungi, and nematodes; and probably promote colonization of plant tissues and disease development in bacteria.

Extrachromosomal inheritance : See cytoplasmic inheritance.

Extrachromosomal inheritance(also cytoplasmic or maternal inheritance) : Inheritance of genes not located in the nucleus, i.e. those in mitochondria and chloroplasts.

Extrachromosomes : Self-replicative genetic elements separate from main chromosome(s) of a cell. This definition usually excludes viruses, but the division is somewhat arbitrary. In bacteria, plasmids are the principal extrachromosomes; they encode functions which are not essential to the growth and division of the host cell. In eukaryotes, extrachromosomes may be either essential or dispensable. They may inhabit (i) the nucleus, e.g., extrachromosomal rDNA molecules, yeast 2?m plasmid; (ii) the cytosol, e.g., dsRNA molecules in fungi; or (iii) the cytoplasmic organelles, e.g., mitochondrial DNA, chloroplast DNA. Eukaryotic extrachromosomal elements may be recognized genetically by their failure to show segregation at meiosis.

Extragenic : An affect due to a second gene. For example, an extragenic suppressor is a mutation in a second gene that repairs the mutant phenotype.

Exudate : Liquid excreted or discharged from diseased tissues, from roots and leaves, or by fungi.

Exude : Slowly discharge; leak liquid material (exudate such as tannins or oxidized polyphenols) through pores or cuts, or by diffusion into the medium. In some woody plant species, exudation is associated with a lethal browning of explants.

F- cell : A cell which does not contain the F-factor, and hence able to act as a recipient (female) in a conjugative DNA transfer in matings with F+ or Hfr strains.

F factor : A bacterial episome that confers the ability to function as a genetic donor in conjugation; the fertility factor in bacteria.

F sp (abbr for forma specialis) : A taxonomic group within a pathogenic species defined in terms of host range, i.e. members of different formae speciales infect different groups of plants.

F+ cell : A cell that carries a F-factor as an autonomous plasmid, which enables the cell acting as a donor (male) to transfer the F-factor to a recipient (female) cell.

F1 : The first filial generation in a cross between any two parents; the first generation of descent from a given mating.

F2 : The second filial generation, produced by crossing two members of the F1, or by self-pollinating the F1. The "grandchildren" of a given mating.

F3 : Progeny obtained by self-fertilization of F2 individuals.

FACS : See fluorescence-activated cell sorting.

Factor : Same as gene.

Factorial mating : A mating scheme in which each male parent is mated with each female parent. Made possible in animals by means of *in vitro* embryo production (q.v.). Such a mating scheme substantially reduces the rate of inbreeding in a selection programme.

Facultative : Organisms that can grow under different conditions. Commonly used to describe bacteria that can grow oxidatively under aerobic or fermentatively under anaerobic conditions.

Facultative parasite : Organism that is normally saprophytic but is capable of being parasitic.

Facultative saprophyte : Organism that is normally parasitic but is capable of being saprophytic.

FAD (flavin adenine dinucleotide) : A co-enzyme important in various biochemical reactions. It comprises a phosphorylated vitamin B_2 (riboflavin) molecule linked to the nucleotide adenine monophosphate (AMP). It functions as a hydrogen acceptor in dehydrogenation reactions, being reduced to $FADH_2$. This in turn is oxidized to FAD by the electron transport chain, thereby generating ATP (two molecules of ATP per molecule of $FADH_2$).

Fallow: Cultivated land kept free from a crop or weeds during the normal growing season.

False fruit : See pseudocarp.

False-negative : A negative assay result that should have been positive.

False-positive : A positive assay result that should have been negative.

Familial : An inherited trait.

Family : A group of individuals directly related by descent from a common ancestor.

Family selection : The selection of progeny families on their mean performance. In addition, the best individuals are usually selected in the best families.

Farm animal genetic resources (AnGR) : In AnGR: Those animal species that are used, or may be used, for the production of food and agriculture, and the populations within each of them. These populations within each species can be classified as wild and feral populations, landraces and primary populations, standardized breeds, selected lines, and any conserved genetic material. (Source: FAO, 1999).

Fasciation : Malformation in shoots or floral organs manifested as enlargements and flattening as if several parts were fused.

Fascicle (adj fasciculate) : Small group, bundle, or cluster.

Fastidious : In reference to prokaryotic organisms, having special growth and nutritional requirements.

Fed-batch fermentation : Culture of cells or micro-organisms where nutrients are added periodically to the bioreactor.

Feedback inhibition : The process by which the accumulated end product of a biochemical pathway stops synthesis of that product. A late metabolite of a synthetic pathway regulates synthesis at an earlier step of the pathway. cf end-product inhibition.

Feeder root : Fine root that absorbs water and dissolved nutrients.

Fermentation : The anaerobic breakdown by micro-organisms of complex organic substances, especially carbohydrates like glucose. The process is energy-yielding. Fermentation is often misused to describe large-scale aerobic cell culture in specialized vessels (fermenters, bioreactors) for secondary product synthesis.

Fermentation substrates : Materials used as food for growing micro-organism. The fermentation substrates and the trace materials needed, together with chemicals added to make the fermentation easier, form the culture medium.

Fermenter : See bioreactor.

Fertile : Of an organism: Capable of breeding and reproduction.

Fertility : Ability to produce viable offspring.

Fertilization (L fertilis, capable of producing fruit) : The union of two gametes from opposite sexes to form a zygote; it involves the fusion of nuclei of gametes (karyogamy) and the fusion of cytoplasm (plasmogamy). Typically, each gamete contains a haploid set of chromosomes. Hence, after fusion of the nuclei, the resulting nucleus contains a diploid set of chromosomes.

Several categories are distinguished : 1. Self-fertilization (selfing) - fusion of male and female gametes from the same euploid organism. 2. Cross-fertilization (crossing) - fusion of male and female gametes from different euploid individuals. 3. Double fertilization - in angiosperms, the fusion of one male gamete with the ovum at about the same time as the second male gamete fuses with the female polar nuclei (or secondary nucleus) to form the endosperm.

Fetal alcohol syndrome : A link between excessive alcohol consumption during pregnancy and birth defects; characteristics include small head and eyes, folds of the skin that obscure the inner juncture of the eyelids, short, upturned nose, and thin lips.

Fetus : See foetus.

Feulgen's test : A histochemical test in which the distribution of DNA in the chromosomes of dividing cell nuclei can be observed. A tissue section is first treated with dilute hydrochloric acid to remove the purine bases of the DNA, thus exposing the aldehyde groups of the sugar deoxyribose. The section is then immersed in Schiff's reagent, which combines with the aldehyde groups to form a magenta-coloured compound.

F-factor : An *E. coli* plasmid that encodes conjugative transfer ("fertility") functions. The F-plasmid may exist as an autonomous plasmid in the cytoplasm or may integrate into specific sites in the chromosome, producing an Hfr cell.

FFDCA (acronym for Federal Food, Drug, and Cosmetic Act) : The federal law which sets legal tolerances of pesticides on food products at harvest.

Fiber : Elongated plant cell with thick walls (in xylem and phloem), dead at maturity.

Fibres : Elongated cells with tapering, pointed ends; the cells interlock to form a strong, rigid tissue; pits in the walls are usually very narrow and not very numerous.

Fibril : Microscopic to sub-microscopic cellulose thread that is part of the cellulose matrix of plant cell walls.

Fibrous root : Root system in which both primary and lateral roots have approximately equal diameters. Opposite is tap root.

Field gene bank : See gene bank (2).

Field trial : A test of a new technology or variety, including biotech-derived varieties, done outside the laboratory but with specific limitations and requirements on location, plot size, methodology, etc.

FIFRA (acronym for Federal Insecticide, Fungicide, and Rodenticide Act) : The federal law which legislates the conditions for use of pesticides.

Filamentous phage : A group of phage with the nucleic acid surrounded by a long, tubular protein coat. Examples of filaments phage include M13 and Ff.

Filamentous(syn filiform) : threadlike

Filial generation (F1, F2) : Each generation of offspring in a breeding program, designated F1, F2, etc.

Filter bioreactor; mesh bioreactor : Cells are grown on an open mesh of an inert material, which allows the culture medium to flow past it but retains the cells. This is similar in idea to membrane and hollow fibre reactors, but can be much easier to set up, being similar to conventional tower bioreactors, but with the meshwork replacing the central reactor space.

Filter mating : A method where conjugation between donor and recipient bacteria is done on a millipore filter. Conjugation is often much more efficient on a solid surface, probably because the donor and recipient bacteria are fixed in close proximity facilitating stable mating bridge formation.

Filter sterilization : Process of sterilizing a liquid by passage through a filter with pores so small that they prevent the passage of micro-organisms and microbial spores.

Filtration : 1. Separation of solids from liquids by using a porous material that only allows passage of the liquid or of solids of a smaller than the pore size of the filter. The material passing the filter forms the filtrate. 2. Removal of cell aggregates to obtain a filtrate of single cells that can be utilized as plating inocula.

Fimbriae : Thin, proteinaceous filaments that extend from the cell surface of microbial cells and facilate adhesion to solid surfaces or other cells.

Fingerprinting : A common term for methods that allow the rapid identification of particular types of bacteria (also known as "typing"). For example, plasmid fingerprinting involves determining the number and types of plasmids in a particular strain of bacteria.

Finished DNA Sequence : High-quality, low error, gap-free DNA sequence of the human genome. Achieving this ultimate 2003 HGP goal requires additional sequencing to close gaps, reduce ambiguities, and allow for only a single error every 10,000 bases, the agreed-upon standard for HGP finished sequence. See also: sequencing, draft sequence.

FISH (fluorescence *in situ* hybridization) : A physical mapping approach that uses fluorescein tags to detect hybridization of probes with metaphase chromosomes and with the less- condensed somatic interphase chromatin.

Fission (L fissilis, easily split) : Asexual reproduction involving the division of a single-celled individual into two new single-celled individuals of equal size.

Fitness : The number of offspring left by an individual, often compared with the average of the population or with some other standard, such as the number of offspring left by a particular genotype.

Fitness, adaptive : The ability of a population to survive, reproduce, and adapt genetically to a changing environment for at least a modest number (usually fewer than 50) generations. (See gene pool and heterozygosity).

Fitness, evolutionary : The ability of a population (usually a whole species or ecotype) to survive, reproduce, and adapt genetically to a changing environment for many (usually more than 50) generations. (See gene pool and heterozygosity).

Fixation : An event that occurs when all the alleles at a locus except one are eliminated from a population. The remaining allele, now occurring with a frequency of 100%, is said to have been fixed.

Flaccid : Wilted, lacking turgor.

Flag : Isolated, wilted or necrotic branch with dead leaves attached.

FLAG : See affinity tag.

Flagella : Long, flexible, helical protein structures that extend from the surface of the cell. Rotation of flagella in one direction results in motility, and rotation of flagella in the opposite direction results in tumbling.

Flagellum (pl flagella; adj flagellate) : A whiplike organelle of locomotion in certain cells; locomotor structures in flagellate protozoa.

Flaming : A technique for sterilizing instruments. The instrument is dipped in alcohol (usually 95% (v/v) ethanol) and then the alcohol on the instrument is ignited, thus heat-sterilizing the tool surface.

Flanking region : For microsatellites, the flanking regions are the stretches of DNA outside the simple sequence tandem repeat (STR). These sequences are used as primer pairs. The flanking regions are usually invariant across a population or species, but mutations in the flanking region can be a cause of null alleles as well as a potentially serious source of homoplasy.

Flanking sequence or flanking region : The DNA sequence located on either side of a specific genetic locus.

Fleck : Minute, discolored spot in green tissue.

Flexuous : Having turns or windings; capable of bending.

Flocculant : Agent that causes small particles to aggregate (flocculate).

Floccule : Micro-organism aggregate or colloidal particle floating in or on a liquid. Contaminated liquid media are usually cloudy, illustrating this flocculation phenomenon.

Flow cytometry : Analysis of biological material by detection of the light-absorbing or fluorescing properties of cells or subcellular fractions (i.e., chromosomes) passing in a narrow stream through a laser beam. An absorbance or fluorescence profile of the sample is produced. Automated sorting devices, used to fractionate samples, sort successive droplets of the analyzed stream into different fractions depending on the fluorescence emitted by each droplet.

Flow karyotyping : Use of flow cytometry to analyze and/or separate chromosomes on the basis of their DNA content.

Flower : Angiosperm reproductive structure bearing pistils, stamens, or both, and usually also sepals and petals. So-called flowers of conifers are the male and female strobili before and during pollination.

Flower break : Break or stripe in flower color (see breaking).

Fluctuation test : An experimental approach designed by Luria and Delbruck to determine whether mutations are random or directed, and to measure mutation rates.

Fluorescence in situ hybridization (FISH) : A physical mapping approach that uses fluorescein tags to detect hybridization of probes with metaphase chromosomes and with the less-condensed somatic interphase chromatin.

Fluorescence-activated cell sorting (FACS) : The use of laser beams to detect differences in fluorescence between different types of cells in a mixture, and the subsequent deflection of cells into separate bins corresponding to each type of cell in the mixture. One of the popular uses of this technology is in sperm sexing (q.v.).

Fluorescent probe : Probe whose response is based on the fluorescence intensity of individual cells or cell components.

Flush end : See blunt end.

Flush-end cut : See blunt-end cut.

Focus (pl foci) : Small area of diseased plants within a population.

Foetus : Pre-natal stage of a viviparous animal, between the embryonic stage and parturition.

Fog : Fine particles of liquid suspended in the air, such as of water in a fog chamber used for acclimatizing recent ex vitro transplants. cf mist propagation.

Fold-back : The structure formed when a double-stranded DNA molecule containing an inverted repeat sequence is denatured and then allowed to re-anneal at low DNA concentrations. The repeated sequence permits the formation of a double-stranded region within each of the separated strands of the original molecule.

Folded genome : The condensed intracellular state of the DNA in the nucleoid of a bacterium. The DNA is segregated into domains, and each domain is independently negatively supercoiled.

Foliar : Pertaining to leaves.

Follicle : Any enclosing cluster of cells that protects and nourishes a cell or structure within. Thus a follicle in the ovary contains a developing egg cell, while a hair follicle envelops the root of hair.

Follicle stimulating hormone (FSH) : A hormone, secreted by the anterior pituitary gland in mammals, that stimulates, in female mammals, ripening of specialized structures in the ovary (Graafian follicles) that produce ova and, in males, the formation of sperm in the testis. It is a major constituent of fertility drugs.

Food processing enzyme : Enzyme used to control food texture, flavour, appearance and, to a certain extent, nutritional value. Amylases break down complex polysaccharides to simplex sugars; proteases tenderize meat proteins. Biotechnology can assist the development of new food enzymes by finding or engineering enzymes which fit better with the other processes which the food must undergo, like cooking or canning.

Foot-candle : An obsolete photometric measure of light intensity. Now superseded by the lux (symbol: lx) (1 fc ? 10.7 lx). See lux; photon.

Footprinting : A method for identification of a protein-binding site on a DNA molecule. For example, DNase I footprinting identifies those phosphodiester bonds which are covered by a specific DNA-binding protein and thus protected from cleavage by DNase I.

Forced cloning : The insertion of foreign DNA into a cloning vector in a predetermined orientation.

Foreign DNA : A DNA molecule that is incorporated into either a cloning vector or a chromosomal site. See exogenous DNA.

Forensics : The use of DNA for identification. Some examples of DNA use are to establish paternity in child support cases; establish the presence of a suspect at a crime scene, and identify accident victims.

Forest decline : Combination of biotic factors (e.g. insect pests, pathogens) and abiotic factors (e.g. drought, air pollution) that contributes to poor growth and death of populations of trees.

Forest genetics : The study of heredity in forest trees.

Forest tree improvement : The application of genetic principles to the improvement and management of forest trees.

Form genus : A genus based on morphology and not on evolutionary relationships, such as used for imperfect fungi.

Forma specialis (abbr fsp; pl formae speciales) : A taxonomic group within a pathogenic species defined in terms of host range, i.e. members of different formae speciales infect different groups of plants.

Formulation : 1. For traditional therapeutic agents, this refers to the method by which a therapeutic agent is delivered to its site of action. 2. For tissue culture, see medium; medium formulation.

Fortify : To add strengthening components or beneficial ingredients to a nutrient medium.

Forward selection : Choosing the best individuals out of a progeny test for use in seed orchards and/or subsequent generations of breeding. (See Backward selection).

Fouling : The coating or plugging (by materials or micro-organisms) of equipment, thus preventing it from functioning properly.

Founder : An individual drawn from a source population who contributes genetically to the derived subpopulation.

Founder animal : In transgenesis research, an organism that carries a transgene in its germ line and can be used in matings to establish a pure-breeding transgenic line, or one that acts as a breeding stock for transgenic animals.

Founder effect : Changes in allele frequencies that occur when a subpopulation is formed from a larger one. Typically many rare and usually undesirable alleles are excluded while a few carried by the founders get a big boost in frequency.

Founder equivalents : The number of hypothetical founders that would have the same diversity as the descendant population. Generally much smaller than the actual number due to unequal use and allele loss (gene dropping).

Founder principle : The possibility that a new, small, isolated population may be genetically different from the "parent" population, because the founding individuals (being a small, random sample from the large, "parent" population) could be quite different from typical members of the "parent" population.

Four-base cutter; four-base-pair-cutter; four-cutter : A type II restriction endonuclease that binds (and subsequently cleaves) DNA at sites that contain a sequence of four nucleotide pairs that is uniquely recognised by that enzyme. Because any sequence of four bases occurs more frequently by chance than any sequence of six bases, four-base cutters cleave more frequently than do six-base cutters. Thus, four-base cutters create smaller fragments than six-base cutters.

F-pilus : A filamentous appendage encoded by the F-plasmid. The pilus promotes adhesion of F+ and F- cells and provides adsorption sites for male-specific phages.

F-prime (F') : An extrachromosomal F-plasmid that carries a fragment of chromosomal DNA.

FQPA (acronym for Food Quality Protection Act) : The federal law which sets a health-based standard governing pesticide tolerance levels in food.

Fractionation : Separation.

Fragile sites : A non-staining gap of variable width that usually involves both chromatids and is always at exactly the same point on a specific chromosome derived from an individual or kindred.

Fragile-X syndrome : X-linked trait; the second most common identifiable cause of genetic mental deficiency.

Fragment : See restriction fragment.

Frameshift : A mutation which adds or deletes one or two base pairs (or any non-multiple of 3) from a coding sequence in a molecule of DNA, so that the genetic code is read out-of-phase. All codons translated downstream of a frameshift mutation will be misread, and frequently an out-of-frame stop codon will prematurely terminate translation.

Frameshift mutation : A mutation (deletion or insertion, never a simple substitution) of one or more nucleotides but never a multiple of 3 nucleotides, which shortens or lengthens a trinucleotide sequence representing a codon; the result is a shift from one reading frame to another reading frame. The amino acid sequence of the protein downstream of the mutation is completely altered, and may even be much shorter or longer due to a change in the location of the first termination (stop) codon: Asn Tyr Thr Asn Leu Gly His Wild-type polypeptide.

AAU UAC ACA AAU UUA GGG CAU mRNA.

Asn Thr Gln Ile STOP Mutant polypeptide | .

Fraternal twin : Siblings born at the same time as the result of fertilization of two ova by two sperm. They share the same genetic relationship to each other as any other siblings. See also: identical twin.

Free water : Water released by a cell when freezing occurs in intercellular spaces. cf bound water.

Free-living conditions : Natural or greenhouse conditions where the plantlets are transferred from *in vitro* conditions to soil mixtures. In such instances, plantlets must manufacture their own food supply for survival. See acclimatization; hardening off.

Freeze preservation : See cryobiological preservation.

Freeze-drying : The process of drying a tissue or an organ in a frozen state under vacuum. Tissues are freeze-dried to measure their dry weight or to preserve them for future analysis. Freeze-drying is the standard way of preserving micro-organisms for long periods of time. cf lyophilization.

Frequency distribution : A graph showing either the relative or absolute incidence of classes in a population. The classes may be defined by either a discrete or a continuous variable; in the latter case, each class represents a different interval on the scale of measurement.

Fresh weight : The weight, including the water content, of a plant or plant part at the time of harvest. cf wet weight.

Friable : A term commonly used to describe a crumbling or fragmenting callus. A friable callus is easily dissected and readily dispersed into single cells or clumps of cells in solution.

Fructification : General term for spore-bearing organs in both macro- and microfungi.

Fruiting body : Any of various complex, spore-bearing fungal structures.

FSH : See follicle stimulating hormone.

Full gene sequence : The complete order of bases in a gene. This order determines which protein a gene will produce.

Fumigant (v fumigate) : A gas or volatile substance used to kill or inhibit the growth of microorganisms or other pests.

Functional domain : A region of a polypeptide chain or RNA that performs a particular function. Functional domains are often folded into independant structures, separated from the rest of the polymer by a flexible hinge region.

Functional gene cloning : See candidate gene strategy.

Functional genomics : The study of genes, their resulting proteins, and the role played by the proteins the body's biochemical processes.

Fungi Imperfecti (syn Deuteromycetes) : Group of fungi without a sexual stage; also, the asexual stage of ascomycetes and basidiomycetes.

Fungicide (adj fungicidal) : Chemical or physical agent that kills or inhibits the growth of fungi.

Fungistasis : Inhibition of fungal growth, sporulation, or spore germination but not death; used to describe the nonspecific phenomenon in natural soils where spore germination is inhibited and often overcome by rhizosphere nutrients.

Fungistat (adj fungistatic) : A chemical or physical agent that inhibits fungal growth, sporulation, or spore geminiation, but does not cause death.

Fungus (pl fungi) : A eukaryotic organism that is usually filamentous (forming a mycelium) and heterotrophic, has cell walls composed of chitin, and reproduces by sexual and/or asexual spores.

Furfural; furfuraldehyde : Used industrially as a solvent and as a raw material for synthetic resin.

Fusiform : Spindle-shaped; tapering at each end.

Fusion biopharmaceuticals : Biopharmaceutical proteins formed as a result of fusion proteins. Their advantages are: - synergistic activities in one molecule; thus when the molecule binds to a cell, it does two things at once; - the adverse effect or poor stability of one molecule may be offset by the properties of the other; and- one molecule acts as a targeting mechanism to bring the other to the site where it is meant to act. See fusion protein: immunotoxin.

Fusion gene : A hybrid gene created by joining portions of two different genes (to produce a new protein) or by joining a gene to a different promoter (to alter or regulate gene transcription).

Fusion protein : A product of recombinant DNA in which the foreign gene product is juxtaposed ("fused") to either the carboxyl-terminal or amino-terminal portion of a polypeptide encoded by the vector itself. Use of fusion proteins often facilitates expression of otherwise lethal products and the purification of recombinant proteins.

G : Guanine residue in either DNA or RNA.

G cap : The 5′-terminal methylated guanine nucleoside that is present on many eukaryotic mRNAs; it is joined, after transcription, to the mRNA. See cap.

Gall (syn tumor) : Abnormal swelling or localized outgrowth, often roughly spherical, produced by a plant as a result of attack by a fungus, bacterium, nematode, insect, or other organism.

Gallery : Insect tunnel in bark and wood.

Gametangium (pl gametangia) : Cell containing gametes or nuclei that act as gametes.

Gamete : Mature male or female reproductive cell (sperm or ovum) with a haploid set of chromosomes (23 for humans).

Gamete and embryo storage : Storage of ova, sperm or fertilized embryos outside their original source. Almost invariably this means cryopreservation in liquid nitrogen.

Gametic (phase) disequilibrium : In relation to any two loci, the occurrence of haplotypes (gametes) with a frequency greater than or less than the product of the frequency of the two relevant alleles. See also gametic (phase) equilibrium.

Gametoclone : A plant regenerated from a tissue culture originating from gametic tissue.

Gametogenesis : The process of the formation of gametes.

Gametophyte : That phase of the plant life cycle that bears the gamete producing organs; the cells have n chromosomes. In angiosperms, the pollen grain is the male gametophyte and the embryo sac is the female gametophyte.

Gametophytic incompatibility : A phenomenon controlled by the complex S locus, in which a pollen grain cannot fertilize an ovule produced by a plant that carries the same S allele as the pollen grain. Thus S1 pollen cannot fertilize an ovule of a plant possessing the S1 allele.

Gap : 1. Period of time, during the cell cycle, between M and S phases. 2. A missing section on one of the strands of double-stranded DNA. The DNA will therefore have a single-stranded region.

Gapped DNA : A duplex DNA molecule with one or more internal single-stranded regions.

Gas transfer : The rate at which gases are transferred from gas into solution. It is an important parameter in fermentation systems because it controls the rate at which the organism can metabolize. Gas transfer can be done by several methods, including use of small gas bubbles, that diffuse faster than larger ones, thanks to their larger surface area per unit of volume; or spreading the liquid out, for example in a thin sheet, or in a thin permeable tube, as in hollow fibre bioreactor.

Gastrula : An early animal embryo consisting of two layers of cells; an embryological stage following the blastula.

GC island : A segment of DNA that is rich in G=C base pairs and often precedes a transcribed gene in the genomes of vertebrate organisms.

GC-rich area : Many DNA sequences carry long stretches of repeated G and C which often indicate a gene-rich region.

gDNA : Shorthand for "genomic DNA".

GDP : Guanosine diphosphate, See guanosine.

Gel : A lyophilic colloid that has coagulated to a rigid or jelly-like solid. It is used for the electrophoretic separation of nucleic acids or proteins, and for encapsulation. See encapsulating agent; gel electrophoresis.

Gel electrophoresis : The process of separating charged species by subjecting them to a voltage gradient. A gel (made of agarose or polyacrylamide) provides mechanical support and prevents mixing of the molecules being separated. Electrophoresis performed in a gel matrix so that molecules of similar electric charge can be separated on the basis of size.

Gel retardation : A technique that identifies a DNA fragment that has a bound protein by virtue of its decreased mobility during gel electrophoresis.

Gel shift assay (gel mobility shift assay, band shift assay) : A method by which one can determine whether a particular protein preparation contains factors which bind to a particular DNA fragment. When a radiolabeled DNA fragment is run on a gel, it shows a characteristic mobility. If it is first incubated with a cellular extract of proteins (or with purified protein), any protein-DNA complexes will migrate slower than the naked DNA - a shifted band.

Gelatin : A glutinous, proteinaceous gelling and solidifying agent. Gelatin is produced by boiling animal connective tissues, which partially hydrolyses the collagen. Gelatin is used to gel or solidify nutrient solutions for tissue culture. cf agar.

Gelatinization : Steam cooking of milled grain, a process that increases the surface area of the starch and converts the original mash to a material with a gel-like consistency.

GelriteT : M The brand name of a Pseudomonas-derived refined polysaccharide used as a gelling agent and agar substitute.

GEM : Genetically engineered micro-organism.

Geminivirus : One of the two groups of DNA viruses that infect plants, the members of which have potential as cloning vectors for some species of higher plants.

Gene (Gr gen, race, offspring) : Conceptually, the unit of heredity transmitted from generation to generation during sexual or asexual reproduction. More generally, the term is used in relation to the transmission and inheritance of particular identifiable traits. Since the molecular revolution, it is now known that a gene is a segment of nucleic acid that encodes peptide or RNA. See polygenes; dominant gene; fusion gene; gene amplification; gene expression; gene flow; gene pool; gene splicing; gene translocation; recessive gene; regulatory gene.

Gene addition : The addition of a functional copy of a gene to the genome of an organism.

Gene amplification : Any process by which specific DNA sequences are replicated disproportionately greater than their representation in the parent molecules; during development, some genes become amplified in specific tissues.

Gene bank : 1. The physical location where collections of genetic material in the form of seeds, tissues or reproductive cells of plants or animals are stored. 2. Field gene bank: A facility established for the ex situ storage and maintenance, using horticultural techniques, of individual plants. Used for species whose seeds are recalcitrant (q.v.), or for clonally propagated species of agricultural importance, such as apple varieties). 3. A collection of cloned DNA fragments from a single genome. Ideally the bank should contain cloned representatives of all the DNA sequences in the genome. 4. A population of micro-organisms, each of which carries a DNA molecule that was inserted into a cloning vector. Ideally, all of the cloned DNA molecules together represent the entire genome of another organism. Also called gene library, clone bank, bank, library. This term is sometimes also used to denote all of the vector molecules, each carrying a piece of the chromosomal DNA of an organism, prior to the insertion of these molecules into a population of host cells.

Gene chip technology : Development of cDNA microarrays from a large number of genes. Used to monitor and measure changes in gene expression for each gene represented on the chip.

Gene cloning : The process of synthesizing multiple copies of a particular DNA sequence using a bacteria cell or another organism as a host. The gene of interest is inserted into a self-replicating DNA molecule (DNA vector, often a plasmid) and the resulting recombinant DNA molecule is amplified in an appropriate host cell. Used in genetic engineering. a.k.a. molecular cloning; cloning. See DNA; host.

Gene conservation; genetic resources conservation : The conservation of species, populations, individuals or parts of individuals, by in situ or ex situ methods, to provide a diversity of genetic materials for present and future generations.

Gene conversion : An genetic event that produces abnormal segregations by non-reciprocal recombination. Gene conversion has been extensively studied in ascomycetes, where it occurs following meiosis as the result of the formation of a mismatched base pair and subseque nt repair by the mismatch repair system.

Gene dropping : Loss of alleles due to genetic drift.

Gene expression : The process by which a a gene product is produced. For genes that encode proteins, the gene must be transcribed into mRNA and then translated into protein. For genes that encode structural RNAs (rRNA, tRNA, etc), the gene must transcribed into RNA.

Gene families : Groups of closely related genes that make similar products.

Gene flow : The spread of genes from one breeding population to another (usually) related populations by migration, possibly leading to changes in allele frequency.

Gene frequency : The proportion in which alternative alleles of a gene occur in a population.

Gene fusion : A construct that joins the coding region of two open reading frames such that expression of the product results in a chimeric protein. For a more detailed description, see Operon and Gene fusions. Sometimes called a translational fusion or protein fusion.

Gene gun : A device invented at Cornell University that allows genetic material to be introduced into a new organism. The genetic material from the donor is "shot" into cells of the recipient and the material is incorporated into its DNA.

Gene imprinting : The differential expression of a single gene according to its parental origin.

Gene insertion : The incorporation of one or more copies of a gene into a chromosome.

Gene library : A random collection of DNA fragments (typically representing the entire genome of an organism) that have been inserted into a cloning vector.

Gene map : The linear arrangement of mutable sites on a chromosome as deduced from genetic recombination experiments.

Gene mapping : Determination of the relative positions of genes on a DNA molecule (chromosome or plasmid) and of the distance, in linkage units or physical units, between them.

Gene modification : Chemical change to a gene's DNA sequence. See DNA.

Gene pool : 1. The total genetic information in all the genes in a breeding population at a given time. 2. In PGR: Use is made of the concept of 1 , 2 and 3 (primary, secondary and tertiary) gene pools. In general, members of a 1 gene pool are inter-fertile; those of the 2 gene pool can cross with the 1 gene pool under special circumstances; with the 3 gene pool, extreme techniques are required to achieve crossing. See germplasm.

Gene pool, primary : The sum total of all the genetic variation in the breeding population of a species and closely related species that commonly interbreed with, or can be routinely crossed with, the species.

Gene pool, secondary : The sum total of all the genetic variation in the breeding populations of related species that can be crossed with the species using mentor pollen, embryo rescue or other unusual measures.

Gene pool, tertiary : The sum total of all the genetic variation in other organisms that cannot be crossed with the species. (With the development of genetic engineering, it is theoretically possible to transfer genes isolated from any organism (plant, animals, virus, or bacterium) into a plant. This makes the line between the secondary and tertiary gene pools somewhat fuzzy).

Gene prediction : Predictions of possible genes made by a computer program based on how well a stretch of DNA sequence matches known gene sequences.

Gene probe : A single-stranded DNA or RNA fragment used in genetic engineering to search for a particular gene or other DNA sequence. The probe has a base sequence complementary to the target sequence and will thus attach to it by basepairing. By labelling the probe, it can be identified after subsequent separation and purification.

Gene product : The biochemical material, either RNA or protein, resulting from expression of a gene. The amount of gene product is used to measure how active a gene is; abnormal amounts can be correlated with disease-causing alleles.

Gene recombination : The appearance of gene combinations in the progeny that differ from the combinations present in the parents.

Gene replacement : The incorporation of a transgene into a chromosome at its normal location by homologous recombination, thus replacing the copy of the gene originally present at the locus.

Gene sequencing : The process of elucidating the nucleotide sequence of a gene. See DNA sequencing.

Gene shears : See ribozyme.

Gene splicing : A stage in the processing of mRNA, occurring only in eukaryote cells, in which intervening sequences (introns) are removed from the primary RNA transcript (hnRNA), and the coding exons are joined together to form the mature mRNA molecule.

Gene superfamily : A large group of genes related (often poorly) by sequence homologies or by the structures of their products, and by their involvement in different aspects of the same larger process (e.g. immunoglobulin gene superfamily).

Gene testing : See: genetic testing, genetic screening.

Gene therapy : The treatment of inherited diseases by introducing into the cells of affected individuals the wild-type copies of the defective gene causing the disorder. If reproductive cells are modified, the procedure is called germ-line or heritable gene therapy. If cells other than reproductive cells are modified, the procedure is called somatic-cell or non-inheritable gene therapy.

Gene tracking : Following the inheritance of a particular gene from generation to generation.

Gene transfer : Incorporation of new DNA into and organism's cells, usually by a vector such as a modified virus. Used in gene therapy. See also: mutation, gene therapy, vector.

Gene transfer agent : A phage-like agent from Rhodobacter capsulates species which can transfer DNA from the donor host to recipient cells.

Gene translocation : The movement of a gene from one chromosomal location to another.

Gene/trait identification : The conducting of high-throughout screening of genetic databases to identify (1) valuable plant traits that can be used in conventional breeding and (2) valuable genes that can be used to improve plants through biotechnology. Screens are applied to broad categories of interest, identifying multiple leads for investigation.

Gene-for-gene hypothesis : The hypothesis that corresponding genes for resistance and virulence exist in the host and pathogen, respectively.

Genera : See genus.

General (homologous) recombination : The common cellular mechanism for genetic recombination between DNA molecules with identical or near identical (homologous) nucleotide sequences. In enteric bacteria, homologous recombination requires the RecA protein.

General resistance (syn horizontal resistance, race non-specific resistance): Resistance that is effective against all biotypes of the pathogen (see durable resistance, specific resistance, vertical resistance).

Generalized transduction : The transfer of any region of the host chromosome into a recipient cell packaged in phage particles. Each phage particle only carries one region of the host chromosome, but within the population of phage every region of the host chromosome is represe nted. Generalized transduction usually results from an error resulting in packaging of host DNA into a phage particle. Typically such transducing particles only carry chromosomal DNA without any phage DNA.

Generally Regarded as Safe (GRAS) : Designation given to foods, drugs, and other materials that have been used for a considerable period of time and have a history of not causing illness to humans, even though extensive toxicity testing has not been conducted. More recently, certain host organisms for recombinant DNA experimentation have been given this status.

Generate : To propagate or (mass) proliferate. The process is generation or regeneration.

Generation time : The amount of time required under specific exponential growth conditions for one cell to divide, forming two daughter cells.

Generative : See somatic cell; somatic embryo.

Generative nucleous : One of the two male gametes in the pollen tube of angiosperms.

Genetic : Relating to heredity; referring to heritable characteristics.

Genetic assimilation : Botany: Eventual extinction of a natural species as massive pollen flow occurs from another related species and the natural species becomes more like the related species. cf gene flow.

Genetic bottleneck : When population numbers are temporarily reduced to a level insufficient to maintain the diversity in the population.

Genetic code : The sequence of nucleotides, coded in triplets (codons) along the mRNA, that determines the sequence of amino acids in protein synthesis. The DNA sequence of a gene can be used to predict the mRNA sequence, and the genetic code can in turn be used to predict the amino acid sequence.

Genetic complementation : When two DNA molecules that are in the same cell together produce a function that neither DNA molecule can supply on its own. a.k.a. complementation.

Genetic counseling : Provides patients and their families with education and information about genetic-related conditions and helps them make informed decisions.

Genetic discrimination : Prejudice against those who have or are likely to develop an inherited disorder.

Genetic disease : A disease that has its origin in changes to the genetic material. Usually refers to diseases that are inherited in a Mendelian fashion, although non-inherited forms may also result from DNA mutation.

Genetic distance : In AnGR: A measure of the genetic similarity between any pair of populations. Such distance may be based on phenotypic traits, allele frequencies or DNA sequences. For example, genetic distance between two populations having the same allele frequencies at a particular locus, and based solely on that locus, is zero. The distance for one locus is maximum when the two populations are fixed for different alleles. When allele frequencies are estimated for many loci, the genetic distance is obtained by averaging over these loci. (Source: FAO, 1999).

Genetic distancing : In AnGR: The collection of the data on phenotypic traits, marker allele frequencies or DNA sequences for two or more populations, and estimation of the genetic distances between each pair of populations. From these distances, the best representation of the relationships among all the populations may be obtained. (Source: FAO, 1999).

Genetic diversity : The heritable variation within and among populations which is created, enhanced or maintained by evolutionary forces. See biodiversity; mutation; genetic drift; gene re-shuffling processes.

Genetic drift : Change in allele frequency from one generation to another within a population, due to the sampling of finite numbers of genes that is inevitable in all real (finite) populations. The smaller the population, the greater is the genetic drift. Sooner or later (depending on the size of the population), genetic drift results in loss of alleles from a population, and hence leads to a loss of genetic variation. Because of this, the minimization of genetic drift is an important consideration for conservation of genetic resources. See genetic diversity.

Genetic engineering : Changes in the genetic constitution of cells (apart from selective breeding) resulting from the introduction or elimination of specific genes through modern molecular biology techniques. This technology is based on the use of a vector for transferring useful genetic information from a donor organism into a cell or organism that does not possess it. See gene cloning. A broader definition of genetic engineering also includes selective breeding and other means of artificial selection.

Genetic engineering technologies : See recombinant DNA technologies.

Genetic equilibrium : Condition in a group of interbreeding organisms in which the allele frequencies remain constant over time.

Genetic fingerprinting : A technique in which an individual's DNA is analysed to reveal the pattern of repetition of particular nucleotide sequences throughout the genome. See DNA fingerprint.

Genetic gain : The change achieved by artificial selection in a specific trait. Gain is usually expressed as the change per generation or the change per year. Gain is influenced by selection intensity, parental variation, and heritability.

Genetic heterogeneity : The situation in which different mutant genes produce the same phenotype.

Genetic illness : Sickness, physical disability, or other disorder resulting from the inheritance of one or more deleterious alleles.

Genetic immunization : Delivery to a host organism of a cloned gene that encodes an antigen. After the cloned gene is expressed, it elicits an antibody response that protects the organism from infection by a virus, bacterium or other disease-causing organism.

Genetic information : Information contained in a nucleotide base sequence in chromosomal DNA or RNA.

Genetic linkage map : A chromosome map showing the relative positions of the known genes on the chromosomes of a given species.

Genetic map : The ordering of genes by the statistical determination of recombination events between them. Genes separated by greater distances are more likely to recombine. (See also linkage map).

Genetic mapping : Determining the linear order of genes and/or DNA markers along a chromosome. a.k.a. mapping.

Genetic markers : Alleles of genes, or DNA polymorphisms, used as experimental probes to keep track of an individual, a tissue, a cell, a nucleus, a chromosome, or a gene. Stated another way, any character that acts as a signpost or signal of the presence or location of a gene or heredity characteristic in an individual in a population. There are 4 chromosome changes that do occur from generation to generation, and these are known as markers: a. indels: these are insertions or deletions of the DNA at particular locations on the chromosome. An example is the YAP (Y chromosome alu polymorphism). b. SNPs: these are single nucleotide polymorphisms in which a particular nucleotide is changed (like A is changed to G). Since SNPs(snips) and indels (stable alus) are very rare, they also are known as unique event polymorphisms (UEPs). c. microsatellites: these are short sequences of nucleotides (typically 2 to 5 core base pairs, example: ATCG) which are repeated multiple times in tandem. Over time changes sometimes do occur, thus the number of repeats may increase or decrease. d. minisatellites: these are longer sequences of nucleotides (typically 9 to 80 core base pairs, example: TAAGGGCCA) which are repeated multiple times in tandem. Over time changes sometimes do occur and the number of repeats may increase or decrease.

Genetic material : See genome.

Genetic modification : The technique of removing, modifying or adding genes to a living organism via genetic engineering or other more traditional methods. Also referred to as gene splicing, recombinant DNA (rDNA) technology or genetic engineering.

Genetic mosaic : An organism in which different cells contain different genetic sequence. This can be the result of a mutation during development or fusion of embryos at an early developmental stage.

Genetic polymorphism : The existence of two or more genetically different classes within a population.

Genetic predisposition : Susceptibility to a genetic disease. May or may not result in actual development of the disease.

Genetic profile : A collection of information about a person's genes.

Genetic recombination : The process by which a fragment of DNA from one molecule (chromosome, plasmid, phage genome) is exchanged with or integrated into another molecule to produce a recombinant molecule(s).

Genetic resources conservation : See gene conservation.

Genetic screening : Testing a group of people to identify individuals at high risk of having or passing on a specific genetic disorder.

Genetic selection : The process of selecting genes, cells, clones, etc., within populations or between populations or species. Genetic selection usually results in differential success rates of different genotypes, reflecting many variables, including selection pressure and genetic variability in populations.

Genetic testing : Analyzing an individual's genetic material to determine predisposition to a particular health condition or to confirm a diagnosis of genetic disease.

Genetic transformation : The transfer of extracellular DNA among and between species by using bacterial or viral vectors.

Genetic variation : A phenotypic variance of a trait in a population attributed to genetic heterogeneity.

Genetically engineered organism (GEO) : A new variety of plant produced using traditional plant breeding techniques, supplemented by the insertion of a specific beneficial gene or genes. Also referred to as genetically modified organism (GMO).

Genetically modified organism (GMO) : A new variety of plant produced using traditional plant breeding techniques, supplemented by the insertion of a specific beneficial gene or genes. Also referred to as genetically engineered organism (GEO).

Genetics : Genetics is the basic science dealing with causes of resemblances and differences among organisms related by descent. It takes into account the effects of genes and the environment.

Genome : The complete genetic content of a cell or organism, including chromosomes, plasmids, and prophages. The total genetic information of an organism. [The Oxford English Dictionary attributes the term genome to Hans Winkler who wrote in 1920 "I propose the expression Genom for the haploid chromosome set, which, together with the pertinent protoplasm, specifies the material foundations of the species.

Genome imprinting : An inheritance pattern in which a gene will have a different effect depending on the gender of the parent from whom it is inherited. Genome imprinting is also known as genetic imprinting.

Genome project : Research and technology development effort aimed at mapping and sequencing some or all of the genome of human beings and other organisms.

Genomic blot : A type of Southern blot specifically used to analyze a mixture of DNA fragments derived from total genomic DNA. Because genomic DNA is very complicated, when it has been digested with restriction enzymes, it produces a complex set of fragments ranging from tens of bp to tens of thousands of bp. However, any specific gene will be reproducibly found on only one or a few specific fragments. A million identical cells will produce a million identical restriction fragments for any given gene, so probing a genomic Southern with a gene-specific probe will produce a pattern of perhaps one or just a few bands.

Genomic clone : A piece of DNA taken from the genome of a cell or animal, and spliced into a bacteriophage or other cloning vector. A genomic clone may contain coding regions, exons, introns, 5' flanking regions, 5' untranslated regions, 3' flanking regions, 3' untranslated regions, or it may contain none of these...it may only contain intergenic DNA (usually not a desired outcome of a cloning experiment!).

Genomic DNA library; genomic library : A collection of clones containing the genomic DNA sequences of an organism. Typically, these molecules are propagated in bacteria or phage. The library is an important tool used in the process of isolating genes. See library.

Genomic sequence : See: DNA.

Genomics : The science that identifies crop traits and accelerate plant breeding. It is one of the tools used to "mine" germplasm, finding the best combinations of characteristics that can be bred or introduced into plants for better products. Genomics allows the "mapping" of a plant's genes to understand its structure and the role it plays in the plant's function.

Genotype : The genetic makeup of an organism or set of DNA variants found at one or more loci in an individual, as characterized by its physical appearance or phenotype. Our external features-what scientists call our phenotypes-are different. We have a wide array of skin color, eye shape and color, hair texture. However our interior profile, or genotype - the organization of our genes on our chromosomes-identifies us all as Homo sapiens.

Genotype-environment interaction : Changes in rank or levels of performance among individuals when tested in different environments. (See progeny tests and clonal tests).

Gentamicin : A type of aminoglycoside antibiotic. Often used to study bacterial entry into eukaryotic cells because gentamicin cannot cross the eukaryotic cell membrane, thus intracellular bacteria are resistant to gentamicin but extracellular bacteria are usually sensitive to gentamicin.

Genus (pl genera) : A somewhat arbitrary group of closely related species, where perceived relationship is typically based on physical resemblance.

GEO : Genetically engineered organism. See GMO.

Geocarp : Fruit that ripens beneath the ground (e.g. peanut pod).

Geographic variation : The phenotypic differences among native trees growing in different portions of a species' range. If the differences are largely genetic rather than environmental, the variation is usually specified as racial, ecotypic, or clinal.

Geotropism (Gr ge, earth + tropos, turning) : A growth curvature induced by gravity. a.k.a. gravitropism.

Germ : 1. In botany, a common name for a plant embryo. 2. Colloquial: A disease-causing micro-organism.

Germ cell : Any cell in the series of cells (the germ line) that eventually produces gametes. In mammals, germ cells are found in the germinal epithelium of the ovaries and testes.

Germ cell : Sperm and egg cells and their precursors. Germ cells are haploid and have only one set of chromosomes (23 in all), while all other cells have two copies (46 in all).

Germ cell gene therapy : The repair or replacement of a defective gene within the gamete-forming tissues, resulting in a heritable change in an organism's genetic constitution.

Germ layers; primary germ layers : The layers of cells in an animal embryo at the gastrula stage, from which the various organs of the animal's body will be derived.

Germ line : A lineage of "generative" cells (= germ track) ancestral to the gametes (sperm and egg cells proper) which, during the development of an organism (animal or plant), are set aside as potential gamete-forming tissues. These ancestral cells, together with the gametes, are called germ cells, as opposed to somatic cells. Location, nature and time of formation of potential gamete-forming tissues are species specific, and may vary greatly from one species to another.

Germ line cells : Cells that produce gametes.

Germ line gene therapy : The delivery of a gene or genes to a fertilized egg or an early embryonic cell. The transferred gene(s) is present in all or some of the nuclei of the cells of the mature individual, including possibly the reproductive cells, and alters the phenotype of the individual that develops.

Germ line genetic mutation : See: mutation.

Germ theory : The theory that infectious or contagious diseases are caused by microorganisms (germs).

Germ tube : Hypha resulting from an outgrowth of the spore wall and cytoplasm after germination.

Germicide : Any chemical agent used to control or kill any pathogenic and non-pathogenic micro-organisms.

Germinal epithelium : 1. A layer of epithelial cells on the surface of the ovary that are continuous with the mesothelium. 2. The layer of epithelial cells lining the seminiferous tubules of the testis, which gives rise to spermatogonia. See spermatogenesis.

Germination (L germinare, to sprout) : 1. The initial stages in the growth of a seed to form a seedling. The embryonic shoot (plumule) and embryonic root (radicle) emerge and grow upwards and downwards respectively. Food reserves for germination come from endosperm tissue within the seed and/or from the seed leaves (cotyledons). 2. The growth of spores (fungal; algal) and pollen grains.

Germplasm : The basic genetic material for any plant, used to develop new seed varieties. Within the germplasm are the basic characteristics that make plants what they are.

Germplasm conservation : Maintenance of the genetic variability of a population. (The term is used in preference to oegermplasm preservation" to reflect the ever-changing nature of living populations).

Germplasm conservation, ex situ : Maintaining the genetic variability of a population in a different environment or geographic location than where it evolved, i.e., plantations of exotic species, breeding orchards, cold storage of seed or pollen. (See gene bank).

Gestation : The period in animals bearing live young (especially mammals) from fertilization of the egg to birth of the young (parturition).

GFP : See Green Flourescent Protein.

GH : See growth hormone.

Giant cell : Enlarged, multinucleate cell formed in roots by repeated nuclear division without cell division induced by secretions of certain sedentary plant-parasitic nematodes.

Gibberellins : Plant growth regulators involved in elongation, enhancement of flower, fruit and leaf size, germination, vernalization and other processes.

Gill : Thin, radial membrane producing basidiospores in the cap of certain basidiomycetous fungi (e.g. mushrooms).

Girdle : To circle and cut through a stem or the bark and outer few rings of wood, disrupting the phloem and xylem.

Glabrous : Smooth; without hairs.

Gland : A group of cells or a single cell in animals or plants that is specialized to secrete a specific substance. In animals, there are two types of glands: endocrine glands discharge their products directly into the blood vessels; exocrine glands secrete through a duct or network of ducts into a body cavity or onto the body surface. In plants, glands are specialized to secrete certain substances produced by the plant. The secretions may be retained within a single cell, or secreted to the outside.

Glaucus : A surface with a waxy, white coating. In most cases, this waxy covering can be rubbed off.

Global warming : The consequences of the greenhouse effect, caused by the rising concentrations of greenhouse gases. The suspicion is that global warming will disrupt weather and climatic patterns. It could lead to droughts in some areas and flooding in others, and is seen as one of the major environmental problems facing the world.

Globulins : Common proteins in blood, eggs and milk, and as a reserve protein in seeds. Globulins are insoluble in water, but soluble in salt solutions. Alpha, beta and gamma globulins can be distinguished in blood serum. Gamma globulins are important in developing immunity to diseases.

GLP (Good Laboratory Practice); GMP (Good Manufacturing Practice) : They are codes of practice designed to reduce to a minimum the chance of accidents which could affect a research project or a manufactured product. The GLP and GMP prescriptions are quite voluminous, but boil down to a few key points. The essential point of both GLP and GMP is that everything is recorded, and that only established procedures are used and by people who have been trained to use them.

Glucocorticoid : A steroid hormone that regulates gene expression in higher animals.

Glucose invertase; glucose isomerase : Enzymes that catalyse the interconversion of the two sugars, glucose and fructose. As fructose is chemically more stable than glucose, a mixture of glucose and fructose with the enzyme will end up almost entirely as fructose. This is valuable for the food industry, as fructose is substantially sweeter than glucose, and so more sweetness per gram is achieved using fructose. The usual use for glucose isomerase is to take glucose made by hydrolysis of corn starch and turn it into a mixture of mostly fructose with some glucose. The corn starch is broken down using amylases. The result is called high-fructose corn syrup (HFCS). Invertase takes sucrose and turns it into glucose and fructose.

Glucoside : A substance that, on decomposition, yields glucose and certain other compounds; some glucosides are defense compounds produced by plants, e.g. cyanogenic or phenolic glucosides.

Gluten : A mixture of two proteins, gliadin and glutenin, occurring in the endosperm of wheat grain. Their amino acid composition varies, but glutamic acid (33%) and proline (12%) predominate. The composition of wheat glutens determines the strength of the flour and whether or not it is suitable for biscuit or bread making. Sensitivity of the lining of the intestine to gluten occurs in coeliac disease, a condition that must be treated by a gluten-free diet.

Glycolysis (Gr glycos, sugar (sweet) + lysis, dissolution) : Sequence of reactions that converts glucose into pyruvate, with the concomitant production of ATP.

Glycosylation : The covalent addition of sugar moities to N or O atoms present in the side chains of certain amino acids of certain proteins, generally occuring within the Golgi apparatus during secretion of a protein. Glycosylation sites are only partially predictable by current computer searches for relevant motifs in protein sequence. Glycosylation may have profound but very unpredictable effects on the folding, stability, and antigenicity of secreted proteins. Glycosylation is a property of eukaryotic cells, and differs among different cell types (i.e., it may be very different in yeast or insect cells used for protein expression, when compared with Chinese hamster ovary (CHO) cells).

Glyphosate : The active ingredient in Roundup herbicide. Roundup is a post-emergent herbicide, and is the world's leading nonselective herbicide brand, used primarily in agricultural settings.

Glyphosate tolerant (Roundup Ready) crops : The most common herbicide-tolerant crops, which provide tolerance to glyphosate, a herbicide effective on many species of grasses, broadleaf weeds and sedges. Roundup Ready crops (cotton, corn, soybeans, canola) contain the Roundup Ready gene, which allows glyphosate to be applied to the crop to provide effective weed control without damaging the crop itself.

GM food : Food that comprises, in whole or in part, material that has been modified by the application of recombinant DNA technology.

GMO : An abbreviation for "genetically modified organism". Commonly used to refer to organisms that have been developed using recombinant DNA approaches, particularly if some foreign DNA remains in the strain.

GMP (good manufacturing practice) : See GLP/GMP.

gobar : See biogas.

Golgi apparatus : A membranous, vesicular structure which is in continuity with the endoplasmic reticulum of eukaryotic cells and generally in close proximity to the nucleus, the Golgi plays an important role in the posttranslational processing and transport of secreted proteins.

Gonad : Any of the usually paired organs in animals that produce reproductive cells (gametes). The most important gonads are the male testis, which produces spermatozoa, and the female ovary, which produces ova (egg cells). The gonads also produce hormones that control secondary sexual characteristics.

Good laboratory practice : See GLP/GMP.

Good manufacturing practice : See GLP/GMP.

Gout : A genetically inherited metabolic disorder in which there is an imbalance of uric acid in the blood which causes a build up of urates in joint areas. Symptoms include painfully inflamed joints, especially of the feet and hands, that can become chronic and result in deformity. Usually, only one joint is involved, most commonly the base of a big toe. Gout is a sex-controlled trait, usually being more severe in men.

G-proteins : Proteins with an important role in relaying signals in mammalian cells. The proteins occur on the inner surface of the plasma membrane and transmit signals from outside the membrane, via transmembrane receptors, to adenylate cyclase, which catalyses the formation of the second messenger, cyclic AMP, inside the cell. G-proteins derive their name from their ability to bind to guanine nucleotides, namely GTP and GDP; the GTP-protein complex is able to activate adenylate cyclase, whereas the GDP-protein complex is unable. G-proteins are activated when the signalling molecule (typically a hormone) binds to the transmembrane receptor.

Gradient of transfer : In an Hfr mating, the decrease in the inheritance of donor markers the farther they are from the origin of transfer.

Graft : Transfer of aerial parts of one plant (e.g. buds or twigs - the scion) into close cambial contact with the root or trunk (the rootstock) of a different plant; a method of plant propagation; the joining of cut surfaces or growing roots of two plants to form a living union.

Graft inoculation test : A test based on the use of a suspected viral carrier which is grafted to an indicator plant. If symptoms appear in the indicator plant, the viral assay is positive.

Graft transmission : Transmission of a pathogen from one host plant to another through fusion of living tissue from the diseased host with living tissue of a healthy host.

Graft union : The point at which a scion from one plant is joined to a stock from another plant.

Gram stain : Procedure used for identification of bacteria in which crystal violet stain, Gram's iodine, ethyl alcohol and safranin stain are applied in succession to cells of the bacteria.

Gram staining : A technique to distinguish between two major bacterial groups, based on stain retention by their cell walls. Bacteria are heat-fixed, then stained with crystal violet, followed by iodine solution, and then rinsed with alcohol or acetone. Gram-positive bacteria are stained bright purple, while Gram-negative bacteria are decolourized.

Gram-negative : Bacteria staining red or pink in the Gram staining procedure after treatment with Gram's stain.

Gram-negative bacteria : A group of bacteria with a cell wall composed of an outer membrane surrounding a thin peptidoglycan layer. When stained with crystal violet then treated with organic solvents, the dye is readily removed by the organic solvent.

Gram-negative organism : Any prokaryotic organism that does not retain the first stain (crystal violet) used in Gram's staining technique. It does retain the second stain (safranin O) and therefore has a pink colour when viewed under a light microscope. Retention of the stain is due to the structure of the cell wall.

Gram-positive : Bacteria staining violet or purple in the Gram staining procedure after treatment with Gram's stain.

Gram-positive bacteria : A group of bacteria with a cell wall composed of thick peptidoglycan layer and no outer membrane. When stained with crystal violet then treated with organic solvents, the dye is retained and thus the cells appear purple colored.

Gram-positive organism : Any prokaryotic organism that retains the first stain used in the Gram technique, which gives a purple-black colour when viewed under a light microscope. Retention of the stain is due to the structure of the cell wall.

Grana (L granum, a seed) : Structures within chloroplasts, seen as green granules with the light microscope and as a series of parallel lamellae with the electron microscope. Disc- or sac-like structures found in chloroplasts composed of stacked membranes and containing the chlorophyll and carotenoid pigments directly involved in photosynthesis (singular: granum).

GRAS : See generally regarded as safe.

Gratuitous inducer : A substance that can induce transcription of a gene or genes, but is not a substrate for the induced enzyme(s).

Gravid : Containing an egg or eggs; capable of depositing eggs

Gravitropism : See geotropism.

GRE : Glucocorticoid Response Element: A binding site in a promoter to which the activated glucocorticoid receptor can bind. The glucocorticoid receptor is essentially a transcription factor which is activated only in the presence of glucocorticoids. The activated receptor will bind to a GRE, and transcription of the adjacent gene will be altered. See also "Response element".

Green Flourescent Protein (GFP) : An intrinsically fluorescent protein from the jellyfish Aequorea victoria. Gene fusions with GFP are commonly used for determining protein localization by fluorescence microscopy.

Green manure : A crop plowed under while still green and growing to improve the soil; sometimes used to enhance populations of antagonistic microorganisms for biological control

Green Revolution : Name given by William Goud to the dramatic increase in crop productivity during the third quarter of the 20th century, as result of integrated advances in genetics and plant breeding, agronomy, and pest and disease control.

Greenhouse effect : Process whereby greenhouse gases released into the atmosphere cause heat to be trapped in the atmosphere rather than escaping into space. The greenhouse gases form an insulating blanket around the planet. This blanket allows light and heat from the sun through, but prevents some of the heat, which radiates back from the earth from escaping. The same effect occurs in greenhouses, hence the name. The effect is natural and vital to current life forms on earth. Without it, the planet would be frozen.

Greenhouse gases : Some gases - such as carbon dioxide, methane and water vapor - trap the sun's heat in the lower atmosphere and prevent it from escaping into space. A major source of the increased concentration of greenhouse gases in the atmosphere is the combustion of fossil fuels.

Gro-luxTM : A wide-spectrum fluorescent lamp suitable for plant growth purposes.

Growth : An increase in cell size or cell number, or both, resulting in an increase in dry weight.

Growth cabinet : A cupboard used for incubating tubes or culture vessels under controlled environmental conditions. The degree of control over temperature, light and humidity is a function of the quality of the cabinet. cf culture room; incubator.

Growth curve : See growth phases.

Growth factor : Any of various chemicals, particularly polypeptides, that have a variety of important roles in the stimulation of new cell growth and cell maintenance. They bind to the cell surface on receptors. Specific growth factors can cause new cell proliferation.

Growth hormone (GH); somatotrophin; somatotropin : A hormone, secreted by the mammalian pituitary gland, that stimulates protein synthesis and growth of the long bones in the legs and arms. It also promotes the breakdown and use of fats as an energy source, rather than glucose. Production of growth hormone is greatest during early life. Its secretion is controlled by the opposing actions of two hormones from the hypothalamus: somatocrinin (growth-hormone-releasing hormone), which promotes its release; and somatostatin (growth-hormone-inhibiting hormone), that inhibits the release of growth hormone from the anterior pituitary gland.

Growth inhibitor : Any substance inhibiting the growth of an organism. The inhibitory effect can range from mild inhibition (growth retardation) to severe inhibition or death (toxic reaction). Two plant growth regulators that may act as inhibitors are ethylene and abscisic acid. The concentration of the inhibitor, the length of exposure to it, and the relative susceptibility of the organisms exposed to the inhibitor, are all important factors which determine the extent of the inhibitory effect.

Growth phases; growth phase curve : The characteristic periods in the growth of a bacterial culture, as indicated by the shape of a graph of viable cell number versus time, namely: lag phase; logarithmic (or exponential) phase; stationary phase; death phase.

Growth rate : Increase in mass per unit of time. See growth.

Growth regulator : A synthetic or natural compound that at low concentrations elicits and controls growth responses in a manner similar to hormones.

Growth retardant : A chemical that selectively interferes with normal hormonal promotion of growth and other physiological processes, but without appreciable toxic effects.

Growth ring : Any of the rings that can be seen in a cross-section of a woody stem, such as a tree trunk. It represents the xylem formed in one year as a result of fluctuating activity of the vascular cambium.

Growth substance : Any organic substance, other than a nutrient, that is synthesized by plants and regulates growth and development. They are usually made in a particular region, such as the shoot tip, and transported to other regions, where they take effect.

GTP : Guanosine triphosphate. See guanosine.

Guanine : One of the four bases in DNA that make up the letters ATGC, guanine is the "G". The others are "A" for adenine, "C" for cytosine, and "T" for thymine. Guanine always pairs with cytosine. Adenine always pairs with thymine. These letters are used as shorthand for the sequences of fragments of DNA e.g. CCAAGTAC. These sequences are the code for genetic information.

Guanosine : A nucleoside consisting of one guanine molecule linked to a D-ribose sugar molecule. The derived nucleotides, guanosine mono-, di-, and triphosphate (GMP, GDP, and GTP, respectively), are important in various metabolic reactions.

Guard-cell : Specialized epidermal cell that occurs as a pair around a stoma and controls opening and closing of the stoma through changes in turgor. See stoma.

Guide RNA : An RNA molecule that contain sequences that function as a template during RNA editing.

Guide sequence : An RNA molecule (or a part of it) which hybridizes with eukaryotic mRNA and aids in the splicing of intron sequences. Guide sequences may be either external (EGS) or internal (IGS) to the RNA being processed and may hybridize with either intron or exon sequences close to the splice junction. See split gene; exon.

Gum : Gelatinous, sugary aggregate that is synthesized and exuded by plant tissues.

Gummosis (pl gummoses) : Pathologic condition characterized by excessive formation of gums; the products of cell degeneration.

GUS : An abbreviation for Beta-glucuronidase (encoded by the uidA gene). Commonly used for constructing operon or gene fusions in organisms with high endogenous Beta-galactosidase activity. A variety of analogs exist that make it easy to detect GUS expression on solid medium and to assay the activity of the enzyme in cells.

Guttation : Exudation of watery, sticky liquid from hydathodes, especially along leaf margins.

Gyandromorph : Organisms that have both male and female cells and therefore express both male and female characteristics.

Gymnosperm : Any plant whose ovules and the seeds into which they develop are borne unprotected, rather than enclosed in ovaries, as are those of the flowering plants (the term gymnosperm means naked seed).

Gynandromorph : An individual in which one part of the body is female and another part is male; a sex mosaic.

Gynogenesis : Female parthenogenesis; after fertilization of the ovum, the male nucleus is eliminated and the haploid individual (described as gynogenetic) so produced possesses the maternal genome only. cf androgenesis; parthenogenesis; anther culture.

Gyrase : An enzymatic activity responsible for maintaining supercoiling in DNA.

Habitat : a place with a particular kind of environment suitable for the growth of an organism.

Habituation : The phenomena that, after a number of sub-cultures, cells can grow, without the addition of specific factors, such as no longer needing exogenous growth regulators in the tissue culture medium. Such cells are autonomous.

Haemoglobin : Conjugated protein compound containing iron, located in erythrocytes of vertebrates; important in the transportation of oxygen to the cells of the body.

Haemolymph : The mixture of blood and other fluids in the body cavity of an invertebrate.

Hair : A single or multicellular, absorptive (root hair) or secretory (glandular hair) and sometimes only a superficial outgrowth (covering hair) of the epidermal cells. The term trichome is often used but includes outgrowths from tissue below the epidermal layer. Distinguishing between hairs and trichomes can be difficult. Trichomes are usually connected to the vascular system, whereas hairs lack a vascular connection.

Hairpin : A region of single-stranded DNA or RNA that can form base pairs between complementary sequences. For example the sequence

5'...GACGTGNNNNNCTGCAC...3' could form a short double-stranded region between the sequences shown while any noncomplementary intervening sequences (shown as N to indicate any nucleotides) would form a single-stranded loop. Also called a "stem and loop" structure or a cruciform structure.

Hairy root culture : A fairly recent development in plant culture, consisting of highly branched roots of a plant. A plant tissue is treated with a culture of the bacterium Agrobacterium rhizogenes, which transfers part of its own plasmid DNA to the cells of an infected plant. This alters the plant's metabolism, including alterations in hormone levels, which in turn cause the explant to grow highly branched roots from the sites of infection. The roots branch much more frequently than the usual root system of that plant, and are also covered with a mass of tiny root hairs. Their most significant feature is that they produce secondary metabolites at levels similar to those made in the original plant. Thus they can be used as replacement plants for making such compounds as food flavours or fragrances.

Hairy root disease : A disease of broad-leaved plants, where a proliferation of root-like tissue is formed from the stem. Hairy root disease is a tumorous state similar to crown-gall, and is induced by the bacterium Agrobacterium rhizogenes, containing an Ri plasmid. See Agrobacterium, crown gall.

Half-site : The DNA sequence recognized by a dimeric protein is composed of two half-sites, with each half-site bound by a monomer. Half-sites are often inverted repeats, and more rarely direct repeats.

Halophyte : A plant that can tolerate a high concentration of salt in the growing medium.

Halothane : A volatile anaesthetic.

Hand-cuffing : A process that leads to replication inhibition in plasmids containing iterons. When a sufficiently high copy number of plasmid molecules accumulate, the replication initiator protein binds to the iteron sequences on two plasmids simultaneously, thereby inactivating both initiation of DNA replication of both plasmids.

Hanging droplet technique : See microdroplet array.

Haplogroup : A collection of closely related haplotypes.

Haploid cell : A cell containing only one set, or half the usual (diploid) number, of chromosomes.

Haploid (Gr haploos, single + oides, like) : A cell or organism containing only one representative from each of the pairs of homologous chromosomes found in the normal diploid cell. Having a single complete set of chromosomes, or referring to an individual or generation containing such a single set of chromosomes per cell. Usually a gamete. See monoploid.

Haplotype : A set of closely linked alleles (genes or DNA polymorphisms) inherited as a unit. A contraction of the phrase "haploid genotype". Different combinations of polymorphisms are known as haplotypes. Collectively the results from several loci could be referred to as a haplotype. "Haplo" comes from the Greek word for "single".

Haplozygous : See hemizygous.

Haptoglobin : A serum protein, alpha globulin, that interacts with haemoglobin during recycling of the iron molecule of haemoglobin.

Hardening off : Adapting plants to outdoor conditions by gradually withholding water, lowering the temperature, increasing light intensity, or reducing the nutrient supply. The hardening-off process conditions plants for survival when transplanted outdoors. The term is also used for gradual acclimatization to *in vivo* conditions of plants grown *in vitro*, e.g., gradual decrease in humidity. cf acclimatization; free-living conditions.

Hardiness : Ability to withstand stress.

Hardy-Weinberg equilibrium : The frequencies of genotypes at a locus resulting from random mating at that locus; for two alleles, A1 and A2, with respective frequencies p and q, the Hardy-Weinberg equilibrium frequencies are p^2 A_1A_1, 2pq A_1A_2, q^2 A_2A_2. Despite the simplifying assumptions required to predict these frequencies, most loci in most populations are in Hardy-Weinberg equilibrium. Thus the Hardy-Weinberg law, which predicts these frequencies, is one of the great unifying themes of biology.

Hardy-Weinberg Law : The concept that both gene frequencies and genotype frequencies will remain constant from generation to generation in an infinitely large, interbreeding population in which mating is at random and there is no selection, migration or mutation.

Haustorium (pl haustoria) : Specialized branch of a parasite formed inside host cells to absorb nutrients.

H-chain : Immunoglobulin heavy chain.

Head : In plants, a type of inflorescence in which numerous small flowers are densely crowded on a disk or around a central stalk.

Headful packaging : The packaging (filling) of a viral capsid with viral nucleic acid until the capsid is full. Headful packaging is based upon the size of nucleic acid stuffed into the capsid rather than the recognition of specific sites on the viral genome. For an example, see the link on P22 packaging.

Heartwood : Central cylinder of nonfunctional xylem in a woody stem (see sapwood).

Heat pump : An apparatus that extracts heat from a fluid or gas that is marginally above ambient temperature. Heat pumps are commonly used to heat (or cool) greenhouses and laboratories.

Heat shock response : A global regulatory response resulting in increased or decreased expression of a number of genes in response to injury by heat, osmotic change and certain other forms of stress.

Heat therapy : See thermotherapy.

Helicases : Proteins that unwind the DNA double helix.

Helix : Any structure with a spiral shape. The Watson and Crick model of DNA is in the form of a double helix.

Helix-turn-helix : A protein structural motif characteristic of certain DNA-binding proteins.

Helminths : Parasitic worms, especially internal parasites of man and animals.

Helper cells : T cells that respond to an antigen displayed by a macrophage by stimulating B and T lymphocytes to develop into antibody-producing plasma cells and killer T cells, respectively.

Helper phage : A phage that is introduced into a host cell in to provide functions needed for replication, morphogenesis, or packaging of a mutant (defective) phage.

Helper plasmid : A plasmid that provides a function or functions to another plasmid in the same cell.

Helper T cells : See helper cells.

Helper T lymphocytes : See helper cells.

Helper virus : A virus that provides a function or functions to another virus in the same cell.

Hemicellulase : An enzyme that degrades hemicellulose to galactose; hemicellulase is available as a commercial product.

Hemicellulose (Gr. hemi, half + cellulose) : Any cellulose-like carbohydrate, but with differing chemical composition. Together with pectin and lignin, hemicelluloses form the cell wall matrix.

Hemimethylation : Double stranded DNA in which only one strand is *E. coli* the A residues on both strands are methylated by DAM methylase but, because DAM methylase lags behind DNA polymerase, immediately after DNA replication only the parental strand is methylated.

Hemiparasite : Parasitic flowering plant that contains chlorophyll when mature (photosynthetic) and obtains water with dissolved nutrients by connecting to the host xylem via the haustorium, e.g. true mistletoe (Phoradendron spp. or Viscum spp.) or witchweed (Striga spp.).

Hemizygous : Having only one copy of a particular gene. For example, in humans, males are hemizygous for genes found on the Y chromosome.

Hemoglobin : See haemoglobin.

Hemolymph : See haemolymph.

Hemophilia : An X-linked genetically inherited recessive disease in which one or more of the normal blood clotting factors is not produced. This results in prolonged bleeding from even minor cuts and injuries. Swollen joints caused by Internal bleeding is a common problem for hemophiliacs. Hemophilia most often afflicts males.

HEPA filter (high efficiency particulate air filter) : A filter capable of screening out particles larger than 0.3 µm. HEPA filters are used in laminar air flow cabinets (hoods) for sterile transfer work. See prefilter.

Herbaceous : Describing primary, soft, nonwoody tissue, as a plant or plant part; having the characteristics of an herb.

Herbicide : Any substance that is toxic to plants; usually applied to agrochemicals intended to kill specific unwanted plants, such as weeds.

Herbicide resistance : The ability of a plant to withstand herbicide. Herbicide resistance has been one of the early targets of plant genetic engineering. If a herbicide is sprayed onto a field planted with such resistant crops, then all the plants except the crop would be killed, thus providing an effective method of weed control without having to develop herbicides specific to each weed type. There is substantial concern in some quarters about the widespread use of this technology, which is essentially giving the plant kingdom the ability to evade man's most effective herbicides. The concern are that, firstly, such engineering will lead to increased use of the herbicides, at a time when it is generally accepted that the use of chemicals should be kept as low as possible, and that, secondly, there is the possibility that resistant crop plants will escape to become weeds, or that their resistant genes could be transferred to other species, including weeds.

Herbicide-tolerant crops : Crops that have been developed to survive application(s) of particular herbicides by the incorporation of certain gene(s) either through genetic engineering or traditional breeding methods. The genes allow the herbicides to be applied to the crop to provide effective weed control without damaging the crop itself.

Hereditary cancer : Cancer that occurs due to the inheritance of an altered gene within a family. See also: sporadic cancer.

Heredity : The handing down of certain traits from parents to their offspring. The process of heredity occurs through the genes.

Heritability : The proportion of observed variability which is due to heredity, the remainder being due to environmental causes. More strictly, the proportion of observed variability due to the additive effects of genes.

Heritability, broad sense : The proportion of the total phenotypic variability for which genetic differences are responsible.

Heritability, narrow sense : The proportion of the total phenotypic variability that is due to additive genetic variability.

Heritable : Passed on from parents to progeny through the chromosomes/ DNA.

Hermaphrodite : 1. Animal that has both male and female reproductive organs, or a mixture of male and female attributes. 2. A plant whose flowers contain both stamen and carpels.

Heteroalleles : Mutations that are functionally allelic but structurally non-allelic; mutations at different sites in a gene.

Heterochromatin : Compact, gene-poor regions of a genome, which are enriched in simple sequence repeats. As it can be impossible to clone, heterochromatin is often ignored when calculating the percentage of a genome that has been sequenced. Heterochromatin was originally identified as regions of the genome that stained differently to euchromatin (gene-rich regions).

Heteroduplex : A double-stranded DNA molecule or DNA-RNA hybrid, where each strand is of a different origin, and consequently containing one or more mismatched (non-complementary) base pairs. A DNA duplex is prepared by the hybridization of single-stranded DNA molecules derived from two different sources. Where the two DNAs have identical or very similar sequences, a double-stranded molecule will be established, whereas where the two DNAs differ in sequence, single-stranded regions will remain. A heterodulex will be revealed as single-strand "bushes" when DNA is observed electron microscopically. A map of homologous and non-homologous regions of the two molecules may thereby be constructed. This process is known as heteroduplex mapping.

Heteroduplex DNA : A double-stranded DNA molecule formed by annealing complementary (or partially complementary) single-stranded DNA from two different sources.

Heteroecious : Pertaining to a rust fungus requiring two unrelated host plants for completion of its life cycle (see autoecious).

Heterogametic sex : Producing unlike gametes with regard to the sex chromosomes. In mammals, the XY male is heterogametic, and the XX female is homogametic.

Heterogeneity : The production of identical or similar phenotypes by different genetic mechanisms.

Heterogeneous nuclear RNA (hnRNA) : Large RNA molecules, which are unedited mRNA transcripts, or pre-mRNAs found in the nucleous of a eukaryotic cell. See RNA.

Heteroimmune phage : A phage that is not sensitive to the repression of lytic development by another prophage in the infected cell.

Heterokaryon : Cell with two or more different nuclei as a result of cell fusion. Opposite: homokaryon.

Heterokaryosis : The process whereby cells fuse to form a multinucleate cell containing nuclei from different parents. Nuclear fusion does not take place.

Heterologous : (i) DNA regions coding for the same function but with differences in their nucleotide sequences; (ii) DNAs which originate from different species.

Heterologous probe : A DNA probe that is derived from one species and used to screen for a similar DNA sequence from another species.

Heteroplasmy : A cellular condition in which two genetically different types of organelles are present. cf homoplasmy.

Heteroploid : Term given to a cell culture when the cells comprising the culture possess nuclei containing chromosome numbers other than the diploid number.

Heteropyknosis (adj heteropyknotic) : Property of certain chromosomes, or of their parts, to remain more dense during the cell cycle and to stain more intensely than other chromosomes or parts.

Heterosis : Hybrid vigor such that an F1 hybrid falls outside the range of the parents with respect to some character or characters. Usually applied to size, rate of growth, or general thriftiness.

Heterothallism (adj heterothallic) : Condition in which sexual reproduction can occur only in the presence of genetically different mycelia (see homothallism).

Heterotroph : An organism that obtains nourishment from outside sources and must obtain its carbon from organic carbon compounds (see autotroph).

Heterotrophic : The ability of an organism to use an organic molecule as carbon source.

Heterozygosity : The presence of different alleles at one or more loci on homologous chromosomes.

Heterozygote : A diploid or partially diploid organism with different alleles of one or more particular genes.

Heterozygous : An individual containing dissimilar alleles for a given gene or locus.

Hfr (high frequency of recombination) : A cell in which the F-factor has integrated into a specific location in the chromosome, causing it to act as a high frequency donor of chromosomal genes in crosses with F- cells.

HFT lysate : A lysate from a lysogenic phage that includes a substantial proportion of transducing fragments. For example, induction of a dilysogen that carries an insertion of both wild-type phage and a defective, specialized transducing phage will often yield about 50% transducing particles.

HGH : Human growth hormone, q.v.

HGP : Human Genome Project.

HHMI : Howard Hughes Medical Institute.

High-input production environment : A production environment where all rate-limiting inputs to animal production can be managed to ensure high levels of survival, reproduction and output. Output and production risks are constrained primarily by managerial decisions. (Source: FAO, 1999).

Highly conserved sequence : DNA sequence that is very similar across several different types of organisms. See also: gene, mutation.

High-throughput sequencing : A fast method of determining the order of bases in DNA. See also: sequencing.

Histocompatibility : The degree to which tissue from one organism will be tolerated by the immune system of another organism.

Histoglobulin : See histocompatibility complex.

Histology (Gr histos, cloth, tissue + logos, discourse) : Science that deals with the microscopic structure of animal and plant tissues.

Histone : A basic protein from eukaryotes (or Archae) which binds to DNA, forming nucleosomes, and packaging the DNA into chromatin.

Histone-like proteins : Proteins from bacteria which bind to DNA, and compact the DNA.

Histones : Highly basic proteins which associate with the chromosomal DNA to package it into a compact, higher order structure.

HIV : Human immunodeficiency virus. The retrovirus that causes AIDS in humans.

HLA : Human-leukocyte-antigen system. See major histocompatibility antigens.

HLA system : The body's genetically inherited system for recognizing and rejecting foreign tissues, such as transplanted organs. The HLA system has the most genes of any other known human multiple-allele series. There are at least 30,000,000 possible HLA genotypes. HLA stands for "human leukocyte antigen".

hnRNA : See heterogeneous nuclear RNA.

hnRNA (heterogeneous nuclear RNA) : Heterogeneous nuclear RNA; refers collectively to the variety of RNAs found in the nucleus, including primary transcripts, partially processed RNAs and snRNA. The term hnRNA is often used just for the unprocessed primary transcripts, however.

Holin : A phage encoded lysis protein that forms pores ("holes") in the cytoplasmic membrane and allows release of an enzyme that degrades peptidoglycan (e.g. lysozyme), facilitating lysis of the host cell.

Holliday junction : The cruciform structure formed as an intermediate in homologous genetic recombination.

Hollow fibre : A tube of a porous material, having an internal diameter of a fraction of a millimetre, and so its ratio of surface area to internal volume is very large. This has had two types of application. Firstly, hollow fibres can be used as filters. Because they have a huge surface area, they take much longer to clog up than normal filters. Secondly, they are used in the hollow fibre bioreactor, in which cells are kept inside the hollow, porous fibres, and the culture medium is circulated outside the reactor. The fibres let nutrients in and products out (as they are in solution), but do not allow the passage of cells. Hollow fibre bioreactors are very effective for maintaining mammalian cells in culture because they have a very large surface area for the cells to grow on without needing a large reactor to hold them, and because the nutrient reaching the cells can be kept fresh. The reactor also provides an easy way of removing the product that the cells are making: such as monoclonal antibodies. Hollow fibre reactors are less use when the cells themselves have to grow, because it is hard to get at the inside of the fibre to remove surplus cells.

Holoenzyme : A complex containing all of the subunits required for a functional enzyme. Used to describe enzymes composed of many different protein subunits. For example, RNA polymerase holoenzyme is composed of six subunits: 2 alpha subunits, beta, beta', omega, and sigma.

Holometabolous : An insect that undergoes complete metamorphosis to the adult from a morphologically distinct larval stage.

Holomorph : The whole fungus in all its states (see anamorph, teleomorph).

Holoparasite : parasitic flowering plant that lacks chlorophyll (nonphotosynthetic) and must rely totally on the contents of the xylem and the phloem of the host plant, e.g. broomrape (Orobanche spp.).

Homeobox : A short stretch of nucleotides whose base sequence is virtually identical in all the genes that contain it. It has been found in many organisms from fruit flies to human beings. In the fruit fly, a homeobox appears to determine when particular groups of genes are expressed during development.

Homeotic mutation : A mutation in a homeotic gene that is responsible for controlling the activities of numerous other genes, usually during embryologic development in higher organisms.

Homoalleles : Mutations that are both functionally and structurally allelic; mutations at the same site in the same gene.

Homodimer : A protein with two identical polypeptide chains.

Homoduplex : A molecule of double-stranded nucleic acid where the two strands have exactly complementary base sequences.

Homogametic sex : Producing similar gametes with regard to the sex chromosomes. In mammals, the XY male is heterogametic, and the XX female is homogametic. See heterogametic sex.

Homogenotization : A genetic technique used to replace one copy of a gene, or other DNA sequence within a genome, with an altered copy of that sequence. The DNA is first cloned and then altered in some way, e.g., a transposon is inserted into a gene. The mutated gene copy can be used to replace the original gene by recombination *in vivo*. The incorporation of the mutated gene is usually selected, for example, by virtue of its containing a transposon-encoded antibiotic resistance. See replacement.

Homoimmune : Expression of a repressor by a lysogenic phage that prevents successful infection by another phage with the same repressor-operator recognition. For example, lambda lysogens are resistant to superinfection by another lambda phage because upon entering the lysogen, cI produced by the prophage turns off gene expression from PL and PR on the incoming phage.

Homokaryon : Cell with two or more identical nuclei as a result of fusion. Opposite: heterokaryon.

Homolog : A member of a chromosome pair in diploid organisms or a gene that has the same origin and functions in two or more species.

Homologies : Similarities in DNA or protein sequences between individuals of the same species or among different species.

Homologous : Nucleic acid molecules with the same base sequence except for minor differences in alleles; (ii) nucleic acid molecules originating from strains of the same species, thus having at least long stretches of identical DNA base sequences; (iii) gene or protein families having a recognizable common evolutionary origin.

Homology : Indicates similarity between two different nucleotide or amino acid sequences, often with potential evolutionary significance. It is probably better to use more quantitative and descriptive terms such as nucleotide "identity" or, in the case of proteins, amino acid "identity" or "relatedness" (the latter refers to the presence of amino acids residues with similar polarity/charge characteristics at the same position within a protein).

Homology of Chromosomes : Applied to whole chromosomes or parts of chromosomes which synapse or pair in meiotic prophase.

Homomultimer : See homopolymer.

Homoplasmy : A cellular condition in which all copies of an organelle are genetically identical. See heteroplasmy.

Homoplasy : Similarity of traits or genes for reasons other than coancestry (e.g., convergent evolution, parallelism, evolutionary reversals, horizontal gene transfer, gene duplications). Homoplasy violates a basic assumption of the analysis of genetic markers -- variants of similar phenotype (e.g., base pair size) are assumed to derive from a common ancestor. [See Sanderson, M., and Hufford. 1996. Homoplasy: The Recurrence of Similarity in Evolution. Academic Press, NY ISBN 618030-X].

Homopolymer : A nucleic acid strand that is composed of one kind of nucleotide, e.g., GGGGGGGG. See polymer.

Homopolymer tailing : Attachment of a sequence of identical nucleotides (e.g. TTTTT) to the end of a nucleic acid molecule. Typically single-stranded homopolymer extensions are added to the ends of a double-stranded DNA molecule.

Homothallism : A process occurring in some yeasts and other fungal strains where haploid spores formed during meiosis undergo self-diploidization due to the switching of mating type by some of the progeny of the spore.

Homozygote; (adj homozygous) (Gr homos, one and the same + zygon, yoke) : An individual that has two copies of the same allele at a particular locus in its two homologous chromosomes. See allele; genotype; heterozygote.

Homozygous : A cell or organism where the same allele is carried by each member of a pair of homologous chromosomes.

Honeydew : Sugary ooze or exudate, often from aphids, and a characteristic symptom of ergot.

Horizontal resistance (syn general resistance, race non-specific resistance) : Resistance which is effective against all biotypes of the pathogen (see durable resistance, specific resistance, vertical resistance).

Hormone (Gr hormaein, to excite) : A specific organic product, produced in one part of a plant or animal body, and transported to another part where, at low concentrations, it promotes, inhibits or quantitatively modifies a biological process.

Horseradish peroxidase : An enzyme that can be complexed to DNA for use in some non-radioactive DNA labeling procedures.

Host : An organism that contains another organism, or a cloning vector.

Host plant : Living plant attacked by or harboring a parasite or pathogen and from which the invader obtains part or all of its nourishment.

Host range : The range of plants on which an organism, particularly a parasite, feeds.

Host resistance : The result of genetic manipulation of the host which renders it less susceptible to pathogens that would or do attack the host.

Host strain (bacterial) : The bacterium used to harbor a plasmid. Typical host strains include HB101 (general purpose *E. coli* strain), DH5a (ditto), JM101 and JM109 (suitable for growing M13 phages), XL1-Blue (general-purpose, good for blue/white lacZ screening). Note that the host strain is available in a form with no plasmids (hence you can put one of your own into it), or it may have plasmids present (especially if you put them there). Hundreds, perhaps thousands, of host strains are available.

Host controlled restriction : A mechanism by which some bacteria degrade unmodified foreign DNA by a restriction endonuclease that recognizes a specific nucleotide sequence.

Host-range : The strain or species specificity of a bacteriophage, plasmid, or pathogen.

Host-specific toxin : A metabolite produced by a pathogen which has a host specificity equivalent to that of the pathogen. Such toxins are utilized for *in vitro* selection experiments to screen for tolerance or resistance to the pathogen.

Hot spot : A region that is especially susceptible to mutations. Mutations occur at such sites at a much higher frequency than most other positions in the genome.

Housekeeping genes : Those genes expressed in all cells because they provide functions needed for sustenance of all cell types.

HUGO : Human Genome Organization.

Human artificial chromosome (HAC) : A vector used to hold large DNA fragments. See also: chromosome, DNA.

Human gene therapy : Insertion of normal DNA directly into cells to correct a genetic defect.

Human genome initiative : Collective name for several projects begun in 1986 by DOE to (1) create an ordered set of DNA segments from known chromosomal locations, (2) develop new computational methods for analyzing genetic map and DNA sequence data, and (3) develop new techniques and instruments for detecting and analyzing DNA. This DOE initiative is now known as the Human Genome Program. The national effort, led by DOE and NIH, is known as the Human Genome Project.

Human genome initiative : Collective name for several projects begun in 1986 by DOE to create an ordered set of DNA segments from known chromosomal locations, develop new computational methods for analyzing genetic map and DNA sequence data, and develop new techniques and instruments for detecting and analyzing DNA. This DOE initiative is now known as the Human Genome Program. The joint national effort, led by DOE and NIH, is known as the Human Genome Project.

Human Genome Organization (HUGO) : The Human Genome Organization (HUGO) is the international organization of scientists involved in the Human Genome Project (HGP), the global initiative to map and sequence the human genome. HUGO was established in 1989 by a group of the world's leading genome scientists to promote international collaboration within the project. HUGO currently has over 1000 members representing over 50 countries. HUGO maintains three regional offices, HUGO Americas, HUGO Europe and HUGO Pacific, which carry out the administrative duties of the organization. Hugo carries out a complex coordinating role within the Human Genome Project. HUGO activities range from support of data collation for constructing genetic and physical maps of the human genome to the organization of workshops to promote the consideration of a wide range of ethical, legal, social and intellectual property issues.

Human Genome Program : This was previously known as the DOE's Human Genome Initiative and is now known as the Human Genome Program.

Human Genome Project (HGP) : The national effort, initially led by DOE and NIH, is known as the Human Genome Project. It is now an international initiative to map and sequence the human genome.

Humoral immune response : The production of antibody by B cells of the immune system in response to the presence of a foreign antigen. See antibody-mediated immune response.

Huntington disease : A disease characterized by irregular, spasmodic involuntary movements of the limbs and facial muscles, mental deterioration and death, usually within 20 years of the onset of symptoms.

Hup+ : See hydrogen-uptake positive.

Hyaline : Transparent or nearly so; translucent; often used in the sense of colorless.

Hybrid : Progeny from a cross among dissimilar genotypes. In forestry, the term is usually used for crosses between species.

Hybrid arrested translation : A method used to identify the proteins encoded by a cloned DNA sequence. A crude cellular mRNA preparation, composed of many individual types of mRNA, is hybridized with cloned DNA. Only mRNA molecules homologous to the cloned DNA will anneal to it. The rest of the mRNA molecules are put into an *in vitro* translation system and the protein products are compared with the proteins obtained by use of the whole mRNA preparation. cf hybrid released translation.

Hybrid cell : The mononucleate cell which results from the fusion of two different cells, leading to the formation of a synkaryon.

Hybrid dysgenesis : A syndrome of abnormal germ-line traits, including mutation, chromosome breakage, and sterility, which results from activity of transposable elements.

Hybrid released translation : A method used to detect the proteins encoded by cloned DNA. The cloned DNA is bound to a nitrocellulose filter and a crude preparation of mRNA is hybridized to the filter-bound DNA. Only mRNA sequences homologous to the cloned DNA will be retained on the filter. These mRNA molecules can then be removed by high temperature or by using formamide. The purified mRNA is then placed in an *in vitro* translation system and the proteins encoded by the message can be analysed by electrophoresis through a polyacrylamide gel. See hybrid arrested translation.

Hybrid seed : 1. Seed produced by crossing genetically dissimilar parents. 2. In plant breeding, used colloquially for seed produced by specific crosses of carefully selected pure lines, such that the F1 crop displays hybrid vigour. As the F1 crop is heterozygous, it does not breed true and so new seed must be purchased each season.

Hybrid selection : The process of choosing individuals possessing desired characteristics from among a hybrid population.

Hybrid vigor (syn heterosis) : Increased vigor of hybrid offspring when compared to either parent.

Hybrid vigour; heterosis : The extent to which the performance of a hybrid in one or more traits is better than the average performance of the two parental populations.

Hybrid (L hybrida) : 1. The offspring of two parents that are genetically different. A cross between two genetically unlike individuals. 2. A heteroduplex DNA or DNA-RNA molecule.

Hybridization : The reaction by which the pairing of complementary strands of nucleic acid occurs. DNA is usually double-stranded, and when the strands are separated they will re-hybridize under the appropriate conditions. Hybrids can form between DNA-DNA, DNA-RNA or RNA-

RNA. They can form between a short strand and a long strand containing a region complementary to the short one. Imperfect hybrids can also form, but the more imperfect they are, the less stable they will be (and the less likely to form). To "anneal" two strands is the same as to "hybridize" them.

Hybridization probe : A labelled nucleic acid molecule that can be used to identify complementary or homologous molecules through the formation of stable base-paired hybrids.

Hybridoma : A clone of plasmacytoma cells which secrete a monoclonal antibody; usually produced by fusion of peripheral or splenic plasma cells taken from an immunized mouse with an immortalized murine plasmacytoma cell line (fusion partner), followed by cloning and selection of appropriate antibody-producing cells.

Hydathode : Epidermal leaf structure specialized for secretion or exudation of water; leaf opening at terminus of vein.

Hydrate : A compound formed by the incorporation of water.

Hydrogen bond : A relatively weak bond formed between a hydrogen atom (which is covalently bound to a nitrogen or oxygen atom) and a nitrogen or oxygen atom with an unshared electron pair. Weak interactions between electro-negative atoms and hydrogen atoms (electro-positive) that are linked to other electro-negative atoms.

Hydrogen fluoride (HF) : A common pollutant from volcanoes and several industries such as brick factories, aluminum smelters, and phosphate fertilizer factories.

Hydrogen-uptake positive (Hup+) : A term describing a micro-organism that is capable of assimilating (or taking up) hydrogen gas.

Hydrolysis : A reaction in which a molecule of water is added at the site of cleavage of a molecule to two products.

Hydrophilicity Plot : A computer plot which examines the relative summed hyrophobicity/hydrophilicity of adjacent amino acid sidechains (usually within a moving window of about 6 amino acid residues) along the primary sequence of a polypeptide chain. Values for the contribution of sidechains of each the 20 common amino acids to hydrophobicity/hydrophilicity have been developed by Hopp & Woods, and Kyte & Doolittle, and these plots are often named after these workers. Generally, hydrophobic regions of proteins are considered likely to be in the interior of the native protein, while hydrophilic domains are likely to be exposed on the surface and thus possibly antigenic sites (epitopes). At best, these are crude predictions.

Hydrophobic interactions : Association of non-polar groups with each other when present in aqueous solutions because of their insolubility in water.

Hydroponics : The growing of plants in aerated water containing all the essential mineral nutrients, with no soil. Also called soilless gardening or cultivation.

Hydroxylamine : NH_2OH -- A mutagen that reacts with C residues, producing a modified base (N4-hydroxyC) that pairs with A instead of G, and hence resulting in GC to AT transition mutations.

Hydroxymethyl cytosine : A modification of cystine residues. This modification is typical of some phage DNA. The modified DNA is resistant to certain restriction endonucleases.

Hymenium : Continuous, spore-bearing layer of a fungus fruiting body.

Hyperparasite : Organism that is parasitic upon another parasite.

Hyperplasia (adj hyperplastic) : Abnormal increase in the number of cells, often resulting in the formation of galls or tumors.

Hyperploid : A genetic condition in which a chromosome or a segment of a chromosome is over-represented in the genotype. Opposite: hypoploid.

Hypersensitive : Extremely or excessively sensitive; often refers to an extreme reaction to a pathogen.

Hypersensitive reaction and pathogenicity (hrp) gene : Gene required for elicitation of the hypersensitive (HR) response in resistant plants and causation of disease in susceptible plants; a type of avirulence gene.

Hypersensitive response (HR) : Rapid and localized cell death at the site of infection in resistant interactions between plants and pathogens.

Hypersensitive sites : Regions in the DNA that are highly susceptible to digestion with endonucleases.

Hypertonic : A solution with an osmotic potential greater than that of living cells, leading to water loss from, shrinkage or plasmolysis of cells in a hypertonic situation. Opposite: hypotonic.

Hypertrophy (adj hypertrophic) : Abnormal increase in the size of cells in a tissue or organ, often resulting in the formation of galls or tumors.

Hypervariability : High degree of variation among individuals within local populations at a given genetic marker. Examples of hypervariable markers include minisatellites and microsatellites.

Hypervariable region : The parts of both the heavy and light chains of an antibody molecule that enable it to bind to a specific site on an antigen.

Hypervariable segment : A region of a protein that varies considerably between strains or individuals.

Hypha (pl hyphae; adj hyphal) : Single, tubular filament of a fungal thallus or mycelium; the basic structural unit of a fungus.

Hyphae : Multicellular filaments formed during the vegetative reproduction of fungi or actinomycetes.

Hyphal sheath (syn mantle) : Dense hyphal mass of ectomycorrhizal fungus enclosing the short feeder roots of plants.

Hyphomycetes : Imperfect fungi that produce conidia freely or on a sporodochium or synnema (see coelomycetes).

Hyphopodium : Stalked, thick-walled, lobed cells that stick to plant surfaces; sometimes used to describe the infection structures produced by ectotrophic hyphae of certain root-infecting fungi such as Gaeumannomyces (take-all pathogen).

Hypochlorite : Generic term for aqueous solutions of sodium hypochlorite, potassium hypochlorite or calcium hypochlorite, which are oxidizing agents and used for disinfecting surfaces and surface-sterilizing tissues, and for bleaching.

Hypocotyl (Gr hypo, under + kotyledon, a cup-shaped hollow) : Portion of an embryo or seedling below the cotyledons, which is a transitional area between stem and root.

Hypomorph : A mutation that reduces but does not completely abolish gene expression.

Hypoplastic : Reduction in plant growth or development (dwarfing, stunting) resulting from an abnormal condition associated with a disease or nutritional stress.

Hypoploid : A genetic condition in which a chromosome or segment of a chromosome is underrepresented in the genotype. Opposite: hyperploid.

Hypostatic; hypostasis : See epistasis.

Hypothalamic peptides : Peptides generated in the vertebrate forebrain and concerned with regulating the body's physiological state.

Hypothesis (Gr hypothesis, foundation) : A tentative theory or supposition provisionally adopted to explain certain facts and to guide in the investigation of other facts. Once proven by rigorous scientific investigation, it becomes a theory or a law.

Hypotonic : Osmotic potential less than that of living cells. Cells placed in a hypotonic solution display swelling and turgidity.

Hypovirulence : Reduced ability to cause disease.

I/E region : See integration-excision region.

Ichthyosis : Any of several hereditary or congenital skin conditions; skin of affected individuals has a dry, scaly appearance.

Icosahedron : A regular geometric polyhedron with 20 equilateral trangular faces and 12 corners. This is a particularly stable structure. The capsid of many phages and viruses are icosahedral.

ICSI : See intracytoplasmic sperm injection.

Identical twin : Twins produced by the division of a single zygote; both have identical genotypes. See also: fraternal twin.

Ideogram : See karyogram.

Ideotype : The ideal type or perfect specimen. A description or illustration of what the final goal of genetic improvement for a species would look like. The ideotype is developed without regard to heritabilities for the individual traits involved. It is not intended to be a practical field guide for selecting "+" trees, but as a starting point for developing a field-selection guide.

Idiotope : Epitopes present in the unique variable sequences of an immunoglobulin molecule. These may or may not coincide with the immunoglobulin's paratope.

IGS : Internal guide sequence See guide sequence.

Illegitimate recombination : An aberrant recombination event between non-homologous sequences that occurs in the absence of a known site-specific recombination system.

Illuminate : To supply or brighten with light. Illumination is an absolute requirement for tissue cultures. Fluorescent lights are commonly employed. The light intensity is dependent on the light source and the requirements of the culture.

Imaginal disc : A mass of cells in the larvae of Drosophila and other holometabolous insects that gives rise to particular adult organs, such as antennae, eyes or wings.

Imbibition (L imbibere, to drink) : 1. The absorption of liquids or vapours into the ultramicroscopic spaces or pores found in materials. 2. The initial water uptake by seeds starting germination.

Immediate early gene : A viral gene that is expressed promptly after infection.

Immobilized cells : Cells entrapped in matrices such as alginate, polyacrylamide and agarose designed for use in membrane and filter bioreactors.

Immortalization : The genetic transformation of a cell type into a cell line which can proliferate indefinitely.

Immortalizing oncogene : A gene that upon transfection enables a primary cell to grow indefinitely in culture.

Immune : Cannot be infected by a given pathogen.

Immune response : The processes, including the synthesis of antibodies, that are used by vertebrates to respond to the presence of a foreign antigen. See primary immune response; secondary immune response.

Immunity : (i) The resistance of a lysogen to superinfection by a phage with a similar regulatory mechanism (also called homoimmunity). For example, expression of the cI repressor by a lambda prophage prevents expression of genes in a secondary infection if the cI protein can bind to the regulatory region of the superinfecting phage. (ii) An adaptive antibody or cellular response against specific microbial infections.

Immunity region : The region of a lysogenic phage that encodes the major regulator of lysogeny and the cognate DNa binding sites (e.g. lambda cI protein and the leftward and rightward operator sites).

Immunization : The production of immunity in an individual by artificial means. Active immunization involves the introduction, either orally or by infection, of specially treated bacteria, viruses or their toxins so as to stimulate the production of antibodies.

Immuno-affinity chromatography : A purification technique in which an antibody is bound to a matrix and is subsequently used to bind and separate a protein from a complex mixture. cf affinity chromatography.

Immunoassay; immunodiagnostics : An assay system which detects proteins by using an antibody specific to that protein. A positive result is seen as a precipitate of an antibody-protein complex. The antibody can be linked to a radioactive atom or to an enzyme which catalyses an easily monitored reaction such as a change in colour.

Immunochemical control : Use of immune agents to combat infections.

Immunodiagnostics : See immunoassay.

Immunogen : See antigen.

Immunogenicity : The ability to elicit an immune response.

Immunoglobulin : One of a group of proteins (globulins) in the body that act as antibodies. They are produced by specialized cells (B lymphocytes) and are present in blood serum and other body fluids.

Immunological screening : The use of an antibody to detect a polypeptide encoded by a cloned gene.

Immunoprecipitation : A process whereby a particular protein of interest is isolated by the addition of a specific antibody, followed by centrifugation to pellet the resulting immune complexes. Often, staphylococcal proteins A or G, bound to sepharose or some other type of macroscopic particle, is added to the reaction mix to increase the size and ease collection of the complexes. Usually, the precipitated protein is subsequently examined by SDS-PAGE.

Immunosensor : A biosensor having an antibody as biological part.

Immunosuppression : The suppression of immune response. Immunosuppression is necessary following organ transplants in order to prevent the host rejecting the grafted organ.

Immunosuppressor : A substance, an agent or a condition that prevents or diminishes the immune response.

Immunosupressant : Chemical or activity that suppresses the natural immune responses in animals, including humans.

Immunotherapy : The use of an antibody or a fusion protein containing the antigen binding site of an antibody to cure a disease or enhance the well-being of a patient.

Immunotoxin : Protein drugs consisting of an antibody joined to a toxin molecule. Immunotoxins can be made by linking toxin and antibody molecules chemically, or by fusing the genes for the toxin and the antibody. The antibody portion of the molecule facilitates binding to a target molecule or cell, and the toxin inactivates or kills the target molecule or cell.

Impeller : An agitator that is used for mixing the contents of a bioreactor.

Imperfect fungi (syn Fungi Imperfecti, deuteromycetes) : Group of fungi without a sexual stage; also, the asexual stage of ascomycetes and basidiomycetes.

Imperfect state (syn anamorph) : The asexual form in the life cycle of a fungus, when asexual spores (such as conidia) or no spores are produced.

Imprinting : A phenomenon in which the disease phenotype depends on which parent passed on the disease gene. For instance, both Prader-Willi and Angelman syndromes are inherited when the same part of chromosome 15 is missing. When the father's complement of 15 is missing, the child has Prader-Willi, but when the mother's complement of 15 is missing, the child has Angelman syndrome.

In planta : In a plant.

***In situ* :** A Latin phrase meaning "in the original place". Commonly used to describe a process that visualizes the position of a biological molecule in a cell.

***In situ* colony; in situ plaque hybridization :** A procedure for screening colonies or plaques growing on plates or membranes for the presence of specific DNA sequences by the hybridization of nucleic acid probes to the DNA molecules present in these colonies or plaques.

***In situ* conservation :** A conservation method that attempts to preserve the integrity of genetic resources by conserving them within the evolutionary dynamic ecosystems of their original habitat or natural environment. cf ex situ conservation.

***In situ* conservation of farm animal genetic diversity :** In AnGR: All measures to maintain live animal breeding populations, including those involved in active breeding programmes in the agro-ecosystem where they either developed or are now normally found, together with husbandry activities that are undertaken to ensure the continued contribution of these resources to sustainable food and agricultural production, now and in the future. (Source: FAO, 1999).

***In situ* hybridization :** Use of a DNA or RNA probe to detect the presence of the complementary DNA sequence in cloned bacterial or cultured eukaryotic cells.

***In situ* plaque hybridization :** See in situ colony.

***In vitro* :** Literally means "in glass", e.g., test tube; now used to mean growth in any type of culture vessel; Outside a living organism.

***In vitro* embryo production (IVEP) :** The combination of ovum pickup (q.v.), *in vitro* maturation (q.v.) of ova, and *in vitro* fertilization (q.v.). A potential means of overcoming the variability between donors in number of ova collected in embryo-transfer programmes.

***In vitro* fertilization (IVF) :** A widely used technique in human and animal science, whereby the egg is fertilized with sperm outside the body. Usually, the fertilized egg is cultured outside the body for a few days (to confirm that fertilization has occurred) before re-implantation into a female.

***In vitro* maturation (IVM) :** Culture of immature ova in the laboratory, usually *in vitro* fertilization.

***In vitro* mutagenesis :** A method for mutating DNA outside of a host cell. Examples include, hydroxylamine mutagenesis of DNA packaged into phage particles or site-directed mutagenesis of plasmid DNA.

***In vitro* packaging :** Synthesis of infective phage particles from a preparation of phage capsid proteins and a concatamer of phage DNA molecules. Commonly used to package DNA cloned onto a lambda vector (separated by cos sites) into infectious lambda particles.

***In vitro* transcription; cell-free transcription :** The specific and accurate synthesis of RNA in the test tube using purified DNA preparations as a template. So-called "coupled systems" may be obtained from *E. coli* which carry out both mRNA synthesis and its translation into protein. For eukaryotes, separate cell-free systems have to be set up to demonstrate the activity of the three functionally distinct RNA polymerase complexes.

***In vitro* translation; cell-free translation :** The synthesis of proteins in the test-tube from purified mRNA molecules using cell extracts containing ribosomal sub-units, the necessary protein factors, tRNA molecules and aminoacyl tRNA synthetases, ATP, GTP, amino acids and an enzyme system for re-generating the nucleoside triphosphates. Prokaryotic translation systems are usually prepared from *E. coli* or the thermophilic bacterium, Bacillus stearothermophilus. Eukaryotic systems usually employ rabbit reticulocyte lysates or wheat germ.

***In vivo* (L for "in living") :** The natural conditions in which organisms reside. Refers to biological processes that take place within a living organism or cell under normal conditions.

Inactivated agent : A virus, bacterium or other organism that has been treated to prevent it from causing a disease.

Inbred line : A line produced by continued inbreeding. In plant breeding, a nearly homozygous line usually originating by continued self-fertilization, accompanied by selection.

Inbreeding : The mating of individuals more closely related than individuals mating at random.

Inbreeding depression : The reduction in vigor often observed in progeny from matings between close relatives. Inbreeding depression is due to the expression of recessive deleterious alleles and is usually severe in open-pollinated outcrossing species (most forest trees) that occur in fairly high population densities.

Inclusion body : Protein that is overproduced in a recombinant bacterium and forms a crystalline array inside the bacterial cell.

Incompatibility : The inability of two plasmids to stably coexist in the same cell.

Incompatibility group : A number of different types of plasmid, often related to each other, that are unable to stably coexist in the same cell.

Incomplete digest : See partial digest.

Incomplete dominance : A type of gene action in which heterozygotes have a phenotype that is distinctly different from, and intermediate to, the homozygous phenotypes. See heterozygote; phenotype.

Incomplete penetrance : When some individuals in a population have a specific genotype that causes an abnormality but are not affected.

Incubation (L incubare, to lie on) : 1. The hatching of eggs by means of heat, either natural or artificial. 2. Period between infection and appearance of symptoms induced by parasitic organisms. 3. The culture of cells and organisms.

Incubator : An apparatus in which environmental conditions (light, photoperiod, temperature, humidity, etc.) are fully controlled, and used for hatching eggs, multiplying micro-organisms, culturing plants, etc. cf culture room; growth cabinet.

Indehiscent : Describing a fruit or fruiting body that does not open to release its seeds or spores when ripe.

Indel : A region of DNA that is present on the chromosome of an organism (Insertion) but absent from closely related organisms (Deletion).

Independence : The relationship between variables when the variation of each is uninfluenced by that of others, that is, correlation of zero.

Independent assortment : The random distribution of alleles (from different loci) to the gametes that occurs when genes are located in different chromosomes or far apart on large chromosomes. The distribution of alleles at one locus is independent of other the distribution of alleles at another locus. cf linkage.

Indeterminate growth : 1. Unlimited growth potential for a definite or indefinite period. Some apical meristems can produce unrestricted numbers of lateral organs. 2. In legumes, used to describe plant architecture.

Indexing : Testing of a plant for infection, often by mechanical transmission or by grafting tissue from it to an indicator plant.

Indicator plant : Plant that reacts to a pathogen or an environmental factor with specific symptoms, used to detect or identify the pathogen or determine the effects of the environmental factor.

Indirect embryogenesis : Embryo formation on callus tissues derived from zygotic or somatic embryos, seedling plant or other tissues in culture. cf direct embryogenesis.

Indirect organogenesis : Organ formation on callus tissues derived from explants.

Indirect penetration : Penetration of plant tissues by a pathogen through natural openings (e.g. stomata) or wounds.

Induced : Produced in response to a stimulus (see constitutive).

Induced systemic resistance (ISR) : Reduced disease symptoms on a portion of a plant distant from the area where the inducing agent is active, caused by the triggering of active plant defenses against a variety of pathogens; used to describe increased resistance in plants induced by certain rhizobacteria (see systemic acquired resistance- SAR).

Inducer : A low-molecular-weight compound or a physical agent that is bound by a repressor so as to produce a complex that can no longer bind to the operator; thus, the presence of the inducer turns on the expression of the gene(s) controlled by the operator.

Inducible : A regulatory system where the genes are only expressed under appropriate conditions (e.g. when the substrate is present or in specific environmental conditions).

Inducible enzyme : An enzyme that is synthesized only in the presence of the substrate that acts as an inducer.

Inducible gene : A gene that is expressed only in the presence of a specific metabolite, the inducer.

Induction media : 1. Media used to induce the formation of organs or other structures. 2. Media causing variation or mutation in the tissues exposed to it.

Induction (L inducere, to lead in) : Act or process of causing to occur; process whereby a cell or tissue influences neighbouring cells or tissues. Turning on transcription of a specific gene or operon. Getting an organism to make a protein by exposing it to some stimulus.

Inembryonation : See artificial inembryonation.

Inert : A support structure that makes no chemical contribution, and whose only function is support. Physiologically it is a neutral or immobile unit.

Infect : Process in which an organism enters, invades, or penetrates and establishes a parasitic relationship with a host plant.

Infection : The invasion of any living organisms by disease-causing microorganisms, which proceed to establish themselves, multiply, and produce various clinical signs in their host.

Infection court : Site in or on a host plant where infection can occur.

Infection cushion : Organized mass of hyphae formed on the surface of a plant from which numerous infective hyphae develop.

Infection focus : Initial site of infection, generally with reference to a population of plants.

Infection peg (syn penetration peg) : The specialized, narrow, hyphal strand on the underside of an appressorium that penetrates host cells.

Infection period : The time required for infection to occur under conducive environmental conditions, usually hours of leaf wetness and temperature.

Infectious : Pertaining to disease, capable of spreading from plant to plant.

Infectious agent : Typically, a proliferating virus, bacterium or parasite that causes a disease in plants or animals.

Infective : Referring to an organism able to attack a host and cause infection; referring to a vector carrying or containing a pathogen and able to transfer it to a host plant.

Infest (n infestation) : To attack as a pest (used especially of insects and nematodes); to contaminate, as with microorganisms; to be present in large numbers.

Infiltrate : To force the passage of liquid into tissue pores or space, such as by applying a vacuum, then releasing it, during the disinfectation procedure.

Inflorescence (L inflorescere, to begin to bloom) : 1. The way the flowers are arranged on the stalk of a plant. 2. The flowers of a plant collectively. 3. The flowering or blooming process.

Informatics : The study of the application of computer and statistical techniques to the management of information. In genome projects, informatics includes the development of methods to search databases quickly, to analyze DNA sequence information, and to predict protein sequence and structure from DNA sequence data.

Informational suppressor : A second mutation that restores function of a primary mutatioin by altering the way the gene is expressed. Examples of this class of suppressors include mutations that alter a tRNA allowing it to misread a nonsense (stop) codon or a frameshift mutation, or mutations in ribosomal genes that affect the fidelity of translation.

Informed consent : An individual willingly agrees to participate in an activity after first being advised of the risks and benefits. See also: privacy.

Inherit : In genetics, to receive genetic material from parents through biological processes.

Inheritance : The transmission of particular characteristics and/or genes from generation to generation.

Inherited : See: inherit.

Inhibitor : Any substance or object that retards a chemical reaction; a major or modifier gene that interferes with a reaction or with the expression of another gene.

Initial : Noun: Cells in a meristem that remain permanently meristematic and form tissues of particular structure and function.

Initial inoculum (syn primary inoculum) : Inoculum, usually from an overwintering source, that initiates disease in the field, as opposed to inoculum that spreads disease during the season.

Initiation : 1. Early steps or stages of a tissue culture process (culture growth, organogenesis, embryogenesis) 2. Early stages of biosynthesis.

Initiation codon : The codon at which translation of a polypeptide chain is initiated. This is usually the first AUG triplet in the mRNA molecule from the 5' end, where the ribosome binds to the cap and begins to scan in a 3' direction. However, the surrounding sequence context is important and may lead to the first AUG being bypassed by the scanning ribosome in favor of an alternative, downstream AUG. Also called a "start codon". Occasionally other codons may serve as initiation codons, e.g. UUG.

Initiation codon; initiator : The codon AUG which specifies the first amino acid methionine of a polypeptide chain. In bacteria, the initiation codon is either AUG, which is translated as n-formyl methionine or, rarely, GUG (valine). In eukaryotes, the initiation codon is always AUG and is translated as methionine. The term is also used to describe the corresponding sequence in DNA, namely ATG. See start codon; initiator codon.

Initiation factors : Proteins that promote the binding of ribosomes and the initiator tRNA to mRNA to begin the process of translation.

Initiator tRNA : A special tRNA responsible for annealing to the start codon, in the P site of the ribosome, to initiate polypeptide synthesis.

Injury : Damage caused by transitory interaction with an agent such as an insect, chemical, or unfavorable environmental condition.

Inner bark : Interior, living portion of the bark; secondary phloem.

Inoculate : Deliberately introduce something into. The process is inoculation. Not the same as contamination. 1. In bacteriology, tissue culture, etc., placing inoculum into (or onto) medium to initiate a culture. 2. In immunology, to immunize.

Inoculation cabinet : Small room or cabinet for inoculation (of tissue or micro-organism cultures) operations, often with a current of sterile air to carry contaminants away from the work area.

Inoculum (pl inocula) : 1. A small piece of tissue cut from callus, or an explant from a tissue or organ, or a small amount of cell material from a suspension culture, transferred into fresh medium for continued growth of the culture. See also minimum inoculum size. 2. Microbial spores or parts (such as mycelium). 3. Vaccine.

Inoculum density : A measure of the number of propagules of a pathogenic organism per unit area or volume.

Inorganic compound : A chemical compound that generally is not derived from living processes; compounds that do not contain carbon.

Inositol (hexahydroxycyclohexane; $C_6H_6(OH)_6$) : 1. A cyclic acid that is constituent of certain cell phosphoglycerides. 2. A water-soluble nutrient frequently referred to as a "vitamin" in plant tissue culture. Also acts as a growth factor in some animals and micro-organisms.

Inositol lipid : A membrane-anchored phospholipid that transduces hormonal signals by stimulating the release of any of several chemical messengers. See phospholipid.

Insect : Member of the class Hexapoda (phylum Arthropoda) possessing three sets of limbs attached to a central body segment.

Insecticide : A substance used to control certain populations of insect pests. In agriculture, insecticides are used to control insect pests that feed on crops or carry plant disease.

Insect-resistant crops : Plants with the ability to withstand, deter or repel insects and thereby prevent them from feeding on the plant. The traits (genes) determining resistance may be selected by plant breeders through cross-pollination with other varieties of this crop or through the introduction of novel genes such as Bt genes through genetic engineering.

Insert : In a complete plasmid clone, there are two types of DNA - the "vector" sequences and the "insert". The vector sequences are those regions necessary for propagation, antibiotic resistance, and all those mundane functions necessary for useful cloning. In contrast, however, the insert is the piece of DNA in which you are really interested.

Insertion : The presence of additional bases within a sequence that are not present in wild-type sequence.

Insertion element : Generic term for DNA sequences found in bacteria capable of genome insertion. Postulated to be responsible for site-specific phage and plasmid integration.

Insertion mutations : Changes in the base sequence of a DNA molecule resulting from the random integration of DNA from another source. See DNA; mutation.

Insertion sequence : A small bacterial transposon carrying only the genetic functions involved in transposition. There are usually inverted repeats at the ends of the insertion sequence.

Insertion site; cloning site : A unique restriction site in a vector DNA molecule into which foreign DNA can be inserted. The term is also used to describe the position of integration of a transposon or insertion site element.

Insertional inactivation : A cloning strategy where insertion of a piece of DNA into a vector inactivates a gene carried by the vector. Vector molecules with an insert can be identified by testing for the phenotype of this gene.

Instability : A random-type variation or a lack of steadiness. Due to genetic instability, cell lines lose certain characteristics or functions in culture.

Instar : One of the stages of growth between molts, in the life history of certain insects.

Insulin : A peptide hormone secreted by the Langerhans islets of the pancreas, and that regulates the level of sugar in the blood.

Integrated pest management (abbr IPM) : A combination of strategies to reduce losses due to pests and pathogens based on environmental and economic considerations.

Integrating vector : A vector that is designed to integrate cloned DNA into the host cell chromosomal DNA.

Integration : The recombination process which inserts a small DNA molecule (usually by homologous recombination) into a larger one. If the molecules are circular, integration involves only a single crossing-over; if linear, then two crossings-over are required. A well known example is the integration of phage λ (lambda) DNA into the *E. coli* genome.

Integration-excision (I/E) region : The portion of bacteriophage lambda (λ) DNA that enables bacteriophage lambda (λ) DNA to be inserted into a specific site in the *E. coli* chromosome and excised from this site.

Integron : DNA fragments that serve as sites for insertion of a variety of other DNA fragments (especially antibiotic resistance genes) and facilitate their transfer into other cells.

Integument : One of the layers that enclosed the ovule, and is the precursor of the seed coat.

Intein : Selfish DNA elements located within coding regions that are translated with the interrupted protein, but then catalyze their own excision and the formation of a peptide bond between the flanking protein regions.

Intellectual property rights : Patents, copyrights, and trademarks.

Intensifying screen : A plastic sheet impregnated with a rare-earth compound, such as calcium tungstate, which absorbs ß radiation and emits light. When placed on one side of a piece of X-ray film with a radioactive sample on the other side, the intensifying screen will capture some of the ß emissions which pass through the film, blackening the X-ray film and so greatly enhancing the sensitivity of the detection. An intensifying screen is used in Southern and northern blotting procedures.

Interaction : In statistics, an effect that cannot be explained by the additive action of contributing factors; a departure from strict additivity.

Interaction suppressor : A mutation at a second site that restores an interaction that was disrupted by a primary mutation. Interaction suppression can occur due to intragenic suppressor mutations (within the same gene that restore interactions of between domains of the protein) or intergenic suppressor mutations (mutations in two different genes that restore protein-protein interactions).

Intercalary (L intercalare, to insert) : Meristematic tissue or growth not restricted to the apex of an organ, i.e., growth at nodes.

Intercalary growth : A pattern of stem elongation typical of grasses. Elongation proceeds from the lower internodes to the upper internodes through thc differentiation of meristematic tissue at the base of each internode.

Intercalating agent : A planer molecule that can insert between two adjacent base pairs in a molecule of double-stranded DNA, distorting the architecture of the double helix. Intercalating agents often cause frameshift mutations. Examples include acridine orange and ICR191.

Intercalating dye : A dye which can insert between the bases of nucleic acids (e.g. ethidium bromide and acridine dyes. Intercalating dyes may be used to stain DNA or to induce frameshift mutation.

Intercalation : The insertion of flat polycyclic molecules between nucleotides in a DNA duplex. See Intercalating agent.

Intercellular : Between or among cells.

Intercellular space : Pore space between cells, especially typical of leaf tissues.

Intercrop : To grow two or more crops simultaneously on the same area of land.

Interfascicular cambium : Cambium that arises between vascular bundles.

Interference : Crossing over at one point that alters the chance of another crossing-over nearby; detected by studying the pattern of crossings-over with three or more linked genes. Interference is positive or negative depending on whether the chance of another crossing-over nearby is reduced or increased, respectively.

Interferon : A family of small proteins that stimulate viral resistance in cells.

Intergeneric : A cross between two different genera.

Intergenic : Between two genes; e.g. intergenic DNA is the DNA found between two genes. The term is often used to mean non-functional DNA (or at least DNA with no known importance to the two genes flanking it). Alternatively, one might speak of the "intergenic distance" between two genes as the number of base pairs from the polyA site of the first gene to the cap site of the second. This usage might therefore include the promoter region of the second gene.

Intergenic regions : DNA sequences located between genes; they comprise a large percentage of a genome and have no known function.

Intergenic suppressor : Restoration of partial or complete wild-type phenotype by second mutation in a different gene than a primary mutation which resulted in a negative phenotype. Intergenic suppression can occur in many ways, for example: A nonsense mutation that inactivates a gene may be suppressed by a second mutation that produces a nonsense suppressor tRNA. A mutation in a gene encoding a protein which must interact with a second protein may disrupt the interaction resulting in a negative phenotype; a compensating mutation in the gene encoding the second protein may restore the protein-protein interactions. Alternatively, the primary mutation may make a protein very sensitive to proteolysis, so there is a not sufficient amount of the protein available to fulfill its function; a mutation that inactivates a specific protease may allow accumulation of the first function, restorying the phenotype.

Interleukin : A group of proteins that transmit signals between immune cells and are necessary for mounting normal immune responses.

Intermediate expression : The situation in which a heterozygous genotype results in a phenotype that is intermediate between those resulting from the homozygous genotypes. The mid-range baritone male voice is an example.

Internal guide sequence (IGS) : See guide sequence.

Internal transcribed spacers : Regions of a DNA sequence that separate genes for certain ribosomal RNAs (rRNAs). These genes appear to be nearly identical among a wide variety of species, but the DNA spacers between them are quite variable and may be species-specific. This makes them good potential targets for "primers" for polymerase chain reaction (PCR) amplification that is used to identify a wide variety of organisms.

Internode (adj internodal) : The portion of a stem between two successive nodes.

Interphase : The stage in the cell cycle when the cell is not dividing; the stage follows telophase of one division and extends to the beginning of prophase in the next division. DNA replication occurs during this stage.

Intersex : An organism displaying a mixture of male and female attributes.

Interspecific : Between two different species, e.g., an interspecific cross is a cross between two species.

Interveinal : Between (leaf) veins.

Intervening sequence : See intron.

Intracellular : Through or within cells.

Intracytoplasmic sperm injection (ICSI) : The injection, using micromanipulation, of a single sperm into the cytoplasm of a mature oocyte.

Intrageneric : Within a genus, such as a hybrid resulting from a cross between species within one genus.

Intragenic complementation : The ability of two mutant forms of a gene, neither of which produce an active gene product, to produce a functional or partially functional gene product. Intragenic complementation is rare, typically only observed for specific missense alleles of a multimeric protein.

Intragenic suppressor : Restoration of partial or complete wild-type phenotype by second mutation in the same gene as a primary mutation which resulted in a negative phenotype. For example, a +1 frameshift mutation located a short distance downstream from a -1 frameshift mutation can restore the reading frame of a protein and thereby allow production of a functional product. Intragenic suppression can also occur via two missense mutations in a single reading frame.

Intramolecular : Within the same molecule. For example, intramolecular transposition is the movement of a transposon from one site on a DNA molecule to a different site on the same DNA molecule.

Intraspecific : Within a species or its populations, including subspecies, such as an intraspecific cross, or variation.

Introgression : The movement of genes from one population into another through hybridization followed by backcrossing. Usually refers to movement of genes from one species to another or among sub-species that have been geographically isolated then brought back together by changes in the species ranges or planting of exotic populations. (See gene flow).

Intron : Introns are portions of genomic DNA which are transcribed (and thus present in the primary transcript) but which are later spliced out. They thus are not present in the mature mRNA. Note that although the 3' flanking region is often transcribed, it is removed by endonucleolytic cleavage and not by splicing. It is not an intron. Compare exons.

Intumescence (syn edema or oedema) : Blisters produced on leaves and other plant parts under conditions of high moisture and restricted transpiration.

Invariant : Constant, unchanging; usually referring to the portion of a molecule that is the same across species.

Invasiveness : Ability of a plant to spread beyond its introduction site and become established in new locations, where it may have a deleterious effect on organisms already existing there.

Inverse PCR : Variation of PCR that makes the amplification of DNA segments of unknown sequence that flank DNA segments of known sequence possible; in brief, total DNA is digested to completion and fragments ligated under conditions that favour circularization of fragments; pair of PCR primers, designed from known sequence known sequence, are used to prime PCR from opposite strands resulting in amplification of fragment of unknown sequence.

Inversion : A DNA rearrangement where a sequence of nucleotides is in the reverse orientation relative to the rest of the molecule.

Inverted repeats : A DNA or RNA sequence where the sequence of nucleotides along one a strand of DNA is repeated in the opposite physical direction along the other strand; inverted repeats are commonly separated by a tract of non-repeated DNA. For example:

3' CTAG . . . GTAC 5'

5' GATC . . . CATG 3'

Ion : A charged particle.

Ion channel : An integral protein within a cell membrane, through which selective ion transport occurs.

Ionic bonds : Attractions between oppositely charged chemical groups.

Ionizing radiation : The portion of the electromagnetic spectrum that results in the production of positive and negative charges (ion pairs) in molecules. X-rays and gamma rays are examples of ionizing radiation.

IPTG : An abbreviation for isopropyl-ß-D-thiogalactoside. A gratuitous inducer of the lac operon. That is, it can induce expression of the lac operon but it is not a substrate for the lac gene products.

Irradiation (L in, into + radius, ray) : 1. Exposure to any form of radiation. 2. Treatment with a ray, such as ultraviolet rays, etc. 3. Apparent enlargement of objects, due to difference in illumination. 4. In food technology: exposure of food to controlled low levels of ionizing radiation to sterilize or disinfest it, without inducing radioactivity in the target. If in sealed containers, the product is sterile and can be stored under ambient conditions. Used for continuous flow disinfestation of grain at unloading facilities.

IS element (insertion sequence) : A short (800-1400 nucleotide pairs) DNA sequence found in bacteria that is capable of transposing to a new genomic location; other DNA sequences that are bounded by IS elements may also be transposed.

Island : A large region of DNA that is present on the chromosome of an organism but absent from closely related organisms.

Islet : A smaller genetic island that only encodes one or a few gene products.

Isoalleles : Different forms of a gene that produce the same phenotype or very similar phenotypes.

Isochromosome : A metacentric chromosome produced during mitosis or meiosis when the centromere splits transversely instead of longitudinally; the arms of such chromosome are equal in length and genetically identical, however, the loci are positioned in reverse sequence in the two arms.

Isodiametric : Term commonly used to describe cells with equal diameters.

Iso-electric focusing gels : Variant of DNA gel electrophoresis, which separates macromolecules on the basis of their iso-electric point rather than their size.

Isoenzyme : An enzyme performing the same function as another enzyme but having a different set of amino acids. The two enzymes may function at different speeds.

Isoform : A member of a family of closely related proteins that have some amino acid sequences in common and some different.

Isogamy : Sexual reproduction involving the production fusion of gametes that are similar in size and structure.

Isogenic : Two strains that are genetically identical except for a single mutation.

Isogenic Lines : Two or more lines differing from each other genetically at one locus only. Distinguished from clones, homozygous lines, identical twins, etc. which are identical at all loci.

Isogenic stocks : Strains of organisms that are genetically identical; completely homozygous.

Isolate : A culture or subpopulation of a microorganism separated from its parent population and maintained in some sort of controlled circumstance; to remove from soil or host material and grow in pure culture.

Isolating mechanism : Any of the biological properties of organisms that prevent interbreeding (and therefore exchange of genetic material) between members of different species that inhabit the same geographical area.

Isolation : The separation of one group from another so that the mating between or among groups is prevented.

Isolation medium : An optimum medium suitable for explant survival, growth and development.

Isoleucine : See amino acid.

Isomerase : Any of a class of enzymes that catalyse the re-arrangement of the atoms within a molecule, thereby converting one isomer into another.

Isometric : Usually used for virus particles to describe those that are icosahedral in structure and appear approximately round.

Isotonic; iso-osmotic : Solutions having the same osmotic potential; the same molar concentration. For protoplasts to survive, the medium they are suspended in must be isotonic with them.

Isotope : One of two or more forms of an element that have the same number of protons (atomic number) but differing numbers of neutrons (mass numbers). Radioactive isotopes are commonly used to make DNA probes and metabolic tracers.

Isozyme : A variant of a particular enzyme. In general, all the isozymes of a particular enzyme have the same function and sometimes the same activity, but they differ in amino-acid sequence. With the help of isozyme analysis, based on electrophoresis techniques that can separate the different variants, related species or cultivars can be distinguished.

Iteron : Repetitive sequences in the origin region of some plasmids that are involved in the control of replication.

ITS (acronym for internal transcribed spacers) : Regions of a DNA sequence that separate genes for certain ribosomal RNAs (rRNAs). These genes appear to be nearly identical among a wide variety of species, but the DNA spacers between them are quite variable and may be species-specific. This makes them good potential targets for "primers" for polymerase chain reaction (PCR) amplification that is used to identify a wide variety of organisms.

IVEP : See *in vitro* embryo production.

IVF : See *in vitro* fertilization.

IVM : See *in vitro* maturation.

J : 1. Joining segment, q.v. 2. Joule, q.v.

J/m2; J m-2 (joules per square metre) : A unit of light measurement.

Jiffy potTM : Pots made from wood pulp and peat, commonly used for transplanting tissue-culture-derived plants into soil medium.

JIVT : See juvenile *in vitro* embryo technology.

Join point : The site where two DNA molecules are connected. Commonly used to describe the junction at the site of a DNA rearrangement such as an inversion mutation.

Joining segment (J) : A small DNA segment that links genes in order to yield a functional gene encoding an immunoglobulin.

Joule (SI symbol J) : The amount of energy needed to apply a force of 1 newton over a distance of 1 metre.

Jumping genes : See transposable elements.

Jumping library : See chromosome jumping.

Junction fragments : DNA fragments adjacent to a particular region of DNA. For example, DNA sequences adjacent to a transposon insertion or DNA sequences adjacent to a DNA fragment that can be detected with a specific probe.

Junk DNA : Regions of DNA that do not encode functional proteins or RNAs and do not function as regulatory sites. Most eukaryotic genomes that have been characterized contain substantial amounts of junk DNA but most prokaryotic genomes that have been characterized contain less than 15% junk DNA. (However, there are exceptions -- for example, the bacterium Mycobacterium leprae has substantial amounts of junk DNA.

Juvenile : An immature form that appears similar to but usually smaller than the adult and is not sexually mature (e.g. insects with gradual metamorphosis, nematodes).

Juvenile hormone : A hormone secreted by insects from a pair of endocrine glands close to the brain. It inhibits metamorphosis and maintains the larval features.

juvenile *in vitro* embryo technology (JIVT; JIVET) : A technology involving collection of immature ova from young animals, *in vitro* maturation of those ova, *in vitro* fertilization, and then transfer of the resultant embryos into recipient females. A method of achieving rapid generation turnover.

Juvenility : Early phase of development in which an organism is juvenile and incapable of sexual reproduction.

Kanamycin (Kan) : An antibiotic that inhibits protein synthesis by binding to the 30s ribosomal subunit and preventing translocation. KanR is usually due to a cytoplasmic aminoglycoside phosphotransferase that inactivates kanamycin by covalently phosphorylating it. Because kanamycin rapidly inhibits protein synthesis, when cells acquire the kanR gene phenotypic expression is required before the cells become kanamycin resistant.

Kanr : Kanamycin-resistance gene. See selectable marker.

Kappa chain : One of two classes of antibody light chains. See lambda chain.

Karyo or Caryo : Greek origin meaning kernel or nut.

Karyogamy : The fusion of nuclei or nuclear material that occurs during sexual reproduction. See fertilization.

Karyogram : A diagram representing the characteristic features of the chromosomes of a species.

Karyokinesis : The division of a cell nucleus. See meiosis; mitosis.

Karyotype : A picture of the chromosomes in a cell that is used to check for abnormalities. A karyotype is created by staining the chromosomes with dye and photographing them through a microscope. The photograph is then cut up and rearranged so that the chromosomes are lined up into corresponding pairs.

kcat : The catalytic rate constant that characterizes an enzyme-catalysed reaction. The larger the kcat value, the faster the conversion of substrate into product.

kcat/Km : The catalytic efficiency of an enzyme-catalysed reaction. The greater the value of kcat/Km, the more rapidly and efficiently the substrate is converted into product.

kd : One thousand Dalton. See Dalton.

Killer T cells : T cells that carry T-cell receptors and kill cells displaying the recognized antigens.

Kilobase (kb) : A length unit equal to 1 000 base pairs of a double-stranded nucleic acid molecule. One kilobase of double-stranded DNA has a mass of about 660 kilodalton.

Kilocalorie (kcal) : It is equal to 1000 cal. See calorie.

Kilodalton (kDa) : A unit of atomic mass equal to 1 000 daltons. See dalton.

Kilojoule (kJ) : It is equal to 1 000 J; 1 kcal = 4.184 kJ. See joule.

Kinase : A kinase is in general an enzyme that catalyzes the transfer of a phosphate group from ATP to something else. In molecular biology, it has acquired the more specific verbal usage for the transfer onto DNA of a radiolabeled phosphate group. This would be done in order to use the resultant "hot" DNA as a probe.

Kinetics : Dynamic processes involving motion.

Kinetin (Gr kinetikos, causing motion) : One of the cytokinins, a group of growth regulators that characteristically promote cell division in plants.

Kinetochore : Spindle attachment. A localized region in each chromosome to which the "spindle fiber" appears to be attached and which seems to determine movement of the chromosomes during mitosis and meiosis.

Kinetochores : Two parallel bushlike filaments situated in the centromere of a chromosome.

Kinetosome : Granular cytoplasmic structure which forms the base of a cilium or flagellum. cf basal body.

Kinin : A substance promoting cell division. In plant systems, the prefix cyto- has been added (cytokinin) to distinguish it from kinin in animal systems.

Kinteochore : A specialized structure found in the centromeric region of the chromosome that is responsible for attaching to the spindle during nuclear division.

Klenow fragment : A proteolytic fragment of DNA polymerase I that contains the DNA polymerase activity and the 3' to 5' proofreading activity but lacks the 5' to 3' exonuclease activity. This enzyme is widely used for DNA sequencing.

Klenow fragment : A product of proteolytic digestion of the DNA polymerase I from *E. coli*; it has both polymerase and 3´-exonuclease activities but not 5´-exonuclease activity.

Klinefelter syndrome : An endocrine condition caused by a an extra X-chromosome (47,XXY); characterized by the lack of normal sexual development and testosterone, leading to infertility and adjustment problems if not detected and treated early.

Km : A dissociation constant that characterizes the binding of an enzyme to a substrate. The smaller the value of Km, the tighter the binding of the enzyme to the substrate. Also called the Michaelis constant.

Knockout : An animal resulting from an embryonic stem cell in which a normal functional gene has been replaced by a non-functional form of the gene. This technique is used extensively in mice - much can be learned about the function of a gene by studying the phenotype of animals that lack the peptide product of that gene.

Knock-out experiment : A technique for deleting, mutating or otherwise inactivating a gene in a mouse. This laborious method involves transfecting a crippled gene into cultured embryonic stem cells, searching through the thousands of resulting clones for one in which the crippled gene exactly replaced the normal one (by homologous recombination), and inserting that cell back into a mouse blastocyst. The resulting mouse will be chimaeric but, if you are lucky (and if you've gotten this far, you obviously are), its germ cells will carry the deleted gene. A few rounds of careful breeding can then produce progeny in which both copies of the gene are inactivated.

Knot : Gall; localized abnormal swelling.

Koch's postulates : The procedure used to prove the pathogenicity of an organism, i.e. its role as the causal agent of a disease.

Label : A compound or atom that is either attached to or incorporated into a macromolecule and is used to detect the presence of a compound, substance, or macromolecule in a sample. a.k.a. tag.

Labeling : The incorporation of a radioactive nucleotide into a nucleic acid molecule.

Lactose : Milk sugar, a disaccharide with one unit each of glucose and galactose.

Lag phase : 1. A state of apparent inactivity preceding a response; called also a latent phase. 2. The initial growth phase, during which cell number remains relatively constant, prior to rapid growth. 3. The first of five growth phases of most batch-propagated cell suspension cultures, being the phase in which inoculated cells in fresh medium adapt to the new environment and prepare to divide. See growth phases.

Lagging strand : The strand of newly replicated DNA that is synthesised discontinuously in the opposite direction from the replication fork. DNA synthesis proceeds in the 5' to 3' direction.

Lambda : A temperate phage that infects *E. coli.* Derivatives of phage lambda are widely used as cloning vectors.

Lambda chain : One of two classes of antibody light chains. cf kappa chain.

Lambdoid : A group of phage that are related to the *E. coli* phage Lambda. Lambdoid phage share a similar gene organization and regulate the lysis-lysogeny decision in a similar way. Examples of lambdoid phage are the *E. coli* phage 434 and the Salmonella phage P22.

Lamella (L diminutive of lamina, a thin blade; pl lamellae) : A double-membrane structure, plate or vesicle that is formed by two membranes lying parallel to each other.

Lamina (L lamina, a thin plate) : Blade or expanded part of a leaf.

Laminar air-flow cabinet; laminar air-flow hood : Cabinet for inoculation of cultures. The working area is kept sterile by a continuous, non-turbulent flow of sterilized air through a HEPA filter. See HEPA filter; prefilter.

Laminarin : A storage polysaccharide of the brown algae.

Lampbrush chromosomes : Large diplotene chromosomes present in oocyte nuclei, and particularly conspicuous in amphibians. These chromosomes have extended regions called loops, which are active sites of transcription.

Land race : A population of trees of a non-native species that has undergone one or more generations of natural selection in a new environment, i.e., Norway spruce in New York State.

Larva (pl larvae) : Immature stage of certain animals (especially insects) that undergo complete metamorphosis; the wormlike or caterpillar stage of the life cycle of such insects.

Latent : Present but not manifested or visible, as a symptomless infection by a pathogen.

Latent agent : Something, usually a virus, that is present in a host organism without producing any symptoms.

Latent bud : An inactive bud not held back by rest or dormant period, but which may start growth if stimulated.

Latent infection : Infection unaccompanied by visible symptoms.

Latent period : The time between the initial infection of a cell with a virus and the production and release of new virions via lysis of the bacterial cell (contrast with eclipse period).

Latent phase : See lag phase.

Lateral bud : A bud produced at the base of a leaf petiole. See axillary bud.

Lateral meristem : A meristem giving rise to secondary plant tissues, such as the vascular and cork cambia. The term is sometimes used to refer to an axillary meristem.

Lawn : In biotechnology: a uniform and uninterrupted layer of bacterial growth, in which individual colonies cannot be observed.

Layering : Technique for vegetative propagation, in which new plants produce adventitious roots before being severed from the parent plant.

L-chain : Immunoglobulin light chain.

LCR : See ligase chain reaction.

LD50 (lethal dose50%) : The amount of a chemical required to kill 50% of the test population. The higher the LD50, the lower the presumed toxicity of the chemical.

Leader peptide : See signal sequence.

Leader sequence : Two usages; 1] N-terminal pre sequence of scretory proteins such as peptide hormones and membrane proteins. 2] The untranslated sequence at the 5'-ends of mRNA molecules.

Leading strand : The strand of newly replicated DNA that is synthesised continuously in the same direction as the replication fork. DNA synthesis proceeds in the 5' to 3' direction.

Leaf blade : The usually flattened portion of the leaf.

Leaf bud cutting : A cutting that includes a short section of stem with attached leaf.

Leaf dip : Inspection of a sample of sap from a suspected virus-infected leaf using an electron microscope.

Leaf margin : The edge of a leaf.

Leaf primordium (L primordium, a beginning) : A lateral outgrowth from the apical meristem, which will become a leaf when fully developed and expanded.

Leaf roll : Virus diseases characterized by symptomatic curling of the host's leaves.

Leaf scar : Mark left on a stem after leaf abscission.

Leaf sheath : Lower, tubular part of a grass leaf that clasps the culm.

Leaf spot : A plant disease lesion typically restricted in development in the leaf after reaching a characteristic size.

Leafhopper : Mobile insects with sucking mouthparts (order Homoptera).

Leaflet : One of the separate blades or divisions of a compound leaf.

Leafroll : Disease symptom in which the edges of the leaf roll or turn up; often a symptom of virus infection.

Leaky mutation : A nucleotide substitution that changes the amino acid sequence of a protein that results in partial loss of its activity.

Lectin : Any of a group of proteins, derived from plants, that can bind to specific oligosaccharides on the surface of cells, causing cells to clump together.

Legume : Simple, dry, dehiscent fruit developed from a simple pistil and splitting at maturity along two seams; a plant of the family Fabaceae (formerly Leguminosae).

Lenticel : A natural opening in the surface of a stem or tuber, fruit or root for gas exchange.

Leptonema (adj leptotene) : Stage in meiosis immediately preceding synapsis, in which the chromosomes appear as single, fine, threadlike structures (but they are really double because DNA replication has already taken place).

Lesion : Localized diseased area or wound.

Lethal allele; lethal gene : A mutant form of a gene that eventually results in the death of an organism if expressed in the phenotype.

Lethal mutation : See lethal allele.

Leucine zipper : A motif found in certain proteins in which Leu residues are evenly spaced through an a-helical region, such that they would end up on the same face of the helix. Dimers can form between two such proteins. The Leu zipper is important in the function of transcription factors such as Fos and Jun and related proteins.

Library : A library might be either a genomic library, or a cDNA library. In either case, the library is just a tube carrying a mixture of thousands of different clones - bacteria or l phages. Each clone carries an "insert" - the cloned DNA. A cDNA library is usually just a mixture of bacteria, where each bacteria carries a different plasmid. Inserted into the plasmids (one per plasmid) are thousands of different pieces of cDNA (each typ. 500-5000 bp) copied from some source of mRNA, for example, total liver mRNA. The basic idea is that if you have a large enough number of different liver-derived cDNAs carried in those bacteria, there is a 99% probability that a cDNA copy of any given liver mRNA exists somewhere in the tube. The real trick is to find the one you want out of that mess - a process called screening (see "Screening"). A genomic library is similar in concept to a cDNA library, but differs in three major ways - 1) the library carries pieces of genomic DNA (and so contains introns and flanking regions, as well as coding and untranslated); 2) you need bacteriophage l or cosmids, rather than plasmids, because... 3) the inserts are usually 5-15 kb long (in a l library) or 20-40 kb (in a cosmid library). Therefore, a genomic library is most commonly a tube containing a mixture of l phages. Enough different phages must be present in the library so that any given piece of DNA from the source genome has a 99% probability of being present.

Library : An unordered collection of clones (i.e., cloned DNA from a particular organism), whose relationship to each other can be established by physical mapping. Compare genomic library, arrayed library.

Life cycle : The complete sequence of events undergone by organisms of a particular species, from the fusion of gametes in one generation to the same stage in the following generation.

Ligand : A molecule that can bind to another molecule in or on cells.

Ligase (DNA ligase) : An enzyme that repairs single-stranded discontinuities in double-stranded DNA molecules in the cell. Ligase joins a 3'-OH residue of a deoxyribonucleotide to the 5'-phosphate residue of an adjacent deoxyribonucleotide. Purified DNA ligase is used in gene cloning to join DNA molecules together.

Ligase chain reaction (LCR) : A technique for determining the presence or absence of a specific nucleotide pair within a target gene.

Ligate (to) : The process of joining two or more DNA fragments.

Ligation : The process of splicing two pieces of DNA together. In practice, a pool of DNA fragments are treated with ligase (see "Ligase") in the presence of ATP, and all possible splicing products are produced, including circularized forms and end-to-end ligation of 2, 3 or more pieces. Usually, only some of these products are useful, and the investigator must have some way of selecting the desirable ones.

Lignification (L lignum, wood + facere, to make) : Impregnation of a cell wall with lignin.

Lignin (L lignum, wood) : A group of high-molecular-weight amorphous materials, comprising polymers of phenylpropanoid compounds, giving strength to certain tissues. Wood is composed of lignified xylem cells (about 15 to 30% by weight).

Lignocellulose : The combination of lignin, hemicellulose and cellulose that forms the structural framework of plant cell walls.

Line breeding : A system of breeding in which a number of genotypes, which have been progeny tested in retrospect to some character or group of characters, are composited to form a variety.

Lineage : Several individuals originating from a common descent, such as the production of a cell line from a single cell plated *in vitro*.

Linear epitope : An epitope formed by a series of amino acids which are adjacent to each other within the primary structure of the protein. Such epitopes can be successfully modelled by synthetic peptides, but comprise only a small proportion of all epitopes. The minimal epitope size is about 5 amino acid residues. Also called a sequential epitope.

Linear phase : When cell numbers constantly increases. The linear phase is located between the exponential growth and the deceleration phases. See growth phases.

Linebreeding : A scheme that attempts to maintain a high contribution of one or two ancestors through successive generations. Often used by breeders for any inbreeding less intensive than between first-degree relatives.

LINEs (long interspersed nuclear elements) : Families of long (average length = 6 500 bp), moderately repetitive (about 10,000 copies). LINEs are cDNA copies of functional genes present in the same genome; also known as processed pseudo-genes.

Linkage : The measure of proximity of two or more markers (e.g., genes, RFLP markers) on a chromosome determined by recombination events. The closer together the markers are, the lower the probability that they will be separated during DNA repair or replication processes (binary fission in prokaryotes, mitosis or meiosis in eukaryotes), and hence the greater the probability that they will be inherited together.

Linkage disequilibrium : Where alleles occur together more often than can be accounted for by chance. Indicates that the two alleles are physically close on the DNA strand. See also: Mendelian inheritance.

Linkage disruption : A genetic rearrangement or insertion that alters the coinheritance of otherwise linked genes.

Linkage equilibrium : See gametic phase equilibrium.

Linkage map : A map of the relative positions of genetic loci on a chromosome, determined on the basis of how often the loci are inherited together. Distance is measured in centimorgans (cM).

Linkage value : Recombination fraction expressing the proportion of crossovers versus parental types in a progeny. The recombination fraction can vary from zero to one half.

Linked : Two genes located close together on the same DNA molecule that are coinherited at detectable frequency during recombination. See Linkage.

Linked genes; linked markers : Genes or markers that show linkage q.v.

Linker : A synthetic, double-stranded oligonucleotide used to attach sticky ends to a blunt-ended molecule.

Lipases : Enzymes which break down lipids into their component fatty acid and "head group" moieties. The lipases used in biotechnology are almost invariably digestive lipases, meant to break down the fats in food. They can be used to break down complex fats into their components, which are then used to make other materials.

Lipid (Gr lipos, fat) : Any of a group of fats or fat-like compounds insoluble in water and soluble in fat solvents.

Lipofectin : A commercially marketed liposome suspension which is mixed with DNA or RNA to facilitate uptake of the nucleic acid by eukaryotic cells (see Transfection).

Lipopolysaccharide (LPS) : A compound containing lipid bound to a polysaccharide; often a component of microbial cells walls.

Liposome : A microscopic artificial membrane vesicle consisting of a spherical phospholipid bilayer. Liposomes can be incorporated into living cells and used to transport relatively toxic drugs into diseased cells, where they can exert their maximum effect. DNA molecules may be entrapped in, or bound to the surface of, the vesicles, and subsequent fusion of the liposome with the cell membrane will deliver the DNA into the cell. Liposomes have been used to develop an efficient transfection procedure for Streptomyces bacteria.

Liquefaction : Enzymatic digestion (often by □ -amylase) of gelatinized starch to form lower molecular weight polysaccharides.

Liquid media : Media without a solidifying agent.

Liquid membranes : Thin films made up of liquids (as opposed to solids) which are stable in another liquid (usually water). Thus the liquid must not dissolve in the water, but nevertheless must be prevented from collapsing into a lot of small droplets.

Liquid nitrogen : Nitrogen gas condensed to a liquid with a boiling point of about -196°C. Very commonly the medium in which containers of genetic material are stored. See cryopreservation.

Litmus : A pH indicator paper (range 4.5-8.3) impregnated with an extracted lichen pigment. It turns red in acidic and blue in alkaline solutions. However, the use of litmus paper as an indicator is not a precise method of pH measurement.

Live vaccine : A living, non-virulent form of a micro-organism or virus that is used to elicit an antibody response that will protect the inoculated organism against infection by a virulent form of the micro-organism or virus. Also a living, non-virulent micro-organism or virus that express a foreign antigenic protein and is used to inoculate humans or animals. The latter organisms are also called a live recombinant vaccines.

Lligase : An enzyme that functions in DNA repair.

Local lesion : Small, restricted lesion, often the characteristic reaction of differential cultivars to specific pathogens, especially in response to mechanical inoculation with a virus.

Localize : Determination of the original position (locus) of a gene or other marker on a chromosome.

Locus (pl loci) : The position on a chromosome of a gene or other chromosome marker; also, the DNA at that position. The use of locus is sometimes restricted to mean regions of DNA that are expressed. See gene expression.

Lod score : The logarithm of the odds of linkage between two loci. Calculated from pedigree data, as the log (to base 10) of the ratio of the probability of the observed pedigree assuming linkage with a specified recombination fraction ?, to the probability of the observed pedigree assuming no linkage, i.e., recombination fraction = 0.5. A lod score (also called a z value) is
□)/Prob(data | 0.5)}. A lod score
of +3.00 (which is odds of 1000:1) or greater is regarded as acceptable evidence for linkage; -2.00 (which is the log of 1:100) or less indicates that no linkage exists.

Lodge : To fall over.

Logarithmic phase(log(arithmic) or exponential growth phase) : The steepest slope of the growth curve; the phase of vigorous growth, during which cell number doubles every 20-30 minutes. See growth phases.

Long interspersed nuclear elements : See LINEs.

Long template : A DNA strand that is synthesized during the polymerase chain reaction and has a primer sequence at one end but is extended beyond the site that is complementary to the second primer at the other end.

Long terminal repeat (LTR) : A string of bases that occurs at each end of the "genome" of a retrovirus that has become integrated into the host genome. Involved in the integration process.

Long(q) and short(p) arms : The regions either side of the centromere, a compact part of a chromosome, are known as arms. As the centromere is not in the centre of the chromosome, one arm is longer than the other.

Long-day plant : Plant requiring short nights before flowering is initiated.

Long-range restriction mapping : Restriction enzymes are proteins that cut DNA at precise locations. Restriction maps depict the chromosomal positions of restriction-enzyme cutting sites. These are used as biochemical "signposts," or markers of specific areas along the chromosomes. The map will detail the positions where the DNA molecule is cut by particular restriction enzymes.

Loop bioreactors : Fermenters in which the fermenting material is cycled between a bulk tank and a smaller tank or loop of pipes. The circulation helps to mix the materials and to ensure that gas injected into the fermenter is well distributed in the liquid. The reactors are also very useful for photosynthetic fermentations, where they allow the photosynthesizing organisms to be passed along a large number of small pipes, where the light can get to them easily, rather than inside a single volume, where only the organisms near the edges get much light.

Low linolenic soybeans : These soybeans produce low linolenic oil which reduces the need for hydrogenation, lowering or eliminating trans fats from foods.

Low-input production environment : A production environment where one or more rate-limiting inputs impose continuous or variable severe pressure on livestock, resulting in low survival, reproductive rate or output. Output and production risks are exposed to major influences which may go beyond human management capacity. (Source: FAO, 1999).

LPS : Lipopolysaccharide. A major componant of the outer layer of the outer membrane of Gram-negative bacteria.

LTR : See long terminal repeat.

Lumen (pl lumina) : Central cavity of a cell or other structure.

Luteinizing hormone : A pituitary hormone which causes growth of the yellow body of the ovary and also stimulates activity of the interstitial cells of the testis.

Lux (SI symbol lx) : The unit of measurement for illuminance (i.e., the amount of illumination) impinging upon a surface. 1 lx is the illuminance impinging upon a surface of 1 m^2, each point of which is at a distance of 1 m away from a uniform point source of light of 1 cd (candela). It supersedes the foot-candle. See foot-candle; photon.

Luxury consumption : Nutrient absorption by an organism in excess of that required for optimum growth and productivity.

Lyase : Any of a class of enzymes that catalyse either the cleavage of a double bond and the addition of new groups to a substrate, or the formation of a double bond.

Lymphocyte : A general class of white blood cells that are important components of the immune system of vertebrates.

Lymphokine : Generic name for proteins that are released by lymphocytes to act on other cells involved in the immune response. The term includes interleukins and interferons.

Lymphoma : Cancer originating in the lymph nodes, spleen and other lympho-reticular sites.

Lyophilize : Rapid freezing followed by a dehydration under high vacuum. The process is lyophilization. cf freeze-drying.

Lysate : A solution containing progeny phage resulting from the lysis of a population of bacterial cells by infecting phage.

Lysis : Disruption of cells with release of the contents.

Lysogen : A bacterial cell carrying a phage genome as a repressed prophage.

Lysogenic : A type or phase of the virus life cycle during which the virus integrates into the host chromosome of the infected cell, remaining essentially dormant for some period of time. See lysogen.

Lysogenic bacteria : Bacteria harbouring temperate (non-virulent, lysogenic) bacteriophages.

Lysogenic conversion : Expression of particular genes by a prophage that confer a novel phenotype on the host (e.g. expression of a phage encoded toxin).

Lysogenic cycle : The pattern of phage infection that involves integration of the phage DNA into the host chromosome.

Lysogeny : The ability of a temperate bacteriophage to maintain itself as a quiescent prophage until induced into the lytic cycle.

Lysosome : A membrane-bound sac within the cytoplasm of animal cells that contains enzymes responsible for the digestion of material in food vacuoles, the dissolution of foreign particles entering the cell and, on the death of the cell, the breaking down of all cell structures. The digestive system of the cell.

Lysozyme : An enzyme that hydrolyzes the peptidoglycan within the cell walls of bacteria.

Lytic : A phase of the virus life cycle during which the virus replicates within the host cell, releasing a new generation of viruses when the infected cell lyses.

Lytic cycle : The development of a bacteriophage, either after infection of a host bacterium or after induction of a prophage, resulting in production and release of free progeny phage particles, and lysis of the host cell.

Lytic phage : A phage that can only enter the lytic cycle when it infects a sensitive bacterial cell.

M_1, M_2, M_3 : Symbols used to designate first, second, third, etc. generations after treatment with a mutagenic agent.

M13 : A filamentous, single-stranded DNA phage that infects *E. coli*. The M13 DNA replicates as a circular double-stranded intermediate (called a replicative form) then single-stranded DNA produced by rolling circle replication is packaged into phage particles. Derivatives of M13 are used as cloning vectors.

M13 strand : The single-stranded DNA molecule that is present in the infective form of bacteriophage M13.

mAB : See monoclonal antibody.

MacConkey plates : MacConkey medium contains pH indicators that can be used to differentiate colonies that can ferment a sugar from colonies unable to ferment a sugar. For example, on MacConkey-Lactose plates Lac+ colonies ferment lactose and the acid turns the pH indicator red colored, but Lac- colonies cannot ferment lactose so they remain white colored.

Macerate : To disintegrate tissues to obtain a cell dissociation. Cutting, soaking or enzymatic actions are commonly used.

Macroconidium (pl macroconidia) : The larger of two kinds of conidia formed by certain fungi (see microconidium).

Macrocyclic : A rust fungus that typically exhibits all five stages of the rust life cycle (see demicyclic, microcyclic).

Macromolecule : Molecule of large molecular weight, such as proteins, nucleic acids and polysaccharides.

Macronutrient : an element needed in relatively large quantities for plant growth, e.g. nitrogen (N), phosphorus (P), and potassium (K); (see micronutrient).

Macrophages : Large, white blood cells that ingest foreign substances and display on their surfaces antigens produced from the foreign substances, to be recognized by other cells of the immune system.

Macropropagation : Production of plant clones from growing parts.

Macrorestriction map : Map depicting the order of and distance between sites at which restriction enzymes cleave chromosomes.

Maintenance methylase : An enzymatic activity responsible for maintaining the patterns of methylation on each strand of a DNA molecule after replication.

Major histocompatibility antigen : A cell-surface macromolecule that allows the immune system to distinguish foreign or "non-self" from "self". A better term is histoglobulin (See histocompatibility antigen). These are the antigens that must be matched between donors and recipients during organ and tissue transplants to prevent rejection.

Major histocompatibility complex : The large cluster of genes that encode the major histocompatibility antigens in mammals.

Male sterility : Absence or non-function of pollen in plants.

Male-specific phage : Phages that only adsorb to receptor sites on the F-pilus.

Malignant : Having the properties of cancerous growth.

Malt extract : A mixture of organic compounds from malt, used as a culture medium adjunct. See organic complex; undefined.

Malting : A process of generating starch-degrading enzymes in grain by allowing it to germinate in a humid atmosphere. See brewing.

Mammary glands : The milk-producing organs of female mammals, which provide food for the young.

Mammary tumours : Tumours of the milk glands.

Mannitol ($C_6H_{14}O_6$; fw 18217) : A sugar alcohol widely distributed in plants. Mannitol is commonly used as a nutrient and osmoticum (q.v.) in suspension medium for plant protoplasts.

Mannose ($C_6H_{12}O_6$; fw 18016) : A hexose component of many polysaccharides and mannitol. Mannose is occasionally used as a carbohydrate source in plant tissue culture media.

Mantle (syn hyphal sheath) : Dense hyphal mass of ectomycorrhizal fungus enclosing the short feeder roots of plants.

Map : 1. Verb: To determine the relative positions of loci on a DNA molecule. Linkage mapping is done by estimating the recombination fraction between loci, from the genotypes of offspring of particular matings. The further apart two loci are on a chromosome, the greater will be the frequency of recombination between them up to a maximum of 50%, the situation observed when they are sufficiently far apart on a chromosome that recombinant gametes are as frequent as non-recombinant gametes, or when they are on different chromosomes. Physical mapping is usually performed by the use of in situ hybridisation of cloned DNA fragments to metaphase chromosomes. 2. Noun: A diagram showing the relative positions of, and distances between, loci.

Map distance: The standard measure of distance between loci, expressed in centiMorgans (cM). Estimated from recombination fraction via a mapping function (q.v.). For small recombination fractions, map distance equals the percentage of recombination (recombination frequency) betwen two genes. 1% recombination = 1 cM. Sometimes called a map unit.

Map unit : An arbitrary unit representing the distance between genes, usually derived from the percentage of recombination, but also defined by the time at which the gene is transferred during conjugation.

Mapping : Determining the location of a locus (gene or genetic marker) on a chromosome. See continuous map; linkage map; physical map.

Mapping function : A mathematical expression relating observed recombination fraction (q.v.) to map distance expressed in centiMorgans. Two common mapping functions are those developed by Haldane (1919; J. Genet., 8: 299-309) and Kosambi (1944; Ann. Eugen., 12: 172-175). In both functions, the relationship between recombination fraction and map distance is approximately linear for recombination fractions less than 10%; as recombination fraction increases above 10% (up to its maximum of 50%), map distance is increasingly greater than recombination fraction.

Mapping population : The group of related organisms used in constructing a genetic map.

Marfan syndrome : Autosomal dominant condition of connective tissue; affects the skeletal, ocular and cardiovascular systems.

Marker : Two typical usages: 1] Molecular weight size marker: a piece of DNA of known size, or a mixture of pieces with known size, used on electrophoresis gels to determine the size of unknown DNA's by comparison. 2] Genetic marker: An identifiable physical location on a chromosome (e.g., restriction enzyme cutting site, gene) whose inheritance can be monitored. Markers can be expressed regions of DNA (genes) or some segment of DNA with no known coding function but whose pattern of inheritance can be determined. See RFLP, restriction fragment length polymorphism.

Marker exchange : Use of recombination to switch allelic forms of a gene with a detectable phenotype. For example, to move a mutation from a plasmid or phage to the chromosome.

Marker gene : A gene of known function and known location on the chromosome. cf genetic marker.

Marker peptide : A portion of fusion protein that facilitates its identification or purification.

Marker rescue : Repair of a mutational defect by recombination. For example, when a cell is co-infected with a mutant phage that is unable to replicate and a wild-type phage, recombination between the two phage can repair the replication defect and allow the recombinant derivative of the mutant phage to reproduce.

Marker-assisted introgression : The use of DNA markers to increase the speed and efficiency of introgression (q.v.) of a new gene or genes into a population. The markers will be closely linked to the gene(s) in question.

Marker-assisted selection (MAS) : The use of DNA markers to increase the response to selection in a population. The markers will be closely linked to one or more quantitative trait loci (q.v.).

Mass selection : As practised in plant and animal breeding, the choosing of individuals for reproduction from the entire population on the basis of individual phenotypes.

Mass spectrometry : An instrument used to identify chemicals in a substance by their mass and charge.

Mass-pedigree method : A system of breeding in which a population is propagated in mass until conditions favorable for selection to occur, after which pedigree selection is practiced.

Maternal effect : An effect attributable to some aspect of performance of the mother of the individual being evaluated.

Maternal inheritance : Inheritance controlled by extrachromosomal (cytoplasmic) factors that are transmitted through the egg.

Mating : See Conjugation.

Mating system : Any number of schemes by which individuals are assorted in pairs leading to sexual reproduction. Random; assortment of pairs is by chance. Genetic assortative mating; mating together of individuals more closely related than individuals mating at random. Genetic disassortative mating; mating together of individuals less closely related than individuals mating at random. Phenotypic assortative mating; mating individuals more alike in appearance than the average. Phenotypic disassortative mating; mating of individuals less alike in appearance than individuals mating at random.

Mating types : Compatible strains, usually designated + and - or A and B, necessary for sexual reproduction in heterothallic fungi.

Matric potential : A water potential component, always of negative value, resulting from capillary, imbibitional and adsorptive forces. See pressure potential.

Maturation : The formation of gametes or spores.

Mb : See megabase.

MCS : See polylinker.

MDA : Multiple drop array. See microdroplet array.

Mean : In statistics, the arithmetic average; the sum of all measurements or values in a sample divided by the sample size.

Mean kinship (mk) : A measure of how related an individual is to the other members of a population. Generally computed as the average IC for the hypothetical progeny of the individual mated to all other members of the population (both sexes). A low average mk for a population indicates that most of the diversity carried by the founders has been retained.

Mechanical injury : Injury of a plant part by abrasion, mutilation, or wounding.

Mechanical transmission : Spread r introduction of inoculum to an infection court (wounding) by human manipulation, accompanied by physical disruption of host tissues.

Media : See culture medium; medium.

Median : In a set of measurements, the central value above and below which there are an equal number of measurements.

Medium (pl media) : 1. In plant tissue culture, a term for the liquid or solidified formulation upon which plant cells, tissues or organs develop. See culture medium. 2. In general terms, it could also means a substrate for plant growth, such as nutrient solution, soil, sand, etc., e.g., potting medium.

Medium formulation : In tissue culture, the particular formula for the culture medium. It commonly contains macro-elements and micro-elements (high and low salt), some vitamins (B vitamins, inositol), plant growth regulators (auxin, cytokinin and sometimes gibberellin), a carbohydrate source (usually sucrose or glucose) and often other substances, such as amino acids or complex growth factors. Media may be liquid or solidified with agar; the pH is adjusted (ca. 5-6) and the solution is sterilized (usually by filtration or autoclaving). Some formulations are very specific in the kind of explant or plant species that can be maintained; some are very general.

Medium-input production environment : A production environment where management of the available resources has the scope to overcome the negative effects of the environment on animal production, although it is common for one or more factors to limit output, survival or reproduction in a serious fashion. (Source: FAO, 1999).

Megabase (abbr Mb) : A length of DNA consisting of 106 base pairs (if double-stranded) or 106 bases (if single-stranded). 1 Mb = 103 kb = 106 bp.

Megabase cloning : The cloning of very large DNA fragments. See cloning.

megadalton (MDa) : One megadalton is equal to 106 daltons. See dalton.

Megaspore; macrospore : A haploid (n) spore developing into a female gametophyte in heterosporous plants.

Meiosis : The process of two consecutive cell divisions in the diploid progenitors of sex cells. Meiosis results in four rather than two daughter cells, each with a haploid set of chromosomes.

Meiotic analysis : A technique used to analyse chromosome-pairing relationships.

Meiotic drive : Any mechanism that causes alleles to be recovered unequally in the gametes of a heterozygote.

Melanin : Pigment, as typically produced by specialised epidermal cells called melanocytes.

Melting : The dissociation of a duplex nucleic acid molecule into single strands, usually by increasing temperature. See Denaturation.

Melting temperature (abbr Tm) : The temperature at which a double-stranded DNA or RNA molecule denatures into separate single strands. The Tm is characteristic of each DNA species and gives an indication of its base composition. DNAs rich in G:C base pairs are more resistant to thermal denaturation than A:T rich DNA since three hydrogen bonds are formed between G and C, but only two between A and T.

Membrane bioreactors : Bioreactors where cells grow on or behind a permeable membrane, which lets the nutrients for the cell through but retains the cells themselves. A variations on this theme is the hollow-fibre reactor.

Memory cells : Long-lived B and T cells that mediate rapid secondary immune responses to a previously encountered antigen.

Mendelian genetics : Inheritance patterns which can be explained by simple rules of dominance and recessiveness of genes.

Mendelian inheritance : One method in which genetic traits are passed from parents to offspring. Named for Gregor Mendel, who first studied and recognized the existence of genes and this method of inheritance. See also: autosomal dominant, recessive gene, sex-linked.

Mendelian population : A natural, interbreeding unit of sexually reproducing plants or animals sharing a common gene pool.

Mendelism : The theory of heredity that forms the basis of classical genetics, proposed by Gregor Mendel in 1866 and formulated in two laws (see Mendel's Laws).

Mendel's laws : Two laws summarizing Gregor Mendel's theory of inheritance. The Law of Segregation states that each hereditary characteristic is controlled by two `factors' (now called alleles), which segregate and pass into separate germ cells. The Law of Independent Assortment states that pairs of `factors' segregate independently of each other when germ cells are formed. See independent assortment; linkage.

Mericlinal : Refers to a chimera with tissue of one genotype partly surrounded by that of another genotype.

Mericloning : A propagation method using shoot tips in culture to proliferate multiple buds, which can then be separated, rooted and planted out.

Meristele : The vascular cylinder tissue in the stem. See stele.

Meristem (Gr meristos, divisible) : Undifferentiated but determined tissue, the cells of which are capable of active cell division and differentiation into specialized and permanent tissue such as shoots and roots.

Meristem tip : An explant comprising the meristem (meristematic dome) and usually one pair of leaf primordia. Also refers to explants originating from apical meristem tip or lateral or axillary meristem tip. Do not confuse the meristem tip with the term "shoot tip," which is much larger and usually has more immature leaves and stem tissue.

Meristem tip culture : Cultures derived from meristem tip explants. The use of meristem tip culture is for virus elimination or axillary shoot proliferation purposes, but less commonly for callus production.

Meristemoid : A localized group of cells in callus tissue, characterized by an accumulation of starch, RNA and protein, and giving rise to adventitious shoots or roots.

Merodiploid : A partially diploid bacterium, carrying both its own chromosome and a chromosome fragment introduced by conjugation, transformation or transduction.

Merozygote : Partial zygote produced by a process of partial genetic exchange, such as transformation in bacteria.

Mesh bioreactor : See filter bioreactor.

Mesoderm : The middle germ layer that forms in the early animal embryo and gives rise to parts such as bone and connective tissue.

Mesophile : A micro-organism able to grow in the temperature range 20 to 50°C; optimal growth often occurs at about 37°C.

Mesophyll (Gr mesos, middle + phyllon, leaf) : Leaf parenchyma tissue occurring between epidermal layers.

Messenger RNA (mRNA) : Proteins are not synthesized directly from genomic DNA. Instead, an RNA template (a precursor mRNA) is constructed from the sequence of the gene. This RNA is then processed in various ways, including splicing. Spliced single-stranded RNAs destined to become templates for protein synthesis are known as mRNAs, The term mRNA is used only for a mature transcript with polyA tail and with all introns removed, rather than the primary transcript in the nucleus. As such, an mRNA will have a 5' untranslated region, a coding region, a 3' untranslated region and (almost always) a poly(A) tail. Typically about 2% of the total cellular RNA is mRNA. See genetic code.

Metabolic cell : A cell that is not dividing.

Metabolism (ML from the Gr metobolos, to change) : In an organism or a single cell, the biochemical process by which nutritive material is built up into living matter, or aids in building living matter, or by which complex substances and food are broken down into simple substances.

Metabolite : 1. A low-molecular-weight biological compound that is usually synthesized by an enzyme. 2. A compound that is essential for a metabolic process. A substance synthesized by the organism, or taken in from the environment. Autotrophic organisms take in inorganic metabolites, such as water, CO_2, nitrates and some trace elements.

Metacentric chromosome : A chromosome with the centromere near the middle and, consequently, two arms of about equal length.

Metallothionein : A protective protein that binds heavy metals such as cadmium and lead.

Metaphase (Gr meta, after + phasis, appearance) : Stage of mitosis during which the chromosomes, or at least the kinetochores, lie in the central plane of the spindle. It is the stage following prophase and preceding anaphase.

Metastasis : The spread of cancer cells to previously unaffected organs.

Methionine : A sulphur-containing amino acid.

Methyl transferase : Enzymes that catalyze the transfer of methyl groups from one molecule to another.

Methylation : The addition of a methyl group (-CH3) to a macromolecule, such as the addition of a methyl group to specific cytosine and, occasionally, adenine residues in DNA.

Methylmalonic acidemia : A group of conditions characterized by the inability to metabolize methylmalonic acid or by a defect in the metabolism of Vitamin B12.

Michaelis constant : See Km.

Microalgal culture : Culture in bioreactors of microalgae; microalgae include seaweeds.

Microarrays : Ordered sets of DNA fragments fixed to solid surfaces. The DNA fragments may represent all the open reading frames in a genome, a particular gene family, or any other subset of genes. Microarrays can be probed with RNA or cDNA derived from an organism to determine which transcripts are expressed under certain conditions. Microarrays are sometimes called gene chips.

Microbe : A general term for a micro-organism.

Microbial : pertaining to or relating to microbes, or microorganisms.

Microbial genetics : The study of genes and gene function in bacteria, archaea, and other microorganisms. Often used in research in the fields of bioremediation, alternative energy, and disease prevention. See also: model organisms, biotechnology, bioremediation.

Microbial mats : Layered groups or communities of microbial populations.

Microbody (Gr mikros, small + body) : A cellular organelle always bound by a single membrane, frequently spherical, from 20 to 60 nm in diameter, containing a variety of enzymes.

Micro-carriers : Small particles used as a support material for cells, and particularly mammalian cells, which are too fragile to be pumped and stirred as bacterial cells are in a large-scale culture.

Microclimate : Weather conditions on a small scale, e.g. at the surface of the plant or within a crop.

Microconidium (pl microconidia) : The smaller of two kinds of conidia formed by certain fungi (see macroconidium).

Microcyclic : Describing a rust fungus that produces only teliospores and basidiospores (see demicyclic, macrocyclic).

Microdroplet array; multiple drop array (MDA); hanging droplet technique : Introduced by Kao and Konstabel (1970), this technique is used to evaluate large numbers of media modifications, employing small quantities of medium into which are placed small numbers of cells. Droplets of liquid are arranged on the lid of a Petri dish, inverted over the bottom half of the dish containing a solution with a lower osmotic pressure, and the dish is sealed. The cells or protoplasts form a monolayer at the droplet meniscus and can easily be examined.

Micro-element : An element required in very small quantities.

Micro-encapsulation : A process of enclosing a substance in very small sealed capsules from which material is released by heat, solution or other means.

Micro-environment (Gr mikros, small + OF environ, about) : The environment close enough to the surface of a living or non-living object to be influenced by it.

Microfibrils (Gr mikros, small + fibrils, diminutive of fibre) : Microfibrils are exceedingly small fibres visible only at the high magnification of the electron microscope.

Microflora : The combination of all microorganisms in a particular environment.

Microgametophytes : See anther.

Micrograft : See shoot-tip graft.

Microinjection : A method of introducing new DNA into a cell by injecting it directly into the nucleus.

Micro-isolating system : Mechanical separation of single cells or protoplasts thus allowing them to proliferate individually.

Micron; micrometre(Gr mikros, small) : A unit of distance: 10-6 m; 0.001 mm. Symbol: □m.

Micronuclei : Chromosome fragments that are not incorporated into the nucleus at cell division.

Micronutrient (Gr mikros, small + L nutrire, to nourish) : For growth media: An essential element normally required in concentrations < 0.5 millimole/ litre.

Microorganism (syn microbe) : An organism of such small size that it can only be seen as an individual organism with the aid of microscope.

Microplasts : Vesicles produced by subdivision and fragmentation of protoplasts or thin-walled cells.

Microprojectile bombardment : A procedure for modifying cells by shooting DNA-coated metal (tungsten or gold) particles into them. See biolistics.

Micropropagation : Miniaturized *in vitro* multiplication and/or regeneration of plant material under aseptic and controlled environmental conditions on specially prepared media that contain substances necessary for growth; used for three general types of tissue: excised embryos (= embryo culture); shoot-tips (= meristem culture or mericloning); and pieces of tissue that range from bits of stems to roots. Four stages of plant tissue culture have been defined by Murashige:

Stage I. Establishment of an aseptic culture.

Stage II. The multiplication of propagules.

Stage III. Preparation of propagules for successful transfer to soil (rooting and hardening).

Stage IV. Establishment in soil.

Micropyle : 1. A small opening in the surface of a plant ovule through which the pollen tube passes prior to fertilization. 2. A small pore in some animal cells or tissues.

Microsatellite : A form of VNTR (q.v.). Specifically, a segment of DNA characterized by the occurrence of a variable number of copies (from a few up to 30 or so) of a sequence of around 5 or fewer bases (called a repeat unit, q.v.). A typical microsatellite is the repeat unit AC, which occurs at approximately 100 000 different sites in a typical mammalian genome. At any one site (locus), there are usually several different "alleles," each identifiable according to the number of repeat units. These alleles can be detected by PCR (q.v.), using primers designed from the unique sequence that is located on either side of the microsatellite. When the PCR product is run on an electrophoretic gel, alleles are seen to differ in length in units equal to the size of the repeat unit, e.g., if the primers correspond to the unique sequence immediately on either side of the microsatellite and are each 20 bases long, and an individual is heterozygous for an AC microsatellite with one allele comprising 5 repeats and the other comprising 6 repeats, the heterozygote will exhibit two bands on the gel, one band being 20 + (2 × 5) + 20 = 50 bases long, and the other allele being 20 + (2 × 6) + 20 = 52 bases long. Microsatellites have been the standard DNA marker: they are easily detectable by PCR, and they tend to be evenly located throughout the genome. Thousands have been mapped in many different species.

Microsclerotium : Microscopic, dense aggregate of darkly pigmented, thick-walled hyphal cells.

Microspore : The smaller of the two kinds of meiospores produced by heterosporous plants in the course of microsporogenesis; in seed plants, microspores give rise to the pollen grain, the male gametophyte.

Microtuber : Cultured tissue capable of growing into tuberous plant.

Microtubules : A minute filament in living cells that is composed of the protein tubulin and occurs singly, in pairs, triplets or bundles. Microtubules help cells to maintain their shape; they also occur in cilia, flagella and the centrioles, and form the spindle during nuclear division.

Middle lamella (L lamella, a thin plate or scale) : Original thin membrane separating two adjacent protoplasts and remaining as a distinct cementing layer between adjacent cell walls.

Mid-parent value : In quantitative genetics, the average of the phenotypes of two mates.

Midrib : Central, thickened vein of a leaf.

Migratory : Migrating from place to place on a plant or from plant to plant when feeding (see sedentary).

Mildew : Thin coating of mycelial growth and spores on the surfaces of infected plant parts.

Minimal essential pairing sequence (MEPS) : The shortest length of DNA homology required for the initiation of homologous recombination.

Minimal medium : A defined medium that provides only the minimum number of different nutrients needed for growth of a particular bacterium.

Minimal tillage practices : Practices that allow farmers to reduce the tilling of the land in order to conserve topsoil and its nutrients.

Minimum effective cell density : The inoculum density below which the culture fails to give reproducible cell growth. The minimum density is a function of the tissue (species, explant, cell line) and the culture phase of the inoculum suspension. Minimum density decreases inversely to the aggregate size and division rate of the stock culture.

Minimum inoculum size : The critical volume of inoculum (q.v.) required to initiate culture growth, due to the diffusive loss of cell materials into the medium. The subsequent culture growth cycle is dependent on the inoculum size, which is determined by the volume of medium and size of the culture vessel.

Mini-prep : A small-scale (mini-) preparation of plasmid or phage DNA. Used to analyse DNA in a cloning vector after a cloning experiment.

Minisatellite : A form of VNTR (q.v.) in which the repeat units (q.v.) typically range from 10 to 100 bases. They are usually detected by Southern hybridization (q.v.), using a probe comprising a clone of the repeat unit. The first DNA fingerprints (q.v.) were minisatellites detected in this way. Minisatellites tend to be located at the ends of chromosomes and in regions with a high frequency of recombination.

Minitubers : Small tubers (5-15 mm in diameter) formed on shoot cultures or cuttings of tuber-forming crops, such as potato.

Mismatch : A defect in the pairing of two complementary DNA sequences where a base in one strand is different from that expected according to complementarity with the other.

Mismatch repair : A mechanism that corrects mismatched base pairs that have escaped correction by the proofreading activities of the DNA polymerases.

Mispairing : Improper alignment of two nucleic acid strands.

Missense mutation : A nucleotide mutation which results in a change in the amino acid sequence of the encoded protein (contrast with Silent Mutation).

Mist propagation : Application of fine droplets of water to leafy cuttings in the rooting stage to reduce transpiration. cf fog.

Mites : Free-living and parasitic animals belonging to the order Acarina, class Arachnida (with spiders). Mites may infest plant crops, reducing their harvest. They may also infest plant tissue culture work areas and incubation facilities in search of sugars, and so contaminate culture vessels and spread bacteria and fungi.

mitochondrial DNA (mtDNA) : A circular ring of DNA found in mitochondria. In mammals, mtDNA makes up less than 1% of the total cellular DNA, but in plants the amount is variable. It codes for ribosomal RNA and transfer RNA, but only some mitochondrial proteins (up to 30 proteins in animals), the nuclear DNA being required for encoding most of these.

Mitochondrion (Gr mitos, thread + chondrion, a grain; pl mitochondria) : A small cytoplasmic organelle that carries out aerobic respiration. Oxidative phosphorylation takes place to produce ATP.

Mitosis : The normal process of nuclear division in a eukaryote, whereby nuclear division occurs on a spindle structure without reduction in the chromosome number in the daughter nuclei.

Mixed bud : A bud containing both rudimentary leaves and flowers.

Mixoploid : Cells with variable (euploid, aneuploid) chromosome numbers. Mosaics or chimeras differ in chromosome number as a result of a variety of mitotic irregularities. 1. The transfer between bacteria of a non-conjugative plasmid by a conjugative plasmid. 2. The transfer between bacteria of chromosomal genes by a conjugative plasmid.

MLO (syn mycoplasmalike organsim) : Archaic term for phytoplasma; plant-parasitic pleomorphic mollicute (prokaryote with no cell wall) found in phloem tissue; cannot yet be grown on artificial nutrient media.

Mobile element : A sequence of DNA that is able to promote its own transposition. An insertion sequence or a transposon.

Mobilizable plasmid : A plasmid that is not self-transmissible byt can be co-transferred with a self-transmissible plasmid, either independantly or as a cointegrate.

Mobilizing functions : The genes on a plasmid that give it the ability to facilitate the transfer of either a non-conjugative or a conjugative plasmid from one bacterium to another.

Modal class; mode : In a frequency distribution, the class having the greatest frequency.

Model : A mathematical description of a biological phenomenon.

Model organisms : A laboratory animal or other organism useful for research.

Modeling : The use of statistical analysis, computer analysis, or model organisms to predict outcomes of research.

Modification (of DNA) : The specific addition of secondary residues (methyl, hydroxymethyl, glucosyl) to deoxyribonucleotides by an organism to differentiate its own DNA from other DNA or for regulatory purposes.

Modifying genes : Genes that affect the expression of a non-allelic gene or genes.

MOET : See multiple ovulation and embryo transfer.

MOI : See Multipilicity Of Infection.

Molarity : The number of moles of a substance contained in a kilogram of solution. See mole.

Mold : Any microfungus with conspicuous, profuse, or woolly superficial growth (mycelium and/or spore masses) on various substrates; especially an economically important saprobe; molds commonly grow on damp or decaying matter and on the surface of plant tissues.

Mole (symbol M) : Amount of substance that has a weight in grams numerically equal to the molecular weight of the substance. Also called gram molecular weight. A mole contains 6.023 × 1023 molecules or atoms of a substance.

Molecular biology : The study of the structure, function, and makeup of biologically important molecules.

Molecular breeding : An enhanced tool that involves the use of DNA markers for genes in combination with physical measurement of traits to accelerate selection in plant breeding programs. Also called marker-assisted breeding.

Molecular cloning : The biological amplification of a specific DNA sequence through mitotic division of a host cell into which it has been transformed or transfected. See cloning.

Molecular farming : The development of transgenic animals to produce human proteins for medical use.

Molecular genetics : The area of knowledge concerned with the genetic aspects of molecular biology, especially with DNA, RNA and protein molecules.

Molecular medicine : The treatment of injury or disease at the molecular level. Examples include the use of DNA-based diagnostic tests or medicine derived from DNA sequence information.

Molecule (L diminutive of moles, a little mass) : A unit of matter, the smallest portion of an element or a compound that retains chemical identity with the substance in mass. The molecule usually consist of a union of two or more atoms; some organic molecules containing a very large number of atoms.

Mollicute : One of a group of prokaryotic organisms bounded by flexuous membranes and lacking cell walls (see phytoplasma and spiroplasma).

Molt : To shed a cuticle or body encasement during a phase of growth.

Monocistronic : A form of gene organization resulting in transcription of an mRNA that contains coding sequence for a single gene or gene product.

Monoclonal antibody : An antibody with very specific and often unique binding specificity which is secreted by a biologically cloned line of plasmacytoma cells in the absence of other related antibodies with different binding specificities. Differs from polyclonal antibodies, which are mixed populations of antibody molecules such as may be present in a serum specimen, within which many different individual antibodies have different binding specificities.

Monocot (syn monocotyledon) : Plant with one cotyledon, such as grasses, grain crops, corn (see dicot).

Monocotyledon (Gr monos, solitary + kotyledon, a cup-shaped hollow) : A plant whose embryo has one seed leaf (cotyledon). Examples are cereal grains (corn, wheat, rice), asparagus, and lily. Colloquially called a monocot. cf dicotyledon.

Monoculture : The agricultural practice of cultivating a single crop on a whole farm or area.

Monocyclic : Having one disease or life cycle per growing season (see polycyclic).

Monoecious : Staminate and pistillate flowers born separately on the same plant.

Monogastric animals : Animals with simple stomachs that do not ruminate. cf ruminant animals.

Monogenic : Determined by a single gene (see polygenic).

Monogenic disorder : A disorder caused by mutation of a single gene. See also: mutation, polygenic disorder.

Monogenic inheritance : See: monogenic disorder.

Monogenic resistance (syn single gene resistance) : Resistance conferred by a single gene (see oligogenic resistance, multigenic resistance, polygenic resistance).

Monohybrid (Gr monos, solitary + L hybrida, a mongrel) : The offspring of two homozygous parents that differ from one another by the alleles present at only one locus.

Monohybrid cross : A cross between parents differing in only one trait or in which only one trait is being considered.

Monolayer : A single layer of cells growing on a surface.

Monomer : A single molecular entity that may combine with others to form more complex structures.

Monomorphic genes : Have only one common allele (rare alleles with frequencies of less than 0.001% may still occur).

Monophyletic : Describing any group of organisms that are assumed to have originated from the same ancestor.

Monophyletic group (clade) : Evolutionary assemblage of taxa that includes a common ancestor and all of its descendants. [See Avise, p. 36].

Monoploid : See haploid.

Monosaccharide : A single sugar. cf polysaccharide.

Monosomic (n monosomy) : Describing a diploid organism lacking one chromosome (2n -1) of its proper (disomic) complement; a form of aneuploidy. See also disomy.

Monosomy : Possessing only one copy of a particular chromosome instead of the normal two copies. See also: cell, chromosome, gene expression, trisomy.

Mono-unsaturates : Oils containing mono-unsaturated fatty acids.

Monoxenic culture : A culture containing one species of organism growing in the presence of one other species of organism.

Monozygotic twins : Identical twins. Twins that come from the same zygote are essentially the same genetically. Differences between monozygotic twins later in life are virtually always the result of environmental influences rather than genetic inheritance. Fraternal twins may look similar but are not genetically identical.

Monozygotic twins : One-egg or identical twins; twins derived from the splitting of a single fertilized ovum.

Morbid map : A diagram showing the chromosomal location of genes associated with disease.

Moron : A short region of DNA that is inserted between a pair of genes in one phage genome but the genes of this same pair are adjacent in a related phage genome. The term (coined by Roger Hendrix) refers to the observation that one phage contains "more" DNA than the other phage at this particular spot. Many virulence factors of pathogenic bacteria are encoded in morons of integrated prophages.

Morphogen : A substance that stimulates the development of form or structure in an organism.

Morphogenesis : The development, through growth and differentiation, of form and structure in an organism.

Morphogenic response : The effect on the developmental history of a plant or its parts exposed to a given set of growth conditions or to a change in the environment.

Morphology (Gr morphe, form + logos, discourse) : 1.The science of studying form and its development. 2. General: Shape, form, external structure or arrangement.

Mosaic : The substitution of portions of a gene with sequences acquired from another site or another organism via genetic recombination. Comparative sequence analysis indicates that many genes are composed of a "patchwork" of sequences derived from multiple recombination events.

Mother plant : See donor plant.

Motif : A recurring pattern of short sequence of DNA, RNA, or protein, that usually serves as a recognition site or active site. The same motif can be found in a variety of types of organisms.

Mottle : Disease symptom comprising light and dark areas in an irregular pattern, usually caused by a virus; often used interchangeably with mosaic.

Mouse model : See: model organisms.

Movable genetic element : See transposon.

Movement protein : For viruses, a protein encoded by the pathogen that is required for spread within an infected plant.

MRCA : Most recent common ancestor.

mRNA : An RNA molecule that includes the coding region(s) and the translation signals for a gene or operon. An intermediate that specifies the amino acid sequence of the encoded polypeptide(s) during translation.

mRNA export : Refers to the movement of spliced mRNA out of the nucleus to the cytoplasm.

mRNA processing : Refers to the processes of polyadenylation, splicing, and addition of a 5' cap structure.

mRNA; messenger RNA : The RNA transcript of a protein-encoding gene. The information encoded in the mRNA molecule is translated into a polypeptide of specific amino acid sequence by the ribosomes. In eukaryotes, mRNAs transfer genetic information from the DNA to ribosomes, where it is translated into protein.

MRUs : Minimum recognition units. See dabs.

mtDNA : Mitochondrial DNA which is passed down from the mother to all her children, males and females. The genetic material of the mitochondria, the organelles that generate energy for the cell.

Mu : A bacteriophage that reproduces by transposition. Mu infects a variety of enteric bacteria.

Mucilaginous : Viscous, slimy.

Mud : A defective derivative of phage Mu. Mud insertions are commonly used to construct operon and gene fusions in enteric bacteria.

Mulch : Layer of material, such as organic matter or plastic, applied to the surface of the soil for purposes such as retention of water and inhibition of weeds.

Multicellular : Composed of many cells (see unicellular).

Multicistronic Message : An mRNA transcript with more than one cistron and thus encoding more than one polypeptide. These generally do not occur in eukaryotic organisms, due to differences in the mechanism of translation initiation.

Multi-copy : Describing plasmids which replicate to produce many plasmid molecules per host genome, e.g., pBR322 is a multi-copy plasmid, there are usually 50 pBR322 molecules (or copies) per *E. coli* genome.

Multicopy Plasmids : Present in bacteria at amounts greater than one per chromosome. Vectors for cloning DNA are usually multicopy; there are sometimes advantages in using a single copy plasmid.

Multifactorial : A characteristic influenced in its expression by many factors, both genetic and environmental.

Multigene family : A number of identical or related genes present in the same organism, usually coding for a family of related polypeptides.

Multigenic : Controlled by several genes, as opposed to monogenic.

Multigenic resistance (syn polygenic resistance) : Resistance conferred by several genes (see monogenic resistance, oligogenic resistance, single gene resistance).

Multiline : mixture of cultivars that differ only in single, major genes for resistance to a pathogen, giving the opportunity to modify the mixture from year to year as needed.

Multi-locus probe : A probe that hybridizes to a number of different sites in the genome of an organism. See probe.

Multimer; multimeric : A protein made up of more than one peptide chain.

Multinucleate : Having more than one nucleus per cell (see uninucleate).

Multiparticulate virus : Plant virus in which the genome is divided into more than one piece of nucleic acid, which are encapsidated in more than one particle.

Multipartite virus : Plant virus in which the genome is divided into more than one piece of nucleic acid.

Multiple alleles : The existence of more than two alleles at a locus in a population.

Multiple cloning site : An artificially constructed region within a vector molecule which contains a number of closely spaced recognition sequences for restriction endonucleases. This serves as a convenient site into which foreign DNA may be inserted.

Multiple drop array (MDA) : See microdroplet array.

Multiple ovulation and embryo transfer (MOET) : A technology by which a single female that usually produces only one or two offspring can produce a litter of offspring. Involves stimulation of a female to shed large numbers of ova; natural mating or artificial insemination; collection of fertilized ova (either surgically, or non-surgically through the cervix); and transfer (usually non-surgical, through the cervix) of these fertilized ova to recipient females.

Multiple sclerosis (MS) : A genetically inherited progressive disease of the central nervous system. MS occurs as a consequence of one's own immune system attacking the insulating sheath that normally protects neurons. Symptoms range from numbness and tingling to paralysis. There is a loss of motor and cognitive functions. The gene(s) responsible for MS are incompletely penetrant in that the onset of the disease is apparently triggered by a virus and possibly other environmental factors. There is a correlation between the amount of sunlight that children are exposed to and the likelihood that they will develop MS later in life. People who spend much of their first 16 years in tropical and subtropical regions of the world are much less likely to develop this disease than those who live in far northern and far southern regions of our planet. It is believed that the lack of abundant sun exposure early in life is somehow responsible for the later onset of MS.

Multiple-allele series : A situation in which a gene has more than two alleles. The ABO blood type system is an example. Multiple-allele series only partly follow simple Mendelian genetics.

Multiplexing : A laboratory approach that performs multiple sets of reactions in parallel (simultaneously); greatly increasing speed and throughput.

Multiplicity of infection (MOI) : The ratio of pathogen to host cells during an infection. For example, the ratio of phage P22 to Salmonella cells, or the ratio of Salmonella to cultured macrophages.

Multiseptate : Having many septa or crosswalls.

Multivalent vaccine : A single vaccine that is designed to elicit an immune response either to more than one infectious agent or to several different epitopes of a molecule.

Mummy : A dried shriveled fruit; plant part or organ partially or completely replaced with fungal structures.

Murine : Organism in the genus Mus. A rat or mouse.

Mutable genes : Genes with an unusually high mutation rate.

Mutagen : A chemical or physical agent that increases the frequency of mutation, usually by directly damaging the DNA.

Mutagenesis : Change(s) in the genetic constitution of a cell through alterations to its DNA.

Mutagenicity : The capacity of a chemical or physical agent to cause permanent genetic alterations. See also: somatic cell genetic mutation.

Mutant : An organism with an altered base sequence in one or several genes. Usually refers to an organism with a mutation that causes a phenotypic difference from the wild-type.

Mutation : A permanent structural alteration in DNA. In most cases, DNA changes either have no effect or cause harm, but occasionally a mutation can improve an organism's chance of surviving and passing the beneficial change on to its descendants.

Mutation breeding : Commonly used practices in plant breeding and other areas in which chemicals or radiation are applied to whole organisms, e.g. plants, or cells so that changes in the organism's DNA will occur. Such changes are then evaluated for their beneficial effects such as disease resistance.

Mutation pressure : A constant mutation rate that adds mutant genes to a population; repeated occurrences of mutations in a population.

Mutation rate : The number of mutations per cell division. The mutation rate can be determined from a Luria-Dulbruck fluctuation analysis. Sometimes the mutant frequency is described instead of mutation rate. The mutant frequency is simply the ratio of mutants / total cells in the population. The mutant frequency is much easier to measure but may show large fluctuations depending upon when the first mutation appeared in the population.

Mutator gene : A mutant gene which increases the frequency of mutation in other genes. Mutations in genes responsible for DNA repair typically have a mutator phenotype. For example, mutations in the genes responsible for methyl-directed mismatch repair, designated mutS, mutL, and mutH, increase the spontaneous mutation frequency about 1000-fold. See Mutator strains.

Mycelia sterilia : A group in the Fungi Imperfecti that does not produce spores, but may produce chlamydospores or sclerotia, e.g. Rhizoctonia and Sclerotium; subsequent research has defined sexual reproduction for some species.

Mycology : The study of fungi.

Mycoparasite : A fungus that attacks another fungus.

Mycoplasmalike organism (syn MLO) : Archaic term for phytoplasma; plant-parasitic pleomorphic mollicute (prokaryote with no cell wall) found in phloem tissue; cannot yet be grown on artificial nutrient media.

Mycoprotein : Fungal protein.

Mycorrhiza (Gr mykos, fungus + riza, root) : Fungi that form an association with or have a symbiotic relationship with roots of more developed plants.

Mycorrhiza (pl mycorrhizae; adj mycorrhizal) : Symbiotic association between a nonpathogenic or weakly pathogenic fungus and the roots of plants (see ectomycorrhiza and endomycorrhiza).

Mycotoxin : Poisonous compound produced by a fungus.

Mycovirus : A virus that infects fungi.

Myeloma : A plasma cell cancer.

Myo inositol : See inositol.

Myosin : See actin.

Myotonic dystrophy : A combination of progressive weakening of the muscles and muscle spasms or rigidity, with difficulty relaxing a contracted muscle; inherited as an autosomal dominant trait.

Myxomycetes (syn slime molds) : Saprophytic organisms that form vegetative amoeboid plasmodia and spores.

N Terminus : See Amino Terminus.

N_2 : Free nitrogen gas. In liquid form, used as a cryopreservant. See cryobiological preservation.

Naked bud : A bud not protected by bud scales.

Nalidixic acid : An antibiotic that inhibits DNA gyrase.

Nanometre : 1×10^{-9} m. One millionth of a millimetre, a.k.a. a millimicron; equals ten angstroms.

Narrow-host-range plasmid : A plasmid that can replicate in one, or at most a few, different bacterial species.

Narrow-sense heritability : In quantitative genetics, the proportion of the phenotypic variance that is due to variation in breeding values.

Native gel : An electrophoresis gel run under conditions which do not denature proteins (i.e., in the absence of SDS, urea, 2-mercaptoethanol, etc.).

Native protein : The naturally occurring form of a protein.

Natural selection : The differential survival and reproduction of organisms because of differences in characteristics that affect their ability to utilize environmental resources.

NDP : Ribonucleoside diphosphate. See nucleotide; ribonucleoside triphosphate (NTP).

Necrosis : Death associated with discoloration and dehydration of all or some parts of organs.

Necrotroph : A parasite that typically kills and obtains its energy from dead host cells (see biotroph).

Nectary : A nectar-secreting gland in a flower.

Needle cast (of conifers) : Disease symptom caused by fungi resulting in premature drop of needles.

Negative autogenous regulation; negative self-regulation : Inhibition of the expression of a gene or set of coordinately regulated genes by the product of the gene or the product of one of the genes.

Negative control system : A mechanism in which the regulatory protein(s) is required to turn off gene expression.

Negative dominance : A gene product that inhibits the function of another gene product.

Negative regulation : A mechanism of genetic regulation in which a protein or RNA molecule inhibits gene expression, e.g. by preventing the transcription of a gene or translation of the mRNA. A negative regulator that inhibits transcription of a gene is commonly called a repressor.

Negative selection : Selection of individuals that do not possess a certain character. A method by which growing cells that do not carry a DNA insert integrated at a specific chromosomal location are selected. See positive selection.

Negative self-regulation : See negative autogenous regulation.

Negative sense RNA : Ribonucleic acid complimentary to the positive or plus sense; not translated into protein (see positive sense RNA).

Neighborhood : The largest population in which mating occurs at random. Generally, a small population within which individuals can be expected to be closely related. Neighborhoods may be created by barriers to pollen and seed movement, or other isolating factors.

Nematicide : Agent, usually a chemical, that kills nematodes.

Nematodes : A class of slender, unsegmented worms, often parasitic. a.k.a. eelworms, especially when phytoparasitic.

Neo-formation : Organogenesis; production of newly formed structures, such as tissues, meristems and embryos.

Neoplasm : Localized cell multiplication. Generally it designates a collection of cells which have undergone genetic transformation, forming a tumour. Neoplasmic cells differ in structure and function from the original cell type.

Neoteny : The retention of juvenile body characters in the adult state, or the occurrence of adult characters in the juvenile state.

Nested Pcr : A very sensitive method for amplfication of DNA, which takes part of the product of a single PCR reaction (after 30-35 cycles), and subjects it to a new round of PCR using a different set of PCR primers which are nested within the region flanked by the original primer pair (see Polymerase Chain Reaction).

Net photosynthesis : Photosynthetic activity minus respiratory activity, as measured by carbon dioxide exchange.

Neurofibromatosis : One of the most common single gene conditions affecting the human nervous system; in most cases, "cafe au lait" spots, are the only symptom; inherited as an autosomal dominant trait, with 50% being new mutations.

Neutral mutation : A mutation that changes the nucleotide sequence of a gene but has negligible effect on the fitness of the organism.

Neutral theory : The theory that much of evolution has been primarily due to random drift of neutral mutations.

NFT : See nutrient film technique.

N-glycosylases : Repair enzymes that are able to excise certain incorrectly paired or damaged nucleotides.

Nick : A single-strand break, involving the absence of one or more nucleotides, in a double-stranded DNA molecule. A break in the phosphodiester bond between adjacent nucleotides in one strand of a DNA duplex.

Nick translation : A method which uses DNA polymerase I to first produce a nick in a DNA duplex, then degrade stretch of single-stranded DNA using its 5'-exonuclease while synthesizing a new strand in its place. By including a labeled dNTP in the reaction, the newly synthesized DNA will be labeled.

Nicked circle; relaxed circle : During extraction of plasmid DNA from the bacterial cell, one strand of the DNA becomes nicked. This relaxes the torsional strain needed to maintain supercoiling, producing the familiar form of plasmid. See plasmid.

NIH : National Institutes of Health.

Nitrate : The only form in which nitrogen can be used directly by plants; a component of chemical fertilizers.

Nitrification : A chemical process in which nitrogen in plant and animal wastes and dead remains is oxidized, first to nitrites and then to nitrates.

Nitrite : A salt or ester of nitrous acid.

Nitrocellulose; cellulose nitrate : A nitrated derivative of cellulose. It is made into membrane filters of defined porosity, used to immobilize DNA, RNA or protein, which can then be probed with a labelled sequence or antibody. These filters have a variety of uses in molecular biology, particularly in nucleic acid hybridization experiments. Used extensively in the Southern and northern blotting procedures involving DNA and RNA.

Nitrogen assimilation : The incorporation of nitrogen into organic cell substances by living organisms.

Nitrogen cycle : The cycling of the element nitrogen from gaseous forms to various inorganic forms, such as ammonium, nitrates and nitrites, and organic forms, such as nucleic acids and proteins and back to elemental nitrogen.

Nitrogen fixation : The conversion of atmospheric nitrogen (N2) into oxidized forms that can be assimilated by plants. Biological nitrogen fixation is catalysed by the enzyme nitrogenase, which is found only in prokaryotes. Certain blue-green algae and some genera of bacteria (e.g., Rhizobium spp.; Azotobacter spp.) are capable of biochemically fixing nitrogen. Such bacteria are very important symbionts for plants growing in nitrogen-poor soils.

Nitrogen oxides : Primary air pollutants produced from internal combustion engines and other industrial processes.

Nitrogenous base : A nitrogen- containing molecule having the chemical properties of a base.

NMP : Ribonucleoside monophosphate. See nucleotide.

NO : See nucleolar organizer.

Nod box : A DNA sequence that controls the transcriptional regulation of Rhizobium nodulation genes.

Nodal culture : The culture of a lateral bud and a section of adjacent stem tissue.

Node (L nodus, a knot) : Slightly enlarged portion of the stem where leaves and buds arise and where branches originate. Stems have nodes but roots do not.

Nodular : Term commonly used to describe a pebbly (rough) texture of a callus.

Nodulation : The formation of nodules by symbiotic bacteria on the roots of plants.

Nodule : The enlargement or swelling on roots of nitrogen-fixing plants. The nodules contain symbiotic nitrogen-fixing bacteria. See nitrogen fixation.

Nomenclature : The scheme by which names are attached to objects, including organisms.

Non-autonomous : A term referring to biological units that cannot function by themselves; such units require the assistance of another unit, or "helper". See autonomous.

Non-coding strand : Anti-sense strand. See "Sense strand" for a discussion of sense strand vs. anti-sense strand.

Non-composite transposon : A transposible element that is NOT flanked by IS elements.

Nonconservative substitution : A mutation which results in the substitution of one amino acid within a polypeptide chain with an amino acid belonging to a different polarity/charge group (see Amino Acids, Conservative Mutation).

Non-disjunction : Failure of disjunction or separation of homologous chromosomes or chromatids in mitosis or meiosis, resulting in too many chromosomes in some daughter cells and too few in others. Examples: In meiosis, both members of a pair of chromosomes go to one pole so that the other pole does not receive either of them; in mitosis, both sister chromatids go to the same pole.

Non-histone chromosomal proteins : In chromosomes, all of the proteins except the histones.

Non-homologous : Nucleic acid molecules that do not share similar nucleotide sequences.

Noninfectious disease : A disease that is caused by an abiotic agent that cannot be transmitted from one plant to another.

Nonpathogenic (syn avirulent) : Unable to cause disease.

Non-permissive conditions : Growth conditions not allowing a conditionally lethal mutant to survive.

Nonpersistent transmission (syn stylet-borne transmission) : A type of virus transmission in which the virus is acquired and transmitted by the vector after short feeding times, and is retained by the vector for only a short period of time.

Nonsense codon : A codon which does not code for any amino acid, but signals a termination of translation, or punctuation. The three nonsense codons are UAG (amber), UAA (ochre), and UAG (opal).

Nonsense mutation : A mutation which replaces a codon for an amino acid with a codon for chain termination (UAG, UAA, or UGA).

Nonsense suppressors : A mutant tRNA that recognizes a nonsense (stop) codon and inserts an amino acid into the growing polypeptide chain. Nonsense suppressors are produced by base substitution mutations in the DNA corresponding to the anticodon of a tRNA that cause the anticodon to pair with one of the terminarion (or "nonsense") codons, UGA (Amber), UAA (Ochre), or UAG (Opal). Note that because of the wobble rules amber suppressing tRNAs can only read UAG codons, but ochre-suppressing tRNAs can read both UAA and UAG codons. Two possible events can occur when a ribosome encounters a nonsense codon in a strain with a nonsense suppressor: termination of peptide elongation can occur if the appropriate release factors associate with the ribosome, or an amino acid can be inserted into the growing peptide chain if the suppressor tRNA associates with the ribosome. The efficiency of suppression depends upon how well the suppressor tRNA is charged with the appropriate amino acid, the concentration of the suppressor tRNA in the cell, and the "context" of the nonsense codon in the mRNA -- especially the base on the 3' side of the codon. Although suppressor tRNAs are often inefficient, the amount of protein produced is often sufficient to repair the mutant phenotype.

Nonseptate : without cross-walls; coenocytic.

Non-target organism : An organism which is affected by an interaction for which it was not the intended recipient.

Non-template strand : In transcription, the non-transcribed strand of DNA. a.k.a. sense strand or coding strand. It will have the same sequence as the RNA transcript, except that T is present at positions where U is present in the RNA transcript.

Nontranslated RNA (NTR) : The segments located at the 5' and 3' ends of a mRNA molecule which do not encode any part of the polyprotein; may contain important translational control elements.

Non-virulent agent : See attenuated vaccine.

Noonan syndrome : A condition characterized by short stature and ovarian or testicular dysfunction, mental deficiency, and lesions of the heart.

NOR : See nucleolar organizer region.

Northern analysis : A technique for transferring electrophoretically resolved RNA segments from an agarose gel to a nitrocellulose filter paper sheet via capillary action.

Northern blot : A technique for analyzing mixtures of RNA by transfer of size-separated RNA fragments to a synthetic membrane, whereby the presence and rough size of one particular type of RNA (usually an mRNA) can be ascertained. See "Blotting" for more information. After Dr. E. M. Southern invented the Southern blot, it was adapted to RNA and named the "Northern" blot.

Northern blotting : Similar to Southern blotting, except that RNA is transferred onto a matrix and the presence of a specific RNA molecule is detected by DNA-RNA hybridization. See northern hybridization.

Northern hybridization : Hybridization of a labelled DNA probe to RNA fragments that have been transferred from an agarose gel to a nitrocellulose filter. See northern blotting; hybridization (3).

Northern transfer : A technique for transferring bands of RNA from an agarose gel to a nitrocellulose or similar membrane.

No-till : A cultural system most often used with annual crops, in which the new crop is seeded or planted directly in a field on which the preceding crop plants were cut down, had the tops harvested, or were destroyed by a nonselective herbicide (the old crop is not removed or incorporated into the soil as is common in preparing a plant bed).

No-till farming : A method of farming that eliminates plowing to prepare the land for planting seeds and weed control. Instead, crop residue is left on fields, and seed and nutrients are placed in narrow rows or in drilled holes. Weed control is accomplished with suitable herbicides. Because the soil is not disturbed or exposed through plowing, it is much less susceptible to erosion from both water and wind.

nt : Abbreviation for nucleotide; i.e. the monomeric unit from which DNA or RNA are built. One can express the size of a nucleic acid strand in terms of the number of nucleotides in its chain; hence 'nt' can be a measure of chain length.

N-terminus : The end of a polypeptide chain that has a free amino acid ($-NH_2$) group.

NTP : Nucleotide triphosphate. Used for the synthesis of RNA or as an energy source.

Nucellar embryo : An embryo which has developed vegetatively from somatic tissue surrounding the embryo sac, rather than by fertilization of the egg cell.

Nucellus (L nucella, a small nut) : Tissue composing the chief part of the young ovule in which the embryo sac develops; megasporangium.

Nuclear run-on : A method used to estimate the relative rate of transcription of a given gene, as opposed to the steady-state level of the mRNA transcript (which is influenced not just by transcription rates, but by the stability of the RNA). This technique is based on the assumption that a highly-transcribed gene should have more molecules of RNA polymerase bound to it than will the same gene in a less-active state. If properly prepared, isolated nuclei will continue to transcribe genes and incorporate 32P into RNA, but only in those transcripts that were in progress at the time the nuclei were isolated. Once the polymerase molecules complete the transcript they have in progress, they should not be able to re-initiate transcription. If that is true, then the amount of radiolabel incorporated into a specific type of mRNA is theoretically proportional to the number of RNA polymerase complexes present on that gene at the time of isolation. A very difficult technique, rarely applied appropriately from what I understand.

Nuclear transfer : A technology by which animals are created by cloning a single diploid somatic cell. It involves taking a single diploid cell from a culture of cells, and inserting it into an enucleated ovum, i.e., an ovum from which the haploid nucleus has been removed. The resultant diploid ovum develops into an embryo that is placed in a recipient female, which gives birth to the cloned animal in the normal manner. Note that the term is somewhat of a misnomer, since it is a whole cell that is transferred, not just the nucleus.

Nuclease : An enzyme which degrades nucleic acids. A nuclease can be DNA-specific (a DNase), RNA-specific (RNase) or non-specific. It may act only on single stranded nucleic acids, or only on double-stranded nucleic acids, or it may be non-specific with respect to strandedness. A nuclease may degrade only from an end (an exonuclease), or may be able to start in the middle of a strand (an endonuclease). To further complicate matters, many enzymes have multiple functions; for example, Bal31 has a 3'-exonuclease activity on double-stranded DNA, and an endonuclease activity specific for single-stranded DNA or RNA.

Nuclease protection assay : See "RNase protection assay".

Nucleic acid : A macromolecule composed of phosphoric acid, pentose sugar, and organic bases. The two nucleic acids, deoxyribonucleic acid (DNA) and ribonucleic acid (RNA), are made up of long chains of molecules called nucleotides (q.v.). They were first isolated as part of a protein complex in 1871, and were separated from the protein moiety in 1889. See DNA; RNA; nucleotides.

Nucleic acid hybridization : Formation of a double-stranded molecule by base pairing between complementary or homologous polynucleotides.

Nucleic acid probe : See DNA probe.

Nuclein : The term used by Friedrich Miescher to describe the nuclear material he discovered in 1869, which today is known as DNA.

Nucleo-cytoplasmic ratio : In a cell, the ratio of nuclear to cytoplasmic volume. This ratio is high in meristematic cells and low in differentiated cells.

Nucleoid : Region in prokaryotes, chloroplasts, and mitochondria where the DNA is concentrated. Unlike a nucleus, it is not bound by a membrane.

Nucleolar organizer (NO); nucleolar organizer region (NOR) : A chromosomal segment containing genes that encode ribosomal RNA; located at the secondary constriction of some chromosomes.

Nucleolar organizing region : A part of the chromosome containing rRNA genes.

Nucleolar zone : Any chromosome region, irrespective of whether or not it is a secondary-constriction, that is associated with the formation of the nucleolus during telophase.

Nucleolus (L nucleolus, a small nucleus) : An RNA-rich intranuclear organelle in the nucleus of eukaryotic cells, produced by a nucleolar organizer. It represents the storage place for ribosomes and ribosome precursors. The nucleolus consists primarily of ribosomal precursor RNA, ribosomal RNA, their associated proteins, and some, perhaps all, of the enzymatic equipment (RNA polymerase, RNA methylase, RNA cleavage enzymes) required for synthesis, conversion and assembly of ribosomes. Subsequently the ribosomes are transported to the cytoplasm.

Nucleoplasm : The non-staining or slightly chromophilic, liquid or semi-liquid, ground substance of the interphase nucleus and which fills the nuclear space around the chromosomes and the nucleoli. Little is known of the chemical composition of this ground substance, which is not easily defined. It may be called "karyoplasm" when it is gel-like, and "karyolymph" when it is a colloidal fluid, but generally the terms are synonymous.

Nucleoprotein : Conjugated protein composed of nucleic acid and protein; the material of which the chromosomes are made.

Nucleoside : A term refering to the combination of adenine, cytosine, guanine, or thymine with a ribose or 2-deoxyribose sugar moiety. A nucleoside is not phosphorylated.

Nucleoside analog : See analog.

Nucleoside analogue : A synthetic molecule that resembles a naturally occurring nucleoside, but that lacks the bond site needed to link it to an adjacent nucleotide. See nucleoside.

Nucleosome : Spherical sub-units of eukaryotic chromatin that are composed of a core particle consisting of an octamer of histones (two molecules each of histones H2a, H2b, H3 and H4) and 146 nucleotide pairs.

Nucleotide : A building block of DNA or RNA consisting of a nitrogenous base (adenine, guanine, thymine, or cytosine in DNA; adenine, guanine, uracil, or cytosine in RNA), a phosphate molecule, and a sugar molecule (deoxyribose in DNA and ribose in RNA). Thousands of nucleotides are linked to form a DNA or RNA molecule. See DNA, base pair, RNA.

Nucleus : The cellular organelle in eukaryotes that contains the genetic material. The center of a cell, where all of the DNA, packaged in chromosomes, is contained.

Null : Completely absent. For example, a null mutation completely disrupts a gene.

Null mutation : See amorph.

Nullisomy (adj nullisomic) : An otherwise diploid cell or organism lacking both members of a chromosome pair (chromosome formula 2n -2). See also disomy.

Nurse culture : Planting a cell from a suspension culture on a raft of filter paper above a callus tissue piece (nurse tissue). The filter paper serves to prevent tissue union but allows the flow of essential substances from the nurse to the isolated cell. In such culture, a piece of callus is first placed on nutrient agar. Over this tissue is laid a strip of filter paper.

Nutrient cycle : The passage of a nutrient or element through an ecosystem, including its assimilation and release by various organisms and its transformation into various organic or inorganic chemical forms.

Nutrient deficiency : Absence or insufficiency of some factor needed for normal growth and development.

Nutrient film technique (NFT) : Hydroponic technique used to grow plants. NFT delivers a film of water or nutrient solution either continuously or through on-off cycles (e.g., on 8 minutes and off 7 minutes).

Nutrient gradient : A diffusion gradient of nutrients and gases that develops in tissues where only a portion of the tissue is in contact with the medium. Gradients are less likely to form in liquid media than in callus cultures.

Nutrient medium (pl nutrient media) : A solid, semi-solid or liquid combination of: major and minor salts; an energy source (sucrose); vitamins; plant growth regulators; and occasionally other defined or undefined supplements. Often made from stock solutions, then sterilized by autoclaving or filtering through a micropore filter.

O-antigen : The arrangement of polysaccharide side chains on the LPS of Gram-negative bacteria. Serves as the receptor for some types of phage.

Obligate : Parasite that cannot multiply in nature without a host.

Obligate parasite (syn biotroph) : Organism that can grow only as a parasite in association with its host plant and cannot be grown in artificial culture media.

Occlusion : Block or plug that stops flow of liquids (as in vessels).

Ochre codon : The UAA stop codon.

Ochre mutation : A change in a codon to create the nonsense codon UAA (the ochre codon).

Octoploid : Cell or organism with eight sets of chromosomes, i.e., chromosome number 2n = 8x.

Oedema (also edema; syn intumescence) : Blisters produced on leaves and other plant parts under conditions of high moisture and restricted transpiration.

Oestrogen; estrogen : The generic term for a group of female sex hormones which control the development of sexual characteristics and control oestrus.

Oestrous cycle (from oestrus) : The cycle of reproductive activity shown by most sexually mature non-pregnant female mammals.

Oestrus (adj oestrous) : In female mammals, the period of sexual excitement and acceptance of the male. a.k.a. rut, heat.

Offset : Young plant produced at the base of a mature plant.

Offshoot : Short, usually horizontal, stem produced near the crown of a plant.

Offspring; progeny (both same in plural) : New individual organisms that result from the process of sexual or asexual reproduction.

Okazaki fragment : Since DNA can be replicated in only one direction, i.e., nucleotides can be added only at the 3′ end, only one of the two strands of a double helix can be replicated continuously. The other strand is replicated in small segments (Okazaki fragments) that are subsequently joined together by DNA ligase. See primosome.

OLA : See oligonucleotide ligation assay.

Oligodeoxyribonucleotide : A short, single-stranded DNA molecule, generally 15-50 nucleotides in length, which may be used as a primer or a hybridization probe. Oligodeoxyribonucleotides are synthesized chemically under automated conditions.

Oligogenic : A phenotypic trait produced by two or more genes working together. See also: polygenic disorder.

Oligogenic resistance : Resistance conferred by a few genes (see monogenic resistance, multigeneic resistance, polygenic resistance, single gene resistance).

Oligomer : A molecule formed from a small number of monomers.

Oligonucleotide : A short sequence of polynucleotide. Oligonucleotides with a defined sequence can synthesized *in vitro* for use for a variety of techniques, including PCR, DNA sequence analysis, hybridization, and site-directed mutagenesis. The length of an oligonucleotide is described as a #-mer -- for example, an oligonucleotide with 20 nucleotides would be called a 20-mer.

Oligonucleotide ligation assay (OLA) : A diagnostic technique for determining the presence or absence of a specific nucleotide pair within a target gene, often indicating whether the gene is wild type (normal) or mutant (defective).

Oligonucleotide-directed mutagenesis : An *in vitro* mutagenesis technique that involves the use of a synthetic oligonucleotide to introduce the predetermined nucleotide alteration into the gene to be mutated.

Oligosaccharide : Carbohydrate consisting of several linked sugar units.

Oliogenic resistance : Resistance determined by one or few genes whose effects are readily detectable.

-Ome : This suffix indicates the complete collection of the subunits described in the word stem. For example, genome indicates the entire haploid complement of genetic material in an organism, pproteome indicates the entire set of proteins expressed by an organism, etc. [Source: J. Lederberg and A. McCray (2001) The Scientist 15:8].

Oncogene : A gene in a tumor virus or in cancerous cells which, when transferred into other cells, can cause transformation (note that only certain cells are susceptible to transformation by any one oncogene). Functional oncogenes are not present in normal cells. A normal cell has many "proto-oncogenes" which serve normal functions, and which under the right circumstances can be activated to become oncogenes. The prefix "v-" indicates that a gene is derived from a virus, and is generally an oncogene (like v-src , v-ras, v-myb , etc). See also "Transformation (with respect to cultured cells)".

Oncogenesis : The progression of cytological, genetic and cellular changes that culminate in a tumour.

Oncogenic : Oncogenic genes are responsible for the transformation of normal cells into tumour cells. See Agrobacterium.

Oncogenic transformation : A change in the behavioral phenotype of a cell to one lacking in normal proliferative control, and often involving invasive characteristics.

Onco-mouse : A mouse that has been genetically modified to incorporate an oncogene; oncogenes cause cells to undergo cancerous transformation.

One-step growth curve : A plot of the production of lytic growth of a virus vs time. An excess of bacteria is infected with virus. There is an initial drop in the number of plaque forming units (PFU) of virus due to entry of the virus into cells and release of the viral nucleic acid. This is followed by a period of time called the eclipse phase where viral proteins and nucleic acid are reproduced but no viable virus is produced. After this period mature viruses are assembled and released from the cell. The "burst size" is the number of virus particles released from the cell divided by the number of virus particles initially added. The one-step growth curve is terminated after the first round of infection.

Ontogeny : Developmental life history of an organism.

Oocyte : The egg mother cell; it undergoes two meiotic divisions (oogenesis) to form the egg cell. The primary oocyte is before completion of the first meiotic division; the secondary oocyte is after completion of the first meiotic division.

Oogenesis : The formation and growth of the egg or ovum in an animal ovary.

Oogonium (pl oogonia) : 1. A germ cell of the female animal, that gives rise to oocytes by mitotic division. 2. The female sex organ of algae and fungi.

Oosphere : The non-motile female gamete in plants and some algae.

Oospore (Gr oion, an egg + spore) : A resistant spore developing from a zygote, resulting from the fusion of heterogametes in certain algae and fungi.

Ooze : Mass of bacterial cells mixed with host fluids.

Opal codon : The UGA stop codon.

Opal mutation : A change in a codon to create the nonsense codon UGA (the opal codon).

Open continuous culture : A continuous culture in which inflow of fresh medium is balanced by outflow of a corresponding volume of culture. Cells are constantly washed out with the outflowing liquid. In a steady state, the rate of cell wash-out equals the rate of formation of new cells in the system. See continuous culture; batch culture; closed continuous culture.

Open pollination : Pollination by wind, insects or other natural mechanisms.

Open reading frame : That segment of a nucleic acid sequence that lies between two stop codons, when translated in a given reading frame. The presence of an open reading frame is necessary to encode a polypeptide sequence (or an exon thereof), but the presence of an open reading frame is not proof that a polypeptide sequence is encoded in that sequence. See "Reading frame" for a simple example.

Open-circular : The non-supercoiled conformation taken up by a circular double-stranded DNA molecule when one or both polynucleotides carry nicks.

Operational definition : An operation or procedure that can be carried out to define or delimit something.

Operator : The DNA sequence where a repressor protein reversibly binds to regulate the activity of one or more closely linked structural genes.

Operon : A sequence of adjacent genes read as a single, polycistronic mRNA. Changes in the level of transcription thus affects all of the genes in an operon, so such genes are often coordinately regulated. Includes both the structural genes, and the operator and promoter sequences that control them.

Operon fusion : A construct that places the coding region of a gene downstream of the promoter for another gene, such that transcription of the downstream gene initiates from this new promoter but translation of the downstream gene occurs from its own translation start site. For a more detailed description, see Operon and Gene fusions. Sometimes called a transcriptional fusion or fusion.

Opine : The condensation product of an amino acid with either a keto-acid or a sugar, produced by the crown gall tissues. Opine synthesis is a unique characteristic of tumour cells.

Ordinate : The vertical axis of a graph. Opposite: absciss; abscissa.

ORF : See open reading frame.

Organ : A tissue or group of tissues that constitute a morphologically and functionally distinct part of an organism.

Organ culture : The aseptic culture of complete living organs of animals and plants outside the body in a suitable culture medium. Animal organs must be small enough to allow the nutrients in the culture medium to penetrate all the cells.

Organellar genes : Genes located on organelles outside the nucleus.

Organelle : A membrane-bounded specialized region within a cell, such as the mitochondrion or dictyosome, that carries out a specialized function in the life of a cell.

Organic : Referring in chemistry to compounds containing carbon, many of which have been in some manner associated with living organisms.

Organic agriculture / Organic farming : To qualify to use recognized organic labels, farmers must conform to a strict agricultural system that minimizes the use of synthetic fertilizers, pesticides, growth regulators and livestock feed additives. As far as practical, organic farming relies on crop rotation, crop residues, animal manure, legumes, green manures, off-farm organic wastes, mechanical cultivation, mineral-bearing rocks, and biological pest control to sustain soil productivity, supply plant nutrients and control crop pests.

Organic complex : A chemically undefined compound added to nutrient media to stimulate growth, e.g., coconut milk; yeast extract; casein hydrolysate.

Organic co-solvent : A compound used to dissolve some neutral organic substances, such as in media preparation. Organic co-solvents include alcohols (usually ethanol), acetone and dimethylsulphoxide (DMSO).

Organic evolution : The process by which changes in the genetic composition of populations of organisms occur in response to environmental changes.

Organism : An individual living system, such as animal, plant or micro-organism, that is capable of reproduction, growth and maintenance.

Organized growth : The development under tissue culture conditions of organized explants (meristem tips or shoot tips, floral buds or organ primordia). See unorganized growth.

Organized tissue : Composed of regularly differentiated cells.

Organizer : An inductor; a chemical substance in a living system that determines the fate in development of certain cells or groups of cells.

Organogenesis : The initiation of adventitious or de novo shoots or roots from callus, meristem or suspension cultures. See micropropagation; regeneration.

Organoid : An organ-like structure produced in culture, such as leaves, roots or callus.

Organoleptic : Having an effect on one of the organs of sense, such as taste or smell.

Origin of replication : A site at which DNA replication is initiated. There is only one in bacterial chromosomes, but numerous origins in eukaryotic chromosomal DNA. (Abbr. "ori").

Origin recognition complex : An organization of protein factors assembled at an origin of replication for the purpose of initiating replication.

Ortet : The plant from which a clone is obtained.

Orthogonal field alternating gel electrophoresis (OFAGE) : A gel electrophoresis technique that employs a pulsed electric field to achieve separation of very large molecules of DNA.

Orthologous : A gene present in multiple species that descended from a unique ancestral gene, when the relative divergence in different species is comparable to the overall difference between the species. Such genes provide useful tools for constructing phylogenetic trees.

Osmic acid (= osmium tetroxide) (OsO4; fw 25420) : A fixing agent commonly used to prepare tissue samples for electron microscopy.

Osmolarity : The total molar concentration of the solutes. Osmolarity affects the osmotic potential of solution or nutrient medium.

Osmosis (Gr osmos, a pushing) : Diffusion from areas of high concentration to areas of lower concentration of a solvent through a differentially permeable membrane.

Osmotic potential [value] : Potential brought about by dissolving a substance, especially in water.

Osmoticum : An agent, such as PEG, mannitol, glucose or sucrose, employed to maintain the osmotic potential of a nutrient medium equivalent to that of the cultured cells (isotonic). Because of this osmotic equilibrium, cells are not damaged *in vitro*.

Osteogenesis imperfecta : A condition also known as brittle bone disease; characterized by a triangular shaped face with yellowish brown teeth, short stature and stunted growth, scoliosis, high pitched voice, excessive sweating and loose joints.

Ostiole (adj ostiolate) : Pore; opening in the papilla or neck of a perithecium, pseudothecium, or pycnidium through which spores are released.

Outbreeding : A mating system characterized by the breeding of genetically unrelated or dissimilar individuals. Since genetic diversity (q.v.) tends to be enhanced and since vigour or fitness of individuals can be increased by this process, it is often used to counter the detrimental effects of continuous inbreeding.

Outcross : A cross, usually natural, to a plant of different genotype.

Outcrossing : Mating two individuals of the same breed that are sufficiently unrelated that the IC of the progeny is lower than the average of the parents.

Outer membrane : The lipopolysaccharide (LPS) containing layer, containing some proteins, which surrounds the cytoplasmic membrane and forms the outermost layer of most Gram-negative bacteria. The outer membrane usually confers resistance to many detergents and contains the receptor sites for certain phage.

Outerbark : Exterior, nonliving portion of bark in woody plants.

Outflow : The volume of growing cells that is removed from a bioreactor during a continuous fermentation process.

Ovary : 1. Enlarged basal portion of the pistil of a plant flower that contains the ovules. 2. The reproduction organ in female animals in which eggs (ova) are produced.

Overdominance : A condition in which heterozygotes are superior (on some scale of measurement) to either of the associated homozygotes.

Overhang : A terminus of a duplex DNA molecule which has one or more unpaired nucleotides in one of the two strands (hence either a 3' or 5' overhang). Cleavage of DNA with many restriction endonucleases leaves such overhangs (see Sticky End).

Overlapping clones : See genomic library.

Overlapping genes : Two genes whose nucleotide sequences partially overlap. The two genes may be translated in different reading frames with the 3' end of one structural gene overlaping with the 5' end of another structural gene, the two genes may be translated in different reading frames with one structural gene located entirely within the other, or the two genes may be translated in the same reading frame with one structural gene located entirely within the other.

Overlapping reading frames : Start triplets in different reading frames generate different polypeptides from the same DNA sequence. See reading frame.

Overproduction suppression : Suppression of a mutant phenotype by expression of a protein at high levels due to a regulatory mutation or expression from a multicopy plasmid. Typically overexpression suppression is used to describe the partial or full restoration of the wild-type phenotype by high levels of a protein with weak activity, either the corresponding leaky mutant protein or another gene product with weak activity.

Overseason : To survive or persist from one planting season to the next.

Overwinter : To survive or persist through the winter period.

Oviposit : To deposit or lay eggs with an ovipositor.

Ovulation : In mammals, the process of escape of the ovum (egg cell) from the ovary.

Ovum (L ovum, egg; pl ova) : 1. A gamete of female animals, produced by the ovary2. The oosphere in plants.

Ovum pickup (OPU) : The non-surgical collection of ova from a female.

Oxygen-electrode-based sensor : Sensor in which an oxygen electrode - a standard electrochemical cell which measures the amount of oxygen in a solution - is coated with a biological material which generates or absorbs oxygen. When the biological coating is active, the amount of oxygen next to the electrode changes and the signal from the electrode changes.

Ozone: A secondary air pollutant that is a highly reactive form of oxygen (O_3) formed when the exhaust of automobiles and other internal combustion engines is in the presence of sunlight.

Ozone layer : A protective layer of ozone in the upper atmosphere that reduces ultraviolet radiation.

P1 : Symbol for the parental generation or parents of a given individual.

P1-derived artificial chromosome (PAC) : One type of vector used to clone DNA fragments (100- to 300-kb insert size; average, 150 kb) in Escherichia coli cells. Based on bacteriophage (a virus) P1 genome. See also: cloning vector.

Pachynema (adj pachytene) : A mid-prophase stage in meiosis, immediately following zygonema and preceding diplonema. In microscopic preparations, the chromosomes are visible as long, paired threads. Rarely, four chromatids are detectable.

Pachytene : The double-thread or four strand stage of meiosis.

Package : In recombinant DNA procedures, refers to the step of incorporation of cosmid or other lambda vector DNA with an insert into a phage head for transduction of DNA into host.

Packaging cell line : A cell line that is designed to produce viral particles that do not contain nucleic acid. After transfection of these cells with a full-size viral genome, fully infective viral particles are assembled and released.

Packed cell volume (PCV) : The volume of cells in a set volume of culture expressed as a percentage of that set volume after sedimentation (packing) by means of low speed centrifugation.

PAGE : Polyacrylamide gel electrophoresis. Separation of molecules through a polyacrylamide gel matrix in an electric field. Separation may depend upon size and charge of the molecules.

Pairing; synapsis : The pairing of homologous chromosomes during the prophase of the first meiotic division, when crossing over occurs.

Pair-rule gene : A gene that influences the formation of body segments in Drosophila.

Palaeontology : The study of the fossil record of past geological periods and of the phylogenetic relationships between extinct and contemporary plant and animal species.

Palindrome : A sequence of DNA that reads the same in the 5' to 3' direction on the complementary strands. For example 5' GATC3' 3' CTAG 5'

Often described as inverted repeats or sequences with dyad symmetry. Many DNA-binding proteins recognize palindromic sequences. Note that the grammatical usage of this term is differnet than the way it is used in genetics and molecular biology - in gramme r it is used to describe a word string that reads the same if read from right to left as if read from left to right.

Palindromic sequence : A segment of duplex DNA whose 5′-to-3′ sequence is identical on each DNA strand. The sequence is the same when one strand is read left-to-right and the other strand is read right-to-left. Typically, recognition sites for type II restriction endonucleases are palindromes. See the entry for inverted repeat for an example of a palindrome.

Palisade parenchyma : Elongated cells found just beneath the upper epidermis of leaves, and containing many chloroplasts.

pAMP : Ampicillin-resistant plasmid. See plasmid.

PAN (abbr for peroxyacetyl nitrates) : A secondary air pollutant formed when the exhaust of automobiles and other internal combustion engines is in the presence of sunlight.

Pandemic : A widespread and destructive outbreak of disease simultaneously in several countries.

Pangenesis : Charles Darwin's incorrect theory about the inheritance of traits. He proposed that hereditary particles in the body are affected by the things an individual does during his or her lifetime. These modified particles were thought to migrate via blood to the reproductive cells and subsequently could be inherited by the next generation. This was a variation of Lamarck's incorrect idea of the "inheritance of acquired characteristics".

Panicle (L panicula, a tuft) : An inflorescence, the main axis of which is branched; the branches bear loose racemose flower clusters.

Panicle culture : Aseptic culture of grain panicle segments to induce microspore germination and development.

Panmictic population : A population in which mating occurs at random.

Panmixis : Random mating in a population. cf apomixis.

Paper raft technique : See nurse culture.

Papilla : nipple-like projection; used to describe the tip of some sporangia and the localized wall thickenings on the inner surface of plant cell walls at sites penetrated by fungi.

Papillae : Small colonies that appear within or at the edge of a colony. For example, blue Lac+ papillae may occur in a white Lac- colony on X-gal plates. (For an excellent example of a papillation assay, see Nghiem et al. 1988. Proc. Natl. Acad. Sci. USA 85: 2709-2713.)

PAR : See photosynthetically active radiation.

Par gene : A gene found in bacterial and plant cells, and involved in partition of plasmids in cells.

Paracentric inversion : An inversion that is entirely within one arm of a chromosome and does not include the centromere.

Paraffin [wax] : A translucent, white, solid hydrocarbon with a low melting point. Paraffin is used as an embedding medium to support tissue for sectioning for light microscopy observation.

ParafilmTM : A stretchable film based on paraffin wax; used to seal tubes and Petri dishes. ParafilmTM is a proprietary name which is applied colloquially to similar products.

Paragynous : Having the antheridium contact the oogonium on the side, as in many Pythium sp. (see amphigynous).

Parahormone : A substance with hormone-like properties that is not a secretory product (e.g., ethylene; carbon dioxide).

Parallel evolution : The development of different organisms along similar evolutionary paths due to similar selection pressures acting on them.

Paralogous : Genes that descended from an original copy of the gene that duplicated within an ancestral genome, and thus each copy of the gene could diverge independently. Because paralogous genes can vary within a species as much as between species, such genes are useful tools for studying protein evolution but not for constructing phylogenetic trees.

Parameter : A numerical quantity which specifies a population in respect to some characteristic.

Parasexual cycle : A sexual cycle involving changes in chromosome number but differing in time and place from the usual sexual cycle; occurring in those fungi in which the normal cycle is suppressed or apparently absent.

Parasexual hybridization : See somatic hybridization.

Parasexualism : A type of recombination, found in certain heterokaryotic fungi, that is based on mitosis rather than meiosis. Genetically distinct haploid nuclei fuse in the heterokaryon. The resulting diploid nuclei multiply by mitotic division, with some crossing-over, and a diploid homokaryon develops. During repeated mitotic divisions, chromosomes may be shed by the diploid nucleus to produce a haploid nucleus.

Parasite (Gr parasites, one who eats at the table of another) : An organism deriving its food from the living body of another organism.

Parasite (adj parasitic) : Organism that lives in intimate association with another organism on which it depends for its nutrition; not necessarily a pathogen.

Parasitism : The close association of two or more dissimilar organisms, where the association is harmful to at least one. cf commensalism; symbiosis.

Parasporal crystal : Tightly packaged insect pro-toxin molecules that are produced by strains of Bacillus thuringiensis during the formation of resting spores.

Paratope : That region within the antigen-binding site of an immunoglobulin molecule responsible for recognition of epitope structure.

Parenchyma : 1. A plant tissue consisting of spherical, undifferentiated cells, frequently with air spaces between them. 2. Loose connective tissue formed of large cells.

Parenchymatous : Adjective used to describe spherical and undifferentiated cells with primary cell walls, capable of both cell division and differentiation.

Parthenocarpy(Gr parthenos, virgin + karpos, fruit) : The development of fruit without fertilization.

Parthenogenesis : Development of an organism from a sex cell in respect to some characteristic.

Partial digest : Addition of a restriction enzyme to a DNA sample under particular conditions or for a limited period, such that only a proportion of the target sites in any individual molecule are cleaved. Partial digests are often performed to give an overlapping collection of DNA fragments for use in the construction of a gene bank. See complete digest; library.

Partial digestion : Treatment of a DNA molecule with a restriction endonuclease under such conditions that only a fraction of all the recognition sites are cleaved. Commonly used to produce a random collection of DNA fragments for construction of gene libraries.

Partial resistance : Resistance in which disease develops more slowly on the host or to a lesser extent, but the host does become diseased (see complete resistance).

Particle radiation : Refers to gamma (□) particles (positively charged) and beta (ß) particles (negatively charged), electrons, protons and neutrons.

In plant tissue culture, these particles are used to produce mutant cells or organisms.

Partitioning : Separation of chromosomes or plasmids into daughter cells following DNA replication and cell division.

Parts per million (ppm) : Units of any given substance per one million equivalent units, such as the weight units of solute per million weight units of solution (i.e., 1 ppm = 1 mg/l).

Parturition : The process of giving birth.

Passage : The transfer or transplantation of cells from one culture vessel to another. cf sub-culture.

Passage number : The number of times the cells in the culture have been sub-cultured. In descriptions of this process, the dilution ratio of the cells should be stated so that the relative cultural "age" can be ascertained.

Passage time : Interval between successive sub-cultures.

Passive immunity : 1. Natural acquisition of antibodies by the foetus or neonate (newborn) from the mother. 2. The artificial introduction of specific antibodies by the injection of serum from an immune animal. In both cases, temporary protection is conferred on the recipient.

Pasteurization : to free a material, usually a liquid, of selected harmful microorganisms using heat.

Patent : In genetics, conferring the right or title to genes, gene variations, or identifiable portions of sequenced genetic material to an individual or organization. See also: gene

Paternal : Pertaining to the father.

Pathogen : A parasite which produces disease in its host.

Pathogen (Gr pathos, suffering + genesis, beginning) : An organism that causes a disease in another organism.

Pathogenecity island (PAI) : A region of a bacterial chromosome that seems to have been acquired via horizontal gene transfer and includes a group of genes encoding virulence factors. Evidence of horizontal gene transfer may be comparison with genomes from related nonpathogenic bacteria, or by flanking genes that suggest transfer via phage, conjugation, or transpositiion. The GC content and codon usage of PAI is often different from the rest of the chromosome.

Pathogenesis : Production and development of disease.

Pathogenesis-related (PR) proteins : Proteins, such as antimicrobial proteins and hydrolytic enzymes, that are synthesized in the early events of the plant defense response.

Pathogen-free : Freedom from disease-causing organisms (bacteria, fungi, viruses, etc.).

Pathogenicity : Ability to cause disease.

Pathology : The study of diseases.

Pathotoxin : Very dilute substance synthesized and released by some pathogen, and which interacts with the host metabolism.

Pathotype : A subdivision of a pathogen species characterized by its pattern of virulence or avirulence to a series of differential host varieties.

Pathovar (abbr pv) : A subdivision of a plant pathogenic bacterial species defined by host range; pathovar for bacteria is equivalent to forma specialis for fungi.

pBR322 : A common plasmid. Along with the obligatory origin of replication, this plasmid has genes which make the *E. coli* host resistant to ampicillin and tetracycline. It also has several restriction sites (BamHI, PstI, EcoRI, HindIII etc.) into which DNA fragments could be spliced in order to clone them.

PCR : Polymerase chain reaction. A method for amplifying a particular region of DNA by a sequence of denaturation, annealing of specific primers, and synthesis. Use of a thermostable DNA polymerase (such as Taq polymerase) allows the reactions to be perfo rmed in a thermocycler that automatically adjusts the temperature to the optimal for each reaction through many cycles. The concentration of the amplified DNA fragment increases exponentially with each cycle.

PCV : See packed cell volume.

Pectin (Gr pektos, congealed) : A white amorphous substance which, when combined with acid and sugar, yields a jelly substance cementing cells together (the middle lamella).

Pectinase (Gr pektos, congealed) : Enzyme catalysing the hydrolysis of pectins.

Pedicel : Small slender stalk; stalk bearing an individual flower, inflorescence, or spore.

Pedigree : A simplified diagram of a family's genealogy that shows family members' relationships to each other and how a particular trait or disease has been inherited.

Pedigree Breeding : A system of breeding in which individual plants are selected in the segregating generations from a cross on the basis of their desirability judged individually and on the basis of a pedigree record.

Pedigree, full- : All parents, grandparents . . . of a particular genotype are known all the way back to some natural population.

Pedigree, partial- : Some, but not all, ancestors of a particular genotype are known, usually the female parents. Partial pedigrees are most common where open-pollinated or polycross seed is used for progeny tests.

Peduncle (L pedunculus, a late form of pediculus, a little foot) : Stalk or stem of a flower that is born singly; the main stem of an inflorescence.

PEG : See polyethylene glycol.

Penetrance : Refers to the proportion of individuals heterozygous for a given dominant allele that express the phenotype of that dominant allele.

Penetrance : The percentage of individuals in population that show a particular phenotype among those capable of showing it, i.e., among those that have the genotype normally associated with that phenotype.

Penetration peg (syn infection peg) : The specialized, narrow, hyphal strand on the underside of an appressorium that penetrates host cells.

Penicillin (pen) : An antibiotic that inhibits crosslinking of peptidoglycan chains in the cell wall of bacteria. Cells growing in the presence of penicillin synthesize weak cell walls, causing them to burst due to the high internal osmotic pressure. PenR usually results from expression of periplasmic ß-lactamase that breaks the Beta-lactam ring of the antibiotic. A large variety of penicillin derivatives are available. Ampicillin is a commonly used derivative of penicillin.

Peptide : A molecule formed by peptide bonds covalently linking two or more amino acids. Short peptides (generally less than 60 amino acid residues, and usually only half that length) can be chemically synthesized by one of several different methods; larger peptides (more correctly, polypeptides) are usually expressed from recombinant DNA.

Peptide bond : A covalent bond between two amino acids, in which the carboxyl group of one amino acid (X1--COOH) and the amino group of an adjacent amino acid (NH_2--X_2) react to form X1-CO-NH-X_2 plus H_2O.

Peptide vaccine : A short chain of amino acids that can induce antibodies against a specific infectious agent.

Peptide-binding groove : That region of a Class I or II MHC molecule that is responsible for binding processed antigen peptides for presentation.

Peptidoglycan : The structural component of most bacterial cell walls.

Peptidyl transferase : An enzymatic activity of the ribosome responsible for formation of the peptide bond between the nascent polypeptide chain and the amino acid carried by the charged tRNA in the A site of the ribosome. In so doing, the ribosome is moved along the mRNA by one codon.

peptidyl-tRNA binding site; P-site : The site on a ribosome that hosts the tRNA to which an amino acid for the growing polypeptide chain is attached.

Perennial : Something that occurs year after year; plant that survives for several to many years (see annual, biennial).

Perfect : Sexual; capable of sexual reproduction (see teleomorph).

Perfect flower : Flower possessing both stamens and pistils.

Perfect state (syn teleomorph) : The sexual form in the life cycle of a fungus (see anamorph).

Pericarp : Fruit wall.

Pericentric inversion : An inversion that includes the centromere, hence involving both arms of a chromosome.

Periclinal : The plane of cell wall orientation or cell division parallel to the surface of the organ. cf anticlinal.

Periclinal chimera : Genotypically or cytoplasmically different tissues arranged in concentric layers.

Pericycle (Gr peri, around + kyklos, circle) : Region of the plant bounded externally by the endodermis and internally by the phloem. Most roots originate from the pericycle.

Periplasm : The space (periplasmic space) between the cell (cytoplasmic) membrane of a bacterium or fungus and the outer membrane or cell wall.

Perithecium (pl perithecia) : Flask-shaped or subglobose, thin-walled fungus fruiting body (ascocarp) containing asci and ascospores; spores are expelled or released through a pore (ostiole) at the apex.

Peritrichous : Having hairs or flagella distributed over the whole surface.

Permanent wilting point (PWP) : The moisture content of soil at which plants wilt to such an extent that they fail to recover even when placed in a humid atmosphere.

Permeable(L permeabilis, that which can be penetrated) : Used of a membrane, cell or cell system through which substances may diffuse.

Permease : An enzyme system concerned with the transport of specific substances, usually nutrients, through the cytoplasm membrane.

Permissive conditions : A particular environmental condition which allows the growth of an organism with a conditional mutation. For example, for temperature sensitive mutations 30 C is typically a permissive temperature and 42 C is the nonpermissive condition.

Peroxyacetal nitrates (PAN) : A secondary air pollutant formed when the exhaust of automobiles and other internal combustion engines is in the presence of sunlight.

Persistence : Ability of an organism to remain in a particular setting for a period of time after it is introduced.

Persistent : 1. Continues to exist or to remain attached. 2. Chemicals with a long inactivation time, such as some pesticides, which may accumulate in the food chain.

Persistent transmission (syn circulative transmission) : A type of virus transmission in which the virus is acquired and transmitted by the vector after relatively long feeding times and remains transmissible for a prolonged period while in association with its vector.

PERV : See porcine endogenous retrovirus.

Pesticide : A substance used to control pests, such as harmful insects, weeds or microorganisms.

Pesticide resistance : A genetic change in response to selection by a pesticide resulting in the development of strains capable of surviving a dose lethal to a majority of individuals in a normal population. Resistance may develop in insects, weeds, or pathogens.

Pest-resistant crops : Plants with the ability to withstand, deter or repel pests and thereby prevent them from damaging the plants. Plant pests may include insects, nematodes, fungi, viruses, bacteria, and weeds, among others.

Petal : One of the parts of the flower that make up the corolla.

Petiole : Stalk portion of a leaf (see lamina).

Petite mutant : A respiration-deficient yeast mutant that produces small colonies when grown on glucose-containing medium.

Petri dish : Flat round dish with a matching lid, made of glass or plastic material, and used for culturing organisms. a.k.a. plates, hence the term to "plate" (q.v.) a culture.

PFGE : See Pulsed field gel electrophoresis.

PFU : Abbreviation for plaque forming unit.

pH : A measure of acidity and alkalinity. Equal to the log of the reciprocal of the hydrogen ion concentration of a solution, expressed in grams per litre. A reading of 7 is neutral (e.g., pure water), whereas below 7 is acid and above 7 is alkaline.

Phage : A virus for which the natural host is a bacterial cell. Used in the laboratory as a cloning vector. (see bacteriophage).

Phage head : The capsid which contains the phage nucleic acid. Often used to describe icosahedral capsid. The pressure inside an icosahedral head from a typical dsDNA phage is about 60 ATM.

Phage induction : A treatment which stimulates prophage to enter; the lytic cycle; eventually the host cell lyses and releases free phage.

Phage tail : Proteins that extend from a phage head and mediate adsorption to receptors on the host cell.

Phagemid : A plasmid that contains a portion of a phage genome. Upon co-infection of the host with a Helper phage, the plasmid can be packaged into phage particles. A commonly used type of phagemid is packaged as ssDNA into phage M13 particles.

Phagocytes : Immune system cells that ingest and destroy viruses, bacteria, fungi and other foreign substances or cells.

Phagocytosis : The process by which minute food particles or foreign particles invading the body are engulfed and broken down by certain animal cells.

Pharmaceutical agent : See therapeutic agent.

Pharmacogenomics : The study of the interaction of an individual's genetic makeup and response to a drug.

Phase state : The coupling or repulsion of two linked genes.

Phase variation : A mechanism which results in variable expression of surface antigens. An example is the switch between H_1 and H_2 flagellar antigens by an invertable switch in Salmonella.

pH-electrode-based sensor : Sensor in which a standard electrochemical pH electrode is coated with a biological material. Many biological processes raise or lower pH, and the changes can be detected by the pH electrode.

Phellem : Cork; a protective tissue composed of nonliving cells with suberized walls produced by the phellogen and accumulating on the surface of stems and roots.

Phelloderm : Tissue formed by and internal to the phellogen; resembles the cortical parenchyma in morphology.

Phellogen : Cork cambium; lateral meristem forming the periderm, a protective tissue in stems and roots; phellem (cork) is produced toward the surface, phelloderm toward the inside.

Phenocopy : An environmental condition that results in a phenotype different that that usually observed for a certain mutant. For example, the surfact properties of F+ bacteria can be altered to mimic F- bacteria when grown to stationary phase in medium with glucose as a carbon source.

Phenol extraction : A method commonly used to remove proteins from aqueous DNA samples. Phenol denatures proteins. The denatured proteins partition into the organic phase or remain at the interphase but the DNA remains in the aqueous phase.

Phenolic oxidation : Many plant species contain phenolic compounds that blacken through oxidation. The process is initiated after plants are wounded. Phenolic oxidation may lead to growth inhibition or, in severe cases, to tissue necrosis and death. Antioxidants are incorporated into the sterilizing solution or isolation medium to prevent or reduce oxidative browning. cf activated charcoal.

Phenological synchrony : The coordinated development of parasites and their hosts based on degree-days and other environmental factors so that parasites can successfully infect their hosts, such as the production and release of ascospores of Venturia inaequalis, the apple scab pathogen, from fallen leaves from the previous season, when new apple leaves are expanding.

Phenols; phenolics : Compounds with hydroxyl group(s) attached to the benzene ring, forming esters, ethers and salts. Phenolic substances are produced from newly explanted tissues, oxidizing to form coloured compounds visible in nutrient media.

Phenotype : The appearance or other observable characteristics of an organism. The phenotype expressed by an organism depends upon the particular forms of its genes (e.g. its wild-type or mutant alleles) and the environmental conditions.

Phenylalanine : See amino acid.

Pheromone : A hormone-like substance that is secreted by an organism into the environment as a specific signal to another organism, usually of the same species.

Phialide : End cell of a conidiophore with one or more open ends through which a basipetal succession of conidia develops.

Phloem : Food-conducting, food-storing tissue in the vascular system of roots, stems, and leaves.

Phloem necrosis : Death of phloem cells, often visible, caused by infection by systemic phloem pathogens such as phytoplasmas, spiroplasmas, and viruses.

PhoA : The gene encoding alkaline phosphatase. Alkaline phosphatase is inactive unless it is exported outside of the cytoplasm where it becomes oxidized. Thus, gene fusions to phoA are commonly used to identify protein domains that are translocated outside of the cytoplasm (including cytoplasmic membrane protein domains exposed to the periplasm, periplasmic proteins, outer membrane proteins, or excreted proteins).

Phosphatase : An enzyme that hydrolyses esters of phosphoric acid, removing a phosphate group from an organic compound.

Phosphodiester bond : The covalent bond between the 3' hydroxyl in the sugar ring of one nucleotide and the 5' phosphate group of the sugar ring of the adjacent nucleotide residue within a nucleic acid:

5'-Ribose- 3' - O - P(O)2 - O - 5' -Ribose - 3' - etc.

Phospholipase A2 : An enzyme which degrades type A2 phospholipids.

Phospholipid : A class of lipid molecules in which a phosphate group is linked to glycerol and two fatty acyl groups. A major component of biological membranes. See inositol lipid.

Phosphorolysis : The cleavage of a bond by orthophosphate; analogous to hydrolysis referring to cleavage by water.

Phosphorylation : The addition of a phosphate monoester to a macromolecule, catalyzed by a specific kinase enzyme. With respect to proteins, certain amino acid side chains (serine, threonine, tyrosine) are subject to phosphorylation catalyzed by protein kinases; altering the phosphorylation status of a protein may have dramatic effects on its biologic properties, and is a common cellular control mechanism. With respect to DNA, 5' ends must be phosphorylated for ligation.

Photoautotroph : See autotroph; heterotroph.

Photo-bioreactor : Bioreactor dependent on sunlight, which is taken up by its content of plant material, usually algae.

Photochemical : Chemical reactions that require sunlight.

Photoheterotroph : See heterotroph.

Photon : A quantum of light; the energy of a photon is proportional to its frequency; E = hv, where E is energy; h is Planck's constant, 6.62 × 10-27 erg-second; and v is the frequency.

Photoperiod (Gr photos, light + period) : The length of day or period of daily illumination provided or required by plants for reaching the reproductive stage.

Photoperiodism : The response of an organism to changes in day length.

Photophosphorylation : The formation of ATP from ADP and inorganic phosphate using light energy in photosynthesis.

Photoreactivation : The repair of ultraviolet irradiated DNA by the cleavage of pyrimidine dimers. Catalyzed by the photoreactivating enzyme, an enzyme that requires post-irradiation exposure to visible light.

Photosynthate : The carbohydrates and other compounds produced in photosynthesis.

Photosynthesis : A chemical process by which green plants synthesize organic compounds from carbon dioxide and water in the presence of sunlight.

Photosynthetic : Able to use sunlight energy to convert atmospheric carbon dioxide into organic compounds. Nearly all plants, most algae and some bacteria are photosynthetic.

Photosynthetic efficiency : Efficiency of converting light energy into organic compounds.

Photosynthetically active radiation (PAR) : Radiant energy captured by the photosynthetic system in the light reactions, usually taken to be the wavelengths between 400 and 700 nm.

Phototropism (Gr photos, light + tropos, turning) : A growth curvature in which light is the stimulus.

Phycomycete : Archaic term for a member of a group of fungi without cross walls (septa) in their mycelium.

Phyllody : Change of floral organs to leaflike structures.

Phylloplane : Leaf surface.

Phylloplane-competent : Used to describe epiphytic microorganisms adapted to living on the leaf surface.

Phyllotaxy : Arrangement of leaves on a stem in relation to one another.

Phylogeny : The classification of organisms according to evolutionary criteria.

Physical map : A map of the locations of identifiable landmarks on DNA (e.g., restriction enzyme cutting sites, genes), regardless of inheritance. Distance is measured in base pairs. For the human genome, the lowest-resolution physical map is the banding patterns on the 24 different chromosomes; the highest- resolution map would be the complete nucleotide sequence of the chromosomes.

Physiological suppression : A second mutation that affects an aspect of cell physiology and thereby restores the function of a primary mutation. For example, a mutation in a protease may allow accumulation of a partially active mutant protein, thereby restoring sufficient function to repair the mutant phenotype; a second mutation that increases accumulation of a limiting substrate may restore function of a primary mutant that has poorer substrate binding; a mutation that increases proline accumulation may suppress a primary mutant with a protein folding defect; etc.

Phytoalexin : Substance produced in higher plants in response to a number of chemical, physical, and biological stimuli that inhibits the growth of certain micro-organisms.

Phytochrome : A reversible pigment system of a protein nature, found in the cytoplasm of green plants. Phytochrome is associated with the absorption of light that affects growth, development and differentiation of a plant, independent of photosynthesis, e.g., in the photoperiodic response.

Phytohormone : A substance that stimulates growth or other processes in plants. They include auxins, abscissic acid, cytokinins, gibberellins and ethylene. Phytohormones are chemical messengers that may pass through cells, tissues and organs and stimulate biochemical, physiological and morphological responses.

Phytokinin : See cytokinin.

Phytolexins : Substances produced or formed by host plants in response to injury, physiological stimuli, infectious agents, or their products that accumulate to levels which inhibit the growth of microorganisms. Some include toxic substances produced to repel insects and nematodes.

Phytoparasite (adj phytoparasitic) : Parasite on plants.

Phytopathogen : An organism that causes disease in plants.

Phytopathogenic : Able to cause disease in plants.

Phytopathology (syn plant pathology) : The study of plant diseases.

Phytoplasma (syn mycoplasmalike organism, MLO) : Plant-parasitic pleomorphic mollicute (prokaryote with no cell wall) found in phloem tissue; cannot yet be grown on artificial nutrient media.

Phytosanitary certificate : Official document that indicates that plant material has been inspected and found to be free of certain pathogens.

Phytostat : The name adopted by Tulecke in 1965 for an apparatus designed for the semi-continuous chemostat culture of plant cells.

Phytotoxic : Harmful to plants (usually used to describe chemicals).

Pigments : Molecules that are coloured by the light they absorb. Some plant pigments are water soluble and are found mainly in the cell vacuole.

Pilus : A filamentous protein polymer protruding from the cell surface. Usually involved in attachment to other cells or surfaces. For example, the F-pilus expressed in *E. coli* cells that carry the F-plasmid is involved in attachment to F- cells prior to conjugation. The plural of pilus is pili.

Pinocytosis : The process by which a living cell engulfs a minute droplet of liquid.

Pipette : A slender graduated tube into which small amounts of liquids are taken up by suction, for measuring and transferring.

Pistil : (L. pistillum, a pestle) Central organ of the flower, typically consisting of ovary, style and stigma. The pistil is usually referred to as the female part of a perfect flower.

Pith : Parenchymatous tissue occupying the center of the stem.

PKU : Phenylketonuria, an enzyme deficiency condition characterized by the inability to convert one amino acid, phenylalanine, to another, tyrosine, resulting in mental deficiency. plasmid double-stranded, circular, bacterial DNA into which a fragment of DNA from another organism can be inserted.

Plant biotechnology : The addition of selected traits (genes) to plants to develop new plant varieties.

Plant breeding : The essential process of constantly developing newer, better and more environmentally suitable plants for the purpose of food production. Plant breeders use cross-pollination, selection and other techniques involving crossing plants to produce varieties with particular desired characteristics (genes) that can be passed on to future plant generations. These become important starting material for further improvements using genetic engineering.

Plant cell culture : Growth of plant cells or roots of plants in bioreactors.

Plant cell immobilization : Entrapment of plant cells in gel matrices; the cells are suspended in small drops of the material, which then is set or allowed to harden to make little carriers. Materials such as alginates, agar or polyacrylamide can be used.

Plant genetic resources (PGR) : Defined in the International Undertaking on Plant Genetic Resources (FAO, 1983) to mean the reproductive or vegetative propagating material of the following categories of plants: (i) cultivated varieties (cultivars) in current use and newly developed varieties; (ii) obsolete cultivars; (iii) primitive cultivars (landraces); (iv) wild and weed species, near relatives of cultivated varieties; and(v) special genetic stocks (including elite and current breeder's lines and mutants).

Plant pathology (syn phytopathology) : The study of plant diseases.

Plant pests : Organisms that may directly or indirectly cause disease, spoilage or damage to plants, plant parts or processed plant materials. Common examples include certain insects, mites, nematodes, fungi, molds, viruses and bacteria.

Plant-incorporated protectants : Formerly referred to as plant-pesticides, plant-incorporated protectants (PIPs) are substances that act like pesticides that are produced and used by a plant to protect it from pests such as insects, viruses, and fungi.

Plantlet : A small rooted shoot developed from seed or from cultured cells either by embryogenesis or organogenesis.

Plaque : A clear spot on an otherwise opaque culture plate of bacteria or cultured bacteria cells, showing where cells have been lysed by viral infection.

Plaque forming unit : A virus particle capable of forming a plaque, indicating that it is able to infect, reproduce, and kill host cells.

Plasma cells : Antibody-producing white blood cells derived from B lymphocytes.

Plasma membrane : See cell membrane; plasmalemma.

Plasmalemma (Gr plasma, anything formed + lema, a husk or shell of a fruit) : A delicate cytoplasmic double membrane found on the outside of the protoplast, adjacent to the cell wall. See cell membrane.

Plasmid : A class of circular, autonomously replicating, extrachromosomal DNA elements found in many bacteria. Contain origins of replication to ensure their maintenance. Some plasmids are capable of integrating into the host genome. A number of artificially constructed plasmids are used as cloning vectors or to alter the characteristics of the bacteria. Common plasmids are pBR322, pGEM, pUC18.

Plasmodesma (Gr plasma, something formed + demos, a bond, a band; pl plasmodesmata) : Fine protoplasmic thread passing through the plant cell wall that separates two protoplasts.

Plasmodiophoromycetes : A fungus-like group of organisms, sometimes called the Plasmodiophoromycota or the endoparasitic slime molds, characterized by the production of zoospores and plasmodia that are restricted to the cells of their host.

Plasmodium (pl plasmodia) : Naked multinucleate mass of protoplasm moving and feeding in amoeboid fashion.

Plasmogamy : The fusion of two sex cells.

Plasmolysis (Gr plasma, something formed + lysis, a loosening) : Separation of the cytoplasm from the cell wall, due to removal of water from the protoplast.

Plastid (Gr plastis, a builder) : A cytoplasmic body found in the cells of plants and some protozoa. Chloroplastids, for example, produce chlorophyll that is involved in photosynthesis.

Plastoquinone : A quinone which is one of a group of compounds involved in the transport of electrons in photosynthesis in chloroplasts.

Plate : 1. Verb: To distribute a thin film of something. Hence micro-organisms or plant cells are plated onto nutrient agar. 2. Noun: Refers to the two segments of a Petri dish or a similar-shaped item.

Platform shaker : See shaker.

Plating efficiency : The percentage of inoculated cells which give rise to cell colonies when seeded into culture vessels.

Pleiotropic : A single mutation that simultaneously changes several apparently unrelated phenotypes.

Pleiotropy : One gene that causes many different physical traits such as multiple disease symptoms.

Pleomorphic : Able to assume various shapes (and perhaps sizes); occurrence of more than one distinct form in the life cycle of an organism.

Ploidy : The number of sets of chromosomes per cell, e.g., haploid, diploid, polyploid.

Plot- : In a field study, a group of trees, all from the same entry (family, clone, provenance . . .) planted together (usually). (See block and randomized-block design).

Plot, contiguous- : In a field study, a group of trees, all from the same entry (family, clone, provenance . . .) planted together. A five-tree row-plot is probably the most common design for forest genetics experiments. (See block and randomized-block design).

Plot, non-contiguous- : In a field study, a group of trees, all from the same entry (family, clone, provenance . . .) planted in the same block, but not planted together. The non-contiguous interlocked-plot design is a recent innovation that allows tests to be thinned systematically and still maintaining uniform spacing. (See block and randomized-block design).

Plumule (L plumula, a small feather) : The first bud of an embryo or that portion of the young shoot above the cotyledons.

Pluripotency : The potential of a cell to develop into more than one type of mature cell, depending on environment.

Pluripotent : See totipotent.

Plus tree- : A phenotypically superior but untested tree. (See elite tree and candidate tree).

Pneumatic reactor : See airlift fermenter.

Point mutation : A single nucleotide substitution within a gene; there may be several point mutations within a single gene. Point mutations do not lead to a shift in reading frames, thus at most cause only a single amino acid substitution (see Frameshift Mutation).

Poisson distribution : A statistical test developed by Simeon Poisson . The Poisson equation describes the probability distribution of random events -- for example, the probability that random mutations in a cell will affect a particular gene, or the average number of viruses to infect a single cell at a particular multiplicity of infection.

Polar : At one end or pole of the cell.

Polar bodies : In female animals, the products of a meiotic division that do not develop into an ovum. The first polar body comprises one of the two products of meiosis I, and it m ay not go through meiosis II. The second polar body comprises one of the two products of meiosis II.

Polar mutation : A mutation which affects not only the gene in which it is located but other genes located immediately downstream on the DNA molecule. See Rho dependent polarity.

Polar nuclei : Two centrally located nuclei in the embryo sac that unite with a second sperm cell in a triple fusion. In certain seeds, the product of this fusion develops into the endosperm.

Polar transport : The directed movement within plants of compounds (usually endogenous plant growth regulators) mostly in one direction; polar transport overcomes the tendency for diffusion in all directions.

Polarity (Gr pol, an axis) : The observed differentiation of an organism, tissue or cell into parts having opposed or contrasted properties or form.

Pole cells : A group of cells in the posterior of Drosophila embryos that are precursors to the adult germ line.

Pollen (L pollen, fine flour) : The mass of germinated microspores or partially developed male gametophytes of seed plants.

Pollen culture : The *in vitro* culture and germination of pollen grains. Callus cultures thus obtained will form shoots or embryoids which develop into monoploid plants. See anther culture; microspore culture.

Pollen grain : A microspore produced in the pollen sac of angiosperms or the microsporangium of gymnosperms. Unicellular, with variable shape and size, and usually ovoid from 25 to 250μm.

Pollination : Deposition of pollen on the receptive part of the female flower.

Poly (A) polymerase : Enzyme that catalyses the addition of adenine residues to the 3′ end of pre-mRNAs to form the poly-(A) tail. See polyadenylation; polymerase.

Poly-(A) tail : See polyadenylation.

PolyA tail : After an mRNA is transcribed from a gene, the cell adds a stretch of A residues (typically 50-200) to its 3' end. It is thought that the presence of this "polyA tail" increases the stability of the mRNA (possibly by protecting it from nucleases). Note that not all mRNAs have a polyA tail; the histone mRNAs in particular do not.

Polyacrylamide gel electrophoresis : See PAGE.

Polyacrylamide gel electrophoresis (PAGE) : A method for separating nucleic acid or protein molecules according to their molecular size. The molecules migrate through the inert gel matrix under the influence of an electric field. In the case of protein PAGE, detergents such as sodium dodecyl sulphate (SDS) are often added to ensure that all molecules have a uniform charge. Secondary structure can often lead to the anomalous migration of molecules. Therefore it is common to denature protein samples by boiling them prior to PAGE. In the case of nucleic acids, denaturing agents such as formamide, urea or methyl mercuric hydroxide are often incorporated into the gel itself, which may also be run at high temperature. PAGE is used to separate the products of DNA-sequencing reactions and the gels employed are highly denaturing, since molecules differing in size by a single nucleotide must be resolved.

Polyadenylation : Post-transcriptional addition of a polyadenylic acid tail to the 3′ end of eukaryotic mRNAs. Also called poly-(A) tailing. The adenine-rich 3′ terminal segments is called a poly (A) tail.

Polycistronic : Refers to a form of gene organization resulting in transcription of an mRNA that contains the coding sequences for multiple gene products, each of which is independently translated from the mRNA.

Polyclonal antibody : A serum sample that contains a mixture of immunoglobulin molecules secreted against a specific antigen, each recognizing a different epitope, some antibodies of which bind to different antigenic determinants of one antigen.

Polycloning site : See polylinker.

Polycross : Open pollination of a group of genotypes (generally selected), in isolation from other compatible genotypes, in such a way as to promote random mating.

Polycross test- : A progeny test to assess general combining ability from crosses among selected parents. Identities can be maintained only for the seed parents. A mixture of pollen is artificially applied to each female parent, or the selections naturally and randomly open-pollinate themselves.

Polycyclic : Having several to many disease cycles in a growing season (see monocyclic).

Polycystic kidney disease (PKD) : A group of conditions characterized by fluid filled sacs that slowly develop in both kidneys, eventually resulting in kidney malfunction.

Polyembryony : In the ordinary course of events, one embryo is formed in each ovule and is derived from the fertilization of the ovum in the solitary embryo sac. However, two or more embryos could start development even though only one may reach maturity.

Polyethylene glycol (PEG); carbowax : A polymer having the general formula: $HOCH_2(CH_2OCH_2)XCH_2OH$ and available in a range of molecular weights from ca 1000 to ca 6000. PEG 4000 and PEG 6000 are commonly used to promote cell or protoplast fusion, and to facilitate DNA uptake in the transformation of organisms such as yeast. PEG is also used to concentrate solutions by withdrawing water from them.

Polyetic : Requiring more than one year to complete one life or disease cycle.

Polygene : One of many genes of small effect influencing the development of a quantitative trait; results in continuous variation and in quantitative inheritance. See gene.

Polygenic : Determined by several genes whose effects are readily detectable.

Polygenic disorders : Genetic disorders resulting from the combined action of alleles of more than one gene (e.g., heart disease, diabetes, and some cancers). Although such disorders are inherited, they depend on the simultaneous presence of several alleles; thus the hereditary patterns are usually more complex than those of single- gene disorders. Compare single- gene disorders.

Polygenic resistance (syn multigenic resistance) : Resistance conferred by several genes (see monogenic resistance, oligogenic resistance, single gene resistance).

Polygenic trait : An inherited trait that is determined by genes at two or more loci. Simple Mendelian rules of dominance do not apply to the complex interaction of these genes. As a result, phenotypes may appear as apparent blends or intermediate expressions. Human skin and hair color are polygenic traits. Many polygenic traits are also influenced by environmental factors.

Polylinker : A short, synthetic DNA sequence containing several restriction enzyme recognition sites. Polylinkers facilitate cloning into plasmid or phage vectors.

Polymer : A compound composed of many identical smaller sub-units; resulting from the process of polymerization.

Polymerase : An enzyme which links individual nucleotides together into a long strand, using another strand as a template. There are two general types of polymerase - DNA polymerases (which synthesize DNA) and RNA polymerase (which makes RNA). Within these two classes, there are numerous sub-types of polymerase, depending on what type of nucleic acid can function as template and what type of nucleic acid is formed. A DNA-dependant DNA polymerase will copy one DNA strand starting from a primer, and the product will be the complementary DNA strand.

A DNA-dependant RNA polymerase will use DNA as a template to synthesize an RNA strand.

Polymerase chain reaction : A technique for replicating a specific piece of DNA *in vitro,* even in the presence of excess non-specific DNA. Primers are added (which initiate the copying of each strand) along with nucleotides and heat stable Taq polymerase. By cycling the temperature, the target DNA is repetitively denatured and copied. Because the newly synthesized DNA strands can subsequently serve as additional templates for the same primer sequences, successive rounds of primer annealing, strand elongation, and dissociation produce rapid and highly specific amplification of the desired sequence. PCR also can be used to detect the existence of the defined sequence in a DNA sample. A single copy of the target DNA, even if mixed in with other undesirable DNA, can be amplified to obtain billions of replicates. PCR can be used to amplify RNA sequences if they are first converted to DNA via reverse transcriptase. This two-phase procedure is known as 'RT-PCR'. Polymerase Chain Reaction (PCR) is the basis for a number of extremely important methods in molecular biology. It can be used to detect and measure vanishingly small amounts of DNA and to create customized pieces of DNA. It has been applied to clinical diagnosis and therapy, to forensics and to vast numbers of research applications. It would be difficult to overstate the importance of PCR to science.

Polymerase, DNA or RNA : Enzymes that catalyze the synthesis of nucleic acids on preexisting nucleic acid templates, assembling RNA from ribonucleotides or DNA from deoxyribonucleotides.

Polymerization : Chemical union of two or more molecules of the same kind such as glucose or nucleotides to form a new compound (starch or nucleic acid) having the same elements in the same proportions but a higher molecular weight and different physical properties.

Polymery : The phenomenon whereby a number of genes at different loci (which may be polygenes) can act together to produce a single effect.

Polymorphic genes : Have 2 or more common alleles in the population.

Polymorphism : Difference in DNA sequence among individuals. To be called a polymorphism, a variant should be present in a significant number of people in the population. Genetic variations occurring in more than 1% of a population would be considered useful polymorphisms for genetic linkage analysis. Compare mutation.

Polynucleotide : A chain of nucleotides in which each nucleotide is linked by a single phospho-diester bond to the next nucleotide in the chain. They can be double- or single-stranded. The term is used to describe DNA or RNA. See nucleotide.

Polynucleotide kinase : Enzyme which catalyzes the transfer of the terminal phosphate of ATP to 5' hydroxyl termini of polynucleotides, either DNAor RNA. Usually derived from T4 bacteriophage.

Polypeptide : A linear molecule composed of two or more amino acids linked by covalent (peptide) bonds. They are called dipeptides, tripeptides and so forth, according to the number of amino acids present.

Polyploid- : A cell, tissue, individual, population, or whole species having more than twice the basic number oex" of chromosomes of the ancestral species. Polyploidy may lead to increased growth rates (Populus) or severe dwarfing (Pinus). Three sets of chromosomes is called triploid, four sets-tetraploid, six sets-hexaploid, etc. (See chromosome number).

Polyprotein : A giant polypeptide that contains multiple individual protein sequences embedded within it and which must be proteolytically cleaved to yield the individual proteins.

Polysaccharide (Gr polys, many + Gr sakcharon, sugar) : Long-chain molecules, such as starch and cellulose, composed of multiple units of a monosaccharide.

Polysaccharide capsules : Carbohydrate coverings with antigenic specificity that are present on some types of bacteria.

Polysome : An mRNA molecule that is in the process of being translated by multiple ribosomes simultaneously.

Polyspermy : The entry of several sperm into the egg during fertilization, although only one sperm nucleus actually fuses with the egg nucleus.

Polytene chromosomes : Giant chromosomes produced by interphase replication without division, and consisting of many identical chromatids arranged side by side in a cablelike pattern.

Polytopic : Commonly used to describe membrane proteins that span the membrane multiple times, such that different regions of the protein are exposed on opposite sides of the membrane. ("Topic" refers to the topology of the protein in the membrane).

Polyunsaturates : Oils containing polyunsaturated fatty acids.

Polyvalent vaccine : A recombinant organism into which antigenic determinants have been cloned from a number of different disease-causing organisms, and used as a vaccine. See vaccine.

Polyvinylpyrrolidone (PVP) (C6H9NO)n) : An occasional constituent of plant tissue culture isolation media. PVP is of variable molecular weight and has antioxidant properties, so is used to prevent oxidative browning of explanted tissues. PVP is less frequently used as an osmoticum in culture media.

Pome (syn pome fruit) : Simple, fleshy, indehiscent fruit derived from several carpels, e.g. apple, pear, quince.

Population : A local group of organisms belonging to the same species and interbreeding.

Population- : A group of individual trees having some characteristics in common, either location, family ancestry, or intended use.

Population density : Number of cells or individuals per unit. The unit could be an area or volume of medium.

Population genetics : The branch of genetics that deals with frequencies of alleles and genotypes in breeding populations. cf. quantitative genetics.

Populations : In genetics, a community of individuals which share a common gene pool. In statistics, a hypothetical and infinitely large series of potential observations among which observations may actually constitute a sample.

Porcine endogenous retrovirus (PERV) : The provirus (q.v.) of a porcine retrovirus. With the increasing interest in the use of pig organs for xenotransplantation to humans, there has been concern that PERVs could be activated after transplantation, creating an infection in the human recipient.

Position effect : The situation in which a change in phenotype results from the change of the position of a gene or group of genes.

Positional candidate gene : A gene known to be located in the same region as a DNA marker that has been shown to be linked to a single-locus trait or to a quantitative trait locus (QTL) (q.v.) and whose function suggests that it could be the source of genetic variation in the trait in question.

Positional cloning : The process that commences with searching for markers linked to a particular inherited trait, then using those markers to identify the approximate location of the gene responsible for the trait, and then using various cloning strategies (including chromosome walking, q.v.) to identify, isolate and characterize the gene. Originally, the strategy was called reverse genetics, a term that some consider to be a misnomer and misleading.

Positive control system : A mechanism in which the regulatory protein(s) is required to turn on gene expression.

Positive selectable marker : See dominant marker selection.

Positive selection : 1. See dominant marker selection. 2. A method by which cells that carry a DNA insert integrated at a specific chromosomal location are selected.

Positive sense RNA : RNA that can serve directly as messenger RNA (see negative sense RNA).

Positive supercoiling : Additional coiling of the circular duplex DNA molecule in the same direction as the winding of the double helix.

Post-emergent herbicide : A herbicide that controls weeds after they emerge from the ground.

Posterior : Toward the back or rear; (see anterior).

Post-meiotic segregation : The type of segregation produced when a recombinant DNA molecule contains an uncorrected mismatched base pair; at the next replication, normal base pairing occurs producing one mutant progeny and one wild-type progeny.

Post-replication repair : A recombination-dependent mechanism for repairing damaged DNA.

Post-transcriptional regulation : Any process occurring after transcription which affects the amount of protein a gene produces. Includes RNA processing efficiency, RNA stability, translation efficiency, protein stability. For example, the rapid degradation of an mRNA will reduce the amount of protein arising from it. Increasing the rate at which an mRNA is translated will increase the amount of protein product.

Post-translational modification : The modification of proteins after the protein has been synthesized. Proteins may be modified in a wide variety of ways, including phosphorylation (addition or a phosphate group), adenylation (addition of an adenine group), glycosylation (addition of a sugar group), acylation (addition of a lipid group), proteolysis (removal of the initiator methionine, cleavage of a signal sequence, or cleavage of a pro-peptide), etc. are subject to a multitude of modification steps. These modifications can have profound, sometimes reversible effects on protein structure and function.

Post-translational processing : The reactions which alter a protein's covalent structure, such as phosphorylation, glycosylation or proteolytic cleavage.

Post-translational regulation : Regulation of gene expression after the gene has been transcribed into mRNA and translated into protein. For example, by regulation of protein activity by covalent modification or proteolysis.

Potentiometric : See enzyme electrode.

Powdery mildew : A white, powdery, superficial ascomycetous fungus that is an obligate parasite and generally is found only on one or a few closely related species of plants.

ppm : See parts per million.

Prader-Willi syndrome : A condition characterized by obesity and insatiable appetite, mental deficiency, small genitals, and short stature. May be deletion of #15 chromosome.

PRE : Progesterone Response Element: A binding site in a promoter to which the activated progesterone receptor can bind. The progesterone receptor is essentially a transcription factor which is activated only in the presence of progesterone. The activated receptor will bind to a PRE, and transcription of the adjacent gene will be altered. See also "Response element".

Precocious germination : Premature germination of the embryo, prior to completion of embryogenesis.

Predispose (n predisposition) : To make prone to infection and disease.

Predisposition : To have a tendency or inclination towards something in advance.

Pre-emergent herbicide : A herbicide that controls weeds before they emerge from the ground.

Pre-filter : A coarse filter used to screen out large particles before air is forced through a much finer filter. See HEPA filter; laminar air-flow cabinet.

Premature chromosome condensation (PCC) : A method of studying chromosomes in the interphase stage of the cell cycle.

Pre-mRNA : An RNA molecule which is transcribed from chromosomal DNA in the nucleus of eukaryotic cells, and subsequently processed through splicing reactions to generate the mRNA which directs protein synthesis in the cytoplasm.

Pressure potential : The pressure generated within a cell; it is the difference between the osmotic potential within the cell and the water potential of the external environment, provided the cell volume is constant.

Presymptomatic diagnosis : Diagnosis of a genetic condition before the appearance of symptoms.

Pre-transplant : Stage III in tissue culture micropropagation; the rooting, hardening stage prior to transfer to soil. See micropropagation.

Preventive immunization; vaccination : Infection with an antigen to elicit an antibody response that will protect the organism against future infections.

Pribnow box : Consensus sequence near the mRNA start-point of prokaryotic genes. See TATA box.

Primary (L primus, first) : First in order of time or development.

Primary antibody : In an ELISA or other immunological assay, the antibody that binds to the target molecule.

Primary cell : A cell or cell line taken directly from a living organism, which is not immortalized.

Primary cell wall : The cell wall layer formed during cell expansion. Plant cells possessing only primary walls may divide or undergo differentiation.

Primary culture : A culture started from cells, tissues or organs taken directly from organisms. A primary culture may be regarded as such until it is sub-cultured for the first time. It then becomes a cell line.

Primary germ layers : See germ layers.

Primary growth : 1. Refers to apical meristem-derived growth; the tissues of a young plant. 2. Refers to explant growth during the initial culture period; such as primary callus growth.

Primary immune response : The immune response that occurs during the first encounter of a mammal with a given antigen. cf secondary immune response.

Primary inoculum (syn initial inoculum) : Inoculum, usually from an overwintering source, that initiates disease in the field, as opposed to inoculum that spreads disease during the season.

Primary leaf : First true leaf that emerges on a plant following the cotyledons.

Primary meristem : Meristem of the shoot or root tip giving rise to the primary plant body.

Primary pollutant : Air pollutant that is released directly into the atmosphere and is harmful to plants, e.g. SO_2 or NOx.

Primary root : Root that develops directly from the radicle of an embryo rather than from a crown or node.

Primary structure : Refers to the sequence of amino acid residues or nucleotides within protein or nucleic acid molecules, respectively (also see Secondary and Tertiary Structure).

Primary tissue : A tissue that has differentiated from a primary meristem.

Primary transcript : When a gene is transcribed in the nucleus, the initial product is the primary transcript, an RNA containing copies of all exons and introns. This primary transcript is then processed by the cell to remove the introns, to cleave off unwanted 3' sequence, and to polyadenylate the 5' end. The mature message thus formed is then exported to the cytoplasm for translation.

Primase : An enzyme that synthesizes ribonucleotide primers for lagging strand DNA synthesis of Okazaki fragments. Polymerizes ribonucleotide triphosphates in the 5' to 3' direction.

Primer : Short, preexisting single-stranded polynucleotide chain to which new deoxyribonucleotides can be added by DNA polymerase (to 'prime' PCR amplification). The primer anneals to a nucleic acid template (DNA of the organism of interest) and promotes copying of the template, starting from the primer site. To amplify microsatellites one uses a forward and reverse primer pair:

[agctcagtccctagtcagtact]acacacacacacacacacacac[ggtacttcggagctatccgaattccct]

In this example the bold, italicized bp are the forward and reverse primers (should not differ among individuals), whereas the unitalicized 'ac' repeat is the microsatellite. By running back and forth across the repeat one can amplify a few copies of the microsatellite region by orders of magnitude, yielding sufficient DNA to allow visualization of the amplified product on an acrylamide gel by staining with ethidium bromide. Some primer sequences may be conserved across wide taxonomic gaps (e.g., across families), while others may differ even among congeners.

Primer DNA polymerase : DNA polymerase that provides primers for the DNA polymerization. Unlike RNA polymerase, DNA polymerase is unable to initiate the de novo synthesis of a polynucleotide chain. DNA polymerase can only add nucleotides to a free 3´ hydroxyl group at the end of a pre-existing chain. A short oligonucleotide, known as a primer, is therefore needed to supply such a hydroxyl group for the initiation of DNA synthesis.

Primer extension : This is a method used to figure out how far upstream from a fixed site the start of an mRNA is. For example, perhaps you have isolated a cDNA clone, but you don't think that the clone has all of the 5' untranslated region. To find out how much is missing, you would first sequence the part you have, and figure out which strand is coding strand (usually the coding strand will have a large open reading frame). Next, you ask the DNA Synthesis Facility to make an oligonucleotide complementary to the 5'-most region of the coding strand (and thus complementary to the mRNA). This "primer" is hybridized to mRNA (say, a mixture of mRNA containing the one in which you are interested), and reverse transcriptase is added to copy the mRNA from the primer out to the 5' end. The size of the resulting DNA fragment shows how far away from the 5' end your primer is.

Primer walking : A method for sequencing long (>1 kb) cloned pieces of DNA. The initial sequencing reaction reveals the sequence of the first few hundred nucleotides of the cloned DNA. On the basis of these data, a primer containing about 20 nucleotides and complementary to a sequence near the end of sequenced DNA is synthesized, and is then used for sequencing the next few hundred nucleotides of the cloned DNA. This procedure is repeated until the complete nucleotide sequence of the cloned DNA is determined.

Primordium : A group of cells which gives rise to an organ.

Primosome : A protein-replication complex that catalyses the initiation of synthesis of Okazaki fragments during discontinuous replication of DNA. It involves DNA primase and DNA helicase activities.

Principle of independent assortment : Gregor Mendel's second principle of genetic inheritance. It states that different pairs of genes are passed to offspring independently so that new combinations of genes, present in neither parent, are possible. In other words, the distribution of one pair of alleles does not influence the distribution of another pair. The genes controlling different traits are inherited independently of one another.

Principle of segregation : Gregor Mendel's first principle of genetic inheritance. It states that, for any particular trait, the pair of genes of each parent separate (during the formation of sex cells) and only one gene from each parent passes on to an offspring. In other words, genes occur in pairs (because chromosomes occur in pairs). During gamete production, the members of each gene pair separate, so that each gamete contains one member of each pair. During fertilization, the full number of chromosomes is restored, and members of gene pairs are reunited.

Prion : An infectious agent thought to be composed solely of protein. Term is derived from "Protein infectious agent".

Privacy : In genetics, the right of people to restrict access to their genetic information.

Pro : Before, in front of, in anticipation of.

Probability : The long term frequency of an event relative to all alternative events, and usually expressed as decimal fraction.

Proband : The individual in a family in whom an inherited trait is first identified.

Probe : Single- stranded DNA or RNA fragment of specific base sequence, labeled either radioactively (often incorporating 32P or 35S) or immunologically, that are used to detect the complementary base sequence by hybridization. For example, if you want to quantitate the levels of alpha subunit mRNA in a preparation of pituitary RNA, you might make a radiolabeled RNA in-vitro which is complementary to the mRNA, and then use it to probe a Northern blot of the pit RNA. A probe can be radiolabeled, or tagged with another functional group such as biotin. A probe can be cloned DNA, or might be a synthetic DNA strand. As an example of the latter, perhaps you have isolated a protein for which you wish to obtain a cDNA or genomic clone. You might (pay to) microsequence a portion of the protein, deduce the nucleic acid sequence, (pay to) synthesize an oligonucleotide carrying that sequence, radiolabel it and use it as a probe to screen a cDNA library or genomic library. A better way is to call up someone who already has the clone.

Probe DNA : A labelled DNA molecule used to detect complementary-sequence nucleic acid molecules by molecular hybridization. To localize the probe DNA sequence and reveal the complementary hybridization sequence, autoradiography or fluorescence is used. See multilocus probe; nucleotide.

Procambium (L pro, before + cambium) : A primary meristem that gives rise to primary vascular tissues and, in most woody plants, to the vascular cambium.

Procaryote : An organism lacking a nuclear membrane and certain organelles such as mitochondria. Refers to both Bacteria and Archae. This term was used synonymously with bacteria before the differences between Bacteria and Archae were understood. Although prokaryotes share some common features because of the lack of membrane bound organelles (e.g., coupled transcription and translation are possible in prokaryotes but not in eukaryotes), overall the Archae are more similar to the Eukarya than they are to Bacteria.

Processed pseudo-gene : A copy of a functional gene which has no promoter, no introns and which, consequently, is not itself transcribed. Pseudo-genes are thought to originate from the integration into the genome of cDNA copies synthesized from mRNA molecules by reverse transcriptase. Pseudo-genes therefore have a poly (dA) sequence at their 5′ ends. Because they are not subject to any evolutionary pressure to maintain their coding potential, pseudo-genes accumulate mutations and often have stop codons in all three reading frames.

Processing : The reactions occurring in the nucleus which convert the primary RNA transcript to a mature mRNA. Processing reactions include capping, splicing and polyadenylation. The term can also refer to the processing of the protein product, including proteolytic cleavages, glycosylation, etc.

Processivity : The ability of an enzyme to continue to act on a polynucleotide for a long distance without dissociating.

Processor preferred high extractable starch (hes) : Hybrids can deliver higher levels of extractable starch, which allows corn wet millers to produce higher starch yields that they can turn into products that include high fructose corn syrup, corn oil, specialty starch products, commodity starches and ethanol.

Processor Preferred High Fermentable Corn (HFC) : Hybrids can deliver higher levels of fermentable starch, which allows dry mill ethanol plants to obtain higher ethanol yields for greater profit potential.

Processor Preferred soybeans : These are soybean varieties proven to have a higher oil or higher protein content, and several varieties offer a combination of high oil and high protein. These are properties valued by soybean processors.

Production environment : In AnGR: All input-output relationships, over time, at a particular location. The relationships will include biological, climatic, economic, social, cultural and political factors, which combine to determine the productive potential of a particular livestock enterprise. (Source: FAO, 1999).

Production traits : In AnGR: Characteristics of animals, such as the quantity or quality of the milk, meat, fibre, eggs, draught, etc., they (or their progeny) produce, which contribute directly to the value of the animals for the farmer, and that are identifiable or measurable at the individual level. Production traits of farm animals ate generally quantitatively inherited, i.e., they are influenced by many genes whose expression in a particular animal also reflects environmental influences. (Source: FAO, 1999).

Productivity : The amount of product that is produced within a given period of time from a specified quantity of resource.

Pro-embryo (L pro, before + embryon, embryo) : A group of cells arising from the division of the fertilized egg cell or somatic embryo before those cells which are to become the embryo are recognizable.

Proflavin : An intercalating agent used as a frameshift mutagen.

Progeny : See offspring.

Progeny test : A test of the value of a genotype based on the performance of its offspring produced in some definite system of mating.

Progeny testing : For a single locus: the practice of ascertaining the genotype of an individual from its offspring, such as by mating it to other individuals and examining the progeny. For a quantitative trait: the collection and use of progeny performance as a clue to the breeding value of the parent of those progeny.

Progesterone : A hormone produced primarily by the corpus luteum of the ovary, but also by the placenta, that prepares the inner lining of the uterus for implantation of a fertilized egg cell.

Prognosis : Prediction of the course and probable outcome of a disease.

Proinhibitin : A plant defense compound that is already present before invasion by a pathogen.

Prokaryote (L pro, before + Gr karyon, a nut, referring in modern biology to the nucleus) : A member of a large group of organisms, including bacteria and blue-green algae, which do not have the DNA separated from the cytoplasm by a membrane in their cells. The DNA is usually in one long strand. Prokaryotes do not undergo meiosis and do not have functional organelles such as mitochondria and chloroplasts. cf eukaryote.

Prolactin : A hormone, produced by the anterior pituitary gland, that stimulates and controls lactation in mammals.

Proliferation (L proles, offspring + ferre, to bear) : Increase by frequent and repeated reproduction; growth by cell division.

Pro-meristem : The embryonic meristem that is the source of organ initials or foundation cells.

Promiscuous plasmid : A plasmid with a broad host-range.

Promoter : A sequence on DNA that functions as the RNA polymerase binding site, thus defining the transcription start site. Promoter recognition is largely determined by the sigma factor associated with RNA polymerase and the interaction of RNA polymerase with positive regulatory proteins. The consensus sequence for a promoter depends upon the sigma factor associated with RNA polymerase.

Promoter sequence : A regulatory DNA sequence that initiates the expression of a gene.

Promoter site : Region of a DNA molecule 5' to a coding sequence that is responsible for assembly of an RNA polymerase complex and the initiation of transcription.

Promycelium (pl promycelia) : In rust and smut fungi, a germ tube issuing from the teliospore and bearing the basidiospores.

Pro-nuclear micro-injection : The initial method of transgenesis in animals, involving injecting many copies of a particular gene into one of the two pro-nuclei of a fertilized ova. Characterised by a very low success rate. As animal cloning (q.v.) becomes more common, pro-nuclear micro-injection will be replaced by micro-injection of cloned genes into a culture of cloned embryos produced by nuclear transfer, which can be tested for expression of the transgene before transfer to recipient females. See transgenesis.

Pro-nucleus : Either of the two haploid gamete nuclei just prior to their fusion in the fertilized ovum.

Proofreading : Removal of mismatched base pairs during DNA replication by the 3' to 5' exonuclease activity of DNA polymerase, followed by resynthesis.

Propagation (L propagare, to propagate) : The multiplication of plants by numerous types of vegetative material; an ancient practice dating from the dawn of agriculture, carried out in a nursery or directly in the field (vegetative propagation), and now in *in vitro* culture (micropropagation).

Propagative transmission (syn circulative propagative transmission) : Pathogen transmission characterized by a long period of acquisition of the pathogen (usually a mollicute, e.g. phytoplasma or spiroplasma, and sometimes a virus) by a vector (typically an insect), a latent period before the vector is able to transmit the pathogen, and retention of the pathogen by the vector for a long period because the pathogen reproduces or replicates in the vector.

Propagule : Any structure capable of giving rise to a new plant by asexual or sexual reproduction, including bulbils, leafbuds, etc.

Prophage : A temperate phage genome whose lytic functions are repressed and which replicates in synchrony with the bacterial chromosome. A prophage may be integrated into the host chromosome (as with phage lambda) or exist as an autonomously replicating plasmid (as with phage P1).

Prophase(Gr pro, before + phasis, appearance) : An early stage in nuclear division, characterized by the shortening and thickening of the chromosomes and their movement to the metaphase plate. It occurs between interphase and metaphase. During this phase, the centriole divides and the two daughter centrioles move apart. Each sister DNA strand from interphase replication becomes coiled, and the chromosome is longitudinally double except in the region of the centromere. Each partially separated chromosome is called a chromatid. The two chromatids of a chromosome are sister chromatids.

Protamines : Small basic proteins that replace the histones in the chromosomes of some sperm cells.

Protandry : Maturation of anthers before pistils.

Protease : An enzyme that hydrolyzes proteins, cleaving the peptide bonds that link amino acids in protein molecules.

Proteasome : A specialized organelle within the cytoplasm of a cell that is responsible for degradation of cytoplasmically situated proteins. The proteasome plays a key role in normal protein turnover and in peptide presentation by MHC Class I antigen.

Protectant : Agent, usually a chemical, applied to a plant surface in advance of a pathogen to prevent infection.

Protectant fungicide (syn contact fungicide) : A fungicide that remains on the surface where it is applied; no after-infection activity (see systemic fungicide).

Protection : A principle of plant disease control in which a barrier is placed between the susceptible plant and pathogen (e.g. the use of protective chemical dusts or sprays).

Protein : A large molecule composed of one or more chains of amino acids in a specific order; the order is determined by the base sequence of nucleotides in the gene coding for the protein. Proteins are required for the structure, function, and regulation of the bodys cells, tissues, and organs, and each protein has unique functions. Examples are hormones, enzymes, and antibodies.

Protein crystallization : Making crystals of a protein, as a key part of most methods used for determining a protein's three-dimensional structure.

Protein drug : See therapeutic agent.

Protein engineering : Generating proteins with modified structures that confer properties such as higher catalytic specificity or thermal stability.

Protein kinase : An enzyme that adds phosphate groups to a protein molecule at the sites of serine, threonine or tyrosine residues.

Protein metabolic step : One step in the chain of reactions that take place in an organism and dictate the composition of that organism.

Protein sequencing : The process of determining the amino-acid sequence of a protein or its component polypeptides.

Protein synthesis : The creation of proteins from their constituent amino acids, in accordance with the genetic information carried in the DNA of the chromosomes.

Proteinaceous infectious particle; prion : An abnormal form of a normal cell protein, with no detectable nucleic acid, found in the brain of mammals and believed to be the agent responsible for the class of diseases called spongiform encephalopathies, including scrapie in sheep and bovine spongiform encephalopathy (BSE; mad cow disease) in cattle.

Proteolysis : The degradation of proteins to peptides or amino acids catalyzed by proteases.

Proteolytic : The ability to break down protein molecules.

Proteome : The full complement of proteins produced by an individual (or a species). It is estimated that each human produces approximately 90,000 types of proteins. See Human Genome Project.

Proteomics : The large-scale study of proteins, using biochemical or physical methods. Approaches include two-dimensional electrophoresis of proteins expressed under particular conditions or in particular cell compartments, Mass spectometry of proteins, use of hybrid approaches to identify protein-protein interactions, large scale crystallography of cellular proteins, etc. This approach complements the analysis of genome sequences (genomics) because it identifies the protein gene products that perform the cellular functions encoded by genes; The analysis of all (or most) of the proteins encoded by a genome. Some of the techniques used to characterize the proteins on this scale include bioinformatics, 2-dimensional gel electrophoresis, mass spectrometry, and protein microarrays.

Proteus syndrome : A condition characterized by distorted asymmetric growth of the body and enlarged head, enlarged feet, multiple nevi on the skin; mode of inheritance is unknown.

Protoclone : Regenerated plant derived from protoplast culture or a single colony derived from protoplasts in culture.

Protocorm : In orchids, seeds contain an unorganized embryo comprising only a few hundred cells. During seed germination this embryo forms a tuberous structure, called a protocorm, from which develops a complete plant. In cultures, vegetative parts of several orchids form round, smooth protocorms which may be multiplied indefinitely or induced to regenerate a whole plant.

Protoderm (Gr protos, first + derma, skin) : A primary meristem tissue that gives rise to epidermis.

Protogyny : The condition in which the female reproductive organs (carpels) of a flower mature before the male ones (stamens), thereby ensuring that self-fertilization does not occur.

Protomeristem : See promeristem.

Proto-oncogene : A gene present in a normal cell which carries out a normal cellular function, but which can become an oncogene under certain circumstances. The prefix "c-" indicates a cellular gene, and is generally used for proto-oncogenes (examples: c-myb , c-myc , c-fos , c-jun , etc).

Protoplasm : The essential, complex living substance of cells, upon which all vital functions of nutrition, secretion, growth and reproduction depend.

Protoplast (Gr protoplastos, formed first) : A bacterial or plant cell for which the relatively rigid wall has been removed either chemically or enzymatically, leaving its cytoplasm enveloped by only a delicate peripheral membrane. Protoplasts are spherical and smaller than the elongate, angular shaped and often vacuolated cells from which they have been released.

Protoplast culture : The isolation and culture of plant protoplasts by mechanical means or by enzymatic digestion of plant tissues or organs, or cultures derived from these. Protoplasts are utilized for selection or hybridization at the cellular level and for a variety of other purposes.

Protoplast fusion : The coalescence of the plasmalemma and cytoplasm of two or more protoplasts in contact with one another. Initial adhesion of two protoplasts is a random process but coalescence can be induced.

Prototroph : A microorganism that is able to grow on minimal medium containing only a carbon source and inorganic salts.

Protozoan (pl protozoa) : A microscopic, single-cell organism.

Protruding end : See extension.

Provenance test- : A test comparing trees grown from seed or cuttings collected in many parts of a species range. (See seed source).

Provirus : A retrovirus in which the single RNA strand has been converted into a double-strand DNA, which has been integrated into a host genome.

Proximal : Nearest to the point of attachment (see distal).

Pseudo-affinity chromatography : See affinity chromatography.

Pseudo-autosomal region : A section at one end of the X and Y chromosomes for which there is sufficient homology that there is pairing (synapsis) during meiosis I.

Pseudocarp; false fruit : A fruit that incorporates, in addition to the ovary wall, other parts of the flower, such as the receptacle (e.g., strawberry).

Pseudogene : An inactive gene derived from an ancestral active gene. Pseudogenes are often recognized by nonsense or frameshift mutations that disrupt an open reading frame that encodes a functional protein in a related genome.

Pseudoknot : A feature of RNA tertiary structure; best visualized as two overlapping stem-loops in which the loop of the first stem-loop participates as half of the stem in the second stem-loop.

Pseudomonas spp : A common Gram-negative bacterial genus that is widely distributed. Many of the soil forms produce a pigment that fluoresces under ultraviolet light, hence the descriptive term fluorescent pseudomonas.

Pseudoreversion : Reversion of a mutant phenotype caused by a second mutation at a different site that partially or fully restores the wild-type phenotype. See suppression.

Pseudorevertant : A mutant virus or organism which has recovered a wildtype phenotype due to a second-site mutation (potentially located in a different region of the genome, or involving a different polypeptide) which has eliminated the effect of the initial mutation.

Pseudothecium (pl pseudothecia) : Perithecium-like fruiting body containing asci and ascospores dispersed rather than in an organized hymenium; an ascostroma with a single locule or cavity and containing bitunicate asci

P-site : See peptidyl-tRNA site.

Psoriasis : An inherited disease characterized by recurring thick, reddish patches of inflamed skin. It affects men and women equally. This trait is subject to genome imprinting.

Psychrophile : A micro-organism that can grow at temperatures below 30°C and as low as 0°C.

Public policy : A set of action guidelines or rules that result from the actions or lack of actions of governmental entities.

PUC : A widely used expression plasmid, containing a galactosidase gene. See plasmid.

Puccinia pathway : The region through which rust urediniospores move from southern areas through all grain-producing areas of the United States to Canada each season.

Pulsed field gel electrophoresis (PFGE) : A gel technique which allows size-separation of very large fragments of DNA, in the range of hundreds of kb to thousands of kb. As in other gel electrophoresis techniques, populations of molecules migrate through the gel at a speed related to their size, producing discrete bands. In normal electrophoresis, DNA fragments greater than a certain size limit all migrate at the same rate through the gel. In PFGE, the electrophoretic voltage is applied alternately along two perpendicular axes, which forces even the larger DNA fragments to separate by size.

Punctuated equilibrium : The occurrence of speciation events in bursts, separated by long intervals of species stability.

Punnett square : A simple graphical method of showing all of the potential combinations of offspring genotypes that can occur and their probability given the parent genotypes. See example below. Punnett squares are commonly used by genetics counselors to predict the odds of a couple passing on particular inherited traits.

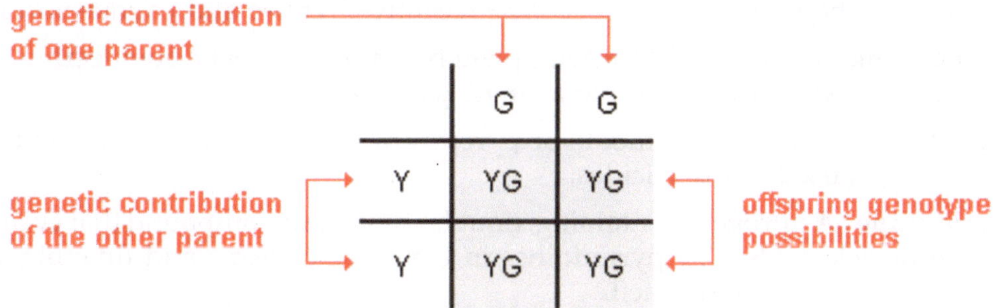

Pupa (pl pupae; v pupate) : Quiescent stage between the larva and the adult of certain insects.

Pure culture : Axenic culture.

Pure line : A strain in which all members have descended by self-fertilization or close inbreeding . A pure line is genetically uniform.

Purebred : Offspring that are the result of mating between genetically similar kinds of parents--the opposite of hybrid. Purebred is the same as true breeding.

Purification tag : See affinity tag.

Purine : A nitrogen- containing, single- ring, basic compound that occurs in nucleic acids. The purines in DNA and RNA are adenine and guanine.

Puromycin : An antibiotic that inhibits protein synthesis. Puromycin is an analog of charged tRNA that binds to the ribosome, and is transferred to the growing peptide chain. The polypeptide is released from the ribosome resulting in premature translation termination. Puromycin inhibits protein synthesis in Bacteria, Eukarya, and Archae.

Pustule : Small blisterlike elevation of epidermis formed as spores emerge.

Pv (abbr for pathovar) : A subdivision of a plant pathogenic bacterial species defined by host range; pathovar for bacteria is equivalent to forma specialis for fungi.

PVP : See polyvinylpyrrolidone.

PWP : See permanent wilting point.

Pycnidiospore : Spore (conidium) produced in a pycnidium.

Pycnidium (pl pycnidia) : Asexual, globose or flask-shaped fruiting body of certain imperfect fungi producing conidia.

Pycniospore (syn spermatium) : Haploid, sexually derived spore formed in a pycnium of rust fungi.

Pycnium (pl pycnia; syn spermagonium) : Globose or flask-shaped haploid fruiting body of rust fungi bearing receptive hyphae and pycniospores.

Pyramiding : The addition, through plant breeding or genetic engineering, of several resistance genes into a single plant cultivar.

Pyrethrins : Active constituents of pyrethrum (Tanacetum cinerariifolium) flowers, used as insecticides.

Pyrimidine : A nitrogen- containing, double- ring, basic compound that occurs in nucleic acids. The pyrimidines in DNA are cytosine and thymine; in RNA, cytosine and uracil.

Pyrimidine : A nitrogen- containing, double- ring, basic compound that occurs in nucleic acids. The pyrimidines in DNA are cytosine and thymine; in RNA, cytosine and uracil.

Pyrimidine dimers : Covalent bonds formed between two adjacent pyrimidines on the same strand of DNA induced by Ultraviolet irradiation.

Pyrogen : Bacterial substance that causes fever in mammals.

QTL : See quantitative trait locus.

Quadrivalent : See tetrad (second definition).

Qualitative character : A character in which variation is discontinuous.

Qualitative resistance : Resistance reactions that can be placed in distinct categories, usually conferred by one or a few genes. (see quantitative resistance).

Quantitative character : A character in which variation is continuous so that classification into.

Quantitative genetics : The area of genetics concerned with the inheritance of continuously-varying traits. Most practical improvement programs involve the application of quantitative genetics.

Quantitative inheritance : Inheritance of measurable traits (height, weight, colour intensity, etc.) that depend on the cumulative action of many genes.

Quantitative resistance : Resistance reactions that have no distinct classes but vary continuously from resistant to susceptible, the result of few to many genes the individual effects of which may be small and difficult to detect. (see qualitative resistance).

Quantitative trait : A measurable trait that shows continuous variation; a trait that can not be classified into a few discrete classes.

Quantitative trait locus (QTL) : A locus that affects a quantitative trait. The plural form (quantitative trait loci) is also abbreviated as QTL.

Quantum (L quantum, how much) : An elemental unit of energy. Its energy value is hv, where h, Planck's constant, is 6.62×10^{-27} erg-second and v is the frequency of the vibrations or waves with which the energy is associated. cf photon.

Quantum speciation : The rapid formation of new species, primarily by genetic drift (q.v.).

Quarantine (It quarantina, from quaranta, forty) : Originally, keeping a person or living organism in isolation for a period (originally 40 days) after arrival to allow disease symptoms to appear, if there was any disease present. Now used for regulations restricting the sale or shipment of living organisms, usually to prevent disease or pest invasion of an area.

Quarternary structure : The folded structure of a protein including all of the polypeptides required for the intact, fuctional protein.

Quiescent : Quiet, at rest, but not necessarily dormant, and having the potential for resumed activity; can apply to non-meristematic cells.

Quinolones : A group of antibiotics that inhibit DNA gyrase. For example, nalidixic acid.

Quorum sensing : The ability of bacteria to interact with each other through a variety of mechanisms; allows a population of bacteria to behave more like a multicellular organism.

Race- : A sub-population, usually fairly large, of a species exhibiting some degree of phenotypic (and presumably genotypic) uniformity among individuals within the population and distinct from the species as a whole.

Race non-specific resistance (syn general resistance, horizontal resistance) : Resistance which is effective against all biotypes of the pathogen. (see durable resistance, specific resistance, vertical resistance).

Raceme : An inflorescence in which the main axis is elongated but the flowers are borne on pedicels that are about equal in length.

Rachilla (Gr rhachis, a backbone + L diminutive suffixilla) : Shortened axis of a spikelet.

Rachis (Gr rhachis, a backbone) : Main axis of a spike; axis of fern leaf (frond) from which pinnae arise; in compound leaves, the extension of the petiole corresponding to the midrib of an entire leaf.

Radiation hybrid : A hybrid cell containing small fragments of irradiated human chromosomes. Maps of irradiation sites on chromosomes for the human, rat, mouse, and other genomes provide important markers, allowing the construction of very precise STS maps indispensable to studying multifactorial diseases. See also: sequence tagged site.

Radicle (L radix, root) : That portion of the plant embryo which develops into the primary or seed root.

Radioactive isotope; radioisotope : An unstable isotope that emits ionizing radiation.

Radioactive suicide : The death of cells due to incorporation of a radioactive substrate with high specific activity. Radioactive suicide can be used as a technique to select for mutants that are unable to incorporate a particular substrate.

Raft culture : See nurse culture.

Ramet- : A vegetatively reproduced copy of a plant. Each ramet will have almost precisely the same genotype as the original parent tree, known as the ortet. (See clone and ortet).

Random : Arrived at by chance without discrimination.

Random amplified polymorphic DNA (RAPD; pronounced 'rapid') : A technique using single, short (usually 10-mer) synthetic oligonucleotide primers for PCR. The primer, whose sequence has been chosen at random, initiates replication at its complementary sites on the DNA, producing fragments up to about 2 kb long, which can be separated by electrophoresis and stained with ethidium bromide. A primer can exhibit polymorphism between individuals, and polymorphic fragments can be used as markers.

Random genetic drift : See genetic drift..

Random mutagenesis : A non-directed change of one or more nucleotide pairs in a DNA molecule.

Random primed synthesis : If you have a DNA clone and you want to produce radioactive copies of it, one way is to denature it (separate the strands), then hybridize to that template a mixture of all possible 6-mer oligonucleotides. Those oligos will act as primers for the synthesis of labeled strands by DNA polymerase (in the presence of radiolabeled precursors).

Random primer method : A protocol for labelling DNA *in vitro*. A sample of random oligonucleotides (6 or 14 nucleotides long) containing all possible combinations of nucleotide sequences is hybridized to a DNA probe. Then, in the presence of a DNA polymerase and the four deoxyribonucleotides - one of which is labelled - the 3′ hydroxy ends of the hybridized oligonucleotides provide initiation sites for DNA synthesis that uses the separated strands of the probe DNA as a template. This reaction produces labelled copies of portions of the probe DNA.

Random primers : A set of short oligonucleotides with variable sequences. Within a population of random oligonucleotides, some will anneal to complementary sequences in a DNA or RNA template. Use of random primers to initiate DNA synthesis in the presence of a radiol abeled (or chemiluminescent) dexoynucleotide (dNTP) will yield probes representing the sequences found in the template DNA. The labeled probes can then be used to identify other sequences homologous to this DNA (e.g. by Southern blotting).

Randomization : Process of making assignments at random.

Randomized complete-block design- : The most common experimental design used in progeny, provenance and clonal field tests. Each entry in the test is replicated once in each block. All the entries are arranged randomly within a block and a new randomization pattern is used for each block. (See replication, block, and plot).

RAPD : See random amplified polymorphic DNA.

Ray : Tissue that extends radially in the secondary xylem and phloem of a woody plant.

Reading frame : When mRNA is translated by the cell, the nucleotides are read three at a time. By starting at different positions, the groupings of three that are produced can be entirely different. The following example shows a DNA sequence and the three reading frames in which it could be read. Not only is an entirely different amino acid sequence specified by the different reading frames, but two of the three frames have stop codons, and thus are not open reading frames (asterisks indicate a stop codon).

A DNA open reading frame: ...ATG ACA TGT AAA GAT AGA CTA ACC TTT TGG...

...Met Thr Cys Lys Asp Arg Leu Thr Phe Trp...

Same bases, different grouping: ...A TGA CAT GTA AAG ATA GAC TAA CCT TTT GG...

... *** His Val Lys Ile Asp *** Pro Phe Gly..

Same bases, another grouping: ...AT GAC ATG TAA AGA TAG ACT AAC CTT TTG G..

... Asp Met *** Arg *** Thr Asn Leu Leu ...

If we shift the grouping again, we will just get the first reading frame again. The reading frame that is actually used is determined by the first methionine codon (the initiation codon). Once that first AUG is recognized, the pattern of triplet groupings follows unambiguously.

Reading frame : A series of triplets beginning from a specific nucleotide. Each triplet is represented by a single amino acid in the protein synthesized. The reading frame defines which sets of three nucleotides are read as triplets in the DNA, and hence as codons in the corresponding mRNA; this is determined by the initiation codon, AUG. Thus the sequence.

AUGGCAAAAUUUCCC would read as

AUG/GCA/AAA/UUU/CCC/ and not as

A/UGC/CAA/AAU/UUC/CC.

Depending on where one begins, each DNA strand contains three different reading frames. See open reading frame, overlapping reading frames.

Read-through : Transcription or translation that proceeds beyond the normal stopping point because of the absence of the transcription or translation termination signal of a gene.

Reca : A protein in most bacteria, and that is essential for DNA repair and DNA recombination.

RecA protein : The protein encoded by the recA gene which is essential for homologous recombination. The RecA protein is also involved in the induction of the SOS response and the induction of lambda prophage in response to DNA damaging agents.

Recalcitrant : Of seeds: unable to survive drying and subsequent storage at low temperature. See field gene bank.

Receptacle (L receptaculum, a reservoir) : Enlarged end of the pedicel or peduncle, to which other flower parts are attached.

Receptive hypha : The part of a rust fungus pycnium (spermogonium) that receives the nucleus of a pycniospore (spermatium).

Receptor : A site that recognizes and binds an elicitor; any organ or molecular site that is sensitive to a distinct (specific) signal molecule.

Receptor-binding screening : One of the biotechnology-based methods for discovering conventional drugs. The method relies on the fact that many drugs act by binding to specific proteins (receptors) on or in cells: these proteins usually bind to hormones or to other cells, and control the cell's behaviour, although they may be enzymes or structural elements of the cell. The drug interferes with the normal role of the protein.

Recessive : Allele that determines phenotype only when homozygous; does not affect phenotype when heterozygous with a dominant allele.

Recessive gene : A gene which will be expressed only if there are 2 identical copies or, for a male, if one copy is present on the X chromosome.

Recessive oncogene; recessive-acting oncogene; anti-oncogene : A single copy of this gene is sufficient to suppress cell proliferation; the loss of both copies of the gene contributes to cancer formation. See oncogene.

Recessive-acting oncogene : See recessive oncogene.

Reciprocal cross : The repetition of a cross where the sexual function of the parents is reversed, i.e., female B x male A is the reciprocal of female A x male B.

Reciprocal recombination : A precise exchange of genetic information like that expected for a simple cut, exchange, and rejoin mechanism. That is, a region of DNA simply moves from the donor molecule to the recipient molecule in exchange for the identical region from the recipient molecule.

Reciprocal translocation : When a pair of chromosomes exchange exactly the same length and area of DNA. Results in a shuffling of genes.

Reciprocating shaker : A platform shaker used for agitating culture flasks, with a back and forth action at variable speeds.

Recognition site : A nucleotide sequence - composed typically of 4, 6 or 8 nucleotides - that is recognized by and to which a restriction endonuclease (restriction enzyme) binds. For type II restriction enzymes (those used in gene-cloning experiments) it is also the sequence within which the enzyme specifically cuts (and their corresponding enzymes methylate) the DNA, i.e., for type II enzymes, the recognition site and the target site are the same sequence. Type I enzymes bind to their recognition site and then cleave the DNA at some more or less random position outside that recognition site. cf restriction site.

Recombinant : A term used in both classical and molecular genetics. 1. In classical genetics: An organism or cell that is the result of recombination (crossing-over), e.g., Parents: AB/ab and ab/ab; recombinant offspring: Ab/ab. 2. In molecular genetics: A molecule containing DNA from different sources. The word is typically used as an adjective, e.g., recombinant DNA.

Recombinant clones : Clones containing recombinant DNA molecules. See recombinant DNA technologies.

Recombinant DNA : DNA molecules in which sequences, not normally contiguous, have been placed next to each other by *in vitro* methods.

Recombinant DNA molecules : A combination of DNA molecules of different origin that are joined using recombinant DNA technologies.

Recombinant DNA technologies : Procedures used to join together DNA segments in a cell- free system (an environment outside a cell or organism). Under appropriate conditions, a recombinant DNA molecule can enter a cell and replicate there, either autonomously or after it has become integrated into a cellular chromosome.

Recombinant frequency (RF) : How often two linked genes are separated by recombination, generally expressed as a percentage of total progeny.

Recombinant protein : A protein whose amino acid sequence is encoded by a cloned gene.

Recombinant RNA : A term used to describe RNA molecules joined *in vitro* by T4 RNA ligase.

Recombinant toxin : A single multifunctional toxic protein that has been created by combining the coding regions of various genes.

Recombinant vaccine : A vaccine produced from a cloned gene.

Recombinase : Enzymatic activity responsible for facilitating intra- or intermolecular recombination of DNA molecules.

Recombination : Genetic exchange resulting from a cross-over between two different DNA molecules or different regions of a DNA molecule. See General recombination, Illegitimate recombination and Site specific recombination.

Recombination frequency : The number of recombinants divided by the total number of progeny. The recombination frequency is proportional to the physical distance between two genetic markers, and thus recombination frequencies can be used to draw genetic maps showing the relat ive distance between genetic markers. Note that often the coinheritance frequency -- the probablity that recombination did NOT occur between two markers is used instead of the recombination frequency.

Recombination-repair : A mode of filling a gap in one strand of duplex DNA by retrieving a homologous single strand from another duplex. Usually the underlying mechanism behind homologous recombination and gene conversion.

Recombineering ("Recombination-mediated genetic engineering") : A term coined by Don Court to describe a method for vector construction based on homologous recombination in *E. coli* using lambda phage recombination proteins exo, bet, and gam. This process allows homologous recombination between much shorter DNA sequences than the *E. coli* homologous recombination system. See http://www.sci.sdsu.edu/~smaloy/MicrobialGenetics/topics/in-vitro-genetics/red-swap.html.

Reconstructed cell : A viable transformed cell resulting from genetic engineering.

Recurrent parent : The parent to which successive backcrosses are made in backcross breeding.

Recurrent selection : A method of breeding designed to concentrate favorable genes scattered among a number of individuals by selecting, each generation, among the progeny produced by matings of the selected individuals (or their selfed progeny) of the previous generation. plants (X0). cell.

Re-differentiation : Cell or tissue reversal from one differentiated type to another differentiated type of cell or tissue. See de-differentiation.

Reduction division : Phase of meiosis in which the maternal and paternal chromosomes of the bivalent separate. See equational division.

Redundant genes : Multiple copies of a gene or different genes with the same function.

Regeneration (L re, again + generate, to beget) : The growth of new tissues or organs to replace those injured or lost. In tissue culture, regeneration is used to define the development of organs or plantlets from a tissue, callus culture or from a bud. See conversion; micropropagation; organogenesis.

Registered seed : The progeny of foundation seed normally grown to produce certified seed.

Regulator : Substance regulating growth and development of cells, organs, etc.

Regulatory gene : A gene whose protein controls the activity of other genes or metabolic pathways.

Regulatory regions or sequences : A DNA base sequence that controls gene expression.

Regulon : A group of genes or operons located at different positions on the chromosome but respond to a common regulatory protein.

Rejuvenation : Reversion from adult to juvenile stage.

Relaxed circle : See nicked circle.

Relaxed circle plasmid : See plasmid.

Relaxed DNA : See Supercoil.

Relaxed plasmid : A plasmid that replicates independently of the main bacterial chromosome and is present in 10-500 copies per cell. See plasmid.

Release factors (RF) : Proteins that facilitate the termination of translation. Translation termination occurs when a ribosome encounters one of the three stop codons (UAA, UAG, or UGA). These codons are not recognized by a tRNA but by protein release factors. RF-1 recognizes UAA and UAG, RF-2 recognizes UAA and UGA. RF-3 stimulates the activity of RF-1 and RF-2. Interaction of the RF with a stop codon causes the peptidyl transferase activity of the ribosome to catalyze the hydrolysis of the ester bond between the tRNA and the polypeptide chain, releasing the polypeptide chain and resulting in dissociation of the ribosome.

Re-naturation : The re-association of two nucleic acid strands after denaturation. The restoration of a molecule to its native form. In nucleic acid biochemistry, this term usually refers to the formation of a double-stranded helix from complementary single-stranded molecules. Some simple proteins can also be re-natured and regain their function.

Reniform : Kidney-shaped.

Rennin : An enzyme secreted by cells lining the stomach in mammals, and that is responsible for clotting milk.

Repeat sequences : The length of a nucleotide sequence that is repeated in a tandem cluster.

Repeat unit : A sequence of bases that occurs repeatedly in the genome, often end-on-end, i.e., tandemly.

Repetitive DNA : A surprising portion of any genome consists not of genes or structural elements, but of frequently repeated simple sequences. These may be short repeats just a few nt long, like CACACA etc. They can also range up to a few hundred nt long. Examples of the latter include Alu repeats, LINEs, SINEs. The function of these elements is often unknown. In shorter repeats like di- and tri-nucleotide repeats, the number of repeating units can occasionally change during evolution and descent. They are thus useful markers for familial relationships and have been used in paternity testing, forensic science and in the identification of human remains.

Replacement therapy : The administration of metabolites, co-factors or hormones that are deficient as the result of a genetic disease.

Replacement; gene replacement : A method of substituting a cloned gene, or part of a gene, which may have been mutated *in vitro*, for the wild-type copy of the gene within the host's chromosome. See homogenotization.

Replica plating : A technique for transferring an identical pattern of bacterial colonies from one petrie plate (the master plate) to a series of other plates containing different media. A plate containing bacterial colonies is pressed against a cylindrical block covered with velveteen, resulting in the transfer of many of the bacteria from each colony onto the corresponding position on the velveteen. Then a series of sterile plates are sequentially pressed against the velveteen to transfer cells from the velveteen to the corresponding position on each petrie plate. After incubation, if the cells can grow on the particular medium in a plate colonies will appear at identical positions on each plate. Typically about 10 replica plates can be made from each master plate with about 200 isolated colonies on the master plate. Thus, it is possible to quickly screen a large number of colonies for a variety of phenotypic characteristics.

Replicase : An enzyme involved in the replication (copying) of nucleic acid.

Replication : The process by which a DNA or RNA molecule makes an exact copy of itself; repetition of an experiment or procedure at the same time and place (one of several identical experiments, procedures, or samples); the process by which a virus particle induces the host cell to reproduce the virus.

Replication fork : The region on a replicating double stranded DNA molecule where synthesis of new DNA is taking place. The replicating fork produces a Y shaped region in the DNA molecule where the two strands have separated and replication is taking place.

Replication restart : An event that occurs when replication forks that collapse or disintegrate spontaneously do not progress to the completion of the chromosome; the replication fork is reasembled at the site of its collapse.

Replicative form (RF) : The molecular configuration of viral nucleic acid that is the template for replication in the host cell.

Replicative transposition : The insertion of a transposable element at a new location without loss from the original location.

Replicon : A DNA molecule that is able to initiate its own replication. A replicon must have an origin of replication and usually also has the necessary regulatory information required for the proper initiation of DNA replication.

Replicon fusion : The integration of one replicon into another to form a single replicon. May occur by recombination between homologous regions of two replicons or by formation of a cointegrate during transposition of a transposable element located on one of the repli cons.

Replisome : The complete replication apparatus present at a replication fork that carries out the semi-conservative replication of DNA.

Reporter gene : A gene that encodes a product that can readily be assayed. Thus reporter genes are used to determinate whether a particular DNA construct has been successfully introduced into a cell, organ or tissue.

Repressible enzyme : An enzyme whose synthesis is diminished by a regulatory molecule.

Repression : A form of gene regulation wherein the promoter is prevented from assembling an RNA polymerase complex, so that transcription does not occur.

Repressor : A gene product that negatively regulates gene expression. Usually refers to a DNA-binding protein that inhibits transcription under certain conditions. Transcriptional repressors typically decrease the transcription of certain other genes by specifically binding to operator sites -- one or more short DNA sequences located upstream of the structural gene. Regulation of repression is usually modulated by a ligand which binds to the repressor protein and alters its DNA-binding properties. DNA-binding may be increased by association with a co-repressor or DNA-binding may be decreased by association with an inducer.

Reproduction : The production of an organism, cell or organelle like itself (self propagation). 1. Sexual reproduction: the regular alternation (in the life-cycle of haplontic, diplontic and diplohaplontic organisms) of meiosis and fertilization (karyogamy) which provides for the production of offspring. The main biological significance of sexual reproduction lies in the fact that it achieves genetic recombination. 2. Asexual or agamic reproduction: the development of a new individual from either a single cell (agamospermy) or from a group of cells (vegetative reproduction) in the absence of any sexual process. See also apomixis.

Repulsion : The phase state in which a dominant (or wild-type) allele at one locus and a recessive (or mutant) allele at a second locus occur on the same chromosome. Also called trans configuration. See coupling.

Residue : As applied to proteins, what remains of an amino acid after its incorporation into a peptide chain, with subsequent loss of a water molecule (see Peptide Bond).

Residue Proven soybeans : These are soybeans suited for conservation-tillage farming practices. They provide disease and pest resistance as well as strong emergence in conservation-tillage conditions.

Resilience- : The ability of a population to persist in a given environment despite disturbance or reduced population size. The resilience of a population is based upon the ability of individuals within the population to survive (fitness) and reproduce (fecundity) in a changed environment and the genetic variability of the population which allows the production of new genotypes.

Resin : Sticky to brittle plant product derived from essential oils; often a defense compound of conifers.

Resinosis : A disease symptom in which there is a copious flow of resin from a wound or infection site of a conifer.

Resistance : The restriction of development of a pathenogenic agent or parasite. Can vary in degree from immunity (no development) to only slight retardation relative to a socalled susceptible reaction.

Resistance factor : A plasmid that confers antibiotic resistance to a bacterium.

Resistance management : Strategies that can be employed to delay the onset of resistance. For insect resistance management, this includes the use of a "refuge" in which the insect will not be challenged by the pesticide used in the rest of the field.

Resistance transfer factor (RTF) : The component of an R-plasmid that encodes the ability to conjugate and to transfer DNA.

Resistant (n resistance) : Possessing properties that prevent or impede disease development (see susceptible).

Resolution : Degree of molecular detail on a physical map of DNA, ranging from low to high.

Resolvase : An enzyme which resolves a co-integrate molecule into its two component replicons.

Respiration : Series of chemical reactions that make energy available through oxidation of carbohydrates and fat.

Response element : By definition, a "response element" is a portion of a gene which must be present in order for that gene to respond to some hormone or other stimulus. Response elements are binding sites for transcription factors. Certain transcription factors are activated by stimuli such as hormones or heat shock. A gene may respond to the presence of that hormone because the gene has in its promoter region a binding site for hormone-activated transcription factor. *Example:* the glucocorticoid response element (GRE).

Response regulator : A regulatory protein that responds to sensor protein of a "two-component" regulatory system. Interaction with the transmitter domain of the sensor protein, leads to the phosphorylation or dephosphorylation of an aspartate residue on the response regulator. The modified protein regulates gene expression in response to the changes in the environment.

Rest period : An endogenous physiological condition of viable seeds, buds or bulbs that prevents growth even in the presence of otherwise favourable environmental conditions. By some seed physiologists, this is referred to as dormancy.

Resting spore : Spore, often thick-walled, that can remain alive in a dormant state for some time, later germinating and capable of initiating infection.

Restitution nucleus : A nucleus with unreduced or doubled chromosome number that results from the failure of a meiotic or mitotic division.

Restriction : The cleavage of double-stranded DNA by an endonuclease (restriction enzyme). The restriction enzyme distinguishes between self and foreign DNA based upon the modification of its DNA-binding site (for example, by methylation).

Restriction analysis : Determination of the number and sizes of the DNA fragments produced when a particular DNA molecule is cut with a particular restriction endonuclease.

Restriction endonuclease [enzyme] : A class of endonucleases that cleaves DNA after recognizing a specific sequence, e.g., BamH1 (5′GGATCC3′), EcoRI (5′GAATTC3′), and HindIII (5′AAGCTT3′). There are three types of restriction endonuclease enzymes.

Type I : Cuts non-specifically a distance greater than 1000 bp from its recognition sequence and contains both restriction and methylation activities.

Type II: Cuts at or near a short, and often palindromic (q.v.) , recognition sequence. A separate enzyme methylates the same recognition sequence. They may make the cuts in the two DNA strands exactly opposite one another and generate blunt ends, or they may make staggered cuts to generate sticky ends. The type II restriction enzymes are the ones commonly exploited in recombinant DNA technology.

Type III : Cuts 24-26 bp downstream from a short, asymmetrical recognition sequence. Requires ATP and contains both restriction and methylation activities.

Restriction enzyme : See restriction nuclease.

Restriction enzyme cutting site : A specific nucleotide sequence of DNA at which a particular restriction enzyme cuts the DNA. Some sites occur frequently in DNA (e.g., every several hundred base pairs), others much less frequently (rare- cutter; e.g., every 10,000 base pairs).

Restriction enzyme, endonuclease : A protein that recognizes specific, short nucleotide sequences and cuts DNA at those sites. Bacteria contain over 400 such enzymes that recognize and cut over 100 different DNA sequences. See restriction enzyme cutting site.

Restriction enzymes : An endonuclease enzyme, isolated from bacteria, that recognizes specific base-pair sequences within DNA and causes endonucleolytic cleavage of the DNA at a site determined by the recognized DNA sequences. The sites vary from frequent to rare cutting, depending upon the length of the restriction site. For example, the restriction enzyme BamHI locates and cuts any occurrence of:

5'-GGATCC-3'

| | | | | |

3'-CCTAGG-5'

Note that both strands contain the sequence GGATCC, but in antiparallel orientation. The recognition site is thus said to be palindromic, which is typical of restriction sites. Every copy of a plasmid is identical in sequence, so if BamHI cuts a particular circular plasmid at three sites producing three "restriction fragments", then a million copies of that plasmid will produce those same restriction fragments a million times over. Bacteria produce restriction enzymes for protection against invasion by foreign DNA such as phages. The bacteria's own DNA is modified in such a way as to prevent it from being clipped. Bacteria contain over 600 such enzymes that recognize and cut over 100 different DNA sequences. See restriction enzyme cutting site.

Restriction fragment : The piece of DNA released after restriction digestion of plasmids or genomic DNA. See "Restriction enzyme". One can digest a plasmid and isolate one particular restriction fragment (actually a set of identical fragments). The term also describes the fragments detected on a genomic blot which carry the gene of interest.

Restriction fragment length polymorphism (RFLP) : The occurrence of variation in the length of DNA fragments that are produced after cleavage with a type II restriction endonuclease. The differences in DNA lengths are due to the presence or absence of recognition site(s) for that particular restriction enzyme. RFLPs were initially detected using hybridization with DNA probes after separation of digested genomic DNA by gel electrophoresis (Southern analysis). Now they are typically detected by electrophoresis of digested PCR product.

Restriction map : A "cartoon" depiction of the locations within a stretch of known DNA where restriction enzymes will cut. The map usually indicates the approximate length of the entire piece (scale on the bottom), as well as the position within the piece at which designated enzymes will cut. This map happens to be of a plasmid, and the two ends are joined together with about 25 nt between the EcoRI and HindIII sites.

Restriction nuclease : A bacterial enzyme that cuts DNA at a specific site.

Restriction site : The specific nucleotide sequence in DNA that is recognized by a type II restriction endonuclease and within which it makes a double-stranded cut. Restriction sites usually comprise four or six base pairs that typically are palindromic (q.v.), e.g., 5′GGCC3′.

3′CCGG5′ The two strands may be cut either opposite to one another, to create blunt ends, or in a staggered manner, giving sticky ends, depending on the enzyme involved. See restriction endonuclease.

Restriction-enzyme cutting site : A specific nucleotide sequence of DNA at which a particular restriction enzyme cuts the DNA. Some sites occur frequently in DNA (e.g., every several hundred base pairs); others much less frequently (rare-cutter; e.g., every 10,000 base pairs).

Restriction-modification : The modification of host DNA to prevent cutting by a restriction endonuclease (see restriction endonuclease). The modification is often via methylation of a specific restriction site sequence.

Restrictive conditions : The particular environmental conditions which do not allow growth of a conditional lethal mutant but result in the expression of the mutant phenotype.

Reticulate : Having netlike markings.

Reticulocyte : A young red blood cell.

Reticulocyte lysate : A lysate of rabbit reticulocytes, which has been extensively digested with micrococcal nuclease to destroy the reticulocyte mRNAs. With the addition of an exogenous, usually synthetic, mRNA, amino acids and a source of energy (ATP), the translational machinery of the reticulocyte (ribosomes, eukaryotic translation factors, etc.) will permit *in vitro* translation of the added mRNA with production of a new polypeptide. This is only one of several available *in vitro* translation systems.

Retinitis pigmentosa : Group of hereditary ocular disorders with progressive retinal degeneration. Autosomal dominant, autosomal recessive, and x-linked forms.

Retinoblastoma : A childhood malignant cancer of the retina of the eye. reverse transcriptase viral enzyme used to make cDNA.

Retro-element : Any of the integrated retroviruses or the transposable elements that resemble them.

Retro-poson; retro-transposon : A transposable element that moves via reverse transcription (i.e., from DNA to RNA to DNA) but lacks the long terminal repeat sequences.

Retroregulation : The effect of a downstream RNA sequence upon the expression of an upstream sequence. First used to describe the regulation of the phage lambda xis gene (see http://www.sci.sdsu.edu/~smaloy/MicrobialGenetics/topics/phage/int-xis.html).

Retrotransposon : A genetic element that transposes to a new location in DNA by first making an RNA copy of itself, then making a DNA copy of this RNA with a reverse transcriptase, and then inserting the DNA copy into the target DNA.

Retroviral infection : The presence of retroviral vectors, such as some viruses, which use their recombinant DNA to insert their genetic material into the chromosomes of the host's cells. The virus is then propogated by the host cell.

Retroviral vectors : Gene transfer systems based on viruses that have RNA as their genetic material.

Retrovirus : Singel stranded RNA virus which is replicated and expressed via a double stranded DNA intermediate. Retroviruses contain enzyme reverse transcriptase (RNA dependent DNA polymerase); this enzyme converts viral RNA into DNA which can combine with DNA of host cell and produce more viral particles.

Reversal transfer : Transfer of a culture from a callus-supporting medium to a shoot-inducing medium.

Reverse genetics : An approach where a cloned gene with an unknown function is used to disrupt the corresponding chromosomal gene to examine the resulting phenotype.

Reverse transcriptase : An enzyme found in the retroviruses which catalyzes the RNA-dependent polymerization of DNA (DNA copy from RNA template). RT is used to make cDNA; one begins by isolating polyadenylated mRNA, providing oligo-dT as a primer, and adding nucleotide triphosphates and RT to copy the RNA into cDNA.

Reverse transcriptase; RNA-dependent DNA polymerase : An enzyme that uses RNA molecule as a template for the synthesis of a complementary DNA strand.

Reverse transcription : The synthesis of DNA on a template of RNA, accomplished by reverse transcriptase.

Reversion; reverse mutation : Restitution of a mutant gene to the wild-type condition, or at least to a form that gives the wild phenotype; more generally, the appearance of a trait expressed by a remote ancestor.

Revertant : See Back Mutation.

RF : See replicative form.

R-factor : A transmissible plasmid that carries genes coding for resistance to several different antibiotics. Also called R-plasmid.

RFLP : See restriction fragment length polymorphism.

Rhizobacterium : A micro-organism whose natural habitat is near, on or in plant roots.

Rhizobium : genus of bacteria that lives symbiotically with roots of leguminous plants; during the symbiosis, atmospheric nitrogen gas is converted into a form useable by the plant.

Rhizobium nodules : Galls on roots caused by Rhizobium spp.

Rhizome : A mostly horizontal, jointed, fleshy, often elongated, usually underground stem.

Rhizomorph : Macroscopic rope-like strand of compacted tissue formed by certain fungi.

Rhizosphere : The soil region in the immediate vicinity of growing plant roots.

Rhizosphere-competent : Used to describe microorganisms adapted to living in the rhizosphere of a plant.

Rho factor : A protein which catalyzes transcription termination at certain sites or when an extended stretch on nontranslated, unstructured RNA is present. See Rho dependent polarity.

Ri plasmid : A class of large conjugative plasmids found in the soil bacterium Agrobacterium rhizogenes. Ri plasmids are responsible for hairy root disease of certain plants. A segment of the Ri plasmid is found in the genome of tumour tissue from plants with hairy root disease.

Ribonuclease : Any enzyme that hydrolyses RNA.

Ribonucleic acid (RNA) : A chemical found in the nucleus and cytoplasm of cells; it plays an important role in protein synthesis and other chemical activities of the cell. The structure of RNA is similar to that of DNA. There are several classes of RNA molecules, including messenger RNA, transfer RNA, ribosomal RNA, and other small RNAs, each serving a different purpose.

Ribonucleotides : See nucleotide.

Riboprobe : A strand of RNA synthesized in-vitro (usually radiolabeled) and used as a probe for hybridization reactions. An RNA probe can be synthesized at very high specific activity, is single stranded (and therefore will not self anneal), and can be used for very sensitive detection of DNA or RNA.

Ribose : A monosaccharide (C5H10O5) rarely occurring free in nature, but important as a component of RNA.

Ribosomal (rRNA) : An RNA molecule that forms part of the structure of a ribosome.

Ribosomal Binding Sequence (Shine-Dalgarno sequence) : In prokaryotic organisms, part or all of the polypurine sequence AGGAGG located on mRNA just upstream of an AUG initiation codon; it is complementary to the sequence at the 3' end of 16S rRNA; and involved in binding of the ribosome to mRNA. The internal ribosomal entry site found in some viruses may be an analogous eukaryotic genetic element.

Ribosomal binding site : A sequence of nucleotides near the 5′ end of a bacterial mRNA molecule that facilitates the binding of the mRNA to the small ribosomal sub-unit. Also called the Shine-Delgarno sequence.

Ribosomal protein : One of the ribonucleoprotein particles that are the sites of translation.

Ribosomal RNA (rRNA) : A class of RNA found in the ribosomes of cells.

Ribosome : An RNA-protein complex responsible for the correct positioning of mRNA and charged tRNAs allowing proper alignment of amino acids during protein synthesis.

Ribosome binding site : A short nucleotide sequence upstream of a gene which forms the site on the mRNA molecule where the ribosome binds. (Also called a Shine-Delgarno sequence).

Ribosomes : Small cellular components composed of specialized ribosomal RNA and protein; site of protein synthesis. See ribonucleic acid (RNA).

Ribozyme : A catalytically active RNA. A good example is the hepatitis delta virus RNA which is capable of self-cleavage and self-ligation in the absence of protein enzymes.

Ribozyme; gene shears : RNA molecule that can catalyse chemical reactions, often cutting other RNAs.

Ribulose : A keto-pentose sugar ($C_5H_{11}O_5$) that is involved in carbon dioxide fixation in photosynthesis.

Ribulose biphosphate (rubp) : A five-carbon sugar that is combined with carbon dioxide to form a six-carbon intermediate in the first stage of the dark reaction of photosynthesis.

Rich medium : A growth medium in which not all the components have been identified. Most "rich" media have a variety of complex, undefined componants.

Rinderpest : Cattle plague; a viral infection of cattle, sheep and goats.

Ringspot : Disease symptom characterized by yellowish or necrotic rings enclosing green tissue, as in some plant diseases caused by viruses.

Risk communication : In genetics, a process in which a genetic counselor or other medical professional interprets genetic test results and advises patients of the consequences for them and their offspring.

R-loops : Single-stranded DNA regions in RNA-DNA hybrids formed *in vitro* under conditions where RNA-DNA duplexes are more stable than DNA-DNA duplexes.

RNA : Ribonucleic acid. An organic acid composed of repeating nucleotide units of adenine, guanine, cytosine and uracil, whose ribose components are linked by phospho-diester bonds. The information-carrying material in some viruses. More generally, a molecule derived from DNA by transcription that may carry information (messenger RNA (mRNA)), provide sub-cellular structure (ribosomal RNA (rRNA)), transport amino acids (transfer RNA (tRNA)) or facilitate the biochemical modification of itself or other RNA molecules. See antigen RNA; gene splicing; heterogeneous nuclear RNA (hnRNA); mRNA; ribosomal RNA; RNA polymerase; small nuclear RNA; transfer RNA.

RNA editing : Post-trancriptional processes that alter the information encoded in gene transcripts (RNAs).

RNA phage : A phage that carries RNA as its genetic material. Some examples include phage MS2 and Qß.

RNA polymerase : Enzymatic activity responsible for DNA-dependent synthesis of RNA. In prokaryotes there is only one RNA polymerase. In eukaryotes, there are three, each of which transcribes a different group of genes.

RNA Splicing : A complex and incompletely understood series of reactions occuring in the nucleus of eukaryotic cells in which pre-mRNA transcribed from chromosomal DNA is processed such that noncoding regions of the pre-mRNA (introns) are excised, and coding regions (exons) are covalently linked to produce an mRNA molecule ready for transport to the cytoplasm. Because of splicing, eukaryotic DNA representing a gene encoding any given protein is usually much larger than the mRNA from which the protein is actually translated.

RNase : Ribonuclease; an enzyme which degrades RNA. It is ubiquitous in living organisms and is exceptionally stable. The prevention of RNase activity is the primary problem in handling RNA. There are many different RNases, some of the more important include: RNase A Cleaves ssRNA 3' of pyrimidines.

RNase T1 Cleaves ssRNA at guanasine nucleotides.

RNase V1 Cleaves dsRNA (helical regions).

RNase H Degrades the RNA part of RNA:DNA hybrids.

Roentgen (symbol r) : Obsolete unit of ionizing radiation. The SI unit is the sievert (symbol: Sv; 1 Sv □ 8.4 r) .

Rogue : A variation from the standard type of a variety or strain. Roguing; removal of undesirable individuals to purify a stock.

Rolling circle (sigma) replication : A type of DNA replication where a replication fork moves around a circular DNA molecule, producing a single-stranded concatamer (much like the way toilet paper peels off the roll). The resulting single-stranded DNA may become double-stranded by the synthesis of a complementary strand.

Root : The descending axis of a plant, normally below ground, which serves to anchor the plant and to absorb and conduct water and mineral nutrients.

Root apex : The apical meristem of a root; very similar to the shoot apical meristem in that it forms the three meristematic areas: the protoderm (developing into the epidermis); the procambium (which develops into the stele); and the growth meristem (which forms the cortex).

Root cap : A thimblelike mass of cells covering and protecting the apical meristem of a root.

Root culture : The culture of isolated root tips of apical or lateral origin to produce *in vitro* root systems with indeterminate growth habits. Root culture was among the first kinds of plant tissue cultures, and is still largely used in the study of developmental phenomena, and mycorrhizal, symbiotic and plant-parasitic relationships.

Root cutting : Cutting made from sections of roots alone.

Root exudates : The various compounds that leak from growing and expanding sections of roots as well as from broken cells at exit points of lateral roots.

Root graft : The fusion of roots of two adjacent plants so that their water and food conducting (vascular) systems become joined.

Root hair : Threadlike, single-celled outgrowths from a root epidermal cell.

Root nodule : A small round mass of cells that is located on the roots of plants and contains nitrogen-fixing bacteria.

Root tuber : Thickened root that stores carbohydrates.

Root zone : The volume of soil or growing medium containing the roots of a plant. In soil science, the depth of the soil profile in which roots are normally found. See rhizosphere.

Rootstock : Portion of the stem (trunk) and associated root system into which a bud or scion is inserted in grafting; fleshy overwintering part of a herbaceous perennial plant with buds and eyes (see scion).

Rosette : Disease symptom characterized by short, bunchy growth habit due to shortened internodes and no comparable reduction in leaf size.

Rot : Softening, discoloration, and often disintegration of plant tissue as a result of fungal or bacterial infection.

Rotary shaker : Rotating apparatus with a platform on which, containers can be shaken, such as Erlenmeyer flasks containing cells in liquid nutrient medium.

Rotation : Growth of different kinds of crops in succession in the same field.

Roundworm : Nematode.

RPM : Revolutions per minute. Often used to describe the velocity of a centrifuge.

rRNA; ribosomal RNA : The RNA molecules which are essential structural and functional components of ribosomes, the organelles responsible for protein synthesis. The different rRNA molecules are known by their sedimentation (Svedberg; symbol S) values. *E. coli* ribosomes contain one 16S rRNA molecule (1541 nucleotides long) in the same (small) sub-unit and a 23S rRNA (2904 nucleotides) and a 5S rRNA (120 nucleotides) in the large sub-unit. These three rRNA molecules are synthesized as part of a large precursor molecule which also contains the sequences of a number of tRNAs. Special processing enzymes cleave this large precursor to generate the functional moieties. See RNA.

RT-PCR : See 'Polymerase Chain Reaction and reverse transcriptase'.

Rubinstein-Taybi syndrome : Condition with multiple congenital anomalies including: mental deficiency, broad thumbs, small head, broad nasal bridge and beaked nose.

RuBP : See ribulose biphosphate.

Rugose : Wrinkled, roughened.

Ruminant animals; ruminants : Animals having a rumen - a large digestive vat in which fibrous plant material is partially broken down by microbial fermentation, prior to digestion in a "true" stomach (the abomasum). There are also two other stomachs - the reticulum and the omasum. Typical ruminants are cattle and sheep.

Runner : A lateral stem that grows horizontally along the ground surface and gives rise to new plants either from axillary or terminal buds. cf stolon.

Runner plants : New plants produced asexually on a runner or stolon.

Russet: Brownish, roughened areas resulting from cork formation.

Rust : A generic descriptor for various plant diseases, especially those caused by a group of parasitic fungi of the phylum Basidiomycota, that attack the leaves and stems of crops.

S phase : The phase in the cell cycle during which DNA synthesis occurs.

S_1 end mapping : A technique to determine where the end of an RNA transcript lies with respect to its template DNA (the gene). Can't be described in a short paragraph. See "RNAse Protection assay" for a closely related technique.

S_1 mapping : See S1 nuclease.

S_1 nuclease : An enzyme that specifically degrades RNA or single-stranded DNA to 5′ mononucleotides. Purified from the filamentous fungus Aspergillus oryzae, S1 nuclease is used in assessing the extent of a hybridization reaction by removing unpaired regions. It is also used to remove the sticky ends of restriction fragments. In S1 mapping, the coding region of a gene is detected by performing mRNA-DNA hybridization and removing unpaired DNA with S1 nuclease.

S1, S2, S3 : Symbols for designating first, second, third, etc. selfed generations from an ancestral plant (S0).

Saccharifaction : Following liquefaction, the hydrolysis of polysaccharides by glucoamylase to maltose and glucose.

Saline resistance : See salt tolerance.

Salmonella : A genus of rod-shaped, Gram-negative bacteria that are a common cause of food poisoning.

Salt tolerance; saline resistance : The ability to withstand a concentration of sodium (Na+ ion), or of any other salt, in the soil (or in culture), which is damaging or lethal to other plants. Breeding and selection for increased tolerance and resistance in crop plants is of great current interest.

Same-sense mutation : A mutation which changes the nucleotide sequence of a codon but does not change the amino acid encoded due to the degeneracy of the genetic code. For example, both CCC and CCA encode proline.

Sanger sequence : DNA is synthesized so it is radioactively labeled and the reaction terminates specifically at the position corresponding to a given base. "plus and minus" or "primed synthesis" method.

Sanitation : Destruction or removal of infected and infested plants or plant parts; decontamination of tools, equipment, containers, work space, hands, etc.

Sap : Fluid content of the xylem and phloem cells of plants. Fluid contents of the vacuole are referred to as cell sap.

Sap transmission : Transmission, usually of viruses, by rubbing sap from an infected plant onto a healthy plant to cause infection.

Saprobe (syn saprophyte) : Organism that obtains nourishment from non-living organic matter.

Saprophyte : A vegetable organism that derives its nutriment from decaying organic matter.

Sapwood : Physiologically active zone of wood contiguous to cambium (see heartwood).

Satellite : A chromosomal segment that branches off from the rest of the chromosome but is still connected by a thin filament or stalk.

Satellite DNA : That portion of the DNA in plant and animal cells consisting of highly repetitious sequences (millions of copies) typically in the range from 5 to 500 bases. Thousands of copies occur tandemly (end-on-end) at each of many sites. It can be isolated from the rest of the DNA by density gradient centrifugation.

Saturates : Noun: Oils containing saturated fatty acids.

SCA (single chain antigen) : Antibody-binding domains in which the two chains are produced by a gene and linked by a short peptide. See dabs.

Scab : Roughened, crustlike diseased area on the surface of a plant organ.

Scaffold : In genomic mapping, a series of contigs that are in the right order but not necessarily connected in one continuous stretch of sequence.

Scaffolding proteins : Proteins which facilitate the assembly of a cell structure but are not included in the final structure. For example, phage encoded proteins that are required for procapsid assembly but are removed after the procapsid is constructed.

Scald : A necrotic condition in which tissue is usually bleached and has the appearance of having been exposed to high temperatures.

Scale up : Conversion of a process, such as fermentation of a micro-organism, from a small scale to a larger scale.

Scanning electron microscope (SEM) : An electron-beam-based microscope used to examine, in a three dimensional screen image, the surface structure of prepared specimens.

Scarification : The chemical or physical treatment given to some seeds (where the seed coats are very hard or contain germination inhibitors) in order to break or weaken the seed coat sufficiently to permit germination.

Scientific name : A unique identifier consisting of a genus and a species name (the specific epithet) in Latin, assigned to each recognized and described species of organism. Based on the Linnean system of classification. Susceptible to considerable blurring at the edges because, as Darwin came to realize, nature has not packaged living things into neatly discrete entities. cf cultivar; pathovar.

Scion : The twig or bud to be grafted onto another plant, the root stock, in a budding or grafting operation.

Scion-stock interaction : The effect of a rootstock on a scion (and vice versa) in which a scion on one kind of rootstock performs differently than it would on its own roots or on a different rootstock.

Sclerenchyma (Gr skleros, hard + echyma, a suffix denoting tissue) : A strengthening tissue in plants, composed of cells with heavily lignified cell walls.

Sclerotium (pl sclerotia) : A vegetative resting body of a fungus, composed of a compact mass of hyphae with or without host tissue, usually with a darkened rind.

Scorch : Any symptom that suggests the action of flame or fire on the affected part, often seen at the margins of leaves.

Score : To determine the presence or absence of a phenotype by testing for growth under different conditions (e.g., plus and minus an auxotrophic supplement or permissive and nonpermissive conditions). Usually done by replica plating or patching colonies onto each type of plate. Synonymous with "screen".

SCP : See single-cell protein.

Scrapie : A disease of sheep; a spongiform encephalopathy See proteinaceous infectious particle.

Screen : To determine the presence or absence of a phenotype by testing for growth under different conditions (e.g., plus and minus an auxotrophic supplement or permissive and nonpermissive conditions). Usually done by replica plating or patching colonies onto each type of plate. Synonymous with "score".

Screening : To screen a library (see "Library") is to select and isolate individual clones out of the mixture of clones. For example, if you needed a cDNA clone of the pituitary glycoprotein hormone alpha subunit, you would need to make (or buy) a pituitary cDNA library, then screen that library in order to detect and isolate those few bacteria carrying alpha subunit cDNA. There are two methods of screening which are particularly worth describing: screening by hybridization, and screening by antibody. Screening by hybridization involves spreading the mixture of bacteria out on a dozen or so agar plates to grow several ten thousand isolated colonies. Membranes are laid onto each plate, and some of the bacteria from each colony stick, producing replicas of each colony in their original growth position). The membranes are lifted and the adherent bacteria are lysed, then hybridized to a radioactive piece of alpha DNA (the source of which is a story in itself - see "Probe"). When X-ray film is laid on the filter, only colonies carrying alpha sequences will "light up". Their position on the membranes show where they grew on the original plates, so you now can go back to the original plate (where the remnants of the colonies are still alive), pick the colony off the plate and grow it up. You now have an unlimited source of alpha cDNA. Screening by antibody is an option if the bacteria and plasmid are designed to express proteins from the cDNA inserts (see "Expression clones"). The principle is similar to hybridization, in that you lift replica filters from bacterial plates, but then you use the antibody (perhaps generated after olde tyme protein purification rituals) to show which colony expresses the desired protein.

SDS : Sodium dodecyl sulphate. See gel electrophoresis.

SDS-PAGE : Denaturing protein gel electrophoresis (see Polyacrylamide Gel Electrophoresis).

Secondary antibody : In an ELISA or other immunological assay system, the antibody that binds to the primary antibody. The secondary antibody is often conjugated with an enzyme such as alkaline phosphatase.

Secondary cell wall : The innermost layer of cell wall, with a highly organized microfibrillar structure, which is formed in certain cells after cell elongation has ceased. It gives rigidity to the cells.

Secondary growth : Type of growth characterized by an increase in thickness of stem and root and resulting from formation of secondary vascular tissues by the vascular cambium.

Secondary immune response : The rapid immune response that occurs during the second (and subsequent) encounters of the immune system of a mammal with a specific antigen. cf primary immune response.

Secondary infection : Infection resulting from the spread of infectious material produced after a primary infection or from secondary infections without an intervening inactive period.

Secondary inoculum : Inoculum produced by infections that took place during the same growing season.

Secondary messenger : A chemical compound within a cell that is responsible for initiating the response to a signal from a chemical messenger (such as a hormone) that cannot enter the target cell itself.

Secondary metabolite : A compound that is not necessary for growth or maintenance of cellular functions but is synthesized, generally, for the protection of a cell or micro-organism, during the stationary phase of the growth cycle.

Secondary oocyte : See oocyte.

Secondary organism : Organism that multiplies in already diseased tissue but is not the primary pathogen.

Secondary phloem : Phloem tissue formed by the vascular cambium during secondary growth in a vascular plant.

Secondary plant products : Metabolic products not having a known functional or structural use in plant cells. They have been extracted from plant tissue cultures for pharmaceutical and food processing purposes (e.g., essential oils, food additives, flavours).

Secondary pollutant : Air pollutant that must be chemically produced from other air pollutants, e.g. ozone (O_3) that is a product of a photochemical reaction of exhaust products from combustion engines in the atmosphere.

Secondary root : Branch from a primary root.

Secondary Structure (also see Primary and Tertiary Structure) : Local structure within a protein which is conferred by the nature of the side chains of adjacent amino acids (e.g., alpha helix, beta sheet, random coil); local structure within an RNA molecule which is conferred by base pairing of nucleotides which are relatively closely positioned within the sequence (e.g., hairpins, stem-loop structures).

Secondary thickening : Deposition of secondary cell wall materials which result in an increase in thickness in stems and roots.

Secondary vascular tissue : Vascular tissue (xylem and phloem) formed by the vascular cambium during secondary growth in a vascular plant.

Secondary xylem : Xylem tissue formed by the vascular cambium during secondary growth in a vascular plant.

Secreted protein : A protein that is exported through the cytoplasmic membrane. Most secreted proteins have specific signal sequences that promote interaction with the export apparatus in the membrane.

Secretion : The passage of a molecule from the inside of a cell through a membrane into the periplasmic space, or the extracellular medium.

Sedentary : Remaining in a fixed location (see migratory).

Seed : Botanically, a seed is the matured ovule without accessory parts. Colloquially, a seed is anything which may be sown; i.e., seed potatoes (which are vegetative tubers); seed of corn, sunflower, etc.

Seed orchard, clonal- : A seed orchard established from vegetatively propagated trees, usually grafts. Usually clonal seed orchards are established with single-tree plots with different ramets of each clone located as far apart as possible to reduce self pollination.

Seed production area- : A stand or plantation not originally planted to produce seed, but rogued of inferior trees and treated in such a manner as to produce large quantities of seed. Usually the stand or plantation is from a known origin or selected seedlot.

Seed source- : The location where a seedlot was collected. If different seedlots of an exotic species are collected and tested, the test is called a seed-source test to distinguish it from a provenance test.

Seed storage proteins : Proteins accumulated in large amounts in seeds not because of their enzymatic or structural properties but simply as a convenient source of amino acid for use when the seed germinates. They are of interest to biotechnologists: 1. As a source of protein. Much of the world's food comes from plant seeds or fruits, and much of the protein in those seeds is storage protein. Thus a substantial amount of the world's food protein comes from plant storage protein. Any improvement of the nutritional content of those proteins could correspondingly improve human diet. 2. As expression systems. Storage proteins are produced in very large amounts relative to other proteins, and are stored in stable, compact bodies in the plant seed. Several workers are seeking to make the plants produce other proteins in similarly large amounts and in as convenient a form, by splicing the gene for a desired protein into the middle of a plant storage protein gene.

Seed transfer guide- : A set of rules for collecting seed and planting seedlings so that genotypes are not moved to microclimates or soils within which they will be mis-adapted. Usually seed transfer guides describe the maximum movement from the point of collection in miles east and west, north and south as well as feet in elevation. (See seed zones).

Seed treatment : Application of a biological agent, chemical substance, or physical treatment to seed, to protect the seed or plant from pathogens or to stimulate germination or plant growth.

Seed zone- : An area within which seed can be collected from any natural stand and planted in any new site without fear of mis-adaptation.

Seedborne : Carried on or in a seed.

Seedlot- : A designation of a group of seeds with some factor in common, i.e., year of collection, stand or seed orchard, individual "+" trees, point of origin in a provenance test, a half- or full-sib family.

Segment-polarity gene : A gene that functions to define the anterior and posterior components of body segments in Drosophila.

Segregant : A hybrid resulting from the crossing of two genetically unlike individuals.

Segregation : The separation of the two members of a chromosome pair from each other at meiosis; the result is seen as the separation of alleles from each other in the gametes of heterozygotes; the occurrence of different phenotypes among offspring, resulting from chromosome or allele separation in their heterozygous parents. Mendel's first principle of inheritance (the Law of Segregation) predicts that heterozygotes will produce equal numbers of gametes containing each allele.

Segregrate : See segregration below.

Selectable : Having a gene product that, when present, enables a researcher to identify and preferentially propagate a particular organism or cell type. See reporter gene.

Selectable marker : A mutation that confers a phenotype on a cell such that when grown on a certain medium only those cells with the mutation will grow. For example, the mutation rpsL1 makes cells resistant to the antibiotic streptomycin -- when grown on medium with streptomycin rpsL1 mutants will grow but rpsL+ cells will die.

Selected marker : An allele that allows growth under a particular selective condition.

Selection : Conditions where only mutant or recombinant cells with a particular phenotype grow and divide.

Selection coefficient, s : The proportion by which the fitness of a genotype is less than the fitness of a standard genotype, which is usually the genotype with the highest fitness. In general, relative fitness = 1 - s.

Selection culture : A selection based on difference(s) in environmental conditions or in culture medium composition, such that preferred variant cells or cell lines (presumptive or putative mutants) are favoured over other variants or the wild-type.

Selection differential, S : The difference between the mean of the individuals selected to be parents and the mean of the overall population.

Selection pressure : The intensity of selection acting on a population of organisms or cells in culture. Its effectiveness is measured in terms of differential survival and reproduction, and consequently in change in the frequency of alleles in a population.

Selection unit : The minimum number of organisms or cells effective in the screening process.

Selection, clonal- : Choosing the best clones from a clonal test.

Selection, index- : Choosing parents on the basis of a weighted score which combines economic value and heritability of several desirable traits.

Selection, recurrent- : Choosing individuals out of families, making crosses, then progeny testing. The process is usually repeated several to many times. Most tree improvement programs are based on recurrent selection.

Selection, tandem- : Selecting for two or more traits sequentially rather than simultaneously. The technique is useful when one trait can be evaluated early and a second trait only after a number of years, or when one trait is much more expensive to measure.

Selective agent : An environmental or chemical agent characterized by its lethal or sub-lethal stress on growing plants, or portion thereof in culture. A selective agent is mainly used when selection of resistant or tolerant individuals is the research aim. See single-cell line.

Selective breeding : Making deliberate crosses or matings of organisms so the offspring will have particular desired characteristics derived from one or both of the parents.

Selective media : Growth media that only allow growth of mutant or recombinant cells with a particular phenotype grow. For example, media containing tetracycline would select for tetracycline resistant bacteria, preventing growth of parental tetracycline sensitive bacteria.

Selective medium : A culture medium containing substances that specifically inhibit or prevent the growth of some species of microorganisms.

SELEX : A method for isolation of high-affinity RNA- or DNA-protein interactions through successive rounds of selection for increased nucleic acid binding.

Self pollination- : The natural or artificial pollination of a female flower with pollen from the same genotype. (See inbreeding, outcrossing, and homozygosity).

Self-fertilization : The process by which pollen of a given plant fertilizes the ovules of the same plant. Plants fertilized in this way are said to have been selfed. An analogous process occurs in some animals, such as nematodes and molluscs.

Self-incompatibility : In plants, the inability of the pollen to fertilize ovules (female gametes) of the same plant.

Selfish DNA : A DNA sequence that does not contribute to the fitness of an organism but is maintained in the genome because it promotes its own replication.

Self-pollination : Pollen of a plant is transferred to the female part of the same plant or another plant with the same genetic makeup. Opposite: cross-pollination.

Self-replicating elements : Extrachromosomal DNA elements that have origins of replication for the initiation of their own DNA synthesis.

Self-sterility : See self-incompatibility.

SEM : See scanning electron microscope.

Semen sexing : See sperm sexing.

Semi-conservative replication : During DNA duplication, each strand of a parent DNA molecule is a template for the synthesis of its new complementary strand. Thus, one half of a pre-existing DNA molecule is conserved during each round of replication. cf replisome.

Semi-continuous culture : Cells in an actively dividing state are maintained in culture by periodically draining off the medium and replenishing it with fresh medium.

Semi-liquid : See semi-solid.

Semi-permeable membrane : A cell or plasma membrane that is partially permeable; certain ions or molecules (water, solvents) can pass through it but others cannot (such as certain solutes).

Semi-solid : Gelled but not firmly so; small amounts of a gelling agent are used to obtain a semi-solid medium; called also semi-liquid.

Semi-sterility : A condition of only partial fertility in plant zygotes; usually associated with translocations.

Senesce (adj senescent, n senescence) : To decline, as with maturation, age, or disease stress.

Senescence : The last stage in the post-embryonic development of multicellular organisms, during which loss of functions and degradation of biological components occur. A physiological ageing process in which cells and tissues deteriorate and finally die.

Sense RNA : A primary transcript (RNA) that contains a coding region (contiguous sequence of codons) that is translated to produce a polypeptide.

Sense strand : The strand of DNA that has the same nucleotide sequence as the mRNA (except the DNA has T where the RNA has U residues). Check out the supplemental resources for a diagram. Note that "sense strand" is sometimes used in different ways in the scientific literature, so it is critical to explicitly indicate what you mean when using this term.

Sense strand : A gene has two strands: the sense strand and the anti-sense strand. The Sense strand is, by definition, the same 'sense' as the mRNA; that is it can be translated exactly as the mRNA sequence can. Given a sense strand with the following sequence:

5' - ATG GGG CCA CGG CTG TGA - 3'

Met Gly Pro Arg Leu stop

The anti-sense strand will read as follows (note that the strand has been reversed and complemented):

5' - TCA CAG CCG TGG CCC CAT - 3'

The duplex DNA will pair as follows:

5' - ATGGGGCCACGGCTGTGA - 3'

| | | | | | | | | | | | | | | | | |

3' - TACCCCGGAGCCGACACT - 5' Note however that when the RNA is transcribed from this sequence, the ANTI-SENSE strand is used as the template for RNA polymerization. After all, the RNA must base-pair with its template strand (see Figure 3), so the process of transcription produces the complement of the anti-sense strand. This introduces some confusion about terminology: Some people use the term 'coding strand' and 'non-coding strand' to refer to the sense and antisense strands, respectively. Unfortunately, many people interpret these terms in exactly the opposite way. I consider the terms 'coding strand' and 'non-coding strand' to be too ambiguous. Some people use the exact opposite definition for 'sense' and 'anti-sense' that I have given here. Be aware of the possibility of a discrepancy. Textbooks I have consulted generally agree with the nomenclature given herein, albeit some avoid defining these terms at all.

Sensitivity : For diagnostic tests, the smallest amount of the target molecule that the assay can detect.

Sepal : One of the modified leaves comprising a calyx.

Sepsis : Destruction of tissue by pathogenic micro-organisms or their toxins, especially through infection of a wound. cf aseptic; axenic; sterile.

Septate (L septum, fence) : Divided by cross walls into cells or compartments.

Septum : The growth of the cell membrane and wall between opposite ends of a growing cell, ultimately separating the cell into two daughter cells.

Sequence : As a noun, the sequence of a DNA is a buzz word for the structure of a DNA molecule, in terms of the sequence of bases it contains. As a verb, "to sequence" is to determine the structure of a piece of DNA; i.e. the sequence of nucleotides it contains.

Sequence assembly : A process whereby the order of multiple sequenced DNA fragments is determined.

Sequence homology : Similarities in nucleic acid sequence and organization, and in their encoded products, that are sufficiently great as to imply common ancestral origins.

Sequence hypothesis : Francis Crick's seminal concept that genetic information exists as a linear DNA code; DNA and protein sequence are collinear.

Sequence polymorphism : See Polymorphism.

Sequence similarity : Similarity in nucleic acid or polypeptide sequences, particularly in shorter segments, that may not be sufficient to imply common ancestral origins.

Sequence Tagged Site (STS) : Short (200 to 500 base pairs) DNA sequence that has a single occurrence in the human genome and whose location and base sequence are known. Detectable by polymerase chain reaction, STSs are useful for localizing and orienting the mapping and sequence data reported from many different laboratories and serve as landmarks on the developing physical map of the human genome. Expressed sequence tags (ESTs) are STSs derived from cDNAs.

Sequencing : Determination of the order of nucleotides (base sequences) in a DNA or RNA molecule or the order of amino acids in a protein.

Sequencing technology : The instrumentation and procedures used to determine the order of nucleotides in DNA.

Sequential epitope : See Linear Epitope.

Serial divisions : Splitting at about monthly intervals of excised shoot-tip material growing on culture medium, in order to induce additional plantlets.

Serial float culture : A technique of floating anthers on liquid medium developed by Sunderland. Anther dehiscence, pollen release and development occur at intervals of several days, and in different nutrient media.

Serology (adj serological) : The study of serum reactions between an antigen and its antibody. Serology is mainly used to identify and distinguish between antigens, such as those specific to micro-organisms or viruses. Serology is also employed as an indicator technique to assay plants suspected of being virus-infected.

Serovars : Different strains of the same species of bacterium that can be distinguished by different reactions to certain antibodies (anti-sera). The name derives from the terms Serological Variety.

Serrate : Edges with teeth, like a saw.

Serum albumin : A globular protein obtained from blood and body fluids.

Sessile : used in reference to a leaf, leaflet, flower, floret, fruit, ascocarp, basidiocarp, etc., without a stalk, petiole, pedicel, stipe or stem; (of nematodes) permanently attached; not capable of moving about.

Seta (pl setae) : Bristle or hair-like structure, usually deep yellow or brown and thick-walled.

Severity values : A means to quantify accumulating opportunities for pathogen infection (e.g. infection periods) to a pre-determined threshold that requires a disease management activity (e.g. a fungicide application); used in disease prediction or forecasting.

Sex cell : A gamete, either a sperm or an ovum. Sex cells are produced by the meiosis process. See somatic cell.

Sex chromosome : The X or Y chromosome in human beings that determines the sex of an individual. Females have two X chromosomes in diploid cells; males have an X and a Y chromosome. The sex chromosomes comprise the 23rd chromosome pair in a karyotype. Compare autosome.

Sex determination : The method by which the distinction between males and females is established in a species.

Sex factor : A bacterial episome (e.g., the F plasmid in *E. coli*) that enables the cell to be a donor of genetic material. The sex factor may be propagated in the cytoplasm, or it may be integrated into the bacterial chromosome.

Sex hormones : Steroid hormones that control sexual development.

Sex linkage : The location of a gene on a sex chromosome, typically on the X chromosome.

Sex mosaic : See gynandromorph.

Sex-controlled gene : A gene that can be inherited by both genders but is usually expressed differently in males and females. See sex-limited gene.

Sexduction : The incorporation of bacterial genes into F factors and their subsequent transfer, by conjugation, to a recipient cell.

Sexed embryos : Embryos separated according to sex.

Sex-influenced dominance : The tendency for the type of gene action to vary between the sexes within a species. Thus the presence of horns in some breeds of sheep appears to be dominant in males and recessive in females.

Sex-limited : Expression of a trait in only one sex; e.g., milk production in mammals; egg production in chickens.

Sex-limited gene : A gene that can be inherited by both genders but is usually expressed in only males or females. See sex-controlled gene.

Sex-linked : A trait transmitted by one of the chromosomes determining sex of an individual.

Sex-linked : Traits or diseases associated with the X or Y chromosome; generally seen in males. See also: gene, mutation, sex chromosome.

Sexual PCR : PCR amplification in which similar, but not identical, DNA sequences are reassembled to obtain novel combinations of related, but variable sequences. When different combinations of a gene sequence are constructed, some of the resulting genes may possess improved characteristics. By repeating many rounds of PCR amplification and selection, an enormous amount of sequence variation can be obtained.

Sexual reproduction : reproduction involving fusion of two haploid nuclei (karyogamy) to form a diploid nucleus followed by meiosis (reduction division) back to haploid nuclei at some point in the life cycle, resulting in genetic recombination.

Sexual spore : Spore produced during the sexual cycle.

Sexually compatible : Able to be cross-mated or cross-fertile.

Shake culture : An agitated suspension in culture providing adequate aeration for cells in the liquid medium. Usually an Erlenmeyer flask containing the culture is attached to a horizontal or platform shaker, or agitated with a magnetic stirrer.

Shear : 1. The sliding of one layer across another, with deformation and fracturing in the direction parallel to the movement. This term usually refers to the forces that cells are subjected to in a bioreactor or a mechanical device used for cell breakage. 2. To fragment DNA molecules into smaller pieces. DNA, as a very long and fairly stiff molecule, is very susceptible to hydrodynamic shear forces. Forcing a DNA solution through a hypodermic needle will fragment it into small pieces. The size of the fragments obtained is inversely proportional to the diameter of the needle's bore. The actual sites at which the shear force breaks a DNA molecule are approximately random. Therefore DNA fragments may be generated by random shear and then cloned (by either tailing their ends or using linkers) so as to create a complete gene library of an organism. This method is little used now, having been replaced by the use of partial digests with four-base-pair cutters, such as Sau3A, as a means of generating random DNA fragments.

Shine-Dalgarno sequence : A conserved sequence in prokaryotic mRNAs that is complementary to a sequence near the 5′ terminus of the 16S ribosomal RNA and is involved in the initiation of translation. See ribosomal binding site.

Shoot : A young branch that grows out from the main stock of a tree, or the young main portion of a plant growing above ground.

Shoot differentiation : The development of growing points, leaf primordia and finally shoots from a shoot tip, axial bud, or even a callus surface.

Shoot tip; shoot apex : The terminal bud (0.1 - 1.0 mm) of a plant, which consists of the apical meristem (0.05 - 0.1 mm) and the immediate surrounding leaf primordia and developing leaves and adjacent stem tissue.

Shoot-tip graft; micrograft : A shoot tip or meristem tip is grafted onto a prepared seedling or micropropagated rootstock in culture. Meristem tip grafting is mainly used for *in vitro* virus elimination with Citrus spp. and other plants.

Short arm (p) : One of the two prominent segments of a chromosome; the long or "q" arm is the other. The arms of a given chromosome join at its centromere.

Short tandem repeats (STR) : Multiple copies of an identical DNA sequence arranged in direct succession in a particular region of a chromosome.

Short template : A DNA strand that is synthesized during the polymerase chain reaction and has a primer sequence at one end and a sequence complementary to the second primer at the other end.

Short-day plant : Plant that requires a night (or dark period) longer than its critical dark period to induce flower formation.

Shotgun cloning : A cloning strategy that involves the insertion of random fragments of a large DNA molecule into a vector, resulting in a large number of different recombinant DNA molecules.

Shotgun cloning : The practice of randomly clipping a larger DNA fragment into various smaller pieces, cloning everything, and then studying the resulting individual clones to figure out what happened. For example, if one was studying a 50 kb gene, it "may" be a bit difficult to figure out the restriction map. By randomly breaking it into smaller fragments and mapping those, a master restriction map could be deduced. See also Shotgun sequencing.

Shotgun method : Sequencing method that involves randomly sequenced cloned pieces of the genome, with no foreknowledge of where the piece originally came from. This can be contrasted with "directed" strategies, in which pieces of DNA from known chromosomal locations are sequenced. Because there are advantages to both strategies, researchers use both random (or shotgun) and directed strategies in combination to sequence the human genome. See also: library, genomic library.

Shotgun sequencing : A way of determining the sequence of a large DNA fragment which requires little brain power but lots of late nights. The large fragment is shotgun cloned (see above), and then each of the resulting smaller clones ("subclones") is sequenced. By finding out where the subclones overlap, the sequence of the larger piece becomes apparent. Note that some of the regions will get sequenced several times just by chance.

Shot-hole : Symptom in which small lesions fall out of leaves, giving the leaf the appearance of being hit by buckshot.

Shuttle vector : A type of cloning vector that contains sequences enabling it to be propagated and maintained in more than one type of host (e.g. *E. coli* and mammalian cells). For this purpose shuttle vectors carry different origin of DNA replication which are characteristic of the desired host systems.

Shuttle vector; bifunctional vector : A plasmid capable of replicating in two different host organisms because it carries two different origins of replication and can therefore be used to 'shuttle' genes from one to the other. For example, the YEp, pJDB219, is a shuttle vector able to replicate in *E. coli* from its pMB9 origin and in Saccaromyces cerevisiae from its 2 μm-plasmid origin.

Sib (sibling) : A term meaning brother or sister. Half sibs have one parent in common. Full sibs have both parents in common.

Sib-mating : Crossing of siblings. Matings involving two individuals of the same parentage; brother-sister matings.

Sickle cell anemia : An hereditary, chronic form of hemolytic anemia characterized by breakdown of the red blood cells; red blood cells undergo a reversible alteration in shape when the oxygen tension of the plasma falls slightly and a sickle-like shape forms.

Sickle-cell trait : A genetically inherited recessive condition in which red blood cells are distorted resulting in severe anemia and related symptoms that are often fatal in childhood. Sickle-cell trait is the result of a pleiotropic gene. Sickle-cell trait is also known as sickle-cell anemia.

Siderophore : A low molecular weight substance that binds very tightly to iron. Siderophores are synthesized by a variety of soil micro-organisms to ensure that the organism is able to obtain sufficient amounts of iron from the environment.

Sieve cell : In vascular plants, a long and slender sieve element with relatively unspecialized sieve areas and with tapering end walls that lack sieve plates.

Sieve element (syn sieve tube element) : A tube-shaped living cell in the phloem functioning in the transport of dissolved organic substances in the plant.

Sieve plate : Perforated wall area in a sieve-tube element through which strands connecting sieve-tube protoplasts pass.

Sieve tube : A tube within the phloem tissue of a plant, and composed of joined sieve elements.

Sievert (symbol Sv) : The SI unit of ionizing radiation.

Sigma factor : A protein that functions as a subunit of bacterial RNA polymerases and is responsible for specificity of recognition of promoters. Different sigma factors allow recognition of different promoter sequences.

Sign : Indication of disease from direct observation of a pathogen or its parts (see symptom).

Signal peptidase : An enzyme present within the lumen of the endoplasmic reticulum which proteolytically cleaves a secreted protein at the site of a signal sequence.

Signal peptide : See signal sequence.

Signal recognition particle (SRP) : A chaperonin complex responsible for arresting polypeptide synthesis and facilitating the docking of a ribosome to the endoplasmic reticulum membrane. Normally, on ribosomes translating polypeptides destined for insertion into or across the endoplasmic reticulum membrane become associated with SRP.

Signal sequence : A segment of about 15 to 30 amino acids at the N terminus of a protein, that enables the protein to be secreted (pass through a cell membrane). The signal sequence is removed as the protein is secreted. Also called signal peptide, leader peptide.

Signal transduction : Conditions that alter the conformation of a protein which regulates expression of other genes. Initially signal transduction was used to refer to extracellular conditions that alter the conformation of a membrane protein and cause it to relay the regulatory signal inside the cell, but more recently the term has been broadly applied to a variety of regulatory cascades.

Silent mutation : A nucleotide substitution (never a single deletion or insertion) which does not alter the amino acid sequence of an encoded protein due to the degeneracy of the genetic code. Such mutations usually involve the third base (wobble position) of codons.

Similarity : Sequence identity between two nucleotide sequences. For example, 85% similarity means that 85 nucleotide positions out of 100 are identical in the two nucleotide sequences. Sequence similarity does not necessarily indicate that two sequences have common ancestry.

Simple sequence repeat (SSR) : See microsatellite.

Simplicity : For diagnostic tests, the ease with which an assay can be implemented.

Sines : Short interspersed nuclear elements. Families of short (150 to 300 bp), moderately repetitive elements of eukaryotes, occurring about 100,000 times in a genome. SINES appear to be DNA copies of certain tRNA molecules, created presumably by the unintended action of reverse transcriptase during retroviral infection.

Single copy : A gene or DNA sequence which occurs only once per (haploid) genome. Most structural genes, those encoding functional proteins, are single-copy genes.

Single cross : A cross between two genotypes, usually two inbred lines, in plant breeding.

Single- gene disorder : Hereditary disorder caused by a mutant allele of a single gene (e.g., Duchenne muscular dystrophy, retinoblastoma, sickle cell disease). Compare polygenic disorders.

Single Nucleotide Polymorphism (SNP) : A variation in the genetic code at a specific point on the DNA. In principle, SNPs could be bi-, tri-, or tetra-allelic polymorphisms. However, in humans, tri- allelic and tetra- allelic SNPs are rare almost to the point of non- existence, and so SNPs are sometimes simply referred to as bi- allelic markers (or di- allelic to be etymologically correct). This is somewhat misleading because SNPs are only a subset of all possible bi- allelic polymorphisms (e.g., multiple base variations). About 30 million SNPs are thought to exist, making them much better markers than alternative markers, such as micro- satellite repeats or short tandem repeats. But it has been the discovery that some SNPs are linked to particular diseases that has fueled the rising interest in this field. see also The SNP Consortium.

Single-cell line; cell strain : A culture initiated from a single cell, usually from suspension cultures of single cells or small aggregates plated on solidified medium. The latter may incorporate a selective agent, from which tolerant or resistant individual cell lines or cell clones can be selected. See selective agent.

Single-cell protein (SCP) : Protein produced by micro-organisms. The dried mass of a pure sample of a protein-rich-micro-organism, which may be used either as feed (for animals) or as a food (for humans).

Single-chain antigen : See SCA.

Single-gene disorder : Hereditary disorder caused by a mutant allele of a single gene (e.g., Duchenne muscular dystrophy, retinoblastoma, sickle cell disease). See also: polygenic disorders.

Single-node culture : Culture of separate lateral buds with each carrying a piece of stem tissue.

Single-nucleotide polymorphism (SNP; pronounced "snip") : A polymorphism (q.v.) at a particular base site in a coding sequence, e.g., at base 306 in a particular gene, one individual could be heterozygous for A and G: the maternal allele could have an A at this site, while the paternal allele has a G at this site. This type of polymorphism is extensive throughout the genome, and has the great advantage of being detectable without the need for gel electrophoresis, which opens the way for large-scale automation of genotyping.

Single-strand conformational polymorphism (SSCP) : Relies on secondary and tertiary structural differences between denatured and rapidly cooled amplified DNA fragments that differ slightly in their DNA sequence; different SSCP alleles are resolved on non-denaturing acrylamide gels, usually at low temperature; ability to resolve alleles depends on conditions of electrophoresis.

Single-strand DNA binding protein (ssb) : The small basic protein that has a high affinity for single-stranded DNA. Ssb protein protects single-stranded DNA from nuclease attack and inhibits it from reannealing into double-stranded DNA.

Single-stranded : A term used to describe nucleic acid molecules consisting of only one polynucleotide chain. The genomes of certain phages, e.g., MI3, are single-stranded DNA molecules; rRNA, mRNA and tRNA are all single-stranded nucleic acids, but they all contain double-stranded regions formed by the intra-strand base-pairing of self-complementary sequences.

Sires : Male animals used for breeding.

Site-directed mutagenesis : The introduction of a mutation, usually a point mutation or an insertion, into a particular location in a cloned DNA fragment. This mutated fragment may be used to "knock out" a gene in the organism of interest by homologous recombination.

Site-specific : A term used to describe any process or enzyme which acts at a defined sequence within a DNA or RNA molecule. Type II restriction enzymes are site-specific endonucleases and the recombination systems encoded by some transposons are site-specific, such as is the integration of phage into the *E. coli* chromosome.

Site-specific mutagenesis : A technique to change one or more specific nucleotides within a cloned gene in order to create an altered form of a protein with one or more specific amino acid changes. a.k.a. oligonucleotide-directed mutagenesis; oligonucleotide-directed site-specific mutagenesis.

Site-specific recombination : Genetic exchange that occurs between particular, short DNA sequences. Site-specific recombination systems do not require RecA. Instead, each site-specific recombination system requires unique enzymes that catalyze the genetic exchange. Unlike RecA mediated recombination, site-specific recombination requires little sequence homology between the two DNA molecules. The best understood example is the integration of lambda into the *E. coli* chromosome by recombination between attP and attB catalyzed by the lambda integrase protein.

Six-base cutter : A type II restriction endonuclease that binds (and subsequently cleaves) DNA at sites that contain a sequence of six nucleotide pairs that is uniquely recognized by that enzyme. Because any sequence of six bases occurs less frequently by chance than any sequence of four bases, six-base cutters cleave less frequently than do four-base cutters. Thus, six-base cutters create larger fragments than four-base cutters.

Slime molds (syn Myxomycetes) : Saprophytic organisms that form vegetative amoeboid plasmodia and spores.

Slippage replication : A mutation process whereby a simple sequence tandem (microsatellite) repeat grows by addition or subtraction of the "beads" of simple units that make up the "necklace". A dinucleotide AC repeat would grow by addition or subtraction of AC units.

Slot blot : Similar to a dot blot, but the analyte is put onto the membrane using a slot-shaped template. The template produces a consistently shaped spot, thus decreasing errors and improving the accuracy of the analysis. See Dot blot.

small nuclear ribonucleoprotein (snRNP; pronounced snurp(s)) : A compound comprising small nuclear RNA (q.v.) and nuclear protein, that is heavily involved in the post-transcriptional processing of mRNA, especially the removal of introns. snRNPs are a major component of spliceosomes (q.v.).

Small nuclear RNA (snRNA) : Short RNA transcripts of 100-300 bp that associate with proteins to form small nuclear ribonucleoprotein particles (snRNPs) (q.v.); most snRNAs are components of the spliceosomes (q.v.) that excise introns from pre-mRNAs in RNA processing. See RNA.

Smut : A disease caused by a smut fungus (Ustilaginales) in the Basidiomycota or the fungus itself; it is characterized by masses of dark brown or black, dusty to greasy masses of teliospores that generally accumulate in black, powdery sori.

SNP : Defined regions of the genome where there are two or more nucleotide variations, each with 1% or greater prevalence in the population. SNPs can be used as genetic or physical markers.

snRNA : Small nuclear RNA; forms complexes with proteins to form snRNPs; involved in RNA splicing, polyadenylation reactions, other unknown functions (probably).

snRNP "snerps", : Small Nuclear RiboNucleoProtein particles, which are complexes between small nuclear RNAs and 7-10 proteins, and which are involved in RNA splicing and polyadenylation reactions.

Sodium dodecyl sulphate (SDS) : See gel electrophoresis.

Soft rot : Softening, discoloration, and often disintegration of plant tissue as a result of fungal or bacterial infection.

Soil amelioration : The improvement of poor soils, usually using bacteria or fungi. This contrasts with bioremediation, which is the cleaning up of toxins, usually in soils. Amelioration includes breaking down organic matter; forming humus; by solubilizing them, making minerals - such as phosphates - in the soil available to plants; fixing nitrogen; and sometimes an element of bioremediation as well.

Soil conservation practices : See minimal tillage practices.

Soil drench : Application of a solution or suspension of a chemical to the soil, especially pesticides to control soilborne pathogens.

Soil inhabitant : An organism that maintains its population in soil over a period of time.

Soil invader : An organism whose population in soil diminishes in several months to years.

Soil pasteurization : Process used to free soil of selected harmful microorganisms using heat.

Soil sterilization : Process used to free soil of all microorganisms.

Soilborne : Carried on or beneath the soil surface.

Soilless culture; soil-free culture : Tissue culture and hydroponics. Growing plants in nutrient solution without soil.

Solarization : Disease control practice in which soil is covered with polyethylene sheeting and exposed to sunlight, thereby heating the soil and controlling soilborne plant pathogens.

Solid media : Nutrient media that has been solidified, such as by addition of agar.

Solution hybridization : A method closely related to RNase protection (see "Rnase protection assay"). Solution hybridization is designed to measure the levels of a specific mRNA species in a complex population of RNA. An excess of radioactive probe is allowed to hybridize to the RNA, then single-strand specific nuclease is used to destroy the remaining unhybridized probe and RNA. The "protected" probe is separated from the degraded fragments, and the amount of radioactivity in it is proportional to the amount of mRNA in the sample which was capable of hybridization. This can be a very sensitive detection method.

Somaclonal variation : Epigenetic or genetic changes, sometimes expressed as a new trait, resulting from *in vitro* culture of higher plants.

Somatic : Refers to non-germline cells. Somatic cells may become terminally differentiated with alterations in their overall genetic compliment, because they are not responsible for passing along the organism's genetic material to the offspring.

Somatic cell : Any cell of a multicellular organism that composes the body of that organism but does not produce gametes. cf gamete; somatic cell gene therapy.

Somatic cell embryogenesis : Embryos are produced either from somatic cells of explants (direct embryogenesis) or by induction on callus formed by explants (indirect embryogenesis). a.k.a. asexual embryogenesis.

Somatic cell gene therapy : The delivery of a gene or genes to a tissue other than reproductive cells of an individual, with the aim of correcting a genetic defect. cf somatic cell.

Somatic cell genetic mutation : A change in the genetic structure that is neither inherited nor passed to offspring. Also called acquired mutations. See also: germ line genetic mutation.

Somatic cell hybrid : Hybrid cell line derived from two different species; contains a complete chromosomal complement of one species and a partial chromosomal complement of the other; human/hamster hybrids grow and divide, losing human chromosomes with each generation until they finally stabilize, the hybrid cell line established is then utilized to detect the presence of genes on the remaining human chromosome.

Somatic cell variant : A somatic cell with unique characters not present in the other cells, such as might be selected for in a screening trial that following a mutation event.

Somatic cells : Any cell in the body except gametes and their precursors.

Somatic hybridization : 1. Asexual fusion of protoplasts from somatic cells of genetically different parents. 2. Hybridization by induced fusion of cells (protoplasts) from two contrasting genotypes for production of hybrids or cybrids which contain various mixtures of nuclear and/or cytoplasmic genomes, respectively. a.k.a. parasexual hybridization.

Somatic hypermutation : A very high frequency of mutational events that occur in specific loci, such as the variable segments of expressed immunoglobulin genes. Somatic hypermutation in immunoglobulin genes occurs after all rearrangements have occurred and provide additional potential for variation in immunoglobulin structure.

Somatic mutation : A mutation occurring in any cell that is not destined to become a germ cell; if the mutant cell continues to divide, the individual will come to contain a patch of tissue of genotype different from the cells of the rest of the body.

Somatic reduction : Halving of the chromosomal number of somatic cells; a possible method of producing "haploids" from somatic cells and calluses by artificial means.

Somatocrinin : Growth-hormone-releasing hormone. See growth hormone.

Somatostatin : Growth-hormone-inhibiting hormone. See growth hormone.

Sonication : Disruption of cells or DNA molecules by high frequency sound waves. a.k.a. ultrasonication.

Sooty mold : Black, nonparasitic, superficial fungal growth on honeydew produced by aphids and other phloem-feeding insects.

Sorus (pl sori) : Compact fruiting compact fruiting structure, especially the erumpent spore mass in the rust fungi (Uredinales) and smut fungi (Ustilaginales); occasionally a group of fruiting bodies as in Synchytriaceae; a cluster of sporangia on a fern sporophyte.

SOS box : The operator sequence recognised by the LexA repressor protein.

SOS response : The coordinate induction of many genes in response to certain types of DNA damage. Many of the induced gene products facilitate repair of the damaged DNA, but the repair processes result in a high frequency of mistakes in the repaired DNA, a process often called error-prone repair. The SOS response in enteric bacteria is initiated by RecA protein, which becomes activated by DNA damage, and stimulates the autoproteolysis of several other proteins, including the LexA protein and the lambda cI repressor protein. The LexA protein is a global repressor that normally turns off expression of many other genes in the SOS regulon.

Source: Agricultural Biotechnology: Informing the Dialogue. Cornell University College of Agriculture and Life Sciences: Ithaca NY. 2003.

Source DNA : The DNA from an organism that contains a target gene; this DNA is used as starting material in a cloning experiment.

Source organism : A bacterium, plant or animal from which DNA is purified and used in a cloning experiment.

Southern blot : The transfer (by absorption) of size-separated (by electrophoresis) DNA fragments to a synthetic membrane whereby the presence and rough size of one particular fragment of DNA can be ascertained by detection of specific base sequences ascertained by radiolabeled complementary probes. Initially described by E.M. Southern.

Southern blotting : Transfer by absorption of DNA fragments separated in electrophoretic gels to membrane filters for detection of specific base sequences by radiolabeled complementary probes.

Southern hybridization : A procedure in which a cloned, labelled segment of DNA is hybridized to DNA restriction fragments on a Southern blot (q.v.).

Southern transfer : See Southern blotting.

Southwestern Blot : The binding of protein to a nucleic acid on a matrix similar to what is done for western, northern, and southern blots. This technique is used to identify DNA binding proteins and the recognition sites for these proteins.

sp (abbr for species; pl spp) : A genus name followed by sp. means that the particular species is undetermined; spp. after a genus name means that several species are being referred to.

SP6 RNA Polymerase : A bacteriophage RNA polymerase which is commonly used to transcribe plasmid DNA into RNA. The plasmid must contain an SP6 promoter upstream of the relevant sequence.

Sparger : A device that introduces into a bioreactor air in the form of separate fine streams.

Specialized : Anatomically or physiologically adapted for particular functions or habitats.

Specialized transduction : A method of gene transfer between bacteria in which a specific region of the bacterial donor DNA is carried by a phage. The host DNA carried by a specialized transducing phage arises by abbarent excision of a prophage. Thus, only regions of DNA adjacent to an integrated phage can be transferred by this method.

Speciation : The development of one or more species from an existing species.

Species : A single, distinct class of living creature with features that distinguish it from others.

Specific epithet : The second word in a Latin binomial.

Specific resistance (syn vertical resistance) : Resistance which is effective against some biotypes or races of the pathogen, but not others, usually inherited monogenically and expressed qualitatively. (see general resistance, horizontal resistance, race-nonspecific resistance).

Specificity : For diagnostic tests, the ability of a probe to react precisely with a specific target molecule.

Spectral karyotype (SKY) : A graphic of all an organism's chromosomes, each labeled with a different color. Useful for identifying chromosomal abnormalities. See also: chromosome.

Spent medium : After each sub-culture, the medium is discarded because it has been depleted of nutrients, dehydrated or accumulated toxic metabolic products.

Sperm : Abbreviation of spermatozoon, q.v.

Sperm competition : Competition between different spermatozoa to reach and fertilize the egg cell of a single female.

Sperm sexing : The separation of sperm into those bearing an X chromosome and those bearing a Y chromosome, in order to be able to produce, via artificial insemination or *in vitro* fertilization, animals of a specified sex. Achieved by means of inactivation of X-bearing or Y-bearing sperm via antibodies directed against sex-specific peptides on the surface of sperm cells, or fluorescence-activated cell sorting (FACS) (q.v.), in which sperm that have been pre-treated with a fluorescent dye that binds to DNA are separated according to the quantity of fluorescence detected by a laser beam, based on the principle that X-bearing sperm contain more DNA than Y-bearing sperm.

Spermagonium (pl spermagonia; syn pycnium for rust fungi) : Structure in which male reproductive cells are produced; in rust fungi, globose or flask-shaped haploid fruiting body composed of receptive hyphae and spermatia (pycniospores).

Spermatid : One of the four cells formed by the meiotic divisions in spermatogenesis. Spermatids become mature spermatozoa (sperm).

Spermatium (pl spermatia; syn pycniospore for rust fungi) : A male sex cell; a nonmotile male gamete; a haploid male gamete.

Spermatocyte : Sperm mother cell. The cell that undergoes two meiotic divisions (spermatogenesis) to form four spermatids; the primary spermatocyte before completion of the first meiotic division; the secondary spermatocyte after completion of the first meiotic division.

Spermatogenesis : The series of cell divisions in the testis by which maturation of the gametes (sperm) of the male takes place.

Spermatogonium (pl spermatogonia) : Primordial male germ cell that may divide by mitosis to produce more spermatogonia. A spermatogonium may enter a growth phase and give rise to a primary spermatocyte.

Spermatozoon (pl spermatozoa; abbr sperm) : The mature, mobile reproductive cell of male animals, produced by the testis.

Spharoblast : Nodule of wood which can give rise to adventitious shoots with juvenile characteristics.

Spicule : Copulatory organ of male nematode.

Spike : Botanical term for an unbranched inflorescence in which flowers (or spikelets of grasses) are attached directly, without petioles, to a central stem. Examples: wheat, gladiolus.

Spikelet : Spike-like appendage comprised of one or more reduced flowers and associated bracts; unit of inflorescence in grasses; a small spike.

Spina bifida : A congenital condition that results from altered fetal development of the spinal cord, part of the neural plate fails to join together and bone and muscle are unable to grow over this open section.

Spindle (AS spinel, a tool for spinning thread by hand) : In mitosis and meiosis, refers to the spindle-shaped intracellular structure in which the chromosomes move.

Spine : Hard, sharp structure on the surface of a plant; usually a modified leaf.

Spirochaete : A non-rigid, corkscrew-shaped bacterium that moves by means of muscular flexions of the cell.

Spiroplasma : Spiral-shaped plant pathogenic mollicute (prokaryote without cell wall).

Splice sequence : A sequence within an mRNA molecule that is the site at which splicing occurs.

Splice site : Location in the DNA sequence where RNA removes the noncoding areas to form a continuous gene transcript for translation into a protein.

Spliceosomes : Organelles responsible for the removal of introns from mRNA by means of splicing (q.v.).

Splicing : The process that removes introns (non-protein-coding portions) from transcribed RNAs. Exons (protein-coding portions) can also be removed. Depending on which exons are removed, different proteins can be made from the same initial RNA or gene. Different proteins created in this way are 'splice variants' or 'alternatively spliced'.

Splicosome : A ribonucleoprotein structure responsible for splicing of primary transcripts.

Split genes : In eukaryotes, structural genes are typically divided up (split) by a number of non-coding regions called introns. cf exon; guide sequence; splicing.

Spontaneous generation, theory : The theory, now known to be invalid, that plants, animals and microorganisms arose suddenly from non-living materials under certain environmental conditions.

Spontaneous mutation : A mutation that occurs without known exposure to a mutagen.

Sporadic cancer : Cancer that occurs randomly and is not inherited from parents. Caused by DNA changes in one cell that grows and divides, spreading throughout the body. See also: hereditary cancer.

Sporangiospore : Non-motile, asexual spore that is borne in a sporangium.

Sporangium (pl sporangia) : Saclike fungal structure in which the entire contents are converted into an indefinite number of asexual spores.

Spore : A highly resistant, resting state of an organism. Under the proper condiditons s spore will germinate to form a viable, reproducing organism. Bacterial spores are often very resistant to heat and dessication.

Spore mother cell : See sporocyte.

Sporidium (plsporidia) : Basidiospore of rust fungi, smut fungi, and other Basidiomycota.

Sporocarp : Spore-bearing fruiting body.

Sporocyte : A diploid cell that gives rise to four haploid spores by meiosis.

Sporodochium (plsporodochia) : Superficial, cushion-shaped asexual fruiting body consisting of a cluster of conidiophores.

Sporophore : A spore-producing or spore-bearing structure such as a conidiophore, ascocarp or basidiocarp.

Sporophyll : A leaf that bears spore producing structures (sporangia).

Sporophyte : The diploid generation in the life cycle of a plant, and that produces haploid spores by meiosis.

Sport : An individual or portion thereof distinguished by a spontaneous mutation. Sports are sometimes of great agricultural worth, but alternatively, they may be disadvantageous and may be rogued during agricultural production.

Sporulate : To produce spores.

Spot : A symptom of disease characterized by a limited necrotic area, as on leaves, flowers, and stems.

Springwood : Early part of the yearly xylem growth ring in woody plants consisting typically of cells that are larger than those formed later in the season (summerwood).

Spririllum : A rigid, spiral-shaped bacterium.

ß-galactosidase : An enzyme that catalyzes the cleavage of lactose into glucose + galactose. In *E. coli* this enzyme is encoded by the lacZ gene. Often used as a reporter for assaying gene expression (see X-gal).

ß-lactam antibiotics : Antibiotics taht contain a ß-lactam ring and act by inhibiting peptidoglycan synthesis. For example, penicillins, cephalosporins, and related antibiotics.

ß-lactamase : An enzyme that cleaves the ß-lactam ring of ß-lactam antibiotics, thus inactivating the antibiotics. The ampicillin resistance encoded by many common plasmids is due to a secreted ß-lactamase.

SSR : Simple sequence repeat. See microsatellite.

Stabilizing selection : The theorized competitive disadvantage of unnecessary virulence genes; races with excess genes would have decreased fitness relative to races with fewer virulence genes, so a "super-race" would be less likely to appear in multiline crops.

Stable transfection : A form of transfection experiment designed to produce permanent lines of cultured cells with a new gene inserted into their genome. Usually this is done by linking the desired gene with a "selectable" gene, i.e. a gene which confers resistance to a toxin (like G418, aka Geneticin). Upon putting the toxin into the culture medium, only those cells which incorporate the resistance gene will survive, and essentially all of those will also have incorporated the experimenter's gene.

Stages of culture (I-IV) : See micropropagation.

Staggered cuts : Symmetrically cleaved phospho-diester bonds that lie on both strands of duplex DNA but are not opposite one another.

Staghead : Defoliated, dead or dying major branches in the crown of a tree, usually resulting from inadequate water uptake or translocation.

Stamen (L stamen, the standing-up things or a tuft of thready things) : Flower structure made up of an anther (pollen-bearing portion) and a stalk or filament. The stamen is the male part of the flower.

Standard deviation : A statistical measure of variability in a population of individuals or in a set of data; the square root of the variance.

Standard error : A statistical measure of variation in a population of means, used to indicate how well sample estimates represent population parameters.

Starch (ME strechen, to stiffen) : A complex insoluble carbohydrate, consisting of various proportions of two glucose polymers, amylose and amylopectin; the chief food storage substance of plants, which is composed of several hundred hexose sugar units and which easily breaks down on hydrolysis into these separate units.

StarLinkTM : An insect-resistant variety of corn that was not labeled for human consumption.

Start codon : That codon at which translation of an mRNA molecule begins. This is always an AUG (encoding methionine) in eukaryotes, and nearly always in prokaryotes. In prokaryotes N-formyl-methionine is used to initiate polypeptide synthesis.

Stationary culture : A culture maintained in the growth chamber with no agitation movement. The antonym is shake culture.

Stationary phase : The plateau of the growth curve after log growth, during which cell number remains constant. New cells are produced at the same rate as older cells die. See growth phases.

Statistic : An estimate based on a sample or samples of a population, providing an indication of the true population parameter.

Steady state : In a continuous fermentation process, the condition when the number of cells that are removed with the outflow is exactly balanced by newly synthesized cells.

Stele (Gr stele, a post) : The central cylinder, inside the cortex, of roots and stems of vascular plants. See meristele.

Stem (OF stemn) : The main body of the above-ground portion of a tree, shrub, herb or other plant; the ascending axis, whether above or below ground, of a plant.

Stem cell : An undifferentiated active somatic cell that undergoes division and gives rise to other stem cells or to cells that differentiate to form specialized cells.

Stem pitting : A viral disease symptom characterized by depressions on the stem.

Stepwise mutation : Microsatellite variation appears to result from slippage in replication, which is most likely to add or delete a single repeat unit (steps of one). As a result, alleles more similar in size will presumably be more closely related. This additional 'phylogenetic' information can be used in assessing genetic differentiation or genetic distance.

Sterigma (pl sterigmata) : Small, usually pointed projection that supports a spore.

Sterile : 1. Medium or object with no perceptible or viable micro-organisms. 2. Incapable of fertilizing or being infertile.

Sterile fungus : A fungus that is not known to produce any kind of spores.

Sterile room : Operation room for performing inoculations under aseptic conditions; usually now replaced by use of laminar air-flow cabinets, in which filtered air is blown from the inside to the outside. See laminar air-flow cabinet.

Sterility : Inability to produce offspring.

Sterilization (adj sterilized) : The total destruction of living organisms by various means, including heat, chemicals or irradiation.

Sterilize : 1. The process of elimination of micro-organisms, such as by chemicals, heat, irradiation or filtration. 2. The operation of making an animal incapable of producing offspring.

Steward bottle : Flask developed by Steward for the growth of cells and tissues in a liquid medium, in which they can be periodically submerged during rotation.

Sticky ends : After digestion of a DNA with certain restriction enzymes, the ends left have one strand overhanging the other to form a short (typically 4 nt) single-stranded segment. This overhang will easily re-attach to other ends like it, and are thus known as "sticky ends". For example, the enzyme BamHI recognizes the sequence GGATCC, and clips after the first G in each strand: The overhangs thus produced can still hybridize ("anneal") with each other, even if they came from different parent DNA molecules, and the enzyme ligase will then covalently link the strands. Sticky ends therefore facilitate the ligation of diverse segments of DNA, and allow the formation of novel DNA constructs.

Stigma : Portion of a flower that receives pollen and on which the pollen germinates.

Stipe : Stalk.

Stippling : Series of small dots or speckles in which chlorophyll is absent.

Stipule : Small leaflike appendage at the base of a leaf petiole, usually occurring in pairs.

Stirred-tank fermenter : A growth vessel in which cells or micro-organisms are mixed by mechanically-driven impellers.

Stochastic : A random process; a process determined by a random distribution of probabilities.

Stock : The lower portion of a graft. See rootstock.

Stock plant : The source plant from which cuttings or explants are made. Stock plants are usually maintained carefully in an optimum state for (sometimes prolonged) explant use. Preferably they are certified-pathogen-free plants.

Stock solutions : Pre-prepared solutions of individual components and used to prepare many different types of media. Certain substances, including Ca and Mg sulphates and phosphates must not be combined until actual medium preparation because insoluble combinations are formed and precipitate.

Stolon (L stolo, a shoot) : A lateral stem that grows horizontally along the ground surface. The runners of white clover, strawberry and bermuda grass are examples of stolons. cf runner.

Stoma (Gr stoma, mouth; pl stomata) : 1. Any of various small openings or pores in an animal body, especially an opening resembling a mouth in various invertebrates. 2. Botany: A minute pore in the epidermis of the leaf or stem of a plant, forming a slit of variable width between two specialized cells (guard-cells), which allows movement of gases, including water vapour, to and from the intercellular spaces. Also, the whole pore with its associated guard-cells.

Stomatal complex : Includes the stoma, together with its guard cells and, when present, the subsidiary cells.

Stomatal index : = (number of stomata per $mm^2 \times 100$)/(number of stomata per mm^2 + number of epidermal cells per mm^2). This value has been found useful in comparing leaves of different sizes. Relative humidity and light intensity during leaf development affect the value of stomata index.

Stone fruit : Fruit with a stony endocarp, e.g. cherry, peach, plum.

Stop (nonsense) codon : A UAG, UAA. or UGA codons which is not representing any cognate aminoacyl tRNA in most organisms. When the ribosome encounters a stop codon in the mRNA, a termination factor interacts with the ribosome causing polypeptide synthesis to stop and the ribosome to dissociate from the mRNA.

STR : See Short Tandem Repeats.

Strain : An organism that is different from other organisms of the same species due to genetic differences. Strain is commonly used in two ways: (i) organisms of the same species that when initially isolated are found to have certain different properties (due to unknown mutations) are called different strains; (ii) derivatives of an organism that have distinct genotypes due to known mutations are called different strains.

Strain isolation : Isolation of any bacterium, animal or plant from the outside world.

Stramenopila (straminipila) : a kingdom including the Oomycota or oomycetes, along with brown and golden algae and diatoms. "Stramenopila" is the originally proposed spelling, whereas "Straminipila" has been proposed as more correct in form on the basis of the Latin root of the term.

Strand slippage : The denaturation and aberrant mispairing of a template strand during DNA replication. Results in frameshift mutations (see http://www.sci.sdsu.edu/~smaloy/MicrobialGenetics/topics/mutations/frame-shift.html).

Stratification : Treatment of moist seeds at low temperature (+2° - +4°C) to break dormancy.

Streptavidin : A bacterial analog of egg white avidin.

Streptomycin : An antibiotic that inhibits protein synthesis in bacteria by binding to the S12 protein of the 30s ribosomal subunit and inhibiting translation. A high level of StrR can result from chromosomal mutations in the gene for the S12 protein (rpsL) which prevent streptomycin from binding to the ribosome. Only mutant ribosomes are StrR, so resistance to streptomycin is recessive to streptomycin sensitivity. Because streptomycin rapidly inhibits protein synthesis, when cells acquire the rpsL gene phenotypic expression is required before the cells become streptomycin resistant.

Stress : See water stress.

Striate (n striations) : Marked with delicate lines, grooves, or ridges.

Stringency : A term used to describe the conditions of hybridization. By varying the conditions (especially salt concentration and temperature) a given probe sequence may be allowed to hybridize only with its exact complement (high stringency), or with any somewhat related sequences (relaxed or low stringency). Increasing the temperature or decreasing the salt concentration will tend to increase the selectivity of a hybridization reaction, and thus will raise the stringency.

Stringent plasmid : A plasmid that only replicates along with the main bacterial chromosome and is present as a single copy, or at most several copies, per cell. See plasmid.

Stringent response : The ability of a bacterium to limit the synthesis of tRNA and rRNA during amino acid starvation. The alarmones ppGpp and pppGpp are at least partially responsible for the stringent response. The concentrations of these alarmones are controlled by the relA and spoT gene products in *E. coli*.

Stroma (pl stromata) : Compact mass of mycelium (with or without host tissue) that supports fruiting bodies or in which fruiting bodies are embedded.

Strong promoter : An efficient promoter that can direct synthesis of RNA transcripts at a relatively fast rate.

Structural gene : The portion of a gene encoding a functional polypeptide or RNA molecule.

Structural genomics : The effort to determine the 3D structures of large numbers of proteins using both experimental techniques and computer simulation.

Structure-functionalism : The scientific tradition that stresses the relationship between a physical structure and its function, e.g., the related disciplines of anatomy and physiology.

STS : See sequence tagged site.

Stunting : Reduction in height of a vertical axis resulting from a progressive reduction in the length of successive internodes or a decrease in their number.

Stuttering alleles : Defective alleles that have segments which are doubled in their transmission from generation to generation. In the case of genetically inherited diseases, the result is increasingly severe symptoms each generation. The myotonic form of muscular dystrophy is an example. Stuttering alleles are also known as unstable alleles.

Style (Gr stylos, a column) : Slender column of tissue that arises from the top of the ovary and through which the pollen tube grows.

Stylet knob (syn basal knob) : Structure at the base of a nematode stylet.

Stylet-borne transmission (syn nonpersistent transmission) : A type of virus transmission in which the virus is acquired and transmitted by the vector after short feeding times, and is retained by the vector for only a short period of time.

Sub-clone : A method in which smaller DNA fragments are cloned from a large insert which has already been cloned in a vector.

Sub-cloning : If you have a cloned piece of DNA (say, inserted into a plasmid) and you need unlimited copies of only a part of it, you might "sub-clone" it. This involves starting with several million copies of the original plasmid, cutting with restriction enzymes, and purifying the desired fragment out of the mixture. That fragment can then be inserted into a new plasmid for replication. It has now been subcloned.

Sub-culture : Division and transfer of a portion or inoculum of a culture to fresh medium. Sometimes used to denote the adding of fresh liquid to a suspension culture. cf inoculum; passage.

Sub-culture interval : The time between subsequent sub-cultures of cells. Sub-culture interval has no relationship to the term cell generation time.

Sub-culture number : The number of times cells, etc., have been sub-cultured, i.e., transplanted by inoculation from one culture vessel to another.

Suberize : To convert into cork tissue.

Subgenomic RNA : A piece of viral RNA, shorter than the entire genome of the virus, found in cells infected by the virus and sometimes encapsidated.

Sublining- : Dividing a breeding population into several to many smaller populations. All controlled crosses for forward selection are made within a subline, leading to inbreeding within sublines. Production seed orchards are planted with clones or seedlings from several to many sublines to minimize inbreeding in the production orchard.

Subspecies : A subpopulation of a species, defined on the basis of more than one character (morphologic for many organisms) that distinguishes the members of the subpopulation from other members of that species.

Substitution : A point mutation in which one base pair in the DNA sequence is replaced by another.

Sub-strain : Derived from a strain by isolating a single cell or groups of cells having properties or markers not shared by all cells of the strain.

Substrate : 1. A compound that is altered by an enzyme. 2. Food source for growing cells or micro-organisms. 3. Material on which a sedentary organism lives and grows.

Substution mutation : A mutation that replaces one nucleotide in a DNA sequence with another nucleotide.

Subtraction library : A cDNA library that only contains those cDNAs that are expressed in a particular type of cell or tissue or under particular growth conditions. The cDNA common to all conditions is removed by hybridization with excess RNA from other cells (i.e. subtraction), leaving only the unique RNA behind.

Sub-unit vaccine : One or more immunogenic proteins either purified from the disease-causing organism or produced from a cloned gene. A vaccine composed of a purified antigenic determinant that is separated from the virulent organism. cf vaccine; enzyme.

Sucker : A shoot that arises from an underground root or stem and grows at the expense of the parent plant. cf turion.

Suckering : Type of vegetative propagation where lateral buds grow out to produce an individual that is a clone of the parent.

Sucrose density gradient centrifugation : A procedure used to fractionate mRNAs or DNA fragments on the basis of size.

Suicide plasmid : A plasmid that cannot replicate in a given host, used to force the; integration into host replicons of genetic determinants carried by the plasmid or phage. It is usually necessary to have a direct selection for the desired recombinants. See Allele exchange.

Sulfur dioxide (SO_2) : A primary air pollutant produced in industrial processes and coal burning that causes interveinal necrosis on broadleaf plants and tip necrosis on conifers.

Summerwood : Part of the yearly xylem growth ring in woody plants formed late in the growing season and consisting of cells smaller than those of springwood.

Sunscald or sunburn : Injury of plant tissues burned or scorched by direct sun or or sunburn.

Superbug : Jargon for the bacterial strain of Pseudomonas developed by Chakrabarty, who combined hydrocarbon-degrading genes carried on different plasmids into one organism. Although this genetically engineered micro-organism is neither "super" nor a "bug", it represents a landmark example because it showed how genetically modified microbial strains could be used in a novel way and because it was the basis for the precedent-setting legal decision that declared that genetically engineered organisms were patentable.

Supercoil : Double-stranded circular DNA which is twisted about itself. Commonly observed with plasmids and circular viral DNA genomes (such as that of hepatitis B virus). A nick in one strand of the plasmid may remove the twist, resulting in a relaxed, circular DNA molecule. A complete break in the DNA puts the plasmid in a linear form. Supercoils, relaxed circular DNA, and linear DNA all have different migration properties in agarose gels, even though they contain the same number of base pairs.

Supercoiled DNA : Double-stranded circular DNA in which either overwinding or underwinding of the duplex makes the circle twist. The conformation of a covalently closed-circular DNA molecule, which is coiled by torsional strain into the shape taken by a wound up elastic band.

Supercoiled plasmid : The predominant *in vivo* form of plasmid, in which the plasmid is coiled around histone-like proteins. Supporting proteins are stripped away during extraction from the bacterial cell, causing the plasmid molecule to supercoil around itself *in vitro*. See Plasmid.

Supergene : A group of neighbouring genes on a chromosome and that tend to be inherited together and sometimes are functionally related.

Superinfection : The secondary infection by a phage in a cell already harbouring the same phage.

Superinfection exclusion : A process whereby the presence of one phage in a cell prevents the secondary infection by other. There are many different mechanisms of superinfection exclusion, including changes in cell surface receptors or the expression of a restriction system.

Supernatant : The soluble liquid fraction of a sample after centrifugation or precipitation of insoluble solids.

Suppression : The restoration (or partial restoration) of a wild-type phenotype by a second mutation. There are many different mechanisms of suppression (see the supplemental resources.

Suppressive soil : Soil in which various diseases are naturally at lower levels than expected due to biological factors in the soil; an example of natural biological control.

Suppressor : Mutations in suppressor genes are able to overcome (suppress) the effects of mutations in other, unlinked, genes. A common, and very useful, kind of suppressor mutation occurs within the gene encoding a tRNA molecule and results in a change in the tRNA's anticodon. Such a mutant tRNA can reverse the effect of chain-terminating mutations, such as amber or ochre, in protein-encoding genes. See stop codon; nonsense mutation.

Suppressor gene : A mutated gene which produces a product which reverses the effect of a previous specific mutation without actually correcting the original mutation in the DNA can be either intergenic or intragenic.

Suppressor gene : A gene that can suppress the action of another gene.

Suppressor mutation : A mutation that restores, partially or completely, the loss of function caused by another mutation. Many suppressor mutations are in genes encoding a transfer RNA species; the altered tRNA can recognise the original mutant codon and, during translation, insert an acceptable substitute amino acid into the polypeptide.

Suppressor tRNA : A mutant tRNA that recognizes a stop codon instead of the codon for the cognate amino acid. This property is sometimes, but not always, due to a base substitution in the anticodon.

Suppressor-sensitive mutant : An organism that can grow when a second genetic factor - a suppressor - is present, but not in the absence of this factor.

Surfactant : A surface-active agent or wetting agent, such as Tween 20TM or Tween 80TM, TeepolTM, Lissapol FTM, AlconoxTM, etc. Surfactants act by lowering the surface tension and are common addenda to solutions used to surface sterilize materials prior to aseptic excision of explants. See detergent; wetting agent.

Suscept : An abbreviated term for a susceptible plant.

Susceptible : The characteristic of a host organism such that it is incapable of suppressing or retarding an injurious pathogen or other factor.

Susceptible (n susceptibility) : Prone to develop disease when infected by a pathogen (see resistance).

Suspension culture : A type of culture in which (single) cells and/or clumps of cells grow and multiply while suspended in a liquid medium. See cell suspension.

Sustainable intensification of animal production systems : In AnGR: The manipulation of inputs to, and outputs from, livestock production systems aimed at increasing production and/or productivity and/or changing product quality, while maintaining the long-term integrity of the systems and their surrounding environment, so as to meet the needs of both present and future human generations. (Source: FAO, 1999) Sustainable agricultural intensification respects the needs and aspirations of local and indigenous people, takes into account the roles and values of their locally adapted genetic resources, and considers the need to achieve long-term environmental sustainability within and beyond the agro-ecosystem. (Source: FAO, 1999).

Symbiont : An organism living in symbiosis with another, dissimilar organism.

Symbiosis (Gr. Syn, with + bios, life) : The close association of two different kinds of living organisms where there is benefit to both or where both receive an advantage from the association. An example is the association of the mycelium of mycorrhizal fungi with roots of seed plants. See commensalism; parasitism.

Symbiotic association : An intimate partnership between two organisms, in which the mutual advantages normally outweigh the disadvantages.

Sympatric speciation : The formation of new species by populations that inhabit the same or overlapping geographic regions.

Symplast : The system of protoplasts in plants, that are interconnected by plasmodesmata.

Sympodial : A type of plant development in which the terminal bud of the stem stops growing due to either its abortion or its development into a flower or an inflorescence, and the uppermost lateral bud takes over the further axial growth of the stem.

Symptom : Indication of disease by reaction of the host, e.g. canker, leaf spot, wilt (see sign).

Symptomless carrier : A plant that, although infected with a pathogen (usually a virus), produces no obvious symptoms.

Synapsis : The pairing of homologous chromosomes or of homologous chromosomal regions.

Synaptonemal complex : A ribbonlike protein structure formed between synapsed homologues at the end of the first meiotic prophase, binding the chromatids along their length and facilitating chromatid exchange.

Synchronous culture : A culture in which the majority of the cells are dividing at the same time or are at a specific phase of the cell cycle.

Syncytium : A group of cells in which cytoplasmic continuity is maintained. cf acellular.

Syndrome : A group of symptoms that occur together and represent a particular disease.

Synergids : The two nuclei within the embryo sac at the upper end in the ovule of the flower, which, with the third (the egg), constitute the egg apparatus.

Synergism (adj synergistic) : Greater than additive effect of interacting factors.

Syngamy : Fertilization.

Syngeneic : Genetically identical members of the same species.

Synkaryon : A nucleus formed by the fusion of nuclei from two different somatic cells during somatic-cell hybridization. See hybrid cell.

Synnema (pl synnemata; syn coremium) : Compact or fused, generally upright conidiophores, with branches and spores forming a headlike cluster.

Synteny : Genes which occur in the same order on the chromosome of different species.

Synthetic variety : A variety produced by crossing a number of genotypes selected for good combining ability in all possible hybrid combinations, with subsequent maintenance of the variety by open pollination.

Systematics : The study of the kinds of organisms and the relationships between them.

Systemic : Pertaining to a disease in which the pathogen (or a single infection) spreads generally throughout the plant; pertaining to chemicals that spread internally through the plant.

Systemic acquired resistance (SAR) : Reduced disease symptoms on a portion of a plant distant from the area where a hypersensitive response occurred or other stimulus was applied; a rapid and coordinated defense response against a variety of pathogens as a signal travels throughout the plant (see induced systemic resistance).

Systemic fungicide : A fungicide that is absorbed into plant tissue and may offer some curative or after-infection activity; includes fungicides that are locally systemic, xylem-mobile (upward moving), and amphimobile (move in phloem upward as well as downward in the plant) (see contact or protectant fungicide).

Sythetic lethal mutations : Two mutations are synthetically lethal if cells with either of the single mutations are viable but cells with both mutations are inviable. As with suppressor analysis, synthetic lethal mutations often indicate that the two mutations affect a single function or pathway.

Sythetic phenotype : A distinct phenotype that requires the presence of two mutations, and either of the mutations alone does not exhibit the same phenotype. Also note the defination of Synthetic lethal mutations.

T : The thymine residue in DNA.

T cell receptor : An antigen-binding protein that is located on the surfaces of killer T cells and mediates the cellular immune response of mammals. The genes that encode T cell antigens are assembled from gene segments by somatic recombination processes that occur during T lymphocyte differentiation.

T lymphocyte : See T cell.

T4 DNA ligase : An enzyme from bacteriophage-T4-infected cells, and that catalyses the joining of duplex DNA molecules and repairs nicks in DNA molecules. The enzyme requires that one of the DNA molecules has a 5´-phosphate group and that the other has a free 3´-hydroxyl group.

T7 RNA Polymerase : A bacteriophage RNA polymerase which is commonly used to transcribe plasmid DNA into RNA. The plasmid must contain a T7 promoter upstream of the relevant sequence.

Tag : See label.

Tailing : The *in vitro* addition of the same nucleotide by the enzyme terminal transferase, to the 3´-hydroxyl ends of a duplex DNA molecule. a.k.a. homopolymeric trailing.

Tandem array : The existence of two or more identical DNA sequences in series, i.e., end -to-end.

Tandem duplication : A DNA sequence that is repeated in direct orientation.

Tandem Repeat Sequences : Multiple copies of the same base sequence on a chromosome; used as a marker in physical mapping.

Tandem repeats : Multiple adjacent copies of the same sequence.

Tank bioreactor : Vessel in which fermentation takes place. A tank bioreactor is a vessel in which a micro-organism is grown in a large volume of liquid. This contrasts with fibre or membrane bioreactors and immobilized cell reactors. The large majority of bioreactors used in biotechnology are tank bioreactors, and most tank bioreactors are stirred-tank bioreactors, because stirring helps to distribute effectively gas and nutrients to the growing organism.

Tap root : Root system in which the primary root has a much larger diameter than the lateral roots. Opposite: fibrous root.

Taq polymerase : A DNA polymerase isolated from the bacterium Thermophilis aquaticus and which is very stable to high temperatures. It is used in PCR procedures and high temperature sequencing.

Target : For diagnostic tests, the molecule or nucleic acid sequence that is being sought in a sample.

Target site duplication : A sequence of DNA that is duplicated when a transposable element inserts; usually found at each end the insertion.

Targeted drug delivery : A method of delivering a drug to the site in the body where it is needed, rather than allowing it to diffuse into many sites.

Targeted mutagenesis : Deliberate change in the genetic structure directed at a specific site on the chromosome. Used in research to determine the targeted region's function. See also: mutation, polymorphism.

Targeting vector : A cloning vector carrying a DNA sequence capable of participating in a crossing-over event at a specified chromosomal location in the host cell.

TATA box : A sequence found in the promoter (part of the 5' flanking region) of many genes. Deletion of this site (the binding site of transcription factor TFIID) causes a marked reduction in transcription, and gives rise to heterogeneous transcription initiation sites.

Tautomeric shift : A reversible change in the position of a hydrogen atom in a molecule which results in the conversion of the molecule between different isomers. A shift between the keto group and a enoyl group in nucleotides can result in altered base-pairing.

tautomerism : A type of isomerism in which the two isomes are in equilibrium.

Taxon (plural taxa) : Group of organisms linked by common ancestry. Taxa can range in scale from populations to kingdoms.

Taxonomy : The classification of organisms.

Tay-Sachs disease : A fatal degenerative disease of the nervous system due to a deficiency of hexosamidase A, causing mental deficiency, paralysis, mental deterioration, and blindness; found primarily but not exclusively among Ashkenazi Jews. Autosomal recessive.

T-cell-mediated (cellular) immune response : The synthesis of antigen-specific T cell receptors and the development of killer T cells in response to an encounter of immune system cells with a foreign immunogen.

T-DNA : The segment of DNA in the Ti plasmid of Agrobacterium tumefaciens that is transferred to plant cells and inserted into the chromosomes of the plant.

Technology transfer : The process of converting scientific findings from research laboratories into useful products by the commercial sector.

Telemeter : The unique structure found at the end of eukaryotic chromosomes containing specialized sequences of DNA that assures the completion of a cycle of DNA replication.

Teleomorph (syn perfect state) : The sexual form in the life cycle of a fungus (see anamorph and holomorph).

Teliospore (sometimes called teleutospore, teleutosporodesm) : Thick-walled resting or overwintering spore produced by the rust fungi (Uredinales) and smut fungi (Ustilaginales) in which karyogamy occurs; it germinates to form a promycelium (basidium) in which meiosis occurs.

Telium (pl telia) : Fruiting body (sorus) of a rust fungus that produces teliospores.

Telomerase : A ribonucleoprotein complex that maintains the repeat sequence structures at the telomeric ends of chromosomes.

Telomere : The terminal part of a linear chromosome. Replication of the ends of linear DNA molecules requires specialized enzymes or structures. Often the telomers have a DNA sequence with a single-stranded end that can fold into a hairpin structure.

Telophase : The last stage in cell division before the nucleus returns to a resting condition.

TEM : See transmission electron microscope.

Temperate phage : A phage that is capable of becoming a prophage in the bacterial host (i.e. maintain itself in a relatively quiescent state). Improperly but frequently called a lysogenic phage.

temperature-sensitive mutant : An organism that can grow at one temperature but not at another.

Temperature-sensitive mutation : A mutation that results in a gene product that is functional within a certain temperature range (e.g. at less than 30°C), but nonfunctional at different temperatures (e.g. at 42°C).

Template : An RNA or single-stranded DNA molecule upon which a complementary nucleotide strand is synthesized. A pattern or mould. DNA stores coded information and acts as a model or template from which information is copied into complementary strands of DNA or transcribed into mRNA.

Template strand : The polynucleotide strand that a polymerase uses for determining the sequence of nucleotides during the synthesis of a new nucleic acid strand.

Temporary wilt : Wilt due to insufficient soil water from which a plant can recover when water is supplied.

Teratogen : A chemical that causes malformations in the fetus.

Teratogenic : Substances such as chemicals or radiation that cause abnormal development of a embryo. See also: mutatgen.

Teratogens : Any agent that raises the incidence of congenital malformations.

Term finalization : Repelling movement of the centromeres of bivalents in the diplotene stage of the meiotic prophase, that tends to move the visible chiasmata toward the ends of the bivalents.

Terminal bud : A branch tip, an undeveloped shoot containing rudimentary floral buds or leaves, enclosed within protective bud scales.

Terminal redundancy : The presence of identical DNA sequences repeated at the two ends of DNA molecule (e.g. phage particles).

Terminal transferase : An enzyme that adds nucleotides to the 3´ terminus of DNA molecules.

Termination codon : See stop codon.

Termination signal : In transcription, a nucleotide sequence that specifies RNA chain termination.

Terminator : A sequence downstream from the 3' end of an open reading frame that serves to halt transcription by the RNA polymerase. In bacteria these are commonly sequences that are palindromic and thus capable of forming hairpins. Sometimes termination requires the action of a protein, such as Rho factor in *E. coli*.

Terminator region : A DNA sequence that signals the end of transcription.

Terminus : The region of DNA sequences where DNA replication terminates.

Tertiary Structure (also see Primary and Secondary Structure) : Refers to higher ordered structures conferred on proteins or nucleic acids by interactions between amino acid residues or nucleotides which are not closely positioned within the sequence (primary structure) of the molecule.

Test tube : Tube in which cells, tissues, etc., can be cultured.

Testa (pltestae) : Seed coat.

Testcross : Backcross to the recessive parental type, or a cross between genetically unknown individuals and a fully recessive tester to determine whether an individual in question is heterozygous or homozygous for a certain allele. It is also used as a test for linkage, i.e., to estimate recombination fraction.

Test-tube fertilization : *In vitro* fertilization.

Tet resistance : See "Antibiotic resistance".

Tetracycline (Tet) : An antibiotic that inhibits protein synthesis by preventing aminoacyl tRNA from binding to ribosomes. There are several possible mechanisms of Tetracycline resistance. TetR encoded by Tn10 and pBR plasmids is due to a membrane protein that actively transports tetracycline out of the cell.

Tetrad : 1. The four cells arising from the second meiotic division in plants (pollen tetrads) or fungi (ascospores). 2. The quadruple group of chromatids that is formed by the association of duplicated homologous chromosomes during synapsis in meiosis I. a.k.a. quadrivalent.

Tetrad analysis : A method for establishing linkage relationships in fungi by analysing the four products from individual meiotic divisions.

Tetraploid : An organism with four basic (x) sets of chromosomes.

Tetrasomic (noun tetrasome) : Pertaining to a nucleus or an organism with four members of one of its chromosomes, whereas the remainder of its chromosome complement is diploid. Chromosome formula: 2n + 2.

Tetratype : In fungi, a tetrad of spores that contains four different types; e.g., AB, aB, Ab and ab.

Thallus (Gr thallos, a sprout) : Plant body without true roots, stems, or leaves.

Thatch : Tightly intertwined layer of plant litter from accumulations of undecomposed or partially decomposed plant residues.

The Bollgard trait : The Bollgard trait offers cotton farmers in-the-seed protection against cotton pests like cotton budworm and bollworm. This protection allows farmers to reduce pesticide applications and get higher yield.

The Roundup Ready trait : Provides farmers the ability to use Roundup agricultural herbicides over the top of their crops, simplifying production and offering the broadest spectrum weed control on the market. The Roundup Ready trait is available in corn, cotton, soybeans, canola and alfalfa.

The SNP Consortium (TSC) : Established (April 1999) to identify SNPs and add them to the public domain rather than patenting them for commercial use. A joint enterprise of pharmaceutical companies (AstraZeneca, Bayer, Bristol-Myers Squibb, F. Hoffmann-La Roche, Glaxo Wellcome, Aventis/ Hoechst Marion Rouse, Novartis, Pfizer, Searle, and SmithKline Beecham) and the Wellcome Trust, with sequencing carried out by three public genomics institutes (Sanger Centre, Washington University School of Medicine, and the Whitehead Institute) with Stanford University Human Genome Center contributing mapping and Cold Spring Harbor Laboratory coordinating bioinformatics activities. [CHI SNPs report] Motorola joined in October 1999 and IBM in March 2000.

The YieldGard Corn Borer trait : Provides farmers with in-the-seed protection against European and southwestern corn borer, reducing the amount of pesticide applications farmers must use to control pest pressure.

The YieldGard Rootworm trait : Provides farmers with in-the-seed insect protection against western, northern and Mexican corn rootworms. Like the other traits in the YieldGard family, YieldGard Rootworm corn reduces the amount of pesticide applications farmers must use to control pest pressure.

Therapeutic agent : A compound that is used for the treatment of a disease or for improving the well-being of an organism. a.k.a. pharmaceutical agent; drug; protein drug.

Thermic shock : Exposure to reduced or increased temperature for several days.

Thermolabile : Not heatproof, e.g., a substance which disintegrates or is unstable upon heating.

Thermophile : An organism which grows at a higher temperature than most other organisms. In general, a wide range of bacteria, fungi and simple plants and animals can grow at temperature up to 50°C; thermophiles are considered to be organisms which can grow at above 50°C. They can be classified according to their optimal growth temperature, into simple thermophiles (50-65°C), thermophiles (65-85°C), and extreme thermophiles (>85°C). Thermophiles and extreme thermophiles are usually found growing in very hot places, such as hot springs and geysers, smoker vents on the sea floor, and domestic hot water pipes.

Thermosensitivity : Loss of activity of a protein at high temperature.

Thermostability : Retention of activity at high temperature.

Thermotherapy : Technique mainly used for virus or mycoplasma elimination. Plants are exposed to elevated temperatures as a treatment. Thermotherapy is used alone or in combination with meristem culture or meristem tip culture.

Theta-replication : Replication of a circular molecule of double-stranded DNA by initiation at a unique origin and proceeding in one or both directions around the molecule. The resulting intermediate looks like the greek letter theta.

Thinning : 1. Removal of older stems to promote new growth. 2. Removal of excess fruits to improve the size and quality of the remaining fruits. 3. Removal of seedlings spaced too closely for optimum growth.

Thorax : Insect body part between the head and abdomen.

Three-factor cross : A method for determining the genetic map position of three linked loci based upon relative frequency of coinheritance of each locus during a cross between two strains with different allelic forms in each gene.

Thymidine : A nucleoside present in DNA but absent in RNA.

Thymidine kinase (tk) : An enzyme that allows a cell to utilize an alternate metabolic pathway for incorporating thymidine into DNA. Used as a selectable marker to identify transfected eukaryotic cells.

Thymine : One of the four bases in DNA that make up the letters ATGC, thymine is the "T". The others are "A" for adenine, "G" for guanine, and "C" for cytosine. Thymine always pairs with adenine. Cytosine always pairs with guanine. These letters are used as shorthand for the sequences of fragments of DNA e.g. CCAAGTAC. These sequences are the code for genetic information.

Thymine dimer : See pyrimidine dimer.

Ti plasmid : Tumour-inducing plasmid. A giant plasmid of Agrobacterium tumefaciens that is responsible for the induction of tumours in infected plants. Ti plasmids are used as vectors to introduce foreign DNA into plant cells. See vector.

Tillage : Cultivation, including hoeing and plowing. In agriculture, tillage is used to fight weeds.

Tiller : A lateral shoot, culm, or stalk arising from a crown bud; common in grasses.

Tissue : Group of cells, usually of similar structure, that perform the same or related functions.

Tissue culture : A technique for cultivation cells, tissues, or organs of plants in a sterile, synthetic medium; includes the tissues excised from a plant and the culture of pollen or seeds.

Tissue-specific expression : Gene function which is restricted to a particular tissue or cell type. For example, the glycoprotein hormone alpha subunit is produced only in certain cell types of the anterior pituitary and placenta, not in lungs or skin; thus expression of the glycoprotein hormone alpha-chain gene is said to be tissue-specific. Tissue specific expression is usually the result of an enhancer which is activated only in the proper cell type.

Titer : Concentration of a virus.

tk : See thymidine kinase.

Tm : The midpoint of the temperature range over which DNA is melted or denatured by heat; the temperature at which a duplex nucleic acid molecule is 50% melted into single strands, it is dependent upon the number and proportion of G-C base pairs as well as the ionic conditions. Often referred to as a measure of the thermal stability of a nucleic acid probe:target sequence hybrid.

Tolerance (adj tolerant) : Ability of a plant to endure an infectious or noninfectious disease, adverse conditions, or chemical injury without serious damage or yield loss; (of pesticides) the amount of chemical reside legally permitted on an agricultural product entering commercial channels and usually measured in parts per million (ppm).

Tonoplast (Gr tonos, stretching tension + plastos, moulded, formed) : The cytoplasmic membrane bordering the vacuole, with a role in regulating the pressure exerted by the cell sap.

Top Cross : A cross between a selection, line, clone, etc., and a common pollen parent which may be a variety, inbred line, single cross, etc. The common pollen parent is called the top cross or tester parent. In corn, a top cross is commonly an inbred-variety.

Topoisomerase : An enzyme which introduces or removes overwinding or underwinding of the DNA circular duplex by causing a nick, rotating the strands, and then ligating them.

Topo-isomerase : An enzyme that introduces or removes supercoils from DNA.

Torr : An obsolete unit of pressure equal to that exerted by a column of mercury 1mm high at 0°C and standard gravity (1mm Hg); named after Evangelista Torricelli (1608-1647), the inventor of the mercury barometer. 1 Torr = 1/760 atm = 133.322 Pa.

Totipotency : Having the potentiality of forming all the types of cells in the body. The property of somatic cells to be induced to undergo regeneration. The diploid zygote formed at fertilization is a single cell which is capable of division and differentiation to give rise to the total range of cell types found in the adult organism.

Totipotent cell; totipotent nucleus : An undifferentiated cell (or nucleus), such as a blastomere, that, when isolated or suitably transplanted, can develop into a complete embryo.

Toxicity : Negative effect of a compound, as shown by altered morphology or physiology. It is meaningful only when the effect itself is also described, such as changes in the rate of cell growth, cell death, etc.

Toxicogenomics : The study of how genomes respond to environmental stressors or toxicants. Combines genome-wide mRNA expression profiling with protein expression patterns using bioinformatics to understand the role of gene-environment interactions in disease and dysfunction.

Toxin (L toxicum, poison) : A compound produced by an organism and poisonous to plants or animals.

Trace element : See microelement.

Tracer : An added or injected substance that can be followed within a reaction or an organism, such as radioactive isotopes and certain dyes.

Tracheid (Gr tracheia, windpipe) : An elongated, tapering xylem cell, with lignified pitted walls, and adapted for conduction and support. Found in conifers, ferns and related plants.

Traditional breeding : Modification of plants and animals through selective breeding. Practices used in traditional plant breeding may include aspects of biotechnology such as tissue culture and mutational breeding (mutagenesis).

Trait : An important characteristic of a crop that is determined by a specific gene or set of genes.

Trait combination or stacking : The combination or "stacking" of traits allows the farmer to have the value of multiple traits, instead of having to choose between them. Combining traits provides tangible cost and yield advantages to farmers. Within each respective crop, farmers can choose to combine different traits in a single seed, including insect-protection and herbicide tolerance. These product offerings are commonly referred to as "stacked" traits.

Trans : Genes located on different DNA molecules present in the same cell (the opposite of cis). For example, one copy of a particular gene may reside on the chromosome and the second copy may reside on a plasmid.

Trans acting factor : A molecule that can diffuse through the cell to act at a distance from where it is made. Often used to distinguish gene products (protein or RNA) that regulate gene expression in trans vs the DNA site where the gene products binds.

Trans configuration : See repulsion.

Trans heterozygote : A double heterozygote that contains two mutations arranged in the trans configuration.

Trans test : See complementation test.

Trans-acting : A term describing substances that are diffusable and that can affect spatially separated entities within cells.

Transcapsidation : The partial or full coating of the nucleic acid of one virus with a coat protein of a differing virus. See coat protein.

Transcript : An RNA molecule that has been synthesized from a specific DNA template. In eukaryotes, the primary transcript produced by RNA polymerase must often be processed or modified in order to form the mature, functional mRNA, rRNA or tRNA.

Transcription : The process of copying a gene into RNA transcript. This is the first step in the expression of any gene. The resulting RNA, if it codes for a protein, will be spliced, polyadenylated, transported to the cytoplasm, and by the process of translation will produce the desired protein molecule, although not all transcripts lead to proteins. Compare translation.

Transcription bubble : A region where the double-stranded DNA is separated while RNA polymerase is actively transcribing RNA. A short region of RNA-DNA duplex is formed between the newly synthesized RNA and the template DNA in this region.

Transcription factor : A protein which controls the transcription of genes. These usually bind to DNA as part of their function (but not necessarily). A transcription factor may be general (i.e. acting on many or all genes in all tissues), or tissue-specific (i.e. present only in a particular cell type, and activating the genes restricted to that cell type). Its activity may be constitutive, or may depend on the presence of some stimulus; for example, the glucocorticoid receptor is a transcription factor which is active only when glucocorticoids are present.

Transcription start site : The point in a DNA sequence at which transcription of a gene into RNA begins.

Transcription terminator : A nucleotide sequence that acts as a signal for termination of transcription. There are two common types of transcription terminators: Rho-independant terminators (usually a stem-loop structure in the transcribed RNA followed by a run of U residues) are typically located at the end of operons, and Rho-dependant terminators (typically an unstructured region of RNA that, when untranslated, is recognized by Rho factor) are typically responsible for translational polarity.

Transcription unit : A segment of DNA that contains signals for the initiation and termination of transcription, and is transcribed into one RNA molecule.

Transcription vector : A cloning vector that allows the foreign gene or DNA sequence to be transcribed *in vitro*.

Transcriptional anti-terminator : A protein that prevents RNA polymerase from terminating transcription at specific transcription termination sequences.

Transcriptome : The complete set of RNA transcripts made by a cell under a particular condition. Typically determined by microarray analysis.

Trans-dimer synthesis : A process which permits nucleotides to be inserted opposite a pyrimidine dimer. Because this process is not based upon complementary base pairing, the wrong base pairs may be inserted, resulting in a mutation.

Transducing phage : See transduction.

Transductant : A genetic recombinant formed by transduction.

Transduction : A method of gene transfer between bacteria in which the bacterial donor DNA is carried by a phage. There are two types of transduction: generalized transduction and specialized transduction. Generalized transduction can transfer of any region of the chromosome from the bacterial host into a recipient cell. Specialized transduction can only transfer regions of DNA adjacent to an integrated phage (prophage).

Transfection : A method by which foreign DNA may be put into a cultured mammalian cell. Such experiments are usually performed using cloned DNA containing coding sequences and control regions (promoters, etc) in order to test whether the DNA will be expressed. Since the cloned DNA may have been extensively modified (for example, protein binding sites on the promoter may have been altered or removed), this procedure is often used to test whether a particular modification affects the function of a gene.

Transfer : The process of moving cultured tissue or cells to a fresh medium.

Transfer RNA (tRNA) : Adaptor molecules which translate the triplet code from the mRNA sequence into the corresponding chain of amino acids. tRNAs are short (about 74-95 bases), single-stranded RNA molecules that contain a high proportion of modified nucleosides. When drawn in two-dimensions, tRNAs can be folded into a characteristic cloverleaf structure with three stem-loop structures. The anticodon at the base of the second loop region. A specific amino acid is are added to the 3' end of each tRNA by a specific aminoacyl tRNA synthetase. The aminoacylated tRNA binds to the ribosome-mRNA complex via interactions with the ribosome and, if the anticodon in the tRNA is properly paired with the complementary bases in the codon at that position, the amino acyl group is transferred to the growing polypeptide chain.

Transferase : Enzyme that catalyses the transfer of a group of atoms from one molecule to another.

Transformant : In prokaryotes, a cell that has been genetically altered through the uptake of foreign DNA. In higher eukaryotes, a cultured cell that has acquired a malignant phenotype. See transformation.

Transformation : 1. The uptake and establishment of DNA in a bacterium or yeast cell, in which the introduced DNA often changes the phenotype of the recipient organism. 2. Conversion by various means of animal cells in tissue culture from controlled to uncontrolled cell growth. Typically through infection by a tumour virus or transfection with an oncogene. See transformant; transformation efficiency.

Transformation (with respect to cultured cells) : A change in cell morphology and behavior which is generally related to carcinogenesis. Transformed cells tend to exhibit characteristics known collectively as the "transformed phenotype" (rounded cell bodies, reduced attachment dependence, increased growth rate, loss of contact inhibition, etc). There are different "degrees" of transformation, and cells may exhibit only a subset of these characteristics. Not well understood, the process of transformation is the subject of intense research.

Transformation efficiency : The number of cells that take up foreign DNA as a function of the amount of added DNA; expressed as transformants per microgram of added DNA. See transformation.

Transformation frequency : The fraction of a cell population that takes up foreign DNA; expressed as the number of transformed cells divided by the total number of cells in a population.

Transformed phenotype : Acquisition of a phenotype characteristic of oncogenic transformation.

Transforming oncogene : A gene that upon transfection converts a previously immortalized cell to the malignant phenotype. See oncogene.

Transgene : A gene from one genome that has been incorporated into the genome of another organism. Often refers to a gene that has been introduced into a multicellular organism.

Transgenesis : The introduction of a gene or genes into animal or plant cells, which leads to the transmission of the input gene (transgene) to successive generations. cf transfection.

Transgenic : Creation of an individual in which genetic modification has occured in the germline tissues and may be transmitted to subsequent generations.

Transgenic mouse : A mouse which carries experimentally introduced DNA. The procedure by which one makes a transgenic mouse involves the injection of DNA into a fertilized embryo at the pro-nuclear stage. The DNA is generally cloned, and may be experimentally altered. It will become incorporated into the genome of the embryo. That embryo is implanted into a foster mother, who gives birth to an animal carrying the new gene. Various experiments are then carried out to test the functionality of the inserted DNA.

Transgenic organism : One into which a cloned genetic material has been experimentally transferred, a subset of these foreign gene express themselves in their offspring.Turner syndrome a chromosomal condition in females (usually 45,XO) due to monosomy of the X- chromosome; characterized by short stature, failure to develop secondary sex characteristics, and infertility.

Transgressive segregation : Appearance in segregating generations of individuals falling outside the parental range in respect to some character.

Transgressive variation : The appearance in the F_2 (or later) generation of individuals showing more extreme development of a trait than either of the original parents.

Transient : Of short duration.

Transient transfection : When DNA is transfected into cultured cells, it is able to stay in those cells for about 2-3 days, but then will be lost (unless steps are taken to ensure that it is retained - see Stable transfection). During those 2-3 days, the DNA is functional, and any functional genes it contains will be expressed. Investigators take advantage of this transient expression period to test gene function.

Transition : A nucleotide substitution in which one pyrimidine is replaced by the other pyrimidine, or one purine replaced by the other purine (e.g., A is changed to G, or C is changed to T) (contrast with Transversion) .

Transition mutation : A base substitution mutation where a purine is replaced by a different purine, or a pyrimidine is replaced by a different pyrimidine.

Transition stage : The integration period of juvenile and reproductive stages of growth.

Transition-state intermediate : In a chemical reaction, an unstable and high-energy configuration assumed by reactants on the way to making products. Enzymes are thought to bind and stabilize the transition state, thus lowering the energy of activation needed to drive the reaction to completion.

Translation : The process of converting the genetic code into polypeptides, catalyzed by the ribosome and a host of soluble factors. mRNA codons are recognized by tRNA anti-codons. Each tRNA codes for a single amino acid, resulting in synthesis of polypeptide wherein the amino acid sequence is dictated by and matches the order of the codons in the mRNA. Sometimes, however, people speak of "translating" the DNA or RNA when they are merely reading the nucleotide sequence and predicting from it the sequence of the encoded protein. This might be more accurately termed "conceptual translation". Compare transcription.

Translational frame shifting : A mechanism used by certain viruses and even higher organisms to change the reading frame used during translation of an mRNA molecule, in a controlled manner, so that the polypeptide product is the result of translation in more than one reading frame.

Translational initiation signal : See initiation codon.

Translational start codon translational stop signal : See initiation codon.

Translesion synthesis : A mechanism that resumes stalled replication due to a damage on the template strand. The stalled replicative polymerase is replaced by translesion polymerase(s) that synthesises a short stretch of DNA across the lesion. Once this occurs, the replicative polymerase resumes DNA synthesis.

Translocation : The process by which a newly synthesized protein is directed toward a specific cellular compartment (i.e, the nucleus, the endoplasmic reticulum).

Translucent : So clear that light may pass through.

Transmission electron microscope (TEM) : A microscope which uses an electron beam to obtain images of objects, with a much greater resolving power than a light microscope. See electron microscope.

Transmit (n transmission) : To spread or transfer, as in spreading an infectious pathogen from plant to plant or from one plant generation to another.

Transorb II technology : A patented technology in Roundup WeatherMAX and other Monsanto herbicides that provides faster herbicide absorption. This gives farmers greater flexibility and better weed control.

Transpiration : Water loss by evaporation from leaf surfaces and through stomata.

Transplant : 1. noun: A plant grown in a cold frame, greenhouse, tissue culture or indoors for later planting outdoors. 2. To dig up and move a plant to another location.

Transposable element : A class of DNA sequences that can move from one chromosomal site to another.

Transposable genetic element : A DNA element that can move from one location in the genome to another. See transposon.

Transposase : An enzyme (or enzyme complex) required for the transposition of a particular transposable element. A transposase must recognize specific sites on the ends of a transposon, cut the transposon out of the original site, and insert the transposon into a new site.

Transposible elements : Genetic elements characterized by their abilities to insert into and withdraw from a given location within the genome, resulting in movement from site to site within the genome over a period of time. Transposable elements may cause epigenetic changes in phenotype.

Transposition : The movement of a discrete segment of DNA from one location in the genome to another.

Transposon : A transposable or movable genetic element. A relatively small DNA segment that has the ability to move (mobile genetic element) from one chromosomal position to another, e.g., Tn 5 is a bacterial transposon that carries the genes for resistance to the antibiotics neomycin and kanamycin and the genetic information for insertion and excision of the transposon.

Transposon tag : Use of a transposon insertion in a gene to follow the inheritance of the gene. Because transposons often express phenotypes that are simple to select and screen for (e.g. antibiotic resistance), it is often much easier to follow inheritance of the transposon insertion than the unmarked gene.

Transposon tagging : The insertion of a transposable element into or nearby a gene, thereby marking that gene with a known DNA sequence.

Transversion : A nucleotide substitution in which a purine replaces a pyrimidine, or vice versa (e.g., A is changed to T, or T is changed to G) (see Transition).

Trap crop : Crop planted around a field to protect the inner crop from diseases transmitted by aerial vectors; host crop of a parasitic plant, such as witchweed (Striga spp.), that is planted to stimulate seed germination, and later sacrified by plowing under before the parasitic plant produces new seeds.

Trenching : Physical separation of soil in a vertical plane to sever grafted roots between trees.

Tribrid protein : A fusion protein that has three segments, each encoded by parts of different genes.

Trichogyne : A female receptive hypha.

Trichome : A plant epidermal hair, of which several types exist.

Tri-hybrid : The offspring from homozygous parents differing in three pairs of genes.

Tri-nucleotide repeats : Tandem repeats of three nucleotides that are present in many genes. In several cases, these trinucleotide repeats have undergone expansions in copy number, and that has resulted in inherited diseases.

Tripartite mating : A process in which conjugation is used to transfer a plasmid vector to a target cell when the plasmid vector is not self-mobilizable. When (1) cells that have a plasmid with conjugative and mobilizing functions are mixed with (2) cells that carry the plasmid vector and (3) target cells, mobilizing plasmids enter the cells with the plasmid vector and mobilize the plasmid vector to enter into the target cells. Following tripartite mating, the target cells with the plasmid vector are separated from the other cell types in the mixture by various selection procedures.

Triplet : A sequence of three nucleotides of DNA which specifies an amino acid. The elucidation of the genetic code involved the binding of charged tRNA species to chemically synthesized ribonucleotide triplets.

Triplet code : A code in which a given amino acid is specified by a set of three nucleotides.

Triplet codon : A set of three nucleotide bases in DNA or RNA that code for an amino acid.

Triploid : A cell or organism containing three times the haploid number of chromosomes.

Trisomy (adj trisomic) (Gr treis, three + soma, body) : An otherwise diploid cell or organism that has an extra chromosome of one pair (chromosome formula: 2n + 1). cf disomy; monosomy.

Triticale : The species formed by crossing wheat and rye.

tRNA; transfer RNA : The class of small RNA molecules that transfer amino acids to the ribosome during protein synthesis. Transfer RNA molecules are folded into a 'clover-leaf' secondary structure by intrastrand base pairing. The anticodon loop contains a nucleotide-triplet complementary to a specific codon within the mRNA molecule. Each tRNA is 'charged' with the correct amino acid molecule, via its 3′ adenosine moiety, by an enzyme called aminoacyl-tRNA synthetase.

Tropism (Gr trope, a turning) : An involuntary plant response to a stimulus, in which a bending, turning or growth occurs, such as phototropism, geotropism or hydrotropism. The response may be positive (towards) or negative (away from) to the stimulus.

True breeding : See purebred.

True-to-type : Applied to a plant or propagation source, this term denotes correct cultivar identification and lack of variation in productivity or performance. Verification is determined visually by an expert or through biochemical, serological or other means.

Truncation : To shorten. For example, a truncated protein results if a premature stop codon interrupts the gene.

Trypsin : A proteolytic enzyme that hydrolyzes peptide bonds on the carboxyl side of the amino acids arginine and lysine.

Trypsin inhibitors : Substances inactivating the enzyme trypsin, which is needed for digestion of peptides.

TSC : See The SNP Consortium.

Tuber : An underground stem adapted for storage, typically produced at the end of a stolon.

Tubulin : The major protein component of the microtubules of eukaryotic cells.

Tumble tube : A glass tube mainly used *in vitro* to agitate and consequently aerate suspension cultures. The tube, which is commonly attached to a slowly revolving platform, is closed at both ends, with a side-neck opening.

Tumor suppressor : A gene that prevents tumor formation until deleted or mutated. The best-known examples of tumor suppressors are the proteins p53 and Rb.

Tumor suppressor gene : Genes that normally function to restrain the growth of tumors; the best understood case is for hereditary retinoblastoma.

Tumor (syn gall) : Abnormal swelling or localized outgrowth, often roughly spherical, produced by a plant as a result of attack by a fungus, bacterium, nematode, insect or other organism.

Tumour virus tumour-inducing plasmid : A virus capable of transforming a cell to a malignant phenotype. See virus.

Tunica : An outer one- to four-cell layered region of the apical meristem, where cell division is anticlinal, i.e., perpendicular to the surface. See apical meristem.

Turbidostat : An open continuous culture in which a pre-selected biomass density is uniformly maintained by automatic removal of excess cells. The fresh medium flows in response to an increase in the turbidity (usually cell density) of the culture.

Turgid (L turgidus, swollen, inflated) : Swollen, distended; referring to a cell that is firm due to water uptake.

Turgidity : State of being rigid or swollen as a result of internal water pressure.

Turgor potential : See pressure potential.

Turgor pressure (L turgor, a swelling) : The pressure within the cell resulting from the absorption of water into the vacuole and the imbibition of water by the protoplasm.

Turion : An underground bud or shoot from which an aerial stem arises. cf sucker.

Turnover : The balance between synthesis and degradation of a product.

Twins : Two individuals originating from the same zygote.

Two μm circle : A plasmid found in the yeast Saccharomyces cerevisiae and used as the basis for a series of cloning vectors.

Two-componant system : A regulatory mechanism that includes at least two functional activities defined as a sensor and a response regulator.

Phosphorylation of the sensor domain is modulated in response to specific physiological stimulus. Phosophotransfer between a histidine residue on the sensor domain and an aspartate residue inÊthe response regulator domain determines the phosphorylation state of the response regulator. The phosphorylation state of the response regulator determines its physiological role (DNA binding, repression, activation, protein-protein interactions, or enzymatic activity). In many cases the two componants are present on separate proteins, but in some cases both componants are present in a single protein, and in other cases there are more than two proteins involved in the phosphotransfer reactions.

Tylosis (pl tyloses) : Balloonlike extrusions of parenchyma cells into lumina of contiguous vessels that partially or completely block them.

Type : The example on which the description of a scientific name is based, and which fixes the application of the name (the type genus of a family, or the type species of a genus).

U : The uracil residue in RNA.

uidA : The gene encoding Beta-glucuronidase (abbreviated GUS). Commonly used for constructing operon or gene fusions in organisms with high endogenous Beta-galactosidase activity. A variety of analogs exist that make it easy to detect GUS expression on solid medium and to assay the activity of the enzyme in cells.

Ultrasonication : See sonication.

Ultrastructure : Submicroscopic structure of a macromolecule, cell, or tissue.

Ultraviolet light; ultraviolet radiation (UV) : The portion of the electromagnetic spectrum with wavelengths from about 100 to 400 nm; between ionizing radiation (X-rays) and visible light. UV is absorbed by DNA and is highly mutagenic to unicellular organisms and to the epidermal cells of multicellular organisms. UV light is used in tissue culture for its mutagenic and bactericidal properties.

Umber : The codon UGA. An less commonly used term for an opal codon.

Unbalanced- : In experimental design, refers to an experiment or set of data in which all treatments or treatment combinations are not equally represented. The most common cause of unbalanced experiments is unequal mortality among entries in a test. (See randomized-block design and replication.) Also used to describe any plant breeder specializing in an organism with nearly as long a life cycle as his or her own.

Undefined : A medium or substance added to medium in which not all of the constituents or their concentrations are chemically defined, such as media containing coconut milk, malt extract, casein hydrolysate, fish emulsion or other complex compounds. cf organic complex.

Undefined medium : A growth medium in which not all the components have been identified. Most "rich" media have a variety of complex, undefined componants.

Understock : Host plant for a grafted scion, a branch or shoot from another plant; an understock may be a fully grown tree or a stump with a living root system.

Undifferentiated : In a meristematic state or resembling a meristem; lacking the specialized or differential gene expression characteristic of specialized cells.

Unequal cross-over : A recombination event that occurs between DNA molecules that are not fully aligned. For example, a crossover may occur between repeated DNA sequences resulting in the deletion or duplication the intervening DNA sequence.

UNESCO : United National Educational, Scientific, and Cultural Organization.

Unicellular : Describing tissues, organs or organisms consisting of a single cell.

Unidirectional replication : DNA replication that proceeds in only one direction along the DNA template.

Uniflagellate : Having one flagellum.

Uninucleate : Having one nucleus (see multinucleate).

Unisexual : Describing animals and plants possessing either male or female reproductive organs, but not both.

Unit inheritance : Gregor Mendel's idea that the characteristics of parents are passed on to descendants unchanged as units. In other words, the hereditary material of any organism is made up of discrete units (now called genes).

Unitunicate : Having a single ascus wall (see bitunicate).

Univalent : An unpaired chromosome at meiosis.

Universal donor cells : Cells that, after introduction into a recipient, will not induce an immune response that leads to their rejection.

Universal genetic code : The "standard" codon usage that is common to most organisms.

Universality : Referring to the genetic code, the codons have the same meaning, with minor exceptions, in virtually all species.

Unorganized growth : *In vitro* formation of tissues with few differentiated cell types and no recognizable structure; many call are unorganized. cf organized growth.

Unselected marker : A genetic marker that may be co-inherited with an adjacent, selected marker but inheritance of the unselected marker is not demanded. For example, in a genetic cross you may select for one gene and screen for inheritance of an adjacent, unselected gene.

Unstable alleles : See stuttering alleles.

Untranslated RNA : See Nontranslated RNA.

Up promoter mutation : A mutation that increasses expression from a promoter.

upstream : 1. In molecular biology, the stretch of DNA base pairs that lie in the 5′ direction from the site of initiation of transcription. Usually the first transcribed base is designated +1 and the upstream nucleotides are marked with minus signs, e.g., -1, -10. Also, to the 5′ side of a particular gene or sequence of nucleotides. 2. In chemical engineering, those phases of a manufacturing process that precede the biotransformation step. Refers to the preparation of raw materials for a fermentation process. Also called upstream processing.

Upstream activator sequence : A binding site for transcription factors, generally part of a promoter region. A UAS may be found upstream of the TATA sequence (if there is one), and its function is (like an enhancer) to increase transcription. Unlike an enhancer, it can not be positioned just anywhere or in any orientation.

Upstream/Downstream : In an RNA, anything towards the 5' end of a reference point is "upstream" of that point. This orientation reflects the direction of both the synthesis of mRNA, and its translation - from the 5' end to the 3' end. In DNA, the situation is a bit more complicated. In the vicinity of a gene (or in a cDNA), the DNA has two strands, but one strand is virtually a duplicate of the RNA, so it's 5' and 3' ends determine upstream and downstream, respectively. NOTE that in genomic DNA, two adjacent genes may be on different strands and thus oriented in opposite directions. Upstream or downstream is only used on conjunction with a given gene.

Uracil : A nitrogenous base normally found in RNA but not DNA; uracil is capable of forming a base pair with adenine.

Uracil-N-glycosylase : A repair enzyme that removes uracil from DNA by cleaving the bond between the base and the sugar backbone. This enzyme ensures that DNA contains T instead of U residues.

Urediniospore (also urediospore, uredospore) : The asexual, dikaryotic, often rusty-colored spore of a rust fungus, produced in a structure called a uredinium; the "repeating stage" of a heteroecious rust fungus, i.e. capable of infecting the host plant on which it is produced.

Uredinium (also uredium; pl uredinia) : Fruiting body (sorus) of rust fungi that produces urediniospores.

Utilization of farm animal genetic resources : In AnGR: The use and development of animal genetic resources for the production of food and agriculture. (Source: FAO, 1999).

UV absorbance spectroscopy : A method for measuring the concentration of a compound by determining the amount of ultraviolet radiation absorbed by a sample.

UV reactivation : A phenomenon in which survival of an ultraviolet-irradiated phage is greater upon infecting a host that has also been irradiated with UV than upon infecting a host that has not been irradiated with UV. The increased survival in the UV irradiated host is due to the induction of the SOS-repair system in the host.

UvrABC : An enzyme complex that functions as an endonuclease, cutting the DNA on both sides of DNA lesions that distort the double-helix (for example, T-T dimers). The DNA fragment is subsequently excised allowing resynthesis and repairing the DNA damage.

V regions : Variable regions in antibodies. See CDR.

v/v : On a volume per volume basis; the percent of the volume of a constituent in 100 units of volume, e.g., (ml/100 ml) × 100.

Vaccination : See preventive immunization.

Vaccine : A preparation of dead or weakened pathogens, or of derived antigenic determinants, that is used to induce formation of antibodies or immunity against the pathogen.

Vaccinia : The cowpox virus used to vaccinate against smallpox and, experimentally, as a carrier of genes for antigenic determinants cloned from other disease organisms.

Vacuole : Generally spherical organelle within a plant cell bound by a membrane and containing dissolved materials such as metabolic precursors, storage materials, or waste products.

Vacuum : Vacuum is created through use of a vacuum pump or aspirator pump, to facilitate specific biological preparations, such as inclusions or disinfection of material for *in vitro* culture, etc.

Variability : Differing from the average value. Absence of uniformity, usually referring to lack of genetic uniformity in a population. (See heterozygosity and resilience).

Variable domains : Regions of antibody chains that have different amino acid sequences in different antibody molecules. These regions are responsible for the antigen-binding specificity of the antibody molecule.

Variable expressivity : Variation in the phenotype caused by different alleles of the same gene and/or by the action of other genes and/or by the action of non-genetic factors. cf expressivity.

Variable number tandem repeat : See VNTR.

Variable surface glycoprotein (VSG) : One of a battery of antigenic determinants expressed by a micro-organism to elude immune detection.

Variance : In statistics: The sum of the squared deviations, divided by one less than the number of observations. A statistical measure of variation in a population.

Variant : An organism that is genetically different from the wild type organism. a.k.a. mutant.

Variation : Differences between individuals within a population or among populations.

Variation : The occurrence of differences among individuals due to differences in their genetic composition and/or the environment in which they were raised.

Variegated : Plants having both green and albino tissues. This difference in colour may result from viral infection, nutritional deficiency, or may be under genetic or physiological control.

Variety : A naturally occurring subdivision of a species, with distinct morphological characters and given a Latin name according to the rules of the International Code of Nomenclature. A taxonomic variety is known by the first validly published name applied to it so that nomenclature tends to be stable. cf cultivar; pathovar.

Vascular (L vasculum, a small vessel) : Referring to any plant tissue or region consisting of or giving rise to conducting tissue, e.g., bundle, cambium, ray.

Vascular bundle : Strand of conductive tissue, usually composed of xylem and phloem (in leaves, small bundles are called veins).

Vascular bundle; fascicle : A strand of tissue containing primary xylem and primary phloem (and procambium if present) and frequently enclosed by a bundle sheath of parenchyma or fibres.

Vascular cambium : In biennials and perennials, cambium giving rise to secondary phloem and secondary xylem.

Vascular cylinder (syn stele) : The cylinder of vascular tissue in stems or roots.

Vascular plants : Plants possessing organized vascular tissues.

Vascular system : 1. A specialized network of vessels for the circulation of fluids throughout the body tissue of an animal. 2. The system of vascular tissue in plants.

Vascular tissue : The tissue that conducts water and nutrients throughout the plant body in higher plants.

Vascular wilt disease : A xylem disease that disrupts normal uptake of water and minerals, resulting in wilting and yellowing of foliage.

Vector (L vehere, to carry) : 1. An organism, usually an insect, that carries and transmits disease-causing organisms. 2. A plasmid or phage that is used to deliver selected foreign DNA for cloning and in gene transfer. See Ti plasmid.

Vegetative : Referring to somatic or asexual parts of a plant, which are not involved in sexual reproduction.

Vegetative propagation : Asexual reproduction; in plants, the use of cuttings, bulbs, tubers, and other vegetative plant parts to grow new plants.

Vehicle : The host organism used for the replication or expression of a cloned gene or other sequence The term is little used and is often confused with vector. See vector.

Vein : Small vascular bundle in a leaf.

Vein banding : Symptom of virus disease in which regions along veins are either darker green or distinctly more yellow than tissue between veins.

Vein clearing : Disappearance of green color in or around leaf veins.

Velocity density gradient centrifugation : A procedure used to separate macromolecules based on their rate of movement through a density gradient.

Velogenetics : The combined use of marker-assisted selection (q.v.) and embryo technologies such as OPU (q.v.), IVM (q.v.) and IVF (q.v.), in order to increase the rate of genetic improvement in animal populations.

Vermiculite : Coarse aggregate material made from expanded mica having a high cation exchange capacity and high water-holding capacity and used as a rooting medium and a soil additive.

Vermiform : Worm-shaped.

Vernalization : The process by which floral induction in some plants is promoted by exposing the plants to chilling for a certain duration.

Vertical resistance (syn specific resistance) : Resistance which is effective against some biotypes or races of the pathogen, but not others, usually inherited monogenically and expressed qualitatively (see general resistance, horizontal resistance, race non-specific resistance).

Vertifolia effect : A term used to describe the loss of general (horizontal, minor gene, multigenic, polygenic) resistance in a cultivar after several generations of selection during which a major gene confers resistance to the dominant race or biotype of the pathogen; first observed in the potato cultivar Vertifolia with late blight resistance.

Vesicle : Thin sac in which zoospores are differentiated and released;the bulbous head terminating the conidiophores of Aspergillus; structure formed by endomycorrhizal fungi within living cells of the root.

Vessel (L vasculum, a small vessel) : 1. A series of xylem elements whose function is to conduct water and nutrients in plants. 2. A container, such as a Petri dish or test tube, used for tissue culture.

Vessel element : A type of cell occurring within the xylem of flowering plants. Many are water-conducting vessels.

Viability : A populations ability to live, grow and develop. It is affected by physical habitat factors (climate, geology, topography, and aquatic features) and by biotic habitat factors (plant and animal populations and communities).

Viability test : Assay of the number or percent of living cells or plants in a population that have been given a specific treatment; such as ascertaining cell viability after cryopreservation treatment, or for seeds.

Viable (n viability) : The state of being alive; able to germinate, as seeds, fungus spores, sclerotia, etc.; capable of growth.

Vibrio : Comma-shaped bacterium.

Vir genes : A set of genes on a Ti plasmid that prepare the T-DNA segment for transfer into a plant cell.

Viral coat protein : Protein present in the outer layer of a virus.

Viral oncogene : A gene in the viral genome that contributes to malignancies in vertebrate hosts. See oncogene.

Viral pathogen : A disease-causing virus.

Viral vaccines : Vaccines consisting of live viruses rather than dead ones or separated parts of viruses. However, as the virus itself cannot be used, because that would simply give the patient the disease, the virus is genetically engineered so that it elicits the immune response to the viral pathogen without causing the disease itself. Two genetic engineering methods can be used. The first is to make the disease virus harmless, but still able to replicate in cultured animal cells. This is similar to producing an `attenuated' virus, i.e., one which has been grown in the laboratory until it loses its ability to cause disease. However, the genetic engineering route seeks to make sure that the attenuated virus has no chance of mutating back to a wild type, pathogenic virus, by deleting whole genes or replacing key regions of genes with completely different genetic material. The second approach is to clone the gene for a protein from the pathogenic virus into another, harmless virus, so that the result `looks' like the pathogenic virus but does not cause disease.

Virescence : State or condition in which normally white or colored tissues (e.g. flower petals) become green.

Virion : An infectious virus particle. A plant pathogen that consists of a naked RNA molecule of approximately 250-350 nucleotides, whose extensive base pairing results in a nearly correct double helix. See satellite RNA.

Viroid : An infectious entity similar to a virus but smaller, consisting only of a strand of nucleic acid without the protein coat characteristic of a virus. See satellite RNA.

Viroplasm : Cellular inclusions that are sites of synthesis of viral components and the assembly of virus particles.

Virulence : Degree or measure of pathogenicity; relative capacity to cause disease.

Virulence (L virulentia, a stench) : The degree of ability of an organism to cause disease. The relative infectiousness of a bacterium or virus, or its ability to overcome the resistance of the host metabolism.

Virulence factor : Any gene product which enhances the ability of an organism to cause disease.

Virulence gene : Slang for a gene encoding a virulence factor.

Virulent : Highly pathogenic; having the capacity to cause severe disease (see avirulent).

Virulent phage : A bacteriophage which always grows lytically.

Viruliferous : Virus-laden, usually applied to insects or nematodes as vectors.

Virus : A small, infectious, obligate intracellular parasite. The virus genome is composed of either DNA or RNA. Within an appropriate host cell, the viral genome is replicated and uses cellular systems to direct the synthesis of other viral components. Progeny viruses are formed by de novo assembly from the newly synthesized components within the host cell. Transmission of the progeny viruses occurs by release from the host cell, and infection of new host cells.

Virus-free : Plant, animal, cell, tissue or meristem which exhibits no viral symptoms or contains no identifiable virus-particles.

Virus-tested : Description of a organism or a cell stock certified as being free of certain specified viruses following recognized procedures of virus diagnosis.

Viscin : Sticky substance produced on the seed coat of parasitic flowering plants in the Viscaceae (mistletoes) that helps attach the seed to the host plant branch.

Visible light : The part of the electromagnetic spectrum with wavelengths between 380 nm and 750 nm and perceived by the human eye.

Vitamin : An organic compound that is required in relatively small amounts for growth of an organism. Vitamins often function as coenzymes.

Vitamin B complex : A large group of water soluble vitamins that function as co-enzymes, including thiamine (B_1); riboflavin or vitamin G (B_2); niacin or nicotinic acid (B_3); pantothenic acid (B_5); pyridoxine (B_6); cyanocobalamin (B_{12}); biotin or vitamin H; folic acid or vitamin M (Bc); inositol; choline; and others.

Vitamin C : See ascorbic acid.

Vitamin H : See biotin; vitamin B complex.

Vitamins (L vita, life + amine) : Naturally occurring organic substances required by living organisms in relatively small amounts to maintain normal health, and which are added to tissue culture media to enhance growth, usually acting as enzyme co-factors.

Vitrified; water soaked : Cultured tissue having leaves and sometimes stems with a glassy, transparent or wet and often swollen appearance. The process of vitrification is a general term for a variety of physiological disorders that lead to shoot tip and leaf necrosis.

Vivipary : 1. A form of reproduction in animals in which the developing embryo obtains its nourishment directly from the mother via a placenta or by other means. 2. A form of asexual reproduction in certain plants, in which the flower develops into a budlike structure that forms a new plant when detached from the parent. 3. The development of young plants in the inflorescence of the parent plant. Vmax The maximal rate of an enzyme-catalysed reaction. Vmax is the product of Eo (the total amount of enzyme) times the value of Keat (the catalytic rate constant).

VLSI : Very large- scale integration allowing over 100,000 transistors on a chip.

VNTR : Variable number tandem repeat. A short DNA sequence that is present as tandem repeats and in highly variable copy number.

Volatilization : The conversion of a solid or liquid into a gas or vapour.

Volunteer : Self-set plant; plant seeded by chance.

Von Hippel-Lindau syndrome : An autosomal dominant condition characterized by the anomalous growth and proliferation of blood vessels on the retina of the eye and the cerebellum of the brain; cysts and cancers in the kidneys, pancreas, and adrenal glands.

VSG : See w/v Weight per volume; the weight of a constituent in 100 cm3 of solution, expressed as a percentage.

VSP repair :

"Very short patch" DNA repair. : A DNA repair mechanism that corrects G-T mismatches arising from deamination of 5-methyl cytosine in enteric bacteria. A short DNA fragment including the mismatch is excised and resynthesized.

Vulva : Exterior opening of a mature female nematode's reproductive system.

Walking : A method for cloning large regions of a chromosome. Starting from a known site, a gene library is screened for clones that hybridize to DNA probes taken from the ends of the first clone. These clones are then isolated, and their ends used to screen the library again. These clones are then isolated and their ends used, and so on. See gene walking; chromosome walking.

Wall pressure : Pressure that a cell wall exerts against the turgor of the cell contents. Wall pressure is equal and opposite to the turgor potential.

Walling-off : Separation of diseased from healthy tissues by barrier tissues produced by a diseased plant.

Wash-out : The loss of the slower growing micro-organism when two organisms are being grown together.

Water potential : Refers to the difference between the activity of water molecules in pure distilled water at atmospheric pressure and 30°C (standard conditions), and the activity of water molecules in any other system. The activity of these water molecules may be greater (positive) or less (negative) than the activity of the water molecules under standard conditions.

Water soaked : See vitrified.

Water sprout : Small, rapidly growing shoot or branch on a large stem, developed from adventitious tissues.

Water stress : The condition when plants are unable to absorb enough water to replace that lost by transpiration. The results may be wilting, cessation of growth, or even death of the plant or plant parts.

Water-soaked : Describing disease symptom of plants or lesions that appear wet, dark, and usually sunken and translucent.

Watson-Crick rules : The normal base pairing rules for DNA and RNA: A pairs with T or U, and G pairs with C.

Wavelength : The distance between two corresponding points on any two consecutive waves. For visible light it is very small and is generally measured in nanometres.

Wax (AS weax, wax) : Esters of alcohol higher than glycerol, which are insoluble in water and difficult to hydrolyze; wax forms protective waterproof layers on leaves, stems, fruits, animal fur and integuments of insects.

Weed : Simply any plant growing where it is not wanted. In agriculture, used for a plant which has good colonizing capability in a disturbed environment, and can usually compete with a cultivated species therein. Weeds are typically considered as unwanted, economically useless or pest species.

Weediness : In agriculture, the ability of a plant to colonize a disturbed habitat and compete with cultivated species.

Weigle reactivation : The increased survival of phage after UV irradiation if they infect cells that have previously been exposed to a lose dose of UV. Due to induction of UV repair systems in the infected cells.

Western blot : A technique for analyzing mixtures of proteins to show the presence, size and abundance of one particular type of protein. Similar to Southern or Northern blotting (see "Blotting"), except that (1) a protein mixture is electrophoresed in an acrylamide gel, and (2) the "probe" is an antibody which recognizes the protein of interest, followed by a radioactive secondary probe (such as 125I-protein A).

Western blotting analysis : A technique used to identify a specific protein; the probe is a radioactively labeled antibody raised against the protein in question.

Wet weight : The gross weight of a product with its full water content, or weight of fully hydrated tissue. See fresh weight.

Wetting agent : A substance that improves surface contact by reducing the surface tension of a liquid: e.g., Triton X-10TM added to disinfecting solutions promotes the disinfestation process. See detergent; surfactant.

White rot (of wood) : Type of wood decay resulting from enzymatic action of fungi; it degrades all components of wood, including lignin, leaving the wood light-colored and spongy.

White rust : Common name usually used for diseases caused by oomycetes in the genus Albugo.

Whorl : Circular arrangement of like parts.

Wild type : An organism as found in nature; the dominant allele usually found in nature, and from which mutations produce other dominants or recessives alleles.

Wilt : Drooping of stems and foliage due to loss of water and decreased turgidity of cells. May be caused by water stress or by disease.

Wilting point (WP) : The moisture content of soil at which plants start to wilt, but not to the extent that they fail to recover when placed in a humid atmosphere. cf. permanent wilting point.

Wind break : A row or other grouping of trees or shrubs used to provide protection against the effects of high velocity winds.

Wind pollination- : Pollination by wind-borne pollen.

Winterburn : Foliar necrosis, often marginal, of plants that retain their leaves in winter due to water deficiency because they cannot take up water from frozen soils.

Witches' broom : Disease symptom characterized by an abnormal, massed, brushlike development of many weak shoots arising at or close to the same point.

Wobble : A hypothesis proposed by Francis Crick to explain how one tRNA may recognize two different codons that differ in the third position. The three bases in the anticodon of each tRNA are antiparallel to the three bases of the codon in the mRNA. Normal Watson-Crick base pairing occurs between the first two bases of the codon with the complementary bases of the anticodon. However, the first (5') base of the anticodon can pair with the third (3') base of the codon in different ways: in this position of the anticodon G can pair with either C or U in the mRNA, U can pair with either A or G in the mRNA, and I can pair with A, C, or U in the mRNA.

Wobble hypothesis : An explanation of how one tRNA may recognize more than one codon. The first two bases of the mRNA codon and anticodon pair properly, but the third base in the anticodon has some wobble that permits it to pair with more than one base.

Wobble position : The third base position within a codon, which can often (but not always) be altered to another nucleotide without changing the encoded amino acid (see Degeneracy).

Wood : Secondary xylem.

Working Draft DNA Sequence : See: Draft DNA Sequence.

Wound : Injuries to plant tissues that often breach barriers (cuticle, bark, cell walls) that might otherwise exclude pathogens; some pathogens (e.g. viruses) can enter plants only through a wound; wounds may occur from natural growth processes, physical and chemical agents, animals (especially insects), and many human agricultural activities, such as pruning.

WP : See wilting point.

X : The basic number of chromosomes in a polyploid series, monoploid = x; diploid = 2x; triploid = 3x; etc.

X chromosome : One of the two sex chromosomes, X and Y. See also: Y chromosome, sex chromosome.

X1, X2, X3 : Symbols denoting first, second, third, etc. generations from and irradiated ancestral.

Xanthophyll (Gr xanthos, yellowish brown + phyllon, leaf) : A yellow chloroplast pigment.

X-chromosome : A chromosome associated with sex determination. In most animals, the female has two, and the male has one X chromosome. The opposite occurs in birds, in which the equivalent is the Z chromosome.

Xenia : The immediate effect of pollen on some characters of the endosperm.

Xenobiotic : A chemical compound that is not produced by living organisms; a manufactured chemical compound.

Xenograft : Tissue or organs from an individual of one species transplanted into or grafted onto an organism of another species, genus, or family. A common example is the use of pig heart valves in humans.

Xeno-immunization : Stimulation of immune response to antigens from different species.

Xenotransplantation : The transplantation of tissue from one species to another species, typically from non-human mammals to humans. This technology has become very important because of a worldwide shortage of human organs for humans requiring a new organ. The most popular non-human species involved in this technology is the pig.

Xerophyte (Gr xeros, dry + phyton, a plant) : A plant very resistant to drought or that lives in very dry places.

X-gal : A common abbreviation for 5-bromo-4-chloro-3-indolyl-ß-D-galactoside. A sensitive, color indicator for ß-galactosidase. X-gal is a colorless compound but upon hydrolysis by beta-galactosidase releases an indolyl moiety. This product is not colored, but dimerizes after oxidation to form an insoluble and highly colored indigo dye. The oxidation is promoted by atmospheric oxygen, hence the color does not form anaerobically. (X-gal is also sometimes written as Xgal, without the hyphen).

X-glu : A common abbreviation for 5-bromo-4-chloro-3-indolyl-glucuronide. A color indicator for beta-Glucuronidase (GUS). X-glu (AKA X-gluc) is a colorless compound but upon hydrolysis by GUS the indolyl moiety is oxidized to form a blue colored product. GUS is often used as a reporter gene in organisms that have high background levels of beta-galactosidase, often due to multiple genes encoding lacZ homologues. (See X-gal).

X-inactivation : The repression of one of the two X-chromosomes in the somatic cells of females as a method of dosage compensation; at an early embryonic stage in the normal female, one of the two X-chromosomes undergoes inactivation, apparently at random, from this point on all descendent cells will have the same X-chromosome inactivated as the cell from which they arose, thus a female is a mosaic composed of two types of cells, one which expresses only the paternal X-chromosome, and another which expresses only the maternal X-chromosome.

XLB (xylem-limited fastidious bacteria) : Bacterial pathogens of plants found only in the xylem, causing wilt, scorch, and stunting symptoms; vectored by insects that feed on xylem fluid; not sap transmissible; require complex culture media.

X-linked : The presence of a gene on the X chromosome. a.k.a. X-linkage.

X-linked disease : A genetic disease caused by an allele at a locus on the X-chromosome. In X-linked recessive conditions, a normal female "carrier" passes on the allele on her X chromosome to an affected son.

X-over : An abbreviation for cross-over.

X-P : A common abbreviation for 5-bromo-4-chloro-3-indolyl phosphate. A sensitive, color indicator for alkaline phosphatase activity. X-P is a colorless compound but upon hydrolysis by a phosphatase the indolyl moiety is released, and upon oxidization it forms a blue colored product. (See X-gal).

X-ray crystallography : The deduction of crystal structure from analysis of the diffraction pattern of X-rays passing through a pure crystal of a substance.

Xylem (Gr xylon, wood) : A complex tissue specialized for efficient conduction of water and mineral nutrients in solution. Xylem may also function as a supporting tissue, particularly secondary xylem.

Xylem-limited fastidious bacteria (XLB) : Bacterial pathogens of plants found only in the xylem, causing wilt, scorch, and stunting symptoms; vectored by insects that feed on xylem fluid; not sap transmissible; require complex culture media.

XYY syndrome : Genetic condition in males with extra Y chromosome (in 1 in 1000 male births). Symptoms: tall stature (over 6'), may including sterility, developmental delay, learning problems.

Y chromosome : One of the two sex chromosomes, X and Y. See also: X chromosome, sex chromosome.

YAC : Yeast artificial chromosome; a linear vector into which a large fragment of DNA can be inserted; the development of YAC's in 1987 has increased the number of nucleotides which can be cloned.

Y-Chromosome : A chromosome that is different in the two sexes and involved in sex determination. The male in our species has one Y and one X chromosome.

Yeast : A unicellular form of fungi. Some fungi (for example Saccharomyces cerevisiae) are only known to reproduce as yeast while other fungi (for example, Candida albicans) are dimorphic, reproducing as yeast under certain conditions and as filamentous mycelia under other growth conditions. Because it is so widely used in studies of Eukaryotic genetics, the term yeast often refers to S. cerevisiae when a specific genus is not indicated.

Yeast artificial chromosome (YAC) : This is a method for cloning very large fragments of DNA. Genomic DNA in fragments of 200-500 kb are linked to sequences which allow them to propagate in yeast as a mini-chromosome (including telomeres, a centromere and an ARS - an autonomous replication sequence). This technique is used to clone large genes and intergenic regions, and for chromosome walking. Compare cloning vector, cosmid.

Yeast episomal plasmid (YEp) : A yeast vector carrying the 2 μm circle origin of replication.

Yeast extract : A water soluble preparation extracted from common yeast. Yeast extract is rich in amino acids, peptides, B vitamins, and trace elements. Commonly used in rich media.

Yeast integrative plasmid (YIp) : A yeast vector that relies on integration into the host chromosome for replication.

Yeast replicative plasmid (YRp) : A yeast vector that carries a chromosomal origin of replication.

Yeast two-hybrid system : An approach developed by Stan Fields and colleagues in 1989 to identify protein-protein interactions *in vivo*. Proteins (aka "prey") that interact with other, known proteins (aka "bait") are identified by using a reporter system that relies upon interaction between the two proteins. A variety of versions of this system are available, but the basic format involves the construction of two distinct hybrid proteins fused to different reporter genes: in the first the "bait" protein is fused to a transcription factor, and in the second the "prey" protein is fused to a related transcription factor. If the bait and prey proteins interact the two reporter genes fused to the proteins are brought into proximity with each other, producing a specific signal.

Yellows : Disease characterized by chlorosis and stunting of the host plant.

Yield : An aggregate of the products resulting from growth or cultivation.

Z-DNA : Alternative left-handed form of the double helix which retains Watson-Crick type base pairing.

Zinc finger : A protein structural motif common in DNA binding proteins. Four Cys residues are found for each "finger" and one finger can bind a molecule of zinc. A typical configuration is: CysXxxXxxCys--(intervening 12 or so aa's)--CysXxxXxxCys.

Zinc-finger protein : A secondary feature of some proteins containing a zinc atom; a DNA-binding protein.

Zonate : Targetlike development of tree canker, characterized by successive, perennial rings of callus; any symptom appearing in concentric rings.

Zoo blot : Northern analysis of mRNA from different organisms.

Zoosporangium : Sporangium, or spore case, that bears zoospores.

Zoospore : Fungal spore with flagella, capable of locomotion in water.

Zygomycetes : A fungus group, also called the Zygomycota, characterized by nonseptate hyphae, sexual zygospores, and sporangiospores produced in a sporangium; common post-harvest pathogens.

Zygospore : Sexual resting spore formed from the union of gametangia in the Zygomycetes.

Zygote : Cell formed by the union of two gametes and the individual developing from this.

Zygotene : A stage in meiotic prophase when the threadlike chromosomes pair.

Zygotic induction : The lytic development of a prophage upon its transfer into a recipient cell which lacks the same prophage. Because the recipient cell does not contain the repressor protein required for maintance of lysogeny, expression of the genes required for lytic growth occurs as soon as the phage enters the recipient.

-DNA : The normal form of DNA found in biological systems. It exists as a right-handed helix.

-galactosidase : An enzyme that catalyses the formation of glucose and galactose from lactose.

-lactamase : An ampicillin resistance gene. See selectable marker.

2µm plasmid : A naturally occurring, double-stranded, circular DNA plasmid (6 318 bp) found in the nuclei of yeast. Many yeast plasmid vectors are derived from the 2µm plasmid. a.k.a. 2µm circle; 2 µ-plasmid; 2-micron plasmid.

3' end : The end of a polynucleotide with a free (or phosphorylated) 3' - hydroxyl group.

3' end/5' end : A nucleic acid strand is inherently directional, and the "5 prime end" has a free hydroxyl (or phosphate) on a 5' carbon and the "3 prime end" has a free hydroxyl (or phosphate) on a 3' carbon (carbon atoms in the sugar ring are numbered from 1' to 5'. That's simple enough for an RNA strand or for single-stranded (ss) DNA. However, for double-stranded (ds) DNA it's not so obvious - each strand has a 5' end and a 3' end, and the 5' end of one strand is paired with the 3' end of the other strand (it is "antiparallel"). One would talk about the 5' end of ds DNA only if there was some reason to emphasize one strand over the other - for example if one strand is the sense strand of a gene. In that case, the orientation of the sense strand establishes the direction.

3' flanking region : A region of DNA which is not copied into the mature mRNA, but which is present adjacent to 3' end of the gene. It was originally thought that the 3' flanking DNA was not transcribed at all, but it was discovered to be transcribed into RNA, but quickly removed during processing of the primary transcript to form the mature mRNA. The 3' flanking region often contains sequences which affect the formation of the 3' end of the message. It may also contain enhancers or other sites to which proteins may bind.

3' untranslated region : A region of the DNA which is transcribed into mRNA and becomes the 3' end or the message, but which does not contain protein coding sequence. Everything between the stop codon and the polyA tail is considered to be 3' untranslated (see Figure 4). The 3' untranslated region may affect the translation efficiency of the mRNA or the stability of the mRNA. It also has sequences which are required for the addition of the poly(A) tail to the message (including one known as the "hexanucleotide", AAUAAA).

3´-extension : A short single-stranded nucleotide sequence on the 3´-hydroxyl end of a double-stranded DNA molecule. a.k.a. 3´ protruding end; 3´ sticky end; 3´ overhang.

3´-hydroxyl end : The hydroxyl group that is attached to the 3´ carbon atom of the sugar (ribose or deoxyribose) of the terminal nucleotide of a nucleic acid molecule.

3'-terminus : The end of a polynucleotide which carries the hydroxyl group attached to the 3' position of the sugar.

5' end : The end of a polynucleotide with a free (or phosphorylated or capped) 5' - hydroxyl group; transcription/translation begins at this end.

5' flanking region : A region of DNA which is not transcribed into RNA, but rather is adjacent to 5' end of the gene. The 5'-flanking region contains the promoter, and may also contain enhancers or other protein binding sites.

5' untranslated region : A region of a gene which IS transcribed into mRNA, becoming the 5' end of the message, but which does not contain protein coding sequence. The 5'-untranslated region is the portion of the DNA starting from the cap site and extending to the base just before the ATG translation initiation codon. While not itself translated, this region may have sequences which alter the translation efficiency of the mRNA, or which affect the stability of the mRNA.

5´ end : The phosphate group that is attached to the 5´ carbon atom of a sugar (ribose or deoxyribose) of the terminal nucleotide of a nucleic acid molecule.

5´-extension : A short single-stranded nucleotide sequence on the 5´-hydroxyl end of a double-stranded DNA molecule. a.k.a. 5´ protruding end; 5´ sticky end; 5´ overhang.

5'-terminus : The end of a polynucleotide which carries the phosphate group attached to the 5' position of the sugar.

7-methyl-guanosine triphosphate : The methylated form of guanosine triphosphate used to cap an mRNA molecule in most eukaryotes.

❑❑❑